The Internal-Combustion Engine
in Theory and Practice

The Internal-Combustion Engine in Theory and Practice

Volume II: Combustion, Fuels, Materials, Design

Revised Edition

by Charles Fayette Taylor
Professor of Automotive Engineering, Emeritus
Massachusetts Institute of Technology

THE M.I.T. PRESS
Massachusetts Institute of Technology
Cambridge, Massachusetts, and London, England

Fifth Printing,1992

Printed and bound by Halliday Lithograph Corporation in the United States of America.

First MIT Press paperback edition, 1977
Revised edition, 1985

Library of Congress Cataloging in Publication Data

Taylor, Charles Fayette, 1894–
 The internal-combustion engine in theory and practice.

 Bibliography: v. 2, p.
 Includes index.
 Contents: v. 2. Combustion, fuels, materials, design.
 1. Internal combustion engines. I. Title.
TJ785.T382 1985   621.43   84-28885
ISBN 0-262-20052-X (hard)
ISBN 0-262-70027-1 (paper)

Preface

As in the case of Volume I, much of the material in this volume derives from the author's work, for many years, as Director of the Sloan Laboratories for Aircraft and Automotive Engines at the Massachusetts Institute of Technology and as a consultant to government and industry.

Since Volume II was published in 1968 there have been no changes in the fundamental principles discussed herein. However, as stated in the preface to Volume I, the petroleum crisis of the 1970s and the adoption of public laws requiring reduced undesirable exhaust emissions have caused changes in emphasis on a number of aspects of engine design, application, and operation. Recent developments in electronic computers and control systems have greatly assisted in the improvement of fuel economy and pollution control. These subjects are discussed in more detail at appropriate points in this volume.

The recent emphasis on fuel economy and pollution control has also stimulated theoretical searches for an automobile power plant better than the conventional spark-ignition or Diesel engine. Such studies (refs. 23.00, 13.01) have found no alternative type that promises to have significant advantages in fuel economy or pollution control, and, above all, none that has nearly the all-around simplicity, safety, and adaptability of present engines. It appears that the conventional types of spark-ignition and Diesel engines will remain in their present predominant position in land and sea transportation and for industrial and portable power for the foreseeable future.

Although great pains have been taken to avoid errors, it is not possible to eliminate them entirely in a work of this magnitude. It is hoped that readers who discover errors will be so kind as to notify the author, in care of The M.I.T. Press, so that they can be corrected in future editions.

Cambridge, Massachusetts   C. FAYETTE TAYLOR
January 1984

DEDICATED TO THE MEMORY OF THE LATE
SIR HARRY R. RICARDO, L.L.D., F.R.S.,
WORLD-RENOWNED PIONEER IN
ENGINE RESEARCH AND ANALYSIS,
VALUED FRIEND AND ADVISER.

Contents

The Internal-Combustion Engine
in Theory and Practice

Introduction

Since it is some years since the first volume of this series was published, it may be well to cite some of the important developments in the fields covered by Volume I since its publication. Very briefly, these are as follows:

Thermodynamic Characteristics of the Fuel-Air Medium. By means of computer techniques, thermodynamic charts similar to those included in Volume I (Charts C-1 through C-4) have been constructed for a wider range of fuel compositions, fuel-air ratios, temperatures, and pressures than hitherto available (0.030–0.035).*

Fuel-Air Cycles. Based on computer programs of the appropriate thermodynamic properties of the charge, the characteristics of fuel-air cycles have also been computed over a much wider range of the important variables than has previously been feasible (0.040–0.045).

Reference 0.040, "The Limits of Engine Performance" by Edson and Taylor, gives the characteristics of constant-volume fuel-air cycles based on conditions at point 1 (beginning of compression). These data are more convenient and more versatile than those incorporating the idealized inlet and exhaust processes. The second edition of Volume I contains data from this reference in place of Fig. 4–5 (p. 82) of the first edition, which was based on cycles with the idealized 4-stroke inlet and exhaust process.

For the convenience of those who have only the first edition of Volume I, Figs. 0–1 through 0–6 herewith give the most important data from ref. 0.040 namely fuel-air cycle efficiencies and ratios of maximum to initial pressure, p_3/p_1. Important conclusions based on ref. 0.040 include the following:

Variations in humidity from 0 to 0.06, mass vapor to mass air, have no effect on fuel-air-cycle efficiency.

Variations in residual-gas content from $f = 0$ to $f = 0.10$ have a negligibly small effect on efficiency.

Efficiency is little affected by the initial pressure p_1, except where $F_R = 1.0$ (Fig. 0–3).

Increasing initial temperature T_1 reduces efficiency (Fig. 0–4) as well as p_3/p_1 (Fig. 0–6).

---

* Numbers in parentheses refer to items in the bibliography, pages 637–761.

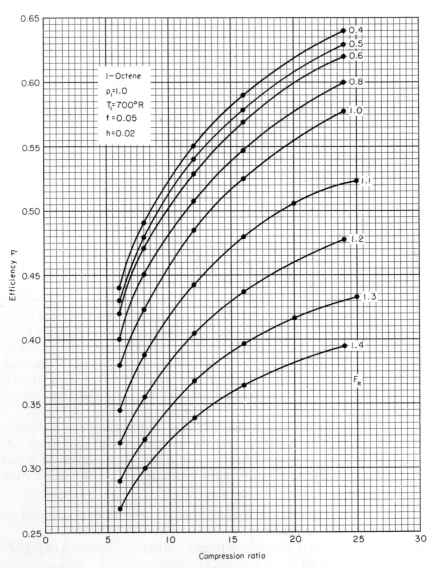

Fig. 0–1. Efficiency versus compression ratio for the constant-volume fuel-air cycle (Edson and Taylor, 0.040).

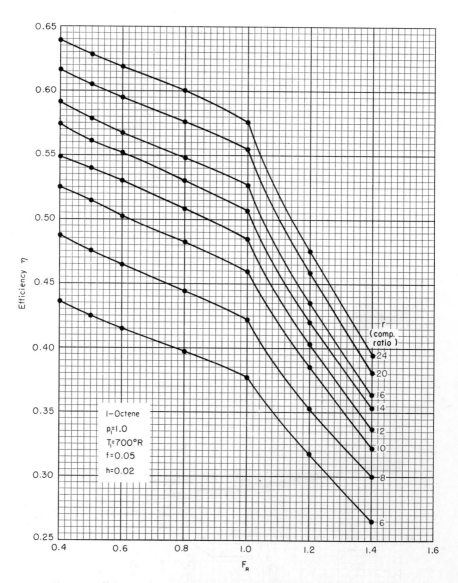

Fig. 0–2. Efficiency versus F_R for the constant-volume fuel-air cycle (0.040).

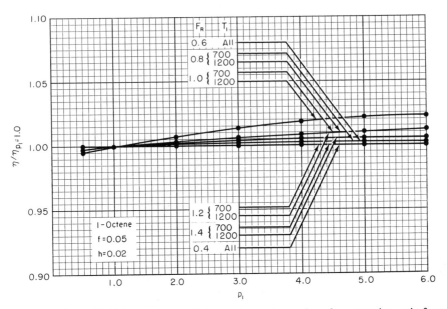

Fig. 0–3. Effect of initial pressure on efficiency at any given value of compression ratio from 6 to 24 (for use with Figs. 0–1 and 0–2) (0.040).

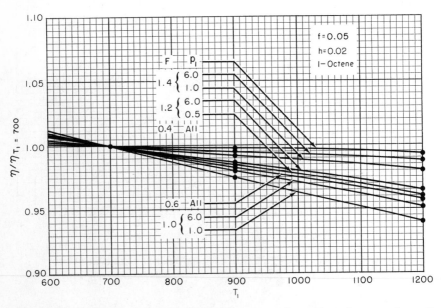

Fig. 0–4. Effect of initial temperature on efficiency at any given value of compression ratio from 6 to 24 (for use with Figs. 0–1 and 0–2 (0.040).

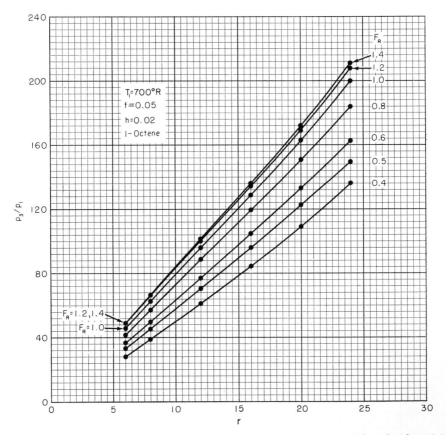

Fig. 0–5. Effect of compression ratio on maximum pressure at any given value of p_1 from 0.5 to 6.0 (for use with Figs. 0–1 and 0–2) (0.040).

In ref. 0.041, Edson shows that the efficiency of constant-volume fuel-air cycles continues to increase with increasing compression ratio up to compression ratio 300. (The efficiency of a constant-volume fuel-air cycle with isooctane at $F_R = 1.0$ and $r = 300$ is 0.80.) However, in practice it has been shown that as the compression ratio is increased, without detonation, the indicated output peaks at about $r = 17$ (Volume I, Fig. 12–15, p. 444, and ref. 12.49, p. 550).

Actual Engine Cycles. Results of cycle calculations which include arbitrary rates of combustion and rates of heat loss have been published (0.121, 0.122). A notable contribution to the measurement and analysis of actual engine cycles will be found in ref. 0.050.

Air Capacity. The appearance of several successful commercial engines with very small stroke-bore ratios confirms the validity of the stroke-bore

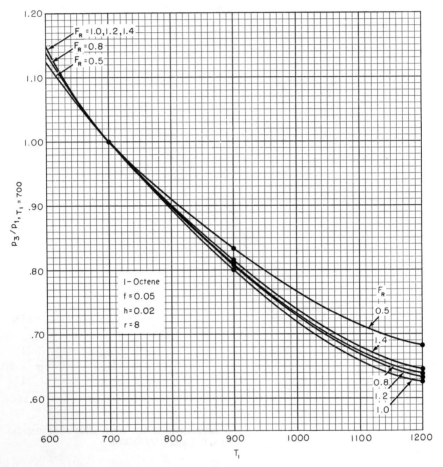

Fig. 0–6. Effect of initial temperature on maximum pressure at any given value of p_1 from 0.5 to 6.0 (for use with Figs. 0–1 and 0–2) (0.040).

ratio effects discussed in Volume I, Chapter 6, pp. 194–195. This question will be examined in detail in Chapter 10 of this volume.

Interest in the dynamic effects of inlet and exhaust pipes has also increased since the publication of the material on inlet-pipe effects in Volume I, pp. 196–200. Some new material on this subject will be found in refs. 12.05–12.092 of this volume.

Heat Losses. Heat-loss research has continued (0.080–0.087, 10.882–10.886). A finding not included in Volume I is the marked effect of valve overlap on the heat rejected to the coolant with supercharged engines (10.884, 10.886).

Miscellaneous. The bibliography also cites important contributions in the following fields, which have appeared since Volume I was published:

Actual-cycle analysis (0.050–0.051)
Friction, lubrication, and wear (0.090–0.097)
Size effects (0.110–0.114)
Engine performance, unsupercharged (0.120–0.122)
Engine performance, supercharged (10.860–10.872).

Material on supercharged engine performance, supplementing that given in Chapters 10 and 13 of Volume I, will be found in Chapters 10 and 11 of this volume and in the bibliographies for those chapters. Recent material on supercharger design, also supplementing Chapters 10 and 13 of Volume I, is here incorporated into Chapter 12 and its bibliography.

The More-Complete-Expansion Engine. Since Volume I was written, there has been some commercial development of engines designed to run on the more-complete-expansion cycle (10.855–10.857). This cycle involves the use of an expansion ratio greater than the effective compression ratio. The reason for this arrangement is that, in practice, maximum cylinder pressure is limited by considerations of stress in the case of Diesel engines and by detonation in the case of spark-ignition engines. In the 4-cycle more-complete-expansion engine the compression pressure, and hence the maximum cylinder pressure, is controlled by early closing of the inlet valve, which gives an effective compression ratio lower than the expansion ratio. In a poppet-valve 2-cycle engine [see Volume I, Fig. 7–1e, p. 212] the same result is obtained by delayed closing of the exhaust valves (10.72). The advantage of this cycle is the possibility of an efficiency higher than could be obtained with an expansion ratio equal to the compression ratio. The disadvantage is a mean effective pressure lower than the conventional arrangement with the same maximum pressure.

The more-complete-expansion cycle is practical only for engines that are not frequently operated at light loads, because in light-load operation the mean cylinder pressure during the latter part of the expansion stroke tends to be near to, or even lower than, the friction mean pressure. Under such circumstances the more-complete-expansion portion of the cycle may involve a net loss rather than a gain in efficiency. Example 10–6 (p. 402) is an illustration of the application of this cycle.

Computer Analysis. Perhaps the most important development in engine research techniques that has occurred since the publication of Volume I is the use of the digital computer to simulate various aspects of engine performance, including the performance of complete engines. As in many other fields, the computer has made possible the solution of complex relations that involve too much labor to be attempted by older methods of calculation.

Computer programs have been very successful in areas where the fundamental relations are known, such as for equilibrium thermodynamic properties of engine gases (0.30–0.035) and for fuel-air cycles (0.040–0.045). The results have improved both the range and accuracy of available data. Other useful applications of computers are for vibration analysis and valve-gear design (see Chapter 12).

Computer techniques have been successfully used for the prediction of 4-cycle volumetric efficiency as a function of valve capacity, valve timing, engine speed, etc. In this case the effects of heat transfer are small enough to be handled by approximate methods (0.060–0.062). Computations of 2-stroke-cycle air-capacity and trapping efficiency are less convincing because of unknown relations in the flow and mixing process during scavenging.

Attempts at predicting heat-transfer rates are also limited by lack of reliable instantaneous local coefficients (0.084). Uncertainties here also handicap computation of over-all engine performance.

A large number of performance computations for complete engines has been published, including comparisons with measured results (0.121, 0.122, 10.864, 10.865). It is obvious that the actual number of variables affecting engine performance is beyond the capacity of present computers and that the knowledge necessary to program many of the variables is inadequate. Ideally, the omitted items should be those that have very small effects. The quality of the program thus depends heavily on the skill of the operator in determining which items should be included and which left out. Of the factors known to be important, many cannot be theoretically programed because of lack of basic data. Thus, the engine programs so far developed have been based partly on theory and partly on assumptions for such unknown factors as instantaneous heat-transfer rates, combustion rates, turbulence, friction, etc. By adjusting the assumptions to agree with measured results, several programs have been made to agree fairly well with measurements from one particular engine size and type. Since most programs published to date ignore cylinder-size effects (as outlined in Volume I, Chapter 11) and the effects of many design details, the quantitative results cannot be taken as applying very far outside the type and size of engine to which these programs apply.

In spite of these limitations, computer technology is already a very valuable tool for the indication of *trends* in engine performance, even though absolute values are not necessarily accurate. Once a program is set up, many important variables can be investigated over very wide ranges, with expenditures of cost and time incomparably less than would be required for actual engine tests. As experience with these techniques accumulates, their accuracy will improve and, hopefully, they will be generalized on a dimensionless basis, so as to cover reasonable ranges of cylinder size and design detail. Computer techniques intelligently programed and interpreted promise to furnish a basis for rapid strides in the improvement of engine performance and engine design. References to the use of computers for various aspects of engine design and performance will be found in many sections of the bibliography.

1 | Combustion in Spark-Ignition Engines I: *Normal Combustion*

The conventional spark-ignition engine is supplied with a mixture of fuel and air which is quite homogeneous and essentially gaseous by the time ignition occurs. Therefore this chapter will be devoted principally to the subject of combustion in homogeneous, gaseous mixtures.

Deliberate use of nonhomogeneous mixtures in spark-ignition engines has been under development for many years but has never attained commercial importance. Since the principal purpose of such charge *stratification* is usually to control detonation, this question will be discussed in Chapter 2.*

BASIC THEORY AND EXPERIMENT

An enormous amount of theoretical and experimental research has been carried out on the subject of combustion of homogeneous, gaseous fuel-air mixtures. Portions of this work that are closely related to the internal-combustion engine are noted in the bibliography applying to this chapter. These investigations include experiments using engines as well as work with other apparatus, such as steady-flow systems and various forms of containers, or *bombs*. This research has shown that combustion in a gaseous fuel-air mixture ignited by a spark is characterized by the more or less rapid development of a flame that starts from the ignition point and spreads in a continuous manner outward from the ignition point. When this spread continues to the end of the chamber without abrupt change in its speed or shape, combustion is called *normal*. When the mixture appears to ignite and burn ahead of the flame, the phenomenon is called *autoignition*. When there is a sudden increase in the reaction rate, accompanied by measurable pressure waves, the phenomenon is called *detonation*. Autoignition and detonation are discussed in the next chapter. Here only normal combustion will be dealt with.

---

* Chapters 1 through 7 include material previously published in Taylor and Taylor, *The Internal Combustion Engine*, International Textbook Co., Scranton, Pa., 1961 and used by permission.

Chemistry of Combustion. Because combustion in fuel-air mixtures occurs with great rapidity and at very high temperatures, observation of the chemical processes involved is very difficult. In spite of continuing research in this field, theories of combustion and flame propagation remain highly speculative (1.10–1.16).

The chemical composition of the unburned gases and that of the products of combustion after cooling can be determined. However, experimental evidence indicates that the transition between these states involves numerous intermediate compounds.

A theory now generally accepted is that combustion of fuel-air mixtures depends on *chain reactions*, in which a few highly active constituents cause reactions which in turn generate additional active constituents in addition to end products, thus multiplying the number of reactions until combustion is complete (to equilibrium) or else until a point is reached where *chain-breaking* reactions overcome the chain-forming ones. In the flame front, the chain-forming reactions can only reach a certain distance into the relatively cool, unburned charge before they are broken, and thus a definite flame boundary is established. However, if the unburned gases become hot enough to sustain chain reactions, the remaining gas will suddenly autoignite.

The chain-reaction theory is discussed more fully in the next chapter, and quite fully in the literature (1.10–1.13). In this chapter subsequent discussion will be concerned with the observable physical aspects of combustion as they affect engine operation.

Flame Propagation. In normal combustion, the forward boundary of the reacting zone is called the *flame front*. This front, together with the burned products behind it, is usually sufficiently luminous for optical or photographic observation. When this is not the case, the flame can be made luminous by the addition of a very small portion of a sodium compound.

In stationary flames the gas moves through the flame, rather than the flame through the gas. This is the type of burning used in gas turbines. If the gas flow is steady and unidirectional, the speed of the flame relative to the gases is equal to the velocity of the gas normal to the flame boundary.

Since the combustion process in engines is complicated by piston motion, by a more or less irregular shape of the chamber, by the presence of residual gases from the previous cycle, and by very turbulent motion of the gases, it will be well to consider first the combustion of gaseous mixtures under less complicated conditions.

Flame Motion and Pressure Development at Constant Volume. Let us first consider the case of combustion of a homogeneous, quiescent fuel-air mixture in a constant-volume container, or *bomb*. Figure 1–1 shows flame position versus time in a fuel-air mixture contained in a spherical bomb having central ignition. The photograph was made by focusing the image of the flame on a film moving at constant velocity. The slope of the edge of the *trace* is thus proportional to flame velocity. It will be noted that the flame proceeds from the ignition point with a well-defined boundary and that the velocity is nearly constant except near the beginning and end of the process.

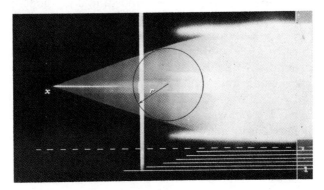

Fig. 1–1. Trace of flame observed in a spherical bomb with central ignition. The sphere of flame starts from the ignition point *x* and makes a "trace" on a film moving at known velocity. The radius *r* of the flame is related to the trace as shown (Fiock *et al.*, 1.27).

The motion of a flame in a mixture confined in a chamber of constant volume is complicated by the fact that expansion of the burned gases compresses the unburned part of the charge. For that reason the boundary of the unburned charge next to the flame front moves relative to the chamber, and the observed flame motion is the sum of two movements: the rate at which the flame moves into the unburned portion of the charge, called the *burning velocity*, and the rate at which the flame front is pushed forward by the expansion of the burned gases, called the *transport velocity*. Figure 1–2 shows burning, transport, and observed velocity as measured in the spherical bomb of Fig. 1–1.

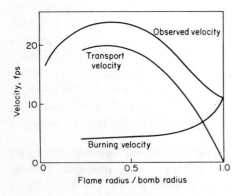

Fig. 1–2. Observed velocity, burning velocity, and transport velocity in spherical bomb with central ignition. $2CO_2$ starting at 1/3 atm, 77°F (1.27).

Figure 1–3 illustrates a container through which a spherical flame front is passing. The relation between flame position and pressure can be deter-

mined by means of the observed fact (1.61, 1.78) that the fraction of the
mass burned is proportional to the fraction of the total pressure rise; that is,

$$\frac{M_b}{M} = \frac{p - p_1}{p_2 - p_1},$$ (1-1)

where M_b = mass burned
$\quad M$ = total mass of the charge
$\quad p_1$ = initial pressure
$\quad p_2$ = pressure at the end of combustion
$\quad p$ = pressure at the instant under consideration.

If it is assumed that the unburned portion of the charge is compressed
adiabatically and is a perfect gas, we can write

$$M_b = M - M_u = M - \frac{p V_u m_u}{R T_u}.$$

Also,

$$T_u = T_1 \left(\frac{p}{p_1}\right)^{(k-1)/k},$$

where the subscript u refers to the unburned portion.

Combining the preceding relation with Eq. 1–1, we get

$$\frac{p - p_1}{p_2 - p_1} = 1 - \frac{p_1 V_u m_u}{M R T_1} \left(\frac{p}{p_1}\right)^{1/k}.$$ (1-2)

It is evident from Fig. 1–3 that, at any given value of the flame radius r,
the value of V_u depends on the shape of the container and the position
of the ignition point.

In order to show the influence of container shape and ignition-point
location, let us examine two extreme cases. The first is a sphere of diameter
D with central ignition, as used for the experiment of Fig. 1–1. In this
case $V_u = (\pi/6)(D^3 - d^3)$. This configuration gives the maximum possible
ratio of burned volume to flame radius; the minimum ratio would be for
a container consisting of a long tube with ignition at one end. Here
$V_u \cong A(L - r)$, where A is the cross-sectional area and L the length of the
tube.

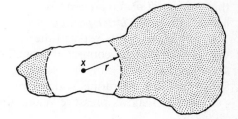

Fig. 1–3. Generalized constant-volume container; x is the ignition point and r the flame radius. Dotted area represents the unburned volume.

Figure 1–4 shows pressure versus flame radius for the above-mentioned sphere and tube, calculated from Eq. 1–2. Experimental measurements for the sphere are also given. These confirm the general validity of Eq. 1–2.

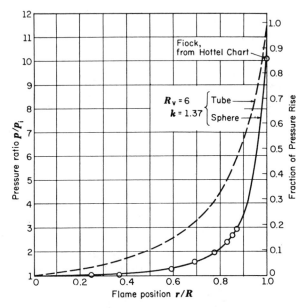

Fig. 1–4. Pressure versus flame radius for constant-volume containers; r = flame radius, R = bomb radius.

———— Sphere with central ignition (Eq. 1-2)
– – – Tube with end ignition (Eq. 1-2)
○ ○ ○ Experimental points for sphere
Fuel C_8H_{18}, $F_R = 1.03$ (Fiock *et al.*, 1.27)

In spark-ignition engines, combustion-chamber geometry generally lies somewhere between the two shapes of Fig. 1–4. An important characteristic of both these shapes is that the pressure rise is very small except near the end of flame travel. For the sphere, only about 2 per cent of the pressure rise has occurred when the flame has traveled halfway across the chamber. Even for the tube, the corresponding fraction is only 11 per cent. From these considerations we can reach the very important practical conclusion that *nearly all of the pressure rise in a constant-volume combustion process occurs during the latter portion of the flame travel across the chamber.*

If the flame radius versus time is known or can be measured, pressure versus time can be computed from Eq. 1–2. If constant observed flame speed versus time can be assumed, an appropriate time scale can be substituted for the radius scale of Fig. 1–4.

Figure 1–5 shows pressure versus time with several fuel-air mixtures as

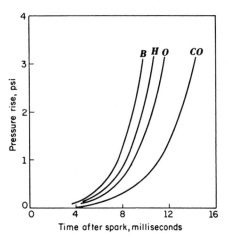

Fig. 1–5. Pressure versus time for flame travel in a spherical bomb with central ignition; initial pressure, 14.7 psia; initial temperature 77°F. Fuel-air mixtures are as follows: B is $C_6H_6 + 7.44O_2 + 12.8N_2$; H is $C_7H_{16} + 11.0O_2 + 19N_2$; O is C_8H_{18} (isooctane) $+ 12.5O_2 + 21.5N_2$; CO is $CO + 0.5O_2 + 0.04H_2O$ (Fiock *et al.*, 1.27).

observed in the bomb of Fig. 1–1. Evidently, changes in mixture composition change the flame speed and hence the pressure-time curve. It is important to note that, except in the case of CO, the major differences in rate are in the early part of the pressure-rise curves, the slopes above pressure rise 1.0 being nearly identical. In engines, the flame speed is greatly influenced by other factors, as we shall see in the following discussion.

FLAME PROPAGATION IN ENGINES

In the light of the foregoing discussion, let us refer to Fig. 1–6, which shows the progress of the flame front in a spark-ignition engine. These photographs show a definite flame boundary, as in the bomb experiments, although this boundary is now quite irregular in shape because of turbulent motion of the gases. The bright spot over the piston edge is probably caused by lubricating oil thrown off by the piston. It may be noted that the brilliancy of the gases near the ignition end of the combustion chamber increases rapidly as the flame approaches the end of its travel, apparently because of the increase in temperature caused by the rising pressure as the rest of the charge burns; this further confirms the existence of the theoretically predicted temperature gradient due to nonsimultaneous burning (see Volume I, pp. 109–112).

Measurements of Flame Velocity in Engines. While photographs such as those shown in Fig. 1–6 are very useful in a study of the nature of flame development, they are not particularly convenient for studying flame velocity. For this purpose the type of flame photograph originally developed by Withrow and Boyd (1.52) and shown in Fig. 1–7a is more useful.

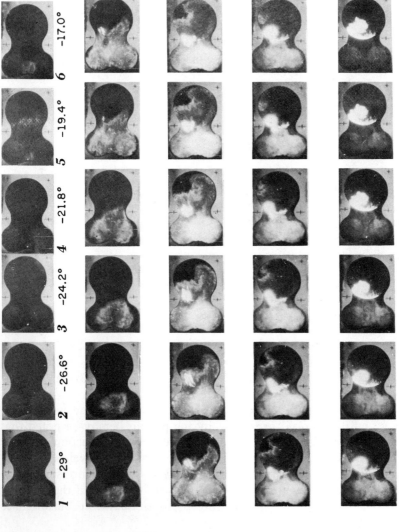

Fig. 1-6. High-speed motion pictures of flame development in a spark-ignition engine, taken through a quartz-glass cylinder head. Crank angle, after top center, is indicated under each exposure (Withrow and Rassweiler, 1.57).

This method employs a narrow glass window running across the cylinder head from a point just over the ignition point to the opposite end of the combustion chamber, as illustrated in Fig. 1–7b. By means of a special type of camera, the image of the flame is focused on a strip of motion-picture film which is moved at constant velocity at right angles to the window. The spark gap is visible through the window, and the passage of the spark makes a bright spot on the film. Exposure is continuous, no shutter being employed. Thus the horizontal axis of Fig. 1–7a represents *time* and the vertical axis, *position* of the flame front. The slope of the flame-front *trace* is a measure of flame velocity.

Because of the irregular shape of the flame front, the flame-front velocity observed through a narrow slot may not be representative of the average velocity. However, measurements of flame velocity from photographs of the type of Fig. 1–7a indicate trends in the average velocity of the front, provided there is no considerable *swirl** of the gases and no systematic change in flame-front shape with the variables being investigated. These conditions held for the tests of the kind referred to here.

Figure 1–7a shows that the shape of the flame position versus time curve in an engine is quite similar to that in the constant-volume bomb of Fig. 1–1. This similarity might have been expected, since the change in volume during the flame-travel period in an engine is not large. However, the mixture in an engine is very turbulent, and thus the absolute values of velocity are much higher than in a bomb with a quiescent mixture.

Effect of Operating Variables on Flame Travel. Since the losses due to combustion time are a function of piston motion during that time, it is convenient to express the results of combustion-time measurements in terms of crank angle rather than time.

Measurements of flame travel have been made from flame-trace photographs such as those shown in Fig. 1–7. Results of these measurements are given in Figs. 1–8 to 1–11 and in Table 1–1.

Since the mass burned in the first 10 per cent of flame travel is very small, variations in this period can be entirely compensated for by suitable adjustment of the spark timing. As we have seen in Volume I, Chapter 5, adjustment to best-power spark advance (bpsa) causes the pressure rise due to combustion to be evenly spread on either side of top center. Here the piston motion is so small that variations in the crank angle occupied by the 10–95 per cent pressure rise as large as 10 crank degrees have a negligible effect on output and efficiency. With reference to Table 1–1, the time losses will be measurably the same for all runs except the one at $F_R = 0.7$, which would show measurably increased time losses (see also Volume I, Fig. 5–20, p. 133).

---

* Swirl is defined as the rotation of the gases in the cylinder, more or less as a whole.

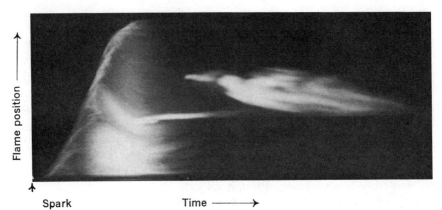

Flame position →

Spark   Time ——→

(a) Engine flame-trace photograph. The flame is photographed through a narrow slot in the cylinder head on a film moving at right angles to the slot.

(a)

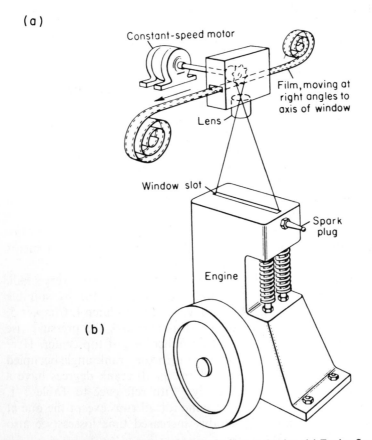

Constant-speed motor

Film, moving at right angles to axis of window

Lens

Window slot

Spark plug

Engine

(b)

Fig. 1–7. Flame-trace photograph and method of its production. (a) Engine flame-trace photograph. (b) The flame is photographed through a narrow slot in the cylinder head on a film moving at right angles to the slot, as indicated (Bouchard *et al.*, 1.58).

Table 1–1. Summary of Flame-Travel Measurements in CFR L-Head Engine

Figure	Variable	Value of Variable	Crank Degrees for Flame Travel		Degrees Change, 10–95% Travel
			0–10%	10–95%	
1–8	Piston speed	450 fpm	15	42	—
		1200 fpm	23	44	+2
1–9	Inlet and exhaust pressure	6 psia	21	48	—
		20 psia	15	39	−9
1–10	p_e/p_i	0.6	18	37	—
		2.0	22	48	+11
1–11	Relative fuel-air ratio, F_R	1.2	13	35	—
		0.7	33	61	+26
		1.5	15	45	+10

Data from Bouchard *et al.* (1.58).

Discussion of Flame-Travel Effects in Engines

The trends shown in Figs. 1–8 through 1–11 and in Table 1–1 may be explained by what is known about the effects of air motion, pressure, temperature, and charge composition on flame velocity in bombs and tubes.

Effect of Engine Speed on Flame Speed. The most significant effect shown is the fact that the variation in crank angle occupied by flame travel as engine speed changes by a factor of nearly 3 : 1 is very small (Fig. 1–8). This means that flame speed must increase nearly in proportion to engine speed. As a matter of experience, the highest-speed engines require only slightly greater spark advance than those running at normal or even quite low speed, and even this is mostly due to the increase in combustion time before 10 per cent flame travel.

The increase of flame speed with increasing engine speed is due to the marked effect of turbulence, as has been noted in many bomb experiments (1.20, 1.21) and in engines where the turbulence was varied independently of speed. This relation will be discussed further in the next section.

The nearly constant crank angle occupied by the major portion of combustion when speed is varied is one of the most important facts in relation to the question of combustion time. Without this characteristic, spark-ignition engines could not run at the very high piston speeds used in some present-day engines.

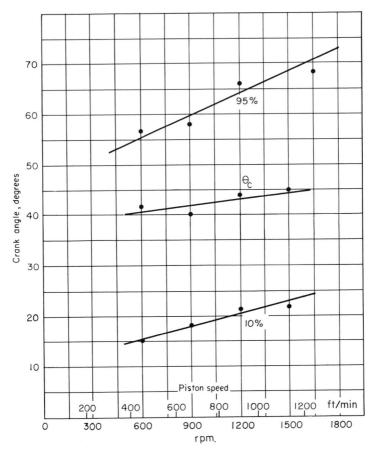

Fig. 1–8. Effect of engine speed on flame travel. Single-cylinder L-head engine, 4.375 × 4.5 in. $F_R \cong 1.2$; θ_c is degrees between 10 and 95 per cent travel (1.58).

Effect of Pressure on Flame Speed. Figure 1–9 shows flame angles versus inlet pressure, with the ratio of exhaust to inlet pressure held constant. Under these circumstances, the average pressure during combustion is nearly proportional to p_i, and there is no change in the residual-gas fraction.

In constant-volume bombs it is found that flame speed increases with increasing initial pressure (1.27, Fig. 9). This corresponds with Fig. 1–9, which shows decreasing crank angles with increasing inlet and exhaust pressure. This trend explains why increased spark advance is required for unsupercharged aircraft engines as altitude increases and why gas-turbine burners tend to blow out at very high altitudes.

Ratio of Exhaust to Inlet Pressure. Bomb experiments show that dilution with inert gas reduces flame speed (1.25). In engines, increasing ratio

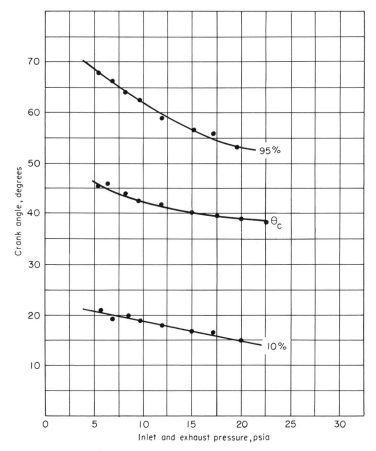

Fig. 1–9. Effect of pressure on flame travel. Engine same as in Fig. 1–8, 900 rpm, $sa = 31°$btc, $F_R = 1.16$, $T_i = 515°$R (1.58).

of exhaust to inlet pressure increases the fraction of residual gas in the charge and thus reduces flame speed, as shown in Fig. 1–10. It is also common experience that throttling an engine calls for advancing the spark timing, because of increasing residual-gas fraction as well as reduction of charge pressure.

Fuel-Air Ratio. Figure 1–11 shows that the fuel-air ratio for minimum time is about the same as that for maximum temperature of the fuel-air cycle (Volume I, Fig. 4–5, pp. 82–87). This similarity is probably due to the effect of flame temperature on the diffusion of active particles and on the transfer of energy from the flame front to the unburned gases.

Adjusting the fuel-air ratio to mixtures slightly leaner than the lowest values shown in Fig. 1–11 causes misfiring. At still leaner mixtures, flame

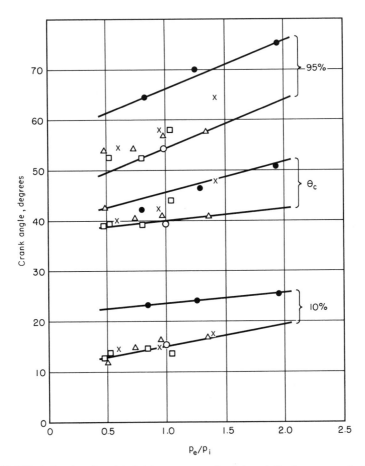

Fig. 1–10. Effect of ratio exhaust-to-inlet pressure on flame travel. Engine same as in Fig. 1–8.

Symbol	●	×	△	□	○
Inlet pressure, psia	4	5.5	7.5	9.2	10

speed is zero; that is, the flame will not propagate. The value of this *lean limit* varies with fuel composition, engine design, and operating conditions, but it is generally in the neighborhood of 60–80 per cent of the chemically correct mixture. Experience shows that the transition from rapid flame speed to zero or near zero speed is very abrupt.

Secondary Effects. The work from which the flame-travel results were taken shows that within the usual limits of engine operation the following operating variables have rather small effects on flame speed: air-inlet temperature, atmospheric humidity, and engine operating temperatures. Measurements of most of these effects are available in refs. 1.60, 1.61.

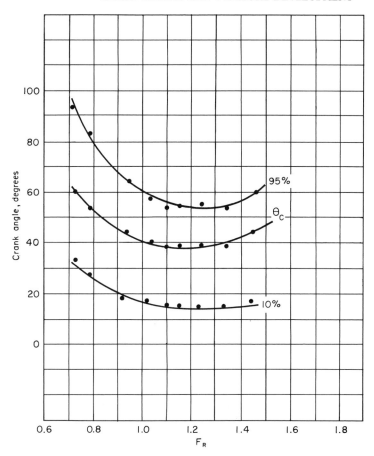

Fig. 1–11. Effect of fuel-air ratio on flame travel. Engine same as in Fig. 1–8, $sa = 30°$ btc, $p_i = 14.7$ psia, $T_i = 510°$R.

FLAME TRAVEL AND PRESSURE DEVELOPMENT IN ENGINE CYLINDERS

As far as engine performance is concerned, the important aspect of flame development is its effect on cylinder pressure. The relation between flame development and pressure is qualitatively similar to that in constant-volume bombs, except for the effects of turbulence and piston position. Figure 1–12 shows the relation of flame position to pressure in an L-head engine cylinder. The fact that in this case the pressure rise coincides with flame position (on a percentage basis) is due to piston motion and to the attenuated form of the L-head chamber. With compact combustion

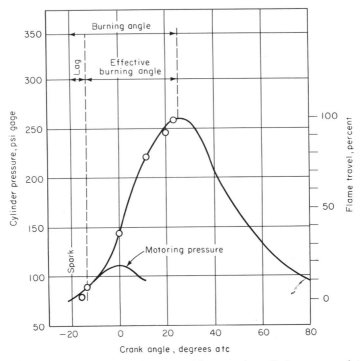

Fig. 1–12. Pressure development and flame travel in an engine cylinder; ○ are points on the flame-travel curve. Single-cylinder L-head engine, 3.75 × 4.0 in., $r = 5$, 1000 rpm (Marvin *et al.*, 1.56).

chambers the relation of pressure to flame position must be more like that shown in Fig. 1–4.

Definitions. For the purpose of analyzing pressure-time or pressure–crank-angle curves, the following definitions, illustrated in Fig. 1–12, will be used:

Burning angle (or burning time). The period between spark and peak pressure.

Lag angle (or lag time). The period between the ignition spark and the appearance of a measurable rise in pressure above the motoring-pressure curve.

Effective burning angle (or time). Burning angle minus lag angle.

Average flame speed. The distance from the spark to the most remote part of the combustion chamber, divided by the burning time.

Effect of Engine Speed. Figure 1–13 shows that the average flame speed remains nearly proportional to the piston speed for three different sizes and two shapes of combustion chamber. This relation is predictable from Fig. 1–8 and the accompanying discussion. The small negative slope of the curves shown in Fig. 1–8 is due almost entirely to lengthening of the lag period. If the spark is advanced as speed increases in such a way as to keep peak pressure at the optimum crank angle (15° to 20° ATC), the apparent time loss (see Volume I, pp. 112–113) will be nearly independent of speed.

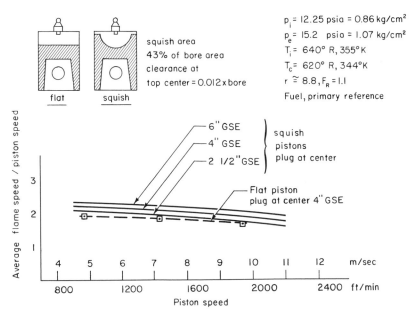

Fig. 1–13. Average flame speed/piston speed for MIT similar engines with flat and squish pistons (Swenson, Jr., and Wilcox, 2.678). Average flame speed = (bore/2)/time, spark to peak pressure.

Effect of Cylinder Size. The ratio of flame speed to piston speed is nearly the same for similar cylinders of different sizes, as shown in Fig. 1–13. The implications of this relation include:

1. At a given piston speed, burning time will be nearly inversely proportional to bore, and, since rpm is inversely proportional to bore, the burning angle will be nearly independent of the bore.

2. At a given rpm, average flame speed will be nearly proportional to bore, and effective burning angle will again be independent of the bore.

Effect of Reynolds Index. Since flame speed increases with both piston speed and bore, and since turbulence is greatly affected by Reynolds number, flame-speed data from the three similar engines at MIT (see Volume I, Chapter 11) have been plotted against inlet Reynolds number in Fig. 1–14. While flame speed always increases with Reynolds number, it is evident that factors other than this number also have a large influence on flame speed.

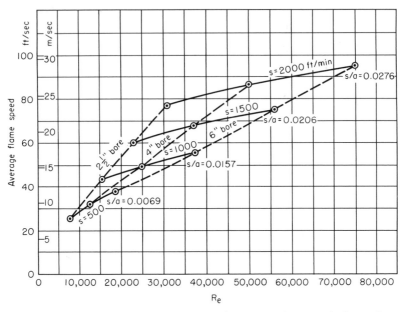

Fig. 1–14. Average flame speed versus inlet Reynolds number for MIT similar engines with flat combustion chambers; $r = 5.53$, $R_e = \rho_i s b / \mu_i g_0$, $F_R = 1.1$, $T_i = 610°R$, $p_i = 13.75$ psia, best-power spark advance, a = inlet sonic velocity (Gaboury et al., 2.672).

Effect of Combustion-Chamber Shape. Figure 1–13 has shown that an open type of combustion chamber tends to have a somewhat lower average flame speed than a chamber with *squish*. This relation is also confirmed by Fig. 2–30. In general, the more compact the chamber, the greater the rate of pressure rise.

Effect of Spark-Plug Position. Flame-position diagrams and pressure–crank-angle curves for different positions of one and two spark plugs are shown in Fig. 1–15. Large differences in burning time are evident, as would be expected from the different distances which the flame must travel. Air motion must also be involved, as indicated by the difference between Runs 2 and 3 and between Runs 4 and 5, which have similar but reversed geometry, with the same flame-travel distance. Dual spark plugs at the sides (Runs 6 and 7) show the same pressure–crank-angle diagrams as does one central plug (Run 1). These tests were made with fixed spark timing. Figure 1–16 shows effects of multiple spark plugs on combustion time with optimum spark timing — for one plug near the edge of the combustion chamber and for seventeen plugs evenly distributed. The ratio of combustion times with one and with seventeen spark plugs is about 2 : 1. The change in flame speed due to deposits in the cylinder is in the opposite direction for the two runs and is not readily explained. It could be due to experimental error. In the absence of detonation, a change from one to two

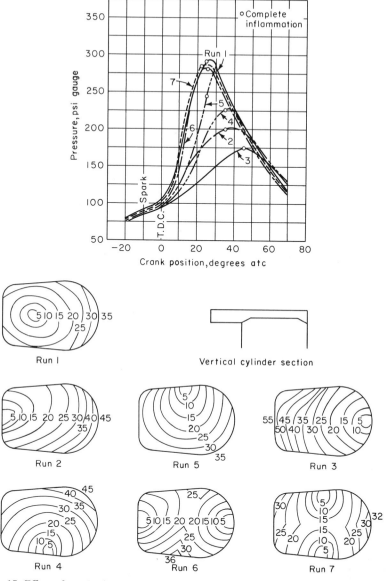

Fig. 1–15. Effect of spark-plug position on flame travel and pressure. Spark timing constant. Numbers in lower diagrams indicate degrees of crank angle after ignition (Marvin *et al.*, 1.56).

spark plugs has only a minor effect on engine performance, provided best-power spark timing is used in each case. The next chapter will deal with detonation effects.

Effect of Stroke-Bore Ratio. According to Fig. 1–17, flame angles tend to be larger as the stroke is reduced, at a given piston speed, with a given

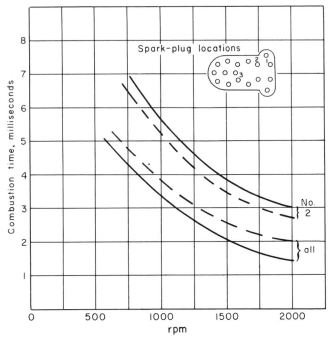

Fig. 1–16. Effect of multiple spark plugs on combustion time, best-power spark advance (Diggs, 2.67).
———— clean combustion chamber   – – – – with 30 hours of deposit formation

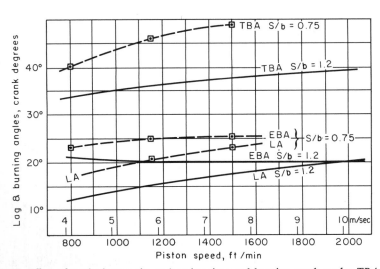

Fig. 1–17. Effect of stroke-bore ratio on burning time and burning crank angle; *TBA*=total burning angle, *LA* = lag angle, *EBA* = effective burning angle, *T* = time (Swenson, Jr., and Wilcox, 2.678). Fuel, iso-octane. Other conditions same as Fig. 1–13.
———— 4-in. bore, 4.8-in. stroke, s/b = 1.2   – – – – 4-in. bore, 3.0-in. stroke, s/b = 0.75

bore. The increase in angle, however, is not as great as the increase in rpm so that, at a given piston speed, flame speeds are actually faster with the shorter stroke.

Cyclic Variation in Pressure Development

The foregoing discussion has been based on average data obtained from a large number of cycles. In practice, one of the prominent characteristics of the spark-ignition combustion process is a wide variation, from cycle to cycle, of the pressure–crank-angle diagram, as illustrated by Fig. 1–18. With spark timing adjusted for peak pressure at the same angle, it is found that the variation in the pressure–crank-angle curve is almost entirely in

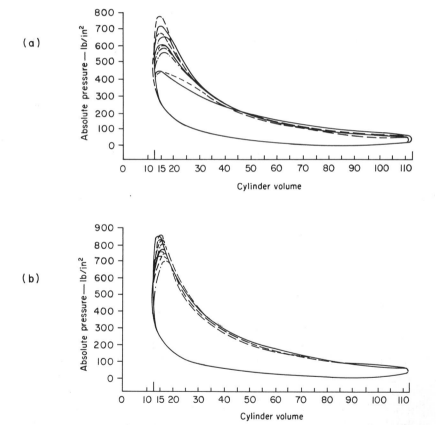

Fig. 1–18. Typical cyclic variation in a spark-ignition engine. (*a*) With a lean mixture, rpm= 2000, *r*= 9.0, $F_R=0.82$. Throttle fully opened. Imep deviation ±4.5 per cent. Peak-pressure deviation ±28 per cent. (*b*) With a rich mixture, rpm=2000, *r*=9.0, $F_R=1.25$. Throttle fully opened. Imep deviation ±3.6 per cent. Peak-pressure deviation ±10 per cent (Soltan, 1.832).

the early stages of combustion, that is, during the first 10 per cent of flame travel.

Research in this field (1.80–1.834), supplemented by general experience in engine operation, indicates that although most commercial spark-ignition systems show some variation in the crank angle at which the spark occurs, this factor is not the primary cause of the observed variation. One obvious cause can be incomplete mixing of fuel, air, and residual gas, with consequent variation in conditions at the ignition point. An extreme example of this is the case of engines in which gaseous fuel is injected directly into the cylinder.

Since a large amount of random turbulent motion exists in all cylinders, it is evident that in the presence of incomplete mixing the spark may occur in mixtures of varying fuel-air ratios and varying quantities of residual gases, and hence with various rates of flame development, possibly including regions where flame will not propagate at all, with consequent misfiring. However, in normal operation of carburated spark-ignition engines, except at low speeds, the mixing process seems to be sufficiently good so that improvements in mixing, such as the use of a premixed gaseous charge, do not appreciably reduce cyclic variation. There is agreement among all students of this subject that even with very complete mixing a large degree of cycle-to-cycle variation remains.

Patterson (1.833) found that with a well-mixed charge the following variables have no measurable influence on cyclic variation: spark energy, spark voltage-rise time, spark-plug gap size and electrode configuration, fraction of residual gas. On the other hand, all investigators have found significant effects due to the fuel-air ratio and to the character and degree of turbulence in the cylinder before ignition.

The degree of variation with fuel-air ratio is minimum at $F_R = 1.0$ and increases as the ratio becomes either greater or smaller than this value (1.833). In view of all the other factors, such as flame temperature, flame speed, etc., that are maximum near $F_R = 1.0$, this relation is not surprising, although no simple theoretical explanation is evident.

To explain the observed variations caused by changes in air motion, the following hypothesis is offered:

Consider a turbulent mixture made up of vortices, as illustrated in Fig. 1–19. If the spark occurs at the center of a vortex a, the flame must spread without the aid of turbulence until it reaches the vortex boundary. On the other hand, ignition at the vortex boundary b will immediately aid the spread of flame because of the shearing action there encountered. Now suppose that the whole charge moves past the ignition point, as in a swirling motion of the cylinder contents. If the spark has a finite duration, as is usual, it will follow the dotted path a–c with relation to the vortices and will soon encounter vortex boundaries.

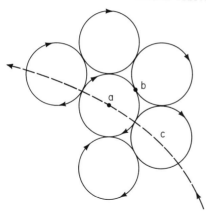

Fig. 1–19. Idealized vortex system.

This theory seems to explain the following observed facts:

1. Multiple ignition points reduce cyclic variation.

2. Cyclic variation is also reduced by increased engine speed or by decreased inlet-valve area. These variables lead to smaller and more intense vortices and to more rapid gross motions of the gas.

3. Cyclic variation is markedly reduced by a tangentially oriented swirl, such as that created by a *shrouded* inlet valve, Fig. 1–20. The same valve arranged so that the gases enter radially instead of tangentially gives much greater cyclic variation. The explanation is probably to be found in the resultant *a-c* type of relative motion indicated in Fig. 1–19.

An experiment that would be useful in evaluating the foregoing hypothesis would be to see if a spark of very short duration leads to more cyclic variation than one that continues for an extended time.

A result published in ref. 1.833 supports the theory here suggested, in its conclusion that " mixture-velocity variations that exist within the cylinder near the spark plug are the major cause of cycle-to-cycle variability."

As a final word on the subject of cycle-to-cycle variation, it should be pointed out that its elimination would be an important contribution to improved engine performance. If all cycles were alike and equal to the average cycle, maximum cylinder pressures would be lower, efficiency would be greater, and, most important of all, the detonation limit, discussed in the next chapter, would be higher, thus allowing appreciable increases in efficiency and/or mep with a given fuel.

Control of Pressure-Rise Rate

It is evident that the least time loss and therefore the highest efficiency will occur with the shortest time of combustion. On the other hand, the shorter the combustion time, the higher must be the time rate of pressure rise in the cylinder.

In some spark-ignition engines a phenomenon known as *roughness* is observed, with otherwise normal combustion. Investigation has shown that roughness is caused by severe vibration of certain engine parts, usually including the crankshaft in bending, and that it may be reduced or eliminated by minimizing the rate of pressure rise during combustion. The control of roughness by mechanical means, that is, by stiffening the engine structure, will be discussed in Chapter 8 and refs. 1.90–1.94, 8.929–8.940.

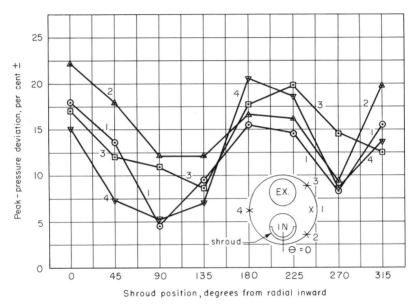

Fig. 1–20. Effect of shrouded inlet valve on peak-pressure deviation. Numbers indicate spark-plug positions (Averette, 1.80).

When the engine structure cannot be stiffened, the only way to cure roughness is to reduce the rate of pressure rise. As we have seen from Figs. 1–3 and 1–4, with a given flame speed, this rate can be controlled, within limits, by the shape of the container.

For practical reasons, engine combustion chambers must lie somewhere between the two shapes of Fig. 1–4. Thus, the more compact the chamber and the more nearly central the ignition, the closer will be the result to that represented by the sphere, and the more rapid the maximum rate of pressure rise. By attenuating the chamber in the direction of the tube with end ignition, a slower maximum rate of pressure rise will be obtained. Another way of saying the same thing is that, for a given volume, the longer the flame path, the lower the maximum rate of pressure rise. However, efficiency will suffer as the chamber is modified in this direction, and, as indicated in the next chapter, the tendency toward detonation may be seriously increased. The modern tendency is to make combustion chambers compact and flame paths short in order to control detonation, and to avoid roughness by a stiff engine structure rather than by attenuating the combustion chamber. With normal combustion in spark-ignition engines, the rate of pressure rise is limited by flame speed to values that make this method of attack generally feasible.

EXHAUST EMISSIONS

Soon after World War II it began to be recognized that the exhaust gases from road-vehicle engines were a major cause of air pollution and city "smog." The causes and the control of this problem have been the objects of an enormous amount of research and development, which is still continuing. Four substances have been recognized as sufficiently objectionable to require legal control. These are unburned hydrocarbons (HC), carbon monoxide (CO), oxides of nitrogen (NO_x), and carbonaceous solids or semisolids ("smoke" or "particulates"). The fraction of these substances found in engine exhaust gases varies with engine type and design and with almost every aspect of engine operating conditions.

Unburned hydrocarbons are caused by cooling of the fuel-air mixture trapped in such pockets as the piston-cylinder space above the top ring—spaces where the cylinder-head gasket is not coterminal with the head wall, poorly fitting spark-plug threads, etc. Cooling of the gases along open cylinder walls, once thought to be important, now appears to be a minor contributor.

Carbon monoxide, a normal component of combustion at fuel-air ratios above stoichiometric (Volume I, p. 223), is formed during the combustion process, and even with lean mixtures some of it fails to burn to CO_2 as would be the case if complete chemical equilibrium was reached before exhaust-valve opening.

Oxides of nitrogen form at the high temperatures reached during combustion, and their concentration increases as the peak combustion temperature goes up. Particulate emissions are not a serious problem with premixed charge engines, but they are a characteristic result of the combustion of liquid-fuel sprays, including those of Diesel engines, stratified-charge engines with spray ignition, and steady-flow burners. Particulates are the result of incomplete combustion of fuel droplets.

These problems are discussed further in Chapters 5, 6 and 7.

2 | Combustion in Spark-Ignition Engines II: *Detonation and Preignition*

DETONATION

By *detonation** is meant the phenomenon which is recognized as "ping," "spark knock," or "carbon knock" and is familiar to all drivers of automobiles. More research has been, and is being, devoted to a study of this phenomenon than to any other aspect of the internal-combustion engine. This chapter will be limited to a discussion of the most relevant aspects of this work to date and to the conclusions that can be drawn therefrom. The bibliography is confined to what seem to be the more important items.

One method of identifying detonation is by its characteristic sound, which is usually audible in the case of engines that are otherwise reasonably quiet, such as automobile engines. The sound is the result of intense pressure waves in the charge, which force the cylinder walls to vibrate and thus communicate the sound to the atmosphere. The pitch of the sound is determined by the natural frequencies of sound waves within the gases. These in turn are governed by the size and shape of the combustion chamber and by the velocity of sound in the charge (2.04, 2.05). A rapid decrease in frequency, caused by the decrease in charge temperature and change in chamber shape accompanying expansion, is one of the identifying characteristics of detonation. The audible frequencies are quite high, ordinarily 5000 cps or more in cylinders of automobile size. The pressure waves in the charge, or the resulting vibration of the cylinder walls, provide the most reliable method so far devised for indicating the presence of detonation.

Various types of instruments are available for detecting pressure waves in the cylinder gases (2.061–2.091). In most cases an electrical signal is produced by the gas pressure on the end of a plug screwed into the cylinder. The end of the plug carries a diaphragm so arranged that its motion generates an electric signal by pressure on a strain-gauge element, by pressure on a piezoelectric crystal, or by changing the geometry of a mag-

---

* The word *knock* is often used interchangeably with the word *detonation*. The author avoids the word "knock" in this sense because the same word is often used to designate a mechanical sound due to loose bearings or pistons. However, the adjective *knock-limited* is used later in this chapter because it is more convenient and more widely used than the equivalent term *detonation-limited*.

34

netic circuit or the gap of a condenser. To ensure a readable signal, the natural frequency of the detector unit must be high in relation to the frequency of the pressure waves to be detected. Figure 2–1 shows the principles of such instruments.

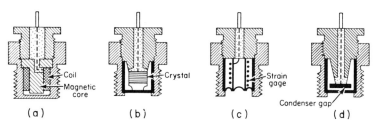

Fig. 2–1. Types of cylinder-pressure pickup units; dashed lines indicate insulated conductors to binding post; dotted areas indicate electrical insulation. (*a*) Electrodynamic: motion of diaphragm generates emf proportional to *dp/dt* (2.061). (*b*) Piezoelectric: pressure on crystal generates emf proportional to *p* (2.09). (*c*) Strain gauge: pressure on diaphragm is transmitted to thin cylinder wound with strain-gauge wire (2.081). (*d*) Capacitance type: motion of diaphragm changes condenser capacity (2.08).

The signal from such *pickup* units is amplified and displayed on an oscilloscope against a time or crank-angle sweep scale. It may appear in terms of pressure against time, as in Figs. 2–2*a* and *b* or a circuit may be used such that voltage is proportional to the rate of pressure change, *dp/dt* or *dp/dθ*. Records of this kind are shown in Figs. 2–2*c* and *d*.

The two sets of records in Fig. 2–2 were taken under identical conditions, except that a nondetonating fuel was used for *a* and *c*. In each set the graphs remain quite similar from the time of ignition to a point near the

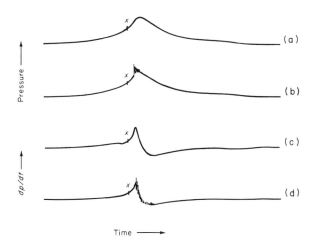

Fig. 2–2. Pressure-versus-time diagrams taken with electric indicator. (*a*) *p* vs. *t* without detonation; (*b*) *p* vs. *t* with detonation; (*c*) *dp/dt* vs. *t* without detonation; (*d*) *dp/dt* vs. *t* with detonation; *x* indicates ignition time. (Sloan Automotive Laboratories).

maximum pressure of the cycle. At this point the normal cycle shows a smooth change in pressure, while the detonating cycle shows severe pressure fluctuations, indicating a vibratory motion of the gases.

Figure 2–3 shows three flame traces taken through a slot window in the cylinder head (see Fig. 1–7). The upper flame trace is typical of normal combustion, while the lower two are typical of detonating combustion. Again, the diagrams are very similar from ignition to a point near the end of flame travel. For the detonating diagram the flame slows down for a brief period near the end of the flame travel; this period is followed by a nearly vertical trace, indicating a very high rate of reaction near the end of combustion. In the lower photograph, vibrations of the gas can be inferred from the motions of incandescent particles near the center of the combustion chamber.

The character of the flame traces of Fig. 2–3 is explained by Fig. 2–4, which shows six series of flame photographs taken through a transparent cylinder head while the engine was detonating. In each case the development of a nucleus of flame ahead of the flame front is evident at a moment when the flame has traversed about two thirds to three fourths of the chamber. It will be noted that after the detonation flame appears, the next photograph shows complete or nearly complete inflammation, indicating the rapidity with which this portion of the charge burns. In series *a*, *b*, and *c* the flame nucleus of detonation appears to be well separated from the flame front, while in *e* and *f* it appears as an excrescence from the front itself. Such variations seem to be characteristic of the process, as is also a wide variation in the sound intensity from cycle to cycle. Often two or more nuclei appear ahead of the flame front, and sometimes the reaction is so rapid that the development of flame nuclei occurs between exposures and the chamber appears completely filled with flame in the next picture.*

Local Pressures with Detonation. There is much evidence to indicate that the peaks of the pressure waves associated with detonation can be very high. Figure 2–5 shows indicator diagrams taken with an averaging or point-by-point indicator, *a* being taken without detonation and *b* with detonation brought about by a change in fuel. Again, the diagrams are nearly identical up a point near the end of flame travel. Since there was no change in power, the average pressure on the piston during expansion must

---

* Some work with photographs taken at ultrahigh speeds [40,000 to 200,000 frames per second (2.26–2.27)] has been interpreted as indicating that the reactions causing detonation may sometimes occur *behind* the flame front, presumably in a part of the charge that has been only partially consumed by normal combustion. Since in these particular tests mixing of the fuel and air was incomplete (because of the use of fuel injection rather than of a premixed charge) the reaction that appeared behind the observed flame front probably took place in portions of the mixture that had been surrounded by the flame but not yet burned. Alternatively, the flame front observed may have been the first stage of the "step reaction" shown in Fig. 2–11, p. 47).

have been the same in both cases. The higher pressure that is recorded in Diagram *b* must have been local in character and due to pressure waves. In this figure the maximum wave pressure recorded with detonation was about 620 psia, or 55 per cent higher than the maximum pressure without

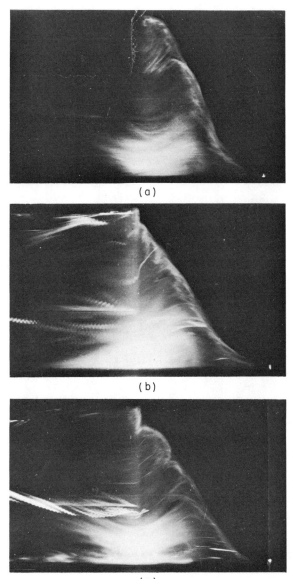

(a)

(b)

(c)

Fig. 2–3. Flame-trace photographs taken by means of the apparatus of Fig. 1–7*b*. (*a*) No detonation (normal combustion); (*b*) detonation; (*c*) detonation. Bright spot is the ignition spark. (Sloan Automotive Laboratories.)

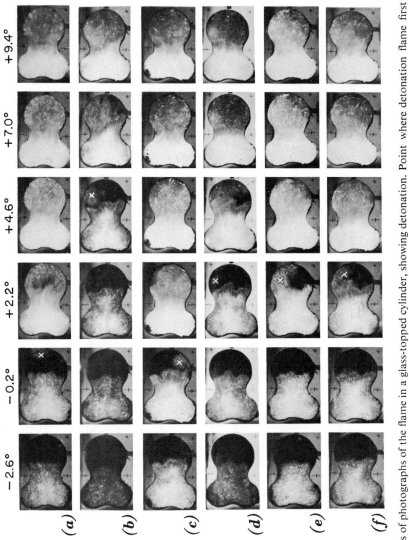

Fig. 2–4. Six series of photographs of the flame in a glass-topped cylinder, showing detonation. Point where detonation flame first appears is indicated by x in each series. Angles are degrees after top center (Withrow and Rassweiler, 1.57).

detonation. A calculation of maximum possible wave pressure, given later in this chapter, shows that actual peak wave pressures can be much higher than this value. The type of indicator used for Fig. 2–5 cannot be expected to record the peak wave pressure because of the extremely short duration and very local character of the wave.

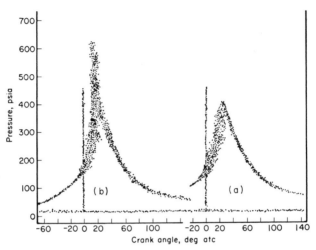

Fig. 2–5. Pressure-crank-angle diagrams taken with MIT point-by-point indicator; CFR engine, 1200 rpm, (*a*) Without detonation. (*b*) With detonation induced by ethyl nitrite. (Sloan Automotive Laboratories.)

Objection to Detonation. In many cases, such as that of passenger-automobile engines, the noise of detonation is objectionable. Detonation may lead to overheating of spark-plug points, with resultant *preignition*, that is, ignition before the spark occurs. Severe preignition causes loss of power and economy, rough and unsatisfactory operation, and often damages the engine (2.100–2.103). Even without preignition, severe detonation sustained over long periods of time often damages aluminum pistons and cylinder heads. Figure 2–6 shows a typical example of this type of damage. Exhaust valves and piston rings also seem to suffer at times from severe and long-sustained detonation. Damage due to detonation may eventually lead to complete failure of the affected parts.

The exact mechanism that causes damage such as that shown in Fig. 2–6 is not known. It seems unlikely that high pressure alone is responsible, since in engines where the susceptible parts run relatively cool, as in unsupercharged liquid-cooled engines, no damage may be evident even after months of operation with severe detonation. A more likely explanation of the damage mechanism is that the pressure waves increase the rate of heat transfer to, and hence the temperature of, the susceptible parts,

Fig. 2–6. Damage to aluminum pistons resulting from extended operation with heavy detonation.

thereby causing either local melting of the material or softening to such a point that the high local pressure causes erosion.

Because of objectionable noise, likelihood of preignition, or the possibility of serious damage, *detonation is an important factor limiting the output and efficiency of carburated spark-ignition engines.* Without detonation, higher compression ratios could be used, giving higher efficiency and output, or else higher inlet pressures could be used in supercharged engines, giving increased output. These facts account for the continued efforts to discover and to produce fuels that have reduced detonation tendencies (Chapter 4), and to develop cylinder and induction-system designs that reduce the tendency to detonate.

Detonation Theory

The Autoignition Theory.* It is now generally accepted that detonation is due to autoignition of the *end gas*, which is that part of the charge which

---

*This theory was proposed by Ricardo about 1922 (2.02) and was based on some work with an early rapid-compression machine (2.01) and on observations with regard to the relation of the autoignition temperature of fuels and their tendency to detonate. At that time it was probably not realized that the end gas could autoignite without giving rise to detonation.

has not yet been consumed in the normal flame-front reaction. When detonation occurs, it is because compression of the end gas by expansion of the burned part of the charge raises its temperature and pressure to the point where the end gas *autoignites*. If the reaction of autoignition is sufficiently rapid and a sufficient amount of end gas is involved, detonation can be observed.

End-Gas Reaction and Pressure Waves. The creation of pressure waves by a rapid reaction in a part of the gases within a closed space is explained by the fact that the reaction, if it takes place with sufficient rapidity, will take place at nearly constant volume. (Because of the inertia of the gas, an instantaneous reaction would evidently take place at exactly constant volume.) Such a reaction will result in a high local pressure, and the portion of the gas in question will subsequently expand rapidly, sending a pressure wave across the chamber. This pressure wave will be reflected from the walls, and a wave pattern of a type predictable by acoustic theory (2.05) will be established quickly.

Calculation of Limiting Local Pressure. In Chapter 5 of Volume I (pp. 109–112) a fuel-air cycle with progressive burning of each element of the charge at *constant pressure* is described and illustrated. Figure 2–7 shows a fuel-air cycle where that element of the charge which burns last, burns at *constant volume*. For this theoretical cycle it is assumed that the end gas is a very small fraction of the charge. This end gas is first compressed by the compression stroke and then by the normal combustion of the main body of the charge until point *2″* is reached, that is, a pressure equal to the maximum pressure of the normal cycle with the temperature of adiabatic compression from point *1*. At point *2″*, let it be assumed, combustion of this small end-gas element occurs instantaneously. Instantaneous combustion also means combustion at *constant volume*. By using the assumptions of the fuel-air cycle, the appropriate thermodynamic chart (C–1 and C–4 of Volume I) gives for conditions at the beginning and end of this combustion process:

$$p_2'' \quad 810 \text{ psia} \qquad T_2'' \quad 1520°\text{R}$$
$$p_3''' \quad 2840 \text{ psia} \qquad T_3''' \quad 5330°\text{R}$$

These values represent high limits for the cycle in question. In an actual detonating cycle, not only is the peak pressure of normal burning lower but also combustion in the end gas can never be quite instantaneous. Both of these effects tend to give a peak pressure lower than that calculated for Fig. 2–7.

The indicator diagrams of Fig. 2–5 were taken under conditions similar to those used in the previous calculation.

In part *b* of that figure, with detonation, the peak pressure is much higher than in part *a*, without detonation. Thus the general relation predicted by the calculation is confirmed. Also, it is evident that the expansion

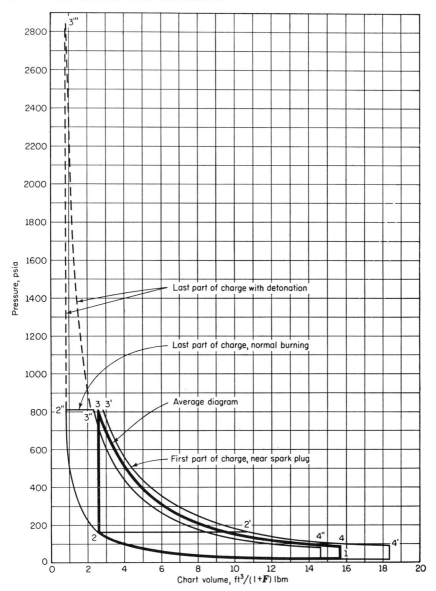

Fig. 2–7. Fuel-air-cycle diagrams showing maximum end-gas pressure (dashed line). See also Fig. 1–1. $F_R = 1.2$; $r = 6$; $T_i = 600°R$; $p_i = 1$ atm; $f = 0$; $h = 0$.

line of the detonation record is lower than that of the record taken with a nondetonating fuel. This agrees with the theoretical detonation cycle shown in Fig. 2–7.

To summarize the discussion, it is evident that high local pressures are to be expected as a result of rapid end-gas reaction and that such pressures will create intense pressure waves throughout the charge. According to acoustic theory, and in actual practice, the frequency of the waves decreases as the piston moves outward, owing to the relationship between speed of sound and gas temperature and owing to the change in chamber shape and size.

In view of the fact that the process of detonation appears closely associated with that of autoignition, it is worthwhile to consider the known facts about autoignition in some detail.

Autoignition

For purposes of this discussion, autoignition in a fuel-air mixture will be defined as a rapid chemical reaction not caused by an external ignition source such as a spark, a flame, or a hot surface. By this definition, ignition in a compression-ignition, or Diesel, engine is autoignition. Autoignition has been observed in various types of vessels, or *bombs*, and in steady-flow systems as well as in engines.

The term *rapid reaction* will be used to indicate a chemical reaction that is completed in a fraction of a second. This term is intended to exclude the extremely slow spontaneous chemical reaction that occurs between fuel and air at normal atmospheric pressure and temperature. Such reaction is normally so slow as to be hardly measurable except over periods of days or weeks. Some authorities use the term " explosion " to refer to the reactions here classified as rapid.

Chemistry of Autoignition. As in the case of combustion with a normal flame front, the chemistry of autoignition is not clearly understood. The most widely accepted theory is that it is a *chain reaction*. As mentioned in Chapter 1, a reaction is called a chain reaction when chemical combinations involving active particles generate additional active particles as well as end products of the reaction. A hypothetical example of this type of reaction, proposed by Lewis and Von Elbe (1.10), is illustrated in Fig. 2–8. Here the active particles are OH, H, and O, and it is seen that they multiply rapidly from the assumed reactions. If each of the reactions requires a finite time, the rate of formation of H_2O will start slowly at first and increase to an extremely rapid one as the reaction proceeds. This principle is the same as that of a nuclear explosion.

The chain-reaction theory postulates that certain conditions of pressure, temperature, container material, etc. may promote chain reactions and that other conditions may inhibit them, or act as *chain breakers*. For example, referring to Fig. 2–8, conditions that would favor the reaction $O + O \rightarrow O_2$, $H + H \rightarrow H_2$, or $OH + H \rightarrow H_2O$ would tend to break the chain and slow down or stop the reaction.

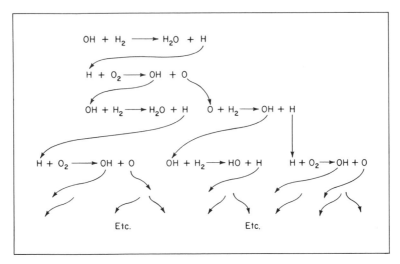

Fig. 2–8. Example of a hypothetical chain reaction involving oxygen and hydrogen.

While the details of chain reaction are not accurately known, this general theory appears to offer a plausible explanation of many of the physical phenomena observed in connection with autoignition.

Rapid-Compression Machine. In order to study the phenomenon of autoignition with compression rates similar to those occurring in the end gas of an engine, the *rapid-compression machine* shown in Fig. 2–9 was designed and developed at MIT under the author's direction. Other machines of this kind had already been used, one of the earliest being that of Tizard and Pye (2.01). However, in most of these the time of compression has been much slower.

In operation, the combustion cylinder of the machine is filled with a carefully prepared homogeneous mixture of air and fuel vapor at a given initial pressure and temperature. This mixture is then suddenly compressed (by introducing gas at high pressure above the driving piston) to a predetermined volume, usually from one eighth to one fifteenth of the original volume. The compression time is about 0.006 sec, or about the same as the time between start of compression and the attainment of maximum pressure in an engine running at 5500 rpm. For each run with the rapid-compression machine, pressure-time records (Fig. 2–10) were made. In addition, high-speed motion pictures of flame development (Fig. 2–11) have been taken for many runs.

Pressure in the combustion cylinder was measured by the electric pressure indicator of Fig. 2–1c. This indicator was connected to a cathode-ray oscillograph in such a manner that a curve was produced on a film, which could be calibrated in terms of pressure as ordinate and time as abscissa. Flame photographs were taken at the rate of 8000 frames per second

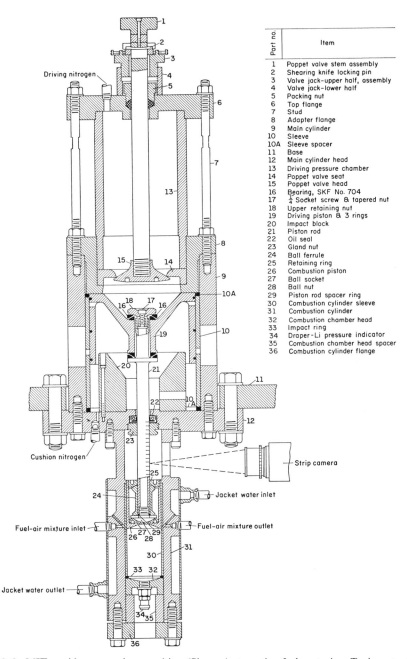

Fig. 2–9. MIT rapid-compression machine (Sloan Automotive Laboratories. Taylor *et al.*, 2.21).

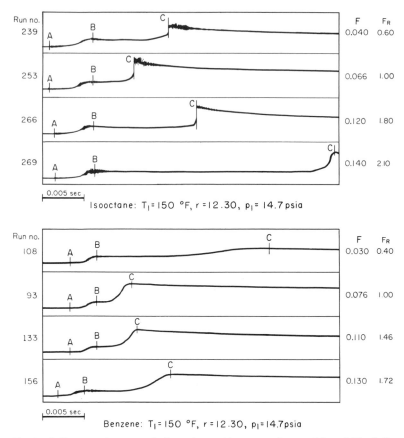

Fig. 2–10. Pressure-time records from the rapid-compression machine of Fig. 2–9.

through a glass window forming the lower end, or head, of the combustion cylinder.

Figure 2–10 shows pressure-time records taken by means of this machine with two different fuels. In each record the piston started to move at *A* and reached the end of its stroke at *B*.* Some time after the end of the piston stroke the pressure in each case starts to rise because of chemical reaction. Presumably the point of maximum pressure is that at which chemical reaction is complete. The rate of pressure rise, or the slope of the curve at any point, is an indication of the rate of chemical reaction.

Although the curves of Fig. 2–10 all indicate a rapid reaction within our definition of the term, they show quite different characteristics. The upper group shows a period of little pressure rise followed by a rapid

---

* The seating of the piston at *B* causes the cylinder head to vibrate, and this causes the small oscillations in the pressure record at this point.

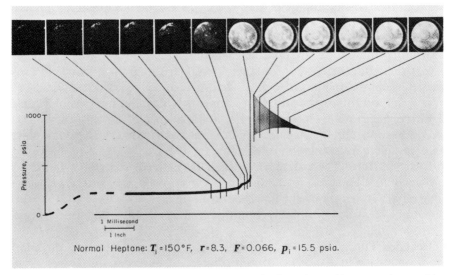

Normal Heptane: $T_i = 150°F$, $r = 8.3$, $F = 0.066$, $p_i = 15.5$ psia.

Fig. 2–11. Pressure-time record with flame photographs from the rapid-compression machine of Fig. 2–9. Note two-stage or step reaction (2.21).

pressure rise completing the process. In the first three curves the reaction is completed by a process so rapid that the recording apparatus is set into rapid vibratory motion. Comparison of the recorded vibration frequency and the computed natural frequency of pressure waves in the chamber indicates that these vibrations are gaseous.

The lower group of curves (for benzene) shows a pressure rise that has a more uniform rate, without any extremely rapid phase. The latter type of reaction has been found to be characteristic of benzene-air mixtures, which are well known to be very resistant to detonation in engines.

All of the pressure-time curves of Fig. 2–10 are of a type that fits the chain-reaction theory; that is, they all show a reaction rate which increases rapidly with time. In the case of isooctane the chains presumably take more time to start than in the case of benzene, but when they do start, they develop much more rapidly.

Flame Photographs of Autoignition. Figure 2–11 represents a series of frames from motion pictures taken of flame development in the combustion chamber of the rapid-compression machine, correlated with simultaneous pressure record. The most striking feature of these photographs is the fact that, even though the mixture is gaseous and as nearly homogeneous as it seems possible to make it, the reaction is extremely nonuniform with respect to position in the chamber. The early stages of reaction are characterized by the appearance of bright spots at various points in the charge. Additional spots appear as the reaction proceeds, while between the spots the gas spontaneously becomes luminous. The luminosity appearing between the large bright spots does not appear to be the spread of a flame front from these spots. Rather, it seems to be formed by the spontaneous appearance of vast numbers of smaller luminous points. Careful tests have shown that these spots are not due to incomplete fuel evaporation or other gross nonhomogeneity of the mixture.* Comparison of these flame patterns with those in the end gas of Fig. 2–4 shows a striking similarity.

It is important to note that the appearance of the first visible flame point coincides quite closely with the first evidence of a pressure rise due to combustion, and that the maximum pressure point in each case coincides with complete inflammation of the charge.

Reaction Rate. The *reaction rate* will be defined, for convenience, as the maximum value of dp/dt, that is, the *maximum slope* of the pressure-time curve. In diagrams such as the top three of Fig. 2–10 this rate is too high to be measured accurately, but the definition is still useful for purposes of discussion.

Reaction Time. The time B–C (Fig. 2–10), between the end of the compression stroke and the end of appreciable pressure rise due to reaction, will be called the *reaction time*. Chemical prereactions probably start at some point *during* the compression process, but since there is no convenient way of identifying this point, the reaction time is defined as beginning at the end of compression.

Delay Period. In the autoignition process, the time during which the rate of pressure rise is relatively slow is often called the *delay period*. In the type of reaction shown in the top three diagrams of Fig. 2–10, the delay period is equal to the reaction time. This type of reaction is undoubtedly the type that causes detonation in an engine; hence, the term "delay period" is frequently used in referring to detonation.

Prereactions. The relatively slow reactions that occur during the delay period are often called *prereactions*.

Results of Rapid-Compression Tests. Using several pure hydrocarbons, runs have been made showing the effects of initial temperature, compression ratio, fuel-air ratio, etc., on the pressure-time curve. Figure 2–10 shows two series of this kind. For each series, plots of the reaction time versus the independent variable have been made, as shown in Figs. 2–12 to 2–14. From the tests made to date it may be concluded that:

1. The autoignition process involves a period of relatively slow reaction followed by a period of rapid reaction. The relative lengths of the slow and rapid reaction periods, and the maximum rate of reaction, depend both on the composition of the fuel and on the test conditions. (The effects of fuel composition, together with the influence of additives, are discussed in detail in Chapter 4.)

---

* There is some evidence that turbulent motion of the mixture before ignition may be the chief cause of the nonhomogeneity (2.21).

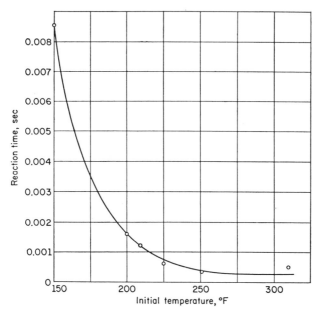

Fig. 2–12. Influence of initial temperature on reaction time in the rapid-compression machine of Fig. 2–9. Compression (volume) ratio, 8.3. Initial pressure, 15.5 psia (1.09 kg/cm^2); normal heptane; $F_R = 1.0$ (2.21).

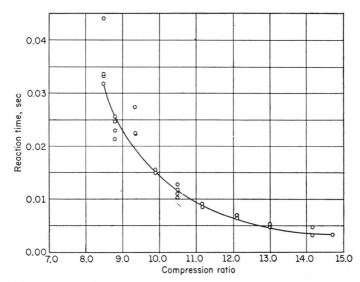

Fig. 2–13. Effect of compression ratio on reaction time of isooctane in the rapid-compression machine of Fig. 2–9. Initial temperature, 150 F (66 C); initial pressure, 14.7 psia (1.04 kg/cm^2), $F_R = 1.0$ (2.21).

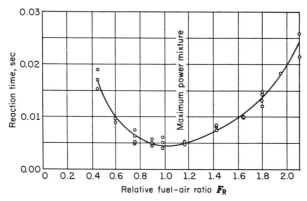

Fig. 2–14. Effect of fuel-air ratio on reaction time for isooctane in the rapid-compression machine of Fig. 2–9. Compression ratio, 12.3; initial temperature, 150°F (66°C); initial pressure, 14.7 psia (1.04 kg/cm²) (2.21).

2. A given fuel tends to retain its characteristic type of pressure-time curve, that is, either the type shown in the upper or in the lower group of Fig. 2–10, over wide ranges of initial temperature, compression ratio, and fuel-air ratio.

3. With a given fuel:

    a. Increasing initial temperature decreases reaction time (Fig. 2–12).

    b. Increasing the compression ratio (ratio of maximum to minimum volume of the working cylinder) decreases reaction time (Fig. 2–13).

    c. When fuel-air ratio is the only variable, a minimum reaction time occurs somewhere near the chemically correct fuel-air ratio (Fig. 2–14). (As will be shown later, this fuel-air ratio is approximately the same as the ratio for maximum detonation in an engine.)

4. A fuel found to have high resistance to detonation in engines has a longer reaction time or a slower reaction rate (or possibly both) than a fuel with less resistance to detonation.

5. If the reactions indicated by the pressure-time curves of Fig. 2–10 are representative of the end-gas reactions in engines, it is clear that such reactions can cause rates of pressure to rise so high as to account for the detonation phenomenon. It is also clear that the autoignition reaction need not always occur with great suddenness. Presumably, reactions of the type of benzene, Fig. 2–10, would not cause the intense pressure waves characteristic of detonation in engines.

Autoignition Map. Figure 2–15 is a plot of temperature at the end of compression versus pressure at the end of compression, with lines of constant reaction time, plotted from rapid-compression machine tests. It indicates that the reaction time of autoignition depends on pressure and

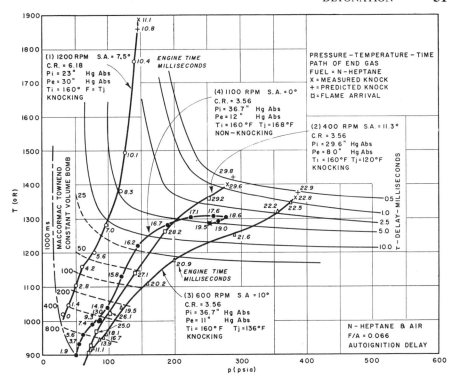

Fig. 2–15. Temperature-pressure-time paths of the end gas in a firing engine superimposed on autoignition-delay "map" for the same fuel-air mixture. Engine data are from tests on CFR engine. Lines of constant delay (reaction time) are from machine of Fig. 2–9 (Livengood and Wu, 2.22).

temperature, and that, in general, reaction time decreases as either the temperature or pressure becomes higher. At low pressures, the influence of a change of pressure on reaction time, at a given temperature, is large. At high pressures, on the other hand, changes in pressure have a smaller influence on reaction time, as evidenced by the near-horizontal nature of the curves.

Correlation of Detonation in Engines and Results of Rapid-Compression-Machine Tests

Superimposed on the curves for the rapid-compression machine, Fig. 2–15 includes several curves that show typical pressure-temperature histories of the end gas in an operating engine. The engine curves are the result of pressure measurements with the MIT balanced-diaphragm indicator (Volume I, pp. 114–119) and temperature measurements by the sound-velocity method (Volume I, pp. 119–122). This figure indicates the basic difference between the compression processes in the engine and in the rapid-compression machine. In the latter, compression is halted early in the reaction process at a measurable temperature and pressure. In the case of a detonating engine, however, the end gas is subjected to a continually rising pressure and temperature up to the completion of the autoignition process.

In order to correlate the two processes, Livengood and Wu (2.22) proposed a hypothesis, applying to reactions of the type shown in the upper three curves of Fig. 2–10, which we may call "the conservation of delay." This theory is based on the assumption that if the rapid-compression-machine process occurred in two steps, such that the compression pressure p_1, with corresponding T_1, was suddenly raised, *during the delay period*, to p_2 and T_2, the fraction of the delay consumed in the first step will be transferred to the second step without change. For example, if the sudden rise in pressure occurred at one third the normal delay time for p_1 and T_1, the autoignition process would be complete after two thirds the normal delay time for p_2 and T_2. The reaction time, t_e, for the process would then be $t_e = \frac{1}{3}\tau_1 + \frac{2}{3}\tau_2$, where τ indicates the normal delay (reaction) time for each step.

Extending this reasoning to an indefinite number of steps makes it evident that the reaction time for any path of T versus p on the ignition-delay map of Fig. 2–15 can be predicted by a step-by-step integration as the curve crosses the lines of constant delay. This operation will give a prediction of t_e for a given fuel-air mixture, provided the foregoing assumptions are valid.

With a plot of T versus p from an actual engine cycle superimposed on the rapid-compression-machine map, as in Fig. 2–15, the integration process indicated by Eq. 2–3 was carried out (with $t = 0$ taken at the start of compression) to the point where the value of t_e was determined for $x/x_c = 1$, which is the presumed point of autoignition. Figure 2–16 shows the integration carried out for one particular cycle, and Fig. 2–17 plots the predicted values of $t_e - t_0$ versus values measured by means of the MIT pressure indicator and the sound-velocity–temperature apparatus. This latter graph includes normal cycles and also motoring cycles with compression ratios high enough to result in autoignition. The fuels used were iso-octane, normal heptane, and a 55–45 blend of these two compounds, each at its stoichiometric mixture, $F_R = 1.0$. Agreement with actual measurements is good enough to validate the assumptions at least for these three fuel-air mixtures. More work would be necessary to establish their validity for other fuels and fuel-air ratios. These results are important for their contribution to the autoignition theory of detonation and for an improved understanding of the detonation process.

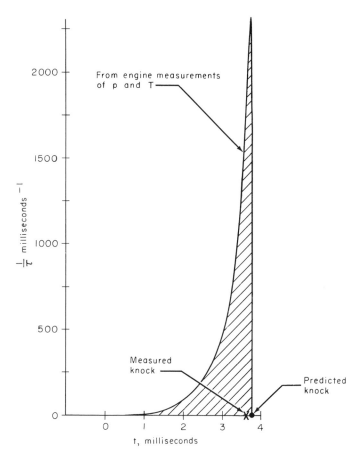

Fig. 2–16. Integration process to predict time of detonation of the end gas by means of the conservation-of-delay theory.

$$\frac{x}{x_c} = \int_{t=0}^{t=t_e} \frac{1}{\tau}\, dt\,,$$

where x/x_c = fraction of prereactions completed in time t.
When this fraction equals 1.0, detonation should occur.

t_e = predicted reaction time.

t_0 = time when compression of end gas starts.

τ = instantaneous delay time determined at the intersection of the end-gas compression curve and the curves of constant delay, Fig. 2–15.

For further details see ref. 2.22.

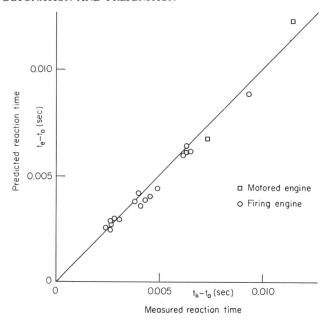

Fig. 2–17. Correlation between prediction and measurement of the time of autoignition. Plot includes both motored and firing engine cycles (2.22).

t_0 = time at which appreciable reaction begins; t_K = time of occurrence of measured autoignition or knock; t_e = time of occurrence of predicted autoignition or knock.

Prereactions in Motored Engines

Another technique which has been used to study autoignition is that of *motoring** a single-cylinder engine with a compression ratio high enough to cause appreciable reaction in the charge (2.35–2.39) without spark ignition. In this case, the whole charge represents the end gas. Figure 2–18 shows some results of this technique. Mixtures of isooctane and normal heptane (see Chapter 4) show decreasing resistance to detonation as the percentage of isooctane decreases. The figure shows a similar trend in reaction time. With more than 50 per cent isooctane, the reaction time was so long that descent of the piston halted the process before the explosive stage.

Such curves show that the rapid phase of autoignition is preceded by slower reactions, which are probably the beginning of chain reactions. When these reactions occupy enough time, the descent of the piston lowers the pressure (and temperature) before autoignition is complete. As the compression ratio increases, the reaction time grows shorter until autoignition occurs before the descent of the piston can stop the process.

---

* By "motoring" is meant driving the engine by means of an electric motor or other source of outside power. The CFR (Cooperative Fuel Research) variable-compression-ratio engine (2.06) is generally used for such tests.

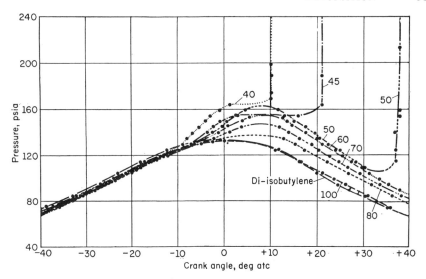

Fig. 2–18. Pressure–versus–crank-angle curves showing prereactions in a motored CFR engine. Numbers on curves indicate per cent of isooctane in normal heptane (Rifkin *et al.*, 2.37).

Work with motored engines shows that fuels that detonate easily will show faster prereactions than those with a higher resistance to detonation. Such work agrees very well with rapid-compression-machine results and gives further support to the chain-reaction theory of detonation.

Cool Flames. In work with bombs and with motored engines it has been observed that at some point in the prereaction process a weak, bluish flame may sweep through the mixture, after which there will be a short delay in the pressure rise before autoignition occurs. The resulting pressure-time curve is a *two-stage* affair, as illustrated for normal heptane in Fig. 2–11. This phenomenon seems to be a normal part of the autoignition process of many fuel-air mixtures. It may be a preliminary chain reaction that involves only a small portion of the molecules in the mixture. It also may account for the fact that some high-speed photographs of detonation seem to show detonation *after* passage of a flame front made visible by the Schlieren technique (2.26).*

End-Gas Reaction in Engines

From the foregoing discussion it is evident that a primary requirement for detonation to be possible in an engine is that the reaction time of the unburned mixture must be shorter than the time for normal flame travel through the mixture. A second requirement is that the reaction be sufficiently intense for sensible pressure waves to be produced.

---

* The Schlieren method can be so sensitive that a "cool" flame front may be mistaken for the principal front.

Intensity of Detonation. In an engine the amount of pressure-wave energy caused by detonation will depend on the mass of end gas that autoignites and the rate of reaction in the autoignition process. Thus, with a given reaction rate, the earlier in the combustion process detonation occurs, the more end gas will participate and the greater will be the intensity of detonation.

In service with an engine operating normally if conditions are changed toward detonation, as by increasing the inlet pressure, the operator usually takes some action to prevent further increase in its severity when detonation becomes audible. In the operation of road vehicles, for example, the operator will not usually open the throttle beyond the point where detonation becomes distinctly audible. In testing for detonation, the usual procedure is to start with conditions that do not produce detonation and then gradually approach the point where detonation can just be detected, in which case it is called *incipient*, or *borderline*, detonation. Here detonation occurs very close to the peak pressure of the cycle.

In order to relate detonation in engines to the data obtained from the rapid-compression machine, it is convenient to consider a hypothetical engine in which compression of the end gas is adiabatic. The instantaneous pressure and temperature are assumed to be uniform throughout the cylinder prior to detonation, and therefore the end-gas pressure is taken as being equal to the cylinder pressure until detonation occurs. Figure 2–19

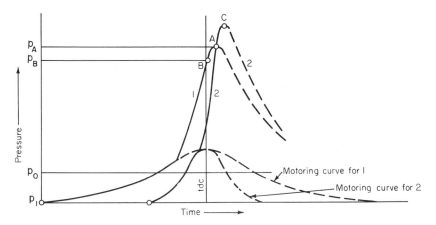

Fig. 2–19. Hypothetical pressure-time curves in an engine.

shows hypothetical pressure-time diagrams for such an engine. In this figure, p_1 is taken as the pressure at the start of the compression stroke, and p_0 as the pressure at which chain reactions start in the end gas.

In the cycle marked 1 let it be assumed that the peak pressure is p_A and that detonation occurs at point A. If the initial temperature is raised, the preliminary reactions will be accelerated and detonation will occur in a shorter time, at a point such as *B*. This change will cause detonation to occur earlier in the cycle, and more end gas will participate. Thus, detonation will be more severe than with the original starting temperature. One way of avoiding detonation with the higher initial temperature is to reduce the maximum cyclic pressure below p_B by reducing the compression ratio or closing the throttle.

Returning to the original inlet temperature, let us suppose that the *speed* of the engine is increased and the time scale so adjusted that the curve passes through point *A*. Curve 2 of Fig. 2–19 represents this condition. By comparing this curve with curve 1, it is evident that, because of the shorter time available for the chain reactions to develop, detonation will now be avoided unless the peak pressure is again raised, this time to a point such as *C*. Thus, a higher inlet pressure or a higher compression ratio could be used at the higher speed. As we shall see, increased engine speed does not always reduce detonation, because factors other than those included in the foregoing assumptions may prevail. But the discussion does serve to show the relation of "delay" to engine conditions.

Detonation Measurements

Before going further into the subject of detonation in engines, it is necessary to discuss methods for detecting the phenomenon and measuring its relative intensity.

The presence or absence of detonation in engines is often judged by ear. A more accurate method is the use of a pressure-sensitive unit similar in principle to those of Fig. 2–1, in which the diaphragm is exposed to the gases in the cylinder. For laboratory use, the output from a unit of this type is generally used to operate a cathode-ray oscillograph, producing curves such as those of Fig. 2–2. A somewhat more convenient but less sensitive method is to connect the amplified output from the pickup unit to a highly damped a-c galvanometer. With this arrangement, the scale reading of the galvanometer increases with increasing amplitude of the signal fluctuations and thus gives a relative measure of detonation intensity.

Octane Requirement. It is, of course, well known that the tendency to detonate depends very much on fuel composition. As will be explained in Chapter 4, the accepted scale for evaluating the resistance to detonation of any fuel is based on the use of two hydrocarbons, namely *isooctane*, which has great resistance to detonation, and *normal heptane*, which detonates readily. The tendency of a given fuel to detonate is measured by its *octane number*, which is the percentage of isooctane in normal heptane that will give borderline detonation under the same conditions as the fuel under test. For fuel tests a specified engine (the CFR, 2.06) is used under a specified set of operating conditions. Conversely, a method of measuring the influence of design and operating conditions on the detonation tendency is to determine the percentage of isooctane in normal heptane that will give incipient detonation under the prevailing conditions. The percentage so found is called the *octane requirement* of the particular engine at the particular set of operating conditions in question. As a given operating variable is changed, by trying different percentages of isooctane in normal heptane a curve of octane requirement versus the variable can be constructed for any engine. The lower the octane requirement, the less the tendency to detonate. This method can be considered an absolute method of measuring the detonation tendency, to the extent that the octane number is accepted as a measure of this tendency. The limitations of the method are discussed in Chapter 4. The great advantage of the octane-requirement method is that it can be used to determine the effect of a single operating or design variable. Its disadvantages are the expense and time involved in the use of reference fuels and the difficulty of defining the octane requirement when it is greater than 100, that is, when something better than pure isooctane is required to suppress detonation.

In order to avoid the disadvantages of the octane-requirement test, several methods of measuring the detonation tendency employ as a dependent variable an operating variable whose effect on detonation is always in the same direction. The effect of the independent variable on detonation is then measured in terms of a dependent variable chosen as a criterion.

There are, in particular, two operating variables that always increase the detonation tendency: increased compression ratio and increased inlet pressure. Thus, one or the other of these variables is often used as a dependent variable to measure the relative effects on detonation of changes in fuel composition or in engine-operating conditions. Since increasing either compression ratio or inlet pressure increases the indicated mean effective pressure, the imep at the point of incipient detonation may also be regarded as the dependent variable.

Knock-Limited Compression Ratio. The knock-limited compression ratio is obtained by increasing the compression ratio (on a variable compression-ratio engine) until incipient detonation is observed. Any change in operating conditions, in fuel composition, or in engine design that increases the knock-limited compression ratio is said to reduce the tendency toward detonation.

Knock-Limited Inlet Pressure. The inlet pressure can be increased (by opening the throttle or increasing the supercharger delivery pressure) until incipient detonation is observed. An increase in knock-limited inlet pressure indicates a reduction in the detonation tendency.

Both of the forementioned methods involve less time and expense than the octane-requirement method, and therefore they are widely used.

Knock-Limited Imep. The indicated mean effective pressure measured at incipient detonation is abbreviated klimep. This parameter and the corresponding fuel consumption are obviously of great practical interest provided the conditions under which they are measured correspond to those which would be used in service.

Performance Number. A useful measure of the detonation tendency, called the *performance number*, has been developed from the conception of knock-limited imep. This number is defined as the ratio of klimep with the fuel in question to klimep with isooctane when inlet pressure is used as the dependent variable. The official Army-Navy (AN) performance number is obtained under specified conditions with a specified test engine (2.093). However, it appears that performance number is reasonably independent of engine design and closely related to the octane number (2.54). One great advantage of the performance number is that it can apply to fuels whose detonation characteristics are superior to those of isooctane; that is, it extends the octane scale beyond 100. For convenience, the initials PN will be used to designate performance number.

End-Gas Temperature

In considering the subject of end-gas temperature, it is well to refer back to Fig. 2–15, where typical pressure-temperature paths for the end gas of a spark-ignition engine are plotted on the pressure-temperature plane, with constant-delay lines (for the same fuel-air mixture) from the MIT rapid-compression machine of Fig. 2–9. An important point to note is that, within the range of engine conditions, the variation of delay time with pressure is relatively small, while the influence of end-gas temperature is great. From this observation it may be concluded that the observed sensitivity of detonation to end-gas pressure is chiefly due to the influence of a near-adiabatic pressure change on the resultant *temperature*. Thus, it may be concluded that whether or not the end gas autoignites is chiefly a question of end-gas temperature and end-gas compression time.

This view is supported by an important contribution to detonation research by Gluckstein and his collaborators (2.33), who measured end-gas temperatures just prior to incipient detonation for many fuels and under a wide variety of operating conditions. The paramount importance of end-gas compression time and *final* end-gas temperature, that is, the temperature of the end gas at the moment of autoignition, is well established by this work.

End-gas temperature is the result not only of initial compression temperature, heat transfer, and mechanical compression of the end gas but also of the temperature rise due to prereactions. The Gluckstein work showed that the end-gas temperature at the detonation limit was nearly constant at $1840 \pm 40°R$ ($1020 \pm 22°K$) for a wide variety of fuels over a considerable range of inlet temperatures with compression ratio as the dependent variable. This aspect of their work will be discussed more fully in Chapter 4. Here we are concerned with detonation with a given fuel composition.

From the foregoing discussion it can be safely concluded that, for a given fuel–air–residual-gas mixture at constant engine speed and with ignition timing set for peak pressure at a fixed crank angle, the end-gas temperature at the point of incipient detonation will be constant. This discovery has led to the ability to predict the detonation-limited operating conditions for a given engine and fuel at a given speed on the basis of $T_{2''}$ (see Fig. 2–7), the predicted temperature of the end gas when adiabatically compressed to the maximum pressure of the cycle. If $T_{2''}$ can be evaluated under one condition of incipient detonation, the detonation-limited operating conditions for other operating conditions, when compression time is constant, can be predicted on the basis of the same value of $T_{2''}$. The following example shows how this procedure can be carried out.

Example 2–1. In the operation of a large gas engine at $F_R = 0.8$, $s = 1200$ ft/min, the detonation limit is found at a maximum cylinder pressure of 650 psia, inlet-pressure 22 psia, inlet temperature 650°R. If the inlet temperature could be reduced by aftercooling to 520°R, estimate the allowable increase in maximum cylinder pressure.

Solution: At the low piston speed of 1200 ft/min, p_1 is taken as equal to p_i (see Volume I, Fig. 6–7, p. 161). Assuming an induction temperature rise of 100°R and using a compression exponent of 1.35, $T_{2''}$ for the original condition is computed as follows:

$$T_{2''} = T_1(p_3/p_1)^{0.35/1.35} = 750(650/22)^{0.26} = 1810°R.$$

If the inlet temperature is reduced to 520°R, $T_i = 520 + 100 = 620°R$; and for the same value of $T_{2''}$:

$$620(p_3/22)^{0.26} = 1810°R \ (1005°K);$$

$$p_3 = 1360 \ \text{psia} \ (96 \ \text{kg/cm}^2).$$

Example 2–2. If the reduction in inlet temperature of Example 2–1 is used to raise the compression ratio, originally 8 in the previous example, what new ratio could be used with the same inlet pressure? F_R for this engine is 0.8.

Solution: For the fuel-air cycle, Figs. 0–5 and 0–6 indicate that at $r = 8$, $T_1 = 750$, $p_3/p_1 = 56(0.95) = 53$. For the engine, p_3/p_1 is $650/22 = 29.6$, or $29.6/53 = 0.56$ times the fuel-air-cycle value. Assuming this same ratio at the higher compression ratio, the value of p_3/p_1 for the corresponding fuel-air cycle will be $(1360/22)/0.56 = 110$. From Fig. 0–6, the correction factor for $T_1 = 520$ is extrapolated at 1.2. Therefore the value of p_3/p_1 to be read from Fig. 0–5 is $110(1.2) = 132$. The corresponding compression ratio is 17.5. From Fig. 0–1 the approximate gain in output and fuel economy will be $(0.534/0.45) = 1.18$, or 18%.

These examples assume that the speed of the engine remains constant and that ignition timing is adjusted to the same crank angle for peak pressure.

Effect of Engine Operating Conditions on Detonation

From the discussion up to this point, it should be evident that, with a given fuel–air–residual-gas mixture, the tendency to detonate will depend chiefly on the temperature of the end gas just before reaction and on the time of compression to this temperature. The influence of end-gas pressure at this time is small, except as it affects the end-gas temperature. Unfortunately it is seldom possible to isolate these variables, because nearly every change in operating conditions which occurs in normal service changes both of them to a greater or lesser extent. However, under laboratory conditions it is often possible to recognize which of them is exerting the principal influence.

Spark Timing. Figure 2–20 plots pressure–versus–crank-angle curves for different spark timings. Within the range shown, peak pressure increases

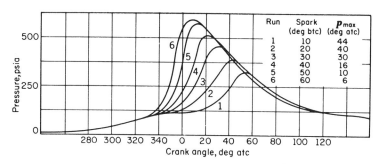

Fig. 2–20. Effect of spark timing on pressure development in an engine cylinder without detonation. CFR engine, 1200 rpm, full throttle. (Sloan Automotive Laboratories.)

and end-gas compression time, measured from the start of compression, decreases as the spark is advanced. The tendency to detonate is promoted by an increase in end-gas temperature (by adiabatic pressure increase) and reduced by a reduction in time. Except for extreme spark advance (not shown in Fig. 2–20 and not of practical interest) the increasing pressure-temperature effect always predominates over the reduced-time effect, and the tendency to detonate invariably increases with advancing spark, within the useful range, as shown in Fig. 2–21.

Because of the fact that it is easily changed, spark timing may be varied in such a way as to help control detonation. The dash-dot line in Fig. 2–21 shows how spark timing is controlled as a function of speed, for a particular engine, in order to reduce octane requirements, particularly at low speeds.

Spark timing is sometimes used as a measure of the detonation tendency. The steep slope of the constant-speed-versus-PN curves of Fig. 2–21 shows that this method furnishes a sensitive measurement but is useful over only

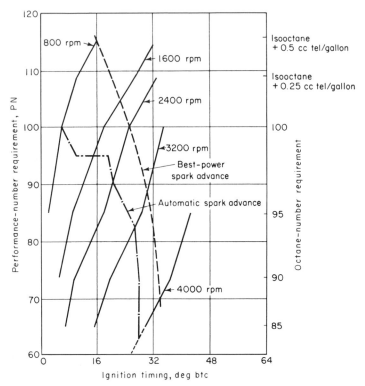

Fig. 2–21. Effect of spark timing on performance number and octane requirement. Automobile engine, full throttle; $r = 10$, $T_i = 100°F$ (33°C); tel = tetraethyl lead (Roensch and Hughes, 2.55).

the small range in which engine performance is not seriously affected (see Volume I, Fig. 12–14, p. 443, for this relation).

With a fixed spark, any variable that increases combustion time will have an influence similar to that of retarding the spark. Thus, where possible, *the effect of other variables on detonation should be determined with the spark so adjusted that peak pressure always occurs at the same crank angle.* Unfortunately, in many cases no data obtained in this manner are available.

Inlet Pressure and Compression Ratio. Increase in either of these variables necessarily increases the detonation tendency, as illustrated in Figs. 2–22 and 2–23. The explanation is obvious: peak pressure, and hence end-gas temperature $T_{2''}$, increases.

Indicator diagrams taken at the knock limit show that, with a given engine at a given speed, and with a given fuel-air mixture, the value of $T_{2''}$ at incipient detonation is the same with either variable. As we have seen, this fact is very useful in predicting the effect on engine performance of changes in compression ratio and inlet pressure.

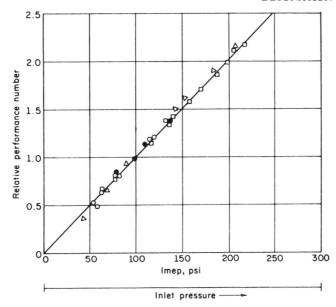

Fig. 2–22. Effect of imep, varied by inlet pressure, on required performance number; relative PN = PN required, divided by PN required at 100 imep (Hesselberg *et al.*, 2.54).

Symbol	Engine	Type	Head	r	RPM
□	1-cyl	Test	I	5.6	900
△	1-cyl	Test	I	7.2	1200
○	6-cyl	Auto	L	6.7	2200
●	6-cyl	Auto	I	8.0	2400

Inlet Temperature. Increasing inlet temperature increases the end-gas temperature at a given pressure, thus increasing the tendency to detonate (Fig. 2–24). It is important to note that different fuels produce quite different slopes of the curve of klimep versus inlet temperature (see also Chapter 4). Those with especially steep slopes, such as di-isobutylene, are called *sensitive* fuels.

Effect of Errors in Distribution. The plots of Fig. 2–24 were made with uniform gaseous fuel-air mixtures. In carbureted engines using liquid fuel, fuel evaporation and distribution are very sensitive to inlet temperature. Thus, in multicylinder engines running with unevaporated fuel in the manifolds (as is the case with most automotive engines) increasing the inlet temperature may sometimes reduce the detonation tendency by reducing the discrepancies in fuel-air ratio between the several cylinders. A similar trend appears with a single-cylinder engine when there are variations in effective fuel-air ratio from cycle to cycle (see further discussion of distribution effects in Chapter 5).

Coolant Temperature. The effect of coolant temperature, Fig. 2–25, is similar to that of inlet temperature, and the explanation is similar, namely, an increase in end-gas temperature. However, in this case this increase must be due to heat transfer. The difference between sensitive and insensitive fuels is again apparent.

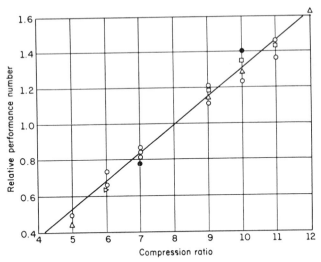

Fig. 2–23. Effect of compression ratio on required performance number; relative PN = PN required, divided by PN required at $r = 8$ (2.54).

Symbol	Engine	Load
△	E	1.0
□	CFR	1.0
●	CFR	0.4
○	G	1.0

Effect of Engine Speed. The effect of changes in engine speed on the fundamental quantities affecting detonation depend on how other operating variables are controlled. When speed effects are under consideration, it is usual to hold inlet and jacket temperatures and fuel-air ratio constant, and they will be assumed constant here. Too often, however, such tests are made with constant spark timing, so that the effective spark advance changes with speed. Tests of this kind should of course be made with the spark adjusted to give peak pressure at the same crank angle. However, even under ideal test circumstances the effects of engine speed on detonation are complex, because a change in engine speed may still affect each of the fundamental variables, namely, end-gas temperature, end-gas pressure, and end-gas compression time.

Increasing engine speed obviously decreases compression time and, as we have seen, this factor tends to reduce the detonation tendency.

On the other hand, as shown in Fig. 2–26, in a 4-stroke engine compression temperature increases with speed even if inlet temperature is held constant. This increase in temperature is due to the fact that as speed increases, less work is done by the gases on the piston during the inlet

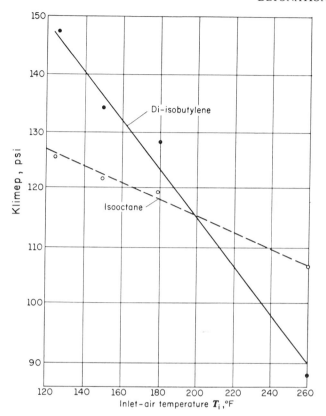

Fig. 2–24. Effect of inlet temperature on knock-limited imep (Heron and Felt, 2.53).

stroke, and the charge temperature at the beginning of compression is higher (see Volume I, pp. 158–164).

A change in engine speed usually changes engine volumetric efficiency (Volume I, Chapter 6). Under most circumstances, peak pressure varies in the same direction as volumetric efficiency.

In Fig. 2–27, the curves marked bpsa were made with spark timing adjusted to give nearly the same peak-pressure angle. Such curves may therefore be considered free from the spark-timing variable. In set *A* of Fig. 2–27 the inlet pressure was kept constant. The combined effects of shorter compression time and reduced volumetric efficiency show improved resistance to detonation as speed increases, in spite of the rising compression temperature.

In curve *C*, the octane-requirement increase with speed from 800 to 1200 rpm is probably due to increasing volumetric efficiency and increasing T_1 sufficient to overcome the shorter time effect. However, the fact that spark timing was varied on an arbitrary schedule introduces an unknown influence.

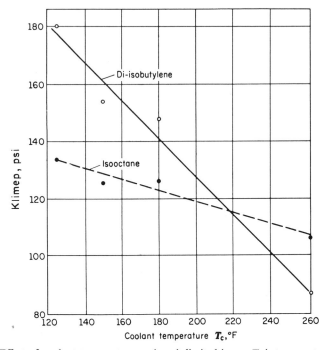

Fig. 2–25. Effect of coolant temperature on knock-limited imep; T_c is temperature of entering coolant (2.53).

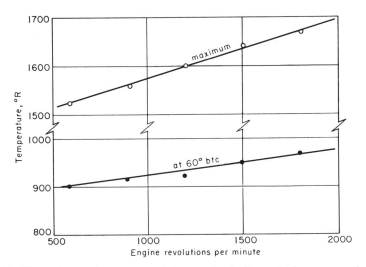

Fig. 2–26. Charge temperature, measured by sound-velocity method, versus engine speed, without detonation (Livengood *et al.*, 1.41.) $p_i = p_e = 14.7$ psia; $T_i = T_c = 160°$F; $r = 6.18$ bpsa; $F_R = 1.07$.

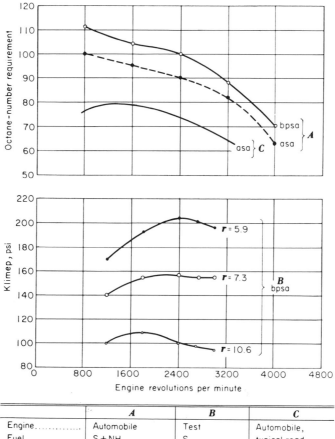

	A	*B*	*C*
Engine.............	Automobile	Test	Automobile,
Fuel................	S + NH	S	typical road
Compression ratio.	10	As indicated	conditions; $r = 7+$
Inlet pressure......	Nearly constant	Variable	
Inlet temp, °F......	100	180	
Fuel-air ratio......	Near 0.08	0.08	
Jacket temp, °F....	....	180	
Reference.........	Roensch and Hughes, 2.55	Heron and Felt, 2.53	Best et al., 2.843

S = isooctane, NH = normal heptane, asa = automatic spark advance, bpsa = best-power spark advance

Fig. 2–27. Effect of engine speed on detonation.

In curves *B* of Fig. 2–27 the inlet pressure was varied to give the knock-limited mep at each speed. A good way of looking at these curves is to regard klimep as measuring relative end-gas pressure at the knock limit. Thus one can say that the time effect predominates over the compression-temperature effect in the range where klimep increases with increasing speed. The reverse is true when klimep decreases with increasing speed. The compression-temperature effect (toward reducing klimep with increasing speed) is evidently stronger as the compression ratio becomes higher. In general, fuels that are especially "temperature sensitive" (see Chapter 4) show a greater tendency for klimep to decrease with increasing speed than do the reference fuels, because of the increasing compression temperature illustrated in Fig. 2–26.

In automobile engines, automatic adjustment of spark timing is used to control detonation (see Fig. 2–21 and the dashed-line curve in Fig. 2–27). The schedule of spark advance versus speed will exert a powerful influence on octane requirement versus speed. It is believed that Fig. 2–21 and curves *A* and *C* of Fig. 2–27 are typical of current practice.

Combined Effects of Speed, Spark Advance, and Imep. Figure 2–28 is

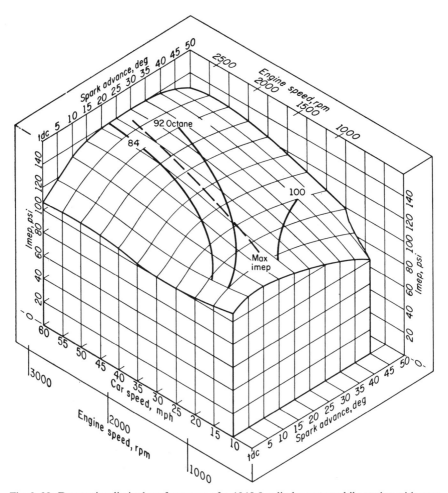

Fig. 2–28. Detonation-limited performance of a 1945 8-cylinder automobile engine with compression ratio 9.0 (Barber, 2.51).

a three-dimensional plot showing the effects of these three variables on octane requirement for an automobile engine. The general shape of these curves is typical, although absolute values are dependent on details of design, engine adjustments, and the character of the fuel.

Fuel-Air Ratio. Changes in fuel-air ratio can cause changes in flame speed, flame and wall temperatures, and the reaction time of the end gas.

Figure 2–29b shows knock-limited compression ratio versus fuel-air ratio for an unsupercharged engine. The maximum tendency to detonate (lowest compression ratio) occurs close to the ratio for minimum reaction time shown in Fig. 2-14. This fact would tend to indicate that reaction time of the mixture is the predominant factor in this case.

The variation of klimep with fuel-air ratio is more complex, since inlet pressure is varied and variations in compression temperature are introduced because of changes in heat transfer, induction work, and end-gas

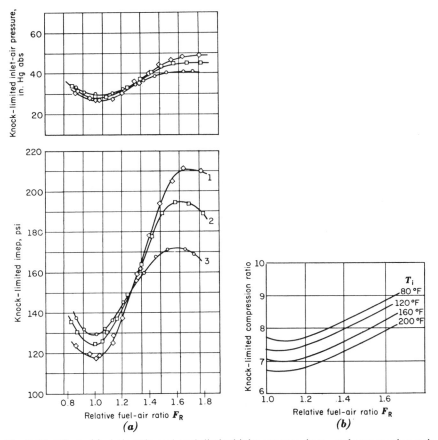

Fig. 2–29. Effect of fuel-air ratio on knock-limited inlet pressure, imep, and compression ratio. (a) Aircraft fuels, $r = 7$, 1800 rpm, $T_c = 212°F$, $T_i = 250°F$ (Cook *et al.*, 2.50); (b) Isooctane fuel, unsupercharged CFR engine, 1200 rpm, $T_c = 211°F$ (Taylor *et al.*, 2.461).

fraction. Figure 2–29a is typical in trend for klimep versus fuel-air ratio with fixed spark timing. Again, the point of maximum detonation tendency (minimum klimep) occurs near the fuel-air ratio having minimum reaction time in Fig. 2–14. In general, except at the rich end, the behavior in the engine generally follows that which would be predicted from the reaction-time curve. Any cooling effect due to fuel evaporation plus the lower maximum temperature at very rich mixtures should also tend toward increasing the klimep as the mixture is enriched. The drop in klimep at the very rich end is due to decreasing thermal efficiency. Note that knock-limited inlet pressure does not fall off.

Dilution of the Charge. References 2.49 and 2.78 indicate that introducing *cooled* exhaust gas with the inlet air reduces the tendency to detonate. Dilution with inert material probably has the same effect as the use of lean mixtures; that is, reaction time is thereby increased.

Exhaust Back Pressure. From studies of the inlet and exhaust process of 4-cycle engines, reported in Chapter 6 of Volume I, it is evident that increasing the ratio p_e/p_i, increases compression temperature, increases the residual fraction f, and lowers maximum pressure p_3. We have seen that the first effect tends to increase the detonation tendency, while the others tend to reduce it. Engine experiments (2.78) show that increasing the exhaust back pressure, with all other operating conditions the same, has only a small effect on the tendency to detonate. Apparently, the opposing factors tend to balance it out in this case.

Atmospheric Humidity. Engine tests invariably show that increasing atmospheric humidity tends to reduce detonation (2.52 and 2.59). It is probable that reduced reaction time is the principal cause of this trend, although chemical factors may also be important.

Water-Alcohol Injection. A very powerful means of reducing detonation is by the injection of water, or water-alcohol mixtures, into the inlet system (2.75–2.77). This method is often used to suppress detonation of supercharged airplane engines during take-off, thus allowing a higher inlet pressure, and hence a higher power output, than would be otherwise possible. Since the quantity of additional fluid needed is relatively large (on the order of 50 per cent of the fuel flow), this method of detonation control is practicable for short periods of time only.

Effect of Cycle-to-Cycle Variation. If an engine is operating with border-line detonation in the cycles with average peak pressure, the cycles with lower peak pressures may not detonate at all, while the cycles with higher peak pressures will detonate with increasing intensity as the peak pressure increases. Thus, increasing the degree of cyclic variation will increase the number of cycles which detonate heavily. Conversely, a series of cycles with little cyclic variation will show a higher klimep than a series with large cyclic variation, the average peak pressure being the same in both cases. Thus, any operating or design factor that reduces cyclic variation will have a tendency to decrease octane requirement and to increase klimep.

Effect of Design on Detonation

Other things being equal, *design*, that is, the choice of shape and materials, can affect detonation chiefly because of its influence on heat transfer to the end gas and on combustion time.

Combustion Time. Under conditions which give the same flame speed, it is evident that the time for combustion is proportional to the distance between spark plug and the end of flame travel. In this chapter we have shown that the shorter the combustion time, the less will be the tendency to detonate, other things being equal. Thus, in general, the more compact the combustion chamber, the lower should be the octane requirement, unless heat-transfer effects interfere.

Heat Transfer to the End Gas. As we have seen, this factor is difficult to predict when speed is the variable. However, in considering the effect of design on detonation, we can consider speed constant, in which case heat-transfer trends, if not absolute values, are easier to predict.

Figure 2–26 shows end-gas temperatures as high as 1700°R with a compression ratio of 6.18. Taking the initial compression temperature at 600°R, the time average of the end-gas temperature will be in the neighborhood of 1000°R, which is hotter than most parts of the combustion chamber. Thus it appears that this part of the charge will lose heat to the walls unless it lies close to a very hot surface, such as the face of a poppet exhaust valve. In the case of combustion chambers of such a shape that the end gas is confined to a specific region, there is some evidence that the presence of a hot exhaust valve in this region tends to promote detonation.* Conversely, if the region in which the end gas is confined is kept cool, and especially if it has a large surface-volume ratio, a reduction in the detonation tendency should result. In this case, however, the cooling effect alone cannot be isolated, because there are other factors in such combustion chambers which operate in the same direction (see later discussion of the *quench area* in combustion chambers).

Effect of Shrouded Inlet Valve. As might be expected, the shape of the combustion chamber, including those portions of the piston and valves that act as chamber wall surfaces, is most important from the point of view of detonation. Figure 2–30 shows best-power spark timing and klimep for a flat-head overhead-valve combustion chamber with two different pistons, with and without a shrouded† inlet valve.

---

* Tests by Heron and others (2.53, 2.54) have indicated that exhaust-valve temperature may have little influence on detonation. However, such tests seem always to have been made with overhead-valve engines in which the charge rotates, with only brief contact between valve and end gas.

† A poppet valve so arranged as to restrict the flow of gas to about 180° of valve circumference and to direct the flow into the cylinder in a tangential direction is called a *shrouded valve*.

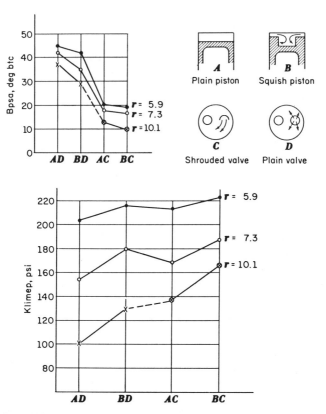

Fig. 2–30. Effect of piston and valve design on knock-limited imep with best-power spark advance. One-cylinder overhead-valve engine, 3.25 × 3.25 in. (83 × 83 mm); $s = 1300$ fpm (6.6 m/sec); $F_R = 1.2$; $T_i = T_c = 180°$ (82°C) (Heron and Felt, 2.53).

As shown by the spark-advance curves, use of the shrouded valve gives a marked reduction in combustion time. This effect in itself is apt to reduce the tendency to detonate. In addition, as shown by Fig. 1–20, the shrouded valve tends to reduce cycle-to-cycle variation, especially when oriented so as to give tangential flow into the cylinder. This orientation was used for Fig. 2–30. Reference 2.631 shows that when the shrouded valve is rotated, the effect is the one which would be predicted from Fig. 1–20, namely, that the minimum tendency toward detonation occurs when the inflow is tangential to the cylinder wall.

Effect of Piston Shape. One of the pistons used for Fig. 2–30 is called a *squish* piston because the charge is squeezed radially inward near top dead center. With the same inlet valve the combustion time is about the same as with the flat-top piston, since the spark advance required is also about the same, but the tendency to detonate is appreciably less.

Figure 2–31 shows results of tests where the squish area was varied in one

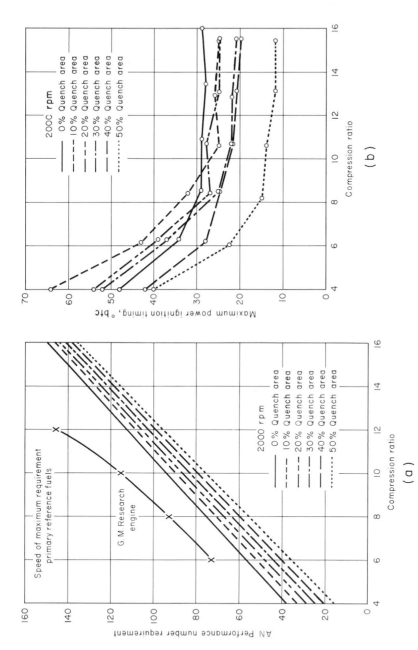

Fig. 2-31. Performance-number requirement and bpsa versus quench (squish) area. COT single-cylinder engine 2.75 × 2.75 in. (70 × 70 mm). Quench-area clearance 0.0065 × bore, $F_R = 1.15$, 800 or 1200 rpm, $T_i = 90°F$, $T_c = 210°F$. (a) Effect of compression ratio and quench area on AN performance number; (b) Effect of compression ratio and quench area on maximum power ignition timing (Hughes, 2.65).

particular type of combustion chamber. In the report from which Fig. 2–31 was taken, it is pointed out that the clearance above the piston in the squish area at top center must be less than 0.005 times the bore (e.g., 0.02 in. for a 4-in. bore) in order to obtain the favorable effects on octane requirement shown.

Results of another series of tests involving squish area are shown in Fig. 2–32. Here again, the squish-type piston shows a marked advantage over the flat piston.

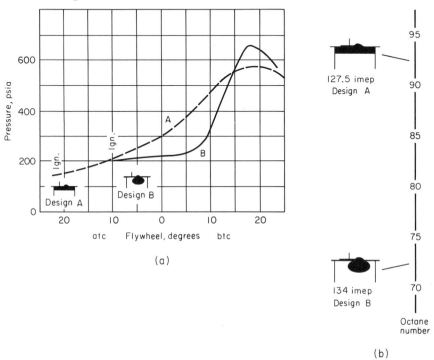

Fig. 2–32. Pressure development and octane requirements of flat and squish combustion chambers. Automotive-type engine, compression ratio 9.0, 1000 rpm, other operating conditions unspecified. (*a*) Effect of chamber design on pressure near top center, (*b*) Effect of chamber design on octane requirement at 99 per cent power (Caris *et al.*, 2.631).

The usual explanation offered for the fact that arrangements involving squish reduce the detonation tendency is based on the assumption of a cooling effect. It is pointed out that near the end of flame travel the end gas is located in a thin space where it makes excellent contact with cylinder walls, which are at a lower temperature than the gas. Because of this theory, the thin space in combustion chambers of this type is often called the *quench* area. While the cooling effect of a quench area may have some influence, there is also the possible influence of increased turbulence. Perhaps even more important is the fact that in nearly every case the combustion chamber becomes effectively more compact as the quench area is increased.

Multiple Spark Plugs. Figure 2–33a shows results of tests with a cylinder head equipped with seventeen spark plugs. Varying the position of a single

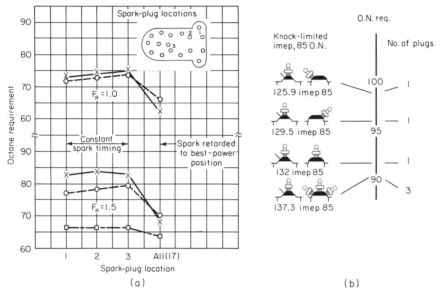

Fig. 2–33. Effect of number and position of spark plugs on octane requirement. Two views of each chamber are shown. The octane-number requirement is at full throttle (Caris *et al.*, 2.631).

    (*a*)   L-head cylinder (Diggs, 2.67);

           □ 900 rpm   ○ 1800 rpm × 2700 rpm

    (*b*)   Automotive-type engine, *r* = 9, 1000 rpm, bpsa.

ignition point seems to have little effect on octane requirement. Probably the reason that the plug with the most central location, and therefore the shortest flame path, does not show the expected reduction in octane requirement is that this position also placed the end gas near the exhaust valve, which was not the case with positions 1 and 2. The use of all seventeen plugs definitely reduces the octane requirement, because of reduced combustion time, provided the spark is retarded to hold peak pressure at the optimum value.

Octane requirements of another type of combustion chamber are shown in Fig. 2–33b as a function of number and position of spark plugs with spark timing adjusted for best power. It is notable that the use of three plugs in this case shows very little advantage over that of one plug in its optimum (central) position.

In most cases the practical alternative is between one spark plug and two, the latter usually at opposite sides of the combustion chamber. Experience shows that the use of two spark plugs instead of one somewhat reduces the tendency to detonate when the appropriate spark timing is used. With two sparks plugs, if one plug fails to fire, the tendency to detonate may be reduced because the effective ignition timing is retarded.

Sleeve Valves. Cylinders equipped with sleeve valves have generally been considered to have a low tendency to detonate, owing to the absence of any high-temperature area such as that of the poppet exhaust valve. There are apparently no definitive measurements that indicate the degree of difference, if any, between poppet and sleeve valves under comparable conditions of operation. There is some evidence (2.48) that the octane requirement of a sleeve-valve engine is reduced by increasing the swirl rate as controlled by the angle of the inlet ports with the cylinder radius. Since port area is reduced, and small-scale turbulence is therefore increased as swirl increases, the observed effect may be due entirely to increased flame speed.

Because of the high cost of manufacturing sleeve-valve cylinders and of the excellent performance of modern poppet valves, sleeve-valve engines are no longer used commercially.

Two-Cycle Octane Requirements. A limited amount of published work on the octane requirements of conventional 2-cycle engines with crankcase compression (2.652) indicates this type to have about the same octane requirements as do 4-cycle engines with cylinders of similar size. The compression ratios used are therefore nearly the same (near 8.0 in 1965 for use with motor gasoline). This relation is somewhat surprising, considering that the 2-cycle engines operate with relatively low maximum cylinder pressures and are not run at low speeds with wide open throttle. Probably the unfavorable effect of large fractions of hot residual gas in the 2-cycle-cylinder charge compensates for the favorable effects.

Summary of Design Effects. In the author's view, the influence of combustion-chamber geometry on detonation appears to be a complicated affair, not yet fully understood. In some cases small differences in geometry or in spark-plug location make larger differences in octane requirement than can be explained by theory. An enormous amount of research has been conducted by engine manufacturers to discover the optimum geometry (1.90–1.94, 2.60–2.67). Most of this work is quite empirical and is difficult to interpret in theoretical terms. Examples of attempts to isolate variables can be found in refs. 2.53, 2.631, and 2.65. Figure 2–34 illustrates current practices in combustion-chamber design. The very fact that so many different designs are used indicates that there are still unknown factors in this area.

Effect of Cylinder Size on Detonation

Figure 2–35 plots available data on the effect of cylinder size on octane requirement and knock-limited compression ratio. It will be noted that, at constant compression ratio, octane requirement increases with bore at either the same rpm or the same piston speed.

When engines of similar design (see Chapter 1) but of different size

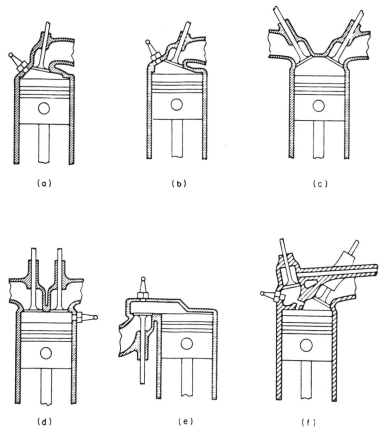

Fig. 2–34. Typical combustion-chamber designs for spark-ignition engines. (Where only one valve is shown, the other is directly behind it.) (a, b) Typical for automobile engines. (c) Typical for sports cars and racing cars, often with four valves. (d) CFR knock-rating engine, also used for engine research (see Chapter 14). (e) L head, used in small, low-cost engines. (f) Honda 3-valve stratified-charge engine. Exhaust valve is behind inlet valve as in a, b, and e.

run at the same piston speed and with the same inlet conditions, fuel-air ratio, and exhaust pressure, the following relations are known:

1. Combustion time increases with bore (2.678);
2. Cylinder-inner-surface temperatures increase with bore (Fig. 2–35d).

Both of these factors tend to increase the octane requirement; this explains the trends shown in Fig. 2–35a for constant piston speed.

At the same rpm, combustion time appears to be nearly independent of bore, as evidenced by the fact that the engines require about the same

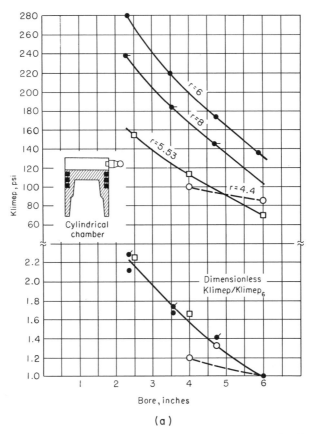

Fig. 2–35. Effects of cylinder size on detonation. Except for Kamm (2.671), all work was done on MIT similar engines (2.672–2.679) with cylinder dimensions 2.5 × 3.0, 4 × 4.8, and 6 × 7.2 in. (63 × 76, 102 × 122, and 153 × 183 mm).

(a) Knock-limited imep versus bore. Constant inlet temperature, exhaust pressure, and coolant temperature, cylindrical combustion chambers.

Symbol	Fuel	Piston Speed	Reference
●——●	aviation	2170 ft/min	2.671
—□—	71.9 O.N.	1200 ft/min	2.672
– – ○ – –	60 O.N.	1200 ft/min	2.676

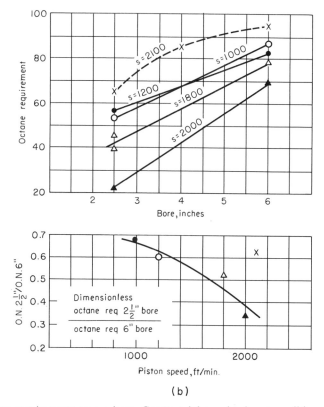

(b) Octane requirements versus bore. Constant inlet and exhaust conditions and coolant temperature. (See part *f* for details of squish-type chambers.)

Symbol	Comp Ratio	Comb Chamber	Reference
O——△	5.53	cylindrical	2.673
●——▲	5.53	cylindrical	2.675
×————×	8.7±	squish	2.678

spark timing. However, under these circumstances there is an even greater increase in cylinder-surface temperature with bore (Fig. 2–35*d*) which probably accounts for the substantial increase in octane requirement as the bore grows larger. Tests with the coolant temperature adjusted to give the same inner-surface cylinder temperature show that octane requirement still increases with bore at the same rpm (Fig. 2–35*e*).

Figure 2–35*c* shows knock-limited compression ratio versus bore. This trend follows from the results obtained with the same compression ratio for all cylinder sizes.

As shown in Fig. 2–35*f*, the use of a squish-type combustion chamber does not alter the fact that increasing bore increases octane requirements.

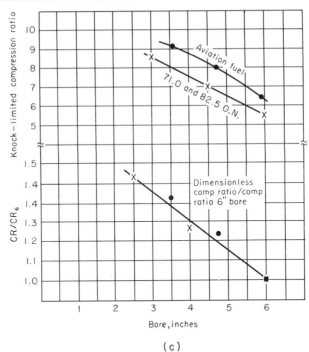

(c)

(c) Knock-limited compression ratio versus bore. Constant inlet and exhaust conditions and coolant temperature, cylindrical combustion chambers.

Symbol	Fuel	Piston Speed	Reference
●	aviation	2170 ft/min	2.671
x	O.N. = 71.0 motor method, 82.5 research method	1000,2000 ft/min	2.675

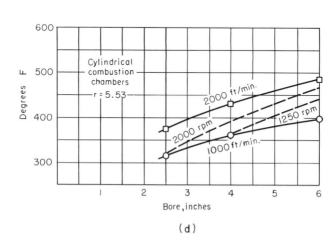

(d)

(d) Cylinder-head surface temperature versus bore. Constant inlet and exhaust conditions and coolant temperature; $r = 5.53$, $T_c = 150°F$ (Taylor, 2.677).

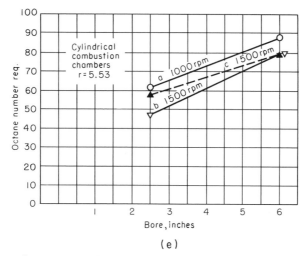

(*e*) Octane requirement versus bore at constant rpm. Same inlet conditions, same coolant temperature —○—▽—. At 1500 rpm with same charge density and same cylinder-head temperature – – ▲ – – (2.677).

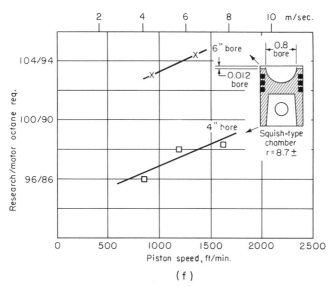

(*f*) Octane requirement versus bore with fuel of sensitivity 10, squish-type combustion chambers (Swenson, Jr., and Wilcox, 2.679).

The results shown by Fig. 2–35 emphasize one of the important drawbacks to using large cylinders on spark-ignition engines. The problem of detonation control is chiefly responsible for the fact that few spark-ignition engines have cylinders of more than 6-in. bore. An exception to this statement is that large cylinder bores are used for some gas-burning spark-ignition engines. Natural gas has such a high octane rating that this practice is possible, although detonation still constitutes a serious limitation on output and efficiency.

Combustion-Chamber Design

Summing up, the following would appear to be the desirable characteristics of combustion chambers for spark-ignition engines, from the point of view of attaining highest resistance to detonation under a given set of service conditions:

1. Small bore
2. High velocity through inlet valves
3. Short ratio of flame path to bore
4. Absence of hot surfaces in the end-gas region.

Most contemporary spark-ignition engines are designed with inlet ports that give a considerable degree of swirl to the cylinder charge.

Figure 2–34 shows examples of contemporary practice in combustion chamber design. Types *a*, *b*, and *c* are typical for spark-ignition vehicle engines. There seems to be no significant difference in the detonation characteristics of these three types in practice. The L head (type *d*) has a large ratio of flame path to bore and a limit on compression ratio due to interference with flow through the inlet valve. This type is used only for small engines where low cost is more important than fuel economy and specific output.

Type *c* chambers are used where high specific output is important, as in sports cars, racing cars, and aircraft. It allows large ratios of valve area to piston area (see pp. 525–527) and seems to be no more prone to detonation than types *a* and *b* under comparable opening conditions. The use of two spark plugs at opposite sides of the chamber, or one in the center where four valves are used, will give improved resistance to detonation.

The CFR fuel-testing and research engine, Fig. 2–34*d*, uses a simple cylindrical chamber, on account of the requirement for a variable compression ratio with a minimum of variation in chamber geometry.

Nondetonating Spark-Ignition Engines

Under the autoignition theory, spark ignition could be employed without the possibility of detonation by rendering the end gas incapable of auto ignition. This can be accomplished by separating or *stratifying* the cylinder contents in such a way that the end gas contains insufficient fuel to support autoignition. Engines using this principle are often called "stratified-charge" engines. In general, charge stratification can be accomplished by fuel injection late in the compression stroke or else by a divided combustion chamber.

An early successful commercial application of the stratified-charge principle was in the Hesselman engine (2.791), now obsolete. This engine took in only air during most of the inlet stroke, and used a shrouded valve to produce tangential swirl. The fuel was injected through a Diesel-type nozzle and ignited by a spark plug suitably located near the edge of the fuel spray. Burning evidently took place in the locality of the spray, and with proper adjustments the end gas had insufficient fuel to support detonation. Fuels of very low volatility and very low octane number were successfully used with a compression ratio of about 7.0. This engine was made obsolete in the early 1940's by the development of excellent small-size Diesel engines, which could run more efficiently on the same fuels.

The principle used in the Hesselman engine has been developed more fully by the Texas Company in what is called the "Texaco" combustion process (2.80, 2.81). In this process the inlet air is made to swirl rapidly by being admitted through a shrouded inlet valve or a tangential inlet port. Fuel is injected near the end of compression. Experimental models of these engines have been run without detonation and with good performance at compression ratios up to 12 and with a wide variety of fuels. Their exhaust emissions are similar to those of Diesel engines, which may be one reason that this system has not yet been put into production.

Divided-Chamber Spark-Ignition Engines

These have been considered for many years, both for the purpose of lean-mixture operation and for detonation control (2.790). The need for lean-mixture combustion for purposes of exhaust emission control (pp. 191–192) has stimulated new interest and new development work on this type. The most successful application is in engines manufactured by Honda. As illustrated in Figure 2.34*f*, a rich mixture is supplied to the small ignition chamber by means of an auxiliary inlet manifold and a small auxiliary inlet valve. A lean mixture is fed to the main chamber by a conventional induction system. This makes possible the successful use of overall mixture ratios leaner than can be used with present conventional chambers (Ref. 2.8111).

Effect of Deposits on Detonation

It is well known that the tendency to detonate increases rapidly with deposit accumulation, at least in the early stages of the operation of automobile engines. This increase is apparently caused by the fact that, as the deposits build up, both the effective compression ratio and the effective temperature of the inner wall surfaces increase. For this reason, compression ratios used in practice are considerably lower than those that could be used if combustion chambers remained clean (2.96–2.994). However, the most troublesome aspect of engine deposits has to do with preignition, the subject of the following section.

PREIGNITION

By preignition is meant ignition of the charge before the ignition spark occurs. This type of ignition is caused by a hot surface. In practice, the chief sources of preignition are overheated spark-plug electrodes or very hot carbon deposits. When preignition occurs, it is the equivalent of an advanced spark and may cause detonation.

Conversely, when detonation is severe and long-continued, it may heat up the spark-plug points or carbon particles to the point where preignition occurs. The heating effect of detonation is due to the high temperature of the burning end gas and to the increased relative velocity of the gases caused by pressure waves (see Volume I, Chapter 8).

As compression ratios in automobile engines have increased, difficulties with preignition have appeared and considerable research into this problem has been undertaken (2.85–2.998). The trouble manifests itself in what is called *rumble* and *wild ping*. These phenomena seem to be due to preignition, which may cause both very high rates of pressure rise due to multiple ignition points and advanced timing, together with irregular detonation. The chief source of ignition in such cases appears to be incandescent particles of hard deposit, which project into the combustion chamber and become overheated.

The tendency of fuels to ignite from a hot surface seems to have little or no relation to the detonation characteristics of the fuels. (See Chapter 4 for a discussion of the effects of fuel composition on the preignition tendency.)

In 1965, automobile gasoline sold in the U.S.A. was in the octane-number range of 85–100, with compression ratios as high as 12. It is quite possible that further increases in compression ratio will be limited by deposit preignition rather than by fuel octane number.

Effect of Fuel Composition

This chapter has been devoted chiefly to the effects of operating conditions, design, and size on the phenomena known as detonation and preignition. It has been evident throughout that fuel composition is a most important factor. This subject will be treated in detail in Chapter 4.

3 | Combustion in Diesel Engines

While the chemical reactions during combustion are undoubtedly very similar in spark-ignition and in Diesel, or compression-ignition, engines, the physical aspects of the two combustion processes are quite different. In the spark-ignition engine in normal operation, the fuel is substantially all in the gaseous state and the fuel, air, and residual gases are rather uniformly mixed at the time of ignition. Ignition occurs at one or more fixed points and at a crank angle that is subject to accurate control. Ignition is followed by the spread of a definite flame front through the mixture at measurable velocities. Except at an ignition point or in a detonating zone, combustion at any given point in the mixture is initiated by means of a *transfer* of energy, or of energized particles, from an adjacent element that is already burning; the time occupied by combustion depends on the rate at which this transfer takes place.

In the Diesel engine air, diluted by a small fraction of residual gas, is compressed through a volume ratio of from 12 to 20, and liquid fuel is sprayed into the cylinder near the top-center position of the piston.

Since both the pressure and temperature of the cylinder contents at the time of injection are very high, some chemical reaction undoubtedly begins as soon as the first droplet of injected fuel enters the cylinder. However, this chemical reaction starts so slowly that the usual manifestations of combustion, namely the appearance of a visible flame or of a measurable pressure rise, occur only after the expiration of an appreciable period of time called the *delay period*. Thus, for Diesel engines, it is convenient to define ignition as the moment when a flame visible to the eye, or to a photographic film, just appears or when the pressure starts to rise appreciably as the result of combustion. Obviously, this point can be detected only in engines equipped with windowed cylinders or with pressure indicators. As Fig. 3–1 shows, the appearance of the first visible flame is also the moment when the pressure begins to rise above the compression pressure. This point can be detected with any one of a number of types of pressure indicator (see Fig. 2–1 and Chapter 5 of Volume I).

In the Diesel engine, the time and place where ignition occurs is not fixed by anything so easily controlled as an ignition spark. Flame normally

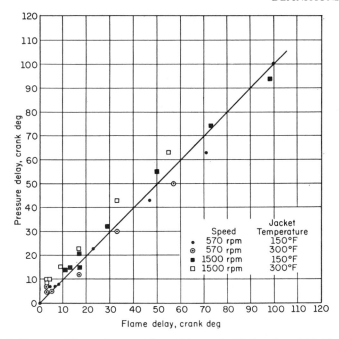

Fig. 3–1. Pressure-delay angle versus flame-delay angle (Rothrock and Waldron, 3.14).

appears while the distribution of fuel in the air-and-residual-gas mixture is extremely nonuniform, and while a considerable part of the fuel is still in liquid form. Because of the high compression ratio, the gases in the cylinder at the time of injection are well above the temperature and pressure required to support chain reactions in a uniform fuel-air mixture. Under these circumstances, the ignition of any element of the charge does not require the transfer of energy from some other portion but will occur when the local circumstances of temperature, pressure, and mixing of fuel and air make combustion possible. Ignition of one part of the charge may ignite adjacent parts of the charge provided they are in a state to support combustion, and combustion of one part may reduce the reaction time in parts of the charge that have not yet burned by raising their pressure and temperature. The point to be emphasized, however, is that combustion in the compression-ignition engine is a question of local conditions in each part of the charge and is not dependent on, although it may be assisted by, the spread of flame from one point to another. Thus, the combustion rate is affected by the state and distribution of the fuel as well as by the pressure and temperature in the chamber.

In Chapter 2 it has been shown that in a well-mixed and evaporated fuel-air mixture the reaction rate following autoignition can be extremely rapid. The fact that the maximum reaction rate in the compression-ignition engine is often slower than that of autoignition in a premixed charge of similar over-all composition is due entirely to the retarding effects of incomplete mixing of fuel with air and of incomplete fuel vaporization.

DEFINITIONS

In order to simplify the discussion, the following terms are defined:

Injection time. The time elapsed between the start of spray into the cylinder and the end of flow from the nozzle.

Injection angle. The crank angle between the start and end of injection.

Delay period. The time between start of injection and first appearance of flame or pressure rise.

Delay angle. The crank angle corresponding to the delay period.

Figure 3–1 shows that the delay angle based on pressure rise is the same as that based on appearance of flame.

PHOTOGRAPHS OF THE COMBUSTION PROCESS

High-speed motion pictures of the combustion process in Diesel engines have been made by the National Advisory Committee for Aeronautics and others (3.11–3.192) through transparent windows in the cylinder. These photographs show the progress of the fuel spray and the development of flame within the combustion chamber. Figure 3–2 reproduces a series of

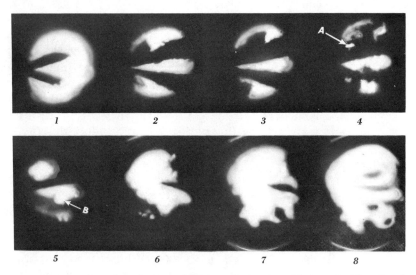

Fig. 3–2. High-speed motion pictures taken through the glass-sided combustion chamber of the NACA compression-ignition apparatus. Delay period shorter than injection period; *A* is first appearance of flame; *B* is second flame nucleus; injection ends in frame 7 (3.14).

such photographs which show the progress of the spray, the formation of ignition points (*A*, *B*) in the spray envelope (where the fuel-air ratio is most favorable to ignition), rapid inflammation immediately after ignition, and subsequent burning as more fuel is injected. Later exposures in these series show the gradual dying out of the flame, which in this case persists during about half of the expansion stroke and which remains longest at scattered points which are probably zones where the fuel is slow in coming into contact with oxygen. In this figure the delay angle is shorter than the injection angle, since fuel is still entering when ignition first becomes evident.

Figure 3–3 shows a series of photographs in which the delay angle was longer than the injection angle. The first flame appears at *A*, eight frames after fuel has ceased to flow from the nozzle and at a time when it appears that evaporation of the fuel is complete, or nearly so. Inflammation is general in the next frame and appears to be complete in the one after that, 1/1250 sec after ignition appears. The observed rate of pressure rise is much higher than that of Fig. 3–2, where inflammation started before injection and evaporation were completed.

Pressure–Crank-Angle Diagram. Figure 3–4 shows pressure–crank-angle diagrams corresponding to Figs. 3–2 and 3–3. Unfortunately the injection timing was not adjusted to bring peak pressure at the same crank angle, so that the two curves are not strictly comparable with regard to pressure development.

THE THREE PHASES OF COMBUSTION

Ricardo (3.02) conceived of the combustion process in the compression-ignition engine as taking place in three stages, the first of which is the delay period. The delay is always long enough that, when ignition occurs, there is an appreciable amount of evaporated and of finely divided fuel well mixed with air. Once ignited, this fuel tends to burn very rapidly, by reason of the multiplicity of ignition points and the high temperature already existing in the combustion chamber. This *period of rapid combustion* is Ricardo's second phase of the process. After the period of rapid combustion, the fuel that has not yet burned, together with any fuel subsequently injected, burns at a rate controlled principally by its ability to find the oxygen necessary for combustion. This period is Ricardo's third stage of combustion.

The first two stages of combustion may be defined on the basis of transition points on the pressure–crank-angle diagram. Referring to Fig. 3–4, the first stage, or delay, has already been defined. The second stage may be taken as beginning at the end of delay and continuing to the point of maximum pressure. The third stage begins at that point, but its end cannot be accurately determined from the pressure–crank-angle diagram.

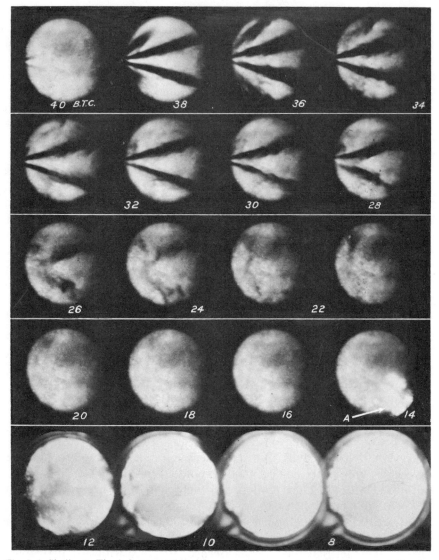

Fig. 3–3. Similar to Fig. 3–2, except that delay period is longer than injection period. Numbers indicate degrees before top center; *A* indicates start of flame 14° btc; injection ends 24° btc (3.14).

The Delay Period. The delay period in Diesel engines, often called the *ignition lag*, apparently corresponds to the period of preliminary reactions that occur prior to the appearance of flame in the autoignition of pre-mixed charges, as discussed in Chapter 2. It has been postulated by some authorities that the delay period in compression-ignition engines may include a period of heating the fuel droplets before any chemical reaction occurs (3.21–3.23, 3.53, 3.55, 3.56). In evaluating this theory, however, it

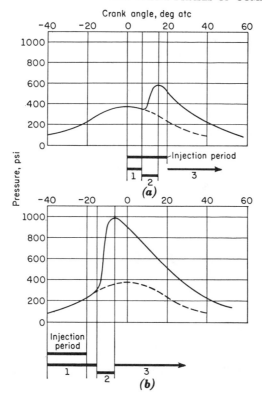

Fig. 3–4. Pressure-crank angle diagrams showing the three stages of combustion. (*a*) corresponds to Fig. 3–2 with delay shorter than injection period; (*b*) corresponds to Fig. 3–3 with delay longer than injection period. 1 = delay period, or first stage; 2 = period of rapid combustion, or second stage: 3 = third stage of combustion. Dashed line is compression and expansion of air only, without combustion. 5 × 7 in. (127 × 178 mm) cylinder, 720 rpm, jacket temp, 150°F (66°C), 0.00025 lbm fuel per stroke (3.14).

must be remembered that each droplet is surrounded by vapor immediately after entering the combustion chamber. It therefore seems to the author that reactions must start in the vapor surrounding the drop surfaces almost simultaneously with the entrance of each drop into the cylinder. This view is supported by the fact that wide variations in those factors that should affect the heating period, such as changes in fineness of the spray or in fuel volatility, have little effect on the delay (see Table 3–3, p. 109).

Studies of the combustion of droplets of fuel in air (3.21–3.23) indicate that ignition starts in the layer of vapor surrounding the drop and that the combustion rate of drops is limited by their evaporation rate. Burning rate decreases as the fraction of oxygen in the surrounding air decreases. These observations help to explain that part of Diesel combustion which occurs prior to complete evaporation of the drops.

Bomb Tests. Some excellent research on the effect of physical and chemical factors on the delay has been done by means of constant-volume bombs equipped with typical fuel-injection systems (3.31–3.36). Figure 3–5, taken from some of this work, shows effects of bomb pressure and bomb temperature on the delay. While the absolute values vary with

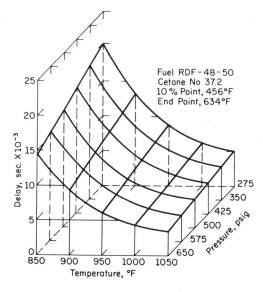

Fuel RDF - 48 - 50
Cetane No 37.2
10 % Point, 456°F
End Point, 634°F

Fig. 3–5. Ignition delay measured in a constant-volume bomb with fuel injection; temperature and pressure are before injection (Hurn and Hughes, 3.34).

the fuel used, the trend of the curves is typical. It is noteworthy that within the range of these tests, which include the usual range of Diesel-engine operation, changes in pressure have a relatively small effect, while changes in temperature have a pronounced effect on the delay. A similar relation for premixed charges was discussed in Chapter 2. Other important observations resulting from this work are as follows:

1. The time lag in establishing temperature equilibrium between injected fuel and bomb atmosphere appeared to be negligible.

2. The effect of volatility changes in otherwise similar fuels was negligible.

3. An important factor influencing delay is fuel composition. This subject will be discussed in Chapter 4.

Ignition Delay in Engines. From the bomb experiments it can be inferred that, with a given fuel, the chief factor influencing delay would be the average temperature of the cylinder contents during the delay period. In engines, however, there is another strong influence on delay, namely, the impingement of the spray on hot surfaces. If these surfaces are sufficiently hot, significant reductions in delay can be obtained by this means.

In premixed charges, we have seen that the time between the completion of the compression process and the appearance of flame varies with fuel-air ratio (see Fig. 2–14). However, in the case of injected fuel, the term " fuel-air ratio" cannot be used in the same sense as in the case of a premixed charge. With fuel injection it is evident that, as long as the fuel is not completely evaporated, the complete range of fuel-air ratios from zero (no fuel) to infinite (no air, within the fuel droplets) must be present, and ignition will occur where the local fuel-air ratio is most favorable. Even the physical character of the spray will not affect the availability of a wide range of fuel-air ratios, and therefore it might be expected that, with a given fuel, the length of the delay period will depend chiefly on the pressures and temperatures that exist in the cylinder gases during that period and on the temperature of the combustion-chamber surfaces against which the spray impinges. That such is the case will appear as the discussion in this chapter proceeds.

Period of Rapid Combustion. Combustion in the period of rapid combustion is due chiefly to the burning of such fuel as has had time to evaporate and mix with air during the delay period. The rate and extent of burning during this period are therefore closely associated with the length of the delay period and its relation to the injection process.

In order to assist our study of these relations, Fig. 3–6 has been prepared.

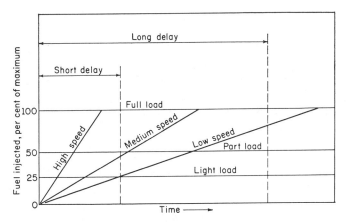

Fig. 3–6. Relation of fuel quantity injected to delay period. Fuel-injection rate is assumed proportional to engine speed.

In this figure the injection rate is assumed to be directly proportional to engine speed. While this characteristic is not achieved completely in practice, the usual injection system follows it closely enough to make the assumption useful. From Fig. 3–6 it is evident that the fraction of full-load fuel quantity injected during the delay period depends on engine speed and load, as well as on the length of the delay period itself. On the other hand, time for evaporation and mixing before ignition depends only on the time of the delay period.

Figures 3–2 and 3–3 constitute a pair where fuel quantity and speed were held constant, but the delay angle was much longer in Fig. 3–3 than in Fig. 3–2. The two figures correspond, qualitatively, to the short- and long-delay diagrams of Fig. 3–6, with full load, intermediate speed.

In Fig. 3–2 only a fraction of the fuel has been injected by the time the delay period ends. Assuming that the period of rapid combustion occupies the first two frames showing flame, it is evident that only a fraction of the chamber is inflamed during this period, indicating that a relatively small quantity of fuel takes part. In Fig. 3–3, on the other hand, the delay is longer than the injection period, and the chamber is completely inflamed in the next frame after ignition. In this case combustion resembles that observed in the autoignition of premixed charges (see Fig. 2–11, top portion).

From the foregoing considerations it might be expected that both the rate and the extent of pressure rise during the second stage of combustion would increase as the delay period increases, since both mixing time and the fraction of fuel taking part in rapid combustion increase. Experience shows that such a relation holds in practice, provided the injection timing* is so adjusted that the period of rapid combustion occurs at substantially the same crank angle. With constant injection timing, piston motion complicates the problem, as we shall see.

The influence of delay on both rate and extent of pressure rise during the second phase is especially strong when the delay period is shorter than the injection period. When the delay is longer than the injection period, it is obvious that the quantity of fuel involved is unaffected by the length of the delay period.

Third Phase of Combustion. The third phase of combustion is the period from maximum pressure to the point where combustion is measurably complete.

When the delay angle is longer than the injection angle, the third period of combustion will involve only such fuel as has not found the necessary oxygen during the period of rapid combustion. In this case, the combustion rate is limited only by the mixing process. This in turn is controlled by the ratio of oxygen to unburned fuel and by the degree to which the two are distributed and mixed at the end of the second period. Even when all the fuel is injected well before the end of the delay period, poor injection characteristics can extend the third period well into the expansion stroke and thus cause low output and poor efficiency.

Under conditions such as those of Figs. 3–2 and 3–4a, where the second phase of combustion has been completed before the end of injection, some portion of the fuel is injected during the third phase, and the rate of burning will be influenced by the rate of injection as well as by the mixing rate. Engines running at low rpm generally operate in this regime. They should

---

* Nominal injection timing is usually taken as the crank angle at the *start* of injection.

be designed to secure rapid mixing of fuel and air during the third period in order to complete the combustion process as early as possible in the expansion stroke.

DETONATION IN THE DIESEL ENGINE

Recalling the study of detonation in Chapter 2, one must be impressed by the apparent similarity between the theory of detonation there presented and the normal process of ignition and combustion in the compression-ignition engine. There is little doubt that the two are fundamentally similar. Both appear to be a process of compression ignition, subject to the ignition-time-lag characteristics of the fuel-air mixture. What happens subsequently to compression ignition, however, is quite different in the two cases. Because under normal conditions the fuel and air are imperfectly mixed in the compression-ignition engine, the rate of pressure rise is normally lower than that in the detonating part of the charge in a spark-ignition engine. However, when the delay angle is as long as or longer than the injection angle, the fuel may be so well evaporated and mixed with air at the time of combustion that the whole fuel quantity burns at a rate similar to that of the end gas in a detonating spark-ignition engine. As might be expected from the discussion of the effects of rapid load application in Chapter 8 (see Fig. 8–39), the result will be high stresses and severe vibration of the cylinder and its associated parts.

In practice, most Diesel engines have a rate of pressure rise sufficiently high to cause audible noise. When such noise becomes excessive in the opinion of the observer, the engine is said to "detonate" or "knock." It is evident that personal judgment is here involved. Thus, in the compression-ignition engine there is no definite distinction between normal and "knocking" combustion. It is merely a question of whether or not the rate of pressure rise is high enough to cause, in someone's judgment, objectionable noise or excessive vibration in the engine structure.

From the foregoing discussion it is evident that knock or detonation in Diesel engines is essentially the same phenomenon as *roughness* in spark-ignition engines (see pp. 32–33 and 296–297). As we shall see, avoidance of knock, or excessive roughness, is one of the important considerations in Diesel-engine design and operation.

Detonation in S-I and C-I Engines Compared

The discussions in Chapter 2 and in this chapter make possible an interesting comparison between detonation in the spark-ignition and detonation, so-called, in the compression-ignition engine.

While detonation in the spark-ignition engine and in the compression-ignition engine have essentially the same basic cause, that is, compression ignition followed by a rapid pressure rise, it is important to note that in the first case the reaction is in the last part of the charge to burn, while in the second it occurs in the first part up to, and sometimes including, the whole charge. In order to avoid detonation in the spark-ignition engine, it is necessary to prevent compression ignition from taking place at all. In the compression-ignition engine the earliest possible ignition is necessary, so that conditions favorable to knocking will not have time to form during the delay period. The result is the rather striking fact that many of the methods of reducing detonation are exactly opposite for the two types. Table 3-1 emphasizes this divergence.

Table 3–1. Characteristics Tending to Reduce Detonation or Knock

Characteristic	Spark-Ignition Engines	Compression-Ignition Engines
Ignition temperature of fuel	High	Low
Time lag of fuel	Long	Short
Compression ratio	Low	High
Inlet temperature	Low	High
Inlet pressure	Low	High
Combustion-chamber wall temperature	Low	High
Revolutions per minute	High	Low
Cylinder size	Small	Large

The Ideal Combustion Process

Up to this point a basis has been provided for establishing an ideal combustion process for the compression-ignition engine. With one exception this process is similar to the combustion process in the limited-pressure fuel-air cycle (see Volume I, pp. 72–74). The exception is that the rate of pressure rise during the period corresponding to constant-volume combustion should be as rapid as possible without, however, exceeding a certain value (usually established experimentally) for the particular size and design in question. The fuel remaining after the period of rapid pressure rise should be burned at a rate such as to hold the cylinder pressure constant, at the maximum allowable value, until all the fuel is burned.

Control of combustion toward this ideal is one of the major problems in the design and operation of compression-ignition engines. Methods of exercising such control will be discussed later in this chapter.

Effect of Operating Conditions on Combustion

In the case of Diesel engines the combustion rate, as well as the rate of pressure rise, depends heavily on the design of the combustion chamber and the injection system. In practice, the design of these elements varies widely, and it is therefore more difficult to generalize trends in this type of engine than in the case of spark-ignition engines, where there is much less variety in design. In view of these facts, the trends noted for Diesel engines of a particular design are not necessarily typical of all types of such engines. For example, the illustrations and figures which follow all apply to engines using *jerk-pump systems* in which injection rate is roughly proportional to engine speed (see pp. 216–224). While this type of system represents nearly all modern practice, there are some engines that have systems so different that the same relations may not apply.

Injection Timing. Pressure–crank-angle curves showing the effect of injection timing at constant speed and fuel-air ratio are shown in Fig. 3–7. Maximum imep, and therefore maximum efficiency, occurs with peak pressure at about 15° after top center (run 409).

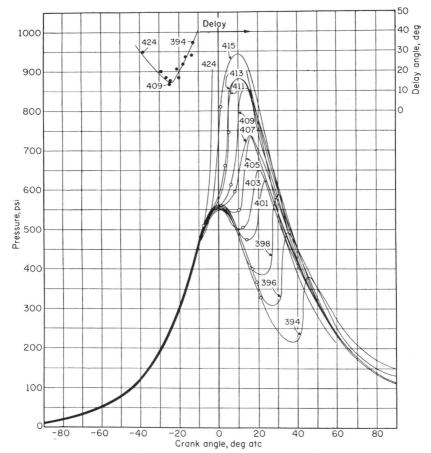

Fig. 3–7. Effect of injection timing on pressure–crank-angle diagram. Numbers indicate run numbers; ○ indicates 30° after start of injection; ● indicates start of injection; single-cylinder 5 × 7 in. (127 × 178 mm) engine with open combustion chamber (Rothrock, 3.03).

Shortest delays occur when the delay period includes top center — in this case when injection starts about 25° before top center (run 409). Earlier or later injection results in longer delays. The longest delay is with the latest injection, where average pressure, and therefore average temperature, during the delay period is lowest.

Rates of pressure rise are affected in this case both by the delay angle and by piston motion. Early injection gives a very high rate. Here a long delay occurs, together with inward piston motion during the subsequent rapid combustion. With late injection, rates of pressure rise are again high, owing to long delays, even though the rapid-combustion period occurs well after top center, with the piston in rapid descent.

Many engines use an injection timing later than that for maximum imep in order to reduce both maximum pressure and rate of pressure rise. Runs 403, 405, and 407 of Fig. 3–7 illustrate this point. Such late timing, of course, involves some sacrifice in output and efficiency.

As in the case of spark-ignition engines, proper timing of the combustion process is an essential element in Diesel-engine operation. Thus, the most satisfactory way of considering the influence of other operating variables would be under conditions of best-power injection timing, that is, timing to give highest mean effective pressure. As in the case of spark-ignition engines, best-power timing brings peak pressure 15° to 20° after top center (run 409). Under such circumstances the period of rapid combustion will occur while piston motion is small, and the maximum rate of pressure rise will not be greatly affected by piston motion. However, Diesel engines are usually operated with fixed injection timing, and therefore a change in the delay period means a change in the piston position and piston velocity at which important events occur. Even under laboratory conditions, very few tests are made with the injection timing adjusted for best power. Thus, most of the effects to be discussed will have to be based on fixed injection timing, that is, a fixed crank angle for the start of injection.

Effect of Rpm. In Chapter 2 it was demonstrated that flame speed in the spark-ignition engine increases nearly in proportion to engine speed, and therefore the number of crank degrees occupied by the combustion process is nearly independent of rpm. Here the pressure–crank-angle diagram changes very little as speed changes, provided the air and fuel quantities per stroke are held constant and the spark timing is properly adjusted.

In the case of Diesel engines an important part of the combustion process, the delay time, tends to be independent of speed. This means that the delay angle will vary with speed and that speed may have large effects on the pressure–crank-angle diagram.

Figures 3–8 and 3–9, together with Table 3–2, show that the delay angle increases with increasing speed but not in direct proportion to speed. The two figures disagree on the effect of increasing speed (and conse-

Table 3–2. Delay Angles and Delay Times for Figs. 3–8 and 3–9

Figure	Rpm	Start of Injection, deg btc	Delay Angle, deg	Delay, seconds × 1000
3–8	570	40	14	4.10
	1500	40	24	2.67
	570	20	7	2.05
	1500	20	12	1.33
	570	10	5	1.47
	1500	10	10	1.12
3–9	500	13	5	1.67
	1000	13.5	7	1.17
	1500	13	11	1.22
	2000	11	10	0.83

quently increasing delay angle) at the second and third stages of combustion. In both cases the delays are generally shorter than the injection angle. Figure 3–8 shows the expected increase in rate of pressure rise and maximum pressure as the delay angle increases. On the other hand, rather opposite trends are shown in Fig. 3–9. Of course many factors other than delay are affected by changes in speed. Among these are spray charac-

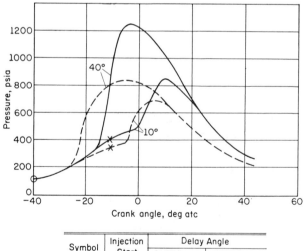

Symbol	Injection Start	Delay Angle	
		At 570 rpm	At 1500 rpm
O	40° btc	14	24
X	10° btc	5	10

Fig. 3–8. Effect of speed on pressure–crank-angle diagram; 5 × 7 in. (127 × 178 mm) open-chamber cylinder; injection angle about 25°; jacket temperature 150 F. Solid line, 1500 rpm; dashed line, 570 rpm (Rothrock and Waldron, 3.14).

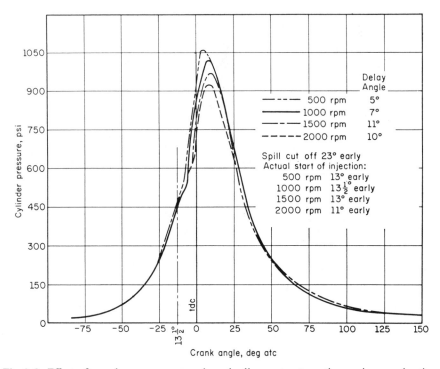

Fig. 3–9. Effect of speed on pressure–crank-angle diagram; automotive engine, combustion chamber of type *e*, Fig. 3–18 (Dicksee, 3.670).

teristics, chamber wall temperatures, and volumetric efficiency. It is not surprising, therefore, to find two very different engines showing different responses to speed changes.

Figure 3–8 is less typical of contemporary Diesel engines than is Fig. 3–9. In the latter it is evident that the second and third stages of combustion respond to engine speed as in the case of spark-ignition engines, that is, with combustion rates nearly proportional to the speed. On this account, Diesel engines can be run successfully at quite high piston speeds,* and some ratings for Diesel engines with small bore are in excess of 3000 rpm. However, to operate satisfactorily at such speeds, fuel of good "ignition quality" (short ignition delay, see Chapter 4) is required.

Turbulence. The response to engine speed of the second and third stages of combustion is probably closely associated with charge turbulence, as in the case of spark-ignition flame propagation. However, in the case of Diesel engines the turbulence effect must be associated more closely with the *mixing* process than with the propagation of chemical reaction. In cases

---

* Some aircraft and racing Diesel engines have been rated at well over 3000 ft/min (15m/sec) piston speed.

where combustion starts early in the injection process, the use of a strong swirl to give a high air velocity across the spray has proved to be very effective in securing short second and third combustion stages. This subject will be discussed at greater length under the heading of Diesel combustion-chamber design.

An investigation of turbulence effects in Diesel combustion is now under way at MIT, using the rapid-compression machine shown in Fig. 3–10.

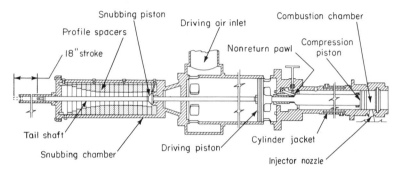

Fig. 3–10. MIT spray-combustion apparatus.

This machine quickly compresses a charge of air to measurable temperature and pressure in the range of Diesel-engine conditions at the point of injection. At the end of the compression stroke the piston is locked in position and the fuel is then injected through a suitable nozzle. In this machine (3.37–3.401) it is easy to give various types of air motion to the charge, either before or during compression. Also, the walls of the combustion chamber do not have to be maintained at the temperature of the air just before injection but can be held at a temperature more nearly representing engine practice.

Work with the MIT machine to date of writing has been preliminary in nature, but the results already show that increasing swirl rate decreases combustion time and gives more complete combustion, as evidenced by higher pressures at a given volume. On the other hand, gentle, small-scale turbulence, induced by placing a screen between piston and combustion chamber at the end of piston travel, appears to have a detrimental effect on combustion, probably because of the heat-absorbing characteristics of the screen. As work with this machine continues, techniques for producing and measuring air motion and for correlating the results with engine observations should improve.

Effect of Fuel-Air Ratio. While it is convenient to define trapped-fuel–air ratio as the ratio of the total mass of fuel injected to the mass of fresh air in the cylinder, we have already seen that for some time after fuel is injected there are local fuel-air ratios varying from infinity at the drop surfaces to zero at points not yet reached by fuel vapor. Thus, as long as evaporation is not complete before ignition, it would be expected that the quantity of fuel injected would have no direct effect on the delay period. This conclusion is confirmed by Fig. 3–11. An indirect effect of fuel quantity on delay is often noted, however, owing to the fact that, as the

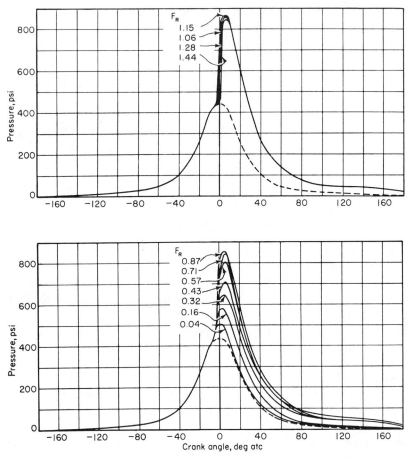

Fig. 3–11. Effect of fuel-air ratio on the pressure–crank-angle diagram; NACA 5 × 7 in. cylinder, 1500 rpm (Rothrock and Waldron, 3.16).

fuel-air ratio is decreased, combustion temperatures are lowered and cylinder-wall temperatures reduced. In many engines, especially those employing hot combustion-chamber surfaces to reduce delay, the length of the delay period increases with decreasing fuel-air ratio because of this reduction in wall temperature.

A most interesting feature of Fig. 3-11 is the fact that there is little reduction in the maximum rate of pressure rise except at the very low fuel-air ratios. On the other hand, maximum pressure falls steadily with the fuel-air ratio, showing that in this case most, if not all, of the fuel was injected during the delay period. Other tests have shown that maximum pressure is little affected by the fuel-air ratio as long as only a fraction of the total fuel is injected during the delay period.

Limits on Fuel-Air Ratio. Figure 3–11 proves that Diesel engines can operate at extremely low fuel-air ratios (F_R min. = 0.04 in the figure). This can be explained by the assumption that burning takes place near and in the spray where local ratios are within the usual inflammable range. However, it is also likely that even with complete evaporation prior to ignition the mixture temperature is high enough for autoignition and burning without dependence on a flame front. A practical difficulty that may appear at very low fuel-air ratios is that of securing spray characteristics adequate for combustion with the minute fuel quantities involved and with relatively low wall temperatures.

Smoke Limit. There is always a practical limit to the maximum fuel-air ratio that can be used in Diesel engines, which is set by incomplete combustion accompanied by a smoky exhaust and rapid build-up of carbon deposits in the cylinders (3.851–3.855). Owing to the extremely short time available for mixing,* as the fuel-air ratio increases beyond a certain value an appreciable fraction of the fuel fails to find the necessary oxygen for combustion and passes through the cylinder unburned or partially burned. The result is a certain amount of smoke. One of the best measures of the quality of a Diesel combustion system is the fuel-air ratio at which smoke becomes pronounced. The higher this ratio, the better the mixing process and, in general, the higher the allowable mean effective pressure. Smoke is objectionable both from the point of view of a public nuisance and because engines that are run for extended periods with smoky exhaust accumulate deposits rather fast and must be overhauled frequently.

Smoke at low fuel-air ratios, that is, at very light loads and at idling, is not uncommon. Smoke under these conditions is due to the difficulties (already mentioned) associated with injection of minute fuel quantities, with relatively cool cylinder-wall surfaces. Fuel composition has important effects on smoke and other exhaust emissions (see Chapter 4 and refs. 4.600–4.608).

Effect of Compression Ratio. The effect of the compression ratio on the pressure–crank-angle diagram when injection timing, speed, and fuel quantity are held constant is plotted in Fig. 3–12. The effect on delay is in the direction that would be predicted from the autoignition experiments reported in Chapter 2. In this case rates of pressure rise are nearly the

---

* Maximum mixing time from the smoke point of view would be the time between injection and exhaust opening, or about 140° of crank travel. Compare this with the corresponding mixing time in a carburetor engine, which includes time in the manifold as well as in the compression and expansion strokes, plus the inlet stroke in the case of 4-cycle engines.

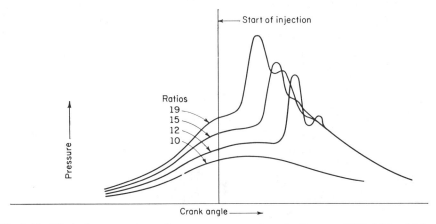

Fig. 3–12. Effect of compression ratio on pressure–crank-angle diagram; CFR 3.25 × 4.5 in. cylinder; 600 rpm; injection advance angle 12°; fuel, kerosene, $F_R = 0.5$ (Pope and Murdock, 3.47).

same. However, if injection timing had been set in a more practical way, that is, so that peak pressure occurred at the same crank angle, it is probable that the rate of pressure rise would have been higher as the compression ratio was reduced and the delay became longer.

Compression Ratios Used in Diesel Engines. Figure 4–6 in Volume I (p. 88) indicates that, unless maximum pressures are allowed to be high, increasing the compression ratio in the Diesel range (14 to 20) gives only a small improvement in efficiency. On the other hand, the higher the compression ratio, the higher the engine friction, leakage, and torque required for starting.

Thus, Diesel-engine designers seek to use the lowest compression ratio consistent with satisfactory starting and operation with the available fuel. Usually it is starting that sets the low limit on compression ratio (3.891–3.896). In practice the lowest compression ratios are used with large cylinders, which are generally located in heated engine rooms. Among the automotive engines, which may have to be started under very cold conditions, the divided-chamber engines generally require higher compression ratios than do open-chamber engines, because of heat losses in the combustion-chamber throat while the engine is cold. Divided-chamber engines are more likely to require glow plugs* or other special starting aids than are open chamber engines. "Multifuel" Diesel engines, discussed later in this chapter, call for relatively high compression ratios.

Increasing Inlet Pressure. Increasing the inlet pressure decreases the

---

* Glow plugs are similar to spark plugs in general construction, but they have a resistance wire in place of the spark gap. For starting purposes this wire is heated to incandescence by means of electric current from a storage battery.

delay, as would be expected from the autoignition experiments of Chapter 2. The tests of Fig. 3–13 were made to show the effect on delay only, and they were made with constant fuel quantity and constant injection timing. The

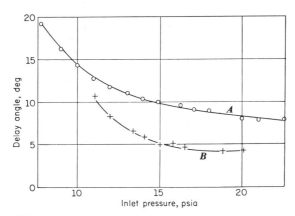

Fig. 3–13. Effect of inlet pressure on delay angle; constant inlet temperature, constant fuel quantity with low fuel-air ratio. A = open combustion chamber; B = divided chamber, type c, Fig. 3–19 (Dicksee, 3.08).

large influence of pressure in this case is associated with its influence on temperature (see Fig. 3–5). As the delay increases, piston motion causes the average temperature during delay to fall. The effect of pressure should be much less with best-power injection timing, where the temperature during delay would be more uniform.

Increasing Inlet and Jacket Temperatures. Increasing the inlet temperature (Fig. 3–14) and increasing the jacket temperature (Fig. 3–15) reduce delay, because the temperature of the air in the cylinder at the time of injection is increased, and in the latter case perhaps also because the fuel spray impinges on a hotter surface. At optimum ignition timing, either temperature increase tends to reduce rate of pressure rise.

Supercharging. Supercharging without aftercooling increases the inlet temperature as well as the inlet pressure, and both of these changes shorten

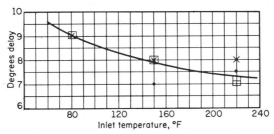

Fig. 3–14. Effect of inlet temperature on delay angle; CFR engine, Comet cylinder head, type c, Fig. 3–19. (Sloan Automotive Laboratories.)
● $F_R = 0.72$ × $F_R = 0.61$ □ $F_R = 0.55$

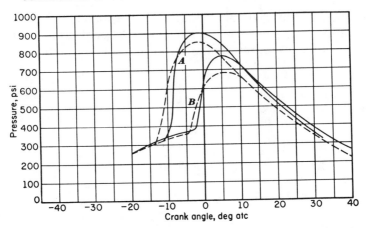

Fig. 3–15. Effect of jacket temperature on pressure–crank angle-diagram. NACA 5×7 in. cylinder; 570 rpm; $r = 14.8$; $F_R = 0.75$. *A*, injection starts, 20° btc; *B*, injection starts, 10° btc; – – – – jacket temperature 300°F (149°C); ——— jacket temperature 150°F (66°C) (Rothrock and Waldron, 3.14).

the delay period. Thus, supercharging is actually favorable to low rates of pressure rise and to maximum pressures lower in proportion to inlet pressure than is the case with the same engines unsupercharged. As we have seen, the reverse is true for spark-ignition engines. In Diesel engines, therefore, no definite limit to supercharging is set by combustion characteristics. In practice, supercharging limits must be set by the less easily determined characteristics of reliability and durability. Excessive supercharging reduces these characteristics because of high maximum cylinder pressures and high rates of heat flow.

Spray Characteristics. The physical characteristics of the fuel spray in relation to the size, shape, and detail design of the combustion chamber can have an enormous effect on the pressure–crank-angle diagram of Diesel engines. One of the most important and most difficult aspects of the development of a new Diesel engine is that of determining the optimum spray characteristics from the point of view of power and efficiency without at the same time incurring excessive rates of pressure rise and high maximum pressures.

The optimum spray characteristics vary with combustion-chamber size and design. Figures 3–16 and 3–17 and Table 3–3 show the effect of a wide variety of spray characteristics with one particular combustion chamber. The start of injection was at 15° before top center in every case. The very poor atomization of the first two nozzles caused a relatively long delay period, probably because of the slow development of very fine droplets, but for the other nozzles the delay tended to remain constant. The maximum rate of pressure rise and the maximum pressure are highest for those nozzles that give the best distribution of the fuel, that is, for the two- and

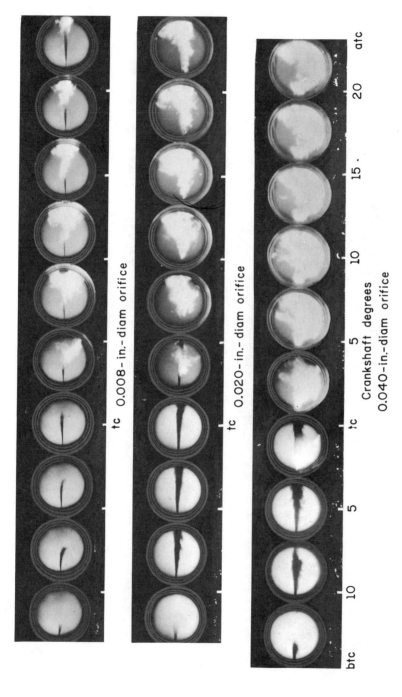

Fig. 3–16. Fuel spray and flame formation with single-orifice nozzles. $F_R = 0.88$; compression ratio 13.2; speed 1500 rpm; jacket temperature 150°F (66°C) (Rothrock and Waldron, 3.17).

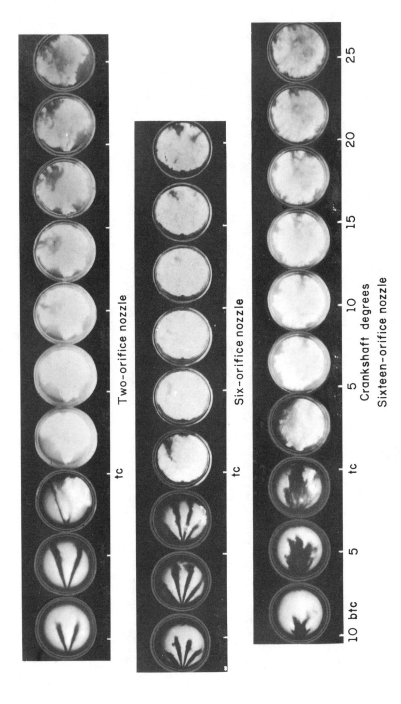

Fig. 3–17. Fuel spray and flame formation with multiorifice nozzles. Same conditions as in Fig. 3–16.

six-orifice nozzles. Distribution of the flame appears to be good in the case of the 16-orifice nozzle, but its poor performance indicates slow mixing, probably because of insufficient spray penetration with consequent slow mixing in the third stage. These results emphasize the enormous influence of mixing and distribution on the combustion process.

Ignition Aids. Because of the advantages of a short delay angle, many devices and methods have been proposed or used to shorten the delay and promote easier starting with a particular fuel and engine. Ignition-quality additives such as amyl nitrate (Chapter 4) are effective delay shorteners.

Table 3–3. Effects of Nozzle Design, from Figs. 3–16 and 3–17

Nozzle Type	Diameter or No. of Orifices	Delay, Crank-shaft Degrees	Max Pressure, psi	BMEP, psi	BSFC, lb/bhp-hr	Approx Max dp/dt psi/deg
Single-orifice	0.008 in.	20	—	—	—	—
	0.020 in.	20	545	66	0.82	30
	0.040 in.	13	690	71	0.79	30
Multiorifice	2	13	745	89	0.63	90
	6	12	800	115	0.47	90
	16	14	650	77	0.72	45

Data from ref. 3.17.

An addition of a few per cent of lubricating oil (which has a relatively low ignition temperature) is said to be similarly effective. Such ignition aids are not yet generally used, probably because a fuel of adequate ignition quality is cheaper than one of poorer quality plus the additive.

Another method of attack is the introduction of part of the same fuel, or of a lower-ignition-point fuel, with the inlet air (3.80–3.83). While apparently effective, the additional complications involved have so far prevented wide commercial application of the method. However, ether or some similar material is sometimes used for cold-weather starting.

Even with low-ignition-point fuels, some heating is usually necessary for starting Diesel engines at extremely low temperatures (3.891–3.896).

Limitations on Maximum Pressure and Rate of Pressure Rise in Diesel Engines

As indicated by the discussions of engine roughness in Chapters 1 and 8, the stress and vibration produced by a given pressure increase as the rate of pressure application increases (Fig. 8–39). Thus, a given maximum pressure following a high rate of pressure rise will always produce higher stresses than the same pressure applied more slowly. Also, vibration will be more severe as the rate of pressure rise becomes greater.

As might be expected, problems associated with high rates of pressure rise are generally more serious in the case of Diesel engines than in that of spark-ignition engines, partly because of the higher maximum pressures involved and partly because, with long delay periods, the rate of pressure rise in Diesel engines can be much more rapid than in the case of normal combustion in a spark-ignition engine. Thus, in Diesel engines both the maximum pressure and the rate of pressure rise during the rapid-combustion period must usually be limited by appropriate design or operating restrictions.

As in the case of spark-ignition engines, because of size effects (see Volume I, Chapter 11), if the indicator diagrams of similar cylinders are identical on a crank-angle basis and if the piston speed is the same, stresses due to rate of pressure rise will also be the same. Thus, for Diesel as well as spark-ignition engines it is appropriate to consider this rate on a crank angle rather than on a time basis.

COMBUSTION-CHAMBER DESIGN IN DIESEL ENGINES

It has now been shown that the combustion process in the Diesel engine should be controlled to avoid both excessive maximum cylinder pressure and an excessive rate of pressure rise, in terms of crank angle. At the same time, the process should be so rapid that substantially all the fuel is burned early in the expansion stroke.

In discussing methods used for attaining these objectives, it is convenient to divide Diesel engines into two types: *open-chamber* engines and *divided-chamber* engines. These two types will here be defined as follows:

An open combustion chamber is one in which the combustion space incorporates no restrictions that are sufficiently small to cause large differences in pressure between different parts of the chamber during the combustion process. The chambers illustrated in Fig. 3–18 come under this classification.

A divided combustion chamber is one in which the combustion space is divided into two or more distinct compartments, between which there are restrictions, or *throats*, small enough so that considerable pressure differences occur between them during the combustion process. The chambers illustrated in Fig. 3–19 are divided combustion chambers. When the burning starts in a chamber separated from the piston by a throat, divided-chamber engines are often called *prechamber* engines.

Open-Chamber Engines. In the open-chamber type, the mixing of fuel and air depends entirely on spray characteristics and on air motion, and it is not vitally affected by the combustion process itself. In this type, once the

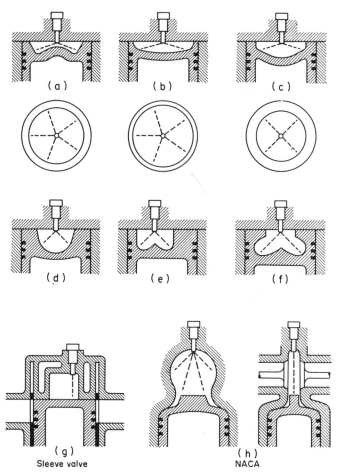

Fig. 3–18. Open combustion chambers. Chambers *a–g* use air swirl from tangential inlet ports or shrouded inlet valves; chamber *h* is NACA type using air motion from displacer on piston. Squish from small clearance over piston increases from *a* to *f*. Dashed lines indicate direction of sprays.

compression ratio, maximum operating speed, and operating temperatures are selected, the delay angle is determined chiefly by fuel characteristics. Engines of this type are very sensitive to spray characteristics, which must be carefully worked out to secure rapid mixing. Subdivision of the spray and the use of high injection pressures are usually required. In the case of high-speed (small-cylinder) engines, mixing is usually assisted by *swirl*, induced by directing the inlet air tangentially, or by *squish*, which is air motion caused by a small clearance space over part of the piston. Chambers *d* to *h* in Fig. 3–18 have notably large squish areas. All except *h* employ swirl induced by a shrouded inlet valve or a tangential inlet port.

The effects of air motion on the Diesel combustion process have already been mentioned in the discussion of the MIT rapid-compression machine. It is evident from Fig. 3–20 that air motion speeds up the combustion process in the second and third stages of combustion. This relation is also confirmed by experience with other engines. Reference 3.191 is an excellent study based on photographs of injection and combustion in typical open- and divided-chamber cylinders, and includes a very complete discussion of squish and swirl in engines of types *d*, *e*, and *f* of Fig. 3–18. This work shows evidence of much smaller radial velocities due to squish than have been assumed; this leads to the question whether the observed beneficial effects of squish-type chambers are due to the squish motion itself or to the increasing swirl velocity in the piston cavity as its diameter is made smaller. The latter increase follows from the conservation of angular momentum in the charge.

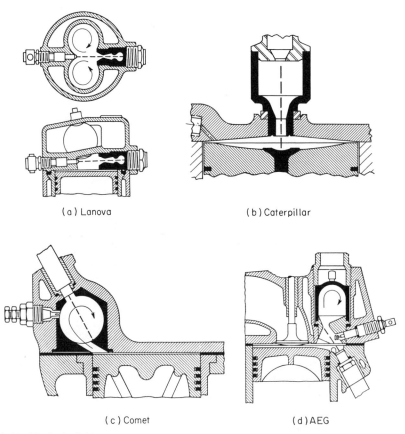

(a) Lanova   (b) Caterpillar

(c) Comet   (d) AEG

Fig. 3–19. Typical divided combustion chambers. Black areas indicate high-temperature inserts, usually of stainless steel. Dashed lines indicate direction of spray; arrows indicate direction of air swirl.

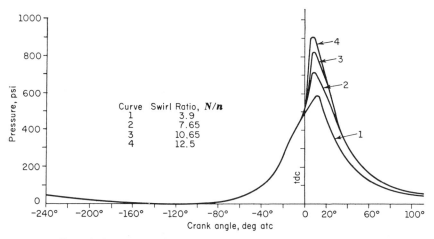

Fig. 3–20. Effect of air swirl on pressure–crank-angle diagram; sleeve-valve cylinder, type *g*, Fig. 3–18

N = air-swirl rpm, n = engine rpm (Alcock, 3.6812).

Figure 3–20 shows the influence of the rate of swirl on the pressure–crank-angle diagram for an engine of type *g* of Fig. 3–18. The swirl rate was varied at constant rpm by changing the angle of some guide vanes that were placed in the inlet ports. It is interesting to note that the delay period is unaffected by the swirl rate, although the effect on the rate of pressure rise and on maximum pressure is large. The improvement in mep and efficiency due to increase in swirl is very marked in this case because the fuel spray was designed in the first place to depend on swirl for proper mixing. However, a considerable gain in performance is usually obtainable by increasing the swirl, even when the optimum spray for use without swirl is used (3.670, 3.680). The beneficial effects of increasing swirl seem to be attributable, to a great extent at least, to faster mixing during injection on account of the increased velocity of the air with respect to the spray. Increasing swirl is accompanied, in nearly all cases, with increased small-scale turbulence, which may assist the mixing process during all stages of combustion.

The M.A.N. "Meurer" Combustion System, Fig. 3–21, shows a combustion system developed about 1954 by the Maschinenfabrik Augsburg-Nürnberg A.G. of Germany for small, high-speed engines. It differs from other open-chamber engines in being designed so that the fuel spray impinges tangentially on, and spreads over, the surface of a spherical cavity in the piston (3.682).

It is known that, except perhaps at light loads, there is some impingement of the spray on the combustion-chamber walls in all successful Diesel engines (see Fig. 3–3, for example). However, until the advent of the Meurer design it had generally been assumed that fuel-spray impingement

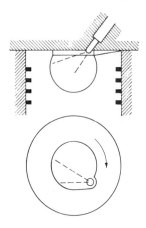

Fig. 3–21. M.A.N. type *M* combustion chamber. Fuel jets impinge tangentially on surface of spherical cavity in piston. Shrouded inlet valve is used to give rapid air swirl in direction of arrow (Meurer, 3.682).

was undesirable. The theory behind this system is that enough of the spray will ignite before impingement so that the delay period will be normal, while the bulk of the spray will have to evaporate from the cavity walls prior to combustion. Thus, the second stage of the combustion process is slowed down, avoiding excessive rates of pressure rise.

In practice this engine gives good performance even with fuels of exceedingly poor ignition quality, such as motor gasoline. (See Chapter 4 for a discussion of ignition quality in relation to fuel structure.) Its fuel economy appears to be extremely good for an engine of small size. From this fact it may be concluded that the third stage of combustion is not unduly extended, and that spray impingement can be employed with useful results. Evidently the advantages of this design are not great compared with those of more conventional designs, since it has been used in only limited quantities although it has been available both in the U.S.A. and in Europe for many years.

Divided Combustion Chambers. Divided types of chamber (Fig. 3–19) have been developed chiefly for use in high-speed (small) engines, in an attempt to overcome some of the limitations of the open-chamber type. In the case of divided combustion chambers the following possibilities may be realized:

1. Extremely high air velocity through the throat during the compression stroke, with resultant intense turbulence and also, in most cases, swirl in the prechamber.

2. The first and second stages of combustion can be forced to take place within a space whose structure is so strong that much higher pressures and higher rates of pressure rise can be tolerated in this space than can be allowed in the space over the piston (Fig. 3–22).

3. The mixing process may be greatly accelerated by the early stages of the combustion process itself. At high fuel-air ratios, combustion is incomplete in the prechamber because of insufficient air, and the high pressure

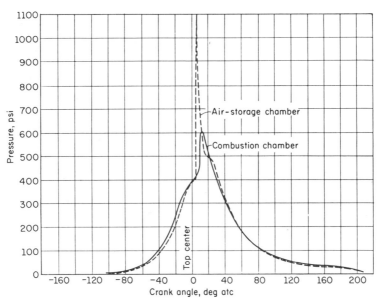

Fig. 3–22. Pressure–crank-angle curves from the combustion chamber and the air-storage chamber of a Lanova type cylinder, type *a*, Fig. 3–19 (Loschge, 3.6813).

developed by the early part of combustion projects the unburned fuel, together with the early combustion products, into the other part of the chamber with very high velocities, thus causing rapid mixing with the air in the space over the piston.

4. It is usually possible to allow all or some part of the walls of the pre-chamber to run at very high temperatures, thereby tending to reduce the delay as compared with that with open-chamber engines using the same fuel. The sections shown in black in Fig. 3–19 are designed to operate at high temperature. These parts usually take the form of stainless steel inserts arranged to have a rather poor thermal connection with the cylinder walls, owing to a loose fit. Such inserts run very hot after the engine is properly warmed up. They do not, of course, assist in the cold-starting process.*

Comparison of Combustion Chambers. Since it has been thoroughly demonstrated that well-designed engines of the open-chamber type give excellent results when properly maintained and when operated with proper fuel, the chief advantages claimed for divided chambers as compared to open chambers in the same engine are the following:

1. Ability to use fuels of poorer ignition quality;
2. Ability to use single-hole injection nozzles and moderate injection pressures and to tolerate greater degrees of nozzle fouling;
3. Ability to employ higher fuel-air ratios without smoke.

---

* A similar result can be achieved in open-chamber engines by using a stainless steel insert in the piston.

4. Lower noise levels.

Disadvantages of the divided chamber are the following:

5. More expensive cylinder construction;
6. Poorer fuel economy due to greater heat losses and pressure losses through the throat, which result in lower thermal efficiency and higher pumping loss.

At the present time (1984), divided-chamber engines* are generally used for passenger automobiles and light trucks, chiefly on account of lower particulate and smoke emissions and lower noise levels. The divided chamber is also used for small industrial engines where fuel quality may be hard to control and where reliability under adverse circumstances is more important than fuel economy. The facts that such engines are less sensitive to spray variations and to nozzle fouling are important considerations for this type of service.

Open chambers are used for heavy road vehicles, railroad locomotives, and marine engines and for nearly all Diesel engines with bores of more than 5 inches.

EFFECTS OF CYLINDER SIZE ON COMBUSTION IN DIESEL ENGINES

The effects of cylinder size on engine design and performance were discussed in Volume I, Chapter 11. There it was shown that, as cylinder size increases, piston speed tends to remain constant, so that rated rpm varies approximately in inverse ratio to the stroke.

For similar cylinders operating at the same piston speed, if the delay time is constant, the delay angle will be inversely proportional to the bore. In practice, delay time does tend to be independent of speed, as was shown in Table 3–2. Therefore, the larger the cylinder, the shorter will be the delay angle and the smaller will be the fraction of fuel taking part in the period of rapid pressure rise. In practice this relation is so marked that fuels of poorer ignition quality can be used as the cylinder size increases. In view of the fact that increase in cylinder size increases combustion problems in the spark-ignition engine (Chapter 2), it is not surprising to find that engines with cylinder bores larger than 6 in. are nearly always of the Diesel type. Exceptions to this rule are found in large spark-ignition engines operating on natural gas, which has high resistance to detonation.

---

* Divided-chamber engines are often designated as "indirect injection" (IDI), and open-chamber engines as "direct injection" (DI).

EXHAUST EMISSIONS FROM DIESEL ENGINES

The most objectionable emissions from Diesel-engine exhaust are finely divided solid or semi-solid particles of carbonaceous matter resulting from incomplete combustion of the fuel spray. The term "particulates" refers to particles not visible but collectable by filtration. The visible particles are classified as "smoke." The concentration of these pollutants increases with fuel-air ratio at a rate that becomes much faster as the overall stoichiometric ratio is approached. The point at which smoke appears is called the "smoke limit" and the corresponding mep or torque the "smoke-limited" values.

The smoke limit tends to be at higher fuel-air ratios with divided than with open combustion chambers. With a given type of chamber, the better the design of injectors and inlet and chamber geometry from the viewpoint of engine performance, the higher will be the smoke limit. At present few Diesel engines of any kind can be operated much above $F_R = 0.75$ without smoke.

Carbon monoxide in Diesel exhaust appears to be low enough to require no special controls. Gaseous unburned hydrocarbons, classified as HC, are emitted in quantities comparable to those of spark-ignition engines of the same size.

Oxides of nitrogen emissions in Diesel engines tend to equal or exceed those from comparable spark-ignition engines. As in S-I engines, they can be reduced by exhaust-gas recirculation, which is now common practice in U.S. Diesel vehicles.

Since present catalytic converters cannot be used for Diesel engines (because of excess exygen and clogging by particulates), this method of controlling NO_x and HC is not yet available.

Considerable progress has been made in reducing Diesel emissions by means of electronic control systems that adjust fuel quantity, injection timing, and EGR to the many operating variables influencing emissions. A typical list of these variables may include engine speed; injection crank angle; temperatures of air, cylinder walls, and fuel; atmospheric and inlet-air pressures; and the injection-pump control's position and rate of movement.

At the present time (1984), new Diesel-powered vehicles can be virtually smoke-free when in good adjustment but only small Diesel passenger cars can meet the limits on HC and NO_x imposed on gasoline cars.

There are as yet no practicable techniques for reducing the emissions of Diesel particulates to acceptable levels in areas of heavy traffic. Continued efforts are being made to develop filters or "traps" for these particles. Because of the high rates of emission, such traps have to be provided with systems to burn out or otherwise dispose of the accumulated material at frequent intervals. As yet, no safe, reliable, and otherwise practicable system of the kind is available.

Heavy vehicles have to be Diesel-powered, but in view of the health dangers associated with particulate emissions it would seem desirable to limit the use of Diesel engines in passenger cars.

DUAL-FUEL DIESELS

A "dual-fuel" Diesel engine is one designed to operate on either Diesel oil or natural gas (3.907–3.909). When operating on gas, this fuel is introduced during the inlet stroke, and ignition is accomplished by injecting a small *pilot charge* of Diesel oil. Appropriate adjustments in inlet pressure and or compression ratio must be used to avoid detonation (see also Chapter 4).

RECOMMENDED REFERENCES ON DIESEL EMISSIONS

SAE Publication P-86, 1980, "Diesel Combustion and Emissions."

SAE Publication SP-502, 1981, "Fuel and Combustion Effects on Particulate Emissions."

SAE Publication SP-484, 1982, "Diesel Combustion and Emissions, Part II,"

"New Directions in Automotive Engineering," Robert Bosch Corp., 2800 South 25th Avenue, Broadview, IL 60153. Excellent description of the mechanical and electronic controls manufactured by this company for both Diesel and spark-ignition engines.

SAE Publication PT-24, 1982, "Passenger Car Diesels." Includes comparison of combustion-chamber types.

"Diesel particulate traps: three approaches." *Automotive Engineering*, January 1984, p. 87.

4 | Fuels for Internal-Combustion Engines

The subject of fuels for internal-combustion engines has been studied intensively ever since such engines have been in existence. More research in this field has been, and is now being, carried on than in any other aspect of engine development. The literature in this field is immense; it includes hundreds of books as well as thousands of technical papers. The bibliography appended to this chapter is only a small, selected sample of the available literature. Many publications referred to in this bibliography include their own extensive bibliographical references.

Under these circumstances it is obvious that a single chapter devoted to fuels can at best indicate only the most important aspects of the subject from a single point of view. The point of view used here is that of the relation of fuel characteristics to engine performance. This subject has already been introduced in Volume I, Chapters 3 to 7, and in the preceding chapters of this volume. Table 4–1 lists important characteristics of the fuels most generally used.*

Internal-combustion engines can be operated on many different kinds of fuel, including liquid, gaseous, and even solid material. The character of the fuel used may have considerable influence on the design, output, efficiency, fuel consumption, and — in many cases — the reliability and durability of the engine.

PETROLEUM FUELS

It is probably safe to say that over 99 per cent of the world's internal-combustion engines use liquid fuel derived from petroleum. In some countries where natural petroleum is scarce, fuels having very similar composition and characteristics are being produced by the hydrogenation of coal (4.117, 4.118). Although the latter are not strictly petroleum fuels, they have characteristics so similar that it is not necessary here to consider them separately. Because of their overwhelming importance as fuels for

---

* Table 4–1 is included in Volume I as Table 3–1, pp. 46–47.

Table 4–1. Properties of Fuels

A. Liquid Fuels; Pure Compounds

Name	Chemical Formula	Liquid Fuel						Fuel Vapor	
		m_f	Specific Gravity (60°F)	H_{lg} at 77°F	C_p 60–80°F	Q_{ch} (High)	Q_c (Low)	C_p at 60°F	k
Normal pentane	C_5H_{12}	72	0.631	−157	0.557	21,070	19,500	0.397	1.07
Normal hexane	C_6H_{14}	86	0.664	−157	0.536	20,770	19,240	0.398	1.06
Normal heptane	C_7H_{16}	100	0.688	−157	0.525	20,670	19,160	0.399	1.05
Normal octane	C_8H_{18}	114	0.707	−156	0.526	20,590	19,100	0.400	1.05
Isooctane	C_8H_{18}	114	0.702	−141	0.515	20,570	19,080	0.400	1.05
Normal decane	$C_{10}H_{22}$	142	0.734	−155	0.523	20,450	19,000	0.400	1.04
Normal dodecane	$C_{12}H_{26}$	170	0.753	−154	0.521	20,420	18,980	0.400	1.03
Octene	C_8H_{16}	112	—	−145	0.525	—	19,035	0.400†	1.05†
Benzene	C_6H_6	78	0.884	−186	0.411	17,990	17,270	0.277	1.08
Methyl alcohol	CH_3OH	32	0.796	−474	—	9,760	8,580	0.41	1.11†
Ethyl alcohol	C_2H_5OH	46	0.794	−361	—	12,780	11,550	0.46	1.13

B. Liquid Fuels; Typical Mixtures

Name	Chemical Formula	m_f	Specific Gravity (60°F)	H_{lg} at 77°F	C_p 60–80°F	Q_{ch} (High)	Q_c (Low)	C_p at 60°F	k
Gasoline	C_8H_{17}	113	0.702	−150	0.50	20,460	19,020		
Gasoline	—	126†	0.739	−142	0.48	20,260	18,900		
Kerosene	—	154†	0.825	−125	0.46	19,750	18,510	~0.4	~1.05
Light Diesel oil	$C_{12}H_{26}$	170	0.876	−115	0.45	19,240	18,250		
Med. Diesel oil	$C_{13}H_{28}$	184	0.920	−105	0.43	19,110	18,000		
Heavy Diesel oil	$C_{14}H_{30}$	198	0.960	−100	0.42	18,830	17,790		

C. Solid Fuels

Name	Chemical Formula	m_f	Specific Gravity	H_{lg}	C_p	Q_{ch} (High)	Q_c (Low)	C_p	k
C to CO_2	C	12	—	—	—	14,520	14,520	—	—
C to CO	C	12	—	—	—	4,340	4,340	—	—

D. Gaseous Fuels; Pure Compounds

Name	Chemical Formula	Fuel Alone				
		m_f	$\dfrac{\rho}{\rho_a}$	C_p	Q_{ch} (High)	Q_c (Low)	k
Hydrogen	H_2	2	0.069	3.41	61,045	51,608	1.41
Methane	CH_4	16	0.552	0.526	23,861	21,480	1.31
Ethane	C_2H_6	30	1.03	0.409	22,304	20,400	1.20
Propane	C_3H_8	44	1.52	0.388	21,646	19,916	1.13
Butane	C_4H_{10}	58	2.00	0.397	21,293	19,643	1.11
Acetylene	C_2H_2	26	0.897	0.383	20,880	20,770	1.26
Carbon monoxide	CO	28	0.996	0.25	4,345	4,345	1.404
Air	—	29	1.00	0.24	—	—	1.40

E. Gaseous Fuels; Typical Mixtures

Name	Fuel Alone										m_f	$\dfrac{\rho_f}{\rho_a}$
	Composition by Volume											
	CO_2	CO	C_2H_6	C_2H_4	H_2	CH_4	N_2	O_2	C_3H_8	C_6H_6		
Blast-furnace gas	11.5	27.5	—	—	1.0	—	60.0	—	—	—	29.6	1.02
Blue water gas	5.4	37.0	—	4.7	47.3	1.3	8.3	0.7	—	—	16.4	0.57
Carb. water gas	3.0	34.0	—	6.1	40.5	10.2	2.9	0.5	—	2.8	18.3	0.63
Coal gas	3.0	10.9	—	1.5	54.5	24.2	4.4	0.2	—	1.3	12.1	0.42
Coke-oven gas	2.2	6.3	—	3.5	46.5	32.1	8.1	0.8	—	0.5	13.7	0.44
Natural gas	—	—	15.8	—	—	83.4	0.8	—	—	—	18.3	0.61
Producer gas	4.5	27.0	—	—	14.0	3.0	50.9	0.6	—	—	24.7	0.86

Sources:
a Maxwell, Volume I, ref. 3.33.
b Taylor and Taylor, ref. 0.010.
Symbols:
† Estimated.
‡ Estimated on basis of zero volume for carbon.
c *Handbook of Physics and Chemistry*, p. 1732.
d *Marks Handbook*, McGraw-Hill, 5 ed., p. 364.
e *Marks Handbook*, McGraw-Hill, 5 ed., p. 820.

f Hottel *et al.*, Volume I, ref. 3.20.
g *Int. Critical Tables*, McGraw-Hill, 1929.
m_f molecular weight fuel
H_{lg} enthalpy of liquid above vapor at 60°F, Btu/lbm
C_p specific heat at constant pressure, Btu/lbm°R
Q_{ch} higher heat of combustion, Btu/lbm
Q_c lower heat of combustion, Btu/lbm
k ratio of specific heats

Used in Internal-Combustion Engines

								1		Source
m_m	F_c	r_m	C_p (60°F)	k	$\dfrac{a}{\sqrt{T}}$	F_cQ_c	$\dfrac{1}{1+F_c\dfrac{ma}{mf}}$		Q_v	of Data

Chemically Correct Mixture Vapor and Air

m_m	F_c	r_m	C_p (60°F)	k	$\dfrac{a}{\sqrt{T}}$	F_cQ_c	$\dfrac{1}{1+F_c\frac{ma}{mf}}$	Q_v	Source of Data
30.1	0.0657	1.05	0.25	1.360	47.3	1280	0.975	95.4	a
30.1	0.0658	1.05	0.25	1.357	47.3	1265	0.976	94.5	a
30.2	0.0660	1.06	0.25	1.355	47.2	1265	0.980	94.7	a
30.3	0.0665	1.06	0.25	1.355	47.2	1270	0.983	95.5	a
30.3	0.0665	1.06	0.25	1.355	47.2	1268	0.983	95.4	a
30.4	0.0668	1.06	0.25	1.353	47.1	1270	0.984	95.6	a
30.5	0.0672	1.06	0.25	1.350	47.0	1275	0.985	96.0	a
30.5	0.0678	1.05	0.25	1.355	47.2	1290	0.983	97.5	g
30.4	0.0755	1.01	0.248	1.350	47.5	1285	0.974	95.7	a
29.4	0.155	1.06	0.245	1.380	47.7	1410	0.876	94.5	b
30.2	0.111	1.06	0.262	1.340	46.9	1320	0.936	94.7	b
30.3	0.0670	1.06				1275	0.984	95.6	b
30.3	0.0668	1.06				1265	0.987	94.8	b
30.4	0.0667	1.06	~0.25	~1.35	~47	1240	0.988	91.0	†
30.5	0.0666	1.06				1220	0.989	88.5	b
30.6	0.0664	1.06				1215	0.989	88.3	†
30.7	0.0670	1.06				1210	0.990	90.5	†
29.0	0.0868	1.00‡	0.24	1.40	49	1260	1.0	96.3	‡
29.0	0.1736	0.825‡	0.24	1.40	49	750	1.0	57.3	‡

Chemically Correct Mixture

m_m	F_c Vol	F_c Mass	r_m	C_p (60°F)	k (60°F)	$\dfrac{a}{\sqrt{T}}$	F_cQ_c	$\dfrac{1}{1+F\frac{ma}{mf}}$	Q_v	Source
21	0.418	0.0292	0.85	0.243	1.64	62.3	1510	0.704	81.2	a
27.7	0.105	0.0581	1.00	0.256	1.39	49.9	1248	0.904	86.4	a
29.0	0.0598	0.0623	1.03	0.249	1.38	48.6	1270	0.943	91.5	a
29.6	0.0419	0.638	1.04	0.250	1.38	48.5	1270	0.961	93.5	a
29.8	0.0323	0.0647	1.05	0.250	1.38	48.4	1270	0.967	94.0	a
28.8	0.0837	0.075	0.96	0.250	1.38	48.7	1555	0.923	110.0	c, d
28.7	0.418	0.404	0.85	0.241	1.43	49.3	1755	0.705	94.7	c, d
28.85	—	—	—	0.240	1.40	49.0	—	1.0	—	

Fuel Alone				Chemically Correct Mixture									
Q_c (Volume)		Q_c (Mass)		m_m	C_p	$\dfrac{a}{\sqrt{T}}$	F_c Vol	F_c Mass	r_m	F_cQ_c	$\dfrac{1}{1+F\frac{ma}{mf}}$	Q_v	Source
High	Low	High	Low										
92	91	1,170	1,160	29.4	0.245†	48.4	1.47	1.50	0.915	1740	0.405	53.9	e
289	262	6,550	5,980	25.0	—	52.6	0.476	0.266	0.864	1580	0.681	83.0	e
550	532	11,350	10,980	27.1	—	51.6	0.218	0.137	0.942	1505	0.821	94.4	e
532	477	16,500	14,800	26.9	—	51.6	0.220	0.092	0.940	1390	0.818	85.6	e
574	514	17,000	15,200	27.3	—	50.3	0.200	0.058	0.957	882	0.890	59.9	e
1129	1021	24,100	21,800	27.3	—	49.6	0.095	0.058	1.00	1265	0.916	88.2	e
163	153	2,470	2,320	27.0	—	—	0.813	0.700	0.913	1620	0.549	68.2	e

m_m molecular weight of stoichiometric mixture
F_c stoichiometric fuel-air ratio
r_m ratio molecular weight before combustion to molecular weight after complete combustion of stoichiometric mixture
a velocity of sound in gas, fps
T temperature, °R
m_a molecular weight of air = 28.95
Q_v heat of combustion of fuel vapor in 1 cu. ft. of stoichiometric mixture at 14.7 psia, 60°F

internal-combustion engines, it seems worthwhile to consider petroleum fuels in some detail at this point.

Crude oil is the term used for the raw petroleum as it comes from the oil wells. (For information on the distribution of the world's supply, and its probable duration, see refs. 4.110–4.115.) It consists chiefly of a mixture of many types of hydrocarbons of many different molecular weights. It also includes a certain fraction, usually small, of organic compounds containing sulfur, nitrogen, etc.

The exact composition of crude oil differs widely according to its source. In general, crudes fall into three classes depending on whether the residue left after distillation is mainly paraffin, asphalt, or a mixture of the two. The corresponding crudes are called paraffin base, naphthenic base, or mixed base. Within each classification, however, the composition of crudes may differ widely.

Refining of crude oil usually starts with distillation at atmospheric pressure, during which process the distillate is separated into various *fractions* according to volatility. The resulting distillates are called *straight-run* products. Distillates or residue may then be subjected to heat treatments and chemical treatments at various pressures and temperatures. Such treatments are defined as *cracking* when they chiefly tend to reduce the average molecular size and as *polymerization* when the reverse predominates. The structure of the resulting products in any of these processes depends both on the details of the process and on the composition of the fraction so treated, which in turn depends on the structure of the basic crude.

The products resulting from the refinement of petroleum are classified by their usage and also according to their specific gravity and their volatility as determined by distillation at sea-level atmospheric pressure. The products of chief interest here are the following:

1. *Natural Gas.* The gaseous hydrocarbons are usually associated with liquid petroleum, either standing above the liquid in the earth or dissolved in it. The dissolved gas is the first product to separate out when distillation is started. Further discussion of natural gas will be found under the heading "Gaseous Fuels."

2. *Gasoline.* This term covers most liquid petroleum fuels intended for use in spark-ignition engines. Gasoline lies in the specific gravity range 0.70 to 0.78 and within the volatility ranges indicated by Fig. 4–1 (see later discussion of volatility). The chemical composition of its constituents varies widely, depending on the base crude and the methods used in refining.

3. *Kerosene* is the next fraction heavier than gasoline. It is intended for use in lamps, heaters, stoves, and similar appliances. It is also an excellent fuel for compression-ignition engines and for aircraft gas turbines. *Jet fuels,* that is, fuels intended primarily for aircraft turbines and jet engines,

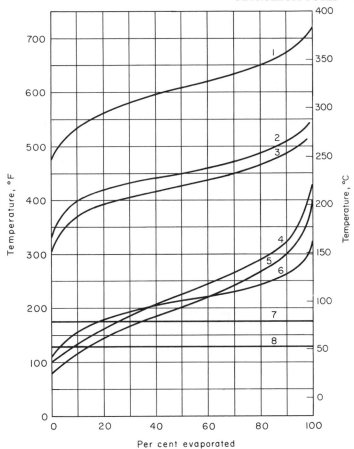

Fig. 4–1. Typical ASTM distillation curves: 1, heavy Diesel oil; 2, distillate; 3, kerosene; 4, summer gasoline; 5, winter gasoline; 6, aviation gasoline; 7, ethyl alcohol; 8, benzene.

are close to kerosene in composition. The specific gravity of kerosene lies between 0.78 to 0.85 and its usual distillation range is near that shown in Fig. 4–1.

4. *Distillate* is slightly heavier than kerosene; it is obtained from some western United States crudes by distillation at atmospheric pressure. It has substantially the same uses as kerosene. Figure 4–1 shows a typical distillation curve for this product.

5. *Diesel oils* are petroleum fractions that lie between kerosene and the lubricating oils. These oils cover a wide range of specific gravity and a very wide distillation range, as shown in Fig. 4–1. Their composition is controlled to make them suitable for use in various types of compression-ignition engines.

6. *Fuel oils* cover a range of specific gravity and distillation similar to that of Diesel oils, but since they are designed for use in continuous burners

their composition does not require such accurate control as is required for Diesel oils.

7. *Lubricating oils* are made up in part from heavy distillates of petroleum and in part from *residual oils*, that is, oils remaining after distillation. *Tar and asphalt* are solid or semisolid products which remain undistilled.

Chemical Structure of Petroleum Products. Each of the fuels and oils mentioned varies widely in chemical composition, depending on the source of the crude and on the methods used in refining. In every case, however, the fuel (or oil) consists almost entirely of a mixture of hydrocarbon compounds having different molecular weights and different types of structure.

The various compounds that make up petroleum fuels are classified according to the number of carbon atoms in the molecules. In the case of many types, a characteristic syllable in the name of each compound refers to the number of carbon atoms, as follows:

Syllable	Number of Carbon Atoms	Syllable	Number of Carbon Atoms
meth	1	hept	7
eth	2	oct	8
prop	3	non	9
but	4	dec	10
pent	5	undec	11
hex	6	dodec	12

The way in which the carbon and hydrogen atoms are arranged is indicated by suffixes to these syllables together with a qualifying adjective where necessary. The following classifications cover the more important constituents in petroleum fuels.

1. Straight chain, or *normal paraffins*. The carbon atoms in this type are connected together as a chain, with hydrogen atoms filling the empty valences. The characteristic name ending for this type is *ane*. Thus, with seven carbon atoms, a compound of this type is called *normal heptane* and its structure is represented diagramatically as follows:

$$
\begin{array}{c}
\text{H} \quad \text{H} \quad \text{H} \quad \text{H} \quad \text{H} \quad \text{H} \quad \text{H} \\
\text{HC—C—C—C—C—C—CH} \\
\text{H} \quad \text{H} \quad \text{H} \quad \text{H} \quad \text{H} \quad \text{H} \quad \text{H}
\end{array}
$$

The general chemical formula for this series is C_nH_{2n+2}, where n is the number of carbon atoms.

2. Branched chain, or *isoparaffins*. In these compounds the chain of carbon atoms is branched. The following are examples:

```
     H          H                          H    H
     HCH        HCH                        HCH  HCH
 H    |    H    |    H                H     |    |     H
 HC——C————C————C————CH     and       HC———C————C————CH
 H    |    H    H    H                H     |    H    H
     HCH                                   HCH
     H                                     H
```

The first compound is popularly called *isooctane*, but this name does not describe it completely, since its characteristics would be different if the relative position of the branches were changed. By a more rigid system of nomenclature this compound is called 2,2,4-trimethyl pentane, indicating five carbon atoms in the straight chain (pentane) with three methyl (CH_3) branches located respectively at stations 2, 2, and 4.

The second compound shown is, strictly speaking, 2,2,3-trimethyl butane. The name *triptane* is more generally used for this compound. Since there is only one possible arrangement of the three methyl branches,[*] the name "triptane" simply indicates trimethyl butane.

The general formula of the isoparaffins is again C_nH_{2n+2}.

3. *Olefines* are compounds with one or more *double-bonded* carbon atoms in a straight chain. Their names are made up with the ending *ene* for one double bond and *adiene* for two. The following examples, respectively, show *hexene* with one double bond and *butadiene* with two:

```
 H  H  H  H  H  H         H  H  H  H
 HC—C—C—C—C=CH            HC=C—C=CH
    H  H  H  H
```

The characteristics of each of these compounds would be different for different locations of the double bond within the chain. Olefines are scarce in straight-run products but are present in considerable quantities in certain highly cracked gasolines. The general formulas are C_nH_{2n} for the *mono-olefines* (one double bond) and C_nH_{2n-2} for the *diolefines* (two double bonds).

4. *Naphthenes* are characterized by a "ring structure" diagramed as follows:

```
   H H               H    H                 H    H
    C              HC————CH                  C————C
   / \              |     |               H /    H \
 HC————CH          HC————CH             HCH          HCH
 H     H           H     H                \ H    H  /
                                           C————C
                                           H    H
```

The three compounds shown are *cyclopropane*, *cyclobutane*, and *cyclohexane*. The general formula is, evidently, C_nH_{2n}.

[*] The "mirror image" of this compound, 2,3,3-trimethyl butane is the same molecule viewed from the other side.

5. Aromatics have as their central structure the *benzene* molecule, C_6H_6, which is represented as follows:

or, more simply

Various other aromatic compounds are formed by replacing one or more of the hydrogen atoms of the benzene molecule with an organic radical; for example, toluene, $C_6H_5CH_3$, is

Two or more benzene rings can also combine with each other. The number of molecular arrangements in the aromatic series is too large to be covered by a simple system of nomenclature.

Aromatic products are found in many crudes and in some of their refined products. This type of structure is most prevalent in the products formed by the distillation of coal. One of the most important of these is *benzol*, made up chiefly of the benzene molecule, C_6H_6, plus a small percentage of other hydrocarbons. For use in engines, it is usually blended with gasoline.

GASEOUS FUELS

Properties of gaseous fuels used in internal-combustion engines are given in Table 4–1 and in refs. 4.106–4.1093. The most important gaseous fuels are natural gas and liquefied petroleum gas. Natural gas is made up chiefly of the paraffinic compound methane, CH_4 (see Table 4–1). It is used for stationary power plants where the supply of gas is abundant, as, for example, near natural-gas fields and pipelines. Natural-gas pipelines are powered by engines or turbines using this gas as fuel (4.447–4.449).

In addition to methane, natural gas may contain ethane, C_2H_6, propane, C_3H_8, and butane, C_4H_{10}. The two latter compounds are extracted from natural gas, or from petroleum in which they are dissolved, and stored as liquids under pressure. In this form they are known as liquefied petroleum gas or bottle gas. Upon release to atmospheric pressure they become gaseous. Bottle gas is used to a limited extent for automotive vehicles in localities where the supply is plentiful (4.400–4.446).

Producer Gas. Producer gas is made by burning carbonaceous material (coal, wood, charcoal, coke, etc.) with a large deficiency of air. The products of this partial combustion contain CO and H_2 in sufficient quantities so that they can be used in an engine as fuel. A typical composition for this type of gas is given in Table 4–1. The composition varies widely with the basic fuel and with the operating conditions of the apparatus. Generated by means of small units that can be carried at the rear of a car or on a light trailer, producer gas has been used in Europe and Asia during war periods when liquid fuels were not generally available (4.400).

Blast-Furnace Gas. As the name implies, blast-furnace gas is the gas issuing from a blast furnace during the operation of reducing iron ore. Since it contains a large amount of combustible material, it is frequently used in internal-combustion engines near blast furnaces, including the engines that drive the air-supply blowers for the furnace itself. A typical composition for this variable fuel is given in Table 4–1.

Artificial Gas. Artificial gas is made from coal or petroleum by various methods, usually involving combustion to CO, together with dissociation of water to yield gaseous hydrogen. Typical compositions are given in Table 4–1 under the headings " blue water gas," " carburated water gas," and " coal gas." Artificial gas is expensive and has been largely displaced by natural gas in many areas. In the past it has been used for engines in standby fire-pump and generator stations. It is a satisfactory fuel from the technical point of view.

Gaseous fuels eliminate most of the starting difficulties associated with liquid fuels, and they never require heating of the inlet system. Distribution of a proper fuel-air mixture to several cylinders is more easily accomplished with gaseous than with liquid fuels.

Disadvantages of gaseous fuels center around storage and handling problems. In general, they are not well suited for use in self-propelled vehicles because of the size or weight of the necessary containers. In carbureted engines, gases displace more air than is the case with liquid fuels and thus tend to reduce maximum power output. (This subject will be fully discussed later in this chapter.)

NONPETROLEUM FUELS

The only nonpetroleum compounds of importance as motor fuels are methyl alcohol, or methanol, CH_3OH, and ethyl alcohol, or ethanol, C_2H_5OH. Where allowed by the rules, methanol is used in racing engines on account of the cooling effect produced by its very large (negative) heat of vaporization (see Table 4–1). Ethanol, blended with gasoline, has been used in locations where it is plentiful as a byproduct of sugar refining, pulp manufacture, or some similar industry. For economic reasons, alcohol is not manufactured specifically for use as a motor fuel. (4.201–4.205).

Animal and vegetable oils have been used experimentally in Diesel engines but are not in commercial use for this purpose except possibly in a few very remote locations where petroleum fuel is not available (4.620–4.624).

Solid Fuels. In isolated cases internal-combustion engines have been made to run on powdered coal, sawdust, or similar materials. The use of such fuels, however, has been limited to experimental installations. The elaborate apparatus required to prepare and inject such fuels, together with difficulties due to the solid residue (ash), have so far prevented successful commercial application (4.125, 4.126).*

Some unusual nonpetroleum fuels, under consideration for special purposes, will be discussed at the end of this chapter (4.127) and in Chapter 14.

FUELS FOR SPARK-IGNITION ENGINES

From the point of view of spark-ignition-engine performance, the following fuel characteristics are of importance:

> Volatility
> Detonation and preignition characteristics
> Heat of combustion per unit mass and volume
> Heat of evaporation
> Chemical stability, neutrality, and cleanliness
> Safety

The cost and availability of fuels are, of course, major economic factors, but discussion of these characteristics lies beyond the scope of this volume. The great preponderance of petroleum products as motor fuel is basically due to their low relative cost and plentiful supply in most parts of the world.

Volatility of Liquid Fuels

Volatility is loosely defined as the tendency of a liquid to evaporate. This quality is of basic importance in carbureted engines, because it has a major influence on the vapor-air ratio in the cylinders at the time of ignition.

---

* Dr. Rudolf Diesel at first tried to operate his newly invented engine on solid fuel (Volume I, ref. 1.05).

In engines burning a uniform mixture of fuel vapor and air, the vapor-air ratio must be not less than about 0.5 times stoichiometric for satisfactory ignition and flame propagation. It is obvious, therefore, that the volatility of the fuel must be sufficient to give at least this vapor-air ratio at the time of ignition under all operating conditions, including starting and warming up a cold engine.

ASTM Distillation. The relative volatility of fuels is ordinarily measured by means of the ASTM distillation test, which is described in detail in ref. 4.05. The apparatus used is illustrated in Fig. 4–2. Since the flask is

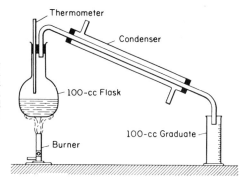

Fig. 4–2. Apparatus used in ASTM distillation. The distillation is started with 100 cc of fuel in the flask; vapor temperatures are recorded for every 10 cc condensate in graduate. Temperature for first drop, end point, and weight of residue in flask are also recorded (ASTM Standards, 4.05).

filled with the fuel vapor and the condenser tube is open to the atmosphere, the process consists of distillation at a vapor pressure of 1 atm. Typical ASTM distillation curves are shown in Fig. 4–1. Between two fuels, the one that has the lower curve has the *greater volatility*.

In fuel-air mixtures, the vapor pressure of the fuel is only a small fraction of 1 atm, and therefore the ASTM results cannot be applied directly. In order to study this problem, consider a given mass of liquid fuel introduced into a given volume of air at a given pressure and temperature. This arrangement could represent a certain portion of the air moving through an inlet manifold, together with the amount of fuel that was introduced to that particular volume of air as it passed through the carburetor. As soon as the fuel is introduced, it will start to evaporate, and, if given enough time, this process will continue to a point of equilibrium dependent on the final temperature and fuel composition.

Using Dalton's law and assuming that $pV = (M/m)RT$ for both fuel vapor and air, the ratio of mass of fuel vapor to mass of air at a given instant may be expressed as follows, since both vapor and air occupy the same volume at the same temperature,

$$F_v = \frac{m_{fv}p_f}{29(p - p_f)}, \qquad (4\text{–}1)$$

where F_v = mass ratio of evaporated fuel to air
$\quad\quad m_{fv}$ = average molecular weight of the fuel vapor
$\quad\quad p_f$ = partial pressure of the fuel vapor
$\quad\quad 29$ = molecular weight of air
$\quad\quad p$ = total pressure of mixture.

If evaporation proceeds to equilibrium with liquid fuel still present and if the fuel consists of a single chemical compound, p_f will be the vapor pressure of that compound and will depend on temperature only. Figure 4–3 gives vapor pressures versus temperature for typical fuel constituents.

If the liquid is a mixture of many constituents, as with most fuels, the molecular weight and vapor pressure in Eq. 4–1 will be average values for the vaporized portion. In that case the composition of the vapor will not be the same as that of the liquid that remains after equilibrium has been reached. Under these circumstances, the pressure relations at *equilibrium* have been formulated in Raoult's law, which may be stated as follows: "Above a solution of normal liquids, the partial pressure exerted by any component is proportional to the product of the normal vapor pressure of that component at the existing temperature and its molecular concentration in the liquid."* Thus, for any constituent of the fuel, at equilibrium with liquid fuel present:

$$p_f = p_v\left(\frac{M}{M_L}\frac{m_L}{m}\right), \tag{4–2}$$

where p_v = normal vapor pressure of constituent in question
$\quad\quad M$ = mass of that constituent remaining in the liquid
$\quad\quad M_L$ = mass of the remaining liquid
$\quad\quad m$ = molecular weight of the constituent
$\quad\quad m_L$ = average molecular weight of the remaining liquid.

The operation of Raoult's law is evident in Fig. 4–1 from the fact that the temperature required to give a vapor pressure of 1 atm depends on the characteristics of the liquid fuel with which the vapor is in equilibrium. For example, the temperature required to maintain a vapor pressure of 1 atm when the liquid has the composition of the original fuel (beginning of evaporation) is quite different from that required when no liquid is left (end point). On the other hand, for a single-constituent fuel, such as alcohol or octane, the temperature required to give a vapor pressure of 1 atm is the same for any fraction evaporated, as shown in the figure.

For a single-constituent fuel the quantity in parentheses in Eq. 4–2 is unity and the equilibrium vapor-air ratio, as a function of temperature, can be computed from its vapor-pressure curve, Fig. 4–3. The results of such a computation for 1 atm total pressure are shown in Fig. 4–4. If

---

* Raoult's law does not hold for certain constant-boiling mixtures, such as 95 per cent ethyl alcohol in water.

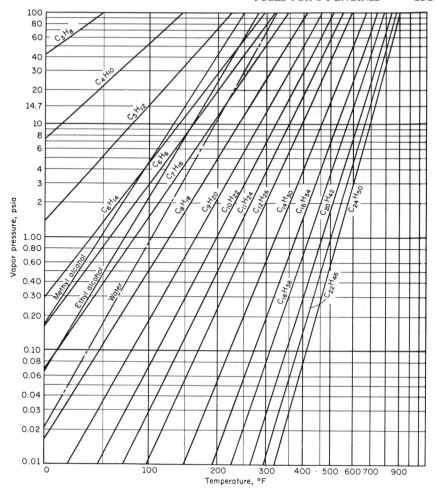

Fig. 4–3. Vapor pressures of some organic compounds (from ref. 4.5421).

the lean firing limit is taken as 0.6 times stoichiometric, or $F_v = 0.04$, it is evident from Fig. 4–4 that, at a total pressure of 1 atm, a firing mixture cannot be obtained below 57°F with this fuel. Lowering the total pressure will increase the vapor-air ratio, as indicated by Eq. 4–1, and will furnish a firing mixture at correspondingly lower temperatures.

Equilibrium Air Distillation. For fuels like gasoline which consist of mixtures of various chemical compounds, computation of equilibrium vapor-air ratios is impracticable, and experiment must be resorted to. Experimental determination of the vapor-air ratio as a function of fuel composition has been conducted by means of a procedure known as the *equilibrium air distillation*, abbreviated EAD (4.554, 4.557, 4.560–4.564).

In the EAD method the fuel is allowed to evaporate into a stream of air

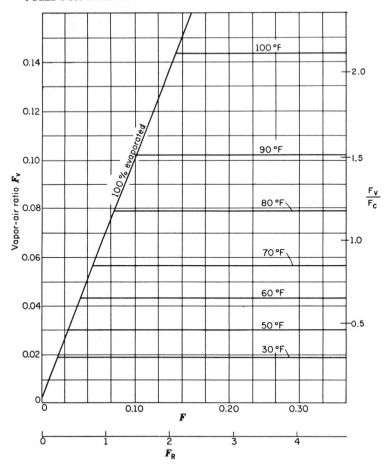

Fig. 4–4. Vapor-air versus fuel-air ratios for octane (C_8H_{18}) at 1 atm. (Computed from vapor pressure given in Fig. 4–3.)

moving through a long, heated tube. The exit vapor-air ratio is measured as a function of the fuel-air ratio supplied, the total pressure, and the exit temperature. The tube must be long enough and the velocity low enough so that equilibrium is attained at the tube outlet. The tube can be regarded as representing an engine manifold in which evaporation reaches equilibrium before delivery to the cylinders.

Figure 4–5 is plotted from the results of EAD tests on a typical gasoline.* These curves are quite different from those of octane in that the over-all fuel-air ratio has a pronounced effect on the vapor-air ratio at any temperature. Because of Raoult's law, even though the gasoline contains many constituents less volatile than octane, a firing mixture can be obtained at

---

* ASTM initial point 140°F, end point 425°F.

30°F (at 1 atm), while octane requires 57°F. However, a very rich over-all fuel-air ratio (4.3 times stoichiometric) is required, with gasoline, since only the more volatile constituents evaporate at this temperature.

An important point not shown by Fig. 4–5 is that, at each temperature, the vapor-air ratio versus fuel-air ratio curve approaches a limiting value as the fuel-air ratio is increased toward the point where evaporation has a negligible effect on the composition of the liquid. The air space over a nearly full gasoline tank is an example of this situation.

Use of EAD Data. EAD results cannot be used to determine, quantitatively, the vapor-air mixtures in an operating engine because of uncertainty as to the mixture temperature at any given location and because equilibrium conditions are seldom achieved. However, curves such as those of Fig. 4–5 are very useful for the study of trends. Most important among

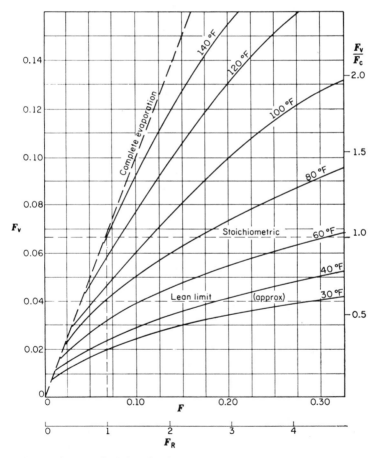

Fig. 4–5. Vapor-air versus fuel-air ratios for gasoline at 1 atm total pressure. (Computed from EAD curves of ref. 4.560.)

these is the fact that with a mixed fuel, the vapor-air ratio at a given temperature and total pressure can be increased by increasing the over-all fuel-air ratio, up to the point where the fuel-air ratio is very large (as in storage tanks).

Conversely, in the presence of the liquid, the vapor-air ratio over a single-constituent fuel is dependent on temperature only.

Relationship Between EAD and ASTM Data. One of the difficulties associated with EAD evaporation is the time and expense required to make

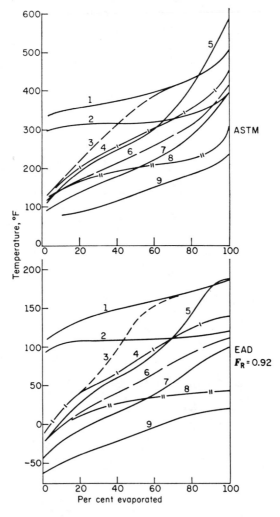

Fig. 4–6. ASTM and EAD volatility curves for nine petroleum fuels (Barber and MacPherson, 4.557). 1, kerosene; 2, low-volatility aviation; 3, tractor fuel; 4, U.S. motor; 5, ANF-58 (used by U.S. Air Force in World War II); 6, premium, summer; 7, premium, winter; 8, 100/130 aviation; 9, cold-starting fuel.

such tests. Separate and rather lengthy runs must be made at each temperature and supply ratio. However, it has been found that for fuels within the range usually classified as gasoline the EAD curves at a given fuel-air ratio tend to lie in the same relative order as the corresponding ASTM curves. This relation is indicated by Fig. 4–6. Thus, the ASTM curves can be used to indicate the *relative* volatility of fuels under actual operating

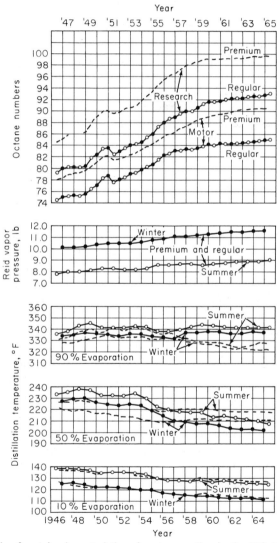

Fig. 4–7. Trends of certain characteristics of motor gasoline in the U.S.A., 1946–1965. Solid lines, regular-price gasolines; dashed lines, premium-price gasolines (U.S. Department of the Interior, 4.105, summer 1965).

conditions. This general relationship is perhaps the most significant result of all the EAD work done to date, since it enables conclusions to be drawn as to the *relative* effects of fuel volatility on engine performance when the only available information consists of ASTM curves for the fuels in question. The trend of ASTM volatility for automobile fuels used in the United States is shown in Fig. 4–7.

Effect of Fuel Volatility on Performance of S-I Engines

Volatility affects engine performance through its influence on the degree of fuel evaporation in intake manifolds and in cylinders prior to and during the combustion process. In the normal type of spark-ignition engine, operation will be satisfactory only if the various cylinders receive approximately the same fuel-air ratio and if nearly all the fuel is evaporated before ignition. Thus, with a given engine at a given set of running conditions, there is a certain minimum volatility below which operation will be unsatisfactory. Conversely, the volatility of the available fuels influences engine design, particularly in regard to size and shape of inlet manifolds and the minimum temperatures of inlet air and inlet manifold required for satisfactory operation.

It is obvious that under conditions where evaporation is complete all fuels will behave alike as far as volatility is concerned. As a corollary to this statement it may be said that the significance of any point on the ASTM volatility curve depends on the temperature range in question. Thus, under low-temperature conditions, as in cold-weather starting, when only a small portion of the fuel is evaporated, the low-temperature end of the curve is of chief importance, while for higher temperature ranges, such as those that exist in the inlet manifold of a warm engine, the shape of the low-temperature end of the curve is of little importance, since all of this part of the fuel is evaporated in any case.

In considering the effects of fuel volatility on engine performance it is necessary to consider the influence of volatility on the behavior of:

1. Cool engines, that is, engines that have not yet reached their normal operating temperatures. A special case of a cool engine is a *cold* engine, that is, one that has been standing idle so long that all its parts are at the temperature of the surrounding atmosphere. *Starting* and *warming up* are cool-engine phenomena.

2. Warm engines, that is, engines that have been in operation long enough so that further operation under the same conditions does not increase operating temperatures.

Except where otherwise noted, the following discussion applies to carburated engines, that is, to engines supplied with fuel and air through a carburetor-manifold system.

Starting. So far as the fuel is concerned, ability to start depends on supplying a combustible mixture in the cylinder at the moment of ignition. Under starting conditions, the pressure and temperature at ignition cannot be predicted with accuracy, and furthermore, vapor equilibrium cannot be assumed. Thus, a quantitative prediction of lowest starting temperature on the basis of Fig. 4–5 is not possible, but this figure does show why the choking process is very effective in starting at low temperatures.

As will be explained in Chapter 5, the choke is a device by means of which the fuel-air ratio may be drastically increased. Figure 4–5 shows that, with gasoline, increasing the fuel-air ratio at any given temperature increases the vapor-air ratio.

Another important point, not shown by Fig. 4–5 but previously mentioned, is the fact that with a given supply of fuel, at a given temperature, the vapor-air ratio is inversely proportional to the air pressure. Thus, the ability of a choke to reduce manifold pressure at cranking speed is helpful in starting. For example, at 1 atm and 30°F the vapor-air ratio at $F = 0.30$ is 0.04. By reducing the manifold pressure to 0.5 atm with the same fuel quantity, the vapor-air ratio becomes 0.08, because the quantity of fuel vapor in a given volume remains constant while the air quantity is reduced by half. This relation probably accounts in a good measure for the fact that ease of starting increases rapidly as cranking speed increases (4.580–4.584).

Starting and the ASTM Curves. As already suggested, the ASTM temperatures can be taken as indicating *relative* EAD temperature at a given fraction evaporated. This relation explains why the ASTM 10 per cent temperature has been accepted as a good indication of the relative starting ability of motor gasoline (4.580–4.582). The lower this temperature, the lower the engine temperature at which starting is easy.

Starting with Single-Component Fuels. Figure 4–4 and Eq. 4–1 show that the equilibrium vapor-air ratio above single-component liquids depends on temperature and air pressure only and cannot be increased by increasing the fuel-air ratio beyond the point of complete evaporation. Thus, with such fuels choking helps only in so far as it reduces air pressure in the induction system.

Minimum Air Temperatures for Starting with Pure Compounds. It is partly because of the difficulty of starting at low temperatures that benzene and the alcohols are seldom used in the pure state.

Warming Up. During the warming-up period, engine temperatures gradually increase to those of normal operation, at which point evaporation is presumably complete in the cylinder before ignition occurs. Thus the relative warming-up performance of fuels should depend on the whole volatility curve. It has even been suggested (4.548) that the time required to warm up a given engine with various fuels be used as a practical measure of the relative volatility of fuels. Figure 4–8 is a graph of warming-up distance versus the ASTM 20 per cent point in automobile engines. The plot is for fuels having similar high-range volatility. That differences in high-range volatility are also important is shown by the crosses on this chart. Evidently, the lower the whole range of temperatures on the ASTM curve, the shorter the warming-up period.

Effect of Volatility on Warm-Engine Performance. Any particular engine type is designed to operate satisfactorily, when warm, on fuel of a certain minimum volatility. This statement implies that, when fuel of this minimum volatility is used in the warm engine, inlet-air and inlet-manifold temperatures, and inlet-system and fuel-system design, are such as to give good distribution and complete fuel evaporation before ignition, with normal fuel-air ratios.

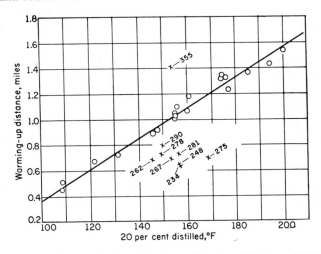

Fig. 4–8. Cold-engine warm-up distance of a passenger automobile versus ASTM 20 per cent temperature. Circles indicate fuels with varying 20 per cent point but with similar high-range volatility; crosses indicate fuels with volatility differing over the whole range. Numbers on these points represent 70 per cent ASTM temperature (Rendel, 4.548).

A good measure of warm-engine performance is acceleration. Figure 4–9 shows that fuel volatility within the range of toleration of a warm engine has little effect on acceleration. In an engine designed to operate with fuels having a certain end point, the use of fuels having lower end points (better over-all volatility) gives no improvement in warm-engine performance unless inlet heating is reduced.

On the other hand, if fuels of relatively high upper-range volatility are available, as for aircraft engines (Fig. 4–1), inlet-system heating can be reduced, with consequent gains in maximum output. Conversely, the use of fuels of lower high-range volatility than the lowest for which the engine is designed will result in slow warm-up, poor distribution, increased deposits, and generally unsatisfactory performance.

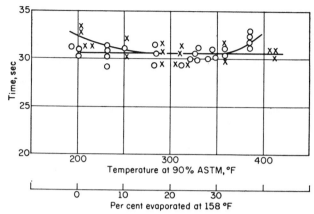

Fig. 4–9. Warm-engine acceleration versus ASTM fuel volatility. Acceleration time was measured between 10 and 38 mph on a 5.5 per cent grade; inlet–manifold temperature was held at 155°F. Crosses indicate 90 per cent point series; Circles indicate 158°F series (Eisinger and Barnard, 4.555).

Low-Volatility Fuels for Spark-Ignition Engines. Fuels as heavy as kerosene and distillate, for which typical ASTM curves are shown in Fig. 4–1, have been used with spark ignition in some farm-tractor engines and in other instances where low fuel cost is very important. When used with a carburetor-manifold system, it is obvious from these curves that much higher mixture temperatures will be required under all conditions than would be required with gasoline, and specific output will thus be lower. Starting and warming up are difficult with such fuels, and an auxiliary supply of a more volatile fuel is generally used for the purpose. The inlet manifold and the engine air supply are kept at high temperature during operation by means of exhaust heat. Unfortunately, the manifold temperatures required for satisfactory distribution may be so high that carbonaceous deposits will accumulate in the inlet manifold and cylinders. Other difficulties often experienced with such fuels include spark-plug fouling and excessive oil dilution. In most cases it has been found economical to use gasoline in spite of its higher cost.

References 2.721–2.81 and 7.900–7.903 indicate that fuels of a volatility somewhat poorer than that of ordinary gasoline can be used with inlet-port injection or with cylinder injection during the inlet stroke. This result is to be expected, since injection avoids the dependence of fuel distribution in the manifold on volatility. However, even in this case the end point of the fuel cannot be too high without resultant trouble from carbonaceous deposits, and to secure good cold starting the low-end volatility must also be nearly as high as with a carburetor engine.

Another and more successful way of utilizing fuels of low volatility with spark ignition is to inject the fuel late in the compression stroke so that ignition occurs within the fuel spray. Fuels of a volatility quite low relative to gasoline can be used with this system, since a combustible mixture can usually be found in some portion of the spray envelope. The principal difficulties encountered with such engines are connected with the problem of securing sufficiently rapid and complete mixing of the fuel and air immediately after ignition, a problem which seems to be little affected by fuel volatility, at least in the range between gasoline and light Diesel oil (2.791–2.81). However, the problem of starting such engines at very low temperatures will probably be affected by volatility. Such systems have already been mentioned in Chapter 2 under the heading "Detonation Control."

Miscellaneous Volatility Effects

Vapor Lock. (4.585–4.590.) Although high volatility at low temperature is desirable for starting, it is quite undesirable from the point of view of *vapor lock*. Vapor lock is defined as a serious restriction of the fuel supply that is due to excessively rapid formation of vapor in the fuel-supply system or carburetor. These parts must be designed to function with the moderate rates of vapor formation that are inevitable under many running conditions. If the vapor formation becomes too rapid, however, serious curtailment or stoppage of the fuel flow often results. As would be expected, it has been found that vapor formation begins to occur in fuel lines, pumps, etc., when the fuel reaches a temperature such that the vapor pressure of the fuel is equal to the pressure in the system. Thus, restrictions to flow in the fuel-supply system, and heating of any part of it, are to be avoided. The temperature of ASTM 10 per cent point has been found to be an approximate indicator of the relative tendency to vapor lock; however, direct measurement of vapor pressure is a more reliable indication. Gasoline specifications usually place an upper limit on vapor pressure, the limit depending on the type of service and the climatic conditions expected. Figure 4–7 shows Reid vapor pressures, defined in refs. 4.05 and 4.06, of motor fuels in the United States from 1945 through 1965.

The danger of vapor lock is especially great in the case of aircraft, where the fall in atmospheric pressure with increasing altitude reduces the boiling temperature of the fuel. But whatever the type of service, the obvious remedy for this trouble lies in eliminating, as far as possible, points of low pressure and high temperature in the fuel system, in proper venting at critical points, and in using a fuel with the lowest vapor pressure consistent with easy starting. Since the danger of vapor lock decreases and the necessity of starting at low temperatures increases as the atmospheric temperature falls, it is desirable to have the 10 per cent ASTM temperature lower (Reid vapor pressure higher) in winter than in summer; this is common practice among fuel manufacturers (Fig. 4–7).

Evaporation Loss. Evaporation loss in tanks, etc., naturally depends on the vapor pressure of the fuel at the storage temperature. In gas-free fuel this is related to the 10 per cent ASTM temperature, but, as in the case of vapor lock, the vapor pressure is a more reliable indicator. Evaporation loss is a reason, in addition to vapor lock, for restricting the low-end volatility of fuels.

Crankcase-Oil Dilution. Dilution of the lubricating oil by unevaporated fuel may reduce the crankcase-oil viscosity in services where frequent starting with low engine temperatures is necessary. As has been seen, very rich mixtures must be supplied under these conditions, and some of the unevaporated fuel leaks past the piston rings and into the oil. It has been found that the relative tendency of fuels to cause dilution of the lubricating oil lies in the order of their 90 per cent ASTM temperatures (4.570).

Control of the 90 per cent point, plus crankcase ventilation, apparently limits the degree of dilution, so that little trouble from this source is experienced in modern motor cars using gasoline. Engines using heavier fuels, such as kerosene and distillate, may suffer from poor lubrication of pistons and rings because of excessive dilution.

Volatility and Fire Hazard. The relationship of volatility and fire hazard has been a subject of much controversy. There is no simple scientific measure for fire hazard, because it depends on particular circumstances. However, the following may be stated as facts which have a bearing on the subject:

1. The atmosphere over fuel in tanks may be explosive with either gasoline, jet fuel, or Diesel oil, depending on temperature. For example, the explosive range for gasoline is from $-80°F$ to $0°F$, for jet fuel (kerosene) $60°F$ to $110°F$, and for light Diesel oil $140°F$ to $180°F$. These are approximate figures, assuming the air space in the tank to be at sea-level pressure (4.122).

2. In partly ventilated enclosed spaces, such as boat bilges and aircraft-wing interiors, gasoline is usually more likely to furnish a combustible atmosphere than are heavier fuels.

3. Gasoline fires develop to great intensity almost immediately after ignition. Fires with heavier fuels develop much more slowly (4.120).

4. The heavier fuels, having lower ignition temperatures, ignite more readily than gasoline when spilled on hot surfaces. On the other hand, the vapors from spilled gasoline are more likely to ignite from electric sparks than is the vapor from heavier fuels, because of the greater likelihood of a combustible mixture at the spark.

From the aforementioned considerations it would appear that, under most circumstances, increasing volatility tends to increase fire hazard, and therefore the theory that the use of Diesel oil in place of gasoline reduces fire hazard is generally correct. However, experience with jet airplanes, which use a fuel less volatile than gasoline, has not shown any noticeably smaller incidence of fire after a crash than has been experienced with gasoline-powered aircraft.

Summary of Volatility Effects

The *relative* performance of gasoline with regard to lowest starting temperature, vapor lock, and evaporation loss can be predicted from the ASTM 10 per cent temperature and the Reid vapor pressure, which should be controlled as to both the upper and lower limits for a given climate and service. The time required for warming up depends on the whole volatility curve.

The inlet temperature required for good warm-engine performance depends chiefly on the upper-end volatility. Increasing upper-end volatility tends to improve maximum engine output and efficiency, provided appropriate design changes can be made. On the other hand, increasing lower-end volatility increases losses in storage and danger of fire and vapor lock.

The ASTM 90 per cent temperature gives a relative indication of the influence of the fuel on oil dilution and on the tendency toward smoky combustion.

Volatility of Commercial Fuels. The increase in vapor pressure and lowering of the 10 per cent temperature, as shown in Fig. 4–7, have been affected to facilitate starting and have been made possible by improved fuel-system design from the point of view of vapor lock. Since 1957, "premium" fuel has had a somewhat lower 90 per cent temperature than "regular" gasoline, but otherwise the differences in volatility between these two grades are insignificant.

DETONATION CHARACTERISTICS OF FUELS FOR S-I ENGINES

The resistance of a fuel to detonation is an extremely important characteristic if the fuel is to be used in spark-ignition engines. As we have seen in Chapter 2, the antiknock property of a fuel apparently depends on its compression-ignition characteristics, and these vary tremendously with its chemical composition.

Knock Rating of Fuels

In Chapter 2 it was shown that, with a given fuel, the presence or absence of detonation and the intensity of detonation when present vary with engine design and with almost every operating variable. Furthermore, it has been known for a long time that even the *relative* knocking tendency of any two fuels is likely to change appreciably, depending on the method used to compare the fuels. Thus the establishment of either an absolute or a relative scale of detonation tendency which will be satisfactory for all applications seems impossible.

Since the two most important applications of the spark-ignition engine are for motor vehicles and for aircraft, most of the work done in the field of knock rating has centered around these two applications. Presumably, the work done in the motor-vehicle field applies reasonably well to marine and industrial applications that use spark-ignition engines of a design quite similar to that of the motor-car engine and also perhaps to unsupercharged aircraft engines of the light-plane class.

Knock Rating of Motor Fuels. By *motor fuels* is meant fuels suitable for general use in unsupercharged engines, and more particularly fuels for road vehicles. Work on this problem in recent years has been done chiefly under the auspices of the Coordinating Research Council (CRC), formerly called the Cooperative Fuel Research (CFR) Committee (4.500). This committee early recognized that four basic requirements were necessary for establishing a comparative scale of the detonation tendency, namely:

1. A standardized engine
2. A set of standard operating conditions
3. A standard method of measuring the intensity of detonation
4. A pair of standard reference fuels.

References 4.210 and 4.500–4.502 afford a rather complete history of the development of these standards. Briefly, the outcome to date has resulted in the following:

The CFR variable-compression single-cylinder engine (4.502)

The ASTM detonation indicator (4.02)

The CFR *motor* method and *research* method of knock testing (4.02, 4.05).

The two standard reference fuels are *isooctane* (page 125), a fuel that has much less knocking tendency than the average motor fuel, and *normal heptane* (page 124), a fuel that has much more tendency to knock than the average motor fuel.

Briefly, the method of test consists of running the fuel to be tested under the standard conditions and setting the compression ratio to give a standard knock intensity as measured by the knock indicator. While holding compression ratio and other test conditions constant, various mixtures of the two reference fuels are tried until a mixture which gives the same knock intensity as the fuel under test is discovered.* The percentage

---

* The final mixture is determined by interpolation between two mixtures, one of which gives slightly more detonation and one slightly less, than the fuel being rated.

of isooctane in the matching mixture is called the *octane number* of the fuel in question.

The test conditions used represent roughly those present in a motor car when detonation is most likely to occur, namely, at full throttle (without supercharging), low rpm, high inlet temperature, and the fuel-air ratio for maximum knock. With motor fuels the mixture for maximum knock is usually close to that for maximum power. The dependent variable is the compression ratio, as seems logical in view of the fact that reduction in the knocking tendency of fuels for unsupercharged engines is most likely to be utilized by raising the compression ratio.

In order that the octane number of motor fuels shall be as representative as possible of their behavior in the average motor-vehicle engine, the CRC has conducted a program of periodic *road tests* (4.515–4.517). As a result of such tests, the standard CFR test conditions have been changed as required to improve the average correlation. Correlation data are available in the literature (4.515–4.524), and the results are good, considering the difficulties involved. In spite of this work, however, with a particular engine operating under a particular set of circumstances, it may be found that the tendency of fuels to knock does not always lie in the order of their octane numbers.

The octane number of a motor fuel thus represents a comparison of the fuel in question with the reference fuels under a set of arbitrary conditions chosen as far as possible to correlate with average service conditions. As might be expected, the more closely the chemical structure of a commercial fuel resembles that of the reference fuels, the more nearly it succeeds in maintaining its octane number constant over a wide range of operating conditions. Thus, gasolines that tend to be largely paraffinic in structure maintain their relative positions, as indicated by the octane number, rather well over a wide range of test conditions. The ratings of gasolines containing a considerable fraction of nonparaffinic material, benzol blends, and alcohol blends, on the other hand, are very sensitive to changes in engine design and operation. Fuel-air ratio, speed, jacket temperature, and intake temperature appear to have especially important influences on the rating of such fuels (4.503–4.510).

Fuel Sensitivity. Two sets of operating conditions, called the *motor method* and the *research method*, are now used in the knock rating of motor fuels (4.02, 4.06). These conditions are as shown in the following table.

Factor	Motor Method	Research Method
Inlet temperature	300°F	125°F
Jacket temperature	212°F	212°F
Speed	900 rpm	600 rpm
Humidity, mass/mass dry air	0.0036–0.0072	0.0036–0.0072

In general, the motor method is considered more severe because of its higher speeds and higher inlet temperature. Because their structure is similar to that of the reference fuels, paraffinic gasolines will have essentially the same octane number with either method.

As the fuel structure departs from paraffinic, the octane number of a fuel will show an increasing difference when determined by the two methods. The research-method octane number minus the motor-method octane number of the fuel is designated as the fuel *sensitivity*. Paraffinic gasolines usually show near-zero sensitivity, while highly cracked fuels give positive sensitivity numbers. Since about 1940, the research-method octane numbers for current gasolines have correlated better with passenger-car road-test octane numbers than have motor-method ratings. Edgar *et al.* (4.515) give data on road-test ratings in commercial vehicles.

Commercial Motor Fuels. Figure 4–7 shows trends of important properties including average octane number, of motor gasoline in the United States from 1946 through 1965. It will be noted that the sensitivity of "premium" gasoline is somewhat greater than that of "regular" gasoline, and that the only important difference between "premium" and "regular" gasoline is in octane number and sensitivity. The fact that the increase in octane number since 1960 has been small appears to indicate that the optimum balance between refining costs and fuel quality is being attained. Also, preignition starts to be a limiting factor when compression ratios are raised to appropriate values for fuels near 100 octane (see Fig. 4–13, p. 151, and later discussion of preignition).

Knock Rating of Aviation Fuels. When aviation engines were mostly unsupercharged, knock ratings obtained by the methods used for motor fuels were a reasonably satisfactory indication of relative performance in flight. However, with the advent of highly supercharged aircraft engines using special fuels of very high resistance to detonation, a special system for knock-rating aviation fuels became necessary. This system takes into account the following differences between aircraft and automotive requirements:

1. In general, improvements in resistance to detonation are used to increase inlet pressure rather than compression ratio.

2. At maximum power, aircraft engines are likely to use fuel-air ratios as high as $F_R = 1.8$, while in motor vehicles F_R seldom exceeds 1.3.

3. Many fuels available for aircraft have a resistance to knocking greater than that of isooctane.

The methods worked out, again largely by the CRC, for knock-rating aviation fuels are covered in refs. 4.511–4.514. These methods differ from those used for motor fuels chiefly in that the engine is run at higher speed, the variable used to produce knock is inlet pressure rather than compression ratio, and the complete useful range of fuel-air ratio is explored. The CFR engine is used.

In rating aviation fuels, curves of knock-limited imep versus fuel-air ratio are plotted for two reference fuels and for the test fuel, as in Fig. 4–10.

Ratings are generally given for the minimum, or *lean-mixture*, and the maximum, or *rich-mixture*, points on the test-fuel curve. Most aviation

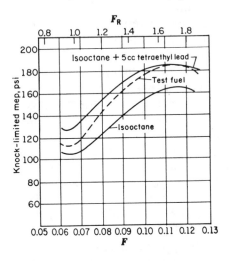

Fig. 4–10. Illustration of ASTM method of knock-rating aviation fuels.

fuels rate better than isooctane (100 ON) and may be rated in terms of isooctane plus tetraethyl lead. By interpolation in Fig. 4–10 the *lean-mixture* rating of the test fuel is "isooctane plus 2 cc lead" and the *rich-mixture* rating is "isooctane plus 5 cc lead." The lead measurements are in terms of cubic centimeters of tetraethyl lead per gallon of fuel.

A more usual and perhaps better method of rating fuels superior to isooctane is in terms of *performance numbers*. Two numbers are given for each fuel, defined, respectively, as

$$\left(\frac{100 \times \text{imep}}{\text{imep}_o}\right)_{\text{lean}} \quad \text{and} \quad \left(\frac{100 \times \text{imep}}{\text{imep}_o}\right)_{\text{rich}},$$

where imep is the knock-limited imep with the test fuel (obtained by varying the inlet pressure at constant compression ratio and inlet temperature) and imep_o is the knock-limited imep with isooctane. "Lean" and "rich" have the meaning previously explained. By this method the test fuel of Fig. 4–10 rates as 110, 113, meaning that its lean performance number is 110 and its rich performance number is 113. Further details of the standard method of rating aviation fuels will be found in refs. 402–4.06. Figure 4–10 illustrates one reason for running aircraft engines with very rich mixtures at times when maximum power is required, such as during take-off.

As in the case of motor fuels, the standard aviation-fuel test conditions have been altered as required to secure good average correlation with service conditions. Also, the correlation between the standardized test and the behavior in a given engine under a given set of circumstances is seldom exact. However, it has been found that the performance number is less dependent on design and test conditions than is the octane number (2.54).

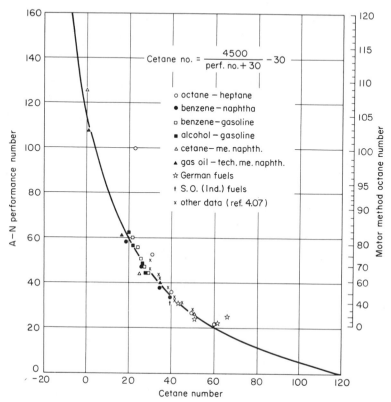

Fig. 4–11. Octane-number, cetane-number, and Army-Navy performance-number relationship for several fuel mixtures (Brewster and Kerley, 4.06, based on unpublished work by Marschner, Standard Oil Co. of Indiana).

Correlation of Detonation Measurements. It has been found that the octane number and performance number are closely related for all common types of fuel (4.07). Figure 4–11 illustrates this relationship, which also affords means for extending the octane-number scale above 100. The cetane number shown as the abscissa of this curve will be explained in the section on Diesel fuels.

Effect of Fuel Structure on Octane Number

Within the limitations of fuel testing outlined in the preceding section, the following general relationships between structure and the detonation tendency of petroleum fuels seem to hold:

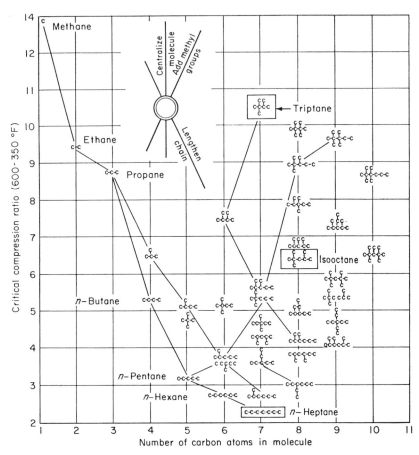

Fig. 4–12. Effect of fuel structure on detonation tendency of paraffinic hydrocarbons. CFR engine, 600 rpm, inlet temp. 350°F (Lovell, 4.221).

Figure 4–12 shows that, for a given number of carbon atoms, branched-chain arrangements are better than the straight-chain arrangement and that antiknock value varies with the number of branches and the position of the branches of the carbon chain.

In general, and always in the case of the straight-chain paraffin compounds, the antiknock value improves as the number of carbon atoms in the molecule decreases. For example, normal pentane has a better antiknock value than normal heptane. The very light paraffins, such as methane, ethane, and propane, have excellent antiknock properties (4.291, 4.292). Naphthenic and aromatic compounds generally have high antiknock characteristics.

Gluckstein *et al.* (2.33) found that, other operating conditions remaining the same, critical end-gas temperature is the same for a number of sensitive and nonsensitive fuels. They attribute the observation that sensitive fuels knock sooner, as inlet temperature is increased, to the fact that sensitive fuels show the greater prereaction activity. Thus, they reach the critical value of $T_{2''}$ (see Chapter 2, p. 60) at a lower inlet temperature than does the corresponding insensitive fuel.

No such systematic relation between fuel structure and octane number has been found for nonpetroleum fuels and fuel additives, as may be seen by reference to Table 4–2.

Table 4–2. Antiknock and Pro-knock Compounds

Compound	Formula	Wt. for Given Effect*	Rel. Mol. Effectiveness†
Aniline	$C_6H_5NH_2$	1	1
Benzene	C_6H_6	9.8	0.085
Toluene	$C_6H_5CH_3$	8.8	0.112
Xylene	$C_6H_4(CH_3)_2$	8.0	0.142
Alcohol	C_2H_5OH	4.75	0.104
Ethyl iodide	C_2H_5I	1.55	1.09
Diethyl selenide	$(C_2H_5)_2Se$	0.214	6.9
Diphenyl selenide	$(C_6H_5)_2Se$	0.49	5.2
Diethyl telluride	$(C_2H_5)_2Te$	0.075	26.6
Diphenyl telluride	$(C_6H_5)_2Te$	0.139	22.0
Triphenylphosphine	$(C_6H_5)_3P$	3.08	0.91
Triphenylarsine	$(C_6H_5)_3As$	2.44	1.35
Triphenylstibine	$(C_6H_5)_3Sb$	1.56	2.42
Tetraethyl tin	$(C_2H_5)_4Sn$	0.66	3.8
Tetraethyl lead	$(C_2H_5)_4Pb$	0.0295	118
Tetraphenyl lead	$(C_6H_5)_4Pb$	0.080	69.5
Diphenyldiethyl lead	$(C_6H_5)_2(C_2H_5)_2Pb$	0.041	110
Triethylbismuthine	$(C_2H_5)_3Bi$	0.135	23.9
Triphenylbismuthine	$(C_6H_5)_3Bi$	0.22	21.5
Nickel carbonyl	$Ni(CO)_4$	0.053	35
Dimethyl cadmium	$(CH_3)_2Cd$	1.23	1.25
Titanium tetrachloride	$TiCl_4$	0.64	3.2
Cumidine	$(CH_3)_3C_6H_2NH_2$	0.96	1.51
Diphenylamine	$(C_6H_5)_2NH$	1.21	1.5
m-Xylidine	$(CH_3)_2C_6H_3NH_2$	0.92	1.4
Monomethylaniline	$C_6H_5NHCH_3$	0.82	1.4
Toluidine	$CH_3C_6H_4NH_2$	0.94	1.22
Amylaminobenzene	$C_5H_{11}C_6H_4NH_2$	1.53	1.15
Ethylaminobenzene	$C_2H_5C_6H_4NH_2$	1.14	1.14
Aminodiphenyl	$C_6H_5C_6H_4NH_2$	1.6	1.14
Methyl-o-toluidine	$CH_3C_6H_4NHCH_3$	1.15	1.13
n-Butylaminobenzene	$C_4H_9C_6H_4NH_2$	1.44	1.11
n-Propylaminobenzene	$C_3H_7C_6H_4NH_2$	1.32	1.10
Monoethylamiline	$C_6H_5NHC_2H_5$	1.27	1.02
Mono-n-propylaniline	$C_6H_5NHC_3H_7$	1.95	0.75

Table 4–2 (*continued*).

Compound	Formula	Wt. for Given Effect*	Rel. Mol. Effective-ness†
Ethyldiphenylanine	$C_2H_5N(C_6H_5)_2$	3.65	0.58
Mono-*n*-butylaniline	$C_6H_5NHC_4H_9$	3.1	0.52
Diethylamine	$(C_2H_5)_2NH$	1.59	0.495
Di-*n*-propylaniline	$C_6H_5N(C_3H_7)_2$	7.15	0.27
Mono-isoamylaniline	$C_6H_5NHC_5H_{11}$	7.1	0.248
Diethylaniline	$C_6H_5N(C_2H_5)_2$	6.7	0.24
Dimethylaniline	$C_6H_5N(CH_3)_2$	6.2	0.21
Ethylamine	$C_2H_5NH_2$	2.4	0.20
Triethylamine	$(C_2H_5)_3N$	7.95	0.14
Triphenylamine	$(C_6H_5)_3N$	30.0	0.09
Ammonia‡	NH_3	−2.0	−0.09
Usopropyl nitrite‡	$C_3H_7NO_2$	−0.085	−11.5
Organic nitrates and nitrites in general‡			

* Amount in grams required to give an antiknock effect equivalent to 1 g of aniline.

† Reciprocal of the number of moles required to give an antiknock effect equivalent to 1 mole of aniline.

‡ Pro-knock compounds.

Aniline in concentrations up to 3 per cent of the fuel by volume was taken as the standard of effect. All measurements made with bouncing-pin apparatus, using kerosene as fuel.

NOTE: The relative effectiveness of these compounds varies with the concentration and with operating conditions.

SOURCE: Boyd, ref. 4.211.

Octane-Number Control

Under the increased demand for higher compression ratios (Fig. 4–13) and therefore for improved antiknock properties, powerful methods of controlling the molecular structure of fuels by means of the refining process have been developed. Thus, while the antiknock value of gasoline of a given volatility was at one time almost entirely controlled by the characteristics of the base crude, it is now subject to a high degree of control in the refining process. However, since it is more expensive to attain a high octane number with certain crudes than with others, it cannot be said that the composition of the base crude has no bearing on the problem. As a result of this situation, the octane numbers of the equivalent commercial fuels produced from various crudes is often brought to the required value by adding the proper amount of antiknock agent.

Antiknock Agents. Deficiencies in the octane number of a gasoline refined for a given service may be made up by the addition of antiknock agents (4.210–4.219). Beginning with the work of Midgeley (4.210), much

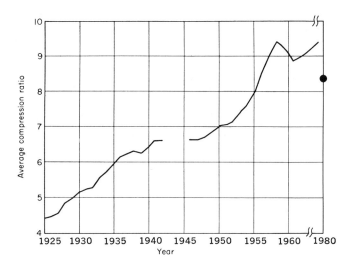

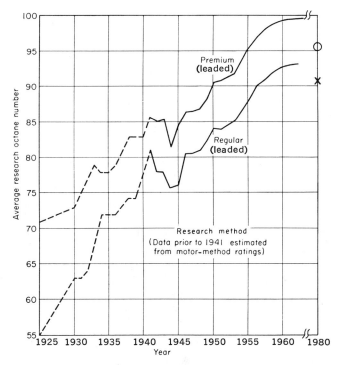

Fig. 4–13. Average compression ratio and gasoline octane number versus year of manufacture. 1935–1957 from Bartholomew *et al.*, ref. 4.518; 1957–1964 comp. ratio from *Auto Industries*, Oct. 15, 1962, p. 63; 1957–1964 ON from Bureau of Mines Surveys, ref. 4.105; 1980— unleaded premium (○); unleaded regular (X).

effort has been expended on the discovery and development of substances which, when mixed with gasoline, reduced its tendency to detonate. In the course of this work substances that increase the detonating tendency have also been discovered. Table 4-2 is a list of some of the chemicals known to have an effect on detonation.

The remarkable effectiveness of tetraethyl lead, $Pb(C_2H_5)_4$, in very small concentrations (Fig. 4–14) has brought this chemical into wide

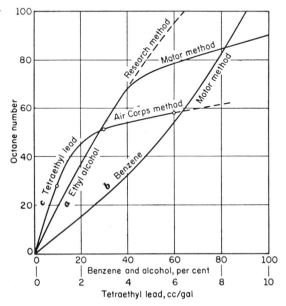

Fig. 4–14. Relative effectiveness of antiknock compounds in normal heptane. The degree of effectiveness and relative effectiveness varies with the nature of the fuel and with test conditions. (*a*) and (*b*) *courtesy National Bureau of Standards*; (*c*) *courtesy Ethyl Corp.*

commercial use as a knock-suppressor. In the United States it has been added in small quantities to most commercial motor and aviation fuels to bring their octane number to standard values for fuels in the various price ranges. The discovery and exploitation of tetraethyl lead has made an enormous contribution toward improving the efficiency and raising the specific output of spark-ignition engines. However, it is nothing more than an antidetonation agent, and its use will not improve the performance of an engine that is not already knocking.

The catalytic reactors used to reduce pollutants in the exhaust systems of cars manufactured since the late 1970s (see p. 211) will not tolerate "leaded" gasoline. Therefore the use of leaded fuel in the United States is decreasing rapidly, and the same trend will probably follow in other countries. Figure 4–13 shows octane numbers of current U.S. unleaded gasoline.

ALCOHOL FUELS

Alcohol mixed with gasoline gives antiknock properties, as indicated in a general way in Figure 4–14. Relative effectiveness varies with fuel-air ratio and with operating conditions. Table 4–4 shows test results at inlet temperatures required for good distribution in a carbureted engine.

As a conservation measure, and in some cases as an economical way of using surplus alcohol (as in Sweden), a mixture of 10–20 percent ethyl or methyl alcohol in gasoline called "gasahol" is marketed for vehicle use. The 10 percent solution seems to be interchangeable with the same gasoline undiluted, as far as engine performance goes. It gives a small increase in octane value and reduces gasoline consumption about 5 percent.

EFFECT OF FUEL COMPOSITION ON PREIGNITION

As stated in Chapter 2, preignition is defined as ignition of the charge prior to passage of the spark, because of contact with a hot surface. The preignition tendency is very sensitive to fuel composition (4.530–4.533) and to test conditions, especially spark-plug electrode temperature, carbon deposits, and engine speed (2.85–2.998).

In a clean engine using gasoline, with proper cooling and choice of spark plugs, it is unusual for preignition, rather than detonation, to set the limit on the allowable manifold pressure or compression ratio. On the other hand, preignition is sometimes encountered in high-compression engines

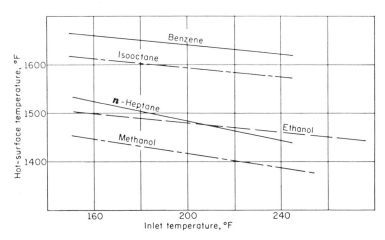

Fig. 4–15. Surface temperatures for ignition of various fuels at their stoichiometric fuel-air ratio; tests with a circular hot surface $\frac{1}{2}$-in. diam flush with inner surface of a CFR cylinder head. 1000 rpm, $r = 5.95$, $F_R = 1.0$, $T_c = 160°$ F (Livengood et al., 4.533).

that have accumulated deposits, especially with leaded fuels. This phenomenon is responsible in good part for the fact (see Fig. 4–13) that compression ratios in U.S. passenger cars are no longer increasing with time at the former rapid rate. Figure 4–15 shows surface temperatures required for preignition of various fuels in a special test engine. These curves may be regarded as relatively significant.

MISCELLANEOUS PROPERTIES OF FUELS FOR S-I ENGINES

Freedom from water, dirt, or other foreign matter is, of course, essential. Any corrosive tendency of the fuel, or of impurities which it may contain, is also to be avoided. Some gasolines have a tendency to deposit gum, a solid oxidation product, in fuel systems and on valve guides. Excessive amounts of gum in gasoline often cause trouble, such as sticking of valves and plugging of fuel passages. Gum is a product of slow oxidation. It is relatively easy to detect and to eliminate gum that is suspended or dissolved in a fuel, but to predict how rapidly gum will form under storage conditions and to control the rate of its formation is more difficult (4.320–4.323). An ASTM accelerated oxidation test (4.05) has been developed; it compares, with a fair degree of accuracy, the tendency of fuels to form gum in storage. Special chemicals, known as *inhibitors*, are mixed with gasoline to reduce the tendency to form gum. A high sulfur content (4.332) is objectional in gasoline because sulfuric acid is formed, with consequent corrosive action on engine parts. Benzol also may contain injurious amounts of sulfur.

Tetraethyl lead may cause troublesome deposits on cylinder walls, spark plugs, valves, etc., and it may promote spark-plug and exhaust-valve corrosion or erosion (4.330–4.336). These difficulties are reduced by adding an organic halide, such as ethylene dibromide, $C_2H_4Br_2$ (4.210). The additive called "ethyl fluid" consists of a mixture of tetraethyl lead and ethylene dibromide. Even with this compound, however, it is necessary to limit the concentration of tetraethyl lead according to the service in which the fuel is to be used.

Sometimes the fuel and lubricating oil seem to work together in building up engine deposits, and it is now believed that certain combinations of fuel and oil are harmful in this respect under circumstances where either one alone would not cause serious difficulty (4.333–4.335). In general, gaseous fuels cause a minimum of trouble from engine deposits, corrosion, etc. The alcohols are also quite free from difficulties of this kind (4.204).

A difficulty with many antiknock agents is their *toxic* character, acting through the skin or when the vapor is absorbed into the lungs. Tetraethyl lead is a dangerous poison of the latter type. On this account, laws exist which require leaded fuels to be colored and which forbid their use except as fuel for internal-combustion engines. Benzene is also somewhat poisonous to the skin and lungs.

EFFECT OF FUEL ON POWER AND EFFICIENCY OF S-I ENGINES

This subject has been touched upon in connection with the discussion of several of the various fuel characteristics. It is now possible to bring together the various effects already mentioned in such a way as to show, at least qualitatively, the total effect of fuel characteristics on the power and efficiency of engines. On the assumption that speed is unaffected by fuel composition, indicated power will be proportional to imep.

For four-stroke engines, from Eq. 12–3, Volume I (p. 426):

$$\frac{P_2}{P_1} = \frac{\text{imep}_2}{\text{imep}_1} = \frac{(\Gamma e_v)_2}{(\Gamma e_v)_1} \frac{\rho_{ai_2}}{\rho_{ai_1}} \frac{(F'Q_c)_2}{(F'Q_c)_1} \frac{\eta_2}{\eta_1}. \tag{4-3}$$

For two-stroke engines, from Eq. 12–4, Volume I (p. 426):

$$\frac{P_2}{P_1} = \frac{\text{imep}_2}{\text{imep}_1} = \frac{(R_s\Gamma)_2}{(R_s\Gamma)_1} \frac{\rho_{s_2}}{\rho_{s_1}} \frac{(F'Q_c)_2}{(F'Q_c)_1} \frac{\eta_2}{\eta_1}, \tag{4-4}$$

where ρ_{ai} = density of air in mixture at inlet

ρ_s = scavenging air density, see Volume I, page 217

R_s = scavenging ratio

e_v = volumetric efficiency

Γ = trapping efficiency

F' = trapped fuel-air ratio

Q_c = heat of combustion (added to the equations in Volume I, which were for one fuel)

η = indicated efficiency based on trapped fuel

and subscripts 1 and 2 refer to two different fuels.

Fuel-Air Ratio. A question that immediately arises with regard to the preceding relations is the fuel-air ratio at which comparisons of different fuels should be made. Our study up to this point (see especially Volume I, Chapters 4 and 5) indicates that the *chemically correct* fuel-air ratio, or some multiple thereof, should furnish a comparable basis.

Effect of Fuel Composition on Efficiency. Under comparable conditions, differences in fuel composition affect the thermodynamic characteristics of the fuel-air mixture, and thus differences in efficiency may result. To determine such differences accurately, thermodynamic data, such as those of charts C–1 to C–4 of Volume I, should be available for each fuel. At present, only a limited number of such data can be found in convenient form (0.030–0.034, and Volume I, 3.20–3.54).

Fuel-air cycles computed for different hydrocarbon and carbohydrate fuels (0.030, 0.034) show little difference in efficiency at identical values of compression ratio, F_R, and conditions at point 1 of the cycle. A study using gaseous mixtures at the same fraction of stoichiometric ratio (4.202) shows that ethyl alcohol gives about 3 per cent higher efficiency than gasoline at the same compression ratio. This difference can be explained by the fact that the heat of combustion of liquid fuels is based on the fuel in the liquid state, while fuel-air cycles are generally constructed on the assumption of a gaseous fuel-air mixture. Under these circumstances the efficiency of liquid fuels with differing enthalpy of evaporation might be expected to be proportional to $Q_c - H_{lg}$ (Table 4–1). In the case of ethyl alcohol and octene this ratio, from the data in Table 4–1, would be

$$\frac{(11,550 + 361)19,035}{(19,035 + 145)11,550} = 1.023$$

Thus, the difference indicated by test is accounted for by this factor alone.

From these considerations it is apparent that basic differences in efficiency are small, as between the most usual liquid fuels. Efficiencies with some unusual fuels are given in refs. 4.126 and 4.127.

Effect of Fuel on Allowable Compression Ratio in Unsupercharged Engines. Unsupercharged engines are designed to carry nearly sea-level atmospheric inlet pressure with wide-open throttle, and the spark timing and compression ratio are adjusted for this condition. Small changes in octane number can usually be handled by adjustment of the spark timing alone. Large changes in octane number, however, involve the necessity for, or opportunity of, changing the compression ratio. Thus, in general, the compression ratios of unsupercharged engines are raised as the antiknock value of available fuels increases. Figure 4–13 shows how the average compression ratio of U.S. motor cars has increased since 1925, largely because of the increase in octane number of the fuel available. The appropriate compression ratio to be used with a given fuel depends on a combination of design and operating factors, and it can be determined only by experiment for a given type of engine and type of service. Thus it is impossible to assign a "highest useful compression ratio" to a given fuel, except for a particular engine under particular circumstances. However, when optimum compression ratios for two fuels are known, an estimate of the effect of this difference on efficiency can easily be made by using relative fuel-air cycle efficiencies (Figs. 0.1–0.6).

Effect of Fuel on Inlet Pressure in Supercharged Engines. In the case of supercharged engines, an increase in octane number may be used to allow an increase in inlet pressure. If no other variables are involved, the performance number of the fuel gives a good measure of the relative imep allowable (see Fig. 4–11). On the other hand, an improvement in octane number (or performance number) gives the choice of raising the compression ratio rather than the inlet pressure, or else raising both by a lesser amount. Thus, when faced with an improvement in the detonation characteristics of the available fuel, it is a matter of choice which combination of inlet pressure and compression ratio is used. Figure 4–16 gives curves of knock-limited imep at various compression ratios and shows clearly

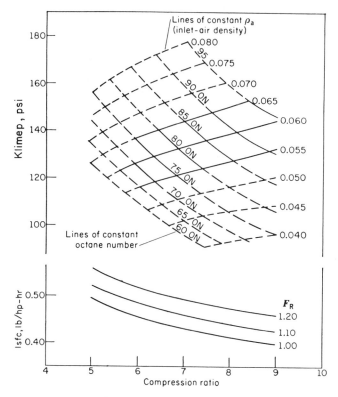

Fig. 4–16. Knock-limited imep and isfc versus compression ratio and inlet air density (Barber, 2.51).

that, with a given fuel, when maximum output is desired in a supercharged engine, a low compression ratio should be used; that is, efficiency must be sacrificed. While the absolute values shown in the figure vary with design and operating conditions, this general relationship holds under all practical conditions. As an example of the factors that must be considered, in the case of aircraft engines the necessity for a high take-off power must be balanced against the requirement for low cruising fuel consumption and a compromise must be decided upon between compression ratio and maximum take-off power.

Effect of Fuel on Inlet Temperature. With gaseous fuels, in any type of engine, there is no fuel characteristic that requires the inlet temperature to be raised above the temperature of the atmosphere. Thus, it can safely be said that with such fuels the inlet temperature is not influenced by the choice of fuel.

With liquid fuels in carbureted engines, as we have seen, heating of the inlet air or of the inlet manifold may be required to secure adequate distribution. In the case of wet mixtures it is difficult to measure T_i, the temperature at the inlet ports; this further complicates the problem. For these reasons it is not easy to treat this problem quantitatively, and we must be satisfied to deal with trends rather than absolute values. From the discussion of volatility, it is apparent that the minimum inlet temperature for satisfactory operation will increase with decreasing fuel volatility.

Under conditions where distribution does not set a low limit on inlet temperature, the use of liquid fuels introduces the possibility of operating with values of T_i below atmospheric temperature. (The temperature drop due to complete evaporation of chemically correct mixtures is given in Table 4–3.) Thus for racing, where output is more important than economy, the alcohols, especially methyl alcohol, are popular fuels or blending agents. In such cases, in order to avoid distribution difficulties, individual carburetors and inlet pipes for each cylinder are often used. An alternative arrangement that accomplishes the same result is injection of fuel at the inlet ports or into the cylinders (see Chapter 7).

Table 4–3. Temperature Drop ΔT due to Complete Evaporation of Liquid Fuels at Stoichiometric Ratio

Fuel and Chemical Formula	H_{lg} at 77°F	ΔT
Gasoline, (C_8H_{16})	−150	40
Light Diesel oil, $(C_{12}H_{26})$	−115	30
Normal heptane, C_7H_{16}	−157	41
Isooctane, C_8H_{18}	−141	38
Benzene, C_6H_6	−186	57
Ethyl alcohol, $C_2H_5(OH)$	−361	153
Methyl alcohol, $CH_3(OH)$	−474	300

H_{lg} is in Btu per lbm.
ΔT is temperature drop, °F, of the stoichiometric mixture of fuel and air due to complete fuel evaporation.

Effect of Fuel on Volumetric and Scavenging Efficiencies. From the discussion of volumetric efficiency in Chapter 6 of Volume I, it is clear that evaporation of liquid fuel in the cylinder of a 4-stroke engine during the inlet stroke will tend to increase e_v. It has been found that a change from carburetion to cylinder injection during the inlet stroke tends to increase volumetric efficiency, because in the latter case more of the heat of evaporation is taken from the gases rather than from the cylinder walls.

In the case of carbureted 2-stroke engines, fuel evaporation during scavenging will tend to increase the scavenging ratio by increasing the fuel-charge density. The effect on power will be proportional to $R_s\Gamma$.

As we have seen in Chapters 6 and 7 of Volume I, the velocity of sound in the inlet system affects the flow of mixture in either 4- or 2-stroke engines.

The influence of fuel composition on this quantity is given in Table 4–1. From these data it is evident that the effect of changes in sonic velocity due to changes in fuel composition will be small except when a change to a fuel containing considerable hydrogen is involved.

In connection with a study of the influence of fuel on volumetric or on scavenging efficiency, it is important to remember that, if changes in fuel composition involve changes in T_i or p_e/p_i, volumetric and scavenging efficiencies will be changed on this account as well as for the reasons already given. Expressions for estimating the effects of p_e/p_i and T_i on volumetric and scavenging efficiencies are given in Chapters 6 and 7 of Volume I.

Summary of Effects of Fuel Composition on Power. Let it be assumed that inlet and exhaust pressures remain constant and that effects on volumetric efficiency and scavenging ratio are confined to the influence of the change in inlet temperature required by, or allowed by, the change in fuel. Let T_i be measured ahead of the carburetor and let the fraction x of the fuel be evaporated before the mixture reaches the inlet port. Let it further be assumed that indicated efficiency is proportional to the fuel-air-cycle efficiencies shown in Fig. 0–1. Equations 4–3 and 4–4 can then be written as follows for both 2-stroke and 4-stroke engines:

$$\frac{\text{imep}_2}{\text{imep}_1} = \frac{\Gamma_2}{\Gamma_1} \sqrt{\frac{(T_i - x\,\Delta T)_1}{(T_i - x\,\Delta T)_2}} \frac{\left[F_c' Q_c \Big/ \left(1 + F' \dfrac{29}{m_f}\right)\right]_2}{\left[F_c' Q_c \Big/ \left(1 + F' \dfrac{29}{m_f}\right)\right]_1} \frac{\eta_{o2}}{\eta_{o1}}, \qquad (4–5)$$

where F' is the trapped fuel-air ratio. Values of FQ_c and $1/[1 + F(29/m_f)]$ are given in Table 4–1. The efficiency η_o is fuel-air-cycle efficiency from Figs. 0–1 through 0–6.

As an example of the use of Eq. 4–5, calculated values of imep are compared with measured values in Table 4–4 for various blends of gasoline and alcohol. The value of x was assumed to be 0.50. Since a 4-stroke engine with small valve overlap was used, Γ was taken as unity. The fact that the calculated imep for the alcohol blends are higher than the measured values indicates that x may have been improperly estimated.

Effect of Fuel on Fuel Consumption

As we have seen, the specific fuel consumption of an engine is inversely proportional to the product of the thermal efficiency and the heat of combustion per unit weight of fuel, that is,

$$\text{sfc} = \frac{2545}{\eta Q_c} \qquad (4–6)$$

Table 4–4. Comparison of Performance with Gasoline and Gasoline–Ethyl Alcohol Blends, Carburated CFR Engine, 1200 rpm

Per Cent of Alcohol	(1)	η (2)	$T_i,°R$ (3)	$x\,\Delta T$ (4)	F_cQ_c (5)	$\dfrac{1}{\left(1+F\dfrac{29}{m_f}\right)}$ (6)	Imep and Isfc			
							Measured (7)		Computed (8)	
0	6	0.32	600	24	1275	0.984	110	0.62	110	0.62
10	6.2	0.33	600	31	1278	0.979	114	0.60	113	0.62
25	8	0.36	600	41	1281	0.972	118	0.61	123	0.615
50	10	0.38	630	57	1288	0.960	123	0.63	129	0.65
100	10*	0.38	640	92	1300	0.936	122	0.85	129	0.89
0 plus 3 cc TEL	7.5	0.345	600	24	1275	0.984	117	0.56	119	0.57

(1) Knock-limited compression ratio.
(2) Fuel-air-cycle efficiency at $F_c = 1.2$.
(3) Required for good distribution at full throttle in a multicylinder engine.
(4) 50 per cent $\times$ 1.2 of ΔT from Table 4–3.
(5)–(6) Interpolated from Table 4–1.
(7) Measured values.
(8) From Eq. 4–5, taking base value for zero alcohol content of 110 imep.
* Highest available value; could have been higher for 100 per cent alcohol.
Test results from Rogowski and Taylor, ref. 4.202.

when expressed in units of pounds per horsepower hour, Btu per pound. Equation 4–6 shows that specific fuel consumption at constant efficiency is inversely proportional to the lower heat of combustion per unit mass of fuel, Q_c. Table 4–4 compares measured isfc with values calculated from Eq. 4–6 for various blends of alcohol and gasoline at knock-limited compression ratio.

FUELS FOR DIESEL ENGINES

As in the case of spark-ignition engines, most Diesel engines are used with liquid petroleum fuels. When "dual-fuel" Diesel engines are operated on gaseous fuel, the gas is introduced during the induction process and is ignited by a small *pilot spray* of Diesel fuel. Under these circumstances, such engines behave essentially as spark-ignition engines (3.907–3.909).

Experimental work with solid fuels has been carried out chiefly with compression-ignition engines (4.123, 4.124), but commercial success with such fuels has never been attained.*

Since the use of any but liquid petroleum fuels is rare in Diesel engines,

* Exhaustive experiments to adapt gas turbines to the use of pulverized coal have also been commercially unsuccessful. See refs. 13.722, 13.723.

the following discussion refers to such fuels only, unless other types are specifically mentioned. For convenience this group of petroleum fuels will be called Diesel oils, which is their customary popular designation.

The general characteristics of Diesel oils have already been listed in Table 4–1, and their volatility range is shown in Fig. 4–1. From the point of view of their use in engines, their important characteristics appear to be ignition quality, density, heat of combustion, volatility, cleanliness, and noncorrosiveness.

Figure 4–17 shows how all but the two last-mentioned properties are interrelated in the case of commercial Diesel fuels. Since density and heat of combustion depend almost entirely on molecular weight, it is impossible to secure appreciable departures from the relationship between these two qualities. On the other hand, at any given density the volatility, viscosity, and ignition quality (cetane number) tend to vary together, since they are sensitive to molecular arrangement as well as to molecular size. These relationships make it very difficult to determine the effect of any one of these qualities alone on engine performance.

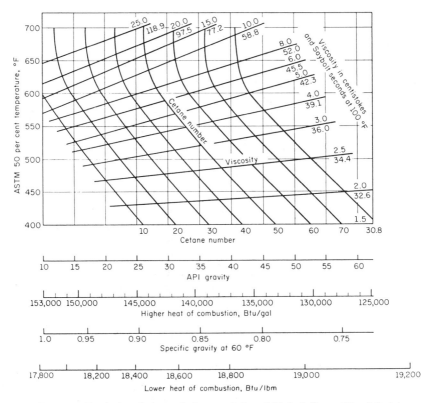

Fig. 4–17. Physical and thermal characteristics of United States Diesel fuels (Burk *et al.*, 4.613).

The most successful method of handling this problem is to plot a given aspect of engine performance against these several variables and by this means discover in which cases a correlation is shown. Such work has been done in a number of instances (4.610–4.618), and the following discussions are based largely on the results of this method of attack.

Ignition Quality of Compression-Ignition Fuels

The term *ignition quality* is used to cover, loosely, the ignition-temperature-versus-delay characteristics of a fuel when used in an engine. Good ignition quality means a short delay angle at a given speed, compression ratio, air inlet, and jacket temperature. The importance of a short delay angle has already been discussed (Chapter 3).

Measurement of Ignition Quality. A great deal of work has been done on the development of methods for rating compression-ignition fuels in respect to their ignition quality. One of the most interesting of these (4.705) uses a constant-volume bomb filled with air at a given pressure and heated to a given temperature. Fuel is then injected by means of an injection system essentially similar to that used in Diesel engines (Chapter 7). Ignition quality is measured in terms of the *delay*, or time between the beginning of ignition and the appearance of an appreciable pressure rise. The shorter the delay, the better the ignition quality. Figure 4–18 gives results of tests in the bomb of ref. 4.705. From the discussion of detonation in Chapter 2 it is apparent that good ignition quality is the opposite of good resistance to detonation; the first requires a short delay and the latter a long delay under conditions of compression ignition.

Engine Tests for Ignition Quality. The current ASTM method of rating Diesel fuels in respect to ignition quality (4.02) depends on engine-test comparisons with reference fuels, as in the case of octane rating. The primary reference fuels are normal cetane, $C_{16}H_{34}$, a straight-chain paraffin having excellent ignition quality, and alpha-methylnaphthalene, $C_{10}H_7CH_3$, a naphthenic compound having poor ignition quality. The percentage of cetane in a blend of these two fuels giving the same delay as the fuel under test is taken as the *cetane number* of the test fuel. A CFR engine with a special compression-ignition cylinder is the generally accepted standard equipment for this test. As in the case of the spark-ignition CFR engine, the compression ratio is variable. A special instrument is used to measure the delay (4.02). Figure 4–11 shows that, as expected, the cetane number decreases as octane and performance numbers increase. A remarkable feature of this relation is its independence of fuel composition.

Effect of Fuel Structure on Ignition Quality. The chemical structure desired in petroleum fuels for compression-ignition engines is opposite to that desirable for spark-ignition engines. For a given number of carbon

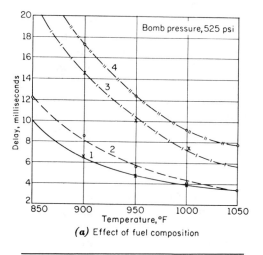

(a) Effect of fuel composition

No.	Type	ASTM distillation temperature		Cetane No.
		10%	90%	
1	Paraffinic	399	617	57.8
2	Paraffinic + aromatics	412	597	48.8
3	Aromatic	....	....	....
4	Highly aromatic....	....	....	....

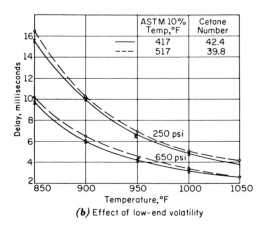

(b) Effect of low-end volatility

Fig. 4–18. Effect of fuel structure and volatility on ignition delay in a constant-volume bomb (Hurn and Hughes, 4.705).

atoms, the straight-chain paraffins have the highest cetane numbers. Also, within this series, the larger the molecule the higher the cetane number. Thus, fuels having very high octane numbers, such as aviation gasoline, aromatic fuels, and alcohols, are not well suited for use in compression-ignition engines.* The best fuels for this type of engine are highly paraffinic, with average molecular weights greater than those of the gasolines. Commercial kerosene is an excellent compression-ignition fuel.

Cetane-Number Control. As in the case of octane number, cetane number is controlled by source of crude, by methods of refining, and by additives, or *ignition accelerators*.

Ignition Accelerators. Just as it is possible to introduce substances that reduce detonation with spark-ignition fuels, so there are additives that will improve the ignition quality of compression-ignition fuels (4.700, 4.701). Substances that increase the tendency to detonate in spark-ignition engines, with few exceptions, facilitate ignition with compression-ignition fuels. Antiknock compounds, such as tetraethyl lead, tend to make compression ignition more difficult. Among the most effective substances for reducing the delay period are amyl nitrate, $CH_3CH_2CH_2CH_2CH_2ONO_2$; ethyl nitrate, $CH_3CH_2ONO_2$; and ethyl nitrite, CH_3CH_2ONO. Figure 4–19

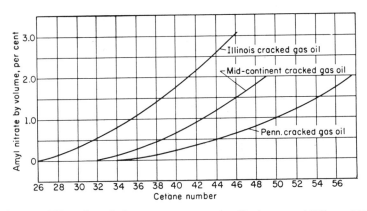

Fig. 4–19. Effect of amyl nitrate on cetane number (Anderson and Wilson, 4.701).

shows that the characteristics of the fuel into which they are introduced affect the behavior of these additives, as is the case with antiknock agents in spark-ignition fuels. It is also possible to reduce the adverse effects of a long delay by injecting a small *pilot spray* a few degrees ahead of the main spray.

---

* In spite of this, some Diesel engines will operate with such fuels, as explained in Chapter 3 and refs. 3.900–3.909.

Effect of Ignition Quality on Engine Performance

As a practical matter of engine performance, the cetane number seems to correlate best with starting ability and with engine roughness.

Starting. Improvement in cold-starting characteristics with increasing cetane number is found with all types of compression-ignition engines; it is one of the most important advantages of a high cetane number (4.610–4.618). Volatile fuels of poor antiknock quality, such as ether, have been found to reduce the ignition delay and to facilitate starting when introduced with the intake air in small amounts.

Engine Roughness. The term *engine roughness* applies to the intensity of vibration of various engine parts, caused by high rates of pressure rise in the cylinders (see Chapter 8, pp. 296–297). Figure 4–20 shows the effect

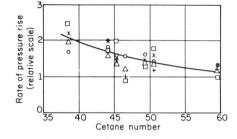

Fig. 4–20. Rate of pressure rise versus cetane number at constant engine speed (Burk *et al.*, 4.613).

   ○ 100 per cent load
   ×   75 per cent load
   △   50 per cent load
   □   25 per cent load
   +   No load

of cetane number on maximum rate of pressure rise in one particular engine. Since the degree of roughness with a given rate of pressure rise varies tremendously with the stiffness of the various engine parts, it is only to be expected that the effect of cetane number on roughness will vary with engine design. This consideration leads to an explanation of the fact that, in general, a high cetane number is usually less important with the divided-combustion-chamber type of cylinder than with open-chamber types. (See discussion of this subject in Chapter 3.) Cetane number is also less important in low-speed (large) cylinders than in high-speed (small) cylinders, because a smaller fraction of the fuel is involved in the rapid pressure rise in the former case.

Influence of Ignition Quality on Compression Ratio. In order to avoid excessively high cylinder pressures, the compression ratio for a Diesel engine is usually the lowest at which good starting can be assured. Other things being equal, the compression ratio required decreases with increasing cetane number. However, when it is desired to employ low-cetane fuels under cold-starting conditions, the designer is more likely to provide auxiliary starting devices, such as heaters, auxiliary fuel, and *glow plugs*, than to raise the compression ratio above 18 to 20. Since correlation of the cetane number with performance of fuels in actual service depends so much on engine design, size, and operating conditions, it is customary for Diesel-engine manufacturers to specify the cetane number required for their particular product on the basis of comprehensive tests on their engines.

Multifuel Diesel Engines

In situations where Diesel fuel may not be available, as in military operations, it is convenient to be able to use other common fuels, such as gasoline or aviation gas-turbine fuels. As stated in Chapter 3 there are some Diesel engines that will operate on fuels in this range. This multifuel capability is achieved principally by raising the compression ratio as required to fire the poorest (highest octane) fuel expected to be used. For example, the General Motors multifuel Diesels have a compression ratio of about 21, whereas the standard model uses 17 to 18 (3.903). The engine structure, of course, must be strong and stiff enough to withstand the resulting increases in maximum pressure and rate of pressure rise.

Figure 4–21 compares the performance of a multifuel engine with

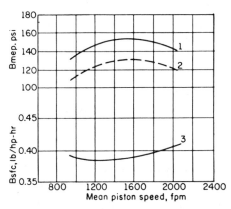

Fig. 4–21. Performance of a multifuel supercharged Diesel engine on gasoline (93 octane) and Diesel oil (cetane number 51.5); (1) maximum bmep with light Diesel oil; (2) bmep with gasoline, same pump setting; (3) bsfc with both fuels at about two thirds rated power (Horiak, 4.618).

gasoline and Diesel oil at maximum fuel-pump delivery rate. The lower power with gasoline is due to its lower heat of combustion per unit volume. The power with gasoline would presumably be equal to that with Diesel oil at the same mass of fuel delivered per stroke. No information on engine roughness, as a function of fuel type, was furnished with these curves.

Volatility and Viscosity of Compression-Ignition Fuels

Apparently the two properties volatility and viscosity cannot be varied independently of each other, and it is therefore difficult to tell which one produces a given effect when fuels of different volatility and viscosity are tested.* In practice, the petroleum fuels used in compression-ignition engines cover a wide range of volatility and viscosity, but, except for multifuel engines using gasoline, the range of these properties lies at lower volatility and greater viscosity than in the case of spark-ignition fuels. High-volatility fuels are excluded from general use in compression-ignition engines, partly because of the great demand for high-volatility fuels for automobiles, and partly because they tend to have poor ignition quality.

The fact that, given a sufficiently high compression ratio, Diesel engines will give acceptable operation on gasoline indicates that high volatility and low viscosity are not serious impediments to good Diesel operation.

In general, however, both volatility and viscosity affect spray character-istics and thus may affect both power and efficiency at a given fuel-air ratio. From the theoretical point of view, it would seem that increasing volatility should increase the rate of evaporation of the fuel, and hence the rate of mixing of fuel and air. With regard to viscosity, it has been shown (Chapter 7) that the tendency of fuel sprays to break up into finely divided particles increases with increasing Reynolds number. This result would indicate that the lower the viscosity, the more rapidly the spray will break up into finely divided particles. From the point of view of both volatility and viscosity, therefore, it would seem that as fuels increase in volatility, at the same time decreasing in viscosity, the mixing process should proceed with increasing rapidity. This theory is supported by tests which show that increasing viscosity results in increasing exhaust smoke (4.602).

One important effect of fuel viscosity is on the leakage of fuel past the injection-pump plunger. At constant pump setting, decreasing viscosity will increase this leakage and thus reduce the volume of fuel delivered.

Effect of C-I Fuels on Power and Efficiency

In the case of compression-ignition engines, changes in fuel within the ordinary range do not usually call for changes in inlet temperature, inlet pressure, or compression ratio. It seems logical, therefore, to test such fuels with these factors constant. As in the case of spark-ignition engines, it appears proper to make such a comparison at the same fraction of the chemically correct fuel-air ratio when it is desired to compare the effect of fuel characteristics on engine performance. Actually, however, most comparisons are made at a *constant fuel-pump-control setting*. Such a setting gives constant fuel-pump displacement volume and tends to give a constant volume rate of fuel delivery per stroke, except as this may be

---

* Volatility and viscosity are also interrelated in spark-ignition fuels, but the range of viscosity is so small in that case that it is not considered an important characteristic.

modified by variations in fuel viscosity. Variation in viscosity affects fuel delivery through its influence on leakage past the pump plunger. If the volume rate of fuel delivery and thermal efficiency remain constant, it is evident that power will vary directly, and specific fuel consumption inversely, with heat of combustion per unit volume.

Figure 4–21 illustrates this point. The relative heats of combustion per unit volume of gasoline and Diesel oil are proportional to fuel density divided by Q_c. From Table 4–1,

$$\frac{Q_{\text{per unit volume gasoline}}}{Q_{\text{per unit volume light Diesel oil}}} = \frac{18{,}250\ (0.876)}{19{,}020\ (0.702)} = 1.19.$$

This is very close to the ratio of bmep's shown in Fig. 4–21, confirming the previous statement that wide differences in volatility and viscosity have only minor effects on the combustion process.

Figure 4–22 shows relative performance versus fuel specific gravity and Btu per gallon when operated in a particular engine at constant fuel-pump setting. The increase in efficiency as the fuel becomes lighter is due to reduced fuel-air ratio. In the range of nearly constant efficiency, the effect on power is due entirely to the changes in heat of combustion per unit volume.

When tests are made at the same fraction of chemically correct fuel-air ratio, $F'Q_c$ for the Diesel oils remains constant (Table 4–1), and no change in output will result at the same efficiency. Under these circumstances specific fuel consumption will of course vary inversely with Q_c. This quantity grows smaller as the specific gravity increases (Fig. 4–17).

Miscellaneous Properties of Fuels for C-I Engines

Engine Deposits. Engine deposits from the fuel and oil constitute a very serious problem in compression-ignition engines. Solid or semisolid carbonaceous matter builds up in all such engines because of incomplete combustion. In most investigations it has been found that the rate of formation of deposits decreases with increasing cetane number. This correlation is probably due not to the change in ignition quality but to the fact that increasing cetane number means reduction in the fraction of nonparaffinic compounds in the fuel. Experience with lamps and burners indicates that the presence of nonparaffinic material tends to increase smoke. It is also well known that deposits tend to increase with increasing specific gravity; this implies increasing viscosity and decreasing volatility. It has been established that sulfur compounds in the fuel tend to promote both deposits and corrosion (4.600–4.605). Thus, low sulfur content is a most desirable characteristic of Diesel fuels.

Cleanliness. Gritty matter in fuel for the compression-ignition engine is extremely objectionable, because it causes rapid wear of the injection

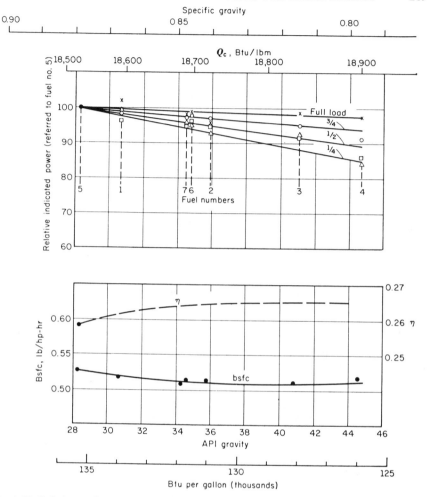

Fig. 4–22. Relative performance of a Diesel engine at constant fuel-pump settings versus specific gravity of fuel; bsfc points are average for all loads (Burk *et al.*, 4.613).

system. The problem of keeping out foreign matter is much more difficult in the case of compression-ignition fuels than in the case of gasoline, because the much higher viscosity of the heavier fuel has a greater tendency to hold solid particles in suspension. Elaborate filtering and centrifuging is necessary for successful use of very heavy fuels in Diesel engines. When such precautions are taken, even *residual* fuels, such as those used for steam generation, can be employed.*

Safety. In connection with fuels for spark-ignition engines it has been pointed out that, in general, the fire hazard increases with increasing

---

* The use of residual fuels in ocean-going Diesel-powered ships is routine as of 1966.

volatility. Thus, general experience indicates that compression-ignition fuels are safer under most circumstances. The chief objection to Diesel oils from the point of view of safety is the fact that light Diesel oils carry an explosive mixture above the fuel in a closed tank at temperatures and pressures often experienced at sea level, while, except at very low temperatures, the fuel-air mixture above gasoline in a closed tank is too rich to be explosive (see page 141). However, this difference appears to be important in military operation only, where the vapor in fuel tanks may be ignited by incendiary bullets.

Nonpetroleum Liquid Fuels for C-I Engines

Compression-ignition engines have been run on various kinds of vegetable and animal oils (4.620–4.624). Satisfactory ignition quality, power, and efficiency are apparently obtainable with many such fuels, and many of them seem to give little trouble from excessive engine deposits or corrosion. Experiments on coal-tar oils, such as creosote oil, have been less successful because of the poor ignition quality of naphthenic compounds.

Gaseous Fuels for Compression-Ignition Engines

Gaseous fuels are used in compression-ignition engines either by introducing gas into the inlet manifold or by injecting it through a special gas-injection system. Since most of the gaseous fuels used have poor ignition quality, a *pilot spray* of fuel oil is used as an ignitor. Engines of this type may be treated as spark-ignition engines where the spark has been replaced by the pilot jet or as compression-ignition engines in which the combustion is augmented by the gaseous fuel. The most useful point of view depends on the relative energy contribution of the liquid and gaseous fuels. If the fuel is entirely liquid, the engine is obviously a true compression-ignition engine, while if the fuel is nearly all gaseous, the engine approaches a spark-ignition engine in behavior. The usual reason for employing gaseous fuel is that it happens to be available at a lower cost than fuel oil at the particular time in question. In some cases engines are so arranged that they can be easily changed over from gas to oil as the relative price of the two fuels fluctuates.

Gas-burning engines have become of considerable importance for pumping gas through pipelines and for power generation in oil-field regions where natural gas is cheap and plentiful. There is a good deal of literature covering the special requirements and characteristics of gas-burning engines with pilot injection of Diesel oil (3.907–3.909). However, except where engines must operate on either gas or oil, spark ignition is more generally used.

UNCONVENTIONAL FUELS

Super Fuels. From time to time attempts have been made to discover fuels that will appreciably increase engine output over that obtainable with a normal fuel-air mixture under similar operating conditions. Table 4–1 shows only acetylene to have a significantly higher value of heat of combustion per standard cubic foot (Q_v) than that of normal fuels, but it has been found impractical both from an economic and technical viewpoint.

While hydrogen has the highest Q_c value per unit mass (see Table 4–1), the volume displaced by this or other gaseous fuels results in a relatively low heat of combustion per unit volume of mixture. It was sometimes burned as a waste product of hydrogen-filled airships (4.128).

With liquid fuels, the limit on Q_c per unit volume, as we have seen, is principally determined by the oxygen content of air. For any significant increase in internal energy per unit volume, some way of increasing the oxygen content per unit volume is necessary. The possibilities include:

1. Increasing the oxygen content per unit volume by supplying gaseous oxygen, instead of air, together with the necessary additional fuel. The heat of combustion of oxygen with gasoline per standard cubic foot with $F_R = 1.1$ would be $Q_v = 0.0845(0.34)(19,020)/(1 + 0.34 \times 32/113) = 500$ Btu per standard cubic foot, or about five times that of gasoline and air. However, the maximum pressures and temperatures would be nearly in proportion, and it is doubtful if an engine that would be durable under such circumstances can be built. In practice, small amounts of oxygen have been successfully added, with resulting small increases in power (4.125). The cost of oxygen and the difficulties of storing and handling it in either liquid or gaseous form make this possibility economically quite unattractive, however.

2. An oxidant to react with the fuel, or a combined fuel and oxidant, could be supplied in solid or liquid form, as in the case of gunpowder and the various rocket fuels. This system has been tried with nitromethane (CH_3ONO_2) as a supplement to a gasoline-air mixture (4.126).

Small amounts of nitromethane are sometimes mixed with methanol for racing purposes (see page 399). The dissociation of hydrogen peroxide, $H_2O_2 \rightarrow H_2O + O$, from liquid to gas, together with sufficient petroleum fuel to consume the surplus oxygen, has been used as a fuel for reciprocating submarine torpedo engines. This is a true "super" fuel, but it is both dangerous and expensive and has not been considered for nonmilitary use. No fuels in this category appear at all attractive for general use.

Ammonia. There has been some interest in the possibility of using anhydrous ammonia (NH_3) mixed with air under circumstances where ordinary fuel is not available. Reference 4.127 indicates that engines can be operated on ammonia-air mixtures, but only at great sacrifice in power and fuel economy. Effects on reliability and durability have not been explored. Only complete unavailability of normal fuels would make the use of ammonia attractive.

For racing purposes where there are no restrictions on type or quantity of fuel, and where maximum power is a major requirement, methyl alcohol, sometimes mixed with small amounts of other fuels thought to be helpful, is often used. Injecting the fuel at each inlet port requires no inlet heating, and full advantage can be taken of the resulting large value of $\triangle T$ (see Table 4–3). Using equation 4–5 and taking T_i as 600°R in both cases gives a gain in imep between 0 and 100 percent methyl alcohol of 38 percent with $x=0.5$, and 75 percent with $x=1.0$.

5 | Mixture Requirements

MIXTURE REQUIREMENTS FOR SPARK-IGNITION ENGINES

It is the purpose of the first part of this chapter to discuss the fuel-air ratios required for various types of spark-ignition engines and for various operating regimes of an individual engine of this type. The second part deals with mixture requirements for Diesel engines.

The profound effect of fuel-air ratio on output and efficiency has been evident from the very beginning of Volume I. The proper fuel-air ratio for each particular set of operating conditions is most conveniently studied under the two headings *steady running* and *transient operation*. Steady running is taken to mean continuous operation at a given speed and power output with normal engine temperatures. Transient operation includes starting, warming up, and the process of changing from one speed or load to another.

Steady-Running Mixture Requirements

In general, the optimum fuel-air ratio at any given engine speed is that which will develop the required torque, or brake mean effective pressure, with the lowest fuel consumption consistent with smooth and reliable operation. This optimum fuel-air ratio is not constant but depends on many factors, as will appear in the following discussion.

One way of determining the optimum fuel-air ratio at a given speed would be to vary the fuel-air ratio while holding the bmep constant by means of the throttle. By repeating this operation over the useful range of speed and bmep, the optimum fuel-air ratios could be determined. It has been found more convenient, however, to cover the range of speed and bmep by making tests at constant throttle opening and allowing the mean effective pressure to vary as the fuel-air ratio varies.

It is convenient to note first the effect of fuel-air ratio on *indicated* performance, thus eliminating the complication of friction effects. Figure 5–1 shows typical curves of indicated mean effective pressure and indicated specific fuel consumption with varying mixture ratio at a fixed throttle opening. The left-hand end of these curves is where the mixture became so lean as to cause missing and explosions in the intake system, usually

173

called *backfiring*. While absolute values differ as between different engines, fuels, and operating conditions, all spark-ignition engines show curves similar in shape to these. The other curves in Fig. 5–1 are for the equivalent fuel-air cycle, that is, for the fuel-air cycle having the same fuel-air ratio and the same temperature and pressure at the start of compression (see Volume I, Chapter 4).

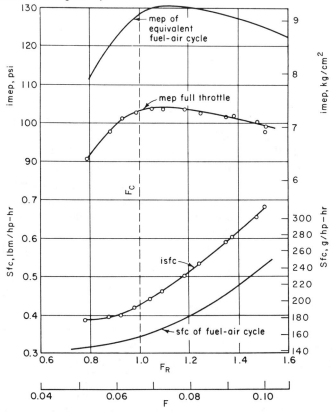

Fig. 5–1. Effect of fuel-air ratio on indicated mep and fuel economy; 221 cu in. (3.6 liters) V-8 engine at 1875-fpm (9.5 m/sec) piston speed (3000 rpm); $r = 6.3$; full throttle, bpsa (Fawkes *et al.*, 5.60).

It will be noted that both the actual and fuel-air imep curves reach a peak between $F_R = 1.0$ and $F_R = 1.2$. The reasons for this peak in the fuel-air cycle were explained in Volume I, Chapter 4. For the actual cycle the explanation is similar, namely that the average pressure during expansion reaches a maximum in this region because of the laws of equilibrium between the various chemical compounds evolved during the chemical reaction.

The specific fuel consumption shown in Fig. 5–1 remains a nearly constant fraction of the corresponding fuel-air cycle consumption. This relation is also shown in Volume I, Fig. 5–20, p. 133. This relation may not hold for lower inlet pressures, as will be shown.

The effect of varying the fuel-air ratio of typical spark-ignition engines over a wide range of speed and load has been widely investigated (5.10–5.65). In most of this work the spark timing was kept fixed while the fuel-air ratio was varied. However, ref. 5.60 makes use of optimum spark advance at all times, and Figs. 5-1 to 5-5 were taken from that reference. The work was done with a typical United States automobile engine in the Sloan Laboratories for Aircraft and Automotive Engines at MIT. The engine was supplied with a gaseous mixture of gasoline vapor and air, and thus it may be assumed that the fuel-air ratio supplied to the engine was also the fuel-air ratio received by each cylinder. In order that the throttle settings be reproducible, orifice plates of different sizes were substituted for the butterfly throttle. The largest orifice plate was called *full throttle*, and orifice plates giving approximately three fourths, one half, and one fourth of the full throttle brake horsepower at 3000 rpm were designated by the terms $\frac{3}{4}$, $\frac{1}{2}$, and $\frac{1}{4}$ throttle, respectively. Runs were made over the useful range of fuel-air ratios at several different speeds and with each of the four orifice plates. Figures 5–2 and 5–3 show results on an indicated basis, and Figs. 5–4 and 5–5 on a brake basis, all at 3000 rpm. Similar curves made at other speeds are not included here, but results of tests made at other speeds are included in the correlation curves, Figs. 5–6 and 5–7.

Curves such as those of Figs. 5–2 to 5–5 are not in a form from which optimum fuel-air ratios can easily be selected. However, a few general observations regarding these curves may be of interest:

Imep. Figure 5–2 shows that the shape of the imep-versus-F_R curve is little affected by throttle position. The highest imep occurs at the same fuel-air ratio in each curve.

Isfc. Figure 5–3 shows that there is a tendency for the isfc curves to show a minimum point at higher fuel-air ratios as the throttle is closed. This tendency can be explained by the reduction in effective flame speed due to increased exhaust-to-inlet pressure, as shown in Fig. 1–10, combined with the reduction in flame speed at lean mixtures shown in Fig. 1–11.

As we have seen in Volume I, Fig. 5–20, p. 133, the effect of lower flame speed at lean mixtures is small when spark timing is adjusted for highest mep at each fuel-air ratio. This procedure was followed for Fig. 5–3 and for all figures in this chapter except where otherwise noted. If, for example, the tests for Fig. 5–3 had been made with a constant spark timing, such as that for best power at $F_R = 1.05$, the minimum point of the isfc curve would have occurred at a higher value of F_R and the minimum isfc would have been greater. The minimum point in the specific fuel-consumption curve is the point at which, as fuel-air ratio is reduced, the rate of increase

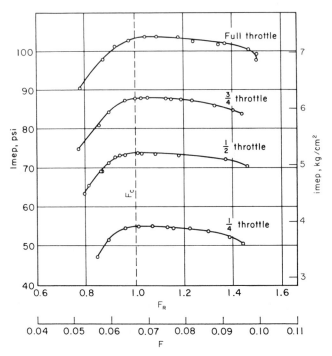

Fig. 5–2. Effect of fuel-air ratio on indicated mean effective pressure at various throttle settings; same test conditions as in Fig. 5–1 except for throttle position (5.60).

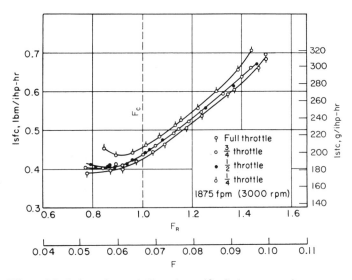

Fig. 5–3. Effect of fuel-air ratio on indicated specific fuel consumption; same tests as in Fig. 5–2.

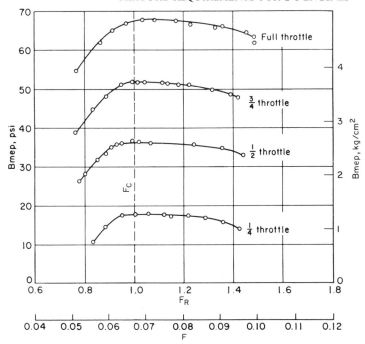

Fig. 5–4. Bmep versus fuel-air ratio; same tests as Fig. 5–2.

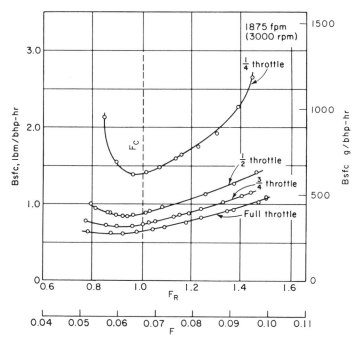

Fig. 5–5. Effect of fuel-air ratio on brake specific fuel consumption at various throttle settings; same tests as in Fig. 5–2.

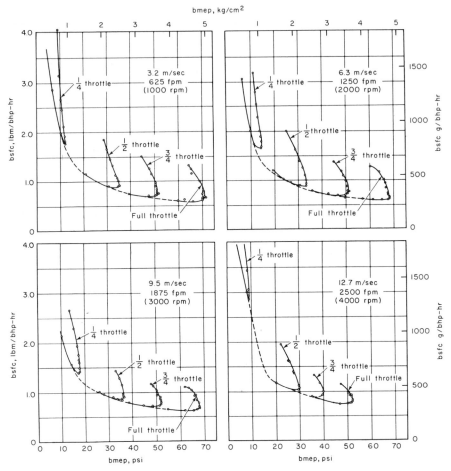

Fig. 5–6. Bsfc versus bmep at various speeds, throttle settings, and fuel-air ratios. Envelope curves show best specific fuel consumption attainable at a given speed and mep; same test series as Figs. 5–1 through 5–5.

in efficiency due to thermodynamic factors (as expressed by the fuel-air cycle) is offset by the rate of decrease in efficiency due to increasing time losses. With optimum spark timing, time losses become large only when the crank angle occupied by flame travel exceeds a certain value. As the mixture is made leaner, this value will be reached at a higher fuel-air ratio when the flame speed has already been slowed by throttling. To generalize on this reasoning, it may be said that any factor that tends to increase the crank angle occupied by combustion will tend to increase the fuel-air ratio for best economy. Thus, increasing the exhaust back pressure (see Fig. 1–10) or decreasing the atmospheric pressure (see Fig. 1–9), will have an effect on the position of the minimum point similar to that of throttling,

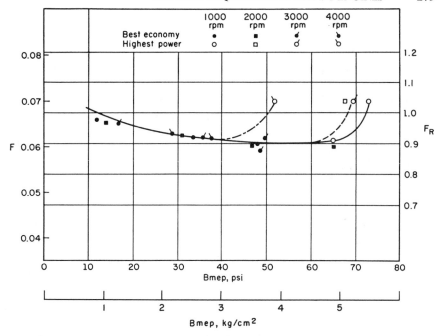

Fig. 5–7. Fuel-air ratio versus bmep; replotted from Figs. 5–5, 5–6, and other tests in same series.

Bmep. Since friction mep is virtually independent of fuel-air ratio (see Volume I, Chapter 9) it is not surprising to find the bmep curves of Fig. 5–4 similar in shape, though not in value, to the imep curves of Fig. 5–2. The bmep values of Fig. 5–4 are nearly equal to the imep in Fig. 5–2 minus 36 psi. For convenience, the fuel-air ratio giving the highest bmep will be called the *best-power* mixture. Tests on many engines under various operating conditions (5.10–5.65) have shown it to be a general rule that the best-power fuel-air ratio is very nearly the same under all operating conditions, provided the distribution of fuel to the various cylinders remains the same.

Bsfc. By definition:

$$\text{bsfc} = \text{isfc} \frac{\text{imep}}{\text{imep} - \text{fmep}}. \tag{5–1}$$

Figure 9–29 of Volume I (p. 352) shows that fmep increases slightly as the throttle is closed. In the range of fmep = 36 psi at full throttle, the fmep would be about 38 psi at $\frac{1}{4}$ throttle. The curves shown in Fig. 5–5 can be quite accurately predicted on the basis of Figs. 5–2, 5–3, 5–4, and Eq. 5–1.

Limitations on Figs. 5–1 through 5–5. It must be kept in mind that Figs. 5–1 through 5–5 were obtained with optimum spark advance for each fuel-air ratio and with uniform *distribution*, that is, with the same fuel-air ratio in each cylinder. The effects of nonuniform distribution will be discussed later.

Optimum Fuel-Air Ratios. In practice, the duty which an engine has to perform at a given instant can always be defined in terms of a required brake mean effective pressure (bmep) at a given speed. In order to determine the optimum fuel-air ratio for a given brake mean effective pressure and speed, the data of Figs. 5–4 and 5–5 can be replotted to show brake sfc versus brake mep (Fig. 5–6). In order to obtain Fig. 5–6, curves similar to those of Figs. 5–4 and 5–5, but for other speeds, were made. The envelope curves (dashed lines) indicate the lowest specific fuel consumption obtainable with this engine at a given speed and mean effective pressure by adjustment of the spark, throttle, and fuel-air ratio. Figure 5–6 shows that the best economy point *at a given throttle setting* does not necessarily fall on the envelope curve. This fact is particularly noticeable at $\frac{1}{4}$ throttle. A study of Fig. 5–6 in relation to Fig. 5–5 reveals that the best economy *at a given bmep* is obtained at mixtures somewhat leaner than those corresponding to best economy at a given throttle opening. This relationship is explained by observing that the bsfc curve for a given throttle position lies below that for the next smaller throttle opening.

From the point of view of fuel economy only, the optimum fuel-air ratio for the engine of Fig. 5–6 is evidently that which will give lowest bsfc at a given bmep. This fuel-air ratio can be found, for each point on the envelope curve by referring back to a figure such as 5–5. For example, let it be required to find the fuel-air ratio for minimum specific fuel consumption for a bmep of 50 psi at 3000 rpm. Figure 5–6 shows this consumption to be 0.67 lbm/bhp-hr at 0.8 throttle. From Fig. 5–5 we see that the fuel-air ratio at this particular point is 0.061. This process has been carried out for the four envelope curves of Fig. 5–6, resulting in the curves shown by Fig. 5–7.

Fuel-Air Ratio versus Load. Inspection of the curves in Fig. 5–7 indicates a similarity of shape between the four speeds, suggesting a possible correlation on the basis of the same fraction of the maximum bmep at a given speed. Let the ratio of bmep to maximum bmep at a given speed be called the *load*. Figure 5–8 shows fuel-air ratios corresponding to the curves of Fig. 5–7 plotted against load. This figure shows no consistent differences between the four speeds at which tests were made, and it may therefore be concluded that best-economy fuel-air ratio, at a given load, is independent of speed, at least down to 20 per cent load, under conditions at which fuel distribution is good.

Idling. Since no useful work is being done at idling (zero load), the optimum fuel-air ratio is that which uses the least fuel at the chosen idling speed, provided firing is steady and reliable. Figure 5–9 shows the fuel-air ratio for lowest idling fuel consumption to be in the neighborhood of the chemically correct ratio for a typical automobile engine. However, this fuel-air ratio is uncomfortably close to that at which misfiring and stalling occur, so that normal idling mixtures usually range between 1.2 and 1.5 times stoichiometric.

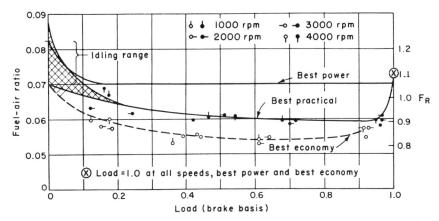

Fig. 5–8. Fuel-air ratio versus load. Replot of Fig. 5–7 on a load basis. "Best practical" is fuel-air ratio at 1.05 × best fuel economy.

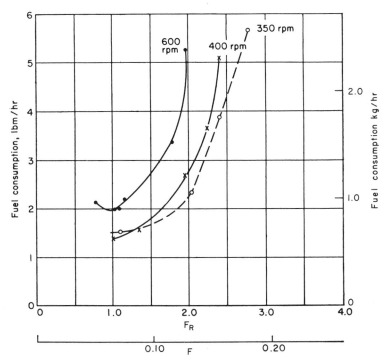

Fig. 5–9. Idling-mixture requirements; 6-cylinder automobile engine; 3.125 × 4.375 in. (80 × 111 mm); $r = 6.7$ (Collmus and Freiberger, 5.61).

Under idling conditions the inlet-manifold pressure is far below atmospheric, and there is usually some leakage of air into the inlet system through the inlet-valve guides. Thus, the combustion fuel-air ratio may be less than that delivered by the carburetor. The leakage rate depends on the inlet valve-guide clearance as well as on the manifold pressure, and will increase as the valve guides wear. Thus, the fuel-air-ratio scale shown in Fig. 5–9 might have been greatly increased if the engine had been an old one with worn guides. The usual range for idling is shown in Fig. 5–8.

Because of uncertainty with regard to idling mixture requirements, carburetors are usually equipped with a hand adjustment for idling (see Chapter 6). The usual criterion of adjustment is regular firing and smooth, reliable operation at the lowest steady speed, since fuel consumption at idling is small in any case.

Effect of Variables Other Than Load on Mixture Requirements. Experience has shown that variables which may have an important effect on curves such as those shown in Fig. 5–8 include spark timing, friction mep, inlet pressure, exhaust pressure, and quality of mixture distribution, that is, the variation of fuel-air ratio from cylinder to cylinder and from cycle to cycle.

Effect of Spark Timing. Any departure from optimum spark timing will tend to increase the best-economy fuel-air ratio, since it tends to increase the time losses.

Effect of Friction. From the foregoing discussion, it is easy to predict that increasing friction mep, with constant indicated mep, will tend to raise the fuel-air ratios for maximum economy.

Effect of Inlet and Exhaust Pressures. Increasing inlet pressure, with exhaust pressure held constant, tends to increase flame speed, as was shown in Fig. 1–9. Increasing pressures during combustion, together with reduced dilution by exhaust gas, contributes to the increase in flame speed. Thus, *supercharging* tends to lower the best-economy fuel-air ratio, just as throttling tends to raise it. The effect on the best-economy ratio is small at pressures above 1 atm, however, so that in supercharged engines the curve of optimum fuel-air ratio versus load is very similar to that shown in Fig. 5–8, provided load is defined as before, namely, bmep at a given speed divided by the maximum (supercharged) bmep at that speed.

As we have already noted, increasing exhaust pressure, at a given inlet pressure, tends to reduce flame speed (see Fig. 1–10) and thus to raise the best-economy fuel-air ratio.

Varying of inlet and exhaust pressures together occurs when the altitude of an engine changes at constant throttle opening. Here the flame speed varies with the pressure level, but at a rate slower than in the preceding examples, because the residual-gas fraction remains substantially constant. However, because of decreasing mechanical efficiency as an engine is carried to a higher altitude, the best-economy fuel-air ratios increase until, at altitudes at which the brake output approaches zero, the maximum-power mixture is optimum at all loads.

Steady-Running Mixture Requirements in Practical Operation. In the absence of severe detonation, and with good distribution at all speeds and loads, spark-ignition engines burning gasoline show best-economy fuel-air ratio curves very close to that of Fig. 5–8.

Departures from the best-brake-economy curve shown in Fig. 5–8 may be required in practice for the following reasons:

1. To allow for possible errors or variations in carburetor metering
2. To compensate for poor distribution
3. To reduce or eliminate detonation
4. To reduce the temperature of hot spots, such as the exhaust valve, spark-plug points, or piston crown; in other words, to assist in cooling.

Allowance for Errors in Metering. In Figs. 5–1 to 5–5, the left-hand end of the curves is the point at which the mixture becomes so lean that engine operation is irregular. It is evident from Fig. 5–5 that, especially at large throttle openings, the best-economy mixture is not far from that below which operation becomes irregular. On the other hand, operation at fuel-air ratios greater than that for best economy is usually smooth all the way up to very high fuel-air ratios. In order to allow for unavoidable variations in carburetors and other types of metering equipment, the normal mixture for which such equipment is set is usually somewhat richer than that for best economy. Figure 5–8 shows that a considerable enrichment above the best-economy ratio is possible with only a 5 per cent increase in bsfc. The reason is evidently that the curves of bsfc versus fuel-air ratio are quite flat near the best-economy point except at light loads, as shown in Fig. 5–5.

Allowance for Poor Distribution. Perfect distribution may be defined as the condition when every cylinder of a multicylinder engine receives the same quantity and quality of mixture for every cycle. This condition, of course, is never reached in practice, although well-designed engines under favorable circumstances often approach it closely. Ordinarily, however, there are irregularities in distribution that are due to variations in mixture quantity and *quality* (fuel-air ratio), both from cylinder to cylinder and from cycle to cycle of the same cylinder.

Variations in the quantity of mixture affect the indicated output of an engine in direct proportion to the total weight of mixture received by all cylinders. This tendency has no effect on the mixture requirements with respect to indicated output, and since variation in quantity is usually small, the effect on the mixture requirements with respect to brake output is generally negligible. The usual irregularities in the *quantity* of fresh mixture, then, leave the mixture requirements substantially unchanged from those with perfect distribution.

In the case of irregularities in the *quality*, that is, in fuel-air ratio, the effect on mixture requirements may be appreciable. If a curve of mixture requirements under conditions of good distribution is available, the effect of known irregularities can be computed. The following is an example:

Let it be assumed that variation in performance with varying fuel-air ratio has been determined for a 6-cylinder engine under conditions of good distribution, with the results shown in Fig. 5–10. Under certain

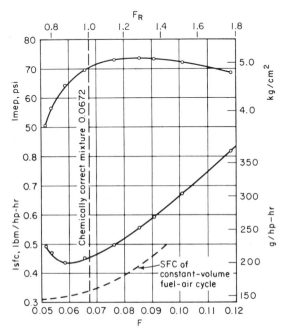

Fig. 5–10. Effect of fuel-air ratio on indicated power and indicated specific fuel consumption at constant spark timing (Sparrow, 5.70).

other operating conditions three of the cylinders run 20 per cent leaner than the rest. If the three normal cylinders receive a 0.08 fuel-air mixture, the three lean cylinders receive a mixture-ratio of $0.08 \times 0.8 = 0.064$. The mixture supplied by the carburetor for this condition must be $[3(0.08) + 3(0.064)]/6 = 0.072$. The indicated mean effective pressure of the engine will be $[3(68) + 3(73)]6 = 70.5$ psi. The specific fuel consumption will be

$$\frac{3(68/70.5)(0.45) + 3(73/70.5)(0.52)}{6} = 0.49 \text{ lbm/hp-hr.}$$

Compare these figures with those for perfect distribution with ratio 0.072, which would be (from Fig. 5–10) 72 psi and 0.475 lbm/hp-hr. By taking various fuel-air ratios and making similar computations it is possible to plot a curve of imep and isfc against mixture ratio furnished by the carburetor. Such curves have been plotted for several assumed irregularities, and the results are shown in Fig. 5–11. It is evident that errors in distribution tend to lower the maximum output, increase the minimum fuel consumption, and narrow the range over which operation is possible. The effect on engine performance is disastrous if the errors in distribution are large.

The curves of Fig. 5–11 would also apply to a single-cylinder engine in

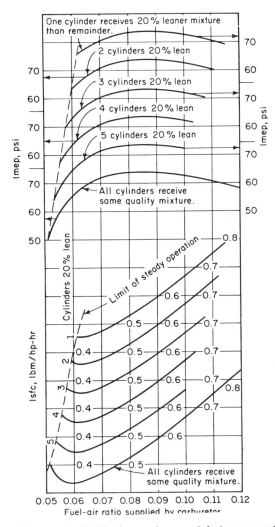

Fig. 5–11. Effect of changes in distribution on imep and fuel consumption of a 6-cylinder engine (5.70).

which the quality of mixture changed from stroke to stroke in such a manner as to give the same proportions of lean or rich cycles to total number of cycles.

In practice, most engines using the carburetor-manifold type of induction system have good distribution only when jacket and inlet temperatures are normal and speed is not too low. Thus, the conclusion that mixture requirements are a function of load alone is likely to break down at low piston speeds or with cold engines. The poorest distribution usually occurs at very low speed and full throttle, and here mixtures richer than normal are often required even at normal engine temperatures in order to secure smooth operation. With closed throttle the low pressure in the inlet manifold assists in evaporating fuel (Chapter 4), and distribution is usually satisfactory as long as the engine is warm. Engines with timed fuel injection to cylinders or inlet ports usually have excellent distribution except at very light loads, at which the mechanical difficulties involved in metering minute quantities of liquid may cause irregularities.

Control of Detonation and Cooling by Fuel-Air Ratio. Figure 2–29 shows that if an engine is operating near $F_R = 1.2$, the tendency to detonate can be reduced by increasing the fuel-air ratio. According to Fig. 8–16 of Volume I (p. 294) the rate of heat transfer to the coolant is also reduced. For most classes of service in which spark-ignition engines are used, it appears practicable to design the engine and cooling system to control detonation and give adequate cooling without depending on fuel-air ratios higher than those called for by a curve such as that of Fig. 5–8. The case may be different, however, where very high specific output is required, as is the case in aircraft engines.

Aircraft-Engine Mixture Requirements. In this type of engine, the compelling need for a high ratio of maximum power to weight and bulk makes it economically defensible to sacrifice some degree of fuel economy to obtain the desired result. Therefore, such engines are usually designed and adjusted so that at high outputs very rich mixtures are used. Referring to Fig. 2–29, with fuel No. 2 for example, the detonation-limited power can be increased by the factor $190/150 = 1.27$ by increasing the fuel-air ratio from 1.2 to 1.6 and increasing the inlet pressure accordingly.

In addition to the reduction in detonation tendency, such rich mixtures result in lower cyclic temperatures, and consequently less heat rejected (see Fig. 8–16, Volume I, p. 294) than would be the case if the same power was produced with the best-power fuel-air ratio. In consequence, less cooling surface in the radiator or on the cylinder is required. Use of this fuel-air ratio also results in lower exhaust-valve and spark-plug temperatures and longer life of these parts.

In place of an excessively rich mixture, an auxiliary cooling fluid, usually a 50-50 water-alcohol mixture, may be introduced along with the fuel. With suitable proportions of water-alcohol to fuel, the allowable power output can be increased by 10 to 25 per cent over that allowable without this auxiliary fluid (2.75–2.77).

As the power of an aircraft engine is reduced from maximum, the fuel-air ratio can be reduced, or the auxiliary injection cut down, until an output is finally reached that is low enough to allow operation at the best-economy mixture without auxiliary injection. It is desirable that this point be reached at normal cruising operation.

The ideal aircraft carburetor or mixture-control device, with or without auxiliary injection, would provide a mixture equal to the richest of the following:

1. The mixture that gives best economy at the particular load in use;
2. The leanest mixture that gives adequate cooling and exhaust valve life;
3. The leanest mixture that gives freedom from detonation.

A typical fuel-air ratio curve fulfilling the aforementioned requirements is shown in Fig. 5–12.

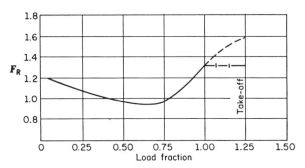

Fig. **5–12.** Fuel-air ratio versus load for supercharged aircraft engines. Normal rated load = 1.00.
– – – Take-off without water-alcohol injection
–⊢–⊢– Take-off with water-alcohol injection

Transient Mixture Requirements

Transient conditions of operation are here defined as conditions at which speed, load, temperatures, or pressures are abnormal or are changing rapidly. The principal transient conditions of operation are starting, warming up, acceleration (increase of load), and deceleration (decrease of load).

Carburetor-Manifold System with Liquid Fuel. Under conditions of steady running an equilibrium is established in the inlet system such that the average fuel-air ratio received by the cylinders is equal to the fuel-air ratio supplied by the carburetor. Also, under normal steady-running conditions, engine temperatures are sufficiently high that practically all the fuel in the mixture is evaporated before ignition occurs.

If we define the mixture ratio as the ratio of evaporated fuel to air in the cylinder at the moment of ignition, the conclusions reached for steady running apply also to any transient condition. But under transient conditions the vapor-air ratio received by a cylinder is not necessarily the same as that which it receives under steady conditions, even though the mixture ratio at the carburetor may be exactly the same in both cases. This is due to one or more of the following factors:

1. Evaporation of the fuel may be incomplete, even when ignition occurs.
2. The quantity of liquid fuel in the inlet manifold may be increasing or decreasing.
3. The distribution of fuel to the various cylinders may differ under the transient condition.

Thus it is apparent that the fuel-air ratio required from the carburetor under transient conditions may be quite different from that required under steady operation, even at exactly the same engine speed and load.

The question of transient mixture requirements has already been discussed at length in Chapter 4 under the heading "Volatility of Liquid Fuels." From that discussion the following general principles, which apply to conditions of *partial evaporation* of the usual liquid petroleum fuels, should be recalled:

1. With a given fuel-air ratio, the vapor-air ratio decreases with decreasing temperature.
2. At a given temperature, the vapor-air ratio at the cylinders can be increased by increasing the fuel-air ratio at the carburetor.

Mixture Requirements for Starting and Warming Up. From the preceding principles it is obvious that a very cold engine will generally require abnormally rich mixtures at the carburetor in order to secure a firing mixture in the cylinders. Thus, the carburetion system must be able to supply *very rich* mixtures for starting, and the fuel-air ratio must be progressively reduced from this point during the warm-up period until the engine will run satisfactorily with the normal steady-running fuel-air ratios.

Mixture Requirements for Acceleration. The term "acceleration," with regard to engines, is generally used to refer to an increase in engine speed resulting from opening the throttle. The immediate purpose of opening the throttle, however, is to secure an increase in torque, and whether or not an increase in speed follows depends on the nature of the load. With governed engines, the throttle is opened in response to a demand for increased torque, and little change in speed occurs. On the other hand, engines driving fixed-pitch propellers or driving vehicles on a level road will respond to opening the throttle with a speed increase.

In carburetor engines using liquid fuel the process of acceleration is complicated by the presence of unevaporated fuel in the inlet manifold. Investigation (5.77) shows that during normal steady operation with gasoline the inlet manifold contains a large amount of liquid fuel which clings to the manifold walls and runs along them to the cylinders at a speed which is very low compared to that of the rest of the mixture, which consists of air, fuel vapor, and entrained fuel droplets. Under steady running conditions at a given speed, the quantity of liquid contained in the manifold at any moment becomes greater as the manifold pressure increases. The principal reason that high manifold pressures result in large quantities of liquid is to be found in the fact that fuel flow must increase with increasing air density, and evaporation is slower as total pressure increases.

When the throttle is opened for acceleration, thus increasing the manifold pressure, fuel must be supplied to increase the liquid content of the manifold. If the carburetor provides a constant fuel-air ratio, the fuel-air ratio reaching the cylinders will be lowered during the time the liquid content of the manifold is being built up to the larger value. With a sudden opening of the throttle, the resulting reduction in the fuel-air ratio received by the cylinders may be such as to cause misfiring, backfiring, or even complete stopping of the engine.

To avoid an abnormally lean mixture in the cylinders as a result of sudden throttle openings, it is usually necessary to increase the supply ratio by injecting into the manifold a quantity of fuel known as the *accelerating charge*. Injection of this charge must take place simultaneously with the opening of the throttle. The optimum amount of accelerating charge is that which will result in the best-power fuel-air ratio in the cylinders. In general, this amount varies with engine speed and with throttle position at the start of acceleration, as well as with fuel volatility, mixture temperature, and rate of throttle opening. Thus, carburetors are designed to furnish the amount required under the most difficult conditions, and when this amount is too large, the error will be on the rich side of best-power mixture, where the sacrifice in output is small, as indicated by Fig. 5–1. Since partial or slow opening of the throttle requires less than the full accelerating charge, the amount of charge delivered is usually made roughly proportional to the rate of throttle opening and the angle through which the throttle moves. Mechanisms that are used to provide the accelerating charge will be described in Chapter 6.

Mixture Requirements with Gaseous Fuels

Since there is no evaporation process associated with gaseous fuels, the transient mixture requirements are substantially the same as the steady-running requirements. Therefore no special enrichment of the fuel-air mixture is required during starting, acceleration, and warming up. Also, distribution in inlet manifolds is generally more uniform with gaseous than with liquid fuels, although experience shows that even here the mixture quality is not necessarily uniform (5.80). Many large gas engines use fuel injection into the cylinders; this will be discussed in Chapter 7.

Mixture Requirements for Fuel-Injection Engines with Spark Ignition

In engines in which the fuel is injected in the form of a metered spray directly into the cylinder or into the inlet port, the mixture requirements for steady operation are essentially those shown in Fig. 5–8, provided the fuel is all evaporated and well mixed with the air before ignition. The use of such a system does not qualitatively change the requirements for richer mixtures when starting and warming up, though it may reduce the amount of enrichment necessary. Direct injection of course eliminates effects due to accumulated liquid fuel in the inlet manifold.

Fuel-injection engines with late injection may have very special mixture requirements because the mixing and evaporation process is incomplete by the time ignition occurs. In the case of the Texas combustion process (2.80, 2.81) control of load is by fuel quantity alone, and the mixture requirements are essentially the same as those for Diesel engines.

Summary of Mixture Requirements for Spark-Ignition Engines

Excluding the late-injection engine, the mixture requirements of a normal spark-ignition engine using gasoline may be summarized as in the following paragraphs.

Steady Operation. The curves shown in Fig. 5–8 are typical for engines at sea level, with good distribution, adequate cooling, and no detonation. Departures from the curves defined by Fig. 5–8 will be required under the following conditions, each of which will tend to increase the required fuel-air ratio in some portion of the operating range:

1. Increased altitude (without supercharging)
2. Increased exhaust pressure
3. Poor distribution (usually a function of rpm and throttle opening)
4. Dependence on fuel-air ratio to control detonation or cooling.

Transient Operation. 1. Starting at low temperatures requires a very rich mixture. The enrichment required decreases as engine and atmospheric temperatures increase.

2. Warming up requires a rich mixture, decreasing to the steady-running requirement as the engine temperatures approach normal.

3. Except with cylinder or inlet-port injection, acceleration requires a temporary enrichment of the mixture to accompany a sudden opening of the throttle. The amount of enrichment required depends on design and

operating conditions, particularly temperature, rate of throttle opening, and extent of throttle opening.

EFFECT OF FUEL-AIR RATIO ON EXHAUST EMISSIONS

As noted in Chapter 1 (p. 33), the substances that have been found objectionable in the exhaust of premixed-charge spark-ignition engines are unburned hydrocarbons (HC), carbon monoxide (CO), and oxides of nitrogen (NO_x).

Figure 5–13 shows, qualitatively, the effect of fuel-air ratio on the relative amounts of these emissions for typical spark-ignition engines without emission controls.

Before the emphasis on fuel economy and emission control, most vehicle engines were operated at fuel-air ratios greater than the ideal requirements shown in Figure 5–8. This was done in order to keep operation safely above

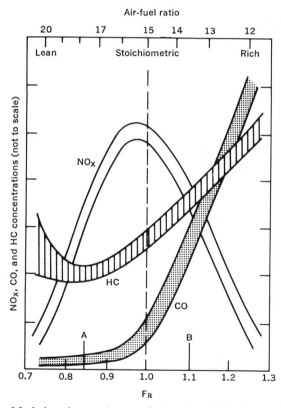

Fig. 5–13. Effect of fuel-air ratio on exhaust emissions in spark-ignition engines. A = normal lean limit; B = normal for maximum bmep (torque).

the lean limit for smooth operation. Since the imposition of government restrictions on fuel economy and exhaust emissions (in 1968 in the United States) much research and development work has been done to extend lean limits. These include improving fuel distribution by means of inlet-manifold design and heating, improving carburetors to provide more accurate fuel-air-ratio control, changing combustion-chamber and inlet-port geometry to provide more favorable air motion, and in many cases shifting from carburetion to fuel injection (see Chapter 7 and references). As a result of this work, contemporary automobile engines run at fuel-air ratios close to the requirements indicated in Fig. 5–8, with modifications for starting, warmup, and acceleration. The mid-range fuel-air ratio used depends on the methods employed for emission control, as explained in the next two chapters. "Lean burn" engines use close to the best-economy ratio indicated by Fig. 5–8. Those using "three-way" converters must control the mid-range fuel-air ratio to near stoichiometric, $F_R = 1.0$.

For future improvements in lean operation there is continuing work on stratified-charge systems such as those mentioned in Chapter 2. With the exception of the Honda divided-chamber type now in successful use (Fig. 2–34), these remain experimental. Those depending on spray combustion have problems with exhaust particulates as well as with the other pollutants.

MIXTURE REQUIREMENTS FOR DIESEL ENGINES

In the Diesel engine the fuel-air ratio is normally the independent variable used to control engine output. Under normal operation conditions and with the usual fixed schedule of injection timing, there is only one fuel-air ratio corresponding to a given speed and bmep.

Diesel engines have very low limits on minimum fuel-air ratio, since burning takes place by compression ignition within the spray envelope or in the evaporated portion of the charge. The practical low limit in any given case is set by mechanical rather than combustion considerations. Poor spray characteristics at very small fuel quantities may set a low limit because of misfiring, but generally the low limit is set by the fuel quantity required to overcome friction during idling. That this limit can be very low is illustrated by Fig. 3–11.

In Diesel engines the practical high limit on fuel-air ratio is set by smoke and particulate emissions, as explained in Chapter 3.

6 | Carburetor Design and Emission Control for Spark-Ignition Engines

The fundamental principles of devices used for introducing fuel into the air supply and for controlling the fuel-air ratio are subjects of this chapter. Systems that inject fuel directly into the cylinders or inlet ports will be discussed in Chapter 7.

The behavior of carburetors may be dealt with in two categories, namely, steady-flow operation and transient operation.

STEADY-FLOW CARBURETION

The mixture requirements of spark-ignition engines were treated in Chapter 5. As indicated by Fig. 5–8, the fuel-air ratio should vary as a function of the load under normal steady-running conditions. For the usual type of spark-ignition engine operating under given atmospheric conditions, the load is determined by the particular values of *any two* of the following variables: 1) torque output, 2) speed, 3) throttle position, 4) rate of air flow, and 5) pressure in inlet manifold.

The conventional carburetor uses the rate of air flow to the engine as the major variable to control fuel-air ratio. Figure 6–1 shows ideal fuel-air ratio requirements versus mass rate of air flow for a typical spark-ignition engine in steady operation. These curves were constructed by measuring air flow at various loads at each speed and plotting the appropriate best practical fuel-air ratio taken from Fig. 5–8. The numbers in circles represent throttle positions measured in angular degrees from the fully closed position.

It is evident that the fuel-air ratio required is a unique function of air flow only at a given speed. Thus, except in constant-speed applications, an additional control, sensitive to at least one of the other variables listed, must be provided if air flow is to be used as one controlling element.

Basic Carburetor. The basic or main element of most carburetors consists of an air passage of fixed geometry containing a venturi-shaped restriction. A fuel nozzle is located in the venturi throat and is supplied with fuel from a constant-level *float chamber* or other constant-pressure device. Air flow is controlled by a *throttle* downstream from the venturi.

193

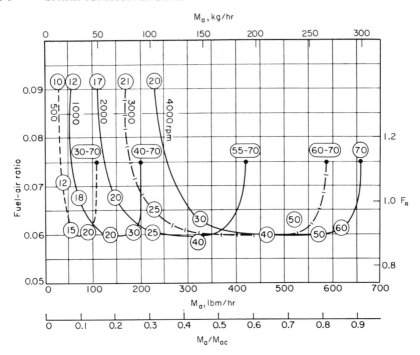

Fig. 6–1. Optimum fuel-air ratio versus air flow; 221 cu in. (3.6 liters) V-8 engine. Numbers in circles give throttle opening in degrees; 70° = full throttle. (Sloan Automotive Laboratories.)

A typical basic carburetor is shown in Fig. 6–2. Carburetors based on this element may be classed as *open-passage* types, in contrast to *air-valve* types which have spring-loaded or gravity-loaded automatic valves in their air passages. The latter type will be described later.

Main Metering System. In the carburetor of Fig. 6–2, the air passes from the air intake through a venturi-shaped passage. This shape is used in order to minimize the pressure loss through the system. The difference between the stagnation pressure at the air intake and the static pressure at the venturi throat is used to create and regulate the flow of fuel. In Fig. 6–2 the air passage is shown as vertical, with downward air flow. Many carburetors are arranged for upward air flow, and some have a horizontal air passage. The principles of operation, however, are the same for any air-flow direction.

Located downstream from the venturi in the air passage is a valve, called the throttle, which in spark-ignition engines is the principal power-control element. It is usually of the butterfly type, as shown in Fig. 6–2.

Fuel is introduced to the air at the throat of the venturi by means of a nozzle which is fed from a constant-level chamber, or float chamber, through a fuel-metering orifice. The fuel level in the float chamber is held constant by a float-operated valve. The pressure above the fuel is held at the air-inlet total pressure by means of an impact tube, which measures the total, or stagnation, pressure at the air inlet.

Carburetor Flow Equations. The behavior of a carburetor such as that shown in Fig. 6–2 can be predicted for steady-flow conditions by means of the usual equations for fluid flow through a fixed passage without heat

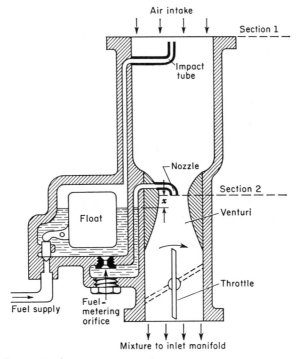

Fig. 6–2 Basic open-passage carburetor. Main metering system.

transfer. These equations, which are given in Volume I, Appendix 3, have already been used in connection with flow through 2-stroke engines (see Volume I, Chapter 7). The same equations are used here, but with some changes in arrangement and symbols for the sake of convenience. For purposes of this analysis, the following symbols and definitions will be used:

A = area of a flow cross section;

p = absolute static pressure;

p_o = total, or stagnation, pressure in absolute units;

ρ = air density at standard sea level;

ρ_o = *stagnation density*, that is, density based on stagnation temperature and pressure; $\rho_o = p_o m / R T_o$;

$\dot{M}$ = mass flow per unit time;

k = specific-heat ratio;

$C = $ *flow coefficient*, that is, the ratio of actual mass flow to ideal mass flow. The ideal mass flow is taken as the mass flow under reversible adiabatic conditions, without shaft work and with uniform velocity across each section.

Subscripts and their significance are as follows:

$_a$ refers to air;

$_f$ refers to fuel;

$_{o1}$ refers to stagnation conditions at the air inlet, Section 1, Fig. 6–2;

$_2$ refers to static conditions in the air at the venturi throat, Section 2, Fig. 6–2;

$_m$ refers to the basic or main system.

Fuel Flow. Since liquid fuel can be considered incompressible, Bernoulli's equation can be used, and

$$\dot{M}_f = A_f C_f \sqrt{2g_o \rho_f \Delta p_f}, \tag{6–1}$$

where $\Delta p_f = $ difference in pressure across the fuel-metering orifice;

$C_f = $ flow coefficient for the fuel-metering orifice.

Air Flow. By using Eqs. A–16, A–18, A–23, and A–25 (pp. 504 and 507) of Volume I for compressible steady flow of air ($k = 1.4$) through the system of Fig. 6–2:

$$\dot{M}_a = \Phi_2[C_2 A_2 \sqrt{2g_o \sigma \rho (\Delta p_a)}], \tag{6–2}$$

where

$$\Phi_2 = 1.87(r)^{0.715} \sqrt{\frac{1 - r^{0.286}}{1 - r}} \tag{6–3}$$

$$\dot{M}_{ac} = 0.484 C_2 A_2 \sqrt{2g_o \sigma \rho p_{o1}} \tag{6–4}$$

$$\frac{\dot{M}_a}{\dot{M}_{ac}} = 2.07 \Phi_2 \sqrt{\frac{\Delta p_a}{p_{o1}}}$$

$$= 3.88(r)^{0.715} \sqrt{1 - r^{0.286}}. \tag{6–5}$$

Here $\dot{M}_a = $ mass flow of air per unit time;

$\dot{M}_{ac} = $ *critical* mass flow, that is, the mass flow when r is equal to or less than its critical value, which is 0.528 when $k = 1.4$;

$C_2 = $ air-flow coefficient at Section 2, Fig. 6–2;

$A_2 = $ air-flow area at Section 2;

$\rho = $ density of air at standard sea-level conditions (0.0765 lb/ft^3 at $p = 14.7$ psia and $T = 520°$R, or 1.23 kg/m^3 at $p = 1.03$ kg/cm^2 and $T = 289°$K);

$\sigma = $ ratio ρ_{o1}/ρ, that is, the ratio of stagnation density at Section 1 to standard sea-level density;

$\Delta p_a = p_{o1} - p_2$;

$p_{o1} = $ stagnation pressure at Section 1;

$r = p_2/p_{o1} = 1 - (\Delta p_a/p_{o1})$.

Values of Φ_2 and $\dot{M}_a/\dot{M}_{ac}$ vs. r and $\Delta p_a/p_{o1}$ are plotted (for $k = 1.4$) in Fig. 6–3.

For the arrangement of Fig. 6–2 it is evident that Δp_f is equal to Δp_a minus the pressure difference necessary to lift the fuel to the top of the

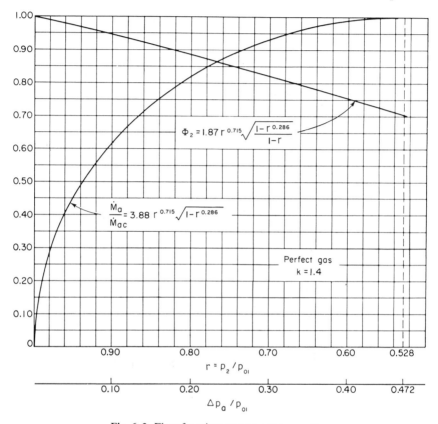

$$\Phi_2 = 1.87\, r^{0.715} \sqrt{\frac{1 - r^{0.286}}{1 - r}}$$

$$\frac{\dot{M}_a}{\dot{M}_{ac}} = 3.88\, r^{0.715} \sqrt{1 - r^{0.286}}$$

Perfect gas
$k = 1.4$

$r = p_2/p_{o1}$

$\Delta p_a/p_{o1}$

Fig. 6–3. Flow functions versus pressure ratio.

nozzle (distance x in Fig. 6–2) and to overcome the surface tension at the nozzle exit. Calling the latter pressure difference Δp_s,

$$\Delta p_f = \Delta p_a - \Delta p_s, \tag{6–6}$$

the fuel-air ratio supplied by such a carburetor will be

$$F_m = \frac{(A_f C_f/A_2 C_2)\sqrt{\rho_f/\sigma\rho}\sqrt{\Delta p_f/\Delta p_a}}{\Phi_2} \tag{6–7}$$

or

$$F_m = \frac{(A_f C_f/A_2 C_2)\sqrt{\rho_f/\sigma\rho}\sqrt{1 - (\Delta p_s/\Delta p_a)}}{\Phi_2}. \tag{6–8}$$

Since both Δp_a and Φ_2 are functions of the critical flow ratio $\dot{M}_a/\dot{M}_{ac}$, F_m can be expressed in terms of this critical flow ratio as soon as σ and Δp_s are known. If $\dot{M}_{aco}$ is the critical flow with ρ_{o1} equal to standard sea-level density, that is, with $\sigma = 1$, F_m becomes a unique function of $\dot{M}_a/\dot{M}_{aco}$, Δp_s, and σ.

In Fig. 6–4, F_m is plotted as a function of $\dot{M}_a/\dot{M}_{aco}$ for a typical value of the ratio $A_f C_f/A_2 C_2$, for $\sigma = 1$, and for $\Delta p_s = 0.4$ in. of water, or about 0.5 in. of gasoline.

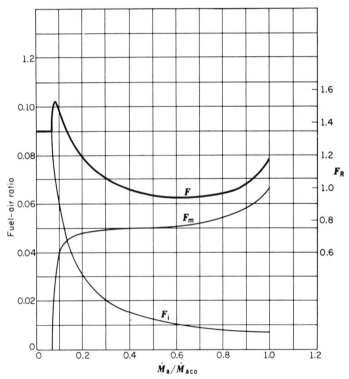

Fig. 6–4. Fuel-air ratio for carburetors of Figs. 6–2 and 6–5; $A_f C_f/A_2 C_2 = 0.0024$; $\Delta p_s = 0.4$ in. H_2O; $F_i = 0.0059/(\dot{M}_a/\dot{M}_{aco})$; $F_m = (0.0493/\Phi_2)\sqrt{1 - \Delta p_s/\Delta p_a}$; $\sigma = 1.0$; $\rho = 0.0765$ lbm/cu ft. (Metric units: $\Delta p_s = 1$ cm H_2O; $\rho = 1.23$ kg/m³.)

As would be expected from inspection of Eq. 6–8, the fuel-air ratio is zero until Δp_a exceeds Δp_s. The value of F_m rises rapidly from this point to the region where Δp_s is very small compared to Δp_a. Beyond this point, and up to the critical pressure ratio,

$$F_m \cong \frac{A_f C_f}{A_2 C_a} \sqrt{\frac{\rho_f}{\sigma\rho}} \left(\frac{1}{\Phi_2}\right). \qquad (6\text{–}9)$$

For the carburetor of Fig. 6–2, assuming that the lean firing limit is $F_R = 0.6$, it is obvious that the metering system will not operate an engine at low rates of air flow. Some device to take care of the low-air-flow regime must therefore be added.

Idling System. The system used to cover the mixture requirement at low rates of air flow is called the *idling system*, although it may influence the fuel-air ratio at loads far above idling, as will appear. One typical arrangement for an idling system is shown in Fig. 6–5. The idle well is a

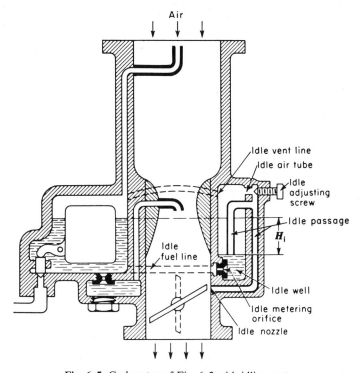

Fig. 6–5. Carburetor of Fig. 6–2 with idling system.

vertical passage connected top and bottom to the float chamber. The bottom connection is restricted by the idle metering orifice with area A_{fi}. The idle passage has a discharge opening located near the edge of the butterfly throttle, when the throttle is in the closed position. The open lower end of the passage is located near the bottom of the idle well. The idle air tube connects into the passage as shown and is controlled by an adjustable restriction called the idle adjusting screw.

For idling at the lowest desired speed, the throttle is set against an adjustable stop so that it remains open just enough to allow the necessary air flow. At this point the edge of the throttle partially overlaps the idle nozzle. By proper location of the nozzle outlet with relation to the throttle, and by proper adjustment of the idle screw, there will be sufficient suction in the idle tube to lift fuel at a rate such as to give the required idling fuel-air ratio, which, as we have seen (Fig. 5–9), is in the neighborhood of $F_R = 1.4$ for the typical automobile engine.

Further opening of the throttle gradually exposes the idle nozzle to the full manifold depression, which may be as much as 10 psi (0.7 kg/cm^2) below atmospheric pressure in a normal idling engine. At this point the pressure difference between the upper and lower ends of the idle passage is so great that it drains the idle well down to the level shown in Fig. 6–5. From this point, as the throttle is opened, there will always be enough suction to keep the fuel level in the idle well at the bottom of the idle tube, and the amount of fuel that flows will be constant and dependent on the area and coefficient of the idling fuel orifice and the gravity head H_i. More or less air will be drawn through the idle passage, depending on the idle suction, but the maximum amount of fuel that can pass through is, from Bernoulli's theorem:

$$\dot{M}_{fi} = A_{fi} C_{fi} \rho_f \sqrt{2gH_i}, \tag{6–10}$$

where $\dot{M}_{fi}$ = idling mass fuel flow;

$\quad A_{fi}$ = idling metering-orifice area;

$\quad C_{fi}$ = idling metering-orifice coefficient;

$\quad H_i$ = height, as shown in Fig. 6–5;

$\quad g$ = acceleration of gravity.

It is obvious that as long as Eq. 6–10 holds, the idling fuel flow is constant and the idling fuel-air ratio F_i can be expressed as

$$F_i = \frac{\dot{M}_{fi}}{\dot{M}_a} = \frac{\text{a constant}}{\dot{M}_a}. \tag{6–11}$$

The fuel-air ratio supplied by the idling system will therefore become richer as the throttle is closed and $\dot{M}_a$ is reduced. In Fig. 6–4, F_i is plotted by assuming dimensions of the idle system in relation to the main system such that $F_i = 0.04$ when $\dot{M}_a/\dot{M}_{aco} = 0.1$. It is further assumed that below $\dot{M}_a/\dot{M}_{aco} = 0.07$, the edge of the throttle blocks off the idle-tube exit in such a way as to hold F_i constant at 0.09.

The fuel-air ratio resulting when both idling and main systems operate together will be

$$F = \frac{\dot{M}_{fi} + \dot{M}_f}{\dot{M}_a} = F_i + F_m. \tag{6–12}$$

Figure 6–4 shows F plotted for the particular values chosen as an example of good practice. The resultant curve shows values that will operate an engine at any value of $\dot{M}_a/\dot{M}_{aco}$ up to critical flow.

Application of Carburetor of Fig. 6–5 to an Engine. In practice the venturi throat area is so chosen that at maximum air flow the ratio $\dot{M}_a/\dot{M}_{aco}$ is less than 1.0. For engines that operate at sea level only, however, the maximum value of this ratio can be nearly 1. For the engine of Fig. 6–1 let the ratio $\dot{M}_a/\dot{M}_{aco}$ at maximum air flow be 0.90, corresponding to a value of $\Delta p_a/p_{o1}$ of 0.264, or $(p_{o1} - p_2) = 3.9$ psia (0.27 kg/cm²) with standard sea-level inlet pressure. Since about 60 per cent of this pressure drop is recovered at the venturi outlet, the manifold pressure at maximum flow would be about 1.6 psi (0.113 kg/cm²) below atmospheric. *

Having chosen the venturi size so that $\dot{M}_a/\dot{M}_{aco}$ is 0.9 at maximum air flow, the curve of fuel-air ratio supplied by the carburetor can be plotted versus $\dot{M}_a/\dot{M}_{aco}$, together with the fuel-air-ratio requirements of the engine which have been shown in Fig. 6–1.

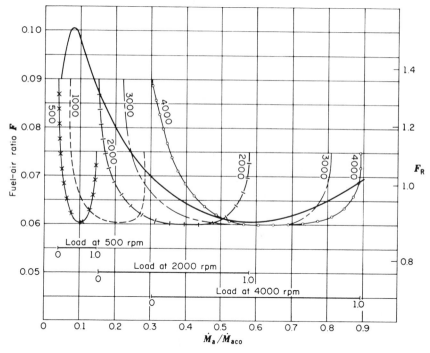

Fig. 6–6. Fuel-air ratio required by engine of Fig. 6–1 and fuel-air ratio delivered by carburetor of Figs. 6–2 and 6–5. Solid line is delivered fuel-air ratio.

Figure 6–6 shows that the mixture supplied is too lean at loads near 1.0 (see Fig. 5–8) for speeds above 1000 rpm and too lean at light loads for speeds above 2000 rpm. The light-load deficiency can be ignored where

---

* In practice this pressure drop may be as low as 0.5 psi (0.035 kg/cm²) where maximum air capacity is important.

light loads are not used above 2000 rpm. The deficiency at loads near 1.0 is corrected by means of a device called a *power jet*, which enriches the mixture when the throttle is opened beyond a certain point. Figure 6–7 shows an arrangement that opens an auxiliary orifice in parallel with the main fuel-metering orifice. Other arrangements to give the same effect are obviously possible.

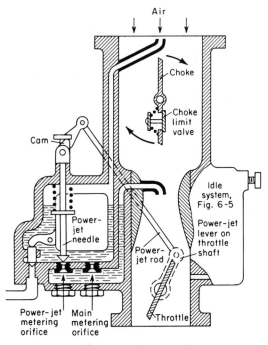

Fig. 6–7. Carburetor of Fig. 6–5 with power jet and choke added.

On referring to the throttle-position numbers on Fig. 6–1, it is evident that the throttle opening at which enrichment should start is a function of speed. A compromise setting in this case would be to have the power jet start to open at 40° throttle angle. Let the power jet have an effective orifice area 10 per cent of that of the main metering orifice and let it be opened as a linear function of throttle angle from 40° to 60°. The results of this modification is shown by the solid-line curves of Fig. 6–8.

The broken-line curves of Fig. 6–8 are replots of the broken-line curves of Fig. 6–6. This plot shows required and delivered fuel-air ratios at three speeds. Similar curves for other speeds could be plotted from Fig. 6–6. The carburetor of Fig. 6–7 approximates the requirements of the engine of Fig. 6–1 quite well except at low speeds with heavy loads and at high speeds with light loads. Since little running is done at heavy loads at low speeds, the fact that an unnecessarily rich mixture is supplied there is

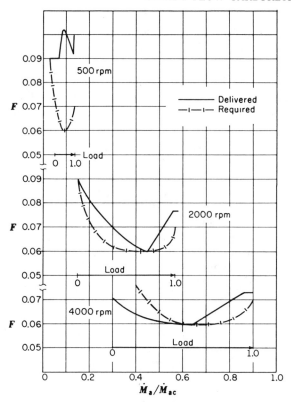

Fig. 6–8. Fuel-air ratios required and delivered. Replotted from Fig. 6–6 with effect of power jet added.

–ı–ı– required, Fig. 6–1;
——— delivered by carburetor of Fig. 6–7

unimportant. Likewise, the deficiency in fuel-air ratio at light loads at high speeds is tolerable, because engines seldom operate in that range. A much closer approximation to engine requirements at all speeds would require a mechanism more elaborate than the carburetor shown in Fig. 6–7, which is typical of modern automotive carburetors.

By adjusting the carburetor for a particular speed, it may be seen that for engines that operate at only one speed, or over a very limited speed range, a very close approximation to optimum requirements can be had by suitable adjustment of a carburetor of the type shown in Fig. 6–7. The curve at 2000 rpm in Fig. 6–8 is an example of this possibility.

Pulsating-Flow Effects

When a carburetor is connected to three cylinders or fewer, the flow is of a strongly pulsating nature, because the suction strokes do not overlap.

Analysis of carburetor behavior under such conditions is not easy, since it involves natural frequencies of the fluid columns in the carburetor, which are difficult to determine and still more difficult to incorporate into suitable flow equations. However, experience shows that the general characteristics shown by Fig. 6–8 remain nearly the same. Some change in design coefficients, such as A_f/A_2, is usually required as compared with a carburetor for steady-flow operation. There is also a tendency toward enrichment at full throttle, where the pulsating effect is strongest, and in some cases this effect eliminates the need for a power jet.

TRANSIENT CARBURETION

Starting. For road-vehicle engines, the very rich mixtures required for cold-weather starting are usually supplied by means of a *choke*. This mechanism consists of a butterfly valve located in the air inlet. Figure 6–7 shows a choke located between the impact tube and the venturi.

To secure the very rich fuel-air ratio necessary for starting, the choke is closed. The choke valve carries an opening with a spring-loaded valve which allows a limited flow of air when the engine is cranked. The operation of cranking with a closed choke creates a large pressure difference across the main metering orifice, which causes fuel to flow rapidly from the main nozzle. At a given cranking speed, the maximum fuel-air ratio is

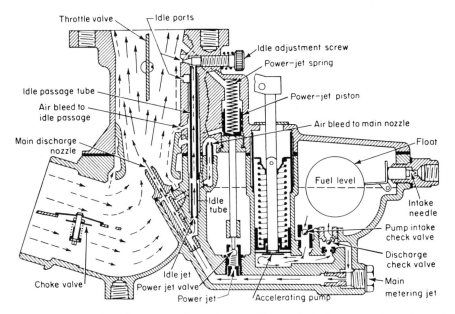

Fig. 6–9. Cross section of an automotive carburetor. The word "jet" corresponds to the word "orifice" in previous figures.

determined by the size of the air hole in the choke valve in relation to the size of the main metering orifice. Smaller fuel-air ratios are obtained by partially opening the choke valve. Gradual opening of the choke as the engine warms up is obviously necessary.

In many cases the choke is operated automatically as a function of exhaust-manifold temperature and inlet-manifold pressure. When the exhaust manifold is cold, the choke is closed. It is opened as the exhaust manifold warms up or as the inlet-manifold pressure becomes so low as to indicate that the engine is running at a speed considerably greater than cranking speed.

Acceleration. The temporarily high fuel-air ratio required for acceleration is furnished by a device called an *accelerating pump*, the usual character of which is shown in Fig. 6–9. The principal feature of this device is a small plunger pump supplied with fuel from the float chamber. The plunger is operated by throttle motion. By a suitable arrangement of passages and check valves, the fuel in the accelerating pump is injected into the air stream when the throttle is opened rapidly. A hole in the plunger, provided with a check valve, regulates the quantity of the accelerating charge so that is roughly proportional to the rate of throttle opening.

COMPLETE AUTOMOTIVE CARBURETOR

Figure 6–9 shows an actual carburetor for road-vehicle and similar service. It will be seen to incorporate the features heretofore described. Minor modifications in the way of auxiliary passages are evident. These are for the purpose of securing flexibility of adjustment or a closer approach to engine requirements than can be obtained with the basic elements alone. The principal differences between this carburetor and the carburetor of Fig. 6–7 are:

1. In this carburetor the space around the removable venturi is used as an impact tube to pick up the inlet pressure p_{o1} and deliver it to the top of the float chamber. This design has the advantages of mechanical convenience and of some compensation for uneven velocity distribution around the periphery of the inlet air passage.

2. The main discharge nozzle is equipped with an *air bleed*, that is, a passage that admits air at the pressure p_{o1} into the fuel passage below the float-chamber fuel level. Analysis of this system shows that it gives somewhat greater fuel flow at low air velocities, thus compensating for the fact that fuel-orifice coefficients tend to be low in this regime.

3. The power jet in this case is operated by a plunger responsive to manifold pressure. At high manifold pressure, which in general corresponds to heavy load, the plunger is forced down by a spring and opens the power-jet passage.

4. The idle well is fed from the main nozzle passage, downstream from the main metering orifice. This arrangement cuts off idle fuel flow when the suction at the main nozzle exceeds the idle suction. This cutoff of the idle flow, however, occurs only at relatively large air-flow rates, where the idle fuel flow is small in comparison to the main fuel flow.

5. The choke is located upstream from the impact-pressure take-off. When the choke is closed, atmospheric pressure bleeds into the float-chamber air space through such leakage points as that around the accelerating-pump plunger, thus furnishing the necessary large pressure differential across the main and idle systems.

In many automobile carburetors, such as the one illustrated, fuel-metering orifices, the main venturi, and some of the air-passage restrictions are removable and replaceable by others of differing size. Thus a carburetor of this type may be adjusted to fit different engines within a given size range.

Other Types of Carburetor

When an engine is to be used in many different positions, as in a chain saw or other hand-held device, a diaphragm-type pressure regulator is substituted for the float chamber. The diaphragm system maintains a pressure upstream of the fuel-metering orifices slightly below atmospheric pressure. The difference between this pressure and atmospheric pressure corresponds to the head x in Fig. 6–2. The principles of operation are otherwise similar to those described for the float-chamber type.

Another type of carburetor employs a spring-loaded or gravity-loaded *air valve* in the main air-supply system. At idling or light-load operation this air valve remains closed. A small passage at the edge of the valve, or else a separate small air passage, supplies the air for idling. An orifice located at the throat of this passage supplies the necessary fuel for idling and light-load operation. As the throttle is opened for more power, the air valve opens and brings the main fuel-supply system into operation. The air valve usually requires a damping system to prevent chattering. With the air valve wide open, this type of carburetor operates according to the principles given for the main metering system of the conventional-type carburetor, as indicated by curve F_m of Fig. 6–4. By using appropriate geometry for the idling system and an appropriate spring or gravitational force on the air valve, the regime between closed and wide-open air valve can be designed to give results similar to those indicated by curve F_i of Fig. 6–4. Carburetors of this type are described in refs. 6.30, 6.31, 6.36–6.38.

Altitude Effects. In many places in the world vehicles have to operate at altitudes as high as 10,000 feet. Aircraft with reciprocating engines are often operated above that level. In either case adjustment of the carburetor (or injection system) is necessary for economical operation.

Equation 6–9 shows that for the main metering system of a carburetor, at a given value of Φ_2 the fuel-air ratio varies as the inverse square root of σ, the inlet stagnation density relative to standard sea-level density. For convenience F_o, the characteristic fuel-air ratio of the carburetor, will be taken as the value of F_m obtained from Eq. 6–9 by setting Φ_2 and σ equal to unity. Thus, F_o is the fuel-air ratio, with standard sea-level inlet density, approached by the main metering system as air flow approaches zero, with $\triangle p_s = 0$. With a real carburetor, F_o is close to the value of F_m at $M_a/M_{aco} = 0.3$ at sea level.

Figure 6–10 shows F_m/F_o as a function of $\dot{M}_a/\dot{M}_{aco}$ at various altitudes for a main metering system with fixed value of $A_f C_f/A_2 C_2$. Obviously, a carburetor based on this arrangement and set for the proper F_m at sea level would be quite unsatisfactory at high altitudes. For this reason, aircraft carburetors usually incorporate a means of correcting F_m for changes in altitude.

In the following discussion the shape of the F_m/F_o curve at any one value of σ will be referred to as the *load compensation* of the carburetor. Satisfactory load compensation will be defined as a maximum of 20 per cent enrichment as a function of $\dot{M}_a/\dot{M}_{aco}$. This limit is indicated by the dot-dash curve of Fig. 6–10.

Altitude Compensation. The characteristics of F_m/F_o as a function of σ will be called the *altitude compensation* of a carburetor. Equation 6–7 indicates that, at given values of σ and Φ_2, F_m can be altered by altering the ratio $A_f C_f/A_2 C_2$ or by altering the ratio $\Delta p_f/\Delta p_a$. The first ratio is most easily altered by substituting an adjustable opening for the main fuel-metering orifice. Figure 6–11 shows F_m/F_o for such an arrangement, assuming that the adjustment is such that $(A_f C_f/A_2 C_2)/\sqrt{\sigma}$ remains constant. This means that $A_f C_f$ is adjusted as a function of air density only; this can be accomplished automatically if desired.

A similar result can be obtained by the second method if $\sqrt{\Delta p_f/\Delta p_a \sigma}$ is held constant by adjustment of $\Delta p_f/\Delta p_a$. This can be accomplished by means of the back-suction system illustrated in Fig. 6–12. The float-chamber air space is so arranged as to be influenced by both p_2 and p_{o1}. The relative influence of these two pressures can be controlled as a function of σ. In the arrangement illustrated, a decrease in inlet density causes the gas-filled bellows to expand and progressively reduce the ratio $\Delta p_f / \Delta p_a$. Proper choice of design coefficients will again result in the characteristics shown in Fig. 6–11.

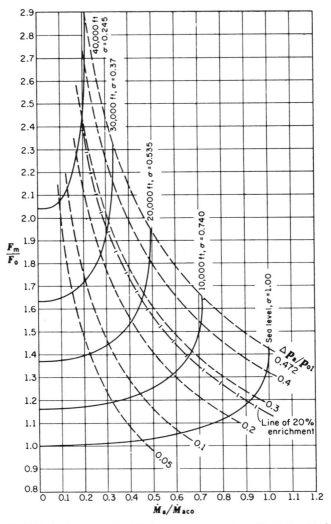

Fig. 6–10. Altitude characteristics of main metering system, Fig. 6–2, with $\Delta p_s = 0$.

As indicated by Fig. 6–10, either of these methods of altitude compensation leaves load compensation inadequate unless $\dot{M}_a / \dot{M}_{aco}$ at sea level is small enough to keep enrichment below 20 per cent at maximum altitude. This question will be dealt with on p. 209 under "Choice of Venturi Size."

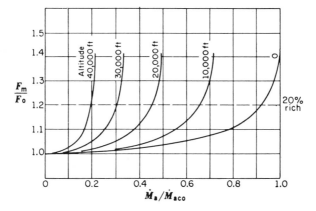

Fig. 6–11. Altitude compensation by back suction or variable fuel orifice.

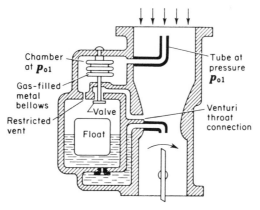

Fig. 6–12. Diagram showing principle of back-suction mixture control. Expansion of bellows reduces $\Delta p_f / \Delta p_a$.

Choice of Venturi Size

For aircraft carburetors that must operate with atmospheric pressure, Fig. 6–10 shows that the venturi size must be chosen on the basis of performance at the highest required altitude.

Let it be assumed that an enrichment of 20 per cent is allowable at the highest altitudes. Unsupercharged aircraft engines will not normally operate above 10,000 feet. Figure 6–10 shows that the maximum value of M_a / M_{aco} at sea level should not exceed 0.65 for satisfactory load compensation up to this altitude. The venturi size should be chosen accordingly.

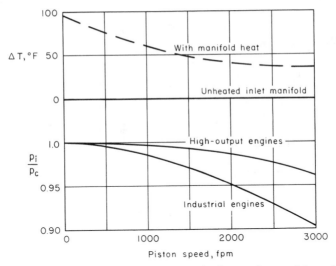

Fig. 6–13. Estimated pressure ratio and temperature rise from carburetor inlet to inlet port at full throttle; based on average values from miscellaneous sources. (This figure supersedes Fig. 6–28, p. 201, of Vol. I.)

p_c = Pressure ahead of carburetor (absolute)
p_i = Pressure at inlet port (absolute)
ΔT = Estimated temperature rise, carburetor inlet to inlet manifold with gasoline fuel.

Pressure Loss in Carburetors

Since the carburetor flow system is not a perfect venturi, some loss in pressure occurs between the carburetor inlet and the outlet, even with the throttle wide open. Figure 6–13 shows typical values of this pressure loss. Estimated typical values ot temperature rise between atmosphere and inlet port are also included in this figure.

Gas Carburetors

Carburetors for gaseous fuels operate on principles similar to those already described. Detail differences include the necessary substitution of a pressure-regulating valve for the float chamber and the provision of larger fuel passages relative to the air passages on account of the low density of gas compared with liquid (6.50, 6.51)

Carburetor Icing

The process of evaporation in a carburetor lowers the temperature of the parts downstream of the fuel nozzle to an extent that, under certain atmospheric conditions, may result in freezing of atmospheric moisture, especially on the throttle valve. This problem is especially critical in aircraft. The remedy lies in adequate heating equipment, suitably controlled (6.71–6.79).

EMISSION CONTROL FOR SPARK-IGNITION ENGINES

Since a major part of air pollution and "smog" in cities all over the world is caused by the exhaust from vehicle engines, the chief emphasis here will be on engines for that type of service.

Many of the measures taken in recent years to reduce the fuel consumption of automobiles have also contributed to reducing air pollution. Included here are reduced size of car and engine (especially in the United States), improved transmissions, and reduced road friction and air drag. These changes, of course, reduce the quantity of exhaust gas delivered to the environment per vehicle. (Unfortunately, the number of vehicles keeps increasing all over the world.)

In Chapter 5 it was shown that control of fuel-air ratio is a key element in the control of emissions. As government restrictions (Table 6–1) have increased, control of fuel-air ratio alone is insufficient to meet these standards. What follows is a summary of present methods of control provided for recent vehicles with spark-ignition engines.

Unburned hydrocarbons and carbon monoxide are reduced by the use of an oxidizing catalytic converter in the exhaust system between the engine and the muffler.

Nitrogen oxides in the exhaust can be reduced by lowering the peak temperature, such as by retarding the spark and by circulating exhaust gas into the inlet system (EGR). The introduction of exhaust gas not only reduces maximum combustion temperatures but also replaces excess oxygen in a lean mixture, thus reducing the oxidation rate of nitrogen during combustion. Since both these methods are detrimental to engine efficiency and car "drivability," their use must be limited in extent.

To avoid excessive economy loss and to meet the latest limitations on NO_x, "three-way" catalytic converters have been developed which are able to reduce NO_x and oxidize HC and CO all in one bed of catalytic material. Satisfactory operation of this type requires controlling the fuel-air ratio very

Table 6–1. Standards (grams per mile) for Exhaust Emissions for Light-Duty Motor Vehicles (U.S. Clean Air Act Amendments, August 1977)

Pollutant	1977	1978	1979	1980	1981	1982	1983	1984
HC	1.5	1.5	1.5	0.41	0.41	0.41	0.41	0.41
CO	15	15	15	7.0	3.4*	3.4*	3.4	3.4
NO_x	2	2	2	2	1.0*	1.0*	1.0	1.0

*subject to appeal

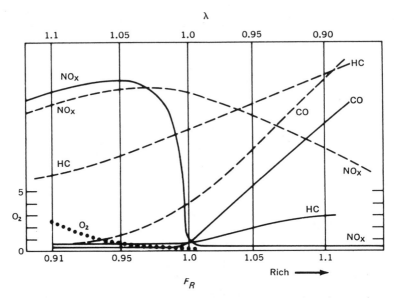

Fig. 6–14. Exhaust emissions vs. fuel-air ratio of spark-ignition engines. (— — —) at exhaust port, (———) after passing through a three-way catalytic converter. (Except for O_2, which is in percent, ordinate scales are qualitative only.)

close to stoichiometric, as shown in Fig. 6–14. The desired fuel-air ratio occurs where the normal oxygen content of the exhaust is about 0.5 per cent. An oxygen sensor can signal this point with a sharp change in voltage (see references, pages 213 and 238). The combination of this sensor with a suitably programmed computer modifies the carburetor fuel flow so as to meet the requirements shown in Fig. 6–14 over as much of the operating range as is possible with satisfactory engine operation (7.932).

It is evident from the above discussion that many of the adjustments required for emission control are detrimental to power and efficiency on account of the reduced compression for unleaded fuel, the retardation of ignition timing from best economy, the use of fuel-air ratios not always optimal for fuel economy, and the increase in backpressure due to cata-lytic converters.

In order to arrive at the best practicable compromise between engine performance and emission control, it has been found necessary to respond to an increasing number of variables as the limits on emissions have become more severe. As this process continues, the desired results have been more easily achieved via fuel injection at the inlet ports than by the use of a carbu-retor or a single-point injection system. At this writing, many manufacturers are switching to fuel injection. This method of fuel supply is discussed further in the next chapter.

Not all the control devices mentioned here are needed on every vehicle. For example, many new light cars with fuel injection need only three-way converters. On the other hand, large cars and light trucks with carburetors may require a combination of several of the systems mentioned.

Pollutants other than those already mentioned include sulfur oxides and sulfuric acid, which are produced partly because of the use of platinum in converters. The need to control these compounds is not yet established.

Further discussion of electronic control systems will be found in the next chapter.

The bibliography on emissions and their control is immense and growing fast. References 1.870–1.8792 are useful as history through 1966. More recent developments will be found in Society of Automotive Engineers reports since that time. These reports continue to cover this field with many issues each year. Recent useful such reports include the following:

SAE P-119, "Automotive Applications of Microprocessors," 20 papers, 1982.
SAE SP-536, "New Developments in Sensor and Actuators," 13 papers, 1983.
SAE SP-511, "Sensors and Actuators," 14 papers, 1982.
SAE SP-505, "Electronic Technologies for Commercial Vehicles," 6 papers, 1981.
SAE SP-486, "Sensors," 9 papers, 1981.
"T_iO_2 Oxygen Sensors, an Update," *Automotive Engineering*, August 1983.
See also reference 7.932 and page 239.

7 | Fuel Injection

Three types of injection system are used with reciprocating internal-combustion engines. They are:

1. *Inlet-port injection.* This type of injection is used with spark ignition only, and the sprays can be either continuous or timed. In timed injection the process occurs over a limited range of crank angle. With timed inlet-port injection the spray usually occurs while the inlet valve is open.

2. *Early cylinder injection.* Fuel is sprayed directly into the cylinder during the inlet or compression stroke, but in any case early enough so that evaporation is complete before ignition. This type of injection is used with spark ignition.

3. *Late cylinder injection.* Fuel is sprayed into the cylinder near top center in such a way that combustion occurs during the mixing and evaporation process. This is the system used both for Diesel engines and for spark-ignition engines as a means of achieving charge stratification and detonation control (see Chapter 2 and refs. 2.790-2.816).

FUEL INJECTION FOR DIESEL ENGINES

Injection systems for Diesel engines are all of type 3. With this type of late injection, the spray must penetrate into air at compression density. This requirement, together with the need for a finely divided spray for rapid mixing, requires high injection pressures. In systems of this type, maximum pressures ahead of the injection nozzle range from about 2000 to over 20,000 psi (140 to over 1400 kg/cm^2).

The technical problems of fuel injection may be divided into the following two categories:

1. *Metering.* By metering is meant the supply of the desired quantity of fuel in each injection, at an acceptable rate and over an acceptable crank angle.

2. *Spray formation.* This term signifies control of the physical characteristics of the spray so as to secure proper mixing of fuel and air, both in time and space.

Although these two aspects of fuel injection are not completely independent, it is convenient to treat them separately, as far as possible.

Fuel Metering for Diesel Engines

For purposes of discussion the following definitions will be used:

Nozzle. The opening through which the fuel emerges into the air.
Nozzle assembly. An assembly of parts including the nozzle, which is removable from the engine as a unit.
Needle. A rod, the inner end of which is used to open and close the nozzle passage. The needle is usually spring-loaded. Most, but not all, high-pressure injection systems use some form of needle.
Opening pressure. The fuel pressure required to lift the needle under test-bench conditions, with the pressure applied gradually.
Injection pressure. The pressure on the upstream side of the nozzle at any particular moment.
Jerk pump. A reciprocating fuel pump that meters the fuel and also furnishes the injection pressure.
Inlet port. The port through which fuel enters the jerk-pump cylinder.
Port closing. The point where the pump ports close, that is, the beginning of the effective pump stroke.
Port opening. The point at which a relief port, communicating with the pump inlet, starts to open.
Effective pump stroke. The length of the pump-plunger stroke between port opening and port closing.
Pump duration. The time or crank angle between start and end of the effective pump stroke.
Primary pump. A pump that supplies fuel, under relatively low pressure, to the injection pump.
Injection quantity. The mass of fuel in one injection.
Injection volume. The volume of fuel in one injection, measured at the injection-pump inlet pressure and temperature.
Spray duration. The time or crank angle between beginning and end of the spray.
Spray penetration. The visible length of the spray, measured from the nozzle, at a given time or crank angle after the start of injection.
Spray length. The maximum spray penetration.
Cone angle. The maximum conical angle of the spray envelope, with its apex at the nozzle opening.

Excluding air injection, which is now obsolete, there are two basic methods of metering in current use; they are called the *common-rail* and the *jerk-pump* systems. The basic principles of these two systems are illustrated diagrammatically in Fig. 7–1. In the common-rail system,

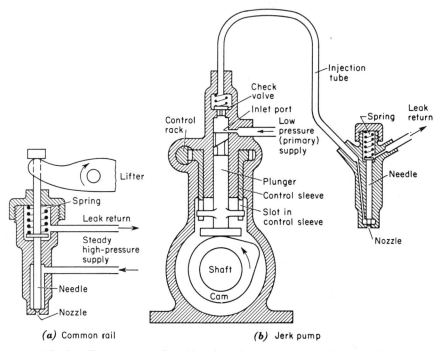

(a) Common rail   *(b)* Jerk pump

Fig. 7–1. The common-rail and jerk-jump injection systems (diagrammatic).

fuel is supplied at a constant high pressure from a reservoir of large volume compared with one injection volume. The reservoir includes the passages connecting several injection valves, which is the "common rail" from which the system derives its name. If all passages are large compared with the nozzle and if there are no serious dynamic disturbances in the reservoir system, the pressure on the upstream side of the nozzle can be considered constant and equal to the supply pressure at all times. Assuming the pressure difference across the nozzle to be in the critical range, and therefore independent of the downstream pressure, we can write, from Bernoulli's theorem:

$$M_f = CAt\sqrt{2\rho p g_o} = CA\sqrt{2\rho p g_o}\,\frac{\theta}{360N}, \qquad (7\text{–}1)$$

where M_f = mass of fuel delivered in one injection
  C = average flow coefficient of the nozzle

$A =$ area of the nozzle cross section

$p =$ upstream pressure

$\rho =$ fluid density

$g_o =$ force-mass-acceleration constant

$t =$ duration time

$\theta =$ duration angle, degrees

$N =$ engine revolutions per unit time.

It is at once evident that the fuel quantity delivered will decrease with increasing speed unless θ is so increased that θ/N remains constant. Since this requirement introduces mechanical complications, the common-rail system is not often used except on engines designed to run at constant speed. For such engines, M_f is controlled either by varying θ or by varying the needle lift, which, when small compared with the nozzle diameter, controls the value of C.

Another serious disadvantage of the common-rail system is that C is sensitive to wear of the nozzle and of the needle point and to deposits of solid matter on the needle and in the nozzle orifice.

Jerk-Pump Systems. In this system the needle is lifted by the fuel pressure, which is supplied by a reciprocating displacement pump. This pump, in turn, is supplied with fuel at relatively low pressure. The quantity of fuel supplied by this system can be written:

$$M_f = AS\rho e_p, \tag{7-2}$$

where $M_f =$ mass of fuel per injection

$A =$ pump-plunger area

$S =$ effective pump stroke

$\rho =$ fuel density at the pump inlet

$e_p =$ pump volumetric efficiency defined, as in the case of reciprocating engines, as the volume of fuel delivered, divided by AS.

On comparing Eq. 7–2 with Eq. 7–1, a basic advantage of the jerk-pump system becomes evident, namely, that the nozzle characteristics and engine speed do not affect the fuel quantity unless they change the pump volumetric efficiency. Actually, small differences in nozzle characteristics, such as those caused by ordinary wear or moderate deposit formation, have little effect on e_p, and thus the jerk-pump system is much less sensitive to such variations than is the common-rail system. Also, with proper design, the effect of speed on e_p can be made small over the usual operating range. The net result is that a well-designed jerk-pump system furnishes a nearly constant fuel quantity across the usual operating range at any given setting of the pump control.

On the other hand, the spray duration and the injection pressure are not fixed in jerk-pump systems as they are in the case of common-rail systems; instead, they depend on nozzle as well as on pump characteristics and on operating conditions, as will appear in the subsequent discussion.

Fuel-Quantity Control. The value of M_f in Eq. 7-2 is usually controlled by varying the effective pump stroke S. A common method of accomplishing this result is to vary the angular position of grooves in the plunger in relation to the fuel-inlet port, as illustrated in Fig. 7-2, which shows the well-known Bosch system.

It is seen from Fig. 7-2 that, at a given angle of the plunger, the inlet port will be closed when the top of the plunger covers it. From this point, as the plunger rises, the fuel is forced through the check valve into the delivery line. Delivery continues until the helical groove in the plunger uncovers the inlet port. Ideally, at this point the pressure in the cylinder drops to delivery pressure, and all flow through the check valve ceases.

Pump Volumetric Efficiency. Possible causes of volumetric deficiency (1 — volumetric efficiency) in pumps of the type shown in Fig. 7-2 include the presence of fuel vapor in the pump cylinder at port closing, backflow through the delivery check valve, fuel and metal elasticity, and leakage.

Fuel Vapor. Vapor in a jerk pump is usually the result of vapor formation in the supply line and can be prevented by avoiding high temperatures in the supply system and supply line and by using adequate supply pressure. This

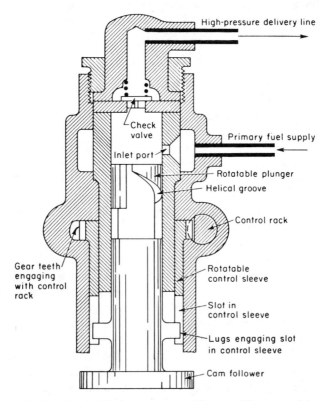

Fig. 7-2. Bosch injection-pump control system (diagrammatic).

pressure, usually 25 to 100 psi (2 to 7 kg/cm^2), is supplied by a primary pump, usually of the gear variety. Vapor in the jerk-pump cylinder during delivery must be avoided. The use of inlet ports of generous dimensions, together with adequate supply pressure and avoidance of vapor in the pump-supply system are necessary to achieve this result. During the return stroke of the pump there will be some vapor in the pump cylinder until the inlet port opens.

Back Flow. At the end of fuel delivery, when the port starts to open, the delivery check valve is open. As the pressure in the pump cylinder falls and the check valve closes, it is evident that there may be an appreciable flow of fuel back into the pump cylinder and out through the port. The extent of this backflow depends on the dynamics of the system, including the mass, lift, and spring constant of the check valve and the phase and amplitude of pressure waves in the fuel line. Obviously, a quick closing of the check valve, without " bouncing," is desirable. In general, the smaller the mass and lift of the valve, the more satisfactory will be its operation.

Compressibility and Deflection. With a large volume of fuel and a flexible line, the stroke of the pump might not result in enough pressure to open the needle. This type of action is of course undesirable, but it can be minimized by minimizing the volume ratio and making the mechanical parts very stiff. Volume ratio is here defined as the ratio of fuel volume under delivery pressure to AS, the effective pump displacement volume. In well-designed systems the deflection of the mechanical parts is small enough to be neglected, but the compressibility of the fuel is always an appreciable factor. The importance of a small volume ratio has lead a number of manufacturers to eliminate the injection line entirely and build the pump and nozzle assembly into single units mounted on each cylinder. The great disadvantage of this design is that the fuel pumps for several cylinders cannot be built into a single compact assembly with a single mount and drive connection but must be separately mounted and driven (7.79).

Leakage. Some degree of leakage in jerk pumps is unavoidable, especially near the beginning and end of the effective stroke, where the leakage distance to the inlet port is short. Leakage can be minimized by using minimum practicable plunger-to-cylinder clearance and maximum feasible length of leakage paths.

Mathematics of Pump Volumetric Efficiency. Assuming mechanical deflections to be negligible, an expression for pump volumetric efficiency can be set up in a manner similar to that used for engine volumetric efficiency in Volume I, Chapter 6, which leads to the expression

$$e_p = \Phi\left(\frac{sA}{aCA_n}, \frac{sb\rho}{\mu g_o}, \frac{p_o g_o}{\rho a^2}, \frac{p_o}{p_i}, \frac{L}{b}, \frac{A}{CA_n}, R_1, \ldots, R_n\right), \qquad (7\text{--}3)$$

where Φ = a function of what follows
A = pump-plunger area

A_n = nozzle area
b = plunger bore
C = nozzle-flow coefficient
L = fuel-delivery-line length
s = pump mean plunger speed
a = speed of sound in the fuel
ρ = fuel density
μ = fuel viscosity
p_o = nozzle opening pressure
p_i = pump inlet pressure
$A/CA_n, L/b$ = very critical design ratios
$R_1, \ldots, R_n$ = remaining design ratios of the system, including pump, line, and nozzle assembly.

The important ratio of fuel volume under delivery pressure to displacement volume of the pump is a function of the design ratios, including especially L/b.

In geometrically similar systems the pressure wave pattern will be a function of the Mach index, sA/aCA_n. The effect of fuel compressibility is included in the Mach index and the ratio $p_o g_o/\rho a^2$. The latter ratio includes the modulus of elasticity of the fuel from the relation $a^2 = E/\rho$, where E is the bulk modulus of the fuel.

Leakage will be a function of pressures and the Reynolds index, $sb\rho/\mu g_o$. When speed or plunger diameter is the only variable, the fraction of fuel that leaks past the piston will decrease as these variables increase, because leakage increases with Reynolds number raised to a power less than one, while fuel flow increases nearly in proportion to s or b^2.

Figure 7–3 shows typical curves of pump volumetric efficiency against four important variables.

The principal factors in each case are probably as shown in the following table:

Curve	Variable in Eq. 7–3	Chief Influence on e_p
1	s	Dynamic effects, including backflow through check valve
2	A/A_n	Leakage decreases as delivery pressure decreases because of larger nozzle
3	p_o	Leakage increases as opening pressure increases
4	L/b	Time for pressure waves to traverse line increases as length increases

Injection Rate. Let the rate at which fuel is delivered from the nozzle be dM/dt. With a constant rate during the effective pump stroke,

$$\left(\frac{dM}{dt}\right)_{\text{ideal}} = \frac{MN}{\theta}, \tag{7–4}$$

where M = mass of fuel delivered in one pump stroke
N = engine speed
θ = injection angle, in radians.

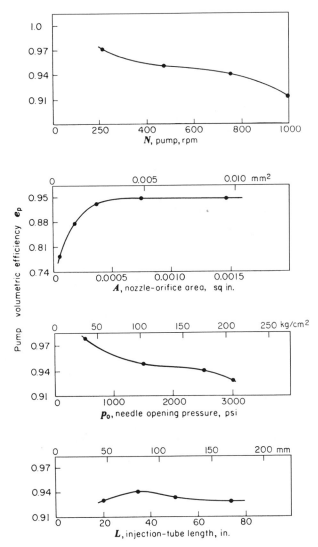

Fig. 7–3. Effect of pump variables on pump volumetric efficiency. When not otherwise noted, $N = 750$, $A = 0.000176$, $p_0 = 2500$ (Gelalles and Marsh, 7.60).

The actual value of the product $(dM/dt)(\theta/MN)$ will be a function of the same set of variables as those which control volumetric efficiency (Eq. 7–3). Actual rates of injection versus speed and nozzle area for one particular system are shown in Fig. 7–4. In general, the type of curve shown for

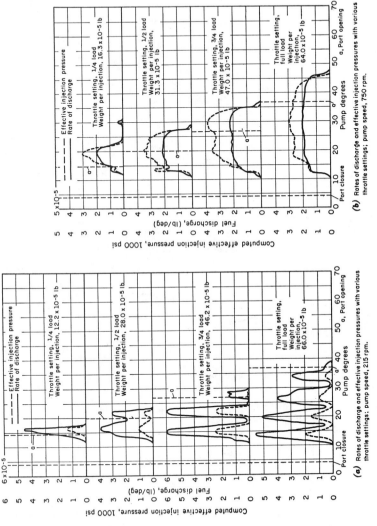

(a) Rates of discharge and effective injection pressures with various throttle settings; pump speed, 215 rpm.

(b) Rates of discharge and effective injection pressures with various throttle settings; pump speed, 750 rpm.

Fig. 7-4. Fuel-discharge rates from a jerk-pump system. Tube length, 34 in. (860 mm); tube inside diameter, 0.125 in. (3.2 mm); nozzle-opening pressure, 2500 psi (176 kg/cm²); nozzle diameter a to c, 0.022 in. (0.56 mm). "Throttle setting" means position of the control rack, cf. Figs. 7-1 and 7-2 (7.60).

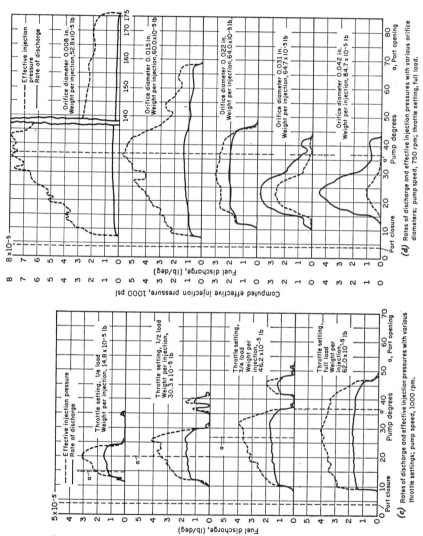

Fig. 7-4 (*continued*)

full-load pump setting at speeds of 750 and 1000 rpm is considered desirable, since the curves are free from needle bouncing and the injection rate is nearly uniform. Secondary discharges, such as those indicated at 1000 rpm, part load, are considered especially undesirable. Certainly freedom from needle bouncing is preferable from the point of view of mechanical durability. Curves giving reasonably uniform rates can usually be obtained over the useful operating range by proper selection of line length and nozzle opening pressure. The most difficult conditions are at low speed, as indicated by Fig. 7–4a.

It is obvious from Fig. 7–4 that the duration angle of the spray may be greater or smaller than that of the pump, and that there is always a considerable lag between the point of port closing and the start of injection. The spray-duration angle can be stretched out indefinitely by reducing the orifice size, as shown in Fig. 7–4d. Diesel engines generally have a full-load injection duration in the range 20° to 40° of crank angle (10° to 20° of pump angle for 4-stroke engines).

Spray Formation

The process of formation of a spray from the discharge through an orifice has been extensively investigated by the NACA (7.10–7.25). Figure 7–5 shows sprays of fuel-oil issuing from a round orifice into air at various densities. To analyze this problem, let us first take the simplified case of steady flow of a liquid fuel through a nozzle under conditions in which evaporation of fuel during spray formation is negligible. In this case it is probable that the factors listed in the following table influence the spray-formation process:

Factor	Symbol	Dimensions
Stream velocity at nozzle exit	u	Lt^{-1}
Fluid viscosity	μ	$FL^{-2}t$
Fluid surface tension	γ	FL^{-1}
Fluid density	ρ	ML^{-3}
Air viscosity	μ_a	$FL^{-2}t$
Air density	ρ_a	ML^{-3}
Nozzle diameter	D	L
Design ratios of nozzle	$R_1, \ldots, R_n$	1

The elastic characteristics of the fluid, as expressed by its sonic velocity, probably have little influence on the character of the free spray under steady-flow conditions.

The physical characteristics of a spray are usually characterized by one or more of the following measures:

Physical Characteristic	Symbol	Dimensions
Cone angle near the nozzle	α	1
Average drop diameter	d	L
Fraction of drops of a given size in the spray envelope	x	1
Length of the visible spray	y	L

By using dimensional analysis, we see that the parameter α can be related to the characteristics of the fluid, air, and nozzle in the following way:

$$\alpha = \Phi_1 \left(\frac{u\rho D}{\mu g_0}, \frac{\gamma}{\mu u}, \frac{\rho_a}{\rho}, \frac{\mu_a}{\mu}, R_1, \ldots, R_n \right). \tag{7-5}$$

Similarly, the ratios x, d/D, and y/D will be functions Φ_2, Φ_3, and Φ_4 of the same group of dimensionless ratios. In this case, in addition to the Reynolds number $u\rho D/\mu g_o$, we have a number covering the influence of surface tension, $\gamma/\mu u$, and also the ratios of air density and viscosity to those of the fuel.

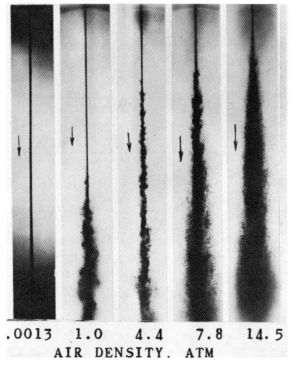

.0013   1.0   4.4   7.8   14.5
AIR DENSITY. ATM

Fig. 7–5. Sprays at various air densities. Injection pressure 250 psi (17.6 kg/cm²), orifice diam 0.020 in. (0.51 mm), No. 1 Diesel oil, see Table 7–1 (Lee and Spencer, 7.23).

It is usually more convenient to measure the fluid pressure p upstream from the nozzle orifice than the velocity u. In this case, from Bernoulli's theorem, $\sqrt{2pg_o/\rho}$ can be substituted for the velocity u.

Although quantitative experimental determination of these functions is far from complete, some quantitative and many qualitative relationships of great interest have been demonstrated (7.11–7.50). Space limitations permit only the more important relations to be mentioned here.

Effect of Air Density. Figure 7–5 shows the great importance of the density ratio ρ_a/ρ on the character of the spray. Apparently the interaction of fuel and air is a most important element in spray formation. With air density near zero, the jet of oil does not break up within the length photographed.

The effect of ρ_a/ρ on drop-size distribution is plotted in Fig. 7–6. Typical values of this ratio in Diesel engines are from 0.015 to 0.05. Within this range the effect on drop size appears to be small.

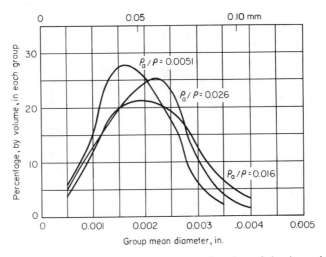

Fig. 7–6. Atomization measured by drop counts as a function of density ratio ρ_a/ρ. Fuel density, 60 lbm/cu ft (960 kg/m³); injection pressure 4140 psi (290 kg/cm²) (Lee, 7.52).

Effect of Reynolds Number. Figures 7–7 and 7–8 show that increasing the Reynolds number, $u\rho D/\mu g_o$, increases the tendency of the fuel jet to break up into small elements. The variable used to change the Reynolds number was the oil pressure upstream from the nozzle, which varies the velocity u. In Fig. 7–7 the air density was only one atmosphere. Figure 7–8 compares spray atomization at different Reynolds numbers at an air density more representative of Diesel-engine conditions.

Effect of Surface Tension. Figure 7–9 shows sprays with different liquids at identical injection pressures and air densities. Since these liquids differ in all their physical properties (Table 7–1) it is difficult to assign the observed

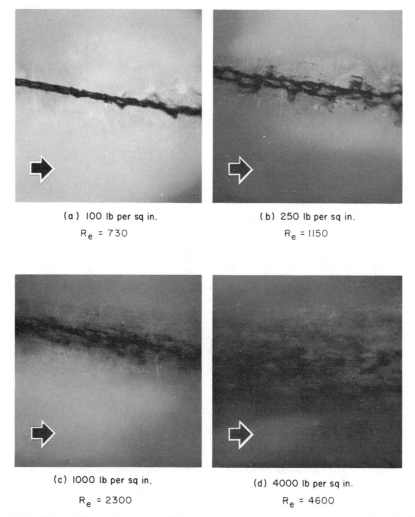

(a) 100 lb per sq in.   (b) 250 lb per sq in.
R_e = 730 R_e = 1150

(c) 1000 lb per sq in.   (d) 4000 lb per sq in.
R_e = 2300 R_e = 4600

Fig. 7–7. Effect of Reynolds number (injection pressure varied) on spray formation. Orifice diam 0.008 in. (0.22 mm); air density 1 atm; distance from nozzle 3 in. (79 mm); No. 1 Diesel oil, see Table 7–1 (Lee and Spencer, 7.23).

spray differences to any one term in Eq. 7–5. However, the alcohol and water sprays have nearly the same density and Reynolds number, so that the difference in their spray patterns are largely attributable to the effect of surface tension on the term $\gamma/\mu u$ of Eq. 7–5. Evidently the fact that the surface tension of alcohol is about one third that of water (see Table 7–1) is responsible for the far finer atomization of the alcohol spray. In the range of Diesel oils used in practice, differences in surface tension are negligible.

Effect of Nozzle Design on Spray Characteristics. Figure 7–10 shows various nozzle designs investigated by the NACA. The plain-orifice and

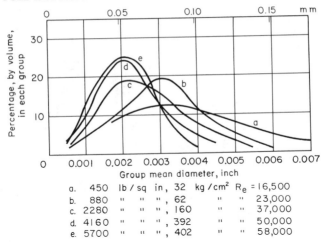

a.	450	lb / sq	in,	32	kg /cm²	R_e =	16,500
b.	880	"	" ",	62	" "		23,000
c.	2280	"	" ",	160	" "		37,000
d.	4160	"	" ",	392	" "		50,000
e.	5700	"	" ",	402	" "		58,000

Fig. 7–8. Effect of Reynolds number on spray atomization. Air density 0.94 lbm/ft³ (1.5 kg/m³); orifice diam 0.020 in. (0.51 mm); Diesel oil near No. 1, see Table 7–1 (Lee, 7.52).

pintle nozzles shown in this figure give so-called *hard* sprays, that is, sprays characterized by a dense core of fuel drops surrounded by an outer layer of well separated drops (Fig. 7–11*a*). The other nozzles give *soft* sprays, such as those illustrated in Fig. 7–12*b*. Soft sprays have no core; they consist wholly of well separated drops.

The drop-distribution curves for hard and soft sprays are plotted in Fig. 7–12, while Fig. 7–13 shows drop-distribution curves for plain nozzles (hard sprays) of various diameters. These figures indicate that hard sprays result in finer atomization than soft sprays and that decreasing the nozzle diameter tends to decrease the drop size.

Relation of Spray Chacteristics to Diesel-Engine Performance

In the case of Diesel engines this subject has already been dealt with in Chapter 3 (see especially Figs. 3–15 and 3–16 and Table 3–3), which illustrated the very close relation existing between spray characteristics and performance in one particular combustion chamber.

In general, every Diesel combustion chamber is a special case for which the optimum spray characteristics must be worked out by patient testing of various nozzle and injection-system components. The following general relations will usually be found:

1. Diesel engines require hard sprays, as from plain-orifice or pintle nozzles. The reason for this is undoubtedly that the soft type of spray does not have adequate penetration into the very dense air contained in the cylinder at the time of injection (14 to 20 times inlet density). The core apparently breaks up soon after ignition occurs; otherwise combustion of a good part of the spray would come too late in the expansion stroke.

Table 7-1. Properties of Liquids of Figs. 7-5 Through 7-9

Fuel	Density, ρ (g/cm³)	(lbm/ft³)	Viscosity, μ (poises)	lbf sec ft⁻² × 10⁶	Surface Tension, γ dynes cm⁻¹	lbf ft⁻¹ × 10⁴	R_e	$\gamma/u\mu$	μ/μ_a	u ft/sec
1 Ethyl alcohol	0.785	49	0.0115	24	24	16.4	56,700	0.128	64	535
2 Diesel #1	0.874	54.6	0.022	46	27	18.5	31,500	0.080	122	505
3 Diesel #2	0.96	60	0.102	212	28	19.1	7,070	0.186	567	483
4 Gasoline	0.738	46	0.0042	8.8	21	14.4	150,000	0.298	23.3	551
5 Water	1.000	62.4	0.0096	20.1	68	46.5	76,400	0.483	53.3	478
6 Lubricating oil	0.915	57	3.07	642	31	21.2	2,280	0.0067	1710	495

R_e is Reynolds number based on liquid properties, jet velocity, and nozzle-orifice diameter. u is jet velocity. (Lee and Spencer, 7.23.)

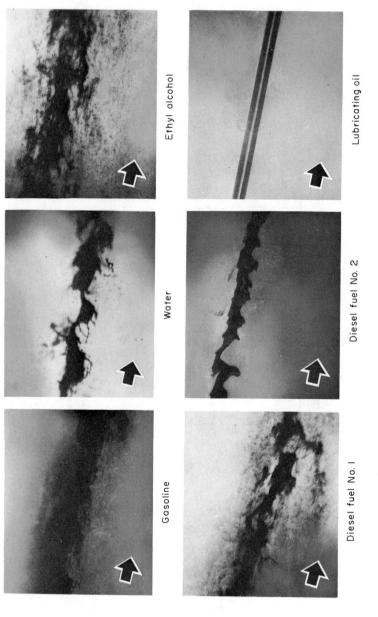

Fig. 7–9. Sprays with various liquids. Injection pressure 1500 psi (106 kg/cm²); orifice diam 0.020 in. (0.51 mm); air density 1 atm; distance from nozzle 5 in. (127 mm). See Table 7–1 for other data (Lee and Spencer, 7.23).

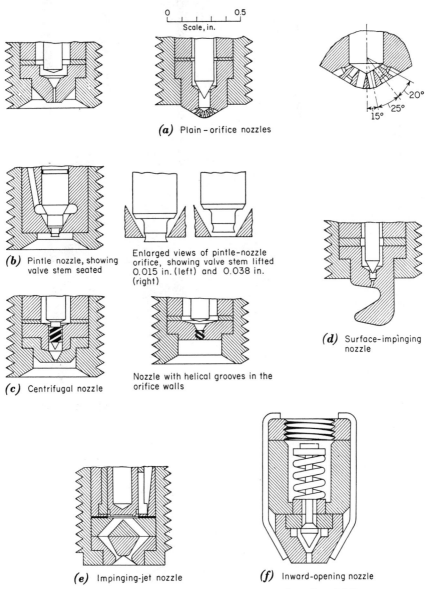

Scale, in. 0 0.5

(a) Plain – orifice nozzles

(b) Pintle nozzle, showing valve stem seated

Enlarged views of pintle-nozzle orifice, showing valve stem lifted 0.015 in. (left) and 0.038 in. (right)

(c) Centrifugal nozzle

Nozzle with helical grooves in the orifice walls

(d) Surface-impinging nozzle

(e) Impinging-jet nozzle

(f) Inward-opening nozzle

Fig. 7–10. Types of injection nozzles (*a* to *e* from Lee, 7.24).

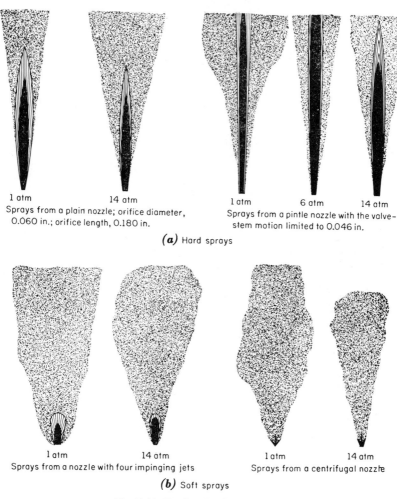

1 atm   14 atm
Sprays from a plain nozzle; orifice diameter,
0.060 in.; orifice length, 0.180 in.

1 atm   6 atm   14 atm
Sprays from a pintle nozzle with the valve-
stem motion limited to 0.046 in.

(a) Hard sprays

1 atm   14 atm
Sprays from a nozzle with four impinging jets

1 atm   14 atm
Sprays from a centrifugal nozzle

(b) Soft sprays

Fig. 7–11. Hard and soft sprays (7.24).

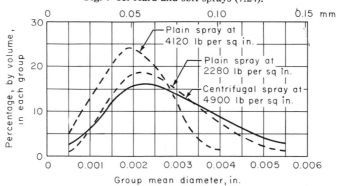

Fig. 7–12. Atomization as recorded by drop-size counts from hard (plain) and soft (centrifugal) sprays (Lee, 7.52).

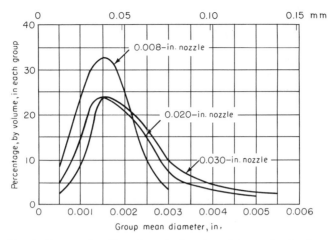

Fig. 7–13. Atomization as a function of nozzle diameter, plain nozzles, hard sprays. Injection pressure 3913 psi (275 kg/cm²). Reynolds number is proportional to nozzle diam (7.52).

2. In open chambers, in the absence of strong air motion, the spray must be directed to various parts of the combustion chamber by means of multiple orifices in the nozzle or by more than one nozzle in the cylinder.

3. Strong air motion, or swirl, is always desirable, because it makes performance less sensitive to spray variations.

4. Divided combustion chambers can usually be made to give satisfactory performance with a single nozzle. Inlet-induced swirl is not necessary with divided chambers.

5. Spray duration at full load should not exceed about 30 crank degrees.

6. Multiple injections, such as those caused by needle bounce, should be avoided as far as possible.

7. As for the effects of different physical characteristics of the fuel on engine performance, the fact that substantially the same performance is obtained with gasoline and Diesel oil, provided the mass of fuel injected is the same (see Fig. 4–21), indicates that when spray characteristics are adjusted for normal Diesel fuel, under the conditions prevailing in a Diesel cylinder during injection, large variations in fuel density, viscosity, and surface tension (see Table 7–1) can be tolerated without serious effects on the resultant mixing and combustion process.

Pilot Injection

In an attempt to secure the advantages of a small fuel quantity during the delay period without sacrifice in over-all performance, numerous systems of pilot injection have been proposed. Pilot injection is defined as injection of a small fraction of the fresh fuel early in the injection period, the remainder of the fuel being injected as in conventional injection systems.

It has even been proposed that the period of pilot injection be so adjusted as to correspond to the delay period. The objective of pilot injection is to have only a small fraction of the fuel take part in the period of rapid combustion.

Experiments along this line have been made with some success (7.791), but the added complication in the injection system, plus the fact that in high-speed engines the whole injection process must be accomplished in an extremely short time, has precluded general application of the principle. In low-speed engines ignition occurs early in the injection process and there is little to be gained by such an arrangement.

FUEL INJECTION FOR SPARK-IGNITION ENGINES

Although a carburetor is usually the simplest and cheapest device available for fuel metering, there may be good reasons for using a separate pumping system for fuel delivery rather than depending on the pressure difference due to air flow. This method of fuel delivery is called *fuel injection*. The usual reasons for using such a system include the following: a need for more accurate fuel metering than is practicable with a carburetor; reducing pressure loss in a venturi; reducing the need for inlet heating; improving fuel distribution when separate injectors are used at each cylinder; improving adaptability to the use of electronic control; reducing fuel loss in two-cycle engines (and in four-cycle engines with large valve overlap) by injecting the fuel during the compression stroke, after the valves have closed; and adapting to stratified-charge systems where Diesel-type fuel injection is required.

Except for stratified-charge engines, the mixture requirements (explained in Chapter 5) are essentially the same for fuel-injection systems as for carburetors, except that when individual injectors are used at each inlet port less enrichment for acceleration will be necessary.

Systems of control may be wholly mechanical, but many include a combination of mechanical, electrical, and electronic elements. In service at the present time are steady-flow systems injecting fuel near the air throttle, steady-flow systems injecting at the inlet ports, and systems with intermittent injection at the inlet ports. High-pressure injection systems with spray ignition are not yet in general use for spark-ignition engines.

Most injection systems use air flow as a primary operating parameter, as in a carburetor. Mechanical control systems usually use motion of an air "flap" valve in the intake system as the primary control element. This moves a control valve to regulate the fuel flow, with metal bellows for temperature and pressure signaling. Figure 7–4 illustrates a typical system of this kind.

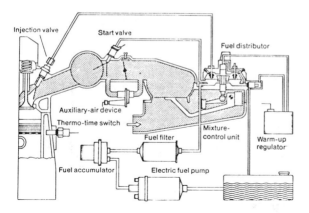

Fig. 7–4. Mechanical system of fuel injection for spark-ignition engines. (Reprinted by permission of Robert Bosch Corporation.)

Mechanical systems are usually assisted by electronics where strict control of emissions is required.

Figure 7–5 shows an electronic control system using individual injectors at each inlet port. From transducers measuring engine speed, crank position, spark timing, fuel-air ratio (*lambda* system), jacket temperature, air-inlet temperature, air flow, and throttle position, signals are processed by the microcomputer so as to give the most economical fuel flow rate for each operating regime for all phases of operation, within the limits of satisfactory engine performance and control of emissions. The system illustrated cuts off fuel flow when the throttle is closed and prevents extended periods of idling. In addition to the functions shown, EGR controls and knock limiting can be added.

In an all electric-electronic system, air flow may be measured by a hot-wire flow meter rather than a flap valve. Since the hot-wire type responds to air density as well as to air velocity, it furnishes an automatic correction for altitude.

Electronic Control for Diesel Engines. Electronic controls are now being used with Diesel engines to improve fuel economy and to reduce air pollution, both in vehicles and in stationary applications. The controlled variables may include injection-pump setting, injection timing, and amount of exhaust-gas recirculation (EGR). Inputs to the control system include pump control position and rate of motion; engine speed; injection angle; temperatures of outside air, engine, and fuel; and pressures of atmosphere and in inlet manifold. Smoke emissions are reduced by limiting the rate of pump-control movement and by limiting the fuel-air ratio to the smoke limit for each operating condition. Injection timing and EGR are controlled for best econ-

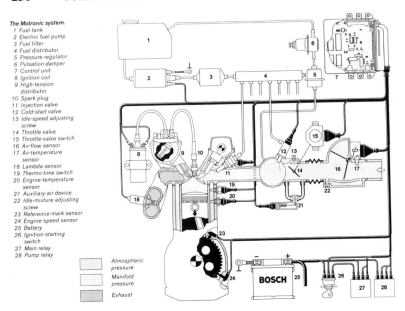

The Motronic system.
1 Fuel tank
2 Electric fuel pump
3 Fuel filter
4 Fuel distributor
5 Pressure regulator
6 Pulsation damper
7 Control unit
8 Ignition coil
9 High-tension
distributor
10 Spark plug
11 Injection valve
12 Cold-start valve
13 Idle-speed adjusting
screw
14 Throttle-valve
15 Throttle-valve switch
16 Air-flow sensor
17 Air-temperature
sensor
18 Lambda sensor
19 Thermo-time switch
20 Engine-temperature
sensor
21 Auxiliary-air device
22 Idle-mixture adjusting
screw
23 Reference-mark sensor
24 Engine-speed sensor
25 Battery
26 Ignition-starting
switch
27 Main relay
28 Pump relay

Atmospheric
pressure

Manifold
pressure

Exhaust

Fig. 7–5. Bosch "Motronic" control system for spark-ignition engines. Knock sensor can be added for use with supercharging. (Reprinted by permission of Robert Bosch Corporation.)

omy consistent with limitations imposed on exhaust pollutants. If practicable filters for particulate emissions are developed, control will be extended to these also.

In the case of stationary Diesel engines, less elaborate systems of operational control may be necessary, and there is usually more opportunity for large-capacity exhaust-treatment systems.

Electronic control is one of the most important developments in the field of internal-combustion engines since the first publication of these volumes. In motor vehicles, its use is being extended into non-engine areas such as controlling transmissions and braking systems, improving instrumentation, and warning of malfunctions of various elements throughout the vehicle. In stationary and marine power plants, the replacement of mechanical by electronic control systems is resulting in cost savings and improved performance.

The reliability and the accuracy of these control systems do not seem to be limited by electronics, but of course depend on the quality of the transducers that translate the quantities being measured into the appropriate electric signals and of the devices that translate the outgoing electric signals into the mechanical or other means required to yield the final result. In spite of their complexity, present systems seem to be quite reliable.

SIZE EFFECTS IN FUEL INJECTION

In Volume I, Chapter 11, it was shown that similar engines used in similar applications theoretically should be operated at the same values of inlet pressure and temperature, fuel-air ratio, and mean piston speed. It was also shown that commercial engines in a given type of service tend to be operated in this manner.

Assuming mechanical similitude (see Volume I, Chapter 11) to include the injection system, and assuming constant values of the engine variables mentioned, with a given fuel and fixed values of p_0/p_i, the injection parameters which will vary with size are fuel-pump volumetric efficiency and spray characteristics.

Pump Volumetric Efficiency. The Reynolds index in Eq. 7–3, $sbp/\mu g_0$, will vary as the bore of the pump. This parameter controls relative leakage past the pump piston, which will fall as size increases. A characteristic of many Reynolds number effects is that they tend to be large only in the range of rather low Reynolds numbers. This probably being the case here, one would expect only fuel pumps of very small-bore to show significant changes in relative leakage as size varies.

Referring to the problem of metering fuel for spark-ignition engines, the fuel-air ratio as expressed by Eq. 7–7 contains both engine and pump volumetric efficiencies and therefore depends on both engine and injection-system Reynolds indices. As far as metering is concerned, one would expect appreciable Reynolds number effects only in the case of very small pump and cylinder sizes. Within the automotive range of sizes, these effects should be small, in which case Eq. 7–8 applies.

Spray Characteristics. It has been shown, as in Fig. 7–7 and Eq. 7–5, that spray characteristics may be strongly influenced by the spray Reynolds number, $up D/\mu g_0$. Thus, for any engine with cylinder injection, a change in cylinder size may require changes in design of the spray nozzles, even though the fuel-air ratio required remains the same.

References for Chapter 7

In addition to references 7.00 through 7.937 showing developments up to 1963, the following references describe practice to 1983:

Bosch, "New Directions in Automotive Engineering," Robert Bosch Corporation, 2800 South 25th Ave., Broadview, IL 60153 (circa 1982).
Ibid. Combined Ignition and Fuel-Injection Syrtem "Motronic," 1983.

SAE Publications

SP505   Electronic Technologies for Commercial Vehicles, 1981.
SP487   Engine Control Modeling, 1981.
SP481   Electronic Management and Driveline Control, 1981.
SP495   Diesel Combustion and Emissions, 1981.
SP477   Implementation of Engine Control, 1980.
P104   Electronic Management and Driveline Controls, 1982.
PT-24   Passenger Car Diesels, 1982.

See also Chapter 6, page 213, and page 636.

8 | Engine Balance and Vibration

The subject of engine balance is taken here to include a study of the internal and external forces and moments associated with engine operation, for various types and cylinder arrangements. Included also is a consideration of special devices for balancing or altering these forces and torques.

" Engine vibration " is here defined to include a study of the motions of engines in their mounts, the transmission of forces and torques between engines and their mounting structures, vibration of parts within engine, and means of controlling the resulting motions and forces.

LITERATURE

There is a large body of literature on these subjects, as evidenced by the bibliography corresponding to this chapter. Unfortunately there is little uniformity in methods of approach, in symbols used, or in presentation of results. The object of this chapter is to present the elements of the subject in a form that can be readily understood and applied to practical problems. Exhaustive mathematical treatment of all aspects is beyond the scope of this chapter but may be found in references listed in the bibliography.

DEFINITIONS

Conventional system refers to engines with cylinder axes and piston-pin axes in the plane of the crankshaft axis, and with all connecting rods hinged around their crankpin axes.

Master-rod system refers to otherwise conventional engines using master rods to which articulated rods are hinged. The words *hinge pin* will be used here.

Offset engine refers to the case where the cylinder axis does not intersect the axis of rotation of the crankshaft.

Offset piston pin refers to the case where the piston-pin axis does not intersect the cylinder axis.

Unless otherwise noted, the conventional system will be assumed.

Moment refers to moments in a plane passing through the axis of rotation of the crankshaft. Counterclockwise moments will be taken to have a positive sign.

Torque refers to moments about the crankshaft-rotation axis. Counterclockwise torques will be taken as being positive.

Forces will be taken as being positive in sign when they are directed away from the crankshaft and toward the cylinder head.

Bank of cylinders is a group of cylinders located on the same side of the crankshaft with their axes in a plane passing through the crankshaft axis.

Row of cylinders is a group of radial cylinders lying in a plane at right angles to the crankshaft axis and operating on a single crank.

In-line engine is an engine with one bank of cylinders.

V engine is an engine with two banks of cylinders, corresponding cylinders in each bank forming a 2-cylinder row.

V angle is the angle between the cylinder planes of a V engine.

W engine is similar to a V engine but with three banks of cylinders. The two V angles are usually equal.

Opposed engine is a V engine with a 180° V angle.

Radial engine is an engine having three or more cylinders in one row with equal angles between their axes.

Multirow radial engine is a radial engine with more than one row of cylinders.

In in-line, V, and W engines, unless stated to the contrary, all cylinders have the same dimensions and the spacing of the cylinders, measured parallel to the crankshaft, is uniform.

SYMBOLS

It is convenient to list here symbols that apply particularly to engine balance and vibration problems. Other symbols are listed on pages 629 to 634. The letters between vertical lines indicate dimensions. In this chapter only three fundamental dimensions are used, namely, Force F, length L, and time t. In all other chapters of Volumes I and II, a fourth, mass M, is assumed. In this chapter mass has the dimensions $FL^{-1}t^2$.

Angles (*dimensionless*)

α	angle between cranks or vectors; angular amplitude of vibration
β	engine V angle, frequency ratio
γ	angle of hinge-pin location (Fig. 8–27)
δ	angle of shaft twist
θ	crank angle, referred to reference-cylinder axis (Fig. 8–1)
ϕ	connecting-rod axis to cylinder axis (Fig. 8–1)
ϕ_n	phase angle for gas torque

Forces $|F|$

F force, general

F_a inertia force acting on frame, opposite to F_p

F_{af} applied force

F_{at} transmitted force

F_{cp} radial inertia force on crankpin

F_{ct} inertia force on frame opposite to F_{cp}

F_{cw} counterweight force

F_p inertia force on piston, along cylinder axis

P gas force on piston $= pA_p$

Z $-M_p\Omega^2 R$

Frequencies $|t^{-1}|$

f vibration frequency $= \omega/2\pi$

N crankshaft revolutions per unit time

Ω crankshaft angular frequency $= 2\pi N$

ω vibration angular frequency

ω_n natural undamped angular vibration frequency

Lengths $|L|$

A_n articulated-rod acceleration coefficient, cosine terms

B_n same as A_n, sine terms

d distance between cylinder axes, measured parallel to crankshaft

h, j rod lengths to rod cg (Fig. 8–3)

L center of crankpin to center of piston pin (Fig. 8–1), shaft length (Fig. 8–28)

R center of crankshaft to center of crankpin (Fig. 8–1)

R_{cw} effective counterweight radius

S distance of piston pin from crankshaft axis

x deflection in an elastic system

Masses $|M/g_0| = |FL^{-1}t^2|$*

M mass, general

M_{cw} mass of counterweight

M_p reciprocating mass

M_R rotating mass

W_1, W_2 conn. rod masses (Fig. 8–3)

Miscellaneous

		Dimensions
A_p	piston area	L^2
a	acceleration	Lt^{-2}
cg	center of gravity	abbreviation
g	acceleration of gravity	Lt^{-2}
g_0	Newton's law constant	$MLt^{-2}F^{-1}$
I	mass moment of inertia (I/g_0)	FLt^2

---

* In this chapter only, all units of mass are taken as equal to mass/g_0 with dimension $FL^{-1}t^2$.

K	stiffness coefficient, linear	FL^{-1}
K_a	stiffness coefficient, angular	FL
N_c	number of cylinders	1
N_{cp}	number of crank positions	1
n	harmonic order	1
p	gas pressure	FL^{-2}
$\dot{S}$	piston velocity	Lt^{-1}
$\ddot{S}$	piston acceleration	Lt^{-2}
V	volume	L^3
V_d	displacement volume	L^3
W	weight	F

Torques and Moments $|FL|$

M_1	first-order moment caused by inertia forces
M_{11}	second-order moment caused by inertia forces
T	torque on crankshaft
T'	torque on frame
T_L	torque on shaft correction for rod inertia
T_L'	torque on frame correction for rod inertia
T_m	mean engine torque
T_p	torque due to gas pressure
T_t	torque due to reciprocating masses

Torque Coefficients (*dimensionless*)

s_n	correction torque on crankshaft for rod moment of inertia
s_n'	correction torque on frame for rod moment of inertia
t_n	torque from inertia of reciprocating mass
U_n	cosine coefficients for gas torque
V_n	sine coefficients for gas torque
W_n	$\sin(\theta + \phi_n)$ coefficients for gas torque

Vibration Coefficients (*dimensionless unless otherwise noted*)

a_n	single-cylinder inertia-force coefficient	1
C	damping coefficient	$FL^{-1}t$
C_c	critical damping coefficient $= 2\sqrt{KM}$	FL^1t
β	relative frequency $\omega_f/\sqrt{K/M}$	1
μ	relative transmitted force $1/\sqrt{(1-\beta)^2 + (2\zeta\beta)^2}$	1
ζ	relative damping, $C/2\sqrt{KM}$	1

SINGLE-CYLINDER GAS FORCES

Under the assumption that the components shown in Fig. 8–1 are all rigid bodies, the forces due to gas pressure in the cylinder can be taken as a force $-P$ on the piston and an equal and opposite force P on the cylinder head.

The force $-P$ is transmitted through the connecting rod to the crankpin and is balanced by an equal and opposite force supplied by the frame through the main bearings. The resultant force on the engine frame is

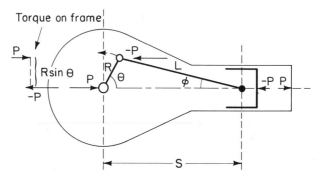

Fig. 8–1. Gas-pressure forces and torques. Torque on crank $= PR \sin \theta$; torque from frame $= -PR \sin \theta$.

obviously zero, but it is also apparent that there is a torque on the frame due to the couple $-PR \sin \theta$, which balances the useful torque from the shaft and which must be absorbed by elements outside the engine frame. This torque will be analyzed in detail later.

INERTIA FORCES AND MOMENTS

In the following discussion it will be assumed that the engine parts behave as rigid bodies. The errors involved in this assumption will be discussed under the heading of engine vibration.

Motion and Acceleration of the Piston. Figure 8–2 represents, diagrammatically, a crank-and-connecting-rod mechanism. It can be seen that

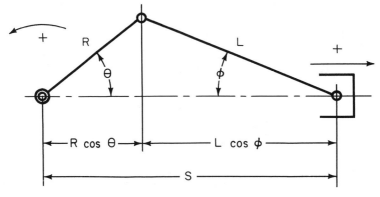

Fig. 8–2. Crank-rod system.

the distance S of the piston pin measured from the crankshaft center, is $R \cos \theta + L \cos \phi$. It follows that:

$$R \sin \theta = L \sin \phi \quad \text{or} \quad \sin \phi = \frac{R}{L} \sin \theta. \tag{8-1}$$

From trigonometry, $\cos \phi = \sqrt{1 - \sin^2 \phi}$, and therefore, from Eq. 8-1,

$$\cos \phi = \sqrt{1 - \left(\frac{R}{L}\right)^2 \sin^2 \theta}.$$

In practical crank-connecting-rod mechanisms R/L is always less than 0.33, so that $(R/L)^2 < 0.11$. For this reason the radical can be expanded by the binomial theorem into a rapidly converging series:

$$\cos \phi = 1 - \frac{1}{2}\left(\frac{R}{L}\right)^2 \sin^2 \theta - \frac{1}{8}\left(\frac{R}{L}\right)^4 \sin^4 \theta - \frac{1}{16}\left(\frac{R}{L}\right)^6 \sin^6 \theta + \cdots. \tag{8-2}$$

The powers of $\sin \theta$ may be replaced by their equivalent multiple angles. From trigonometry,

$$\sin^2 \theta = 1/2 - 1/2 \cos 2\theta$$

$$\sin^4 \theta = 3/8 - 1/2 \cos 2\theta + 1/8 \cos 4\theta$$

$$\sin^6 \theta = 5/16 - 15/32 \cos 2\theta + 3/16 \cos 4\theta - 1/32 \cos 6\theta$$

etc.

Substituting in Eq. 8-2, we obtain

$$\cos \phi = a_0' + a_2' \cos 2\theta + a_4' \cos 4\theta + a_6' \cos 6\theta + \cdots, \tag{8-3}$$

where

$$a_0' = 1 - \left[\frac{1}{4}\left(\frac{R}{L}\right)^2 + \frac{3}{64}\left(\frac{R}{L}\right)^4 + \frac{5}{256}\left(\frac{R}{L}\right)^6 + \cdots\right]$$

$$a_2' = + \left[\frac{1}{4}\left(\frac{R}{L}\right)^2 + \frac{1}{16}\left(\frac{R}{L}\right)^4 + \frac{15}{512}\left(\frac{R}{L}\right)^6 + \cdots\right]$$

$$a_4' = - \left[\frac{1}{64}\left(\frac{R}{L}\right)^4 + \frac{3}{256}\left(\frac{R}{L}\right)^6 + \cdots\right]$$

$$a_6' = + \left[\frac{1}{512}\left(\frac{R}{L}\right)^6 + \cdots\right].$$

Since

$$S = R(\cos \theta + L/R \cos \phi),$$

$$S = R[\cos \theta + L/R(a_0' + a_2' \cos 2\theta + a_4' \cos 4\theta + a_6' \cos 6\theta + \cdots)].$$

Let

$$a_0 = \frac{L}{R} a_0' \qquad a_2 = \frac{L}{R} a_2' \qquad a_4 = \frac{L}{R} a_4' \qquad \text{etc.}$$

Then:

$$S = R(a_0 + \cos\theta + a_2 \cos 2\theta + a_4 \cos 4\theta + a_6 \cos 6\theta + \cdots) \qquad (8\text{–}4)$$

Note that the second term of Eq. 8–4, $R \cos\theta$, is the projection of the crank position along the cylinder axis. The remainder of the terms are functions of R/L and are due to the angularity of the connecting rod. After the first three terms the coefficients become very small. If the crank is assumed to rotate at the constant angular velocity Ω, expressions for the velocity and acceleration of the piston may be obtained by differentiating Eq. 8–4 with respect to time as follows:

Piston velocity:

$$\frac{dS}{dt} = \dot{S} = -\Omega R(\sin\theta + 2a_2 \sin 2\theta + 4a_4 \sin 4\theta + \cdots) \qquad (8\text{–}5)$$

Piston acceleration:

$$\frac{d^2 S}{dt^2} = \ddot{S} = -\Omega^2 R(\cos\theta + 4a_2 \cos 2\theta + 16a_4 \cos 4\theta + \cdots). \qquad (8\text{–}6)$$

The reasons for expressing the piston acceleration in this form are, first, that it greatly facilitates the addition of forces with varying phase relations such as we have in multicylinder engines and, second, that this form is helpful in solving vibration problems.

At this point it is convenient to consider the connecting rod as equivalent to two masses concentrated at its ends, such that the sum of the masses is equal to the mass of the rod, and so proportioned that the center of gravity of the two masses is at the center of gravity of the rod, as shown in Fig. 8–3. Since the mass substituted for the connecting rod is equal to the actual mass of the rod, and the center of gravity is at the same point, the magnitude and direction of the resultant forces given by this assumption are equal to the magnitude and direction of the resultant of the actual forces on the connecting rod. Couples which may exist will be considered later. This substitution of a hypothetical connecting rod of two masses results in great simplification, since the substitute mass at the piston end of the rod reciprocates with the piston and may be considered a part of the piston assembly, while the substitute mass at the crank end of the rod simply rotates at constant speed with the crankpin.

Unbalanced Forces on the Single-Cylinder Engine. From the foregoing analysis it can be seen that a single crank–connecting-rod mechanism

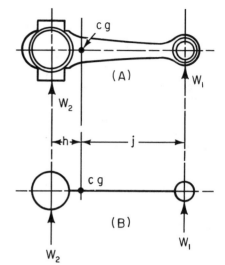

Fig. 8–3. Connecting-rod masses.

operating at a constant rotational speed produces forces as follows (see Fig. 8–4):

1. A force F_p along the cylinder axis, acting on the piston assembly to produce acceleration of the piston assembly and the reciprocating part of the connecting rod.

2. A force F_{cp}, acting on the crank pin and lower end of the connecting rod, directed radially inward toward the center of the crankshaft, to produce the centripetal acceleration of the parts which are considered to revolve with the crank pin.

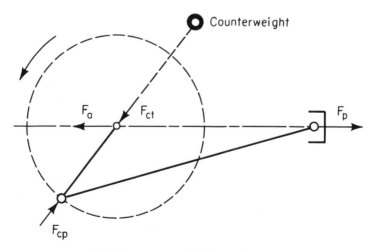

Fig. 8–4. Counterweight for rotating mass.

The reactions to these two forces are the two major forces acting *on* the engine frame tending to produce translation of the whole engine. Other forces, which may be due to the acceleration of exhaust gas during the exhaust blowdown process or the acceleration of various miscellaneous moving parts, such as valves, etc., are usually small compared with the unbalanced forces due to the crank train and will not be considered here.

The reaction *on* the engine frame due to F_p is a force acting along the cylinder axis, equal and opposite to F_p (i.e., $-F_p$) which will be called F_a. The reaction *on* the engine frame due to F_{cp} is a force acting at the crankshaft center, directed radially outward, and passing through the crank pin. It is equal and opposite to F_{cp} and will be called F_{ct}. The force F_{ct} is usually balanced by the arrangement of cranks in a multicylinder engine. In any case it may be balanced by suitable counterweights on the crankshaft,* as shown in Fig. 8–4. If F_{ct} is so balanced, the *only* unbalanced force acting on the engine is F_a. If the engine is mounted rigidly, the force acting *on* the engine from the rigid support must be F_a and the reaction *on* the support is $-F_a$. From Eq. 8–6 the force acting on the piston and the upper end of the connecting rod is

$$F_p = M_p \ddot{S} = -M_p \Omega^2 R(\cos\theta + 4a_2 \cos 2\theta + 16a_4 \cos 4\theta + \cdots), \qquad (8\text{–}7)$$

where M_p is the mass (M/g_0) which is considered to reciprocate with the piston, including the piston, piston rings, piston pin, and the equivalent mass of the upper end of the connecting rod.

The coefficients $4a_2$, $16a_4$, etc. are functions of R/L and are plotted in Fig. 8–5. The convergence of the series is so rapid for ordinary values of R/L (<0.3) that $16a_4$ and higher terms can be considered negligible.

Let $Z = -M_p \Omega^2 R$. From Fig. 8–5 it will be seen that $4a_2$ is numerically almost equal to R/L. Therefore to a good approximation,

$$F_a = -F_p \cong Z\left(\cos\theta + \frac{R}{L}\cos 2\theta\right). \qquad (8\text{–}8)$$

The coefficient of θ in Expression 8–8 will be referred to hereafter as the *order* of the harmonic. Thus, the first term of the expression represents the first-order force, or the one that varies periodically once per shaft revolution. The second term represents the second-order force, or the one that varies periodically *twice* per shaft revolution.

Representation of the Unbalanced Forces in a Single-Cylinder Engine. The unbalanced inertia force F_a acts along the cylinder axis only. It is due to the inertia of the reciprocating assembly and has *no side component*. It may be expressed by Eq. 8–8 and is the sum of two sinusoidal forces

---

* The crankshaft itself is assumed to be in static and dynamic balance. It may obviously be made so by means of proper counterweights.

of different amplitudes and frequencies but having maximum amplitude at $\theta = 0$. These forces are plotted in Fig. 8–6, which shows first- and second-order forces and their sum F_a as functions of the crank angle θ.

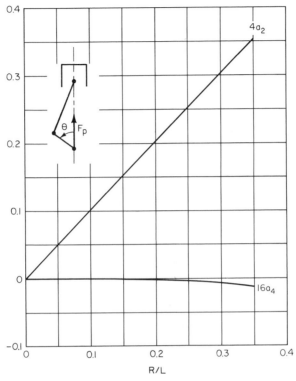

Fig. 8–5. Coefficients of F_p, the force acting on the engine frame due to acceleration of the piston and upper end of rod.

$$F_p = Z\left(\cos\theta + \sum_{n=1}^{\infty}(2n)^2 a_{2n}\cos 2n\theta\right)$$

$$Z = M_p\Omega^2 R$$

M_p = mass of reciprocating parts divided by g_0

n = harmonic order
R = crank radius
L = length of connecting rod
$\Omega = d\theta/dt$ (radians per second).

When the inertia forces of several cylinders must be added together to find the total inertia force in a multicylinder engine, it is convenient to use a vector system. Figure 8–7 shows the first- and second-order inertia-force vectors for a single-cylinder engine. The first-order vector has a length proportional to Z. It rotates in the direction of the crankshaft at crankshaft speed, and its projection on the vertical (cylinder) axis is the first-order inertia force, $Z\cos\theta$. The second-order vector has a length represented by $Z(R/L)$ and rotates at twice crank speed in the crank direction. Its projection on the cylinder axis is thus $Z(R/L)\cos 2\theta$, which is the second-order inertia force. In each case only the axial component of the vector, that is, its cosine projection, represents the force in question.

Summation of Forces $\sum F_a$ for Multicylinder In-Line Engines. In this type of engine the unbalanced reaction forces F_a from the individual pistons add algebraically, since their lines of action are parallel. Thus

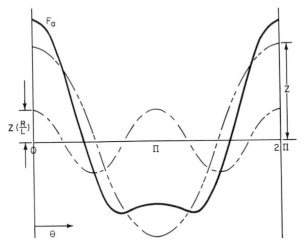

Fig. 8–6. Unbalanced forces versus θ.

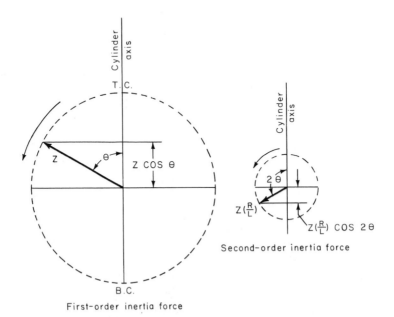

Fig. 8–7. First- and second-order vectors.

the resultant unbalanced reaction force acting on the engine frame is simply

$$\sum F_a = F_{a_1} + F_{a_2} + F_{a_3} + \cdots, \tag{8-9}$$

where the symbols F_{a_1}, F_{a_2}, etc. represent the individual reactions from the separate cylinders. These separate forces may be obtained by the addition of forces represented by Expression 8–8, using appropriate values of θ for each cylinder. Ordinarily the stroke, the reciprocating mass, and the length of the crank and connecting rod will be the same for all cylinders. Therefore:

$$\sum F_a = Z\,[(\cos\theta_1 + \cos\theta_2 + \cos\theta_3 + \cdots)$$
$$+ R/L(\cos 2\theta_1 + \cos 2\theta_2 + \cos 2\theta_3 + \cdots)]. \tag{8-10}$$

The angles θ_1, θ_2, θ_3, etc., for the various cylinders may all be expressed in terms of θ_1. When the cranks are evenly spaced, which is the usual arrangement, and α is the smallest angle between two cranks,

$$\alpha = \frac{2\pi}{N_{cp}} \text{ when } N_{cp} = \text{total number of crank positions.}$$

For example in a 3-cylinder in-line engine with cranks evenly spaced (Fig. 8–8):

$$N_{cp} = 3 \text{ and } \alpha = 2\pi/3 \text{ or } 120°$$
$$\theta_2 = (\theta_1 + \alpha)$$
$$\theta_3 = (\theta_1 + 2\alpha)$$
$$\text{and } 2\theta_2 = 2(\theta_1 + \alpha) = 2\theta_1 + 2\alpha$$
$$2\theta_3 = 2(\theta_1 + 2\alpha) = 2\theta_1 + 4\alpha.$$

Hence, from Eq. 8–10 the first- and second-order inertia forces are

$$\sum F_a = Z\bigg\{[\cos\theta_1 + \cos(\theta_1 + \alpha) + \cos(\theta_1 + 2\alpha)]$$
$$+ \frac{R}{L}\,[\cos 2\theta_1 + \cos(2\theta_1 + 2\alpha) + \cos(2\theta_1 + 4\alpha)]\bigg\}. \tag{8-11}$$

The first- and second-order parts of this expression are shown as separate vectors in Fig. 8–8. Note that the angle between successive first-order vectors is α and that between the second-order vectors, 2α. Thus the vectors are evenly spaced and form a "star diagram" with a resultant inertia force of zero.

Since the cranks are at $120°$ for the 3-cylinder engine of Fig. 8–8, the centrifugal forces F_{cp} will be balanced. However, since the cranks lie in different planes, these forces would have a resultant moment that would "wobble" the engine unless counterweights were added to counteract this wobble. In the subsequent discussion, it will be assumed that the F_{cp} forces and moments are balanced, either by the arrangement of the cranks or by suitable counterweights attached to the crankshaft.

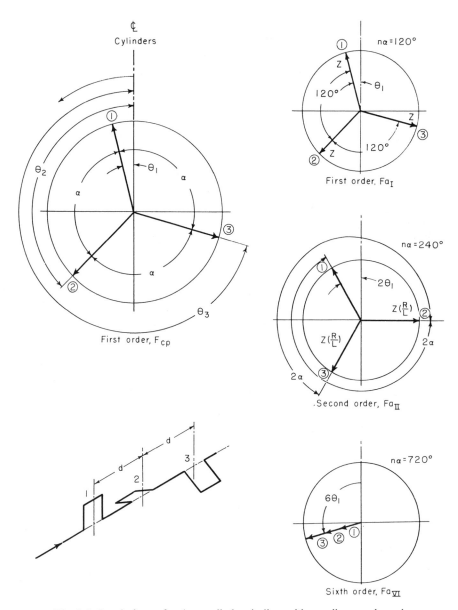

Fig. 8–8. Inertia forces for three cylinders in line, with equally spaced cranks.

The foregoing example illustrates a general rule for in-line engines with (more than one) evenly spaced cranks. This rule states that the resultant forces for any order will be zero (star diagram) except for those orders that are an even multiple of the number of separate crank positions. When n is an even multiple of the number of separate crank positions, the vectors coincide. Where each crank is at a different angle, the resultant unbalanced force is N_{cp} times that of a single cylinder. Thus, for the 3-cylinder example, the lowest order which adds (see Fig. 8–8) is the sixth. The sixth-order vector diagram is plotted in Fig. 8–8 on a greatly expanded scale. When two cranks are at the same angle, as in 4-stroke in-line engines, the resultant force is $2N_{cp}$ times that of a single cylinder.

In the case of unbalanced forces, the magnitude of orders above the second is usually considered negligibly small, although such orders may have to be considered in vibration problems, as will appear later.

Applying the above rule to in-line engines with evenly spaced cranks gives the following tabulation:

Number of cylinders	Cycle	N_{cp}	Orders which add
2	4	1	1, 2, 4, 6 $\cdots$
2	2	2	2, 4, 6 $\cdots$
3	2 or 4	3	6, 12, 18 $\cdots$
4	4	2	2, 4, 6 $\cdots$
4	2	4	4, 8, 12 $\cdots$
5	2 or 4	5	10, 20, 30 $\cdots$
6	4	3	6, 12, 18 $\cdots$
6	2	6	6, 12, 18 $\cdots$
etc.			

In the case of 4-stroke engines with even numbers of cylinders, for evenly spaced firing there must always be two cranks at the same angle θ, so that N_{cp} will be $N_c/2$, where N_c is the number of cylinders. In the case of two cylinders on each crank, all unbalanced forces are equal to twice those of one cylinder.

If coefficients above the second order are neglected, it is evident that, with one exception, all in-line engines of three cylinders or more, with evenly spaced firing, are free of large unbalanced inertia forces (but not necessarily free of important unbalanced *moments*). The exception is the 4-cylinder, 4-cycle engine with cranks at 180°, which has an unbalanced second-order force of $4Z(R/L) \sin 2\theta$.

Unbalanced Moments Resulting from Summation of Unbalanced Forces. Although the forces due to acceleration of the pistons in an in-line engine are in the same plane, they do not have the same line of action. This may give rise to unbalanced moments in the plane of the cylinders.

When the forces are balanced, it is convenient to take moments about the intersection of the crankshaft centerline and the axis of one cylinder. A positive pitching moment is then defined as one which would be caused by positive unbalanced forces from the other individual cylinders.

Take for example the 3-cylinder engine of Fig. 8–8. As in the case of forces, we shall consider the moments due to various orders separately. The distance between cylinder axes is assumed uniform and is taken as d. Starting with the first order, and taking moments about an axis normal to the axis of cylinder No. 1, the resultant moment is

$$M_1 = Z(d \cos \theta_2 + 2d \cos \theta_3). \tag{8–12}$$

For this engine, as before,

$$\alpha = 120°;$$

$$\theta_2 = \theta_1 + 120°;$$

$$\theta_3 = \theta_1 + 240°.$$

$$M_1 = Z[d \cos(\theta_1 + 120°) + 2d \cos(\theta_1 + 240°)]. \tag{8–13}$$

Similarly the second-order moments will be

$$M_{11} = Z\left(\frac{R}{L}\right) [d \cos (2\theta_1 + 240°) + 2d \cos(2\theta_1 + 480°)]. \tag{8–14}$$

The vector diagram for Eq. 8–13 is shown in Fig. 8–9, and that for Eq. 8–14 in Fig. 8–10.

Cylinder No. 1 of Fig. 8–9 has no moment. The vectors for Nos. 2 and 3 add to give a resultant vector of value $\sqrt{3}\,Zd$ acting at an angle $\theta_1 + 210°$. The actual moment is the projection of this vector on the cylinder-axis plane, or,

$$M_1 = \sqrt{3}\,Zd \cos(\theta_1 + 210°). \tag{8–15}$$

The same method is used to find M_{11} in Fig. 8–10. The second-order moment is seen to be

$$M_{11} = \sqrt{3}\,Z\left(\frac{R}{L}\right) d \cos(2\theta_1 + 150°). \tag{8–16}$$

Similar vector diagrams may be used for any in-line engine.

In order to eliminate all moments, in-line and V engines can be made with even numbers of cylinders and with cranks arranged *symmetrically* about a plane perpendicular to the crankshaft axis. Figures 8–11a and 8–12a illustrate symmetrical shafts. The unbalanced force at each crank is balanced by an equal and opposite force on another crank at 180°, so placed that the moment arms around the plane of symmetry are equal.

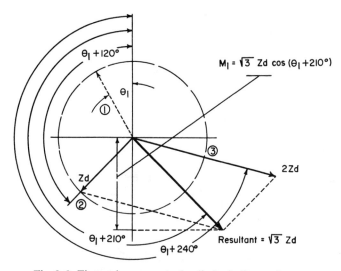

Fig. 8–9. First-order moments, 3-cylinder in-line engine.

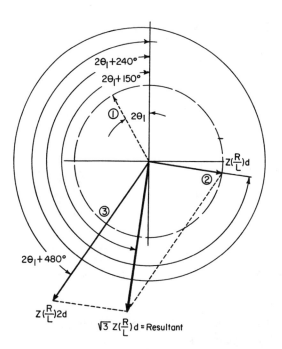

Fig. 8–10. Second-order moments, 3-cylinder in-line engine.

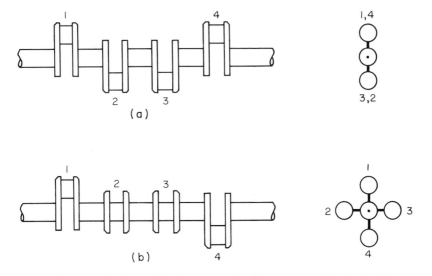

Fig. 8–11. (*a*) Symmetrical and (*b*) antisymmetrical 4-throw crankshafts. Shaft (*a*) is conventional for 4-cycle, 4-cylinder engines. Shaft (*b*) is conventional for 4-cycle V-8 engines with cylinder banks at 90°, since both first- and second-order forces and moments are balanced in this case. The antisymmetrical shaft must be counterweighted with respect to rotating masses.

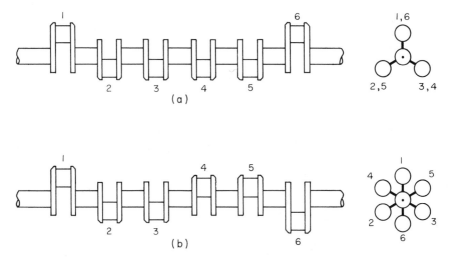

Fig. 8–12. (*a*) Symmetrical and (*b*) antisymmetrical 6-throw crankshafts. Symmetrical shaft is conventional for 4-cycle in-line and V-12 engines. Antisymmetrical shaft is appropriate for 2-cycle in-line 6 and V-12 engines. The antisymmetrical shaft should be counterweighted with respect to rotating masses.

Conventional four-cycle 4-, 6-, and 8-cylinder in-line engines as well as conventional four-cycle V-12 and V-16 engines have symmetrical crankshafts. The symmetrical arrangement is especially adapted to 4-cycle engines since, with an even number of cylinders, it is necessary to have two pistons at top center at the same crank position in order to get evenly spaced firing impulses (one cylinder starting the power stroke while the other starts the suction stroke). For 2-cycle engines, however, it is undesirable to have two cylinders fire at the same time. Consequently, the symmetrical crank arrangement is seldom used for 2-stroke engines (see Table 8–3 on pp. 304–305).

An *antisymmetrical* crankshaft is one in which the crank of the last cylinder is 180° from cylinder No. 1, that for the next-to-last cylinder is 180° from cylinder No. 2, etc., as in Figs. 8–11*b* and 8–12*b*. In engines with antisymmetrical crankshafts, the moment about the center of the engine due to secondary forces is zero. In cases where the primary moment can be offset by counterweights (as in the conventional 4-cycle V-8 engine), all important moments are thus eliminated, provided the rotating masses are suitably counterweighted.

Unbalanced Forces and Moments in V Engines with Simple Connecting Rods. In many V engines the forces due to piston acceleration and the resulting moments are balanced for each bank of cylinders. This is the case in conventional 4-stroke 12- and 16-cylinder engines. In some cases, however, notably in the case of 90° V engines, it is possible to obtain excellent balance, even though the separate banks of cylinders may not in themselves be balanced.

The best method for analyzing the inertia forces of V engines is to consider a unit consisting of the cylinders that act on the same crank (Fig. 8–13). If YOY is the axis of symmetry and XOX the perpendicular to it through the crankshaft center, the unbalanced forces from each cylinder may be projected on the XY axes and added to find the unbalanced forces of the unit.

With the V angle β, let the crank angle for the right-bank cylinder be taken as θ; that for the left-bank cylinder will then be $2\pi - \beta + \theta$, which is trigonometrically equivalent to $\theta - \beta$.

For the right cylinder:

$$F_{a_R} = Z(\cos \theta + R/L \cos 2\theta), \qquad (8\text{--}17)$$

and for the left cylinder:

$$F_{a_L} = Z[\cos(\theta - \beta) + R/L \cos 2(\theta - \beta)]. \qquad (8\text{--}18)$$

The Y projection of these unbalanced axial inertia forces will be $F_{a_Y} = (F_{a_R} + F_{a_L}) \cos(\beta/2)$, so that

$$F_{a_Y} = Z \cos \frac{\beta}{2} \left\{ [\cos \theta + \cos(\theta - \beta)] + \frac{R}{L} [\cos 2\theta + \cos 2(\theta - \beta)] \right\}. \quad (8\text{--}19)$$

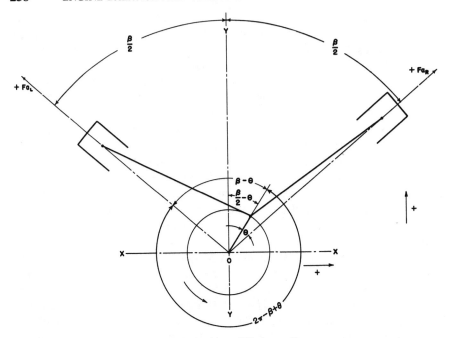

Fig. 8–13. Ninety-degree V balance diagram.

The X projection of F_{a_R} and F_{a_L} will be

$$F_{a_X} = Z \sin \frac{\beta}{2} \left\{ [\cos \theta - \cos(\theta - \beta)] + \frac{R}{L} [\cos 2\theta - \cos 2(\theta - \beta)] \right\}. \quad (8\text{–}20)$$

The first-order terms

$$F_{a_{Y_1}} = Z \cos \frac{\beta}{2} [\cos \theta + \cos(\theta - \beta)], \quad (8\text{–}21)$$

and

$$F_{a_{X_1}} = Z \sin \frac{\beta}{2} [\cos \theta - \cos(\theta - \beta)] \quad (8\text{–}22)$$

may be simplified by using the relationship:

$$\cos X + \cos Y = 2 \cos \tfrac{1}{2}(X + Y) \cos \tfrac{1}{2}(X - Y);$$
$$\cos X - \cos Y = -2 \sin \tfrac{1}{2}(X + Y) \sin \tfrac{1}{2}(X - Y).$$

If $X = \theta$ and $Y = (\theta - \beta)$, then

$$F_{a_{Y_1}} = 2Z \cos^2 \frac{\beta}{2} \left[\cos\left(\theta - \frac{\beta}{2}\right) \right]; \quad (8\text{–}23)$$

$$F_{a_{X_1}} = -2Z \sin^2 \frac{\beta}{2} \left[\sin\left(\theta - \frac{\beta}{2}\right) \right]. \quad (8\text{–}24)$$

vector, and components of them may be taken on any set of axes. The length of each vector is half that of the conventional single vector. In order to lend reality and to avoid confusion, the two vectors may be represented as weights revolving in opposite directions about the crankshaft center line, as in Fig. 8–14b. The weights may be considered as having the mass $M_p/2$ revolving at crankshaft speed at radius R for the primary forces, and the mass $M_pR/2L$ revolving at twice crankshaft speed at radius R for the secondary forces. Evidently the unbalanced force created by primary weights is $Z \cos \theta$, and that by secondary weights is $Z(R/L) \cos 2\theta$; they are thus equal to the unbalanced primary and secondary forces of a single-cylinder engine.

To illustrate the utility of this representation, let us solve the balance of a 90° V pair of cylinders by this method. With reference to Fig. 8–15, if the crank is at angle θ for the right bank, the two primary force weights

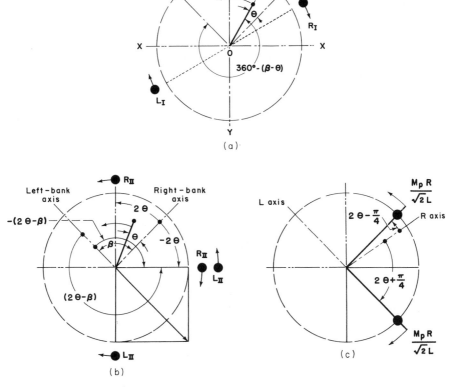

Fig. 8–15. Ninety-degree V engine, two cylinders only. Representation of first- and second-order forces by rotating masses. (*a*) First-order forces; (*b*) second-order forces separate; (*c*) second-order forces combined.

R_1 will have moved θ and $-\theta$ from the right-bank axis. The two primary force weights L_1 for the left bank will have moved in both directions from the left-hand axis, the same number of degrees as has the crank, that is, $360° - (\beta - \theta)$. Thus, one will be rotating counterclockwise with the crank, the other clockwise at the same angle from the left-bank axis. The result is that two weights are rotating counterclockwise *with* the crank while the other two are 180° apart and rotate clockwise at crank speed. The latter two cancel out, while the former two give a resultant force of magnitude Z, acting outward, which can be balanced by a counterweight such that $M_{cw} R_{cw} = M_p R$.

The second-order vectors, also shown in Fig. 8–15, are located by moving the two second-order weights to an angle from the axis of their cylinders twice as great as the crank has moved. The diagram shows that the two clockwise weights will always be 90° from each other and may be combined into a single clockwise mass of magnitude $M_p R/\sqrt{2} L$. The two counterclockwise vectors may be similarly combined, Fig. 8–15c. It will be seen that the resultant force will always act along the X (horizontal) axis and will have a maximum of twice the value of the combined vector, or $\sqrt{2} Z(R/L)$, whenever the combined vectors cross each other.*

The "weight" method is particularly useful in connection with radial engines. Figure 8–16 shows the first- and second-order analysis for a 3-cylinder radial engine. The crank is arbitrarily shown at $\theta = 0°$. The crank has moved 240° from top center for cylinder No. 2, the No. 2 primary vectors have therefore moved 240° counterclockwise and clockwise from axis No. 2. For cylinder No. 3 the crank has moved 120° from top center, and the primary weights have moved 120° counterclockwise and 120° clockwise. Evidently the three clockwise weights cancel each other. The three counterclockwise weights add up to a force of $3Z/2$ rotating with the crank, and may be balanced by a counterweight opposite the crank.

For the second order, the three counterclockwise weights cancel out, leaving three second-order weights rotating in a direction opposite to the crank at twice crankshaft speed. The unbalanced second-order force is evidently $(3Z/2)(R/L)$. Single-crank radial engines with more than three cylinders, with all connecting rods pivoted on the crankpin, have no unbalanced primary or secondary forces provided the primary unbalanced force is properly counterbalanced (see Table 8–2). With the usual master-rod system, they have a small secondary unbalanced force that can be balanced by weights rotated at twice engine speed (see later discussion of the master-rod system).

---

* The second-order unbalanced forces are neutralized in the conventional 4-cycle V-8 engine, as already explained.

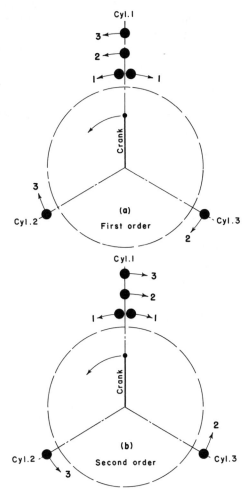

Fig. 8–16. Three-cylinder radial engine. First- and second-order forces, represented by rotating masses.

ENGINE TORQUE

Moments around the crankshaft axis will be here classified as engine torque.

We have seen that the gas pressure on the piston gives a resultant torque on the crankshaft of $PR \sin \theta$. This torque varies because both the cylinder pressure and $\sin \theta$ vary with crank angle. In addition to the torque due to gas pressure, a torque that varies with crank angle is produced as a result of the inertia of the piston and connecting rod.

So far we have taken the connecting rod to be dynamically equivalent to two masses concentrated at its ends and having the same total mass and center of gravity as the actual rod. While this assumption introduced no error in computing forces and moments, it is not correct in the case of inertia torque, because the moment of inertia of the actual rod referred to its cg will not be quite as large as the moment of inertia of the assumed equivalent.

If we denote by T the torque applied to the crankshaft by one of the cylinders at any given crank angle θ, we can write

$$T = T_t + T_L + T_p. \tag{8-32}$$

Here T_t is the torque reaction due to the change in momentum of the piston and the piston end of the connecting rod, with the mass distribution assumed previously; T_L is a correction term arising from the fact that the moment of inertia of the assumed rod is not equal to the moment of inertia of the 'actual rod. Finally, T_p is the torque at any crank angle due to gas pressure in the cylinder.

Torque Due to Inertia of the Moving Parts. The instantaneous torque T_t may be found by noting that the change in kinetic energy of the piston must be equal to the work done by the crankshaft on the connecting rod; hence,

$$d\left(\frac{M_p}{2}\dot{S}^2\right) = -T_t \, d\theta, \tag{8-33}$$

$$\frac{d}{dt}\left(\frac{M_p}{2}\dot{S}^2\right) = M_p\dot{S}\ddot{S} = -T_t\frac{d\theta}{dt} = -\Omega T_t, \tag{8-34}$$

$$T_t = -M_p\ddot{S}\left(\frac{\dot{S}}{\Omega}\right). \tag{8-35}$$

The quantity $\dot{S}/\Omega$ is a function of the crank angle and may be found in the form of a Fourier series from Eq. 8–5. Since S itself is a Fourier series in terms of θ, the quantity T_t can be represented as the product of two such series. This may be reduced to a single series by the use of the identities

$$\sin n\theta = \frac{e^{in\theta} - e^{-in\theta}}{2i}, \tag{8-36}$$

$$\cos n\theta = \frac{e^{in\theta} + e^{-in\theta}}{2}, \tag{8-37}$$

where $i = \sqrt{-1}$ and n is the harmonic order.

By substituting these exponentials for the sines and cosines of the two series, combining terms and resubstituting sines and cosines for the exponentials it may be shown that

$$T_t = M_p\Omega^2 R^2[t_1(\sin\theta_1 + \sin\theta_2 + \sin\theta_3 + \cdots)$$

$$+ t_2(\sin 2\theta_1 + \sin 2\theta_2 + \sin 2\theta_3 + \cdots)$$

$$+ t_3(\sin 3\theta_1 + \sin 3\theta_2 + \sin 3\theta_3 + \cdots) + \cdots], \quad (8\text{--}38)$$

where $t_n = f(a_1, a_2, a_3 \cdots)$.

Since t_n is a function of the $a\dot{S}$ it must also be a function of R/L. Figure 8–17 is a plot of t_1, t_2, t_3 and t_4 vs. R/L. Other terms are too small to show on the plot.

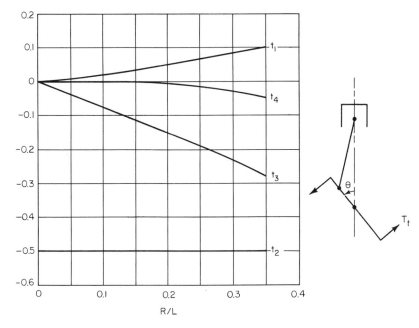

Fig. 8–17. First-, second-, third-, and fourth-order inertia-torque coefficients.

$$T_t = M_p\Omega^2 R^2 \sum_{n=1}^{\infty} t_n \sin n\theta = ZR \sum_{n=1}^{\infty} t_n \sin n\theta$$

T_t = torque applied to crankshaft due to acceleration of reciprocating parts
$-T_t$ = torque reaction on engine due to acceleration of reciprocating parts

Note that θ_1, θ_2, etc., in the above expression are the angles measured from the top-center position of each cylinder under consideration. The torque applied by the inertia of the reciprocating parts to the shaft depends only upon θ for the individual cylinders and is independent of the angle between cylinder axes.

It will be observed from Fig. 8–17 that the second-order inertia torque is large and not appreciably affected by R/L, in fact, the second order torque is strongly present even when $R/L = 0$, that is, when the piston motion is harmonic.

Torque on the Crankshaft Due to Angular Acceleration of the Connecting Rod. We have already considered all the forces (and torques) acting on a connecting rod which is equivalent to two masses concentrated at its ends. The motion of the center of gravity of such a fictitious rod is the same as that of the actual rod, and the forces acting are therefore the same. However, the moment of inertia I'_L of the fictitious connecting rod is not in general equal to that of the actual rod.

The moment of inertia of the fictitious connecting rod (see Fig. 8–3) is

$$I'_L = (W_1 j^2 + W_2 h^2)/g. \tag{8–39}$$

From the definition of the center of gravity, $W_1 j = W_2 h$, $W_1 j^2 = W_2 hj$, and $W_2 h^2 = W_1 hj$. Thus,

$$I'_L = (W_1 + W_2)hj/g = hjM_L, \tag{8–40}$$

where M_L is the mass of the connecting rod.

If the moment of inertia of the actual connecting rod is I_L, the difference between the torque acting on the actual rod and that acting on the substitute rod is, by Newton's law,

$$(I_L - hjM_L)\ddot{\phi}. \tag{8–41}$$

Let us now consider the torque that must act on the connecting rod in order to impart angular acceleration to it. With reference to Fig. 8–18, the torque acting on the connecting rod is $fL \cos \phi$, and, by Newton's law,

$$I_L \ddot{\phi} = fL \cos \phi \quad \text{or,}$$

$$f = \frac{I_L \ddot{\phi}}{L \cos \phi}. \tag{8–42}$$

The torque acting on the crankshaft will be

$$fR \cos \theta = \frac{I_L \ddot{\phi} R \cos \theta}{L \cos \phi}. \tag{8–43}$$

By a process similar to that used to find the series representing the acceleration of the piston, we find that the torque acting on the crankshaft to accelerate the connecting rod is

$$I_L \Omega^2 (s_2 \sin 2\theta + s_4 \sin 4\theta + \cdots), \tag{8–44}$$

where $s_2 = + \left[\frac{1}{2}\left(\frac{R}{L}\right)^2 - \frac{1}{32}\left(\frac{R}{L}\right)^6 + \cdots \right]$

$s_4 = - \left[\frac{1}{4}\left(\frac{R}{L}\right)^4 + \frac{1}{8}\left(\frac{R}{L}\right)^6 + \cdots \right].$

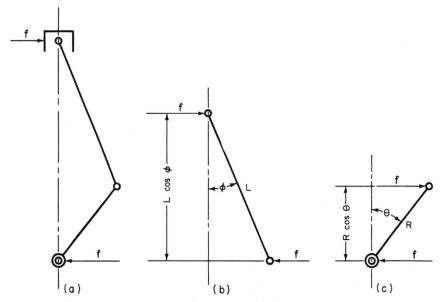

Fig. 8–18. Inertia torques on rod and frame. (*a*) Forces that engine frame applies to moving parts in order to produce angular acceleration of the connecting rod; (*b*) forces on the connecting rod; (*c*) forces on the crank.

A plot of s_2 vs. R/L is shown in Fig. 8–19. The quantity s_4 is too small to be of any importance. However, we have already accounted for a torque of

$$hjM_L \Omega^2(s_2 \sin 2\theta + s_4 \sin 4\theta + \cdots)$$

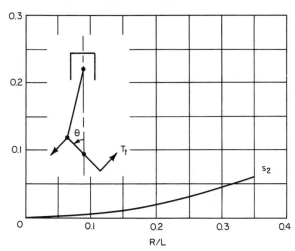

Fig. 8–19. Coefficient for correction of inertia torque on crankshaft for true moment of inertia of connecting rod. Torque on crankshaft due to correction to moment of inertia of connecting rod is $T_L = (I_L - hjM_L)\Omega^2 \sum s_{2n} \sin 2n\theta$.

from forces acting on the fictitious rod. We must therefore apply only the correction torque T_L, equal to the difference of these quantities

$$T_L = (I_L - hjM_L)\Omega^2(s_2 \sin 2\theta + s_4 \sin 4\theta + \cdots). \qquad (8\text{-}45)$$

The torque on the crankshaft due to the inertia of the moving parts is then

$$T_t + T_L = M_p\Omega^2 R^2 \sum_1^\infty t_n \sin n\theta + (I_L - hjM_L)\Omega^2 \sum_1^\infty s_{2n} \sin 2n\theta. \quad (8\text{-}46)$$

Unbalanced Torque Acting on the Engine Frame. It might be assumed that the torque reaction of the parts on the engine frame would be equal and opposite to the torque applied by these parts to the crankshaft. This assumption would be allowable were it not for the fact that the connecting rod, alone of all engine parts, changes in angular momentum as the crank turns. Thus we may write

$$T' = -T_t - T_p + T'_L, \qquad (8\text{-}47)$$

where T_t and T_p are, respectively, the torque acting on the crankshaft due to inertia of the reciprocating parts and the torque on the crankshaft due to gas pressure; T'_L is the correction to torque acting on the engine frame due to the angular motion of the connecting rod. It may be seen from Fig. 8–18 that this torque *is not equal and opposite* to the correction torque T_L that acts on the crankshaft:

$$T'_L = (I_L - hjM_L)\ddot{\phi}\left(\frac{R \cos \theta}{L \cos \phi} + 1\right). \qquad (8\text{-}48)$$

The quantity T'_L may be reduced to a trigonometric series

$$T'_L = (I_L - hjM_L)\Omega^2 \sum_{n=1}^\infty s'_{2n} \sin 2n\theta. \qquad (8\text{-}49)$$

The values of the coefficients s' of this series are plotted in Fig. 8–20.

Torque on the Crankshaft Due to Gas Pressure. The instantaneous torque T_p acting upon the crank due to gas pressure may be found by noting that, neglecting friction, the work done on piston is equal to the work done on crankshaft:

$$p\,dV = T_p\,d\theta,$$

where p is the pressure on the piston, and V the cylinder volume. Since $dV = A_p\,dS$, where A_p is the area of the piston,

$$T_p = pA_p \frac{dS}{d\theta}, \qquad (8\text{-}50)$$

but

$$dS/d\theta = \frac{dS/dt}{d\theta/dt} = \dot{S}/\Omega.$$

Therefore

$$T_p = pA_p \dot{S}/\Omega.$$

The value of $\dot{S}/\Omega$, as given by Eq. 8–5, is

$$\dot{S}/\Omega = -R(\sin\theta + 2a_2 \sin 2\theta + 4a_4 \sin 4\theta + \cdots). \qquad (8\text{–}51)$$

The value of p must be found from an indicator diagram. Multiplication of the instantaneous pressure, the piston area, and the ratio of piston

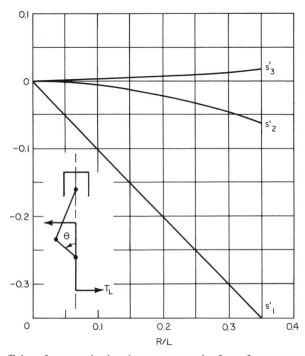

Fig. 8–20. Coefficients for correcting inertia torque on engine frame for true moment of inertia of connecting rod. Torque reaction on engine frame due to corrected moment of inertia of the connecting rod is $T_L' = (I_L - hjM_L)\Omega^2 \sum\limits_{n=1}^{\infty} s_{2n}' \sin 2n\theta.$

velocity to crank velocity gives T_p. A typical plot of pressure versus crank angle and the resultant torque-crank-angle diagram are shown in Fig. 8–21 for a 4-cycle engine.

The instantaneous torque for each cylinder may be analyzed and expressed as a Fourier series as follows:

$$\frac{T_p}{T_m} = 1 + U_{\frac{1}{2}} \sin \tfrac{1}{2}\theta + U_1 \sin\theta + U_{1\frac{1}{2}} \sin 1\tfrac{1}{2}\theta + U_2 \cdots$$

$$+ V_{\frac{1}{2}} \cos \tfrac{1}{2}\theta + V_1 \cos\theta + V_{1\frac{1}{2}} \cos 1\tfrac{1}{2}\theta + \cdots, \qquad (8\text{–}52)$$

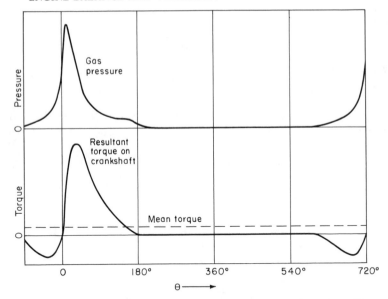

Fig. 8–21. Typical curve of gas pressure versus crank angle for a 4-cycle engine, with resultant torque on crankshaft.

where θ is the crank angle measured from firing top center of any one of the cylinders, and T_m the mean torque of the engine. For a 4-cycle engine,

$$T_m(4\pi) = \text{work done in two revolutions}$$
$$= \text{mep} \times A_p \times 2R$$
$$= \text{mep} \times V_d$$
$$T_m = (\text{mep} \times V_d)/4\pi,$$

where A_p is the piston area and V_d the piston displacement.

Values of the coefficients U and V for a 4-cycle spark-ignition engine are plotted in Fig. 8–22. In this case the indicated mean effective pressure was 155, the ratio R/L was 0.291 and the compression ratio was 6.

From Fig. 8–23 it is seen that $U_n \sin n\theta + V_n \cos n\theta$ may be expressed as

$$W_n \sin(n\theta + \phi_n),$$

where $W_n = \sqrt{U_n^2 + V_n^2}$

$$\phi_n = \tan^{-1} \frac{V_n}{U_n};$$

therefore,

$$\frac{T_p}{T_m} = 1 + \sum_{n=\frac{1}{4}, 1, 1\frac{1}{4}, \text{etc.}}^{n=\infty} W_n \sin(n\theta + \phi_n). \qquad (8\text{--}53)$$

Values of W_n are given in Fig. 8–24, and values of ϕ_n in Fig. 8–25.

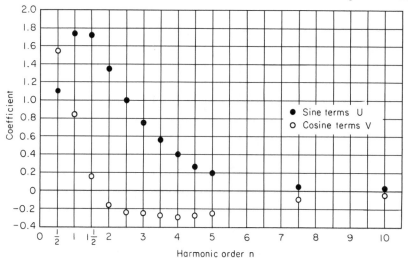

Fig. 8–22. Harmonic coefficients of gas-pressure torque for a single-cylinder 4-cycle engine, compression ratio 6.0 (Taylor and Morris, 8.20).

$$T_p/T_m = 1 + U_{\frac{1}{2}} \sin \frac{\theta}{2} + U_1 \sin \theta + U_{1\frac{1}{2}} \sin 1\tfrac{1}{2}\theta + U_2 \sin 2\theta + \cdots$$

$$+ V_{\frac{1}{2}} \cos \frac{\theta}{2} + V_1 \cos \theta + V_{1\frac{1}{2}} \cos 1\tfrac{1}{2}\theta + V_2 \cos 2\theta + \cdots,$$

where T_p is the instantaneous torque on crankshaft and T_m the mean torque on crankshaft.

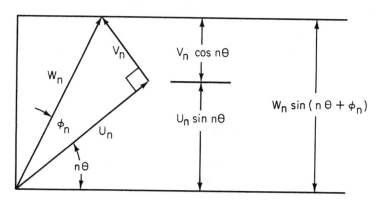

Fig. 8–23. Geometric proof that $U_n \sin n\theta + V_n \cos n\theta = W_n \sin (n\theta + \phi_n)$.

Effect of Engine Variables on Torque Coefficients. The coefficients given in Figs. 8–22, 8–24, and 8–25 were taken from indicator diagrams from a 4-cycle CFR engine, bore = 3.25 in. (83 mm), stroke = 4.5 in. (114 mm), $r = 6.0$ (ref. 8.20). It was found that these coefficients were little changed by the variables indicated in Fig. 8–24, i.e., spark timing, higher compression ratio, lowered fuel-air ratio (from $F_R = 1.2$). From these observations it

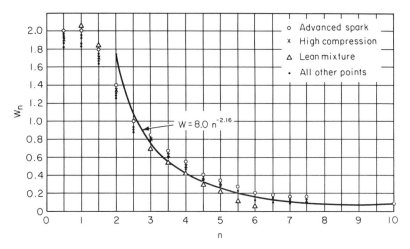

Fig. 8–24. W_n vs. n corresponding to Fig. 8–22.

$$\frac{T_p}{T_m} = 1 + \sum_{n=\frac{1}{2},1,1\frac{1}{2}\cdots}^{n=\infty} W_n \sin(n\theta + \phi_n),$$

where

$$W_n = (U_n^2 + V_n^2)^{\frac{1}{2}}$$
$$\phi_n = \tan^{-1}(V_n/U_n).$$

may be concluded that the coefficients shown are reasonably representative for 4-cycle spark-ignition engines. The coefficients are taken from an average indicator diagram in each case and do not take into account the fact that there is considerable cycle-to-cycle variation in the p-θ diagram, especially in the case of spark-ignition engines (see Chapter 1). This cycle-to-cycle variation will to some extent modify the torque coefficients shown and will introduce new coefficients which are multiples of the lowest frequency of the variability. Thus, an exact prediction of engine-torque coefficients is not possible.

Two-cycle engines with no cycle-to-cycle variation would obviously have no half-order coefficients. This approximation applies pretty well to 2-cycle Diesel engines, in which the cyclic variation is usually small, as is also the case with 4-cycle Diesel engines.

Torque coefficients for various types of engines under various simplifying assumptions are available in the literature (8.20–8.24).

Even though exact prediction of engine-torque coefficients is impossible, the coefficients based on cyclic uniformity are generally the most important ones, and analysis based on such data is valuable, as will appear in the subsequent discussion, which is based on the assumption of cyclic uniformity unless otherwise indicated.

Summation of Torques Acting on the Crankshaft in Multicylinder Engines. From Eq. 8–52 or 8–53 it is evident that if all the coefficients, U, V, and W, are zero, the torque on the crankshaft is equal to the mean

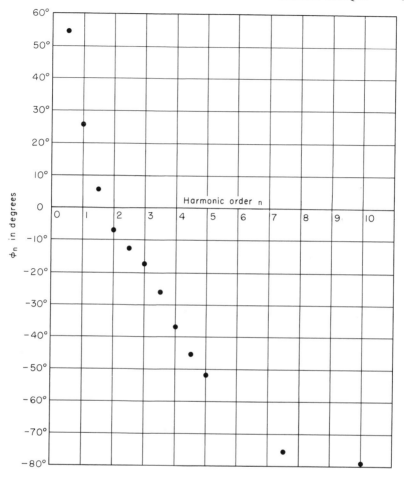

Fig. 8–25. Values of ϕ_n for Fig. 8–24.

torque and is a constant quantity. This condition is desirable in the interest of smooth engine operation and is approached as the coefficients (U and V, or W) cancel out with increasing numbers of cylinders. Since the value of the coefficients is small above the eighth order (see Fig. 8–24), it is particularly desirable that the orders below eight cancel in multicylinder engines.

The process of summing torque in a multicylinder engine is simple, since all torques act about the same axis and are applied to the same body and thus add directly.

In the expressions for gas torque, the angle α between the first-order vectors is equal to the number of crank degrees between successive arrivals at *firing* top center. Two examples of torque-vector addition are shown in Fig. 8–26.

These examples illustrate the general principle that, with evenly spaced firing, the torque vectors for 4-cycle engines add when n is a multiple of $N_c/2$ and cancel for other values of n. In the case of 2-cycle engines the

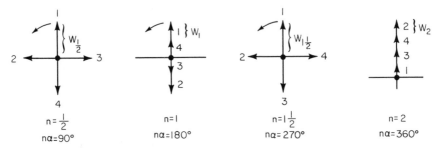

Fourth-order torque component is four times that of one cylinder.

(a)   4-cylinder 4-cycle engine $\alpha = 720/4 = 180°$

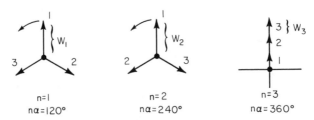

Third-order torque is three times that of one cylinder.

(b)   3-cylinder 2-cycle engine $\alpha = 360/3 = 120°$
(no half-order terms)

Fig. 8-26. Vector diagrams of torque output for two engines with cyclic uniformity and equal firing intervals; α is the angle between torque vectors and is equal to $720/N_c$ degrees for 4-cycle engines when N_c is an even number. For all other engines $\alpha = 360/N_c$ degrees.

vectors add when n is a multiple of N_c and cancel for other values. Table 8-1 summarizes these relations for various numbers of cylinders with even firing. Note that the arrangement of cylinders does not make any difference so long as the firing intervals are equal.

It should be remembered that the preceding discussion assumed identical indicator diagrams for all cylinders of a given engine. In practice, of course, the diagrams vary to an appreciable extent between strokes of a given cylinder and also between cylinders during a given revolution of the crankshaft. On account of this nonuniformity, all engines will show some degree of torque at all harmonics of the cyclic variation as well as of the crank speed. However, since only small differences are involved, the magnitude of the torque variations in orders that theoretically cancel will usually be small compared with those that theoretically add in the lower (<8) orders.

Table 8–1. Orders of Gas-Pressure Torque Which Appear in Multicylinder Engines with Even Firing Intervals and Uniform Pressure-Crank-Angle Diagrams

Number of Cylinders	Cycle	Orders Appearing, n
1	2	1, 2, 3, 4, etc.
1	4	$\frac{1}{2}$, 1, $1\frac{1}{2}$, 2, $2\frac{1}{2}$, etc.
2	2	2, 4, 6, 8, etc.
2	4	1, 2, 3, 4, etc.
3	2	3, 6, 9, etc.
3	4	$1\frac{1}{2}$, 3, $4\frac{1}{2}$, 6, $7\frac{1}{2}$, etc.
4	2	4, 8, 12, etc.
4	4	2, 4, 6, 8, etc.
5	2	5, 10, 15, etc.
5	4	$2\frac{1}{2}$, 5, $7\frac{1}{2}$, etc.
6	2	6, 12, 18, etc.
6	4	3, 6, 9, 12, etc.
etc.		

The coefficient of each torque order which appears is, in each case, n times the value for one cylinder.

ENGINES WITH NONUNIFORM FIRING

In practice, a good many engines are made with nonuniform firing intervals. The reason for such an arrangement may be for mechanical simplicity (for example, to reduce the number of cranks), for engine compactness (for example, small V angle) or to avoid a serious vibration resonance at a critical frequency (see examples in Table 8–3). For such cases, all normal orders are present in the torque diagram, even assuming cyclic uniformity. The magnitude and phase of each order can be determined by the methods previously described.

ENGINES WITH ARTICULATED CONNECTING RODS

Figure 8–27 shows the geometry of a master-articulated connecting-rod system. When the hinge-pin angle γ is equal to the angle between the cylinder axes β, the articulated piston has the same length of stroke as that

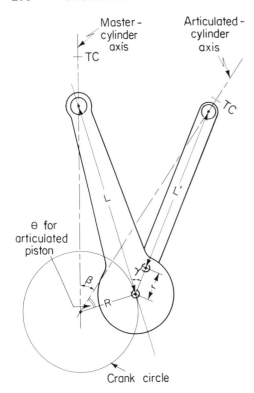

Fig. 8–27. Geometry of the master-articulated connecting-rod system.

of the master-rod piston. This arrangement is generally used in radial engines.

Analysis of the articulated rod motion is a difficult task, for which space is not here available. Tanaka (ref. 8.10) gives a particularly complete analysis of the articulated system as applied to radial engines and includes an excellent analysis of articulated-rod effects on balance and torque. A comparison with V-type engines is included.

When $\gamma = \beta$, the acceleration of the articulated piston may be expressed as follows:

$$a = -\Omega^2[(-R\cos\theta + 4A_2\cos 2\theta + 16A_4\cos 4\theta\cdots)$$
$$+ (4B_2\sin 2\theta + 16B_4\sin 4\theta\cdots)], \qquad (8\text{--}54)$$

where each A and B is equal to R multiplied by a function of R/L, R/L', r/L', and the angles β and γ. It will be noted that the $\cos\theta$ term is the same as for the plain or master connecting rod. As in the case of the plain rod, only even harmonics appear after the first, but the appearance of sine terms indicates that the acceleration is not exactly in phase with that of the master piston. This difference involves a difference in the arrival of the piston at top center. For $R/L = 0.25$, $R/L' = 0.308$, $r/L' = 0.21$, $\gamma = \beta$, (reasonably typical values) the crank positions for top and bottom centers are:

β	0°	45°	60°	90°	120°	150°	180°
degrees atc	0	0.53	1.09	2.66	3.23	1.99	0
degrees bbc	0	0.8	1.73	4.35	6.36	6.88	0

(atc and bbc refer to the crank position of the cylinder in question).

This table (from ref. 8.10) is for the case where crank rotation is from the master cylinder toward the articulated cylinder. In general, the departures shown have no effect on performance with best valve and ignition timing for the master cylinder.

For the case cited, the values of A/R and B/R are as follows:

β	0	45°	60°	90°	120°	150°	180°
A_1/R	-1	-1	-1	-1	-1	-1	-1
A_2/R	-0.625	-0.565	-0.585	-0.631	-0.853	-1.11	-1.225
A_4/R	0.00003	0.00003	0.00006	0.00006	0.00009	0.00009	0.00009
B_2/R	0	-0.0021	-0.0041	-0.0098	-0.0127	-0.0092	0
B_4/R	0	←———————— 0.00003 ————————→					0

It is evident that the first-order term is the same as for the master rod ($\beta = 0$) and that only the second-order A and B terms are large enough to be important.

Effect of Articulated Rods on Engine Balance. From the foregoing, the following general characteristics of the articulated-rod system are evident:

1. Only the first- and second-order terms are important, as with plain-rod arrangements.

2. The second-order articulated-rod forces are not quite in phase with the crank.

3. The master rods behave qualitatively like plain rods.

4. The connecting-rod masses and moments of inertia will, in general, be different for both master and articulated rods as compared with the corresponding plain-rod assembly.

V Engines with Articulated Rods. The V-8 engine with master and articulated rods will require first-order counterweights different from those of the plain-rod engine, and will have a small second-order unbalance because the second-order forces of the articulated rods are not quite in phase with those of the master rods. It will be remembered that in this engine the second-order forces balanced out with plain connecting rods.

In the case of V engines in which each bank is balanced for certain orders with plain rods, the same will be true with the master-rod system. Also, the orders that add will be the same, since the same crank angles are involved. However, with articulated rods the orders that add will be slightly different in value and have a phase relation slightly different from those in the case of plain rods. In general, the differences in balance between plain-rod and articulated-rod V engines of otherwise similar design are too small to be of practical importance.

Radial Engines with Articulated Rods. Complete analysis of radial-engine balance with the master-rod system is too lengthy for consideration here, but is well covered in ref. 8–10. Since there is only one master rod in each row (ring) of radial cylinders, and since this rod is much heavier than the articulated rods, the single-row radial engine might be expected to have balance characteristics similar to those of a single-cylinder engine. Fortunately, however, the first-order forces have a constant radial component and can be balanced by a counterweight opposite the crank. The value of this weight is very nearly*

$$M_{cw} = \frac{R}{R_{cw}} \left(M_R + \frac{\Sigma M_p}{2} \right). \tag{8–55}$$

The rotating and reciprocating masses are computed with the assumption that the connecting-rod masses are concentrated at their ends, as was assumed for plain connecting rods earlier in this chapter. With the first-order term balanced, the only serious unbalance in a conventional single-row radial engine is a second-order force, which can be represented by a constant vector leading the crank by $180°$ at top center of the master-rod cylinder and revolving at twice crank speed in the direction of crank rotation (see ref. 8.10, pp. 254–256). In large radial engines this force is often balanced by fore-and-aft weights revolving about the crankshaft axis at twice engine speed.

Moments in Engines with Articulated Rods. The moments due to the unbalanced forces in articulated-rod systems can be handled in the way that is used with plain rods. Again only the first and second orders are usually important. The second-order moments will have small sine components (see Eq. 8–54) and will therefore be somewhat out of phase with 2θ.

Inertia Torque from Articulated-Rod Systems. An analysis for one pair of cylinders will be found in ref. 8.10. In general it is a Fourier series containing sines and cosines (except $\cos 2\theta$) for all orders of θ. Orders above 2 are

---

* Reference 8.16 gives a small correction term to be added in Eq. 8–55.

small, and for ordinary designs the over-all result is not greatly different from that of a plain-rod system of equal total mass.

Gas-Pressure Torque with Articulated-Rod Systems. The torque from a master cylinder is obviously that of a plain-rod system. The articulated rod has a lever arm around the crankpin which varies with crank angle and results in a somewhat different torque-versus-crank-angle curve. As with plain rods, all half orders from the 1/2 to the nth are involved for 4-stroke engines, and all integer orders with 2-stroke engines. In multicylinder arrangements, orders cancel and add as in the case of plain rods, but with slightly different values at most crank angles.

The differences in the gas torque-versus-crank-angle curves between articulated and plain rods have very little practical importance. However, the torque on the master rod produced by the articulated rods is very important from the point of view of master-rod stresses, master-rod design, and side forces on the master-rod piston (see Chapter 11).

BALANCE OF TYPICAL ENGINES

Table 8–2 gives balance characteristics for 4-cycle engines with various cylinder arrangements; and Table 8–3 lists similar data for 2-cycle engines. Both tables assume evenly spaced firing, except as noted. Both tables will be found at the end of this chapter.

ENGINE VIBRATION

The subject of engine vibration can be divided into two categories: vibration of the engine parts relative to each other, called *internal vibration*, and movement of the engine as a whole, called here *external vibration*.

Internal Engine Vibration. Within the engine structure, the forces created by the inertia of the moving parts and by the varying gas pressure in the cylinders must result in deflections of the structural members of the engine, since these parts are elastic. Thus, vibrations of varying frequencies and amplitudes are set up throughout the engine structure.

In any machine in which the applied forces vary with time, there is always vibratory motion. In practice, this motion must be controlled so as to avoid malfunction, mechanical failure, or excessive noise. Experience shows that the problems in this category most often include crankshaft vibration, both torsional and bending, and vibrations in the valve-operating mechanism.

Crankshaft Torsional Vibration. This subject is treated exhaustively in refs. 8.400–8.694. The scope of this volume allows only to summarize the important general relationships involved. It is assumed that the reader is familiar with the general characteristics of vibrating systems.

In general, any engine speed in which torsional vibration is markedly severe is called a *critical* speed. The order of this speed is the ratio of the observed vibratory frequency to the crankshaft rotational frequency.

The torsional vibration characteristics of a crankshaft are of course heavily influenced by the connected equipment. Reduction gears, drive shafts, generators, propellers, couplings, etc., all become part of the torsional system and must be considered in connection therewith (8.06). Particularly complex is a vibration system consisting of an engine with an air propeller mounted on its output shaft, since the propeller blades usually have important modes of vibration which tie in with crankshaft torsional vibration (see ref. 8.492).

In predicting critical speeds and other aspects of torsional vibration, the crankshaft-rod-piston system is usually considered to be equivalent to a shaft carrying a disc at each crank position: so chosen that its moment of inertia is equal to the moment of inertia of the rotating mass at that crank plus R^2 times a mass equal to half the total reciprocating mass. The rotating and reciprocating masses are considered to include the concentrated connecting-rod masses indicated in Fig. 8–3. The remaining system is simplified in a similar manner by assuming a system of shafts carrying discs whose moments of inertia are equivalent to those of the flywheel, couplings, gears, propellers, etc., which form part of the torsional system. The sizes and lengths of the shafts between the various inertia elements are chosen so as to be equivalent in stiffness to the actual shafts, as nearly as that can be estimated. Methods of estimating crankshaft stiffness are given in refs. 8.400–8.490 (see especially ref. 8.400).

In general a system consisting of inertia elements connected by shafts has as many modes of vibration, or " degrees of freedom," as there are inertia elements, and the equations representing vibratory motion will be in degree (powers) equal to the number of such elements. Solutions for the higher-degree equations are generally found by trial, using some such method as that proposed by Holzer (8.400, 8.481, 8.488).

In multicrank engines the problem of torsional-vibration analysis is complicated by the fact that the torsional impulses originate at different crank positions at different times. Thus, even if the indicator diagrams were exactly the same in all cylinders and the firing impulses exactly evenly spaced in time, no torsional order would completely cancel out because of angular deflection of the crankshaft. In the actual case, where indicator diagrams are never exactly alike, it is apparent that all orders of the Fourier series representing engine torque may cause some torsional vibration.

In many practical cases, engines are so connected to the load that their crankshafts behave essentially as isolated vibration systems. A familiar example is the conventional automobile engine, where the external system (consisting of gear box, propeller shaft, rear axle, and wheels) has great torsional flexibility compared with that of the crankshaft and is excited in torsional vibration at very low engine speeds only. The use of a hydraulic clutch or a torque converter between load and flywheel is also an effective way of isolating the engine's torsional system. In such cases, crankshaft torsional vibration is essentially an internal engine problem, and critical speeds can be predicted with a good degree of approximation (8.400–8.490). Calculation of torsional amplitudes is more difficult because damping coefficients are not easy to predict. Some work on this problem (8.50–8.52) indicates that a typical figure for the engine torsional system is 0.02 times critical damping.

Modes of Crankshaft Vibration. Let us consider an isolated torsional system consisting of a crankshaft with a flywheel at one end that has a relatively large moment of inertia. The first mode of vibration will be with a node at the flywheel and an antinode at the free end. The higher modes will have a node and antinode at these positions, plus intermediate nodes from 2 to n integer numbers. It is seldom that modes higher than the second are of practical importance, except in the case of in-line engines of more than 8 cylinders (8.400).

For exactly similar designs, frequency in any mode is proportional to $1/L$ (see Volume I, Chapter 11). Figure 8–28 gives data on the first-mode torsional frequency of crankshafts torsionally isolated from the external driven system and carrying a heavy flywheel at one end. In spite of large differences in detail design, this natural frequency shows the paramount influence of shaft length on the torsional frequency. This figure can be used for a rough prediction of first-order critical speed. For example, estimate the first-order critical speed for a 6-cylinder in-line 4-stroke engine of 6 in. bore. From Fig. 8–28, $L = 6/12$ (1.1) $6 = 3.3$ feet. The reciprocal of this is 0.3, and from the figure, $f = 140$ cps. Since the lowest large torque component is the third order, the lowest critical speed is estimated as follows:

$$\left(\frac{\text{rpm}}{60}\right)3 = 140, \text{ rpm} = 2800.$$

External-System Effects. An example where the external system has a major effect on crankshaft torsional behavior is that of marine engines with long propeller shafts. Where such a shaft is long compared with the crankshaft, the lowest critical speed is likely to be engine against propeller, with the major angular deflections in the long propeller shaft. However, when long propeller shafts are used with engines having 8 to 12 cylinders in line, as in many large Diesel-engined ships, all the important modes of vibration are likely to involve considerable crankshaft angular deflection. Such arrangements are likely to have one or more critical speeds in the operating range. These speeds must be passed through quickly in operating the power plant. Sometimes troublesome critical speeds are changed or suppressed by using unevenly spaced crank positions, with consequent

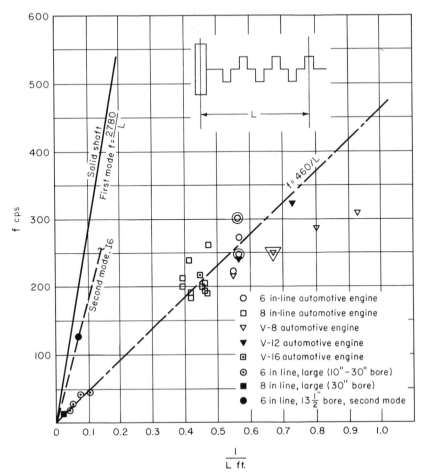

Fig. 8–28. First-mode torsional frequency of crankshafts with flywheel at one end. Where the actual value of L was not known, it was taken as $b(1.1N_c)$ for in-line engines and as $b(0.55\ N_c + 0.5)$ for V engines, where b is the bore in feet and N_c the number of cylinders.

acceptance of uneven firing intervals and larger-than-normal ratios of maximum to mean torque.

An example of an especially complex torsional system is illustrated by Fig. 8–29, which shows the spectrum of crankshaft torsional vibration for a 12-cylinder V aircraft engine operating an air propeller through a 1 : 2 reduction gear. With the assumptions of uniform cycles and fixed crank angles, the lowest expected order would be the sixth. This order does not even appear in the range of speeds investigated, probably because no element in the system had the corresponding natural frequency. It is notable that a change of only two degrees in propeller-blade angle has a large effect on the observed amplitudes and relative importance of the various

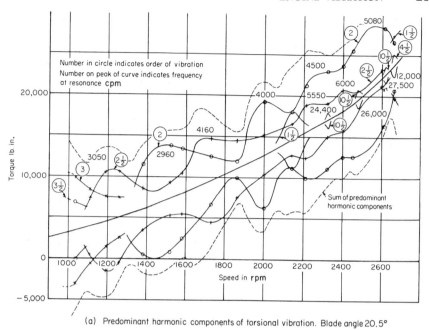

(a) Predominant harmonic components of torsional vibration. Blade angle 20.5°

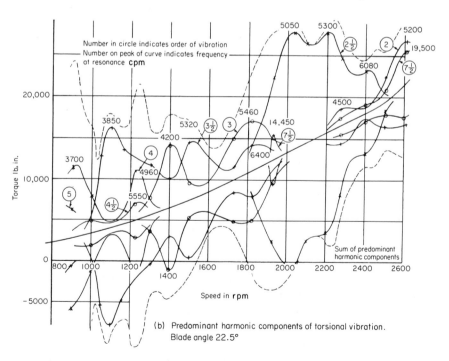

(b) Predominant harmonic components of torsional vibration.
Blade angle 22.5°

Fig. 8–29. Torque at drive end of crankshaft of Rolls-Royce *Merlin* II aircraft engine with variable-pitch propeller, 1-to-2 reduction gear (Carter and Forshaw, 8.492).

orders observed as well as on the stresses in the crankshaft itself. In this type of system, stresses in the propeller are likely to be a more serious problem than those in the crankshaft. For this reason the problem of designing a satisfactory air propeller involves exhaustive investigations of its vibration characteristics while mounted on each engine type with which it may be used.

Torsional Vibration Dampers and Absorbers. Vibration amplitudes and the resulting stresses in the torsional system may be modified and often greatly reduced by means of *dampers* or *absorbers*. In current practice these are of three types:

1. Frictional dampers
2. Tuned absorbers
3. Pendulous absorbers

The usual form of frictional damper, Fig. 8–30a, consists of a flywheel-type mass mounted on the crankshaft through a set of closely spaced discs arranged like the plates of a disc clutch, the space between discs being filled with a viscous fluid (8.692–8.694). Such a device increases the system damping. Like all dampers, it is most effective when located at the antinode of the particular mode whose amplitude it is designed to reduce, or at the free end of the crankshaft. This type of damper can function at any mode or frequency. Its effectiveness depends on its moment of inertia, plate area, and fluid viscosity. It is a true damper and must have sufficient heat capacity and heat-dissipation characteristics to avoid destructive over-heating.

A tuned absorber consists of an inertia element added to the crankshaft, as shown in Fig. 8–30b. Its rotation is restrained by springs in such a way that its undamped natural frequency is the same as the frequency at which damping is required. Its basic principle is illustrated by the linear system shown in Fig. 8–31, where the absorber is represented by the small mass M_2 attached by a spring to the large mass M_1 which represents the crankshaft. This system has two natural frequencies, one higher and one lower than the natural frequency of the original system.

The equation given in the figure caption shows that the amplitude of the large mass will be zero when the natural frequency of the absorber, $\omega_2 = (K_2/M_2)^{\frac{1}{2}}$, is equal to the frequency ω of the applied force. At this point the absorber will be 180° out of phase with the applied force and will therefore act to nullify the amplitude which M_1 would otherwise have at this particular applied frequency. For given values of K_2 and M_2, it is evident that this type of absorber is effective at one frequency only. Applied to a crankshaft, as shown in Fig. 8–30b, the system must be analyzed in terms of torsional stiffnesses and torsional moments of inertia. The natural frequency of the absorber mass is $(K_{a2}/I_2)^{\frac{1}{2}}$, where I_2 is the mass moment of inertia of the absorber around its pivot point and K_{a2} the spring stiffness in terms of torque per unit angle. As with any damper or

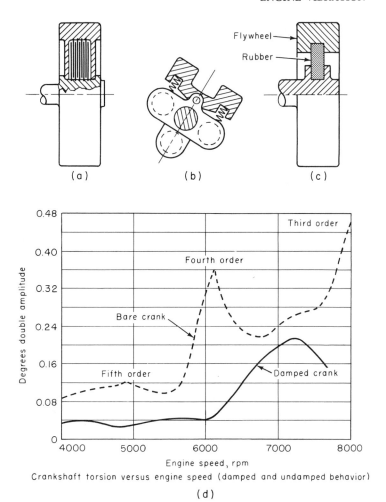

(a)   (b)   (c)

Crankshaft torsion versus engine speed (damped and undamped behavior)

(d)

Fig. 8–30. Crankshaft vibration dampers and absorbers: (*a*) Viscous damper (space between plates is filled with viscous fluid); (*b*) tuned damper; (*c*) rubber-mounted damper-absorber; (*d*) effect of a rubber-mounted damper-absorber in Ford V-8 engine (Scussel, 10,803).

absorber, such a unit should be located as near the antinode as practicable. Also, it must have sufficient mass so that it does not "bottom" (hit the stops) during normal operation. Since its motion must furnish a force equal and opposite to the applied force when $\omega_2 = \omega$, it follows from the equation under Fig. 8–31 that when $\omega_2 = \omega$,

$$F \cos \omega t - I_2 \frac{d^2}{dt^2} (\alpha_2 \cos \omega t) = 0, \qquad (8\text{–}56)$$

or

$$\alpha_2(\text{maximum}) = F/\omega^2 I_2. \qquad (8\text{–}57)$$

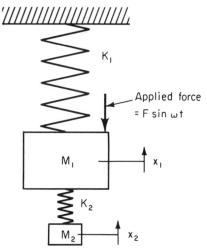

Fig. 8–31. Principle of the tuned vibration absorber.

$$\frac{X_1}{X_{st}} = \frac{1 - (\omega/\omega_2)^2}{\left[1 - \left(\dfrac{\omega}{\omega_2}\right)^2\right]\left[1 + \dfrac{K_2}{K_1} - \left(\dfrac{\omega}{\omega_1}\right)^2\right] - \dfrac{K_2}{K_1}};$$

K = spring-stiffness coefficient $|F/L|$; t = time; $X_{st} = F/K_1$; ω = applied frequency; $\omega_1 = (K_1/M_1)^{\frac{1}{2}}$; $\omega_2 = (K_2/M_2)^{\frac{1}{2}}$. When $\omega = \omega_2$, the amplitude X_1 is zero.

Thus, I_2 must be large enough so that the amplitude does not exceed the space allowed for the absorber's motion.

Most automotive-size engines now use rubber-mounted absorbers, Fig. 8–30c, mounted on the antiflywheel end of the crankshaft. This type acts as a combination frictional damper and tuned absorber (8.69, 8.691, 10.803). While it will be most effective when the applied frequency is equal to the natural frequency of the absorber itself, it is also effective in reducing amplitude at other frequencies because of internal friction of the rubber. Figure 8–30d shows the effect of such a device on a V-8 automobile engine. Apparently the natural frequency of the absorber is tuned to the fourth order. The damping effects at other orders seem to be considerable.

Pendulous Absorber. This type of absorber, although earlier patented in France, was independently conceived and put into practice by E. S. Taylor in 1935 in order to overcome serious torsional vibrations encountered in geared radial aircraft-engine-propeller systems (8.60).

A pendulous absorber is essentially a tuned absorber whose natural frequency varies in direct proportion to its rotational speed. For small displacements a pendulum in a rotating field (see Fig. 8.32a) has a natural frequency expressed by

$$\frac{\omega}{\Omega} = \sqrt{\frac{r}{L}},$$

where ω is the pendulum frequency, Ω the field rotative frequency, r the distance from the pendulum pivot to the center of field rotation, and L the pendulum length. It is evident that the ratio of pendulum natural frequency to rotational frequency is fixed by the ratio r/L and is independent of the rotative frequency Ω. This relation is very convenient for use in variable-speed machinery, because the absorber can be tuned to any

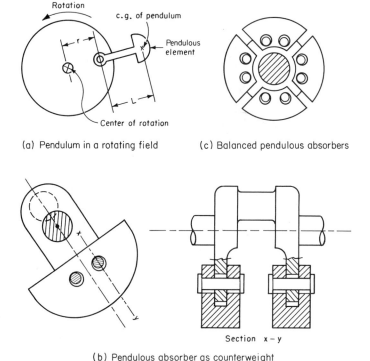

(a) Pendulum in a rotating field

(c) Balanced pendulous absorbers

(b) Pendulous absorber as counterweight

Fig. 8–32. Pendulous vibration absorbers.

order of torsional vibration. For example, if it is desired to suppress the 4.5 order of a 9-cylinder engine, ω/Ω is 4.5 and $r/L = (4.5)^2 = 20.25$. Thus, the pendulum length L must be $1/20.25$ of the radius of rotation of the pivot. Except perhaps in very large engines, this is an inconveniently short length for a conventional pendulum. This problem was very neatly solved by Sarazen in France and Roland Chilton in the U.S.A. by the arrangement shown in Fig. 8–32b, where the counterweights are mounted to serve also as pendulous absorbers (8.63). The effective radius of the pendulum is here the difference between the hole radius and the pin radius and can be made as small as necessary. Where counterweights are not used, balanced absorbers, similar in principle to that shown in Fig. 8–32c, may be mounted on the crankshaft (8.61, 8.67, and 8.68).

The two masses shown in Fig. 8–32*b* can be tuned to different orders if desired, as is also the case with the different elements shown in Fig. 8–32*c*.

As in the case of the simple tuned absorber, the pendulous absorber must have sufficient mass to keep the amplitude small. When large counterweights are used, this problem is usually automatically solved.

The result of adding a 4.5-order pendulous absorber to a 9-cylinder radial engine is shown in Fig. 8–33. Without the absorber, a large 4.5-order

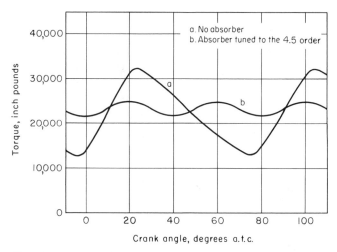

Fig. 8–33. Effect of a pendulous absorber on the torque on the crankshaft of a 9-cylinder radial aircraft engine (Taylor, 8.60).

component (80° of crank angle between peaks) is apparent. Addition of the absorber suppresses this component completely, leaving only the 9th order as an important one. This could be suppressed by tuning one of the counterweights to the 9th order.

Internal Engine Vibrations Other Than Crankshaft Torsional. It is fully apparent to anyone listening to an internal-combustion engine in operation that the engine structure vibrates in an indefinite number of modes and frequencies. The higher frequencies generate noise, which will be discussed briefly later in this chapter. In general the vibrations that cause ordinary noise do not generate objectionable stresses, because of their small amplitude.

Vibrations that result in excessive stress, and even failure, have been found in a great many components of engine structures. In the author's experience these have included the following:

1. *Crankshaft bending.* A dangerous example is where heavy counterweights engage in a "tuning-fork" mode. This type is especially prevalent in radial engines (8.494). Another type of vibration which may lead to

bending failure involves flywheel wobble. On the other hand, many crank-shaft failures in bending result from bearing misalignment or excessive wear or failure of main bearings.

2. *Torsional vibration of auxiliary-driven systems, such as superchargers, water pumps, generators, etc.* Gear-driven supercharger systems are especially prone to this kind of trouble. Various types of flexible couplings have been used to reduce critical speeds below operating speeds (Volume I, refs. 19.21, 19.22).

3. *Rotational vibration of pistons.* This type of vibration can occur with connecting rods that are too flexible in torsion. The cure is to use torsionally stiffer connecting rods.

4. *Diaphragm-type vibration of flat, thin sections of crankcases, gear cases, etc.* The best way to overcome this is to use curved or dished shapes.

5. *Gear-tooth-excited vibrations.* With accurate geometry, this kind of vibration is more apt to be noisy than destructive. Modern methods of gear manufacture and quality control have made destructive vibration from this source infrequent.

6. *Cantilever-beam type of vibration of auxiliaries such as carburetors, generators, etc.* The cure is, obviously, stiffer mounting.

7. *Vibration of pipes and tubes, including fuel lines, exhaust pipes, etc.* Proper design of mountings and supports can control this very common type of vibration.

8. *High-frequency vibration caused by combustion, detonation, gear teeth, etc.* This is classed as noise and will be discussed later in this chapter.

EXTERNAL ENGINE VIBRATION AND VIBRATION ISOLATION

By external vibration is meant vibration of the engine structure as a whole. It is caused by unbalanced forces, moments, and inertia torques, and by variations in the output torque transmitted to the crankshaft. Space is not available here to discuss the complete theory and practice of flexible engine mounts. Only a general outline of the principles involved will be attempted. For a fuller treatment of the subject see refs. 8.70–8.893.

It will be of assistance in this brief discussion to have reference to the familiar curves of amplitude and transmitted force versus applied frequency for a system with one degree of freedom (Figs. 8–34 and 8–35). Such curves apply to either torsional or linear vibration in any particular mode, provided the system conforms to Hooke's law, and provided the damping force is proportional to velocity.

The mass M in Fig. 8–34 may be taken as representing, in one of its degrees of freedom, an engine attached by a suspension system, consisting of a spring and damper, to a foundation whose mass is so large compared with that of the engine that it can be considered as fixed in space. While it is never completely realized in practice, the concept of a fixed foundation is a useful one and is approximated in many actual cases.

For torsional systems, substitute I (mass moment of inertia) for M, and K_a (torque/deflection angle) for K in Figs. 8–34 and 8–35.

If the suspension system is rigid, a case represented by $\beta = 0$ in Figs. 8–34 and 8–35, the inertia forces, moments, and torques, together with the

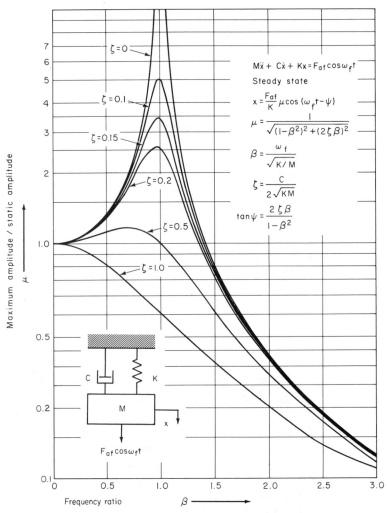

$$M\ddot{x} + C\dot{x} + Kx = F_{af}\cos\omega_f t$$

Steady state

$$x = \frac{F_{af}}{K}\mu\cos(\omega_f t - \psi)$$

$$\mu = \frac{1}{\sqrt{(1-\beta^2)^2 + (2\zeta\beta)^2}}$$

$$\beta = \frac{\omega_f}{\sqrt{K/M}}$$

$$\zeta = \frac{C}{2\sqrt{KM}}$$

$$\tan\psi = \frac{2\zeta\beta}{1-\beta^2}$$

Fig. 8–34. Steady-state amplitude versus frequency ratio, single degree of freedom; vibrating system with damping.

$M = \text{mass}/g_0$
$K = \text{spring rate, force/deflection}$
$C = \text{damping coefficient, critical value} = 2\sqrt{KM}$

torque on the crankshaft, are transmitted in full to a rigid base structure, which opposes them with equal and opposite forces and couples. Stationary engines rigidly bolted to very heavy foundations approach this condition unless the natural frequency of the engine-foundation mass on its earthen "spring" has a natural frequency near some frequency coming from the engine. With such mounts, the foundation is never completely rigid, so that appreciable vibration is transmitted to the surroundings, and can usually be felt in the adjacent structure.

Another example of engine mounting with low values of β is the case of reciprocating engines in boats and ships where the engine is bolted as rigidly as possible to a heavy ship's structure. That β is not actually zero in

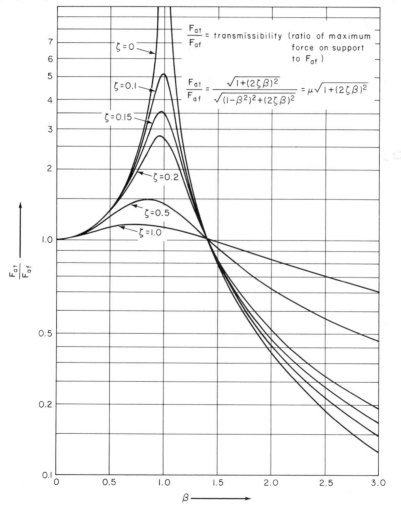

Fig. 8–35. Force transmissibility for the system of Fig. 8–34.

such cases is indicated by appreciable vibration of the ship's structure at engine frequencies. It should be remembered, however, that much of the observed hull vibration may originate from the propeller rather than from the engine.

Vibration Isolation. In many cases, including especially those of aircraft and road vehicles, it is desirable to minimize the vibratory forces transmitted to the surrounding structure. In general, this end is achieved by providing for operation at large values of β and small values of the relative damping (see Fig. 8–35). Operation at large values of β is achieved by using flexible engine mountings. By the term *flexible* is meant a mounting structure such that the natural frequencies of the engine mass in its mount are much lower than any vibratory frequencies generated by the engine in normal operation.

If the mount is flexible in all directions, the engine has six degrees of freedom to vibrate: linear displacement along three axes and rotation about each of these axes. In the most general case (8.87, pp. 44, 45) such a system has six different natural frequencies, and any force or torque applied to the engine will cause motion in all six modes. In this case the problem of designing the mount so as to keep all important natural frequencies outside the operating range may be difficult.

Where feasible, the problem of vibration control can be greatly simplified by means of the principle illustrated in Fig. 8–36. The body shown is suspended in such a way that a force applied along any one of the reference axes will excite motion along that axis only, and a couple applied around any reference axis will excite motion around that axis only. To accomplish

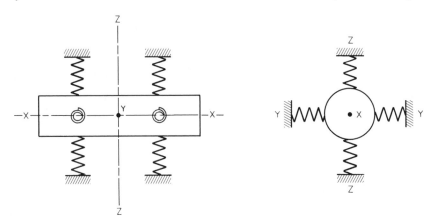

Fig. 8–36. Decoupled flexible mounting of a cylindrical mass. Reference axes are chosen through the center of gravity and along the principal axes of inertia. All springs are alike and are symmetrically mounted around the center of gravity and the principal axes of inertia. A force along any reference axis excites motion only along that axis. A couple in a plane perpendicular to a reference axis excites rotation around that axis only.

FLOATING POWER

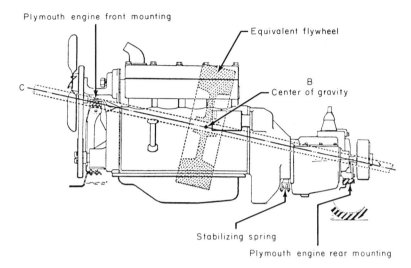

Plymouth engine front mounting

Equivalent flywheel

C

B
Center of gravity

Stabilizing spring

Plymouth engine rear mounting

Plymouth engine front mounting

Plymouth engine stabilizing spring

Fig. 8–37. Plymouth "Floating Power," a torsionally decoupled mount for a 4-cylinder automobile engine, 1931. Engine is effectively mounted along its principal axis of inertia (Denham, 8.81).

this result, the suspension forces must act at the center of gravity and the suspension couples must act around the principal axes of inertia. Such a suspension may be called *decoupled* because, within the restrictions noted, only one mode of vibration will be excited by any one force or couple. As a further simplification, the spring rates and positions may be chosen so that all six natural frequencies are identical. If this principle could be completely applied to an engine mount, there would be only one speed where the natural frequency of the engine in its mount corresponded to a given order of the engine-exciting forces and couples.

This principle was used by the Chrysler Corporation (8.81) in 1931 for mounting the 4-cylinder Plymouth engine so as to allow rotation around its principal axis of inertia, Fig. 8–37. Except at very low speeds, this mounting effectively eliminated transmission of inertia-torque and gas-torque variation, both of which have large second-order components in this type of engine (8.12).

Another notable example of the use of this principle is the E. S. Taylor-Browne "Dynamic Suspension" for radial aircraft engines, shown in Fig. 8–38. This mounting proved so effective in reducing the transmission

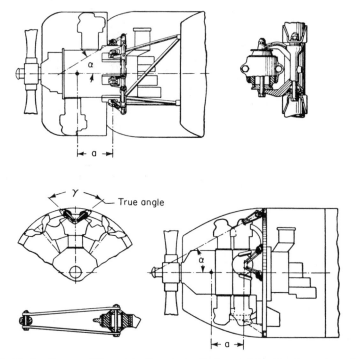

Fig. 8–38. Taylor-Browne "Dynamic Suspension," for radial aircraft engines. In upper diagram mount is applied to crankcase and in lower diagram to cylinder heads. Engine is effectively mounted at its center of gravity and torsionally around its principal axis of inertia (Browne, 8.87).

of vibratory forces and couples to the airframe that it has been universally used ever since its development (8.87, 8.89).

Except in the case of radial airplane engines, most contemporary flexible engine mounts make no attempt to achieve decoupling. Instead, rubber or spring mountings sufficiently "soft" to keep all natural frequencies relatively low are generally used. When properly applied, this method proves quite effective, especially for engines of six or more cylinders, where vibrating forces and couples below the third order are generally very small.

ENGINE NOISE

Perhaps the best definition of noise is "disagreeable sound." The subject is too large to discuss in detail here. For more complete information see refs. 8.900–8.959.

The factors which cause engine noise can be classified as follows:

1. Mechanical noise, due to the impact of one engine part against another.
2. Noise due to vibrations resulting from combustion.
3. Intake and exhaust noise.

Mechanical Noise. Important sources of mechanical noise may be the vibrations set up by impact of pistons (8.911), valve-gear parts, bearings, gear teeth, etc. Aircraft engines use large clearances for pistons and bearings, with consequent high mechanical noise levels. Gears can be another important source of mechanical noise.

Noise control is especially important for passenger automobiles. Here mechanical noise is minimized by small bearing clearances, flexible piston skirts that operate with virtually no clearance, hydraulic valve lifters, very accurate gear-tooth geometry, and effective soundproofing between engine and body interior.

Another application where low noise level is very critical is for Diesel submarine engines (refs. 8.950–8.957). Here one of the most difficult problems appears to be gear noise.

In Diesel engines an important source of noise may be the injection system, especially the seating of the injector needle and check valve.

Combustion Noise. It can be shown theoretically that when a force is quickly applied to an elastic structure, vibrations will be set up in the structure with an amplitude depending on the time of force application and on the natural periods of structure vibration. Figure 8–39 shows this relation for a simple undamped system with one degree of freedom. The vibration amplitude is small as long as t_n/t does not exceed 0.4. As the ratio increases from that value, the amplitude increases rapidly toward a value approaching twice the static amplitude as defined in the figure. From analogy with Fig. 8–39, it is evident that engine vibration (and consequent noise) due to a high rate of pressure rise must be due to natural periods of vibration that are considerably longer than the period of rapid pressure rise in the cylinder.

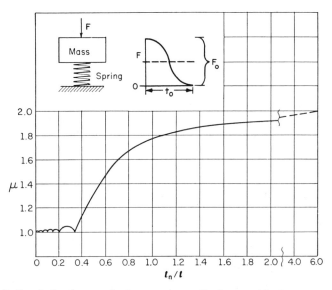

Fig. 8–39. Ratio of vibration amplitude to static amplitude caused by force $F = (F_0/2) \cos \omega t$ applied over time t_0 in the manner shown. The natural period of the spring-mass system, $t_n = 2\pi \sqrt{M/K}$. μ is amplitude of vibration divided by F_0/K, where K is spring stiffness.

Noise caused by mechanical vibrations set up by rapid rates of combustion has been mentioned in Chapters 1, 2, and 3. In spark-ignition engines, noise due to rapid rates of pressure rise in the absence of detonation is called engine "roughness." (See Chapter 1, pp. 32 to 33 and refs. 1.90–1.94 and 8.929–8.940.) Research has shown that engine roughness usually involves vibration of the crankshaft-flywheel system in bending. As discussed in Chapter 2, detonation involves high-frequency vibration of the engine structure with a characteristic noise called "knock" or "ping."

Rates of pressure rise in Diesel engines tend to be higher than in non-detonating spark-ignition engines; consequently many Diesel engines produce noise that would be classified as roughness in spark-ignition engines. This characteristic is one of those which make Diesel engines less attractive than spark-ignition engines for use in passenger automobiles. (8.950–8.959, 10.21, 10.28–10.291).

From the relation shown in Fig. 8–39 it is evident that noise due to combustion can be reduced either by reducing the rate of pressure rise, or by increasing the natural frequencies of the vibrations involved. The rate of pressure rise with detonation is so high that only the elimination of that form of combustion is effective in eliminating this source of noise. Roughness in spark-ignition engines can be controlled to a limited extent by combustion chamber design (8.940). As stated in Chapter 3, control of the rate of pressure rise is an important objective in the design of Diesel engines and their injection systems.

In order to secure high natural frequencies, a stiff engine structure is necessary, particularly with regard to the crankshaft in bending. From this point of view short, compact engines with generous crankshaft dimensions are called for. For example, the V-8 engine lends itself particularly well to rigid design, including a crankshaft which can be short and stiff compared with that required for a 6- or an 8-cylinder in-line engine of equal power.

Exhaust and Intake Noise. Silencers for exhaust and intake are available for all services. The basic problem is to achieve the required noise suppression with a minimum pressure difference across the silencer. In the case of U.S. passenger automobiles, the requirements for exhaust-noise suppression are such that considerable exhaust back pressures (2–3 psi) are accepted at high engine output. The problem of exhaust and intake silencing is an acoustical one on which there exists a large body of literature. For further discussion see refs. 8.920–8.928 and the bibliographies contained in these references.

Noise Control

The vibration and noise control achieved in the best contemporary passenger automobiles is remarkably good, especially with automatic transmissions which do not allow the engine to run with high torque at low speed. Submarine and large marine Diesel engines have also been made relatively quiet. To date, however, insufficient attention has been given to noise reduction in Diesel engines used in road vehicles.

Table 8–2. Four-Cycle Engines with Even Cylinder Spacing
and Even Firing Intervals

Arrangement of Crankshaft and Cylinders	No. of Cyl.	Inertia Balance and Firing Intervals				Firing Interval, Deg.
		Primary		Secondary		
		Shaking Force	Moment	Shaking Force	Moment	
Four-Cycle In-Line Engines						
	1	$F_c = 0$ $V = Z \cos \theta$ $H = 0$	$M = 0$	$F_c = 0$ $V = \lambda Z \cos 2\theta$ $H = 0$	$M = 0$	720
	2	$F_c = 0$ $V = 0$ $H = 0$	$M_v = aZ \cos \theta$	$F_c = 0$ $V = 2\lambda Z \cos 2\theta$ $H = 0$	$M = 0$	180 540
	3	$F_c = 0$ $V = 0$ $H = 0$	$M_v = a\sqrt{3}Z \sin \theta$	$F_c = 0$ $V = 0$ $H = 0$	$M_v = a\sqrt{3}\lambda Z \sin 2\theta$	240
	4	$F_c = 0$ $V = 0$ $H = 0$	$M = 0$	$F_c = 0$ $V = 4\lambda Z \cos 2\theta$ $H = 0$	$M = 0$	180
	6	$F_c = 0$ $V = 0$ $H = 0$	$M = 0$	$F_c = 0$ $V = 0$ $H = 0$	$M = 0$	120
	8	$F_c = 0$ $V = 0$ $H = 0$	$M = 0$	$F_c = 0$ $V = 0$ $H = 0$	$M = 0$	90
	8	$F_c = 0$ $V = 0$ $H = 0$	$M = 0$	$F_c = 0$ $V = 0$ $H = 0$	$M_v = 4a\lambda Z \cos 2\theta$	90

Table 8–2 (*continued*)

Inertia Balance and Firing Intervals

Arrangement of Crankshaft and Cylinders	No. of Cyl.	Primary Shaking Force	Primary Moment	Secondary Shaking Force	Secondary Moment	Firing Interval, Deg.
Four-Cycle Opposed Engines						
	2	$F_c = 0$ $V = 2Z\cos\theta$ $H = 0$	$M = 0$	$F_c = 0$ $V = 0$ $H = 0$	$M = 0$	180 540
	2	$F_c = 0$ $V = 0$ $H = 0$	$M_v = aZ\cos\theta$	$F_c = 0$ $V = 0$ $H = 0$	$M_v = a\lambda Z\cos 2\theta$	360
	4	$F_c = 0$ $V = 0$ $H = 0$	$M = 0$	$F_c = 0$ $V = 0$ $H = 0$	$M_v = 2a\lambda Z\cos 2\theta$	180
	4	$F_c = 0$ $V = 0$ $H = 0$	$M_v = 2aZ\cos\theta$	$F_c = 0$ $V = 0$ $H = 0$	$M = 0$	180
	8	$F_c = 0$ $V = 0$ $H = 0$	$M = 0$	$F_c = 0$ $V = 0$ $H = 0$	$M = 0$	180 2 cyls. simult.
	8	$F_c = 0$ $V = 0$ $H = 0$	$M_v = 2(2a+b) \times Z\cos\theta - 2bZ\sin\theta$	$F_c = 0$ $V = 0$ $H = 0$	$M = 0$	90
	12	$F_c = 0$ $V = 0$ $H = 0$	$M = 0$	$F_c = 0$ $V = 0$ $H = 0$	$M = 0$	60

Table 8–2 (*continued*)

Inertia Balance and Firing Intervals

Arrangement of Crankshaft and Cylinders	No. of Cyl.	Primary		Secondary		Firing Interval, Deg.
		Shaking Force	Moment	Shaking Force	Moment	
Four-Cycle Opposed Engines (*continued*)						
(diagram)	16	$F_c=0$ $V=0$ $H=0$	$M=0$	$F_c=0$ $V=0$ $H=0$	$M=0$	90 2 cyls. simult.
Four-Cycle V-Type Engines						
(diagram)	2	$F_c=0$ $V=2Z\cos\theta\cos^2\dfrac{\alpha}{2}$ $H=2Z\sin\theta\sin^2\dfrac{\alpha}{2}$	$M=0$	$F_c=0$ $V=2\lambda Z\cos2\theta\cos\dfrac{\alpha}{2}\cos\alpha$ $H=2\lambda Z\sin2\theta\sin\dfrac{\alpha}{2}\sin\alpha$	$M=0$	
(diagram 90°)	2	$F_c=Z$ $V=0$ $H=0$	$M=0$	$F_c=0$ $V=0$ $H=\sqrt{2}\lambda Z\times\cos2\theta$	$M=0$	450 270
(diagram 60°)	2	$F_c=\tfrac{1}{2}Z$ $V=Z\cos\theta$ $H=0$	$M=0$	$F_c=\tfrac{1}{2}\sqrt{3}\lambda Z$ $V=0$ $H=0$	$M=0$	420 300
(diagram 45°)	2	$F_c=0.293Z$ $V=\sqrt{2}Z\times\cos\theta$ $H=0$	$M=0$	$F_c=0.541\lambda Z$ $H=0$ $V=0.765\lambda Z\times\cos2\theta$	$M=0$	405 315
(diagram 120°)	6	$F_c=0$ $V=0$ $H=0$	$M_v=\tfrac{1}{2}a\sqrt{3}\times Z\sin\theta$ $M_h=\tfrac{3}{2}a\sqrt{3}\times Z\cos\theta$	$F_c=0$ $V=0$ $H=0$	$M_v=\tfrac{1}{2}a\sqrt{3}\times\lambda Z\sin2\theta$ $M_h=\tfrac{3}{2}a\sqrt{3}\times\lambda Z\cos2\theta$	120
(diagram 90°)	8	$F_c=Z^*$ $V=0$ $H=0$	$M=0$	$F_c=0$ $V=0$ $H=4\sqrt{2}\lambda Z\times\cos2\theta$	$M=0$	90

* Opposite each crank

Table 8–2 (*continued*)

Inertia Balance and Firing Intervals

Arrangement of Crankshaft and Cylinders	No. of Cyl.	Primary		Secondary		Firing Interval, Deg.
		Shaking Force	Moment	Shaking Force	Moment	
Four-Cycle V-Type Engines						
(diagram, 90°, 1 2 3 4)	8	$F_c = Z*$ $V = 0$ $H = 0$	$M = 0$	$F_c = 0$ $V = 0$ $H = 0$	$M = 0$	90
(diagram, 60°, 1 2 3 4)	8	$F_c = 0$ $V = 0$ $H = 0$	$M = 0$	$F_c = 2\sqrt{3}\lambda Z$ $V = 0$ $H = 0$	$M = 0$	60 120 60 120 etc.
(diagram, 60°, 1 2 3 4 5 6)	12	$F_c = 0$ $V = 0$ $H = 0$	$M = 0$	$F_c = 0$ $V = 0$ $H = 0$	$M = 0$	60
(diagram, 45°, 1 2 3 4 5 6)	12	$F_c = 0$ $V = 0$ $H = 0$	$M = 0$	$F_c = 0$ $V = 0$ $H = 0$	$M = 0$	45 75 45 75 etc.
(diagram, 45°, 1 2 34 56 78)	16	$F_c = 0$ $V = 0$ $H = 0$	$M = 0$	$F_c = 0$ $V = 0$ $H = 0$	$M = 0$	45
(diagram, 135°, 1 234 5678)	16	$F_c = 0$ $V = 0$ $H = 0$	$M = 0$	$F_c = 0$ $V = 0$ $H = 0$	$M = 0$	45

* Opposite each crank

Table 8–2 (*continued*)

Inertia Balance and Firing Intervals

Arrangement of Crankshaft and Cylinders	No. of Cyl.	Primary		Secondary		Firing Interval, Deg.
		Shaking Force	Moment	Shaking Force	Moment	

Radial Engines, Properly Counterweighted, with All Connecting Rods Operating Around Crankpin Center

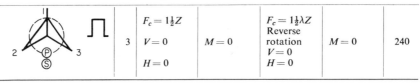	3	$F_c = 1\tfrac{1}{2}Z$ $V = 0$ $H = 0$	$M = 0$	$F_c = 1\tfrac{1}{2}\lambda Z$ Reverse rotation $V = 0$ $H = 0$	$M = 0$	240

Radial engines of this kind with more than 3 cylinders in each row have no primary or secondary unbalanced forces or moments. Four-cycle radials must have an odd number of cylinders in each row for even firing at intervals of $720°/N_c$: Radial engines with master and link-rod systems have a small second-order unbalance in each row. For details of these, see pp. 275–279.

Four-Cycle Radial Engines; with Link Rods

| Single row Single crank n cylinders $n > 3$ $r =$ distance crankpin to knuckle pin $l =$ length of linkrod | n | $F_c = \dfrac{n}{2}Z$ $V = 0$ $H = 0$ | $M = 0$ | $F_c = \dfrac{r}{l}\lambda Z$ same side as crankpin when master rod is in top dead center $V = 0$ $H = 0$ | $M = 0$ | $\dfrac{720}{n}$ |
| Conditions: 1. $r =$ constant 2. $L = r + l$ 3. Angle between knuckle pins is same as angle between cylinders | | | | | | |

Table 8–2 (*continued*)

Arrangement of Crankshaft and Cylinders	No. of Cyl.	Primary		Secondary		Firing Interval, Deg.
		Shaking Force	Moment	Shaking Force	Moment	
Inertia Balance and Firing Intervals						

Four-Cycle *W*-Type Engines

Arrangement of Crankshaft and Cylinders	No. of Cyl.	Shaking Force	Moment	Shaking Force	Moment	Firing Interval, Deg.
(diagram) 60° 60° P S	3	$F_c = 1\frac{1}{2}Z$ $V = 0$ $H = 0$	$M = 0$	$F_c = \frac{1}{2}\lambda Z$ $V = 0$ $H = \lambda Z \sin 2\theta$	$M = 0$	120 300 300
(diagram) 60° 60°	6	$F_c = 1\frac{1}{2}Z^*$ $V = 0$ $H = 0$	$M = 0$	$F_c = \lambda Z$ $V = 0$ $H = 2\lambda Z \times \sin 2\theta$	$M = 0$	120
(diagram) 60° 60°	12	$F_c = 0$ $V = 0$ $H = 0$	$M = 0$	$F_c = 2\lambda Z$ $V = 0$ $H = 4\lambda Z \times \cos 2\theta$	$M = 0$	60
(diagram) 45° 90° 90° MISC.	16	$F_c = 0$ $V = 0$ $H = 0$	$M = 0$	$F_c = 0$ $V = 0$ $H = 0$	$M = 0$	45

* Opposite each crank

$Z = MpR\Omega^2$
$F_c =$ rotating force, balanceable by suitable counterweights
$H =$ the horizontal unbalanced inertia resultant after the counterweight indicated has been incorporated
$V =$ the vertical unbalanced inertia resultant after the counterweight indicated has been incorporated
$\theta =$ crank angle, measured from the crank position as shown, in a clockwise direction
$a =$ distance as shown
$b =$ distance as shown
$c =$ distance as shown
$M_v =$ unbalanced moment in vertical plane, after counterweights are incorporated
$M_h =$ unbalanced moment in horizontal plane after counterweights are incorporated
$\lambda = R/L$
(From Kalb, ref. 8.15.)

Table 8–3. Two-Cycle Engines with Even Cylinder

No. of Cylinders	In-Line Engines		Unbalanced Forces*		Unbalanced Moments*		
	Crank Diagram	Firing Order	1st order Z	2nd order ZR/L	1st order Rotating dM_R	1st order Vertical dZ	2nd order dZR/L
1		1	1.0†	1.0	0	0	0
2		1–2	0	2.0	1.0	1.0	0
3		1–2–3	0	0	$\sqrt{3}$	$\sqrt{3}$	$\sqrt{3}$
4		1–3–2–4	0	0	$\sqrt{2}$	$\sqrt{2}$	4.0
5		1–5–2–3–4	0	0	0.449	0.449	4.98
6		1, 6–3, 4–2, 5 Double firing	0	0	0	0	0
6		1–5–2–4–3–6 Uneven firing	0	0	0.89	0.89	1.74
6		1–6–2–4–3–5	0	0	0	0	3.464
6		1–6–4–2–5–3	0	0	2.0	2.0	6.928
6		1–5–3–6–2–4	0	0	3.46	3.46	0

* Multiply the quantities Z, ZR/L, dM_R, etc., by the numbers tabulated.
† Rotating mass assumed counterbalanced. d = distance between adjacent cylinder axes.
Note on other cylinder arrangements:
Two-cycle radial engines have the same balance characteristics as 4-cycle radials, see Table 8–2. All radials with evenly spaced cylinders above three in number have both primary

Spacing and Even Firing, Except as Noted

No. of Cylinders	In-Line Engines		Unbalanced Forces*		Unbalanced Moments*		
	Crank Diagram	Firing Order	1st order Z	2nd order ZR/L	1st order Rotating dM_R	1st order Vertical dZ	2nd order dZR/L
7	(crank diagram)	1–7–2–5–4–3–6	0	0	0.267	0.267	1.0
7	(crank diagram)	1–7–4–2–6–3–5	0	0	0.85	0.85	5.53
8	(crank diagram)	1–8–2–6–4–5–3–7 $X = 36°\ 50'$ $O = 53°\ 10'$	0	0	0	0	0
8	(crank diagram)	1–8–2–6–4–5–3–7	0	0	0.448	0.448	0
9	(crank diagram)	1–9–2–7–4–5–6–3–8	0	0	0.194	0.194	0.548
9	(crank diagram)	1–9–4–3–7–5–2–8–6	0	0	0.92	0.92	1.28
10	(crank diagram)	1, 10–5, 6–2, 9–3, 8–4, 7 Double firing	0	0	0	0	0
10	(crank diagram)	1–10–2–8–4–6–5–7–3–9	0	0	0	0	0.896
11	(crank diagram)	1–11–2–9–4–7–6–5–8–3–10	0	0	0.153	0.153	0.382
12	(crank diagram)	1–6–8–10–3–5–7–12–2–4–9–11	0	0	0	0	0

and secondary balance except for the master-rod effect. All 2-cycle V engines with even firing and with less than twelve cylinders have some degree of unbalance with regard to either primary or secondary moments, or both. Two-cycle V–12 and W–12 engines can have complete primary and secondary balance if cranks and V angle are properly arranged. (See Nakanishi, 8.13.)

9 | Engine Materials

In a volume of this kind it is practicable to treat only the most relevant aspects of engine materials. The point of view taken will be that of the engine designer, builder, and user rather than that of the metallurgist or materials specialist. A more intensive study of the subject will be assisted by reference to the bibliography for this chapter.

The number and variety of materials now available to the designer are so vast that the problem of making a choice is indeed formidable. Fortunately, in the case of internal-combustion engines, the available materials have been reasonably well classified and standardized and are presented in a well-organized form in the *SAE Handbook*, ref. 9.00, which constitutes the best guide available in the English language, in this field. This handbook is revised and issued annually and is therefore not subject to serious obsolescence. All the usual materials used in engines, both metallic and nonmetallic, are included. In most cases advice is given as to the appropriate application of each material. Other valuable sources of similar information are refs. 9.01–9.042.

It is not within the scope of this volume to attempt to duplicate or even to summarize the contents of the handbooks. It is expected that the serious reader or the active designer will have access to these documents for ready reference. This chapter will be devoted rather to outlining a basic philosophy of material selection as related to engine design, in an attempt to establish a background for understanding the fundamental questions involved.

The choice of material for any machine part can be said to depend on the following considerations:

1. General function: structural, bearing, sealing, heat-conducting, space-filling, etc., or combinations thereof
2. Environment: loading, temperature and temperature range, exposure to corrosive conditions or to abrasion, wear, etc.
3. Life expectancy
4. Space and weight limitations
5. Cost of the finished part and of its maintenance and replacement
6. Special considerations, such as appearance, customer prejudices, etc.

While every part must be designed to function satisfactorily in its operating environment for a suitable period of time, the relative importance of items 3 to 6 varies greatly with the type of service. For example, in aircraft the importance of light weight and small space justifies higher cost and shorter life than might otherwise be desirable. In the case of passenger-car engines low initial cost is so important that considerable sacrifice in life of the parts is often justified. On the other hand, large stationary and marine engines are expected to have such long service lives that space, weight, and initial cost may be relatively great in order to attain these objectives. Table 10-1, in the next chapter, indicates the characteristics emphasized for various types of service.

Except in the case of certain small parts,* the materials used must be plentiful, easily available, and readily manufactured.

As would be expected, after some seventy years of development of the internal-combustion engine, the types of materials used have become rather standardized, as indicated by Table 9-1. However, the detailed composition of these materials is subject to continuous improvement, so that such a list must be considered subject to change with time.

STRUCTURAL MATERIALS

Materials whose essential function is to carry relatively high stresses will here be classed as *structural*. The heavily stressed materials include those which carry and transmit the forces and torques developed by cylinder pressure and by the inertia of the moving parts in the power train† and valve gear.

The success of structural materials is measured by their resistance to structural failure. In engines the most prevalent type of structural failure is that due to *fatigue*.‡ (Failures due to wear, seizure, corrosion, oxidation, etc., are not here classified as structural.)

Fatigue Failure

The fatigue of materials has been the subject of extensive research and publication for many years. Here it is feasible to cite only the most important references (9.190–9.303) and to point out the most significant factors from the point of view under discussion.

---

* Examples of small parts where relatively expensive (rare) materials are often used include ignition breaker points, spark-plug electrodes (platinum has been used in aircraft plugs) and special alloying elements and materials used in exhaust valves and exhaust turbines.

† The power train is taken to consist of cylinders, pistons, piston pins, connecting rods, crankshaft, rod and main bearings, and their supporting structure and fastenings.

‡ Strangely enough, the question of fatigue is not mentioned in the SAE Handbook.

Table 9–1. Materials Typically Used in Principal Engine Parts

Part	Material Type	SAE No.	BHN	Remarks	Reasons
Cylinder Heads	Gray cast iron J431a*	G3000 G3500 G4500	170–269	Usual	4, 7
	Cast aluminum J465	39, 322		Aircraft (some others)	1, 4
	Forged aluminum J454c	A2218		Aircraft engines	2
Cylinder barrels	Gray cast iron J431a	G3000 G3500 G4500	170–269	Usual	4, 5, 7
	Steel**	4130	300±	Aircraft engines, often nitrided	2
	Cast aluminum J465	34 39		Small engines, plated bore	1, 4
Pistons	Sand-cast aluminum or Die-cast aluminum	34, 39 309, 314, 321, 328		Usual for engines of less than 10-in. bore	2, 3, 4
	Forged aluminum J454c	AA2018		Aircraft and some Diesel	1, 3
	Gray iron J431a	G3000 G3500 G4500	170–269	Small engines and most engines of more than 10-in. bore	4, 7
Piston pins	Steel	4140 4340	250–300	Usual	2
	Steel, c.h.†	5015 4119 4320	670–720	Hard-surfaced	1, 10
Piston rings	Special cast iron		150–250	Usual material	5, 10
	Steel, chrome plated			Heavy-duty	1, 10
Connect-ing rods	Steel	1041 4130 4137	190–230	Small rods	1
		4340 9840	190–230	Large rods	2, 13
	m- or n-iron	J433, J434	200–300	Small engines	4, 7
Bolts, studs, nuts	Steel	4137 4340	300–340	Highly stressed	2
		1137	200–250	Minor fastenings	1, 7
Crank-shafts	Steel	1046 4140 4340	230–275††	Usual	1 2 2, 13
	Cast steel J435a	0150 0175	311–363†† 363–415	Frequent	1, 4, 7
	n-Iron J433	Grade III Grade IV	240–290 300–350	Rare	1, 4, 7
	m-Iron J434	53004	197–241	Rare	4, 7

Table 9–1 (*continued*)

Part	Material		BHN	Remarks	Reasons
	Type	SAE No.			
Crank-cases	Gray iron J431a	G3500 G4500	187–269	Automotive engines	4, 7
	Cast aluminum J465	310, 311 312, 322		Aircraft and some automotive engines	1, 4
	Forged aluminum J454c	AA2218		Aircraft engines only	2
	Welded steel J410b	950C 945C	150±	Many large engines	1
Main and rod bearings (SAE Handbook J459, Table 1)	Tin-base babbit Lead-base babbit	11, 12 13–15		Light-duty non-automotive	5, 7, 10
	Lead-tin overlay Copper-lead Aluminum	19, 190 49, 48, 480, 481 770–781		Heavy-duty	5, 10, 11
Other plain bearings	Bronze	J459a, Table 1		For all lightly loaded bearings	5, 10
	Aluminum	J461a, Table 10		Many lightly loaded bearings are directly in the material of the casting	5, 7, 10
	Cast Iron	J431a	187–269		
	Sintered	J471b		Special purpose	5, 10
Anti-friction bearings	Steel	51100 52100	670–739	Small bearings	
	Steel, c.h.†	5115 5120		Large bearings	2, 10
Camshafts	Special cast iron	G4000 d, e, f	241–321	Automotive practice	7, 10
	Steel, c.h.†	4119 4317	650–700	Heavy duty	10
Cam followers	Same as camshafts				10
Push rods	Steel tubing	0.20–0.30 carbon steel		Usual	1, 7
Rocker arms	Steel	4140 4340	250–300	Usual	1
	n-Iron J433	III or IV			
	Sintered steel	J471b		Rare	4, 7
	m-Iron	J434			
Valves and valve seats	Special steels	J775			8, 9, 10

Table 9–1 (*continued*)

Part	Material		BHN	Remarks	Reasons
	Type	SAE No.			
Valve springs	Alloy steel	4150 4350 6150	400–445	Often shot peened	11
Gears	Steel, c.h.†	4119 4320 9317	455–525	Heavy duty	2, 10
	Steel	4150	375–425	Medium duty	1, 7
	Carbon steel		375–425		
	Bronze	J461*a*		Light duty	7
	Sintered	J471*b*			
Gear cases	Cast iron	G2000 G3000	187–229	Usual	4, 7
	Cast aluminum J452	33		Common	4, 1
	Cast magnesium J465	50, 500, 502, 504, 505		Aircraft	
Cylinder-head gaskets and spark-plug gaskets	Copper or aluminum			Diesel engines, heavy-duty engines, large gas engines	11
	Copper-asbestos			Automotive spark-ignition engines and some automotive Diesels	7
Water seals	Rubber O-rings	J14		Best water seal where appropriate	12
	Cork Vellum Fiber	J90*a*		For flat surfaces	7
Low-pressure gaskets	Cork Vellum Fiber	J90*a*		For flat surfaces	7

* Numbers starting with **J** refer to sections in the SAE Handbook (9.00). Other numbers are for specific compositions.

** "Steel" means forged or rolled steel.

† c.h. means casehardened.

†† Bearing surfaces of iron and steel shafts are often flame-hardened to much higher values.

m-iron = malleable cast iron n-iron = nodular cast iron

Reasons:

1. High strength/weight ratio
2. Very high strength/weight ratio
3. High heat conductivity
4. Can be cast in intricate shapes
5. Good bearing properties
6. Best bearing properties
7. Low cost, adequate
8. High hot strength
9. Resistance to corrosion
10. Resistance to wear
11. Strength and resilience
12. Water-tightness and durability
13. Good heat treatability

Fatigue failure is fracture due to the repeated application of many cycles of varying stress. Its basic physical and chemical causes are still not completely understood (9.2082). The typical appearance of a fatigue failure is shown in Fig. 9–1. It is progressive in nature, starting as a crack which usually begins at the surface. The crack grows as the stress cycle is repeated, until a point is reached where the remaining material is stressed beyond its ultimate strength and fails suddenly, in tension or shear or a combination of both. The boundary between the fatigued area (ABC in Fig. 9–1) and the area where failure was sudden is usually clearly distinguishable. Other characteristics of fatigue failure are:

1. The fatigue crack starts at the point of maximum tensile or shear stress. This point is generally, but not always, at the surface of the part.*

2. The point of maximum stress is usually a point of "stress concentration," such as an original crack in the surface, a notch created by corrosion or in the manufacturing process, or a point where stress is high because of the shape of the part and the method of loading.

3. The fatigue crack occurs at a maximum stress usually well below the elastic limit and hence without significant distortion of the surrounding material.

The comparative resistance to fatigue of materials is measured by laboratory tests such as the rotating beam, the vibrating cantilever beam, or the repeated-torsion test (9.211–9.214). In such tests the load is usually reversed in direction in a sinusoidal pattern, with the average stress equal to zero. Such tests are often designated by the terms *reversed bending* or *reversed torsion*, respectively. In homogeneous materials fatigue failure occurs in the region of maximum stress in the test specimen. The stress at this point is naturally the one of principal interest.

The results of such tests can be plotted, as illustrated for example by Fig. 9–2. Each point on this plot represents a test to failure of a single specimen. Such a figure is called an S-N plot because it records *stress* at failure against *number* of cycles to failure. Important characteristics are:

1. There is a general tendency for life to increase as stress is reduced.

2. In the case of the materials of Fig. 9–2, no failures seem to occur when stress is less than 48 per cent of the ultimate tensile strength.†

3. Dispersion of the test points increases with decreasing stress.

---

* The most common example of failure below the surface is that of loaded curved surfaces, such as gear teeth and tappet rollers, where maximum stress occurs below the surface. This tendency is exaggerated when the surfaces are casehardened, since here the thin, hardened outer layer is much stronger than the interior material.

† Ultimate strength is the strength determined by the standard ASTM tensile test, ref. 9.02, part 3.

Fig. 9–1. Fatigue failure of automobile crankshaft. A = start of fracture in fillet; B = "beach marks" typical of crack progress under intermittent operation; C = line of final sudden failure. (*Courtesy Lessells and Associates, Inc., Waltham, Mass.*)

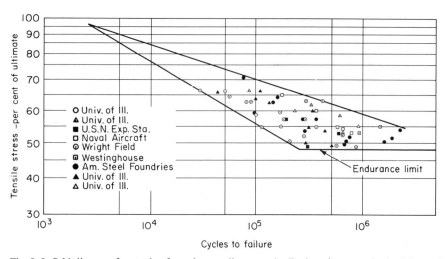

Fig. 9–2. S-N diagram for steels of varying tensile strength. Each point records the failure of one specimen in the rotating-beam test (Almen, 9.205).

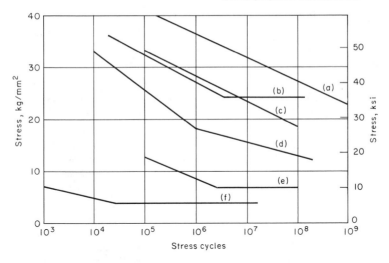

(a) Monel metal, $\sigma_B = 63$ kg/mm^2 (b) 0.18%C-steel, $\sigma_B = 47$ kg/mm^2
(c) Nickel, $\sigma_B = 53$ kg/mm^2 (d) Refined Dural, $\sigma_B = 46$ kg/mm^2
(e) Pure aluminum, $\sigma_B = 17$ kg/mm^2 (f) Pine wood, $\sigma_B = 17$ kg/mm^2

Fig. 9–3. S-N diagrams (lower limit of test points) for various materials in reversed bending. σ_B is the ultimate tensile stress (Matthaes, 9.202).

Figure 9–3 shows lower boundaries of S-N plots for various materials. It will be noted that three of these materials (steel, pure aluminum, and wood) show a definite *endurance limit* or *fatigue limit,** that is, a stress below which failure does not occur even after 10^8 or 10^9 cycles. This type of curve is characteristic of most ferrous metals but is not typical for the common nonferrous metals, including the aluminum alloys. For materials that do not show a definite fatigue limit of this kind, the nominal fatigue limit is taken as the no-failure stress at an arbitrary number of cycles, usually 10^8. Thus, according to Fig. 9–3 the fatigue limits for these particular specimens are as follows:

		Fatigue Limit	
	Material	kg/mm^2	ksi*
a.	Monel metal	27	38.2
b.	Steel (low carbon)	25	35.5
c.	Nickel	19	27.0
d.	Aluminum (Dural)	12	17.0
e.	Aluminum (pure)	7.5	10.6
f.	Wood (pine)	4	5.7

* The symbol ksi stands for thousands of pounds per square inch, and will be so used throughout this chapter.

* These two terms are used interchangeably throughout the literature in English.

It is obvious that S-N curves require a considerable outlay in time and expense compared with simple one-time tests, such as elastic limit, ultimate strength, impact value, etc. Therefore, much effort has been expended on finding a correlation between endurance limit and one of the simpler tests. The only reasonably good, though still very approximate, correlation appears to be with ultimate tensile strength, and hence it is customary to plot endurance limit versus ultimate tensile stress. Figure 9–4 illustrates such a plot for steel.

In the case of steel, the ultimate tensile strength (UTS) is nearly proportional to the Brinell hardness number (BHN), as shown in Fig. 9–5. Therefore, in the case of steel, a scale of BHN can be added to, or substituted for, the UTS scale.

Figure 9–6 shows torsional (shear) endurance limits for steel, and Figs. 9–7 and 9–8 show endurance limit versus UTS for aluminum and magnesium alloys, respectively. Similar curves for titanium (9.200) show endurance limit averaging 50 per cent of UTS, with occasional instances of failure at 35 per cent.

Forrest (9.200) gives the following approximate ratios for torsional endurance limit σ_t divided by reversed-bending endurance limit σ_b:

Material	Ratio σ_t/σ_b
Forged steel	0.52–0.69
Forged aluminum alloys	0.43–0.74
Forged copper alloys	0.41–0.67
Forged magnesium	0.49–0.60
Titanium	0.37–0.57
Cast iron	0.79–1.01
Cast aluminum-magnesium alloys	0.71–0.91

The characteristically higher values for cast materials are notable.

It should be pointed out that plots such as those of Figs. 9–2 through 9–8 are based on an arbitrary method of testing, at room temperature (about 70°F, or 21°C), through a sinusoidal stress pattern with average stress zero, and with highly polished specimens of arbitrary size and shape. Such tests presumably furnish a good comparison of the endurance qualities of materials only under these particular circumstances. It is therefore pertinent to inquire into the effects of different test conditions.

Such an inquiry should include, as a minimum, the effects of other values of frequency of test cycle, temperature, stress-cycle pattern, shape of specimen, surface finish and surface treatment, corrosion, size of specimen, and direction of grain. The following sections summarize what is known of these effects up to the present time (1966).

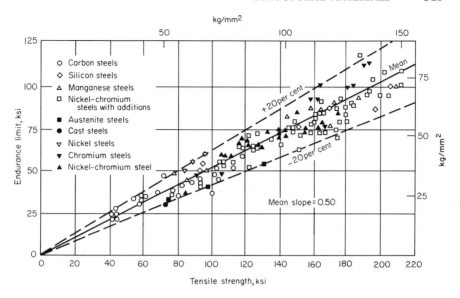

Fig. 9–4. Endurance limit (reversed bending) versus tensile strength for steel (9.202).

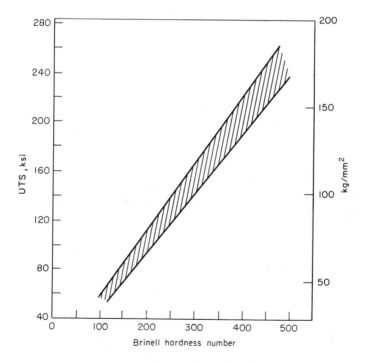

Fig. 9–5. Relation of ultimate tensile strength to Brinell hardness number for steel. UTS $\cong$ 0.5 (BHN) ksi or 0.35 (BHN) kg/mm^2.

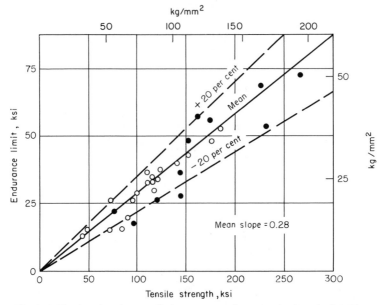

Fig. 9–6. Torsional endurance limit versus tensile strength of steels (9.202).

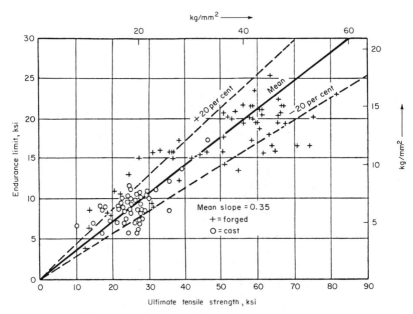

Fig. 9–7. Endurance limit in reversed bending versus tensile strength of aluminum alloys (9.202).

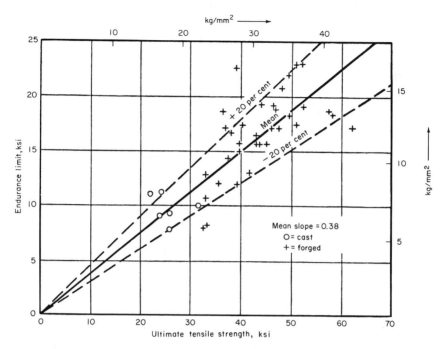

Fig 9–8. Endurance limit of magnesium alloys in reversed bending (9.202).

Frequency. All investigators seem to agree that, within practical limits, the frequency of load reversal has little effect on endurance-limit test results.

Temperature Effects. Figure 9–9 is a summary of the effects of temperature on endurance limit. From a practical point of view these data would indicate the following approximate temperatures, above which endurance limit is adversely affected:

Material	Temperature	
Aluminum alloys	150°C	300°F
Low-carbon steel	350°C	660°F
Medium-carbon steel	430°C	800°F
Austenitic steel	500°C	930°F
Cast iron	450°C	840°F
Magnesium alloy	40°C	100°F
Titanium	400°C	750°F
Nimonic alloy (a special tu.bine-blade material)	750°C	1380°F

Above 1000°F the austenitic steels and special nickel alloys, such as

Nimonic, are indicated. Titanium* and magnesium alloys seem to be adversely affected above room temperature, but titanium still has excellent endurance strength up to 900°F.

Stress-Cycle Effects. Reversed-bending tests can be considered as tests to determine the no-failure stress range when the average stress is zero. The allowable stress range is then twice the endurance limit. Endurance tests when the average stress is not zero can be regarded as an alternating stress superimposed on an average or mean steady stress. Results of tests where the mean stress is not zero are shown in Fig. 9–10. This type of graph is called an R-M diagram, because it shows allowable *range of stress* plotted against *mean stress*. In this figure the average stress was varied from the compressive yield stress (-1.0) to the tensile yield stress ($+1.0$), and the maximum value of the alternating stress at which no failure occurred up to 10^8 cycles was determined by endurance tests.

To illustrate the use of Fig. 9–10, assume a steel with yield point of 90 ksi and an endurance limit of 50 ksi. Find the allowable maximum and minimum stress when the average stress is 45 ksi in tension. Figure 9–10 indicates that the allowable stress range at mean stress $= 45/90$ or 0.50 times the yield stress will be 80 per cent of twice the endurance limit, or $0.80(2 \times 50) = 80$ ksi. The allowable high stress will thus be $45 + 40 = 85$ ksi in tension, and $45 - 40 = 5$ ksi in compression.

Figure 9–11 shows R-M test results for cast iron. These are based on ultimate tensile stress, since cast iron has no measurable yield point. In the case of cast iron, torsion (shear) tests give results similar to tension-compression tests. When the average stress on cast iron is compressive, the allowable range increases as with steel or at an even higher rate.

The effect of stress range in torsion is summarized in Fig. 9–12. For these materials it appears that the allowable high stress is independent of the mean stress, at least up to a mean stress of 0.8 times the elastic limit in shear (see discussion of valve-spring design in Chapter 11).

Combined Stresses. Much experimental work has been done on the problem of fatigue failure under combined stresses (9.207, also 9.200, pp. 107–114), but the results are difficult to correlate, and vary with test conditions. As a practical matter, values of combined stresses in any machine part are seldom known with sufficient accuracy to allow meaningful interpretations on the basis of existing data or theories. In the case of combined bending and torsion, when the two stresses are known, the following equation may be used (9.200):

$$\sigma_n = \sqrt{\sigma_b^2 + 4\sigma_s^2},$$

where σ_n is the equivalent rotating-beam stress, σ_b the maximum tensile

---

* Because of its light weight (60 per cent that of steel) and high strength below 900°F, titanium is used in supersonic-aircraft structures with working temperatures in the range 600° to 800°F.

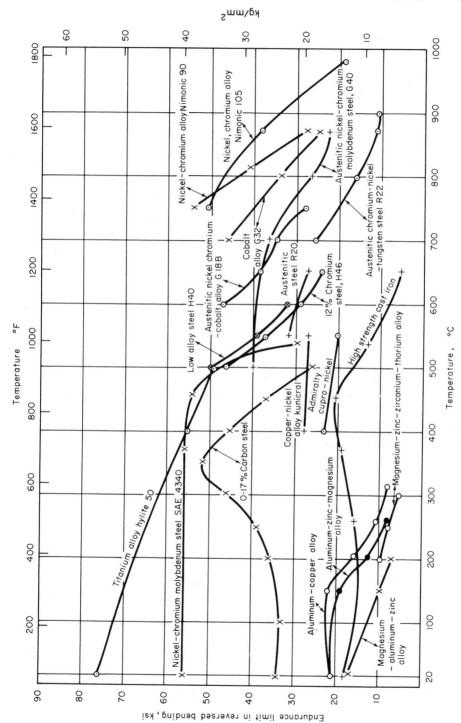

Fig. 9-9. The influence of temperature on the fatigue strength of metals (Forrest, 9.200; *courtesy Pergamon Press*).

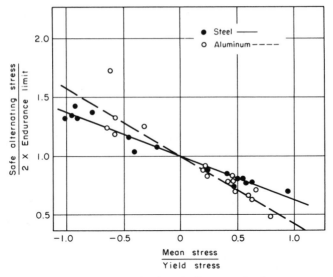

Fig 9–10. R-M diagram for forged alloys, rotating-beam tests.

Steel from refs. 9.2086 and 9.2092
Aluminum from refs. 9.2086 and 9.2091

stress, in bending, and σ_s the maximum shear stress. This equation assumes the shear endurance limit to be one half the tensile endurance limit, a conservative estimate for most materials.

Effects of Shape on Endurance Limit. The usual standard rotating-beam endurance-test piece is shown in Fig. 9–13. The maximum stress in this specimen is probably very close to that calculated by the simple beam theory for the smallest section. Important points to note are that there

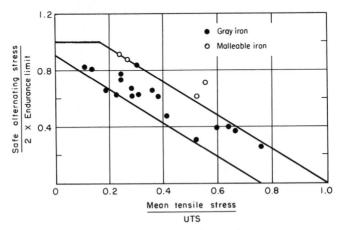

Fig. 9–11. R-M diagram for cast iron
(● Smith, 9.2088; ○ Pomp and Hempel, 9.2089).

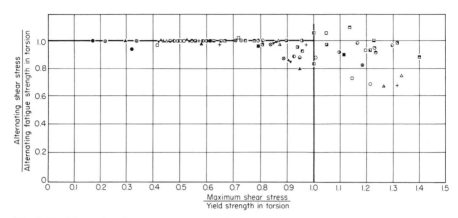

Fig. 9–12. Alternating shear stress versus maximum shear-stress. Graph includes a wide variety of forged and cast steels, malleable iron, aluminum alloys, brass and bronze alloys. (From Forrest, 9.200; *courtesy Pergamon Press*.)

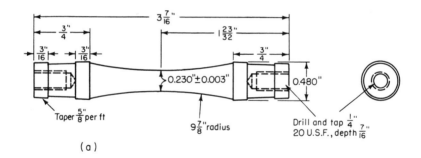

(a)

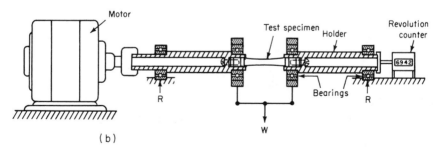

(b)

Fig. 9–13. ASTM rotating-beam endurance-test equipment. (*a*) Standard fatigue specimen; (*b*) arrangement of apparatus (W = weight suspended by equalizing linkage; R = support points).

is no sharp change of shape in the test section, that the test-section surface is highly polished, and that the diameter at the critical section is small (0.23 in., 0.9 cm).

Stress Concentration. Figure 9–14 shows a section of a plate with circular notches, under a tensile load. By means of elastic theory (9.300) it can be shown that the stress pattern in the vicinity of the notches has both vertical and horizontal components. The maximum stress, which occurs at the center of the notch, is a tensile stress equal to twice the average stress in the smallest section. The notches cause what is defined as a *stress*

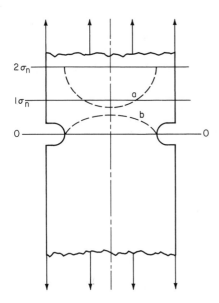

Fig. 9–14. Theoretical stresses in notched plate in tension. Plate is of indefinite length normal to page.

σ_n = load/area at section O-O
a = axial stress at section O-O
b = transverse stress at section O-O

concentration. If a circular groove is cut around the minimum section of a rotating-beam specimen, the maximum stress at the bottom of the groove is K_t times the stress at the bottom of the groove calculated by the simple beam theory. The value of K_t depends on the geometry of the groove, and can be computed for simple shapes by elastic theory (see ref. 9.301 for values of K_t).

Notch Sensitivity. Assuming that K_t is known, it might be expected that the endurance-limit load which could be carried by a notched specimen would be $1/K_t$ times the endurance-limit load of the standard specimen. Experiments show, however, that this relation does not necessarily hold in practice, but that in general materials are less affected by notches than the relation would indicate. A convenient index of the effect of notches on endurance limit is given by the relation

$$q = \frac{K_f - 1}{K_t - 1},$$

where q = notch sensitivity

K_t = stress-concentration factor

$$K_f = \frac{\text{endurance-limit stress of unnotched specimen}}{\text{endurance-limit stress of notched specimen}}.$$

The endurance-limit stress of the notched specimen at the bottom of the notch is usually calculated by means of the simple beam (or tension) theory.

It is evident that if the notch has the effect predicted by elastic theory, $K_f = K_t$, and $q = 1.0$. On the other hand, if the notched specimen fails at the same calculated stress as the unnotched specimen, $K_f = 1.0$ and $q = 0$; that is, the material has zero notch sensitivity.

Figure 9–15 and Table 9–2 show values of K_f and q for different shapes and materials. It is evident that these values vary with material and with geometry and that there is a wide scatter. It is also evident that, except for relatively hard steels, q tends to be less than 1.0, which means that the stress concentration has less effect than would be predicted on the basis

Table 9–2. Observed Notch Sensitivity of Various Materials

(Data for 5×10^6 cycles in rotating beam tests)

Material	UTS		K_t	K_f	q
	ksi	kg/mm²			
Soft steel	48.5	34	1.6	1.3	0.5
Hard steel	142.0	100	1.6	1.6	1.0
Gray cast iron	16.6	117	1.6	1.0	0
18-8 stainless steel	96.4	68	1.6	1.0	0
Forged aluminum	59.0	43	1.6	1.0	0
Magnesium alloy	44.6	31	1.6	1.1	0.17
Bronze	81.0	57	1.6	1.0	0

From Grover, ref. 9.201, p. 70.

of elastic theory. This may be due in a large measure to the fact that actual materials are nonuniform in structure, with small imperfections such as cracks and inclusions. Thus the unnotched specimen already has a certain degree of stress concentration due to these imperfections. The fact that cast materials generally show low values of q reinforces this supposition.

Sharp Notches. The effect of notch sharpness on the fatigue load-carrying ability of specimens all of which have the same minimum diameter is illustrated in Fig. 9–16. The disastrous effect of sharp notches on strength is apparent. The theoretical stress at the base of sharp notches is very high, being infinite when the notch is completely sharp. Of course a completely sharp notch is impossible in practice, although the radius

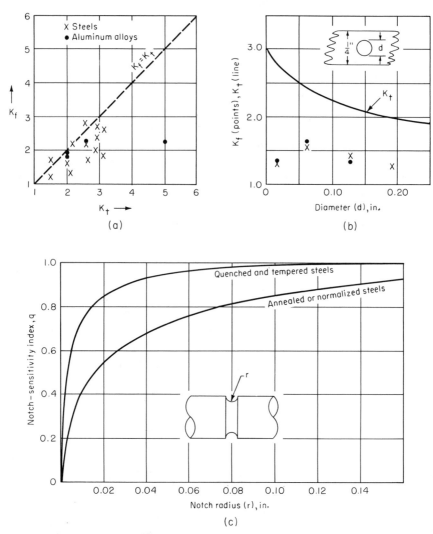

Fig. 9–15. Notch-sensitivity data. (*a*) and (*c*) Grover, 9.201; (*b*) Kuhn and Hardrath, 9.2087.

at the bottom of the notch can be made extremely small. The larger the specimen, the greater the effective sharpness of a given notch.

More will be said in subsequent chapters about the question of stress concentration as related to shape. The point to be discussed here is the relative behavior of various materials with notches under stress.

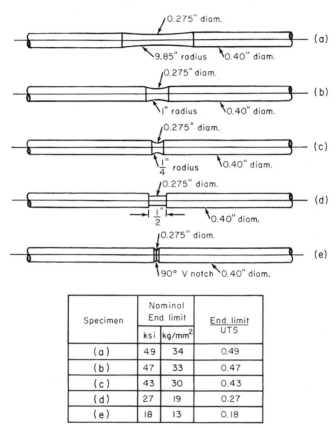

Specimen	Nominal End. limit		End. limit UTS
	ksi	kg/mm²	
(a)	49	34	0.49
(b)	47	33	0.47
(c)	43	30	0.43
(d)	27	19	0.27
(e)	18	13	0.18

Fig. 9–16. Effect of sharp notches on endurance–limit stress as computed by simple beam theory at smallest cross section of 0.49 per cent carbon steel, water quenched and drawn at 1200°F (Almen, 9.205).

Figure 9–17 shows endurance limits of standard and notched steel specimens versus ultimate tensile strength. It is evident that the notch effect increases with increasing UTS (which also means increasing hardness and brittleness). Above about 180 ksi tensile strength (350–400 BHN), the increase in notch sensitivity offsets the effect of increasing UTS, and the endurance limit falls off with further increase in hardness. Thus, for steel parts that are effectively notched, the maximum endurance strength occurs at 350–400 BHN.

Surface Finish. A rough surface is obviously one form of notched surface. In the case of steel the effects of surface irregularities are especially great, as indicated by Fig. 9–18. It is evident that for maximum strength a polished surface is important (see also Fig. 9–17). Grinding may introduce unfavorable surface conditions which should be polished or peened for maximum strength (Fig. 9–18*c*).

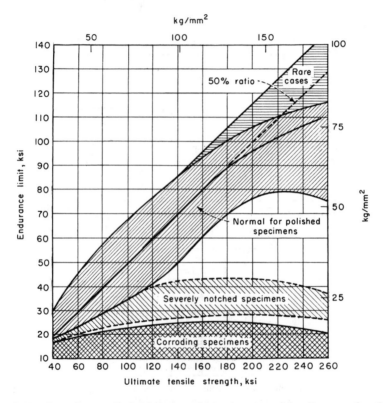

Fig. 9–17. Tensile endurance limit of steel specimens in reversed bending as a function of ultimate tensile strength (Bullens, 9.4101).

Effects of Corrosion on Endurance Limit. This question can be considered as one of surface finish but deserves special consideration on account of its importance. As Fig. 9–19 shows, the effect of severe corrosion on endurance limit can be very large, even though the depth of the corroded layer may appear small. Apparently corrosion has the effect of producing

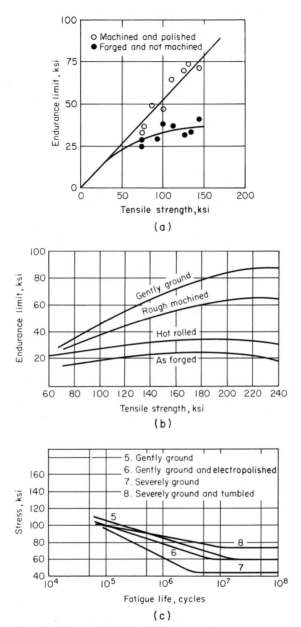

Fig. 9–18. Effects of surface finish on endurance limit of steel (rotating-beam, unnotched specimens). (*a*) Polished versus forged surface (Matthaes, 9.202); (*b*) effect of surface condition (Noll and Lipsin, 9.209); (*c*) effect of degree of grinding. Curve 7 indicates damage due to excessive depth and feed of grind (Tarasov and Grover, 9.229).

sharp cracks, or notches that readily develop into cracks (9.200, Chapter VII), and accelerates crack growth.

Fretting corrosion, which occurs when there is relative motion between two metallic surfaces held together under high pressure, is also damaging to the endurance limit. This type of corrosion is most often observed in

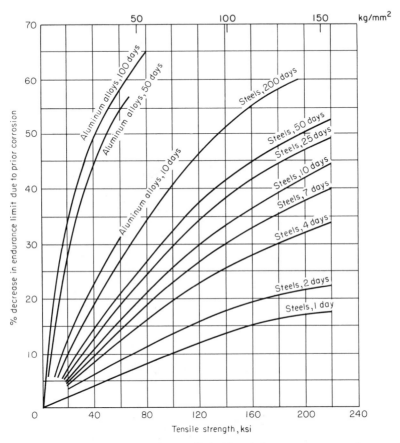

Fig. 9–19. Effect of corrosion on endurance limit. Time is for exposure to a salt atmosphere prior to rotating-beam tests (McAdam and Clyne, 9.051).

the case of parts joined together by means of press fitting or shrink fitting, or in tapered or splined joints, when the loads are such as to cause some degree of slippage.

The obvious remedy for corrosion fatigue is to prevent corrosion by the use of appropriate materials or special coatings (9.050). However, some coatings have a tendency to reduce the endurance limit because of their effect on surface geometry (see Fig. 9–20d).

Summary of Notch Effects. From the foregoing results, the following practical conclusions may be drawn:

1. Notches or other sharp changes in section are to be avoided as far as possible in all machine parts.

2. In forged materials stress concentrations are more dangerous as the material becomes more brittle.

3. Stress concentrations are more harmful in forged than in cast materials.

4. Rough surfaces have strong adverse effects on endurance strength, especially in forged materials.

5. Corrosion, including fretting corrosion, may cause serious reductions in endurance strength, probably because of sharp notches or cracks caused by the corrosion process.

Surface Treatments Influencing Endurance Limit. Since fatigue cracks nearly always start at points of high tensile stress at the surface, any treatment that increases surface strength or produces a high compressive stress at the surface tends to increase the endurance strength of a given part. The treatments found effective include those listed below, most of which have the combined effect of strengthening the surface and introducing some surface compressive stress as well. Among the more effective treatments are casehardening, Fig. 9–20a, rolling, Fig. 9–20b, shot peening, Fig. 9–20c, and nitride hardening. On the other hand, platings can reduce endurance strength, sometimes by considerable factors (Fig. 9–20d). A more complete discussion of surface-treatment effects can be found in refs. 9.101, 9.200–9.208, and 9.220–9.229. Treatments considered include:

1. Surface heat treatments, such as casehardening, flame hardening, quenching, and nitriding.

2. Cold working of the surface, as by cold rolling, cold drawing, shot peening, stretching, tumbling, and bending.

In the case of engines, shot peening has been found especially effective in prolonging the life of valve springs. Rolling of fillets and threads is also very effective (Fig. 9–20b). The various hardening processes are employed chiefly for the purpose of reducing wear but may also have a strengthening effect, as indicated in Fig. 9–20a and the references quoted. Nitriding, which is essentially a chemical treatment resulting in a wear-resistant surface, may also increase endurance strength appreciably.

Effect of Size on Endurance Limit. Research on the effect of size on endurance limit has, with numerous exceptions, indicated that the endurance limit falls with increasing size of specimen, particularly in the range of diameters less than one inch, as shown in Fig. 9–21. The tendency toward decreasing endurance limit with increasing size seems especially marked with notched specimens, even when the notches are geometrically similar. References 9.200, pp. 135–45, and 9.201, pp. 111–19, give excellent

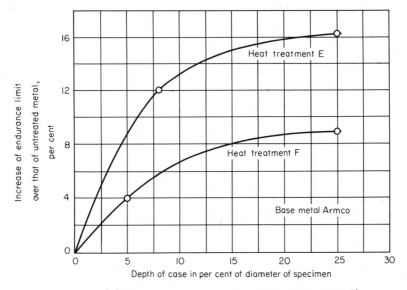

(a) Effect of casehardening (Moore & Kommers, ref. 9.190)

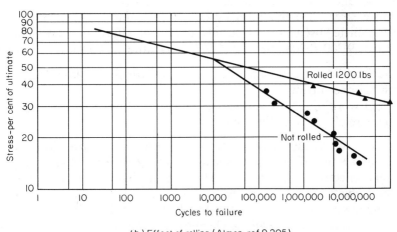

(b) Effect of rolling (Almen, ref. 9.205)

Fig. 9–20. Effect of various surface treatments on endurance limit of steel.

summaries of work on size effects. Grover *et al.* (9.201) recommend a 10 to 15 per cent reduction in standard endurance-limit values for diameters greater than one inch.

Effect of Direction of Grain. Rolled and drawn materials, when etched, show a directional structure or *grain* caused by elongation of the crystalline

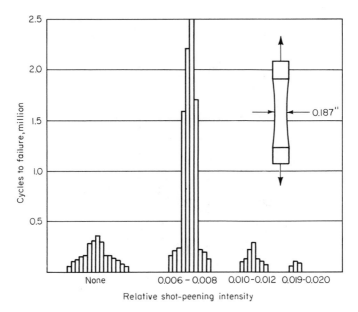

(c) Effect of shot peening, SAE 4340 steel 320 BHN (Almen, ref. 9.2054)

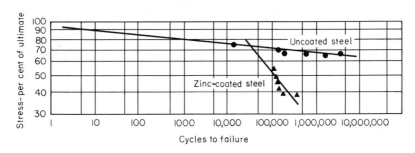

(d) Effect of galvanizing (Almen, ref. 9.205)
Fig. 9–20 (*continued*)

structure and the impurities in the direction of rolling or drawing. Reference 9.200, pp. 154–155, indicates reductions in endurance limit for steel up to 30 per cent and for magnesium forgings up to 20 per cent when the rotating-beam specimen is cut with the grain at right angles to its axis, as compared with the usual longitudinal orientation. Such results indicate that it is important to control the forging process for engine parts in such a way as to orient the grain in the direction most favorable to endurance strength. Thus, parts such as crankshafts, connecting rods, and gears should never be machined from straight bar stock but should be

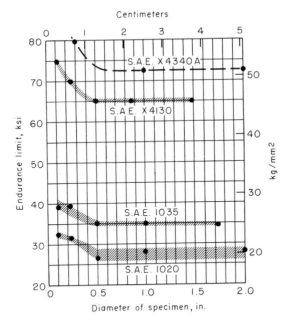

Fig. 9–21. Size effect in fatigue testing (Gadd *et al.*, 9.2085).

made from individual forgings fabricated, as far as possible, so that the grain will run parallel to the principal stresses. For example, crankshaft forgings should be bent from straight stock, and connecting-rod ends should be *upset*, that is, hammered out in the direction of the rod axis.

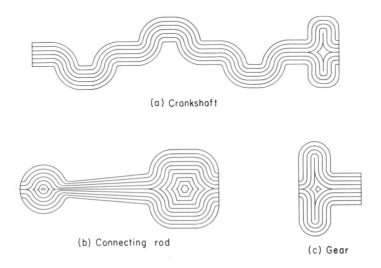

(a) Crankshaft

(b) Connecting rod

(c) Gear

Fig. 9–22. Preferred grain direction in forgings.

Gear forgings should also be upset, in the direction of the gear axis. Figure 9–22 indicates the desired grain direction in such forgings.

Creep Failure

When subjected to high stresses at high temperatures for long periods of time, gradual deflection of a material may take place, even though the stress is below the nominal elastic limit (9.101, 9.35, 9.36). This type of failure is limiting in the case of turbine wheels and turbine blades, which run at high stresses at high temperature. Creep is also observed in the case of valve springs, which may gradually lose load with time, even at moderate temperatures.

The creep characteristic of gas-turbine-blade materials is one of the important factors limiting turbine inlet temperature. In the case of reciprocating engines, this limitation applies to the allowable temperature at the inlet of an exhaust-driven turbine. In turbo-supercharged 4-cycle engines a large valve overlap is often used in order to limit turbine inlet temperature.

In the case of valve springs, creep is extremely slow and is usually handled by spring replacement at suitable intervals of operation.

The question of creep will be discussed further in connection with exhaust-turbine design, Chapter 12.

NONSTRUCTURAL PROPERTIES OF MATERIALS

Other than structural strength, as measured by endurance limit or creep limit for most engine parts, the important properties of materials include cost of raw material, availability, cost of fabrication, density, hardness, heat conductivity, thermal expansion, bearing properties, resistance to corrosion, etc. Typical values of many of these properties for the most important engine materials are given in Tables 9–3 and 9–4. Such properties are often decisive in the choice of a material for a particular use, as will appear in subsequent discussions.

STEEL

Steel is relatively low in cost (except for stainless and other special alloys), has the highest endurance strength of any available material, and has a naturally hard surface. Its strength and hardness can be controlled over a wide range by heat treatment. Compared with other structural materials it has the following disadvantages:

It is not easy to cast and for most purposes must be forged and machined.
It is subject to rapid corrosion (except in stainless form).
It has a relatively low thermal conductivity.

Table 9–3. General Properties of Wrought Metals

	Carbon Steel	Alloy Steel	Stainless Steel	Aluminum Alloys	Magnesium Alloys	Copper-Base Alloys	Titanium
UTS, ksi	45–120*	75–300	100–170	15–77	29–47	32–200	90–250‡
Elastic modulus, psi $\times 10^{-6}$	30	30	28	10–11	—	10–22	15.5‡
BHN	85–250*	100–600+	160–180	23–135†	very soft	25–210†	—
Endurance ratio EL/UTS	0.35–0.60	0.4–0.6	0.3–0.6	0.35–0.50	0.25–0.50	0.25–0.50	0.35–0.60
Elongation, per cent	0–50	0–50	10–55	1–30	4–12	4–48	8–27
Specific gravity	7.6–7.85	7.6–7.85	7.6–8.1	2.2–3.0	1.77–1.82	7.8–8.9	4.5
Heat conductivity { cal/cm °C hr	0.108–0.115	0.11±	0.06–0.10	0.37–0.53	0.17–0.38	0.06–0.94	0.04±
Btu/ft °F hr	25–26	25±	10–23	90–120	39–87	14–226	9±
Coeff. of expansion per °F $\times 10^6$	6.08–7.34	6.4±	5.7–9.6	12.7–13.3	16±	9.4–11.8	5.0‡
Relative machinability	good	good to impossible	poor	excellent	excellent	good to excellent	very poor

* Usual limit for forgings. Carbon steel can be hardened to very high UTS and hardness.
† 500-kg ball measure.
‡ Pure metal.

Table 9-4. General Properties of Cast Metals

	Gray Iron	Malleable Iron	Nodular Iron	Cast Steel	Aluminum Alloys	Magnesium Alloys	Copper Base
UTS, ksi	20–50	50–90	60–160	60–175	17–45	18–40	30–110
Elastic modulus, psi $\times 10^{-6}$	15–20	23	25	28–30	10	—	10–12
BHN	170–269	163–285	160–360	131–415	40–125*	very soft	wide range
Endurance ratio	0.35–0.50	0.50–0.60	0.40–0.60	0.40	0.30	0.40	0.40
Elongation, per cent	0	2–18	0–19	6–32	0–12	1–7	8–22
Specific gravity	7.01–7.315	7.4±	7.1–7.3	7.7±	2.6–2.9	1.8±	7.8–8.9
Heat conductivity { cal/cm °C hr	0.096–0.149	0.10±	0.10±	0.11±	0.26–0.40	0.33–0.38	0.06–0.94
Btu/ft °F hr	22–35	23±	23±	26±	60–95	77–90	14–226
Coeff. of expansion per °F $\times 10^6$	6.0±	6.6±	6.5±	6.5±	12.1–13.8	16±†	9.0–9.8
Relative machinability	very good	good	good	fair to good	excellent	excellent	good to excellent

* 500-kg ball.
† Pure metal.

Because of its low cost, great strength, and high hardness, it is the preferred material for most moving parts, such as crankshafts, connecting rods, gears, auxiliary shafts, etc., and for bolts, screws, and other fastenings.

In general, the steels used in engines can be divided into five categories, namely

1. Low-carbon steels (C = 0.10 to 0.20 per cent)
2. Medium-carbon steels (C = 0.30 to 0.50 per cent)
3. High-carbon steels (C = 1.0 per cent ±)
4. Stainless steels
5. Special steels (for valves, turbine blades, etc.)
6. Cast steels.

A further division is that between so-called *carbon* steels, that is, steels with no special alloying elements,* and *alloy* steels, which contain special alloying elements, principal among which are nickel, chromium, and molybdenum. All the stainless and special steels are alloy steels.

SAE Classification. In the early part of the twentieth century the principal incentive for the classification, control, and improvement of steel came from the developing automobile industry, where the structural requirements of light weight and resistance to fatigue called for better structural characteristics than those available in the ordinary carbon steels used in buildings, bridges, and stationary power machinery. The leading organization in this development was the (U.S.) Society of Automotive Engineers, and the classification of steels now used in the U.S.A. and many other parts of the world is the one developed by that organization and published annually in the SAE Handbook. Table 9–5 gives the key to this classification, which will be used hereafter in referring to various types of steel. Steels beginning with digits 10 and 11 are called carbon steels, while the rest are known as alloy steels. The last two digits indicate the approximate carbon content in hundredths of one per cent.

Carbon Steels. At a given hardness, carbon steel can be just as strong as alloy steel. Some of the strongest material known is in the form of carbon-steel "piano wire," where the drawing process gives this material UTS values of more than 300 ksi. Figure 9–4 shows endurance-limit values of carbon-steel rotating-beam specimens equal to any of the alloy steels up to over 200 ksi UTS. In considering the figure, however, it must be remembered that the results were obtained with specimens of small cross section (0.23 in., 5.8 mm, diameter), longitudinal grain structure, high polish, and without notches.

In general, carbon steel is used wherever sections are small and stresses are low. Such uses include sheet stock for oil pans, covers, etc., screw

---

* Reference 9.410, part III, p. 51, defines carbon steels as those containing less than 1.65 per cent manganese, 0.60 per cent silicon, and 0.60 per cent copper, with no special alloying elements such as nickel, vanadium, chromium, molybdenum, etc.

Table 9–5. SAE Numbering System for Wrought or Rolled Steel

(From SAE Handbook 1964, ref. 9.00)

Type of Steel*	Identifying Number†
Carbon steels	
Plain carbon	10xx
Free cutting	11xx
Manganese steels	
Mn 1.75	13xx
Nickel steels	
Ni 3.50	23xx
Ni 5.00	25xx
Nickel-chromium steels	
Ni 1.25; Cr 0.65	31xx
Ni 3.50; Cr 1.57	33xx
Corrosion and heat resisting	302xx, 303xx
Molybdenum steels	
Mo 0.25	40xx
Chromium-molybdenum steels	
Cr 0.50 and 0.95; Mo 0.25, 0.20, and 0.12	41xx
Nickel-chromium-molybdenum steels	
Ni 1.82; Cr 0.50 and 0.80; Mo 0.25	43xx
Ni 1.05; Cr 0.45; Mo 0.20	47xx
Ni 0.55; Cr 0.50 and 0.65; Mo 0.20	86xx
Ni 0.55; Cr 0.50; Mo 0.25	87xx
Ni 3.25; Cr 1.20; Mo 0.12	93xx
Ni 1.00; Cr 0.80; Mo 0.25	98xx
Nickel-molybdenum steels	
Ni 1.57 and 1.82; Mo 0.20 and 0.25	46xx
Ni 3.50; Mo 0.25	48xx
Chromium steels	
Low Cr—Cr 0.27, 0.40, and 0.50	50xx
Low Cr—Cr 0.80, 0.87, 0.92, 0.95, 1.00, and 1.05	51xx
Low Cr (bearing) Cr 0.50	501xx
Medium Cr (bearing) Cr 1.02	511xx
High Cr (bearing) Cr 1.45	521xx
Corrosion and Heat Resisting	514xx, 515xx
Chromium-vanadium steel	
Cr 0.80 and 0.95; V 0.10 and 0.15 min.	61xx
Silicon-manganese steels	
Mn 0.65, 0.82, 0.87, and 0.85; Si 1.40 and 2.00; Cr None,	
0.17, 0.32, and 0.65	92xx
Low alloy, high-tensile steel	950
Boron Intensified Steels, B denotes boron steel	YYBxx
Leaded Steels, L denotes leaded steel	YYLxx

* Numbers give the average percentage of the principal alloying elements.

† Carbon content in hundredths of one per cent is given in the spaces xx. For boron- or lead-intensified steels, the identifying numbers appear in spaces YY.

fastenings not under heavy load (outside the power train), some small casehardened parts, and similar applications. A special category of carbon steels, SAE numbers 11xx and the closely related manganese steels, 13xx are made to have especially "free cutting" properties. Such steels are therefore particularly appropriate for low-stressed bolts, studs, nuts, and other small parts made on automatic screw machines.

Another use of carbon steel is for welded crankcases, where good welding properties are the decisive factor. In some cases important structural parts are made of carbon steel, usually for the sake of economy. Typical examples include those crankshafts which are sized and designed to avoid high stresses and thus obviate the need for high-endurance-limit material.

Alloy Steel. From the structural point of view, the advantage of alloy steel over carbon steel is almost entirely a question of "heat treatability." This property is usually measured by the *Jominy* test (9.00, 9.410), which consists of heating a 1-in. (25.4-mm) diameter bar above its transition temperature and quenching one end in a stream of cold water. By measuring the resultant hardness as a function of distance from the quenched end, a hardenability "band" is constructed, as shown in Fig. 9–23. Such curves show that the heat treatment penetrates much farther into an alloy-steel bar than into a carbon-steel bar. In practice this means that parts made of alloy steel can be heat treated with a slower cooling rate than similar parts made of carbon steel, resulting in more uniformity of physical properties and less danger of unfavorable residual stresses, cracks, or deformation. This in turn allows treatment to greater strength and hardness. The difference does not show up in Fig. 9–4 because of the small size of the test pieces. The advantage of alloy steels becomes more important

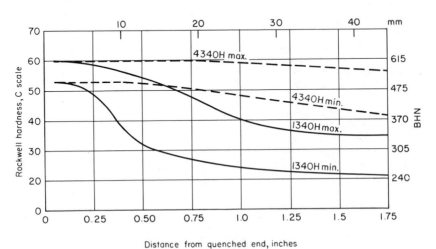

Fig. 9–23. Hardenability curves ("bands") for two steels having equal carbon content. 1340H is a carbon steel; 4340H is a chrome-molybdenum-nickel steel. (See Table 9–5 for compositions.)

as size increases and shape becomes more complex. The result in practice is that alloy-steel parts can be made with greater hardness and therefore greater endurance limits than would be practicable with carbon steels. In practice, most engine parts treated to UTS values exceeding 100 ksi (200 BHN, 70 kg/cm²) are made of alloy steel.

Reference 9.410 is especially helpful in selecting the alloy steel appropriate for a given duty. Part III of this reference groups the various alloy steels in terms of their relative hardenability for a given carbon content. As parts are made larger, it is advisable to choose steel of higher hardenability so as to ensure the attainment of the required physical properties throughout the structure. Of the steels most commonly used, the following are rated as having high hardenability:

Carbon Content, per cent	SAE Number
0.20	3310, 4320, 4820, 9317
0.30–0.37	2330, 4137, 5135, 8635, 9437
0.40–0.42	4340, 9840
0.45–0.50	4150, 9850
0.50–0.62	8653, 8655, 8660, 9662
1.00–1.02	51100, 52100

In this list, steels with carbon content of 0.50 per cent or less are used for general structural purposes. Numbers 51100 and 52100 have been especially developed for ball and roller bearings.

Referring to the alloy steels not listed here (see Table 9–5), series 92xx is for vehicle springs and steel 950 is for vehicle-body sheets. Series 302xx, 303xx, 514xx, and 515xx are stainless-type steels. The other alloy steels are for structural purposes where very high hardenability is not required. In this group, hardenability is generally greater as the fraction of the principal alloying elements increases. At the same hardenability level, the choice between them is largely one of cost, availability, and individual preference based on experience in heat treating, quality control, etc. Table 9–1 indicates the most popular choices in engine practice as of 1965.

Alloys versus Carbon Steels. For structural purposes, the only disadvantage of alloy steel as compared with carbon steel is its greater cost, including that of heat treatment, and the fact that at times there may be shortages of some of the alloying elements. For example, during World War II, nickel and chromium (both imported by the U.S.A.) were in short supply, and there was pressure to avoid their use as far as practicable.

Stainless Steels. These steels are characterized by high chromium content, usually between 11.5 and 26 per cent, which makes them corrosionproof under most circumstances. The hardness obtainable by heat treatment of such steels, including even the heat-treatable varieties, is limited, and they are therefore unsuitable where very high surface hardness is required. In general, stainless steels are not required in internal-combustion engines except for parts exposed to hot exhaust gases. Most of the special exhaust-valve steels and some of the inlet-valve steels are stainless. Ordinary stainless steels, SAE series 302xx, 303xx, may be used for exhaust piping, Diesel-engine combustion-chamber inserts, and sheet-metal work exposed to the atmosphere. The heat-treatable stainless steels in the group SAE 514xx are used where wear resistance is important, as for water-pump shafts, injection-pump parts, and valve stems. Exhaust-valve heads, one-piece exhaust valves, and most inlet valves are made from the special SAE valve steels discussed in the next section.

Special Alloys. The most important special alloys are those used for highly stressed parts that run at very high temperature, including particularly valves, and exhaust-turbine nozzles, blades, and rotors. Steels for these purposes must have nonoxidizing (stainless) properties and also good endurance and creep strength at their working temperatures. SAE series NV, HNV, EV, and HEV (9.00, Section J41a) are special valve steels. All but series NV are stainless. Materials used for exhaust-turbine and gas-turbine nozzles, blades, and rotors are usually classed as *superalloys* (ref. 9.00, Section J467a). These range from essentially stainless steels to alloys composed largely of nickel and chromium and containing very little iron. The choice of steel for valves and turbine parts will be considered in more detail in Chapter 12.

Surface Hardening of Steel

Where a structural member also acts as a bearing, it is often desirable to secure a very hard surface without affecting the interior structure of the part. Methods employed for this purpose include casehardening, surface heat treatment, nitriding, and plating.

Casehardening. Casehardening is applied only to steels with low (0.1 to 0.2 per cent) carbon content. It consists of a process of adding carbon to the surface layers only by baking the part in carbonous material. The result is a high carbon content (up to 1 per cent or more) near the surface while the original low carbon content remains in the interior. Subsequent heat treatment then hardens the surface far more than the interior. Without alloying elements, the core remains practically unaffected. On the other hand, certain alloy steels, notably SAE 3310, 4320, 4815, 4817, 4820, 8822, and 931 yield especially strong core characteristics when properly case-hardened (9.00, Section J41b).

Surface Heat Treatments. Surface heat treatments are applied to medium (0.3 to 0.5 per cent) carbon steels by heating only the surface areas to be hardened. This objective is accomplished either by electrical induction heating at the surface or by some form of high-temperature flame applied so quickly that the interior structure is not affected (see ref. 9.04 for details). In engines this type of surface hardening is used mainly for crankshaft and camshaft bearing surfaces.

Nitriding. Nitriding is a process of heating the part in ammonia gas, which produces a very hard, wear-resistant surface. This is an expensive process, but has been much used for aircraft-engine parts, including cylinder bores.

Plating. Plating for increased wear properties is usually done with chromium. Chromium plating has been found very effective in reducing the wear of piston rings and cylinder bores. This process is used on cast-iron as well as on steel parts and will be discussed in greater detail in Chapters 11 and 12.

Hardness for Various Engine Parts. Figure 9–24 shows hardness values usually specified for steel parts in automotive vehicles and their engines. (The largest use of steel in vehicles is the sheet stock for bodies, which is low-carbon steel, not shown in the figure.) The scale at the top of Fig. 9–24 gives the approximate carbon content. These hardness values are the result of long experience rather than of theoretical considerations.

Cast Steel

(See ref. 9.00, Section J435a.) The casting of steel requires sophisticated techniques and is far more difficult and more expensive than that of iron or aluminum. The characteristics of high-grade cast steel, including endurance limits and hardenability, are nearly equal to those of the forged material. The most frequent use of cast steel in engines is for crankshafts. Recent developments in nodular iron indicate that it may replace steel for this application on account of its lower cost and its almost equal physical properties.

High-chromium, that is stainless steel, can also be cast. Section J436b of ref. 9.00 gives specifications for this type of material. Cast stainless steel has been used to a limited extent for exhaust valves. Many of the special alloys used for turbine wheels and blades are suitable for casting.

The question of choice of steel will be dealt with further in Chapters 11 and 12, in connection with the discussion of the design of individual parts.

CAST IRON

Gray Iron. Except for aircraft engines and other engines where light weight is a primary requirement, the main structure of the engine, that is the crankcase and cylinder structure, is made of gray cast iron. Typical compositions and physical properties for this material are given in Section

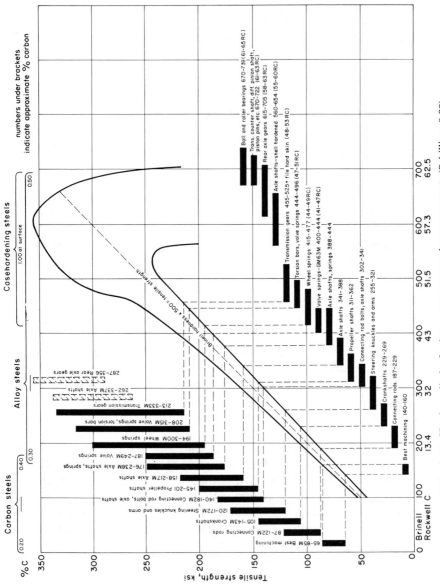

Fig. 9-24. Hardness limits for important automotive components (Schilling, 9.08).

J431*a* of the SAE Handbook. Table 9–4 shows the range of properties available. The outstanding characteristic of gray iron is the ease with which it can be cast into thin and intricate shapes.

The endurance limit of gray iron (Fig. 9–11) is lower than that of steel, but it is remarkably free from notch sensitivity (Table 9–2), probably because the structure is basically notched by particles of free carbon. However, as with steel, large sizes are more sensitive to notches than are small sizes.

In addition to being relatively easy to cast in thin, complicated shapes, gray iron is an excellent bearing material. It is therefore the most popular material for cylinder barrels, which may be either integral with the main casting or separately inserted. Other uses for gray iron are piston rings, gear cases, and miscellaneous nonstructural parts of complex shape, such as water- and oil-pump casings, covers, etc. The structural qualities of gray iron are improved by the addition of nickel (see ref. 9.00, Part J431*a*).

Chilled Cast Iron. A very hard surface can be obtained on gray iron by "chilling," that is, by using a metal insert in the mold which achieves quick local cooling of the surface. The composition of the iron and the technique of chilling have developed this material to the point where it is generally used for camshafts and tappets of automobile and other low-cost engines (see Chapter 2). An alternative to chilling is flame hardening of the bearing surfaces (see ref. 9.00, Part J431*d*, Section 7.1).

Malleable Iron. This iron (see ref. 4.00, Sections J432, J433), sometimes called "white" iron, is made so as to be without free graphite. It is annealed after casting, which gives it remarkable strength and ductility. For small-sized parts the result is a structure comparable in strength to a mild-steel forging (Table 9–4).

The advantage of malleable iron parts over steel forgings is the resultant saving in manufacturing cost in cases where the shape would involve expensive machine operations. Malleable iron is more frequently used in various automobile-chassis parts than in engines. However, in a limited number of cases it has been used for such items as rocker arms, rocker-arm brackets, and other small, highly stressed parts of complicated shape.

Nodular Iron. A development since World War II is the so-called *nodular* iron which is cast in such a way that the free-carbon granules are spherical rather than stringy in form. This iron has a much higher tensile and endurance strength than gray iron, but is somewhat more expensive. Table 9–4 shows its physical properties as compared with other cast materials (see ref. 4.00, Section J434).

Nodular iron is now (1966) beginning to be used to replace steel forgings, including crankshafts (see Chapter 11), where a cost saving can be achieved. Increasing use of this interesting material may be expected as experience is accumulated.

Figure 9–25 shows typical microstructures of the various types of cast iron. Maintenance of the appropriate microstructure is an important element in quality control of this material.

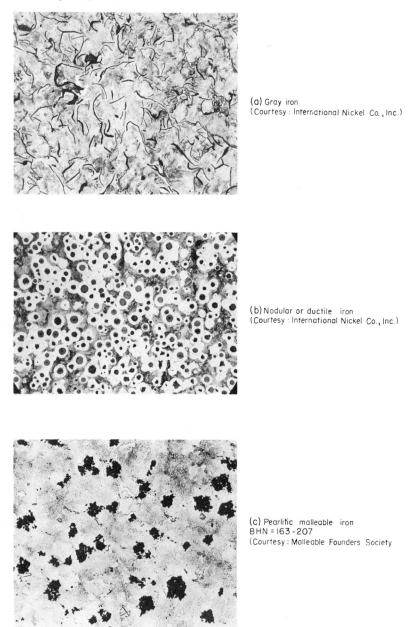

(a) Gray iron
(Courtesy : International Nickel Co., Inc.)

(b) Nodular or ductile iron
(Courtesy : International Nickel Co., Inc.)

(c) Pearlitic malleable iron
BHN = 163-207
(Courtesy : Malleable Founders Society

Fig. 9–25. Typical cast-iron microstructures.

ALUMINUM

Specifications and physical and chemical characteristics of aluminum alloys suitable for use in internal-combustion engines are given in ref. 9.00, Section J452–460, together with recommendations as to application. Outstanding characteristics of aluminum are (Tables 9–3 and 9–4):

1. Low density (one third that of iron)
2. Ease of casting (including die casting), forging, and machining
3. High heat conductivity
4. Good resistance to corrosion
5. Good bearing characteristics against steel or iron.

Disadvantages, for most purposes, include:

6. High thermal expansion coefficient
7. Low hardness
8. Adverse effects of high temperature on strength (Fig. 9–9)
9. Cost of raw material (somewhat higher than iron or carbon steel).

In engines, characteristics 1–5 make aluminum especially suitable for pistons, cylinder heads, and bearing surfaces, in spite of disadvantages 6 and 8.

Most engines with cylinders of less than 6-in. (152-mm) bore use cast-aluminum pistons. Many larger-bore engines also use this material. While the sections of an aluminum piston must be somewhat thicker than those of the corresponding iron piston, the over-all weight is less and the high heat conductivity of aluminum tends to reduce piston temperatures. This advantage is especially important when no special cooling system (using oil or water circulation) is used for the piston. For heavy-duty operation, as in aircraft and Diesel engines, aluminum pistons may be forged (see Chapter 11).

Where light engine weight is sufficiently important to offset higher costs, aluminum is used for cylinder heads and crankcase–water-jacket structures. These are usually cast, but some radial aircraft engines have been built with forged aluminum heads and crankcases (see Chapter 11). Most examples of the use of forged aluminum are found in aircraft and military engines. Aluminum is also gradually replacing other materials for crankpin and crankshaft bearings (see Chapter 12). The use of aluminum will be discussed further in Chapters 11 and 12. Physical properties are indicated in Tables 9–3 and 9–4 and usual applications in Tables 9–6 and 9–7.

MAGNESIUM

Magnesium is considerably lighter than aluminum, more expensive, and much softer. Its physical properties are given in Tables 9–3 and 9–4.

Table 9–6. Applications of SAE Aluminum Casting Alloys—SAE J452

Report of Nonferrous Metals Division, approved January 1934 and last revised by Nonferrous Metals Committee June 1961.
The SAE Standards for aluminum casting alloys cover a wide range of castings for general and special use, but do not include all the alloys in commercial use. The castings are made principally by sand-cast, permanent-mold, or die-cast methods; however, shell molding, investment casting, plaster cast, or other less common methods may be used.

Alloy Designations				Similar Specifications			Typical Uses and General Data
SAE No.	ASTM Designation	Commercial* Designation	Type of Casting	ASTM	Federal	AMS	
33	CS72A	113	Sand and permanent mold	B26 B108	QQ-A-601 Class 9 QQ-A-596 Class 1	— —	General-purpose alloy with fair strength and resistance to corrosion; often used for oil pans, crankcases, camshaft housings, and other parts not highly stressed.
34	CG100A	122	Sand and permanent mold	B26 B108	QQ-A-601 Class 7 QQ-A-596 Class 2	— —	Primarily a piston alloy, but also used for aircooled cylinder heads and valve-tappet guides.
35	S5B	43	Sand and permanent mold	B26 B108	QQ-A-601 Class 2 QQ-A-596 Class 7	— —	Used for intricate castings having thin sections; good resistance to corrosion; fair strength but good ductility.
38	C4A	195	Sand	B26	QQ-A-601 Class 4	4230, 4231	General structural castings requiring high strength and shock resistance.
39	CN42A	142	Sand and permanent mold	B26 B108	QQ-A-601 Class 6 QQ-A-596 Class 3	4222	Used primarily for aircooled cylinder heads, but also used for pistons in high-performance gasoline engines.
300	CS66A	152	Permanent mold	B108	—	—	Pistons primarily.
303	SC114A	384	Die	B85	QQ-A-591 Alloy 384	—	General-purpose alloy with high fluidity; used for thin-walled castings or castings with large areas.
304	S5C	43	Die	B85	QQ-A-591 Alloy 43	—	Good casting characteristics and resistance to corrosion.

No.	Designation	Alloy	Casting method	ASTM	Federal	SAE	Characteristics
305	S12A	A13	Die	B85	QQ-A-591 Alloy A13	—	Highly resistant to corrosion; excellent casting characteristics; used for complicated castings of thin section.
306	SC84A	A380	Die	B85	QQ-A-591 Alloy A380	4291 Comp. 2	Good casting characteristics and fair resistance to corrosion; not especially suited for thin sections; limited to cold-chamber machines.
308	SC84B	380	Die	B85	QQ-A-591 Alloy 380	—	Same as SAE 306 but suitable for use in either cold-chamber or gooseneck machines.
309	SG100A	A360	Die	B85	QQ-A-591 Alloy A360	—	Excellent casting characteristics; suited for use in thin-walled or intricate castings produced in cold-chamber casting machine; corrosion resistance high.
310	ZG61A	40E	Sand	B26	QQ-A-601 Class 17	—	General-purpose structural castings developing strengths equivalent to SAE 38 without requiring heat treatment.
311	ZG32A	Ternalloy 5	Sand and permanent mold	B26, B108	QQ-A-601 Class 24, QQ-A-596 Class 13	—	High-strength general-purpose alloy; excellent machinability and dimensional stability; very good corrosion resistance; can be anodized.
312	ZG42A	Ternalloy 7	Sand and permanent mold	B26, B108	QQ-A-601 Class 25, QQ-A-596 Class 14	—	High-strength general-purpose alloy; excellent machinability and dimensional stability; very good corrosion resistance; can be anodized.
313	ZG61B	A612	Sand	B26	QQ-A-601 Class 23	—	High-strength general-purpose alloy; excellent machinability; easily polished; very good corrosion resistance; can be anodized.
314	ZC60A	C612	Permanent mold	B108	—	—	High-strength general-purpose alloy; excellent machinability; easily polished; very good corrosion resistance; can be anodized.

* Most commonly used commercial alloy designation listed for information only.
† SC64C was in ASTM B26-57T and B108, but has not been in subsequent issues.

Table 9–6 (*continued*)

						SAE	Characteristics and uses
315	ZC81A ZC81B	Tenzalloy Tenzalloy	Sand and permanent mold	B26 B108	QQ-A-601 Class 22 QQ-A-596 Class 12	— —	High-strength general-purpose alloy; excellent machinability; easily polished; very good corrosion resistance; can be anodized
320	G4A	214	Sand	B26	QQ-A-601 Class 5	—	Moderate-strength high-resistance to corrosion.
321	SN122A	A132	Permanent mold	B108	QQ-A-596 Class 9	—	Pistons, low expansion.
322	SC51A	355	Sand and permanent mold	B26 B108	QQ-A-601 Class 10 QQ-A-596 Class 6	4210, 4212, 4214 4280, 4281	General use where high strength and pressure tightness are required, such as in pump bodies and liquid-cooled cylinder heads.
323	SG70A	356	Sand and permanent mold	B26 B108	QQ-A-601 Class 3 QQ-A-596 Class 8	4217 4284, 4286	Alternate for SAE 38 in intricate castings.
324	G10A	220	Sand	B26	QQ-A-601 Class 16	4240	High strength and ductility; structural use; required special foundry practice.
326	SC64D	Allcast	Sand and permanent mold	B26 B108	QQ-A-601 Class 18 QQ-A-596 Class 11	—	General purpose alloy.
327	SC82A	Red X-8	Sand	B26	QQ-A-601 Class 20	—	Similar to SAE 322 and 323.
328	SC122A	B132	Permanent mold	B108	QQ-A-596 Class 10	—	Pistons.
329	SC64C†	319	Sand and permanent mold	B26 B108	QQ-A-601 Class 18	—	General-purpose low-cost alloy; good foundry characteristics; good machinability; similar to SAE 326.
332	SC103A	F132	Permanent mold	B108	—	—	Primarily intended for automotive use in pistons.
334	—	E132	Permanent mold	—	—	—	Pistons.
335	SC51B	C355	Permanent mold	B108	—	—	Similar to SAE 322, but has greater strength.
336	SG70B	A356	Permanent mold	B108	—	—	Similar to SAE 323, but has greater strength.
380	CS42A	B195	Permanent mold	B108	QQ-A-596 Class 4	4282, 4283	Modification of SAE 38 suitable for use in permanent molds.

Table 9–7. Applications of SAE Wrought-Aluminum Alloys

SAE No.	Former SAE No.	General Data
AA1100	25	Where forming and resistance to corrosion are required and low stresses are involved.
AA2011	202	Screw-machine products and fittings.
AA2014	260	Highly stressed forgings; truck, frame rails, and cross members.
Alclad AA2014	246	High-strength structural applications where high resistance to corrosion is required.
AA2017	26	Screw-machine products and ordinary structural applications.
AA2018	270	Forged pistons.
AA2024	24	High-strength structural applications; truck and trailer forged wheels.
Alclad AA2024	240	High-strength structural applications where high resistance to corrosion is required.
AA2117	204	Rivets for fastening truck and body panels.
AA2218	203	Forged cylinder heads and muffs.
AA3003	29	Higher strength and slightly less workable than AA1100; similar uses.
Alclad AA3003	245	Combines good forming characteristics with high resistance to corrosion; mechanical properties similar to AA3003.
AA3004	20	Miscellaneous body parts, fuel and brake lines.
Alclad AA3004	242	Body parts where high resistance to corrosion is required; mechanical properties similar to AA3004.
AA4032	290	Forged pistons.
AA4043	205	Welding wire for welded assemblies.
AA5050	207	Higher strength and slightly less workable than AA1100; similar uses.
AA5052	201	Truck- and bus-panel construction.
AA5083	—	Cargo tanks, storage tanks, armor plate, high-strength welded parts, and missiles.
AA5086	—	Storage tanks, stressed skin applications, and welded parts.
AA5154	208	Fuel tanks for trucks and tractors; truck cab parts.
AA5252	—	Automotive trim.
AA5257	—	Automotive trim.
AA5357	209	Automotive trim.
AA5454	251	Fuel tanks and hot cargo tanks for trucks and tractors.
AA5456	—	Cargo tanks, storage tanks, high-strength welded automotive and missile parts, and armor plate.
AA5457	252	Automotive trim.
AA5557	—	Automotive trim.
AA5657	—	Automotive trim.
AA6061	281	Combines good forming characteristics with high resistance to corrosion and moderately high strength; truck, trailer, and body chassis parts, frame rails.
Alclad AA6061	244	High-strength structural applications where high resistance to corrosion is required; mechanical properties similar to AA6061.
AA6062	211	Combines good forming characteristics with high resistance to corrosion and moderately high strength; uses similar to AA6061.

Table 9–7 (*continued*)

AA6063	212	Truck-body and trailer-flooring construction.
AA6151	280	Complicated forgings of moderate strength; fittings.
AA6463	253	Decorative trim for automotive-vehicle construction.
AA6951	213	For fabrication of assemblies by brazing.
AA7072	214	Wire for metallizing applications.
AA7075	215	High-strength structural applications.
Alclad AA7075	241	High-strength structural applications where high resistance to corrosion is required.
AA11 Brazing Sheet	247	Special one-side-clad sheet suitable for fabrication of assemblies by furnace brazing operations.
AA12 Brazing Sheet	248	Similar to AA11 Brazing Sheet, except both surfaces of the sheet are clad.
AA21 Brazing Sheet	249	Special one-side-clad sheet for production of stronger and more rigid assemblies, since core alloy is heat-treatable, and the product can be quenched from the brazing temperature.
AA22 Brazing Sheet	250	Similar to AA21 Brazing Sheet except both surfaces of the sheet are clad.

(From 1964 SAE Handbook.)

Magnesium is used for covers and other lightly loaded parts where weight saving is very important, as in aircraft and military engines. It can be cast in sand or dies, forged, and is available in bar stock. Reference 9.00, Sections J464–466, gives chemical and physical properties of commerical magnesium alloys and discusses their manufacture and application.

BEARING AND BUSHING ALLOYS

Alloys useful in internal-combustion engines are covered by ref. 9.00, Section J459. Included are alloys based on tin, lead, copper, and aluminum. Their applications will be discussed in greater detail in Chapter 11.

The range of their physical properties is listed in Tables 9–3 and 9–4.

Copper-Base Alloys

An enormous number of so-called brasses and bronzes are included in this category, quite aside from copper for electrical applications (see ref. 9.00, Section J46/9; see also Tables 9–3 and 9–4). Especially notable is their very wide range of properties, depending on composition and method of fabrication. By suitable control of these factors extremely high values of strength, resistance to corrosion, easy casting and machining, high heat and electrical conductivity, and excellent bearing properties against steel can be obtained. In engines they are used in the electrical systems, for bushings, for water-pump rotors and water-system fittings, for oil and water radiators, and for fuel lines, carburetor parts, etc.

Sintered Materials

These are mixtures of metal powder, formed into parts in dies under extremely high pressure (9.00, Section J471a). The resulting somewhat porous structure has good oil-absorbing properties. Most so-called oilless bearings are made of sintered bronze impregnated with oil. Bearings of this type are used in engine water pumps, generators, distributors, and for other lightly loaded bearings where it is inconvenient or undesirable to provide any other lubricant. Other uses for sintered material are oil-pump gears and water-pump rotors. The forming of sintered materials in dies can give finished dimensions so accurate that little or no machining is subsequently required. Thus, for large-quantity production this method of manufacture may be very economical, especially for small parts. Iron-base sintered materials can have tensile strengths comparable to those of steel and have been used to some extent for gears (9.70, 9.71). Further developments in this field are to be expected.

MISCELLANEOUS MATERIALS

Reference 9.00 contains useful information and specifications on the following types of materials used in internal-combustion engines:

Material	Section
Electrical resistance alloys	J470a
Zinc die-casting alloys	J468
Brazing alloys	J472
Solders	J473a
Elastomer compounds	J14
Elastomeric materials	J200a
Elastomer compounds for engine mounts	J16

SPECIFIC CHOICE OF MATERIALS

Table 9–1 shows materials typically used in internal-combustion-engine parts as of 1966. The question of the most appropriate materials for each type of engine part will be dealt with further in connection with Engine Detail Design, Chapters 11 and 12. See also Tables 9–6 and 9–7.

10 | Engine Design I: *Preliminary Analysis. Cylinder Number, Size, and Arrangement*

INTRODUCTION

The internal-combustion engine has been in use for approximately 100 years, during which time it has gone through an astonishing amount of detail development, but without basic changes since the invention of the Otto "silent" engine about 1876 and the Diesel engine about 1897 (see Volume I, refs. 1.01–1.061 for historical material). Internal-combustion engines have reached the stage where improvement in detail design is more likely than in over-all concept. This is not to say that further important improvement is unlikely. On the contrary, there is enough variety in successful contemporary design to indicate that the optimum is still unknown.

All basic indices, such as fuel economy, size, weight, and cost in proportion to output, reliability, and durability, are being improved each year. Even such important choices as 2-cycle versus 4-cycle and supercharged versus naturally aspirated engines are still open for many types of service. Thus, the design of a successful internal-combustion engine presents many challenging problems, many of which still remain unanswered or incompletely understood. It is with the hope of assisting in the continuing evolution of this type of power plant that the following three chapters are presented.

BASIC DECISIONS AND PRELIMINARY ANALYSIS

The discussion in this chapter will be confined to the conventional type of piston–connecting-rod–crankshaft engine. Unconventional types will be discussed in Chapter 13.

In view of the enormous expense of developing a new engine, the decision to design and build one should be taken only after the most careful consideration, which should result in answers to the following questions:

1. Reasons for a new design?
2. Type of service for which engine is intended?
3. Type of fuel to be used?
4. Power and fuel-economy requirements?

5. Best type to meet these requirements:
   Diesel or spark-ignition?
   Two- or 4-cycle?
   Supercharged or naturally aspirated?
   Number and arrangement of cylinders?
6. Estimated cost of development?
7. Estimated time of development?
8. Estimated manufacturing cost?
9. Assuming successful on-schedule completion of the development, will the new design be able to compete successfully against other equipment expected to be available for the same purpose at that time?

Reasons for a New Design. The reasons for a new design may be very definite, as in the case of a government or private contract for such an engine, or a vehicle with power requirements not satisfied by engines currently available. On the other hand, the reason for the new design may be the hope of competing successfully with existing engines used for the same purpose; in this case question 9 will always involve some degree of uncertainty.

Type of Service. The requirements of different types of service differ so widely that every engine must be designed with the intended type of major service in view. Success is very unlikely for designs not specifically oriented toward a particular service or group of services. Table 10–1 gives a classification of engines by types of service, based on present (1966) use, together with the approximate range of rated power per engine.

Type of Fuel. Except in very special cases, the fuel to be used must be one that is readily available in suitable quantities and at reasonable cost. In the case of spark-ignition road vehicles, the choice narrows to one or more of the gasolines available in the region. In the case of Diesel engines for road vehicles, the type of Diesel oil available at the roadside must be used. This is usually a light or medium grade (see Table 4–1). On the other hand, large marine Diesels are forced to use very heavy oils for economic reasons. Industrial applications may have special fuel limitations, such as natural gas or crude oil from nearby wells. " Bottle gas," that is, commercial propane-butane gas, may be appropriate where the cost of such fuel is low compared with that of other available fuels (5.443–4.447). Characteristics of available fuels change with time, as indicated by Fig. 4–7 and ref. 4.105.

General Service Requirements. Every successful engine must have to a reasonable degree the general characteristics of light weight and small bulk for a given power, good fuel economy, low initial cost, reliability, low maintenance requirements, long life, etc. However, these qualities are a matter of degree, and extreme values of any one are obtained only at the expense of some or all of the others. Thus, each type of service has priorities in regard to such qualities, as suggested in Table 10–2.

Table 10–1. Classification of Reciprocating Engines by Types of Service

Class	Service	Approximate hp Range One Engine	D or S-I	Cycle	Cooling
Road vehicles	Motorcycles, scooters	1–100	S-I	2,4	A
	Automobiles (ordinary)	20–150	S-I	4	W
	Automobiles (sport)	100–500	S-I	4	W
	Light commercial	50–200	S-I-D	4	W
	Heavy (long distance) commercial	150–500	D	2,4	W
Off-road vehicles	Light vehicles (factory, airport, etc.)	2–50	S-I	2,4	A–W
	Agricultural	4–200	S-I-D	2,4	A–W
	Earth moving	50–2000	D	2,4	W
	Military	50–2500	D	2,4	A–W
Railroad	Rail cars	200–1000	D	2,4	W
	Locomotives	500–4000	D	2,4	W
Marine	Outboard	0.5–200	S-I	2	W
	Inboard motorboats	5–1000	S-I-D	4	W
	Medium-size vessels	1000–4000	D	2,4	W
	Ships over 2000 tons	4000–50,000	D	2,4	W
Airborne vehicles	Airplanes	65–3500	S-I	4	A
	Helicopters	65–2000	S-I	4	A
Home use	Lawn mowers	1–4	S-I	2,4	A
	Snow blowers	3–6	S-I	2,4	A
	Light tractors	3–10	S-I	4	A
Stationary	Building service	10–1000	D	2,4	W
	Electric power	20–30,000	D	2,4	W
	Gas pipe line	1000–5000	S-I	2,4	W
Special for racing	Vehicles and boats	100–2000	S-I	4	W
Toys	Model airplanes, autos, etc.	0.01–0.50	HW	2	A

S-I = spark-ignition; D = Diesel; HW = hot-wire ignition of carburated mixture.

Service Overlapping. Many of the services indicated in Tables 10–1 and 10–2 are sufficiently similar so that engines designed for one can be used in another. Wherever this overlapping is possible, manufacturing costs are lowered because of the consequent larger production rate.

Power Requirements. Your author has suggested the following definition of an engine: "A machine with insufficient power." The validity of this definition is confirmed not only by the continuous upward trend of power ratings for almost all types of services, but also by the fact that in nearly fifty years of experience he has never heard a complaint about an engine having too much power!

Table 10–2. Engine Characteristics Emphasized by Type of Service

Service	Very Important	Moderately Important	Less Important
Small engines for home use, Outboard engines,	Light weight* Small bulk* Low first cost	Low noise level Reliability Low maintenance	Fuel economy Long life Vibration
Passenger automobile engines	Low noise and vibration Low first cost Reliability Flexibility† Low maintenance	Fuel economy Weight Bulk	Long life
Engines for commercial vehicles, light marine, and industrial use	Reliability Fuel economy Low maintenance	Weight Bulk Low noise and vibration Long life	First cost
Locomotive engines	Small bulk Fuel economy Low maintenance Reliability	Long life	First cost Weight Noise Vibration
Aircraft engines	Light weight Small bulk High take-off power Fuel economy** Reliability	Low vibration Low maintenance	First cost Long life Noise
Racing engines	High output within established rules	Reliability	All other
Large engines (over 12-in., 300-mm bore)	Fuel economy Long life Reliability Low maintenance	Low noise and vibration Bulk	First cost Weight

* Weight and bulk are considered in proportion to power
† Flexibility means smooth and efficient operation over a wide range of speed and load
** In relation to ton miles rather than horsepower

A dramatic illustration of the continuing demand for increased power is the fact that airplane-engine ratings have increased from the 12 hp of the Wright brothers' first engine in 1903 to a maximum of 400 hp in World War I, 3000 hp in World War II, and the present huge power ratings of turbojet engines.* The average power of U.S. automobile engines increased rapidly from 1951 to 1965, as shown by the curves of Fig. 10–1. The dramatic decrease in average power since that time, indicated by the 1984 data, was caused by the 1973 petroleum crisis and the legal restrictions on the fuel consumption of road vehicles put into effect in 1978 (Table 10–3).

Table 10–3. U.S. Department of Energy corporate average fuel economy (CAFE) standards (Code of Federal Regulations, Title 49, part 531). Mileage determined by standard test procedures.

	1978	1979	1980	1981	1982	1983	1984	1985 and after
Miles/gal.	18	19	20	22	24	26	27	27.5

These regulations have been met chiefly by means of drastic reductions in the size and weight of American automobiles, and by increases in the proportion of small cars (many imported from Europe and Japan).

In the process of producing and importing smaller cars, the six- and eight-cylinder engines of the period before 1973 have been largely replaced by the now-dominant four-cylinder type. Again, even for these smaller vehicles, there is a constant trend to increase engine power per unit weight of vehicle, so as to be able to advertise increased acceleration. Methods of improving engine power/weight ratio include the use of aluminum in place of cast iron, the improvement of air capacity by revised valve and port design, and the supercharging of some spark-ignition and many Diesel engines (See Vol. I, Chapter 13).

It must be evident that the maximum engine power required for a given service depends not only on technical requirements but also on custom, tradition, and the existing state of the art. In the case of engines for personal transportation the nontechnical factors are especially important, whereas in the case of engines for commercial use the technical factors tend to predominate, but even here other considerations are often very important.

In large aircraft, marine, and stationary installations the problem of engine ratings is complicated by the possibility, and often the necessity, of using multiple units to achieve a given total power. In spite of the many theoretical advantages of multiple small units, discussed in Chapter 11 of

---

* 20,000-lbf thrust is equivalent to 32,000 thrust hp at 600 mph and 85,000 hp at 1600 mph (supersonic).

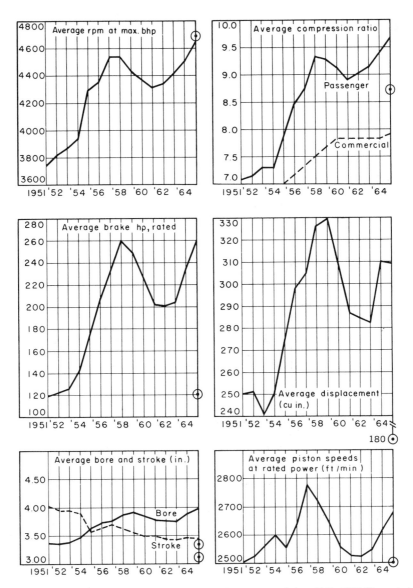

Fig. 10–1. Trend of U.S. passenger-automobile engine characteristics, 1951–1965 (from Automotive Industries, March 15, 1965). (●) 1984 values, estimated, spark-ignition engines only.

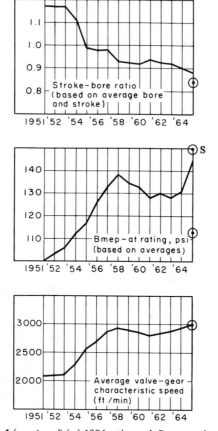

Fig. 10–1 (*continued*) (●) 1984 estimated. S = supercharged.

Volume I, the general trend has been toward larger power ratings even for engines used in multiple.

It must be evident, then, that the choice of maximum rated power for a new design is a critical one, and one for which no precise formula is available. However, the following general rules are in accordance with the author's experience:

1. Never, without special reasons, plan an engine with less power than competitive engines designed for the same use.

2. Design the engine so that increased ratings will be possible through further development. (An important item here is to use the largest valves that can be readily accommodated in the cylinder head, even though smaller valves might be sufficient for the initial power rating.)

3. In many cases it will be well to plan on the use of a given cylinder design in a range of engines having different numbers of cylinders (Table

10–4) ("cylinder design" is here taken to include pistons, rods, valves, and valve-gear parts).

4. The power rating should be based not on present competitive power ratings, but on the ratings of competitive engines estimated for the time the new design is ready for production.

Table 10–4. Number and Arrangement of Cylinders
(Usual Practice)

0–3	1	I
3–30	2–3	I
30–50	4	I
20–100	4–6	I, O, V
100–200	6–8	I, O, V
200–1000	6–12	I, O, V, OP
1000 +	12–24	I, V, OP

I = in-line, vertical or horizontal
O = opposed cylinders (in pairs)
V = V arrangement
R = radial
OP = opposed-piston (2-cycle)

Figure 10–2 shows the trend of power ratings of several airplane engines for the period 1935–1945. Experience showed that it took at least four years to design and develop a large aircraft engine to the point of service use. Thus, an engine design started in 1940 to compete with the Pratt and Whitney R-2800 engine should have been planned for a rating of not less than 3000 hp, or 65 per cent more power than that of the competitive engine at the time the design was started. As a matter of interest, the manufacturer of this engine made at least one attempt at a competitive design, which however did not measure up to the improved performance of the existing design when the preliminary development period of the new design was completed. The new design was therefore abandoned.

Fuel Economy. While good fuel economy is always desirable, its importance relative to other desirable characteristics varies according to the type of service. As we have seen in Volume I, the attainment of maximum fuel economy usually comes at the expense of specific output. For a given power, therefore, the most efficient engine will be larger, heavier, and probably more expensive than one where fuel economy can be sacrificed, to a greater or lesser degree, in favor of high specific output.

Fuel economy becomes more important as the *use factor* of the engine increases. By use factor is meant the fraction of time in use multiplied by

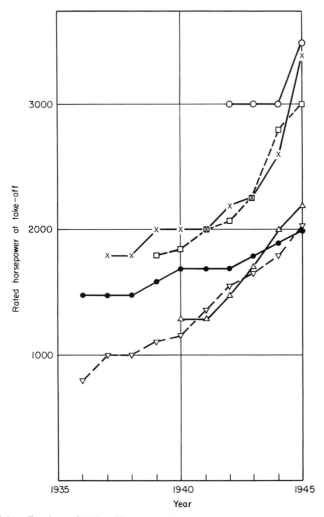

Fig. 10–2. Take-off ratings of U.S. military aircraft engines, 1936–1945 (from Army Air Forces Letter TSEPP: EAW jds, Feb. 11, 1956).

×	Wright	R-3350	△ Packard Rolls-Royce	V-1650
○	P and W	R-4360	▽ Allison	V-1710
□	P and W	R-2800	● Wright	R-2600

the average power developed and divided by the rated power. The use factor of a home lawn-mower engine is near zero, especially in climates with a long winter. Obviously, fuel economy has but little importance here in comparison with factors such as low cost, light weight, and easy starting.

The use factor of a passenger automobile is generally low, especially for high-powered U.S. automobiles. For example a car rated at 200 hp, used on the average of 2 hours per day (20,000 miles per year) at an average of

20 hp has a use factor of $(2/24) \times (20/200) = 0.0083$, or less than 1 per cent. Under U.S. conditions the combined cost of depreciation, maintenance, garaging, and insurance is many times the cost of fuel.

Diesel Versus Spark Ignition

In many cases this question is decided by the type of fuel to be used. For example, only fuel suitable for Diesel engines is available at dockside in quantities appropriate for large vessels, and in many parts of the world only gasoline is available at the roadside. In view of the adverse effects of large cylinder size on detonation (see Chapter 2), the larger the bore, the more difficult it becomes to design for spark ignition. With the exception of engines designed for natural gas, few spark-ignition engines are built with a bore greater than 6 in. (152 mm).

For engines where fuel supply is not restricted and bore is less than about 6 in., the chief reason for considering a Diesel engine as an alternative to a spark-ignition engine is fuel economy and fuel cost. Unsupercharged, the Diesel engine is heavier, larger, considerably more expensive, rougher and noisier than a spark-ignition engine of equal power, and it generally requires more sophisticated maintenance. With appropriate supercharging the size and weight of the Diesel can be made reasonably competitive, but the other disadvantages remain.

As we have seen, the economic value of low fuel comsumption and low fuel cost depends heavily on the *use factor*. Therefore services with low use factors tend toward spark-ignition engines, while the Diesel engine tends to be preferred where the use factor is high. Thus, nearly all engines used for personal service, that is, for noncommercial vehicles, small motorboats, lawn mowers, etc., use spark ignition. In most such services, the saving in fuel cost achieved by using a Diesel would not even be enough to offset the increased first cost alone, during the useful life of the apparatus.

For passenger automobiles the use of Diesel engines has been fairly common abroad (17 percent in Europe in 1984), chiefly on account of very high fuel prices. In the United States, Diesel automobiles were rare until after the fuel crisis of 1973. In 1984 about 5 percent of the automobiles in the United States were Diesels, most of them imported. Considering the now excellent fuel economy of gasoline automobiles, the fact that the retail prices of gasoline and Diesel oil are nearly the same, and the higher first cost and service requirements of Diesel automobiles, this use of Diesel engines seems not to be justified by economic reasons alone. Noise, difficult cold starting, and emissions are also handicaps in automotive applications. These problems are discussed more fully in chapter 13 of volume I and in chapters 3 and 7 of this volume.

Aircraft engines of the reciprocating type are all spark ignition because of weight considerations. Intensive development of Diesel engines for aircraft service during the 1930's (see Volume I, refs. 7.32, 12.58, and 13.22) did not produce really competitive types. Except for small sizes (less than 500 hp), the development of reciprocating engines for aircraft was completely abandoned upon the advent of the jet engine.

The question as to which type is appropriate comes up chiefly in the case of commercial engines for road, industrial, marine, and stationary service where both gasoline and Diesel fuels are available and the power requirement per engine does not exceed about 500 hp. Wherever this question is an open one, the final decision should be based on the advantages of the Diesel engine in regard to fuel economy, balanced against its disadvantages in weight, initial cost, maintenance cost, and noise in operation. In some cases, notably in marine service, the reduced fire hazard with Diesel fuel may be the decisive factor.

Nonhighway engines designed for more or less continuous use are likely to be Diesel, even in small sizes. For nonhighway use, the absence of a fuel tax makes the cost ratio of Diesel fuel to gasoline much lower than when a substantial tax is applied to both fuels. This fact furnishes a further incentive to use Diesel engines for many nonhighway purposes.

Two-Stroke or 4-Stroke Cycle

Perhaps the most convenient way of approaching this controversial subject is to consider the two applications in which 2-cycle engines are now predominant, namely, small spark-ignition engines and medium to large Diesel and gas engines.

The small spark-ignition 2-cycle engine is generally used for motorcycles, outboard motorboat engines (where it dominates the field) and as a light, portable engine for lawn mowers, chain saws, etc. These applications have the following features in common:

1. Low first cost
2. Low use factor
3. Low weight/power ratio (see Table 10–5 at end of this chapter).

All such engines are of the carbureted crankcase-compression type (see Volume I, Chapter 7) which constitutes the simplest prime mover now available. Besides the ignition system there are only three moving parts per cylinder: the piston, the crankshaft, and the connecting rod. The specific output is generally somewhat higher than that of the competing 4-stroke type, so that cost and weight are basically lower for a given power. On the other hand, the fuel economy is poorer by at least 25 per cent on account of wastage of carbureted mixture during scavenging. For this reason the type predominates only where the use factor is low and fuel economy is not critical.

Disadvantages, in addition to poor fuel economy, are irregular idling and light-load operation, and relatively high oil consumption, especially when the lubricating oil is mixed with the fuel, as is generally the case. This feature also has a nuisance aspect. The 2-cycle spark-ignition engine long ago acquired a reputation for difficult starting. This disadvantage has been largely overcome by technical advances but still remains a barrier to complete public acceptance, except in the outboard-motor field. For example, most lawn mowers use 4-cycle engines on this account alone.

Application of the 2-cycle engine to passenger automobiles is confined to a very few manufacturers (10.272). To this author, the 4-cycle engine seems much more appropriate on account of its better idling and light-load operation and its better fuel economy, achieved at the expense of very small, if any, increase in first cost under mass-production conditions.

In the case of Diesel engines the simple crankcase scavenging system is less appropriate, because the relatively low fuel-air ratios at which Diesel engines operate gives rather low mean effective pressures with this system. Also, crankcase scavenging becomes less attractive structurally as the number of cylinders is increased.

In adopting a separate scavenging blower, much or all of the cost advantage of the crankcase-scavenged engine is lost as compared with an unsupercharged 4-cycle engine. On the other hand, when using a separate scavenging system the 2-cycle Diesel engine suffers very little in fuel economy or light-load operation and tends to have a higher specific output than a 4-cycle engine of the same size (see later examples in this chapter) and also Tables 10–6 through 10–11).

For Diesel engines up to about 12-in. (30.5-mm) bore, both types are used, with 4-cycle engines predominating. With larger cylinders the 2-stroke cycle is predominant. The principal reason for using the 2-stroke cycle with large bores is that, with cylinders of equal size at or near the same piston speed and power rating, cylinder pressures are lower with 2-stroke operation than with 4-stroke operation, and maximum stresses in the engine structure are correspondingly less. As we have seen in Volume I, Chapter 11, it becomes increasingly difficult to avoid high thermal stresses as cylinders get larger, and the allowable maximum cylinder pressure decreases as the cylinder bore increases.

Since the possibility of higher specific output with the 2-cycle Diesel engine is present for all cylinder sizes, the question arises why it is not more generally used in the smaller cylinder sizes. The answer is probably to be found in the following facts:

1. There is a greater background of experience in 4-cycle-engine design.
2. The development work to secure good scavenging in a new 2-cycle design is likely to be greater than that required for good air capacity in a 4-cycle design.

3. Unless simple loop-scavenged cylinders are used, the 2-cycle design is no less complex than its 4-cycle counterpart.

4. Most 2-cycle Diesels have slightly poorer fuel economy than their 4-cycle competitors. (See examples 10-2, page 387, and 10-4, page 395.)

An interesting side light on the 2-versus 4-cycle question is the fact that the General Motors Corporation built none but 2-cycle Diesel engines until 1962, at which time they started to market a 4-cycle design (10.38) competitive with their 2-cycle automotive line (10.37). However, their larger Diesel engines ($5\frac{3}{4}$-in. and 9-in. bore) are still exclusively 2-cycle.

In the case of spark-ignited natural-gas engines, many of which are converted Diesels, there are many 2-cycle types even in the larger cylinder sizes. Since the gas is injected after scavenging, there is no loss of fuel during the scavenging process. For engines of equal size, the 2-cycle examples generally have somewhat higher fuel consumption (Table 10–12).

Air-Cooling Versus Water-Cooling

The fundamentals of engine cooling are explained in Volume I, Chapter 8. In that chapter (Eq. 8–1, p. 269) it is shown that in systems of similar geometry the heat transfer between a surface and a moving fluid is approximately proportional to kR^nP^m/L, where k is the coefficient of thermal conductivity of the fluid, L a characteristic length, P the fluid Prandtl number, and R the fluid Reynolds number $\rho u L/\mu g_0$. Here ρ is the density, u the velocity, μ the viscosity of the fluid, and g_0 is Newton's law coefficient. At given values of u and L, the heat-transfer coefficient varies with $k(\rho/\mu)^nP^m$. Measurements indicate that $n \cong 0.8$ and $m \cong 0.4$ for the usual system (Volume I, Fig. 8–2, p. 270). For water at 160°F,

$$k(\rho/\mu)^{0.8}(P)^{0.4} = 0.415(62.4/0.4)^{0.8}(2.50)^{0.4} = 5.4.$$

For air at the same temperature and at standard sea-level pressure, this quantity is:

$$0.0129(0.065/0.02)^{0.8}(0.70)^{0.4} = 0.031.$$

In these equations the conductivity is in Btu/hr ft°F, the density in lbm/ft³, and the viscosity in centipoises.

The inherent advantage of water as a coolant is in the ratio $5.4/0.031 = 175$. In practice this advantage can be largely offset by using air at much higher velocity and at lower temperature than in the usual water-cooling system, together with the addition of finning to increase the area of the outside surface of the cylinders.

Typically, the finned area of an air-cooled cylinder is from 10 to 125 times larger than the area of the unfinned cylinder, and the air velocities used are four to eight times as great as the corresponding water velocities. The smaller of these numbers correspond to small engines of low specific output, while the larger ones apply to aircraft and military engines of high specific output. Another factor offsetting the fundamental advantage of water as a coolant is that air is seldom hotter than 100°F (38°C), while with water cooling the usual coolant-in temperature is 160–180°F (71–82°C). However, with comparable quality of design, air-cooled cylinders generally show higher temperatures at the critical areas (exhaust valves, seats and ports, and spark-plug bosses) than water-cooled cylinders under similar circumstances.

Water-cooled engines in land vehicles and in locations without a convenient supply of cool water must of course give up their heat to the air through heat-transfer devices which have large surface areas compared to the cylinder surfaces. The fundamentals of this problem are discussed in Volume I, Chapter 8, and some further design considerations are discussed in Chapter 12 of this volume.

Applications of Air-Cooling. For reasons already given, the difficulties associated with cooling increase as cylinders become larger. On this account few, if any, engines with cylinder bores greater than 6 in. (152 mm) are air-cooled. Above this cylinder size, air-cooling is inappropriate in any service. Another category where air-cooling seems inappropriate is for watercraft (with the exception of those driven by air propellers, in which case aircraft-type engines are a logical choice).

In addition to the above categories, air-cooling is particularly attractive for engines with one or two cylinders. In these there is plenty of room for cooling fins, and such engines are most often used under conditions where a radiator and water system would hardly be justified. Thus air-cooling seems the logical choice for small portable and stationary engines and motorcycle engines (10.60–10.650, 10.66–10.691).

Another place where air-cooling is attractive is in opposed-cylinder engines. In these, the cylinder spacing required by the crankshaft is usually large enough to allow for fins between the cylinders (10.22, 10.26, 10.816, 10.850–10.854).

For in-line and V engines of four or more cylinders, a comparison of air-cooled and water-cooled engines of equal power will usually indicate that

1. Including the cooling system, the over-all weight will be nearly the same, provided both use aluminum cylinders and crankcases. The water-cooled installation will be heavier than one with an aluminum air-cooled engine if it uses the conventional cast-iron structure.

2. The air-cooled engine will be longer because of the wider cylinder spacing. The over-all dimensions of the two installations will depend on the design and placement of radiators, fans, and air ducts. With best design

for each, the over-all volume may not be very different. However, the water-cooled installation has the advantage that the radiator can be remote from the engine.

3. If the water-cooled engine is of the conventional en-bloc design, the cost of engine plus cooling system will be less than that of the air-cooled installation with its separate cylinders and more elaborate fan-and-duct system.

4. The air-cooled installation will be noisier because of the higher air velocities and the absence of the sound-absorbing properties of water jackets.

5. Formerly, water-cooling had serious disadvantages due to leakage, corrosion, wear of pump shafts and seals, and loss of antifreeze due to boiling. These difficulties have been overcome by means of better pump and seal design, improved antifreeze and antirust additives, more durable rubber hose, and the use of closed water systems at pressures above ambient. Well-designed water-cooling systems now appear to be as reliable and trouble-free as air-cooling systems.

In practice only a small fraction of automotive and industrial engines with four or more cylinders in line are air-cooled (10.391, 10.398, 12.920). The prevailing considerations favorable to water-cooling are lower cost, shorter engine length, and less noise.

The U.S. military services have favored air-cooling for most of their special vehicle engines because, compared to water-cooled engines, they are less vulnerable to small-arms fire and to shell fragments, more tolerant of wide extremes of ambient temperature, and because they require no water or antifreeze material (10.820, 10.828, 12.920).

For civilian aircraft engines, air-cooling appears the logical choice because of the high air velocities available during operation, and the elimination of water-system elements that are subject to failure.

Supercharging or Not?

Spark-Ignition Engines. For reasons already given, the supercharging of spark-ignition engines involves a compromise with efficiency which is only justified in special cases, including

1. *Aircraft Engines.* Here supercharging is used both to provide high specific output for takeoff and to compensate for lowered air density at operational altitudes. All but small engines for light airplanes are supercharged. The problem of detonation is solved by using high-octane fuel.

2. *Automobile Engines.* With the advent of small, efficient turbosuper-

chargers and reliable electronic controls, many luxury and sports-type automobiles are being supercharged. As explained in Chapter 13 of Volume I, this usually lowers fuel economy in comparison with the same engine naturally aspirated. The decision to use supercharging in this way is more one of marketing than one of utility.

3. *Racing Engines.* Here high specific output has exaggerated importance and supercharging must be used wherever it is allowed. See example 10–5 on p. 398.

4. *Large Natural-Gas Engines.* In this case the saving in weight and bulk by means of supercharging is great, and the fuel used has a high resistance to detonation. Nearly all gas engines rated above 500 hp are supercharged. (10.70–10.77).

Diesel-Engine Supercharging. As already mentioned in this volume and in Volume I, no limit on supercharging is imposed on Diesel engines by combustion. The decision whether or not to use supercharging, and if so how much, depends on a balance between the question of the relative mechanical simplicity of the unsupercharged engine together with its generally lower mechanical and thermal stresses, and the smaller size and weight of a supercharged engine with the same rating. The initial cost may be in favor of either type, depending on the cost of the supercharging equipment and the reduction in cost due to decreased engine size. In general, the larger the engine, the greater the likelihood of the supercharged design being cheaper. Actually, the use of supercharging for Diesel engines seems to be increasing in all categories, except for small industrial or utility engines where low cost is more important than small size and light weight. As indicated by Table 10–7 (page 413), small automobile Diesels may be supercharged at ratings of 50 hp or less. Diesel engines for trucks, buses, and locomotives and medium and large marine engines are almost always supercharged. The same is true for stationary applications, such as electric generating, except in the smallest sizes. In view of the general availability of relatively efficient turbosuperchargers of all sizes, it is becoming difficult to justify no supercharging for most Diesel-engine applications (see Tables 10–7 through 10–11).

Limits on Supercharging. As stated in Volume I and again in this volume, the allowable amount of supercharging in the case of Diesel engines depends only on questions of reliability and durability, plus economic factors such as the cost of supercharging equipment as a function of its capacity and pressure ratio.

The limits based on reliability and durability are difficult to determine on a theoretical basis. In practice they have to be determined experimentally and by service experience. The limits on engine rating also depend on the type of service and especially on the percentage of the running time that is at rated conditions. At one extreme are large marine and stationary installations designed to run at rating with great reliability for long periods of time. At the other extreme is the racing engine, where the requirement is to run at rating for short periods only, and reliability may be sacrificed for high specific output.

There is a large body of literature on the effects of supercharging of Diesel engines on engine temperatures, thermal and mechanical stresses, and durability (10.45, 10.492–10.494, 10.880–10.900). A study of this material should be helpful to the designer in selecting appropriate output limits for any given design.

With a given cylinder size and a given engine design, thermal stresses will be proportional to the temperature differences across a given section, which in turn will be dependent on the *heat flux*, or rate of heat flow through the wall per unit time per unit area.

Equation 8–16 (page 290) in Volume I expresses heat flux to the coolant as follows:

$$\frac{\dot{Q}}{A_p} = KG_g^{0.75}(T_g - T_c),$$

where $\dot{Q}$ is the flow of heat to the coolant per unit time, A_p the piston area and K a coefficient depending on units and detail design, G_g the flow of air and fuel through the engine per unit piston area per unit time, T_g the mean effective gas temperature, and T_c the mean coolant temperature.

For a given engine operating at a given fuel-air ratio with constant values of T_g and T_c, the indicated power output will be proportional to mass flow of air. In this case the heat flux $\dot{Q}/A_p$ will be proportional to $(P_i/A_p)^{0.75}$, where P_i is the indicated power. At maximum output, where mechanical efficiency is high, heat flux will be nearly proportional to the specific brake output, P/A_p raised to the 0.75 power. From this relation we may conclude that for a given engine size and design, if heat flux is the limiting factor, there will be a limit on specific output. French and Lilly (10.45) state that with 4-cycle locomotive Diesel engines of about 10-in. (254-mm) bore, the limit found practical in 1965 is about 3.55 hp per sq in. piston area (0.55 hp/cm²). Tables 10–9 through 10–11 show a number of recently designed Diesel engines exceeding this limit. Improved turbo-supercharger efficiency and improved heat flow and structural design are responsible.

Figure 13–9 of Volume I (p. 474) explores the possibility of achieving a given specific output (in this case 6.15 ihp per sq in. for a 2-cycle engine) by increasing the supercharger pressure ratio as the fuel-air ratio is reduced. Evidently this procedure can be used effectively down to fuel-air ratios as low as $F_R = 0.3$, provided appropriate supercharger pressure ratios can be provided. In actual practice, when Diesel engines have been converted to supercharging, the rated fuel-air ratio has often been reduced, with a consequent reduction in T_g and improvement in reliability, and also a cleaner

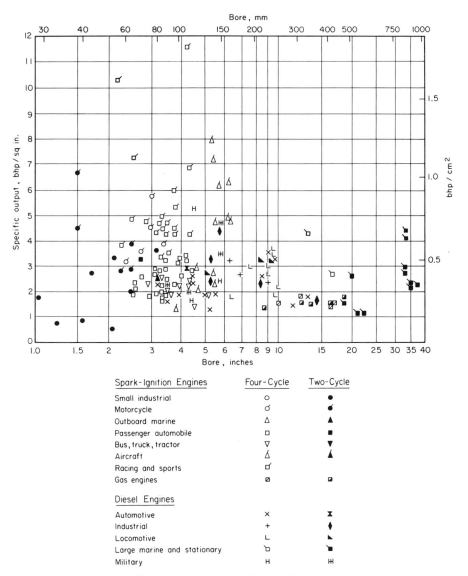

Fig. 10–3. Specific output versus bore. From Tables 10-5 through 10-15.

Spark-Ignition Engines	Four-Cycle	Two-Cycle
Small industrial	o	●
Motorcycle	♂	♂
Outboard marine	△	▲
Passenger automobile	□	■
Bus, truck, tractor	▽	▼
Aircraft	△	▲
Racing and sports	♂	
Gas engines	⊘	▣

Diesel Engines		
Automotive	×	✕
Industrial	+	◆
Locomotive	L	◣
Large marine and stationary	ʋ	◪
Military	H	Ⱨ

exhaust. Usually, however, fuel-air ratios at rating are no lower than $F_R = 0.5$.

Effect of Cylinder Size on Specific Output. In Volume I, Chapter 8, it was shown that thermal stresses tend to increase as cylinder bore increases. In order to minimize this effect, design changes such as those suggested by Volume I, Fig. 8–7 (p. 282) may be used. Examination of current successful engine designs shows an increasing use of composite pistons and increasing attention to piston, port, and valve cooling as the bore increases. In spite of these measures, however, Fig. 10–3 shows that the maximum rated specific output decreases with bore. This relation is attributable both to thermal-stress limitations, and to the fact that engine life and reliability tend to be of greater importance as cylinder size increases. Note that the highest specific outputs are for racing engines of small bore, in a service where durability and reliability are of minimum importance. Another factor influencing the trend toward decreasing specific output with increasing bore is that, as unit size increases, endurance testing becomes more difficult and much more expensive. Low values of rated specific output in the small-bore range are for applications where low first cost is more important than light weight, or where piston speed is held to low values to avoid gearing.

Figure 10–4 is presented to facilitate rapid computation of specific output.

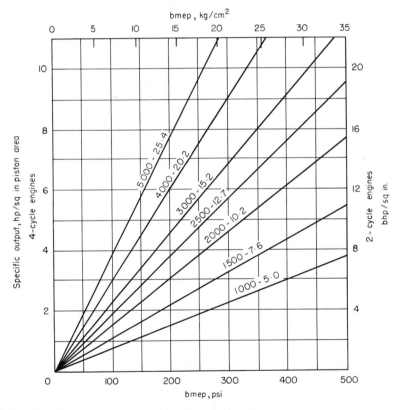

Fig. 10–4. Specific output versus bmep and piston speed. Numbers on curves are piston speed in ft/min and m/sec.

Relation of Engine Performance to Supercharger Characteristics. This question is fully discussed in Volume I, Chapters 10 and 13. Figures 13–9 and 13–10 of Volume I show Diesel-engine performance versus compressor pressure ratio for various supercharger arrangements, based on realistic assumptions as to the state of the art in 1960.

Figure 10–5 in this volume compares some test results on 4-cycle Diesel engines with the corresponding curves of Fig. 13–10 in Volume I. The curve for the engine with variable-compression-ratio pistons shows that gains in output, at a sacrifice in fuel economy, can be made by this means. For constant-compression-ratio engines, it appears that a compressor pressure ratio of 3 will yield a bmep for 4-cycle engines in the range of 250 psi (17.5 kg/cm²), which corresponds to a specific output of 3.8 bhp/in.² at a piston speed of 2000 ft/min (10.1 m/sec). This pressure ratio is also near the practical maximum for a single-stage centrifugal supercharger.

In conclusion, the choice of natural aspiration or supercharging, and the inlet-pressure ratio to be used with supercharging, must depend on the particular application, including especially the relative importance of weight, bulk, first cost, and maintenance costs.

Improvement in engine design since World War II, together with the never-ceasing demand for increased engine power, has resulted in a steady increase in the proportion of Diesel engines offered with supercharging and also in specific output of supercharged engines. There is good reason to believe that this increase will continue as design methods improve.

Number and Arrangement of Cylinders

This choice will first of all depend on the power output required. For engines below 3 or 4 hp there are few circumstances that would justify more than a single cylinder, partly because of the usual emphasis on low first cost in this class and partly because the theoretical benefits of small cylinders, discussed in Volume I, Chapter 11, are difficult to realize with cylinders smaller than about 2 in. (50 mm) bore.

As rated power increases, the advantages of small cylinders in regard to size, weight, and improved engine balance, point toward increasing the number of cylinders per engine. As we have seen (Volume I, Chapter 11), engine size and weight decrease as the number of cylinders increases, as long as similitude in design is maintained and provided there are no restrictions on rpm. As a matter of actual experience, however, it is impracticable to maintain exact similitude for very small cylinder sizes, and therefore the gain in weight and size diminishes to near zero when the cylinder bore drops below about 2 in. However, reduction in displacement volume for a given power continues down to quite small cylinder sizes (see example 10–5, page 398).

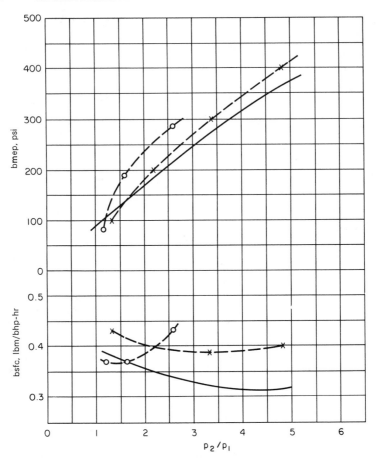

Fig. 10–5. Performance of turbosupercharged 4-cycle Diesel engines versus compressor pressure ratio, p_2/p_1.

Curve	Comb. Chamber	Bore in.	Piston Speed ft/min	F_R	r	Maximum Cyl. Press. psia	After-cooled	Reference
——————	Open	—	2000	0.6	14	1760	Yes	Volume I, Fig. 13–10 # 1
— — × —	Divided	4.5	2200	0.68 0.60	—	3200	Yes	10.880
— — ○ —	Open, variable comp. ratio	8.5	2000*	0.7	15–8	1250	Yes	10.881

* Tests were made at 1375 ft/min. Bmep and bsfc corrected to 2000 ft/min by subtracting 10 psi for increased friction mep (see Volume I, Fig. 9–27, p. 350).

It is evident, then, that the choice of the number of cylinders to be used will be a compromise between low specific weight and volume, tolerable vibration levels, costs of manufacture and maintenance, life expectancy, and over-all shape of the engine package.

For installations where rpm is fixed by outside considerations, stroke is determined by rpm and piston speed, and bore by an acceptable ratio to stroke. Here the number of cylinders will be governed by bmep and power output (see example 10–3, page 391).

Present practice in regard to number of cylinders is summarized in Table 10–4.

Cylinder Arrangement. One of the most important factors in the choice of cylinder arrangement is the appropriateness of the resultant shape of the engine to the space available, including considerations of accessibility for service and repairs. For example, cylinders located below the crankshaft are extremely inappropriate for marine use and for automobiles. On the other hand, radial and inverted-V engines have proved both appropriate and accessible in aircraft. The following discussion should be read with these considerations in mind.

In-Line Engines. With 2 or more cylinders, the in-line type is used where the shape is especially suitable or where some sacrifice in light weight and compactness for the sake of mechanical simplicity and easy maintenance seems justified. The in-line six, as we have seen, is especially attractive from the point of view of balance and vibration. In-line engines of more than 6 cylinders tend to be very long in relation to their other dimensions and usually have problems involved with crankshaft torsional vibration. In spite of these disadvantages, in-line engines of up to 12 cylinders are still used for large ships because their shape fits well into hull design, especially in the case of ships with 2 or more propellers.

V Engines. V engines of 8 cylinders or more have excellent balance and relative freedom from vibrational problems. The V-8 engine fits very well into the space available in front-engine automobiles, especially when the stroke-bore ratio is less than 1.0 (see Tables 10–6 and 11–2). This characteristic, together with its good balance and relatively simple design explains the great popularity of the V-8 engine for higher-powered automobiles and for the many other applications where it proves to be a good compromise between compactness, high specific output, and reasonably low cost. The V-12 engine is also an excellent type from all points of view and is widely used for engines above about 300 hp. V engines of more than 12 cylinders are used for large power outputs when it is necessary to limit cylinder size or in order to use a standard cylinder design for a wide range of power ratings. Two-, four-, and six-cylinder V engines have balance problems (see Chapter 8). They are used chiefly in small sizes where the resultant shape is appropriate, such as in motorcycles and small automobiles.

Opposed-Cylinder Engines. Such engines might be classed as engines with a V angle of 180°, except that they usually require a separate crank for each cylinder, while normal V engines have 2 cylinders working on each crank. This cylinder arrangement lends itself particularly well to air-cooling, since the individual cranks require a relatively wide cylinder spacing which provides room for cooling fins between the cylinders. Examples of this type will be found in Fig. 11–8 and in refs. 10.22, 10.26, 10.850–10.854. Opposed-cylinder engines are most often used where light weight and short length are important, as in outboard marine engines, motorcycles, rear-engined automobiles, and light airplanes. They are also appropriate for under-the-floor installation in buses and trucks.

Radial Engines. Radial engines are used mostly for aircraft, because the shape is convenient for this service and lends itself readily to air-cooling and to cylinder accessibility. Basically, the radial arrangement gives the lowest weight per unit displacement because the material in crankshaft and crankcase is at a minimum for a given number of cylinders. Such engines are used only rarely for other types of service (10.58, 10.75, 10.76).

The Opposed-Piston Engine. This type is used for 2-cycle Diesels in many types of service. Included in this category are locomotive engines (10.48), large marine and stationary engines (10.57, 10.592), and military-vehicle engines (10.823, 10.824, and Fig. 11–24). The now obsolete Junkers aircraft Diesel engine still holds the record for specific output of Diesel engines in actual service (Volume I, Fig. 13–11). The advantages of this engine type include the absence of cylinder heads, which should result in reduced heat loss, and the opportunity to design a very effective through-scavenging system with piston-controlled ports (see Volume I, Chapter 7). Disadvantages include the requirement for two crankshafts geared together, and a shape that may be awkward in many installations. Opposed-piston is also the usual arrangement for free-piston engines (see Volume I, pp. 478–482 and refs. 13.50–13.392, and the discussion of this type of engine in Chapter 13 of this volume).

Choice of Engine Type

Where there is any doubt about the best number and arrangement of cylinders, the final decision should be made only after a study made by means of actual "layouts" (preliminary assembly drawings) of the various alternatives. Such layouts will also be necessary for estimates of manufacturing costs, accessibility factors, etc.

Development Time

The time for development will depend on the intensity of the effort, as measured by funds, personnel, and equipment to be devoted to it; the experience of the personnel assigned to the work; and the degree in which the projected engine incorporates new and untried features.

Experience has shown that the time between start of the design and first service use of *conventional** engines, when carried out by large organizations which can devote very intense effort to the work, is from two to five years. As a measure of the effort expended by large organizations, it may be mentioned that in the development of a very conventional U.S. automobile engine of 1962, six hundred engines are said to have been used up in preproduction testing of the engine and vehicle. Other examples are:

1. The development time for the (unconventional) sleeve-valve cylinders of the Bristol air-cooled radial aircraft engines used in World War II extended over a period of more than ten years.

2. The very unconventional rotary "Wankel" engine (see Chapter 13) has been under development for more than ten years and is not yet (1966) fully serviceable. Many competent engineers doubt that it ever will be.

3. Untold time and funds have been spent on various types of "crankless" engines and other unconventional types, without success and without present promise of success. (See Chapter 13.)

Cost of Development

To predict this cost, estimates of the time, materials, personnel, and equipment to be devoted to development must be made. The more conventional the design and the more experienced the organization, the more accurate will these estimates be.

History has shown that organizations without a background of engine-development experience generally greatly underestimate time and costs of an engine development. In any case, the high cost of engine development makes it necessary to consider very carefully whether a new design is really justified.

Manufacturing Cost. This cost, of course, depends not only on design details but also on the rate and methods of production. After the assembly drawings are made, experienced engine manufacturers can predict production costs quite accurately. Of course production costs include not only the actual labor and materials but also the appropriate overhead charges.

The tremendous influence of rate and methods of production is illustrated by the case of U.S. passenger-automobile engines. The large quantities in which they are produced has made them very much lower in cost per horsepower than any other type. Passenger-car engines in the U.S.A. retail for about $2.00 per rated horsepower. Other high-production

---

* The word "conventional" denotes engines that do not depart in their essential features from those made successfully in the past by the same organization or its close competitors.

engines include those for commercial vehicles, small motorboat engines, small engines for home use, and certain industrial categories. These types retail for $10.00 to $20.00 per horsepower. Large engines, built to order, may cost $50.00 or more per horsepower.

In considering any new design, possible competition from existing or converted low-cost engines should be taken into serious consideration. For example, the low cost of quantity-produced engines connected to separate air compressors makes it unlikely that a basically simpler device, such as the free-piston air compressor, could compete successfully in the small-air-compressor market. The free-piston compressor is a single-purpose machine and therefore would never be produced in quantities comparable to those of conventional industrial engines used for many other purposes (see Table 10–3).

Market for a New Design

Except for designs completed under contract for government or other outside sources, the potential market of the completed product is obviously the most important consideration of all. While "market research" is outside the scope of this volume, it should be pointed out here that the

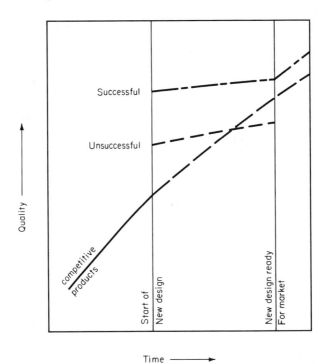

Fig. 10–6. Design-forecast curves.

designing engineer can contribute to a meaningful answer by insuring that the technical aspects of the problem are understood by management. One of the principal contributions the engineer can make is an estimate of the probable development of competitive equipment during the development period of the new design. Figures 10–1 and 10–2, together with the discussion of power rating, have already indicated the nature of this problem.

Figure 10–6, a graph similar to Fig. 10–1, can be adapted for any given qualitative measurement, such as fuel economy, specific weight, specific volume, service life, cost per rated horsepower, etc. In reaching a final decision on whether or not to proceed with a new design, curves similar to those of Fig. 10–6 should be prepared, showing each projected quality of competitive equipment as compared with the estimated quality of the new design at the end of its development period. Intelligent preparation of such curves would have saved many millions expended in the past on designs which could have been shown in advance to have a very poor promise of success.

In general, the success of a design will be measured by its accomplishment of the required objectives with the minimum feasible engine size, weight, cost, fuel consumption, and maintenance requirements.

DETERMINATION OF CYLINDER NUMBER, DIMENSIONS, AND ARRANGEMENT

At this point it will be assumed that major design decisions have been made and that the designer possesses the following information:

1. Type of service
2. Type of fuel
3. Rated power
4. Ambient conditions at rating
5. Diesel or spark ignition
6. Supercharged or not
7. Limit on rpm, if any
8. Special qualifications to be emphasized relative to competitive engines for the same service.

For a given power requirement, since engine size, weight, and to a large extent cost, go down as bmep and piston speed increase, these latter parameters become of major importance in the design procedure. The relation of power to bmep and piston speed, already given in Volume I, may be expressed as follows:

$$P = \frac{A_p(\text{bmep})s}{(2 \text{ or } 4)K}, \tag{10–1}$$

where P is the power required, in horsepower

A_p is the piston area

s is the mean piston speed

2 is used for 2-cycle engines

4 is used for 4-cycle engines.

Values of K are as follows:

A_p	bmep	s	K
in²	psi	ft/min	33,000
cm²	kg/cm²	m/sec	76

Choice of Piston Speed. Piston speed is limited by one or both of the following factors:

1. *Mechanical stresses as they influence reliability and durability.* As we have seen, inertia stresses increase as the square of the piston speed. As will be pointed out in the next chapter, stresses cannot be computed accurately in complicated engine parts. Reliability and durability are even less subject to mathematical analysis. In practice, therefore, the limits on piston speed for good reliability and durability are set by reference to previous experience with engines in similar service.

2. *Fuel economy.* In services where good fuel economy is important, the choice of piston speed will be influenced by this factor as well as by those previously mentioned. The effect of piston speed on fuel economy is considered in Volume I, Chapter 12 (pp. 445–449). Where maximum economy is desired, piston speed should be in the range 1000 to 1200 ft/min (5 to 6 m/sec). Piston speeds above that range involve a compromise between fuel economy and specific output. Tables 10–5 through 10–13 and Figs 10–1 and 10–2 give piston speeds used in current practice for various types of service.

Fuel-Economy Estimates. Brake specific fuel consumption may be expressed as follows:

$$\text{bsfc} = (K_p/J)\eta_b Q_c, \tag{10-2}$$

where

$$\eta_b = \eta_i \frac{\text{imep} - \text{fmep}}{\text{imep}} \tag{10-3}$$

η_b = brake thermal efficiency

η_i = indicated thermal efficiency

$K_p/J = 2545$ when Q_c is in Btu/lbm and bsfc in lbm/hp-hr

$K_p/J = 642$ with Q_c in kg cal/g and bsfc in gm/hp-hr.

The heat of combustion Q_c is determined by the type of fuel to be used (see Table 4–1). Brake thermal efficiency may be written:

$$\eta_i = \left(\frac{1}{\Gamma}\right)\eta'_r\eta_0, \tag{10-4}$$

where η'_r is the ratio of actual indicated efficiency based on trapped fuel-air ratio, to the corresponding fuel-air-cycle efficiency, η_0. The multiple $(1/\Gamma)$ is used only for carbureted engines. To determine η_0 it is necessary to estimate F'_R, the trapped relative fuel-air ratio. The question of suitable values for F'_R is discussed in Volume I, Chapter 12 (pp. 437–441) and in Chapter 5 of this volume.

Values for η'_r are given in Volume I, Chapter 5, Figs. 5–21 and 5–27. Figures 10–7, 10–8, and 10–9 of this present volume give more recent and more complete data for spark-ignition and automotive-Diesel engines, respectively. Figures 0–1 through 0–6 in this volume plot the latest data available on the efficiency η_0 of the constant-volume fuel-air cycle. Figure 4–6 (p. 89) and ref. 0.004 of Volume I give values of η_0 for limited-pressure fuel-air cycles.

Limitations on the compression ratio r have been discussed both in Volume I and Chapters 2 to 4 of this volume. In unsupercharged spark-ignition engines, r is limited by detonation, which responds to a complex function of fuel, engine design, and operating conditions. As the octane number of available fuel has improved (see Fig. 4–7), compression ratios have been increased, as indicated by Fig. 10–1, where it will be noted that from 1958 to 1961 the curve of average compression ratio did not respond to the increase in fuel octane number shown in Fig. 4–7. This is because at that time a second limitation on compression ratio appeared, namely, "rumble" and "wild ping" (see Chapter 2 and refs. 2.96 et seq.), caused by deposit-induced preignition. Since 1958 much research has been conducted on the problem of deposit preignition, as a result of which the curve of average compression ratio has been increasing since 1961, along with the fuel octane number.

In supercharged spark-ignition engines the usable compression ratio is limited by the supercharger pressure ratio as well as by fuel characteristics. This question is discussed in some detail in Volume I, Chapter 13. There are few commercial applications of supercharging to spark-ignition engines except in the cases of aircraft engines and large engines operating on natural gas.

In the case of Diesel engines, since efficiency is but little affected by compression ratio (see Volume I, Fig. 4–6, p. 88), the lowest value of r which gives satisfactory starting is usually chosen in order to limit maximum cylinder pressures. This value varies with the type of combustion chamber, with the type of fuel to be used (see Chapters 3 and 4), and with the lowest expected starting temperature. A number of schemes have been proposed for varying the compression ratio, but none of these has so far come into wide commercial use. The most promising system is a hydraulic one (see Fig. 10–5 and ref. 10.881) now in use on a military engine. As specific outputs increase to the point where maximum cylinder pressures must be strictly limited, this device appears to be worth further development.

From the foregoing discussion it must be evident that the designer's choice of compression ratio must be based on previous experience with similar engines in the same service, plus a rational estimate of expected improvements in design and in fuel during the development time.

Indicated Mean Effective Pressure. The problem of computing an appropriate value of imep was discussed by implication in Volume I, Chapters

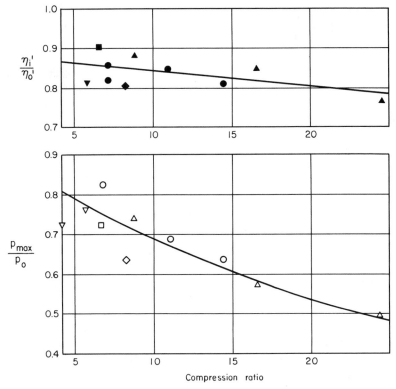

Fig. 10–7. Indicated efficiency and maximum pressure of spark-ignition engines compared with values for equivalent constant-volume fuel-air cycle.

▲	△	Caris and Nelson, automobile engines (0.123)
●	○	Kerley and Thurston, automobile engines (0.050)
◆	◇	Taylor, CFR engine (Volume I, p. 134)
■	□	Taylor, aircraft engine (Volume I, p. 124)
▼	▽	Taylor, CFR engine (Volume, I p. 134)

Subscript 0 indicates equivalent fuel-air-cycle value. (Best-power spark advance in all cases.)

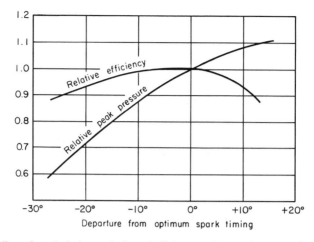

Fig. 10–8. Effect of spark timing on indicated efficiency and on peak pressure (based on Volume I, Fig. 5–15, p. 128).

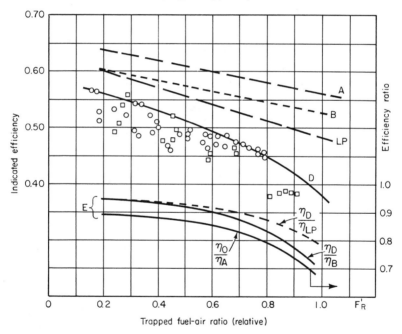

Trapped fuel-air ratio (relative)

Fig. 10–9. Diesel-engine indicated efficiency compared with fuel-air cycle efficiency, automotive-type engines.

> A constant-volume fuel-air cycle, $r = 20$
> B constant-volume fuel-air cycle, $r = 15$
> LP limited-pressure fuel-air cycle, $r = 15$, $p_3/p_1 = 70$
> D maximum measured indicated efficiency of Diesel engines
> E efficiency ratios
> □ 4-cycle divided-chamber engines
> ○ 4-cycle open-chamber engines

For the above engines, average compression ratio is 18 and average p_3/p_1 is about 70.

12 and 13. In order to amplify that discussion, let us begin by recalling the following relationships. For 4-cycle engines,

$$\text{imep} = J\rho_a e_v \Gamma(F'Q_c \eta_i'). \tag{10-5}$$

For 2-cycle engines,

$$\text{imep} = J\rho_s \Gamma R_s(r/r - 1)(F'Q_c \eta_i'), \tag{10-6}$$

where the primes indicate values for the trapped cylinder charge.

In the case of unsupercharged engines, ρ_a and ρ_s are limited by atmospheric pressure and temperature and by the changes of pressure and temperature as the inlet air passes from the atmosphere into the inlet ports (see Fig. 6–14).

In the case of supercharged engines, ρ_a or ρ_s is a function of:

atmospheric conditions
supercharger pressure ratio
supercharger efficiency
aftercooler effectiveness
aftercooler coolant temperature.

In spark-ignition engines, detonation sets a high limit on ρ_a or ρ_s. A method of arriving at an appropriate value is illustrated in example 10–6, page 402.

In the case of supercharged Diesel engines, the appropriate imep should

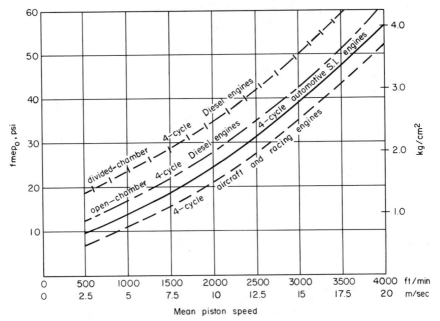

Fig. 10–10. Curves for estimating friction mep of 4-cycle engines; fmep$_0$ is friction mep at $p_e - p_i = 0$, bmep $= 100$ psi. Correct for other conditions by means of Volume I, Eq. 9–16, p. 354 (see Volume I, Chapter 9, for details).

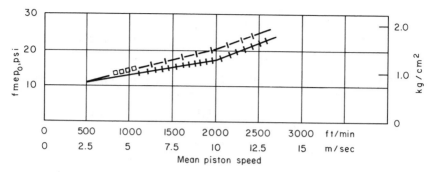

Fig. 10–11. Curves for estimating friction mep of 2-cycle engines. $fmep_0$ is friction mep at bmep = 100 psi and with no scavenging pump connected. Correct for other values of bmep by means of Eq. 9–16, Volume I, p. 354. See Volume I, Chapter 9, for details.

—|—|—|—|—	miniature S-I engine, 1.01 × 0.75 in.
—|–|–|–|—	General Motors 6-71 Diesel, 4.25 × 5.0 in.
□ □ □ □ □ □	Sulzer Marine engine, 28.4 × 47.3 in.

be chosen on the basis of the maximum values of mean and maximum cylinder pressures considered feasible with appropriate requirements regarding reliability and durability, and on supercharger limitations. The resulting value of ρ_a or ρ_s is then calculated from Eq. 10–5 or 10–6, and from this value the required supercharger characteristics are computed. Examples 10–3 and 10–4 (pp. 391 and 395) illustrate this method of approach.

The estimate of the limiting imep and cylinder maximum pressure must be based primarily on previous successful practice, plus appropriate allowances for possible improvements in design. Determination of e_v, R_s, and Γ has been discussed at length in Volume I, Chapters 6 and 7. By definition, $e_s = R_s \Gamma$.

Friction mep. This quantity is estimated on the basis of Volume I, Chapter 9, and bmep is obtained by subtracting this value from the imep. Figures 10–10 and 10–11 based on the data of Volume I, are included here for convenient reference.

With imep and piston speed determined, the piston area is computed from Eq. 10–1, and cylinder dimensions follow from the choice of number of cylinders and bore-stroke ratio.

EXAMPLES

Perhaps the most effective way of illustrating the recommended procedure for arriving at suitable values for cylinder dimensions is by means of a few typical examples. These will be chosen to illustrate the following cases:

10–1 Unsupercharged 4-cycle spark-ignition automobile engine.
10–2 Unsupercharged Diesel engine for motor trucks.

10–3 Large marine Diesel engine, rpm limited.

10–4 Supercharged Diesel locomotive engine.

10–5 Unsupercharged limited-displacement automobile-racing engine.

10–6 Supercharged 4-cycle spark-ignition gas engine.

Example 10–1. Unsupercharged   4-Cycle   Spark-Ignition   Automobile Engine. Establish preliminary specifications for a passenger-automobile engine with a nominal rating of 200 bhp, primarily for use in the U.S.A. Specifications are to include number and arrangement of cylinders, bore, stroke, fuel economy at rating, speed, and compression ratio.

Development Time. Experience shows that the development time for a new conventional design by the large U.S. manufacturers is between two and three years.

Current Trends. Figure 10–1 shows trends in engine design in the U.S.A. for this type of engine since 1951. These trends should be extrapolated, so far as possible, from the starting time (1965 assumed here) to the conclusion of developments (1967–1968). Table 10–6 gives basic characteristics of current (1965) engines for this service.

Fuel. Figure 4–7 shows the trend of octane number for automobile gasoline in the U.S.A. Projecting these data to 1968 indicates an O.N. number for " regular " gasoline of 94. From Table 4–1 for gasoline, $F_c = 0.0670$ and $Q_c = 19,020$ Btu/lb (2180 kgCal/kg).

Compression Ratio. Projecting Fig. 10–1 to 1968, a compression ratio of 10 is selected for the new design.

Indicated Efficiency. According to Fig. 0–1 the fuel-air-cycle efficiency at $r = 10$, $F_R' = 1.2$ is 0.385. It may be assumed that under test conditions where the engine is rated, p_1 and T_1 will be close to 1 atm and 700°R, so that no corrections for fuel-air-cycle efficiency at other values will be necessary (see Figs. 0–2 through 0–6 for such corrections). Figure 10–7 justifies taking the ratio of actual to fuel-air-cycle efficiency at 0.85 so that the estimated value of $\eta_i' = 0.85(0.385) = 0.32$.

Piston Speed. In order to keep the engine size to a minimum, the highest practicable rated piston speed will be used. In the case of passenger-automobile engines, high rated piston speeds can be used, since maximum engine speed is used only a small fraction of the time. This fact also reduces objections to high rated piston speed on the basis of its adverse effects on fuel economy.

When short stroke-bore ratios are used, the limiting piston speed is likely to be imposed by the valve gear. Since valve dimensions are proportional to bore, with similar valve-gear designs, valve-gear linear velocities will be proportional to piston speed multiplied by bore-stroke ratio. Assuming similitude in valve-gear details, a *valve-gear characteristic speed* can be defined as follows:

$$s_v = s(\text{bore}/\text{stroke})/\sqrt{N_i}, \tag{10–7}$$

where s_v = valve-gear characteristic speed

   s = mean piston speed

   N_i = number of inlet or exhaust valves per cylinder, whichever is less.

Thus, for the usual 2-valve head, the valve-gear characteristic speed is s (bore/stroke). The average trend for this factor for U.S. automobile engines is shown in Fig. 10–1. Projecting this trend to 1968 indicates an average value of 3200 ft/min (16 m/sec) for the conventional push-rod type valve gear. That this figure can already be considerably exceeded is illustrated in Table 10–6 by the Ford 289-CID-PF which has a characteristic valve-gear speed in 1966 of 3950 ft/min (20 m/sec). This accomplishment is the result of experience carried over from Ford developments in racing engines (see Table 10–13 and refs. 10.803, 10.804). Table 10–13 shows that this value can be as high as 4250 ft/min (21.6 m/sec) for racing engines with overhead camshaft and excellent design. For the engine in question it should be possible to design a 2-valve head and push-rod gear for a characteristic speed of 3500 ft/min (17.8 m/sec) without going to high-cost design or materials. With a stroke-bore ratio of 0.8, the corresponding piston speed will be 2800 ft/min (14.2 m/sec).

Inlet Air Density. In the U.S.A. it is customary to rate automobile engines without air cleaners or mufflers, and at standard sea-level ambient* conditions (14.7 psia, 520°R or 1.035 kg/cm² and 287°K). Allowing for the influence of the carburetor as shown in Fig. 6–14 (superseding Fig. 6–28 in Volume I),

$$p_i = 0.96(14.7) = 14 \text{ psia}(0.99 \text{ kg/cm}^2);$$

$$T_i = 60 + 15 = 75°F = 535°R = 297°K.$$

We have seen from Volume I and from Chapter 5 of this volume that the fuel-air ratio for maximum power with gasoline is $F' = 1.2(0.067) = 0.08$. According to Volume I, Fig. 6–2 (p. 152), the fuel vapor at $F_R = 1.2$ (maximum power) reduces inlet-air density by the factor 0.98, so that

$$\rho_a = 0.98(2.7) \ 14/535 = 0.0692 \text{ lbm/ft}^3 = 1.11 \text{kg/m}^3.$$

Exhaust Pressure. With no muffler or tail pipe, p_e is taken as 14.7 psia, and p_e/p_i then is $14.7/14.0 = 1.05$.

Volumetric Efficiency. We may assume at this point, to be verified and corrected later, that the inlet valves can be designed to give $Z = 0.5$ at the rated piston speed and that inlet pipes 4 × stroke can be accommodated. From Volume I, Fig. 6–27 (p. 198) by using cam C it appears that a base volumetric efficiency of 0.83 $(1.08) = 0.90$ can be attained. Comparing the base conditions with those of the present example,

---

* The word "ambient" means the atmospheric conditions surrounding the engine.

	Volume I, Figure 6–27 Base Conditions	Example 10–1
r	6	10
T_i	560°R (311°K)	535°R (297°K)
p_e/p_i	1.145	1.05
F'	0.08	0.08

Correcting the base volumetric efficiency by means of Volume I, Eq. 6–26 (p. 205), $K_F = 1.0$, $K_{LC} = 1.0$, $K_{LP} = 1.0$, T_c correction $= 1.0$ gives:

$$e_v = (0.90)\sqrt{\frac{535}{560}}\left(\frac{0.285 + \dfrac{10 - 1.05}{1.4(9.0)}}{0.285 + \dfrac{6 - 1.145}{1.4(5.0)}}\right) = 0.872.$$

Trapping Efficiency. In a normal spark-ignition engine, valve overlap will be small enough so that $\Gamma = 1.0$ can be assumed.

Imep. From Eq. 10–5,

$$\text{imep} = \frac{778}{144}(0.0692)(0.872)(0.08)(19,020)0.32 = 159 \text{ psi}(11.5 \text{ kg/cm}^2).$$

Friction Mep. From Fig. 10–9, fmep at 2800 ft/min (14.2 m/sec) is estimated at 37 psi. Correcting this by means of Volume I, Eq. 9–16 (p. 354);

$$\text{fmep} = 37 + 0.13(14.7 - 14.0) + 0.028(159 - 100) = 39 \text{ psi};$$

$$\text{bmep} = 159 - 39 = 120 \text{ psi } (8.24 \text{ kg/cm}^2).$$

Piston Area. From Eq. 10–1,

$$A_p = \frac{(200)132,000}{(120)(2800)} = 78.7 \text{ sq in.} = 508 \text{ cm}^2.$$

Bore. For 6 cylinders $b = 4.08$ in. (104 mm), and for 8 cylinders $b = 3.54$ in. (90 mm). With cylinder spacing $1.20 \times$ bore, the approximate length of the 6-cylinder block would be $1.20(6)4.08 = 29.4$ in. (75 cm), and that for the 8-cylinder block $1.20(4)3.54 = 17$ in. (43 cm).

Number of Cylinders. The 8-cylinder engine is the logical choice not only because of its shorter length and lighter weight but also because of the reduced octane requirement of the smaller bore (see Fig. 2–34).

Stroke. In order to keep engine size, especially height and width, to a minimum, U.S. passenger-car engines have been using relatively small stroke-bore ratios with excellent results (see Fig. 10–1 and Table 10–6). Taking this ratio as 0.8, the stroke is $3.60(0.8) = 2.88$ in. (73 mm), and the rpm at rating is therefore $2800(6/2.88) = 5820$.

Fuel Economy at Rated Load. Brake thermal efficiency is $0.32(120/159) = 0.242$, and bsfc from Eq. 10–4 is $2545/19,020(0.242) = 0.553$ lbm $= 251$ grams per bhp-hr at rated power.

Weight. Displacement is $78.7 \times 2.88 = 227$ cu in. (3.70 liters). For some reason, manufacturers of passenger-automobile engines do not publish the weight of their engines. Table 10–7 for spark-ignition truck and bus engines shows weights ranging from 1.36 to 2.92 lbm per cu in. displacement. Passenger-car engines probably approach the lower of these figures. Taking the weight of the new engine as 1.5 lbm/cu in. gives an estimated engine weight of $227(1.5) = 341$ lbm (155 kg). A 6-cylinder engine would have a displacement of $78.7(0.8)4.08 = 257$ cu in. (4.2 liters) and would weigh $257 (1.5) = 386$ lbm (175 kg).

Discussion. The result depends on the following "experience" factors taken from knowledge of previous good practice:

1. Maximum allowable piston speed and valve-gear speed for good reliability under service conditions. These speeds can be high for passenger-car service since they are used only a small fraction of the time. As indicated by Fig. 10–1, the average rated piston speed has been increasing with time as the quality of engine design has improved.

2. Maximum practical compression ratio, allowing for trend of fuel quality with time (Figs. 4–7 and 10–1).

3. Pressure drop and temperature rise through carburetor (Fig. 6–14).

4. Allowable length of inlet pipes. This is decided by space available in the expected installation. Study by means of assembly drawings will be necessary for final decision.

5. Stroke-bore ratio (see Volume I, pp. 194–195). In passenger automobiles there is a great premium on low engine height and small width. In a V engine with a given bore, the shorter the stroke, the smaller these dimensions, while the power remains unaffected provided piston speed can be kept the same. The limit on piston speed, then, becomes that which is imposed by the valve gear, since valve size remains constant and rpm increases as stroke is shortened. A stroke-bore ratio of 0.8 appears to be a good compromise, but even shorter strokes are possible, provided valve-gear design is improved accordingly (see Chapter 12).

In addition there are the more precise factors, including
Maximum-power fuel-air ratio, Fig. 5–1
Friction imep (Volume I, Chapter 9)
Volumetric efficiency (Volume I, Chapter 6)
η_i'/η_0, Fig. 10–7.

Example 10–2, Unsupercharged Diesel Truck Engine. Compare cylinder size and performance of a 4-stroke and a 2-stroke-loop-scavenged engine, 300 bhp for use in motor trucks. Both engines to be V-12, unsupercharged.

Rating is to be maintained up to 5000 ft (1525 m) altitude, 70°F ambient temperature (530°R, 295°K).

Solution. For clean combustion, F_R' at rating is taken as 0.7. From Table 4–1 for light Diesel oil, $F_c = 0.0666$; $Q_c = 18,250$ Btu/lbm (1012 kcal/g); $F' = 0.07 (0.0666) = 0.0465$.

At the specified altitude, from Volume I, Table 12–1 (p. 434), pressure at 5000 ft is 12.22 psia (0.86 kg/cm²).

Piston and valve-gear speeds for current engines in this type of service are given in Table 10–8. The two engines designed for passenger-automobiles, Mercedes and Mitsubishi, are already rated at 2300 and 2480 ft/min, respectively. One truck engine, the Mack, is rated at 2100 ft/min.

During the period of development of a new engine, which may be taken as three years, these ratings will undoubtedly increase. For this example a piston speed at rating of 2500 ft/min (12.7 m/sec) for the 4-cycle engine will be assumed. Using four valves will give a moderate valve-gear characteristic speed. For the 2-cycle engine, in order to avoid too high a scavenging mep, a piston speed of 2000 ft/min (10.1 m/sec) will be chosen.

Compression Ratio. Table 10–8 indicates that with an open combustion chamber a compression ratio of 17 should give adequate starting characteristics. Divided chambers require higher compression ratios as indicated in the table. This ratio can be changed without appreciable effect on performance if a different value is found necessary during field trials of the engine.

Four-Cycle Volumetric Efficiency. To keep the engine compact, short inlet pipes will be used. Assume that valves can be designed for $Z = 0.5$ and $p_e/p_i \cong 1.0$, $T_c = 640°R$. The volumetric efficiency is estimated by means of Volume I, Eq. 6–26, p. 205, as follows: Take base volumetric efficiency from Volume I, Fig. 6–16, which gives $e_b = 0.83$ at $T_i = 610°R$, $F_R' = 1.10$, $p_e/p_i = 1.14$, $T_c = 640°R$.

Correction for p_e/p_i from Volume I, Fig. 6–4, is $1.0/0.99 = 1.01$
Correction for $T_i = \sqrt{530/580} = 0.956$
K_F from Volume I, Fig. 6–18, is 1.01
$K_{IC} = K_{IP} = 1.0$
Correction for coolant temperature $= 1.0$
Then: $e_v = 0.83(1.01)(0.956)(1.01) = 0.81$.

Four-Cycle Imep. From Fig. 10–8 the indicated efficiency of an open-chamber 4-cycle engine at $F_R' = 0.7$ can be taken as 0.47, assuming proper combustion-chamber and injection-system development. The inlet (dry) air density is $2.7(12.2/530) = 0.0521$. From Eq. 10–3,

$$\text{imep} = \frac{778}{144}(0.0521)0.81(1.0)(0.0465 \times 18,250 \times 0.47) = 91.5 \text{ psi (6.4 kg/cm²)}.$$

Four-Cycle Friction Mep. According to Fig. 10–10 the base friction mep is 34 psi at sea level. From Volume I, Fig. 9–16 (p. 338), the pumping mep in a 4-cycle motoring test at 1800 ft/min (9.1 m/sec) piston speed, is 2.8-20/28 = 0.287 parts of the total motoring fmep. Extrapolating to 2500 ft/min (12.7 m/sec) this fraction is estimated at 0.30. The pumping mep at 5000-ft altitude should be reduced in proportion to the inlet pressure, or in the ratio 12.2/14.7 = 0.83. Therefore, the base pumping loss is 37(0.30) = 11 psi which is reduced at 5000 ft to 11(0.83) = 9 psi. The corrected base fmep is therefore 37 − 11 + 9 = 35 psi (2.46 kg/cm²).

Four-Cycle Performance and Cylinder Size.

bmep = 91.5 − 35 = 56.5 psi (4 kg/cm²)
isfc = 2545/18,250(0.47) = 0.30 lbm (136 g) per ihph
bsfc = 0.30(91.5/56.5) = 0.485 lbm (220 g) per bhph

$$A_P = \frac{300 \times 132,000}{2,500 \times 56.5} = 280 \text{ sq in.}$$

bore for 12 cylinders = 5.45(138 mm).

If the stroke is taken as 0.8 × bore, the valve-gear characteristic speed for four valves will be $2500/0.8 \times \sqrt{2} = 2200$ ft/min (11.2 m/sec) which is within the range of current practice as shown in Table 10–8.

stroke = 0.8(5.45) = 4.37 in. (111 mm);
displacement = 280(4.37) = 1220 cu in. (20 liters);
rpm = 2500(6/4.37) = 3440.

Two-Cycle Loop Scavenged Engine for Example 10–2.

As a first trial, let the scavenging system to be designed for $R_s = 1.2$, with cylinders of configuration II of Volume I, Fig. 7–15 (p. 240). From this figure at $R_s = 1.2$, 2000 ft/min, e_s is estimated at 0.65. Therefore $\Gamma = 0.65/1.2 = 0.54$. C from the same figure is 0.022. To use Volume I, Fig. 7–10 (p. 229), compute:

$$R_s s/C\left(\frac{r-1}{r}\right) = 1.2(2000/60)/0.022(17/18) = 1940.$$

Reading from this figure, $p_i/p_1 = 1.52$ at $a_i = 1200$ ft/sec (370 m/sec).

At this pressure the ratio Y for the scavenging pump (from Volume I, Fig. 10–2, p. 364) is 0.125. Assuming a Roots blower with efficiency 0.70 (see Volume I, Fig. 10–7, p. 372), $T_i = 530$ (1 + 0.14/0.70) = 636°R (352°K), $a_i = 49\sqrt{636} = 1230$ ft/sec (375 m/sec), close enough to the original assumption. Then:

$$\rho_s = 2.7(12.22/636) = 0.0521.$$

From Eq. 7–21 of Volume I (p. 235):

$$\text{cmep} = \frac{778}{144}(1.2)(0.0521)0.24(530)(0.125/0.70)\frac{17}{16} = 8.2 \text{ psi } (0.58 \text{ kg/cm}^2).$$

Two-Cycle Indicated Efficiency.

From Volume I, Figs. 5–7 (p. 118) and 7–3 (p. 214), the indicated efficiency for the 2-cycle engine is estimated at 0.95 times that of the corresponding 4-cycle engine, or 0.47(0.95) = 0.445. From Eq. 10–4:

$$\text{imep} = \frac{778}{144}(0.0521)0.65(17/16)(0.0465 \times 18{,}250 \times 0.445)$$

$$= 73.3 \text{ psi } (5.16 \text{ kg/cm}^2).$$

Two-Cycle Fmep.

From Fig. 10–11, base fmep = 20 psi. From Volume I, Eq. 9–16 (p. 354), fmep = 20 + 0.08 (81 − 100) = 19.3 psi (1.36 kg/cm^2).

Two-Cycle Performance and Cylinder Size.

bmep = 73.3 − 19.3 − 8.2 = 45.8 psi (3.2 kg/cm^2)
isfc = 2545/18,250(0.465) = 0.314 lbm (143 g) per ihph
bsfc = 0.314(73.3/45.8) = 0.50 lbm (227 g) per ghph
$$A_p = \frac{300 \times 66{,}000}{2000 \times 45.8} = 216 \text{ sq in. } (1400 \text{ cm}^2)$$
bore for 12 cylinders = 4.8 in. (122 mm)
let stroke = 0.8(bore) = 3.84 in. (96.5 mm)
displacement = 216(3.84) = 830 cu in. (13.5 liters)
rpm = 2000(6/3.84) = 3130.

Comparison of 2-Cycle and 4-Cycle Automotive Diesel Engines at 5000-ft Altitude, Unsupercharged*

	Two-Cycle		Four-Cycle	
	English	Metric	English	Metric
bore	4.80	122	5.45	138
stroke	3.84	96.5	4.37	111
displacement	830	13.5	1220	20
rpm	3130	3130	3440	3440
bsfc	0.50	227	0.485	220
approximate weight	2900	1320	3180	1440
specific output	1.39	0.276	1.07	0.166

* For 4-cycle engines, weight is taken at 2.6 lbm/cu in., and for 2-cycle engines at 3.5 lbm/cu in. (see Table 10–8). The 2-cycle engine is heavier in proportion to its displacement on account of the Roots blower, large inlet and exhaust headers, and wider cylinder spacing. Units are in inches, mm, cu in., liters, lbm/bhp hr, g/bhp hr, lbm, and kg. Specific output is in bhp/sq in. and bhp/cm^2.

Discussion. The 2-cycle engine is smaller and has slightly poorer fuel economy under the conditions assumed for this problem. Because of the higher piston speed at which it is rated, the mechanical efficiency of the 4-cycle engine is reduced enough to nearly offset its higher indicated efficiency. The fuel economy of the 4-cycle engine would still be better than that of the 2-cycle engine if both ran at the same piston speed, but such an engine would be still larger and heavier.

Experience factors in this problem are used to establish:

1. Rated piston speed, in this case the highest estimated for good life and reliability, and in the case of the 2-cycle engine, competitive fuel economy as well.

2. Compression ratio, lowest for cold starting in this particular service.

3. Fuel-air ratio (trapped), highest without excessive smoke at maximum rating.

These engines will give nearly $(14.7/12.22)(300) = 360$ bhp at sea level, same air temperature, with somewhat better fuel economy on account of improved mechanical efficiency. They could be much smaller if supercharged.

Example 10–3: Large Marine Diesel Engine. It is desired to design a line of marine engines using cylinders giving 1200 hp each, for direct drive to propellers which are limited to 150 rpm. Maximum cylinder pressure is to be limited to 1200 psia (84.5 kg/cm^2) and, for long life and best fuel economy, piston speed will be limited to 1200 ft/min, 20 ft/sec (6.1 m/sec). Choose cycle, cylinder type, bore stroke, and compute fuel economy. Ambient conditions at rating are 14.7 psia (1.033 kg/cm^2) 560°R (311°K). Sea water at rating is at 560°R.

Solution. Table 10–11 indicates 1965 practices for engines of this class. Because both piston speed and rpm have been specified, the stroke is determined as $1200 \times 6/150 = 48$ in. (122 cm). For such large cylinders the 2-stroke cycle is the obvious choice in order to limit their bore with a given limit on maximum pressure. The compression ratio can be low, since this type of engine is started in a heated engine room. Referring to Volume I, Fig. 4–6 (p. 88), a good compromise between high efficiency and a definite limit on maximum cylinder pressure would seem to be at $r = 14$, $p_{max}/p_i = 0.60$.

As a first trial, assume $F'_R = 0.6$ with heavy Diesel oil. From Table 4–1, $F' = 0.6(0.0666) = 0.04$ and $Q_c = 17{,}790$ Btu/lbm. Let us also assume the following for the first trial:

Loop-scavenged cylinder, Volume I, Fig. 7–15, similar to configuration II, but designed so as to give best scavenging at 1200 ft/min, so that: At $R_s = 1.2$, $e_s = 0.65$, $\Gamma = 0.54$, $C = 0.022$. With this combination, the overall fuel-air ratio $F_R = 0.6(0.54) = 0.325$; $F = 0.325(0.0666) = 0.0216$. From Volume I, Fig. 10–11 (p. 384), $T_e = (550 + T_i)°R = (306 + T_i)°K$.

Indicated Efficiency. Since this engine is far from an automotive type, rather than use Fig. 10–6, refer to the basic limited-pressure cycle, Volume

I, Fig. 4–6 (pp. 88–89). The fuel-air-cycle efficiency at $F_R' = 0.6$ and $r = 15$ is 0.52. Converting this to $r = 14$ gives $0.52(0.475/0148) = 0.515$. In a large cylinder, carefully designed and developed, the heat losses will be low, and an efficiency ratio of 0.90 should be attainable even with the 2-stroke cycle.* Thus the estimated indicated efficiency is, $0.515(0.90) = 0.464$.

Scavenging. At $p_{max} = 1200$ psia, $p_{max}/p_1 = 60$, $p_1 = p_e\,1200/60 = 20.0$ psia (1.21 kg/cm²). In order to use Fig. 7–10 of Volume I (p. 229) compute

$$\frac{R_s s}{C(r-1)/r} = \frac{1.2(20)(14)}{0.022(13)} = 1170.$$

From that figure, at an estimated value of $a_i = 1150$ ft/sec (350 m/sec) the required pressure ratio is 1.17 and the scavenging pressure will be 1.17(20.0) 23.4 psia (1.65 kg/cm²). From Volume I, Fig. 10–2 (p. 364), Y_c at $23.4/14.7 = 1.59$ is 0.13. With a compressor efficiency of 0.80, the compressor outlet temperature (with 560°R ambient temperature) will be $560(1 + 0.13/0.80) = 650$°R. With sea water at 560°R and an aftercooler effectiveness of 0.95, $T_i = 650 - 0.95(650 - 560) = 565$°R (314°K). Thus, $a_i = 49\sqrt{565} = 1160$ ft/sec (354 m/sec), and $T_e = 550 + 565 = 1115$°K. Assuming humidity $h = 0.02$, the scavenging density according to Volume I, Fig. 6–2 (p. 152), based on exhaust pressure is $(0.97)2.7$ $(20.0)/565 = 0.093$ lbm/cu ft. (1.51 kg/m³).

It is now necessary to see if a free turbosupercharger can operate the scavenging pump. From Volume I, Fig. 10–11 (p. 384), at over-all fuel-air ratio, $F_R = 0.325$; $T_e = 1115$°R (620°K). From Volume I, Fig. 10–2 (p. 364), Y_t at pressure ratio $14.7/20.0 = 0.78$ is 0.08. From Volume I, Eq. 10–38 (p. 391), with a steady-flow exhaust turbine,

$$\eta_{ts}\eta_c = \frac{0.24(560)(0.13)}{1.0216(0.27)1115(0.08)} = 0.71.$$

With a blower efficiency of 0.80, the turbine efficiency would have to be 0.89, higher than is practical for a steady-flow turbine. However, by using a mixed-flow system with multiple exhaust pipes, enough blowdown energy can be added to make the supercharger selfsustaining. For example, if the steady-flow efficiency is 0.79, Fig. 10–15 of Volume I (p. 390) will show that at $p_2/p_1 = 20.0/14.7 = 1.36$ the output of a mixed-flow turbine can be 30 per cent higher than with steady flow, increasing the apparent turbine efficiency to $0.75(1.3) = 0.98$.

---

* Figure 10–8 shows that a 4-cycle automotive-size engine can attain an efficiency ratio of 0.91 at $F_R' = 0.60$.

Imep. From Eq. 10–4:

$$\text{imep} = \frac{778}{144}(0.093)0.65(14/13)(0.04 \times 17,790 \times 0.464)$$

$$= 116 \text{ psi } (8.2 \text{ kg/cm}^2).$$

Fmep. From Fig. 10–10, base fmep $= 13$, and from Volume I, Eq. 9–16 (p. 354):

fmep $= 13 + 0.06 (116 - 100) = 14$ psi (0.99 kg/cm^2).

Brake Mean Effective Pressure.

   With compressor driven by a free exhaust turbine:

bmep $= 116 - 14 = 102$ psi (7.2 kg/cm^2)

Indicated Specific Fuel Consumption.

isfc $= 2545/17,790(0.464) = 0.31$ lbm/ihp-hr (141 g/ihp-hr)

Brake Specific Fuel Consumption.

bsfc $= 0.31(116/102) = 0.35$ lbm/bhp-hr (159 g/bhp-hr)

Cylinder Size.

$$A_p = \frac{1200 \times 66,000}{102 \times 1200} = 648 \text{ sq in. } (4200 \text{ cm}^2)$$

Bore $= 28.7$ in. (73 cm)

Stroke $= 48$ in. (122 cm)

Displacement per cylinder $= 648 \times 48 = 31,100$ cu in. (5070 liters)

   Poppet-valve cylinders. As an alternative to a loop-scavenged cylinder, let us try a poppet-valve cylinder similar to the configuration of Volume I, Fig. 7–21 (p. 249). At 1200 ft/min (20 ft/sec) the scavenging efficiency at $R_s = 1.2$ is estimated at 0.75, with C practically the same as before.

   With the same load on the engine structure, the piston area will be reduced about in the ratio $0.65/0.75 = 0.87$, and the maximum cylinder pressure can be increased to $1200/0.87 = 1380$ psia (97 kg/cm^2). With the exhaust pressure still at 20.0 psia, p_{max}/p_i will be 69. Interpolating from Volume I, Fig. 4–6 (p. 89) part 1, this will increase the indicated thermal efficiency by one percentage point to 0.474. Thus, the new performance is calculated as follows:

imep $= 116(0.75/0.65)(0.474/0.464) = 136$ psi (9.6 kg/cm^2)

fmep $= 13 + 0.06(136 - 100) = 15$ psi (1.06 kg/cm^2)

bmep $= 136 - 15 = 121$ psi (8.5 kg/cm^2)

   isfc $= 2545/17,790(0.474) = 0.302$ lbm $(137 \text{ g})/\text{bhp hr}$

bsfc $= 0.302(136/121) = 0.340(154 \text{ g})/\text{bhp hr}$

   $A_p = 648(102/121) = 546$ sq in. (3520 cm^2)

bore $= 26.4$ in. (67 cm)

Stroke, same as before = 48 in. (122 cm)
Displacement per cylinder = 546 × 48 = 26,200 cu in. (4280 liters)

The following tabulation summarizes these results:
For 1200-hp cylinder, 48-in. stroke, 150 rpm:

| Type | R_s | e_s | F'_R | Bmep | | Bsfc | | Bore | | Stroke | Specific Output bhp/sq in. |
				Eng.	Metric	Eng.	Metric	in.	cm		
Loop scav.	1.2	0.65	0.6	102	7.2	0.35	159	28.7	73	48 in. (122 cm)	1.85
Poppet valve	1.2	0.75	0.6	121	8.5	0.34	158	26.4	67	48 in. (122 cm)	2.20

Experience factors include

1. Rated piston speed 1200 ft/min (20 ft/sec) for long life, low maintenance, and best fuel economy. (See also Volume I, table on page 447.)
2. Maximum cylinder pressure (about 1200 psi, 8.45 kg/cm²) for long life and low maintenance.

A next step should be to repeat the calculation with various values of the scavenging ratio and fuel-air ratio, so as to find the best compromise between cylinder size and fuel economy.

Discussion. The loop-scavenged cylinder has a 9 per cent larger bore but is lower in height because of the absence of poppet valves in the cylinder head (compare Figs. 11–20 and 11–21). The absence of valve gear gives an important advantage in first and maintenance costs.

A second alternative would be to use a rotating valve in the exhaust port, as shown in Fig. 11–20 and in Volume I, Fig. 7–1a (p. 212). (See also ref. 10.594.) By means of this device, unsymmetrical timing can be obtained without adding to the height of the engine. Presumably performance (and cylinder size) more nearly equal to that of the poppet-valve type could be obtained in this way without the increase in height. (See Volume I, Chapter 7 for a discussion of 2-cycle cylinder design.)

Optimum values for fuel-air ratio and scavenging ratio should yield a performance equal to or exceeding that shown for the large crosshead-type marine engines of Table 10–11 (last 5 items). Referring to that table, the highest specific output and the lowest fuel consumption are attained by a poppet-valve-type cylinder (Stork). There is little to choose in performance between the other large engines, that is, the Burmeister and Wain with poppet valves, the Fiat with nonreturn valves in the inlet system (Volume I, Fig. 7–17, p. 242), the M.A.N. which is simply loop-scavenged, and the Sulzer with auxiliary rotating exhaust valves (Volume I, Fig. 7–18,

p. 244). Since all these engines are turbosupercharged, the bmep can be adjusted in each case to the maximum where reliability and durability are considered satisfactory. Thus, the scavenging system is only one factor in the total problem, which is mainly that of securing adequate life and reliability with acceptable cylinder size and fuel economy.

Current practice calls for crosshead-type engines in this class, principally because this design has made possible the burning of residual (very viscous but very cheap) fuel. References 10.52, 10.53, 10.55 discuss this problem and state that rapid progress is being made in adapting trunk-piston engines to the use of residual fuel. By the time a new design is ready for the market, it is quite possible that this problem will have been solved for trunk-piston engines. However, with the long stroke-bore ratios used (in order to limit rpm), the saving in height by omitting the crosshead is not as great as might at first be expected, because the connecting rod for the long-stroke trunk-piston engine has to be abnormally long in order to avoid interference with the cylinder bore.

If the smaller, higher-rpm, trunk-piston engines prove satisfactory when using residual fuel, the advantages of multiple engines geared to the propeller shaft are such that the huge direct-drive marine engines may soon be obsolescent.

Example 10–4. Diesel Locomotive Engine. Determine number of cylinders, bore, stroke, and fuel economy at rated power for a 2000-hp Diesel engine for railroad-locomotive service. Engine is to use medium Diesel oil and rating is at ambient conditions of 14.7 psia, 560°R (1.035 kg/cm², 311°K), dry air.

Solution. For railroad-locomotive service high specific output, especially in regard to engine size, is extremely important. Therefore the highest maximum cylinder pressure and piston speed compatible with good reliability and freedom from excessive maintenance costs will be used. Preliminary computations will be made for both 2-cycle and 4-cycle operation. Table 10–9 shows the status of this type of engine in the U.S.A. as of 1965. Allowing a four-year period for development, reasonable goals would appear to be as follows:

Type	p_{max}* psia	p_{max}* kg/cm²	r	Piston Speed ft/min	Piston Speed m/sec
Four-cycle	1800	127	14	2200	11.2
Two-cycle	1500	106	14	1800	9.15

* Present maximum cylinder pressures (1965) are in the range of 1500 for 4-cycle and 1200 for 2-cycle engines.

Indicated Efficiency. From locomotive experience, the rating of these engines may be taken at $F_R' = 0.75$. From Table 4–1, for medium Diesel oil $F' = 0.75(0.067) = 0.05$; $Q_c = 18{,}000$ Btu/lbm. With a fixed limit on maximum pressure, it is evident that p_i and therefore specific output will increase as p_{max}/p_1 decreases, although fuel economy will suffer. Because of the need for high specific output for this service, let $p_3/p_1 = 50$ for both engines. Figure 10–9 for 4-cycle automotive engines (average $p_3/p_1 = 70$) shows the best indicated efficiency at $F_R = 0.75$ to be 0.46. Correct this for $p_3/p_1 = 50$ and $r = 14$ by means of Volume I, Fig. 4–6 (pp. 88, 89) as follows:

$$\eta_i = 0.46 \left(\frac{0.48}{0.50}\right)\left(\frac{0.47}{0.51}\right) = 0.417,$$

which is the estimated indicated efficiency for the 4-cycle type. For the 2-cycle engine:

$$\eta_i = 0.417(0.95) = 0.40.$$

Four-Cycle Inlet Conditions. With an air-cooled aftercooler the inlet temperature can probably be held to $560 + 20 = 580°$R. The inlet pressure will be $1800/50 = 36$ psia, and therefore

$$\rho_i = 2.7(36)/580 = 0.167 \text{ lbm/cu ft.}$$

Estimating exhaust pressure to be $0.80 \times$ inlet pressure, and pressure loss in aftercooler to be 3%,

$$p_2/p_1 = 36/14.7(0.97) = 2.52 \text{ for the compressor}$$

and

$$p_e = 0.80(36) = 28.8 \text{ psia.}$$

Four-Cycle Performance. Assume inlet valves can be designed for $Z = 0.50$, valve overlap 140°, short inlet pipes. e_{vb} from Volume I, Fig. 6–13, is 0.83. Correcting for valve overlap by Volume I, Fig. 6–22: $e_{vb} = 0.83$ $(1.2) = 1.0$, $\Gamma = 0.9$ and $e'_v = 1.0$ $(0.9) = 9.0$. Other corrections are within the accuracy of the estimate. Therefore, from Eq. 10–5,

$$\text{imep} = \frac{778}{144}\, 0.167(0.90)(0.05)(18{,}000)0.417 = 306.$$

From Fig. 10–10 and Volume I, Eq. 9–16 (p. 354),

$$\text{fmep} = 30 + 0.023\,(306 - 100) + 0.13\,(28.8 - 36) = 34$$
$$\text{bmep} = 306 - 34 = 272$$
$$\text{bsfc} = [2545/0.417(18{,}000)]\frac{306}{272} = 0.382 \text{ lbm/bhph}$$
$$\text{Piston area} = \frac{2000 \times 132{,}000}{272 \times 2200} = 441 \text{ cu in. } (2840 \text{ cm}^2).$$

Two-Cycle Scavenging Density. From the limitation on p_{max} with $p_{max}/p_1 = 50$, $p_1 = 1500/50 = 30$ psia, which is assumed equal to the exhaust pressure, p_e. Taking $C = 0.022$ (Volume I, Fig. 7–15, p. 240) and $R_s = 1.4$, in order to use Volume I, Fig. 7–10 (p. 229), compute

$$\frac{R_s s}{C[(r-1)r]} = \frac{1.4(30)}{0.022(13/14)} = 2050, \text{ and at } a_i = 49\sqrt{580} = 1180 \text{ ft/sec,}$$

$$p_i/p_e = 1.55.$$

Therefore, $p_i = 1.55\,(30) = 46.5$ psia, and $p_2/p_1 = 46.5/14.7\,(0.97) = 3.26$. With aftercooling to $580°R$, $\rho_s = 2.7\,(30)/580 = 0.140$ lbm/ft^3.

Two-Cycle Scavenging Efficiency. Assume loop-scavenged cylinders, configuration II of Volume I, Fig. 7–15. At $R_s = 1.4$, $s = 1800$; $e_s = 0.71$; $\Gamma = 0.51$.

Two-Cycle Performance. From Eq. 10–6,

$$\text{imep} = \frac{778}{144}(0.14)(0.71)0.05(18,000)0.40 = 193.$$

From Fig. 10–11 and from Volume I, Eq. 9–16, p. 354,
fmep $= 20 + 0.075(193 - 100) = 27$ psi
bmep $= 193 - 27 = 166$

$$\text{bsfc} = [2545/0.40(18,000)]\frac{193}{166} = 0.41$$

$$\text{piston area} = \frac{2000(66,000)}{1800(166)} = 441 \text{ cu in., or exactly equal to that of the}$$
2-cycle engine under the assumed limitations.

Number of Cylinders. For locomotive service in this size, the V engine appears to be the best choice. Assuming this arrangement and a stroke/bore ratio of 0.83, the following table compares the two types:

2000-hp Locomotive Engines	English Units (inches, pounds)		Metric Units (cm, kg)	
Cylinders	12	8	12	8
Cycles	2—4	2—4	2—4	2—4
Piston areas	441	441	2840	2840
Bore	6.85	8.45	17.4	21.4
Stroke	5.5	6.75	14.00	17.2
Displ.	2420	2980	39.6*	48.8*
Bmep	166—272	166—272	11.7—19.1	11.7—19.1
Bsfc	.41—.38	.41—.38	186‡—173‡	186‡—173‡
Weight†	8700	10,700	3950	4860
Rpm	1970—2400	1600—1960	1970—2400	1600—1960
Wgt/bhp†	4.35	5.35	1.98	2.43
Specific output	4.53	4.53	.70	.70

* liters
† Estimated at 3.6 lbm/cu in., see Table 10–9
‡ grams

Discussion. The 4-cycle engine would seem the logical choice here because of its better fuel economy. The advantages of the 2-cycle engine are its lower maximum cylinder pressure, lower rpm, and the absence of valves and valve gear. Both engines have the same " thermal loading," as expressed in specific output. In practice, the 4-stroke cycle is predominant except in the U.S.A., where a large fraction of the locomotive engines are Electro-motive (General Motors) 2-cycle poppet-valve engines.

In this example, power per unit of weight and displacement is in advance of engines in Table 10–9, due to shorter stroke-bore ratio and higher bmep and piston speeds. The specific output is higher than the limit of 3.5 bhp/cu in. (0.54 bhp/cm^2) suggested in reference 10.45; this means that improved piston, cylinder-head, and exhaust-valve design will be required, particularly from the point of view of control of thermal stresses.

The 8-cylinder engines have cylinders of acceptable size, and over-all dimensions smaller than any in Table 10–9. By using 8 cylinders for 2000 hp, a 3000-hp V-12 and a 4000-hp V-16 engine could be added to the line, using identical cylinders. Alternatively, opposed-piston configurations could be used with somewhat smaller cylinders, because of better scavenging (see Volume I, Fig. 7–9, p. 227).

In this example, an appreciable sacrifice in fuel economy in favor of high mep has been made by the choice of F_R' 0.75 and $p_{max}/p_1 = 50$. This bias toward high specific output seems to be justified in locomotive practice.

Example 10–5. Limited-Displacement Automobile Racing Engine.* Determine the basic characteristics of an unsupercharged racing engine where the rules specify a maximum displacement of 91 cubic inches (1.5 liters), no supercharger, and no limitations on fuel. It has been decided that the design should have the highest practicable power output and acceleration and that cost is a minor consideration.

Solution. The basic problem in this case is to achieve the highest possible power output consistent with enough reliability and durability so that the chances of completing the intended races without breakdown are at least very good. Since piston area equals piston displacement divided by stroke, Eq. 10–1 can be written:

$$\frac{P}{V_d} = \frac{\text{bmep} \times s}{4KS}, \tag{10–8}$$

where $P = $ power, $V_d = $ piston displacement, $s = $ piston speed, $S = $ stroke, and K is a constant depending on units of measurement.

---

*For an excellent account of racing-engine development see J. Blunsden, *The Power to Win*, Motor Racing Publications, Ltd. 28 Devonshire Rd., London W42HD, England

From this relation it is obvious that with a given bmep and piston speed, the shorter the stroke, the greater the power for a given displacement. The shortest stroke will obviously be obtained with the smallest feasible bore and the smallest feasible stroke-bore ratio.

Table 10–13 gives statistics on a number of successful racing engines, in most cases built under limited-displacement rules and all unsupercharged except one. The highest output per unit of displacement is for the engine with the shortest stroke.

Use of the 2-stroke cycle for this class of engine is ruled out by the excessive power required to scavenge at the very high piston speeds characteristic of racing engines. Figure 7–10 of Volume I (p. 229) indicates how the pressure ratio, and hence the power required, increases as a power function of the piston speed. All the engines of Table 10–13 are 4-cycle engines. Let it be assumed that, after studying this table, the designer feels confident of securing adequate racing reliability at 4400 ft/min (22.4 m/sec) piston speed.

Stroke-Bore Ratio. As already mentioned, in 4-cycle engines speed is likely to be limited by the valve mechanism. The highest valve-gear characteristic speed given in Table 10–13 is just over 4500 ft/min (22.8 m/sec). If this value is assumed for the engine in question, the stroke-bore ratio would be 4400/4500 = 0.98 for the new design with two valves per cylinder. By use of four valves per cylinder the stroke-bore ratio is reduced to $0.98/\sqrt{2} = 0.69$. A more conservative figure would be 0.75, giving a valve-gear characteristic speed of $4400/0.75\sqrt{2} = 4150$ ft/min (21 m/sec). Let the design be fixed, then, at piston speed 4400 ft/min (22.3 m/sec) stroke-bore ratio 0.75, with four valves per cylinder.

In choosing such high rated speeds it must be realized that reliability and durability may be marginal, even with the best materials and detail design. In some races the rated speed may even be exceeded for short periods of time. The engine can always be run more slowly, at a corresponding sacrifice in output, if experience indicates this to be necessary.

Fuel. Where no restrictions on fuel are imposed it is customary to use fuel consisting of a mixture of 80 per cent methanol (methyl alcohol), 5 per cent nitrobenzene, and 15 per cent acetone (10.801). For such a mixture $Q_c = 9000$ Btu/lbm, $F_c = 0.150$, and $m_f = 34$. The high value of H_{lg} (−474 Btu per lbm, Table 4–1) and the high resistance of methanol to detonation are its great advantages where high specific output is at a premium. However, the low value of Q_c will entail high specific fuel consumption. Nitromethane (CH_3NO_2) contains some oxygen, which gives it a Q_v value (see Table 4–1) of about 130 Btu per standard cubic foot of fuel-air mixture, as compared with 90 to 95 for nonoxygen-bearing fuels. Acetone is added to secure solubility of the mixture. This formula has been worked out largely on an empirical basis, by the racing fraternity. Most of the ratings given in Table 10–13 are with this type of fuel.

Compression Ratio. Table 10–13 shows a compression ratio as high as 15 with a bore of 4.28 in. (109 mm). With smaller cylinders at the same piston speed the compression ratio using the above fuel is probably not limited by detonation. Volume I, Fig. 12–15 (p. 444) shows peak bmep in an unsupercharged engine at $r = 17$. A choice of $r = 16$ would therefore seem to be logical.

Inlet Density. From Vol. I p. 185 the temperature drop due to complete evaporation at the stoichiometric mixture for methanol is 300°F (166°C). Perhaps half the fuel will evaporate during induction and one third of the heat will be contributed by the inlet system. In this case, a rough estimate of the inlet temperature is $530 - [(300/2)2/3] = 430°R$ (239°K). Volume I, Eq. 6–6, p. 151, indicates that with

$$h = 0.02, \ F_i = 0.150/2 = 0.075, \text{ and } m_f = 34,$$

$$\rho_a = \frac{2.7(14.6)}{430} \left(\frac{1}{1 + 1.2[0.075(29/34)] + 0.032} \right)$$

$$= 0.0826 \text{ lbm/ft}^3 = 1.33 \text{ kg/m}^3.$$

In this computation a loss of 0.1 psia has been assumed in the carburetor.

Inlet-Valve Mach Index. The cylinder would be designed with slanted overhead valves (see Fig. 11–22), and the largest feasible inlet-valve size would be determined. Let it be assumed that the design layout with four valves allows an inlet-valve diameter of 0.4 times the bore. A_p/A_i is then $1/(0.4)^2 \times 2 = 3.1$.

With reference to Volume I, Table 6–1 (p. 176) C_i is estimated at 1.60 $(0.25) = 0.40$, with L/D taken as 0.25. From Volume I, Eq. 6–22, p. 173:

$$Z = \frac{(A_p/A_i)s}{a_i C_i} = \frac{3.1(4400)}{49\sqrt{430}(0.40)60} = 0.58.$$

Volumetric Efficiency. In order to secure the highest possible volumetric efficiency, individual inlet pipes of optimum length and diameter can be used. In Volume I, Fig. 6–26 (pp. 196-197) a maximum volumetric efficiency of 0.99 is shown at $Z = 0.35$ with Cam C. It is quite probable that by suitable alterations of valve timing and pipe dimensions this value could also be obtained at $Z = 0.58$. Assuming this possibility, the "base" volumetric efficiency for the engine in question will be taken as 0.99 under the operating conditions of the figure. These conditions include $p_e/p_i = 1.15$ and $T_i = 560°R$. Allowing for a loss of 0.1 psi in the carburetors, p_e/p_i for the new engine will be $14.7/14.6 = 1.01$. From Volume I, Fig. 6–4 (p. 156) the correction factor for p_e/p_i is $0.999/0.990 = 1.01$. Reference 10.801, p. 48, states that use of individual exhaust pipes developed for racing purposes can increase volumetric efficiency by 5 to 7 per cent. Using the lower figure and correcting for the reduced inlet temperature, we find that

$$e_v = 0.99(1.01)1.05\sqrt{\frac{430}{560}} = 0.92.$$

Indicated Mean Effective Pressure. From Fig. 0–1 the fuel-air-cycle efficiency for $r = 16$, $F_R = 1.2$, and $T_1 = 700°R$, is 0.437. In the engine in question, T_1 is probably 50°F higher than inlet temperature, or $430 + 50 = 480°R$. Correcting the fuel-air-cycle efficiency to this value from Fig. 0–4 by extrapolation $\eta_0 = 0.437(1.015) = 0.443$. From Fig. 10–7, the efficiency ratio is predicted at 0.83, so that $\eta_i = 0.443(0.83) = 0.370$. Then:

$$\text{imep} = \frac{778}{144}(0.0826)0.92[0.150(1.2)9000(0.370)] = 246 \text{ psi.}$$

Friction. For a racing engine, aircraft-engine friction characteristics can be assumed. From Fig. 10–10, base fmep is estimated at 60 psi and from Volume I, Eq. 9–16 (p. 354):

fmep $= 60 + 0.04 (246 - 100) = 66$ psi
bmep $= 246 - 66 = 180$
   isfc $= 2545/9000 (0.375) = 0.755$
   bsfc $= 0.755 (246/180) = 1.03$ lbm/bhp hr (470 g/bhp-hr).

Piston Area. With displacement limited to 91 cu in. (1.48 liters) and a stroke-bore ratio of 0.75:

$$A_pS = A_pb(0.75) = 91 \text{ cu in.}$$

$$A_p = 121.3/b \text{ sq in.}$$

Number of Cylinders.

$$N_c\frac{\pi b^3 \times 0.75}{4} = 91 \text{ cu in.}$$

$$b^3 = 154.5/N_c \text{ cu in.}$$

The following table has been prepared for a piston displacement of 91 cu in. (1.5 liters), bmep 180 psi (12.6 kg/cm²), piston speed 4400 ft/min (22.3 m/sec), stroke-bore ratio 0.75:

No. of Cyls.	Bore		Stroke		Piston Area		bhp	rpm 1000's	bhp/D		Weight/bhp*	
	in.	mm	in.	mm	sq in.	cm²			per cu in.	per liter	lbm	kg
4	3.38	86.0	2.54	64.2	35.8	231	225	10.4	2.47	150	1.22	0.56
6	2.95	75.0	2.22	56.4	41.0	264	246	11.9	2.70	164	1.11	0.51
8	2.68	68.0	2.01	51.0	45.2	292	271	13.1	2.98	181	1.01	0.46
12	2.34	59.5	1.75	44.5	52.0	335	312	15.1	3.43	204	0.88	0.40
16	2.13	54.1	1.60	40.6	56.8	366	335	16.5	3.68	224	0.82	0.38
24	1.86	47.2	1.40	35.3	65.0	419	390	18.9	4.28	260	0.70	0.32

* Based on assumed 3.0 lbm/cu in. displacement, or 273 lbm for all engines.

In practice, power would increase with cylinder number at a rate slightly lower than indicated, on account of increasing heat losses as cylinders get smaller and the inlet Reynolds number decreases (see Volume I, Chapter 8). Difficulties in detail design also increase as cylinders get smaller, as do those of reduction-gear design, assuming a given vehicle wheel speed. Successful racing engines of this displacement have been built in 12 and 16 cylinders. A good compromise would be the 12-cylinder engine giving 312 bhp at 15,100 rpm. This performance is well above that of any 91-cu in. (1.5 liters) engine shown in Table 10–13, but seems to be a very real possibility. Note that the Offenhauser engine of Table 10–13 has already exceeded these design goals in bmep at practically the same piston speed, but because of using only four cylinders with a stroke-bore ratio 1.02 it has a much lower power output.

Discussion. This example illustrates requirements exactly opposite to those of example 10–3. Fuel economy, engine life, and engine reliability are all sacrificed in order to obtain the highest possible (unsupercharged) specific output. This example also illustrates the possibilities inherent in using small cylinders, small stroke-bore ratio, very high piston speed, "tuned" inlet and exhaust pipes, and special fuel. It is obvious that this end is achieved at the expense of very high fuel consumption (due to high piston speed and to the low Q_c of methyl alcohol). At the designed piston speed the design details and materials would have to be of the highest quality in order to achieve even the modest degree of reliability required for racing. A difficult feature of the design would be in the valve gear, which would have to be of the overhead-camshaft type (see Fig. 11–22). Details of valve-gear design are discussed in Chapter 12. A long development period, starting with operation at lower speeds, would be required to achieve even racing reliability. The 4.11-liter Offenhauser engine of Table 10–13 has been under development for many years, but is now (1966) being outperformed by engines with more and smaller cylinders for the same displacement, such as the 1966 Ford.

Example 10–6. Large Gas Engine. Determine the characteristics of an engine to drive directly a natural-gas compressor unit requiring 5000 hp at 500 rpm at standard sea-level conditions. The engine is to run on natural gas (methane). It is to be used in pipeline pumping stations where great reliability, long life, and good fuel economy are required. Cooling water at a temperature not exceeding 550°R (305°K) is to be available.

Solution. (Characteristics of 1965 gas engines are given in Table 10–12.) Since no restrictions other than rpm are imposed on the designer, he is free to decide the following:

1. Type of cycle to be employed
2. Supercharged or not
3. Aftercooled or not

4. Piston speed
5. Number and arrangement of cylinders.

The objective will be to design the lowest-cost engine consistent with the stated requirements. Experience shows that gas engines should be spark-ignited, except where alternative operation on Diesel oil is required. For this example, spark ignition will be used.

Cycle. From previous discussions and examples it is evident that where extremely high fuel economy is desired, the 4-cycle engine has a small advantage over the 2-cycle one when both are equally well designed (see Table 10–12). Since fuel economy is paramount, the 4-stroke cycle will be used here.

Supercharging leads to reduced engine size and improved fuel economy on account of better mechanical efficiency. It also makes possible the use of the "more-complete-expansion" cycle, that is, a cycle where the expansion ratio is greater than the compression ratio. An idealized cycle of this type is illustrated by Fig. 10–12 (see also Introduction, p. 7, and also Volume I, example 13–4, p. 488). Here the inlet valve is closed before the end of the inlet stroke, point x on the diagram. From this point the gases in the cylinder are expanded to maximum cylinder volume, V_1, and then compressed from that point as in the normal cycle. References 10.72, 10.73, and 10.77 describe engines using this cycle. For this cycle let r be the compression ratio = V_x/V_2, and ξ the expansion ratio = V_1/V_2, where 1 and 2 indicate volumes at bottom and top center, respectively (Fig. 10–12).

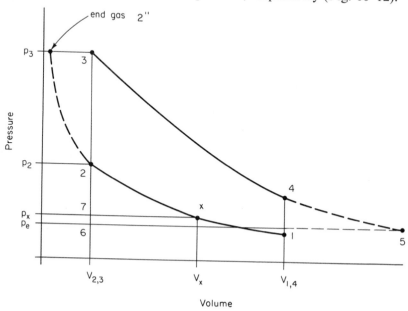

Fig. 10–12. Pressure-volume diagram of more-complete-expansion fuel-air cycle.

This type of cycle may be used to advantage where maximum cylinder pressure is limited by detonation or stress considerations and where a sacrifice of specific output (an increase in cylinder size) is permissible in order to achieve the best practicable fuel economy. The cycle is suitable only for engines that operate most of the time under conditions of high mechanical efficiency, that is, at relatively low piston speeds and near full load. All these conditions hold in the present case.

In addition to the use of this cycle, other devices that may be used to achieve high fuel economy are low piston speed and low fuel-air ratio.

Piston Speed. Since fuel economy and durability are of primary importance, piston speed will be limited to 1200 ft/min (see Volume I, pp. 445–449).

Fuel. From Table 4-1 the characteristics for methane are $Q_c = 21,480$ Btu/lbm (1.195 kgcal/g), $F_c = 0.0581$.

Fuel-Air Ratio. Experience with natural-gas engines indicates a lean limit for regular firing at about $F'_R = 0.60$. For purposes of this example $F'_R = 0.7$ will be chosen so that $F' = 0.7 (0.0581) = 0.0406$. In actual practice these calculations should be repeated at assumed fuel-air ratios from $F'_R = 0.6$ to 1.0, and the effects on cylinder bore and fuel economy plotted. Optimum fuel-air ratio can then be determined on the basis of a suitable compromise between cylinder dimensions and fuel economy.

Inlet Temperature. In order to use the more-complete-expansion cycle, aftercooling is necessary. With water temperature not exceeding 550°R, let it be assumed that the aftercooler effectiveness is such that the inlet-air temperature can be held to 565°R (314°K). For this type of engine it has been found that the rise in temperature during induction is about 150°F, so that T_x is estimated at 715°R (397°K).

Maximum Cylinder Pressure. For large cylinders, in order to secure the necessary durability and reliability, maximum cylinder pressure is usually limited to a certain specified value. For this example let it be assumed that experience has shown a limit of 100 kg/cm² (1420 psia) to be appropriate.

Detonation Limit. In Chapter 2 it was shown that, with a given engine at a given speed, the detonation limit may be defined as a nearly unique function of T''_2, the temperature of the end gas compressed adiabatically to the peak pressure of the cycle. For purposes of this example, let it be assumed that the designer's experience with similar engines at 500 rpm using methane gas indicates that the value of T''_2 at which detonation starts will be about 1950°R. To allow a margin of safety in operation, the limit for T''_2 will be taken as 1865°R (1036°K). With $T_x = 715°$ for adiabatic compression of the end gas, with $k = 1.35$,

$$\left(\frac{1420}{p_x}\right)^{0.26} = \frac{1865}{715} = 2.61; \quad p_x = 36.4 \text{ psia } (2.56 \text{ kg/cm}^2).$$

Thus, the conditions at point x will be 715°R (397°K) and 36.4 psia (2.5 kg/cm²).

Compression Ratio. Figures 0–5 and 0–6 show ratios of p_3/p_1 for the fuel-air cycle, and Fig. 10–7 shows the most recent data on the relation of actual to fuel-air cycles. The appropriate value of r is arrived at by trial from these figures as follows:

r	p_3/p_1 Fuel-Air Cycle	$\dfrac{p_{max}\ \text{Actual}}{p_3\ \text{Fuel-Air}}$	$\dfrac{p_{max}}{p_x}$	p_x psia	p_x kg/cm^2
8	53	0.72	38.1	37.2	2.62
9	60	0.70	42.0	33.9	2.49
8.5	57	0.71	40.5	35.2	2.48
8.3	55	0.71	39	36.4	2.50
Column No.	2	3	4	5	6

Column 2 is from Fig. 0–5 at $F_R = 0.7$. Column 3 is from Fig. 10–7. Column 4 is Column 2 × Column 3. Column 5 is 1420/Column 4.

Thus, the compression ratio is established at 8.3.

Expansion Ratio. An engine operating on the normal cycle beginning at conditions x would be limited to an expansion equal to the compression ratio of 8.3. For the more-complete-expansion cycle, the expansion ratio can be taken at any suitable value greater than the compression ratio, the practical limit being where the improvement in indicated efficiency is offset by reduced mechanical efficiency. A range of expansion ratios will be tried, and the engine performance calculated for each.

Effect of Expansion Ratio on Efficiency. The indicated efficiency of the compression-expansion part of the cycle will depend on the expansion ratio. This efficiency is estimated from fuel-air-cycle efficiency (Figs. 0–1 through 0–6) and the ratio of actual to fuel-air-cycle efficiency, Fig. 10–7. The results are tabulated in Table 10–16.

Inlet Density. Assuming that the moisture content $h = 0.02$ and that the gas is injected after inlet valve closing, from Volume I, Eq. 6–6 (p. 151):

$$\rho_i = \frac{2.7(36.4)}{565} \times \frac{1}{1 + 1.6(0.02)} = 0.169\ \text{lbm/ft}^3 (2.74\ \text{kg/m}^3).$$

Volumetric Efficiency. At point x, the inlet valve is timed so that $p_i = p_x = 36.4$ psia. The temperature of the cylinder contents at that point has been estimated at 715°R. Engines of this type are usually operated with considerable valve overlap, so that residual gas content is small. Assuming this content to be 3 per cent, the volumetric efficiency based on cylinder volume at point x will be $(565/715)0.97 = 0.765$.

Indicated Mean Effective Pressure. For a normal cycle with expansion ratio equal to compression ratio $= 8.3$ from Eq. 10–5,

$$\text{imep} = \frac{778}{144}(0.169)(0.765)(0.0406 \times 21{,}480)0.395 = 241\ \text{psi}(1.7\ \text{kg/cm}^2).$$

For other expansion ratios, imep is in proportion to $\eta_i'/(V_1 - V_2)$ which results in the relation

$$\text{imep} = \frac{241(7.3)}{0.395(\xi - 1)} \times \eta_i'.$$

Results of this computation are given in Table 10–16.

Pumping Mean Effective Pressure. Referring to Vol. I p. 155, the ideal mep of the exhaust stroke is evidently p_e. Assuming that a turbosupercharger can be operated at $p_e/p_x = 0.75$, the exhaust stroke mep is $0.75 (36.4) = 27.2$ psi (1.92 kg/cm^2).

For the inlet stroke, referring to Fig. 10–11,

$$\text{mep}_i = w_i/(V_1 - V_2) = \frac{p_x(V_x - V_2) + JM(E_x - E_1)}{V_1 - V_2},$$

using the relations $r = V_x/V_2$, $\xi = V_1/V_2$, $M/V_1 = p_x r/\xi$ and (for a perfect gas) $E_x - E_1 = C_v T_x[1 - (V_x/V_1)^{k-1}]$, the preceding expression, in English units, reduces to

$$\text{mep}_i = p_x\left(\frac{r-1}{\xi-1}\right) + JC_v T_x\left(\frac{r}{\xi}\right)\left(\frac{1 - (r/\xi)^{0.35}}{1 - 1/\xi}\right) = \frac{266 + 926[1 - (8.3/\xi)^{0.35}]}{\xi - 1}.$$

Subtracting the results of this relation from the exhaust-stroke mep gives the pmep of the ideal inlet process. The results are given in Table 10–16. For piston speed 1200 ft/min and with normal size of inlet valves, Z will be about 0.25. Figure 10–13 gives the relation between ideal and actual pmep, from which the actual pmep is estimated as in Column 7 of Table 10–16.

Friction. The mechanical friction is estimated from Volume I, Fig. 9–8 (p. 329) as 13 psi. Adding this to the pmep gives the estimated friction mep, Column 8 of Table 10–16.

Brake Mean Effective Pressure. The brake mep in Table 10–16 is imep − fmep.

Fuel Economy. The fuel economy for gas engines is usually expressed in units of heat. Therefore isfc $= 2545/\eta_i$ Btu per bhp-hr and $642/\eta_i$ kgcal/bhp-hr.

From Table 10–16 it is obvious that expansion ratios higher than 14 give diminishing or negative returns in fuel economy. The logical choice appears to be 14, at which point the specific consumption is 6200 Btu/hp-hr (1560 kgcal) and the piston area 3920 sq in. (25,300 cm^2).

Piston Area. The piston area, computed from Eq. 10–1 is given in Table 10–16.

Stroke. The stroke at 500 rpm and 1200 ft/min is $1200 \times 6/500 = 14.4$ in. (366 mm).

Bore. The bore will depend on the number of cylinders, as follows:

Number of Cylinders	Bore in.	mm	Stroke in.	mm
6	28.8	730		
8	25.0	635		
12	20.4	518	14.4	366
16	17.7	450		
20	15.8	401		

Sixteen cylinders, 17.7-in. bore, and 14.4-in. stroke appear to be a good choice, since the cylinders are well proportioned; in case a higher-powered engine is required in the future, a 20-cylinder model at 6250 bhp is possible. The V arrangement would ordinarily be used, with the alternative of opposed pistons in the case of a 2-cycle engine.

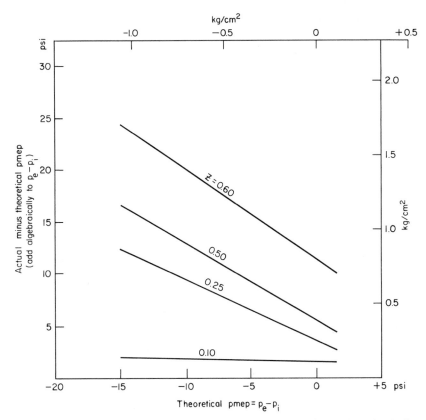

Fig. 10–13. Actual versus theoretical pumping mep of 4-stroke engines. Positive sign indicates *lost* mep. Z = inlet-valve Mach index as defined in Volume I, Chapter 6 (from Volume I, ref. 6.44).

In practice, the above calculations would be repeated over the suggested range of fuel-air ratios, and the required ratio of exhaust to inlet pressure would be computed on the basis of turbosupercharger characteristics, discussed in Volume I, Chapter 10.

Remarks. This example is intended to illustrate another case where fuel economy and long life are primary considerations. These requirements also apply to many large stationary and marine Diesel engines, as in example 10–3. The more-complete-expansion cycle is used by at least one U.S. manufacturer of Diesel engines (10.77) and also by at least two manufacturers of large gas engines (10.72, 10.73). In one case the 2-stroke cycle is used, and the compression ratio is made lower than the expansion ratio by closing the poppet exhaust valves abnormally late in the compression stroke (10.72). As in the other examples, the 2-stroke engine will be smaller for the same output and have a somewhat higher fuel consumption (see Table 10–12).

Experience factors used in this example include evaluation of T_2'' from experience with similar engines operating under similar conditions, piston speed for long life and high fuel economy, maximum cylinder pressure consistent with long life and great reliability, and relation of actual engine cycles to their equivalent fuel-air cycles.

GENERAL DISCUSSION ON EXAMPLES

The foregoing examples indicate a basic method of approach to engine design. The computed characteristics are first approximations, to be modified by subsequent computations* which should include, where necessary and appropriate, plots of performance versus the following:

Fuel-air ratio (Volume I, Chapters 12 and 13, and Volume II, Chapter 5)
Speed and load (Volume I, Chapters 9, 12, and 13)
Scavenging ratio for 2-cycle engines (Volume I, Chapter 7)
Inlet conditions as a function of supercharger pressure ratio
Ambient conditions including altitude (Volume I, Chapters 12 and 13).

In the case of supercharged engines it is essential to verify or modify preliminary assumptions as to supercharger performance (Volume I, Chapter 10).

Uncertainties in the assumptions and computations should always be resolved on the conservative side, that is, by allowing for somewhat less than the highest possible efficiencies of the various units involved. To state this axiom in another way, the cylinder bore chosen should always be at least a little larger than the most optimistic assumptions would call for. An engine hardly ever has too much power!

---

* Such computations can be carried through by means of computers where facilities permit.

In many cases in the past too little time and effort have been devoted to preliminary analysis, with the result that much time and expense have been wasted on designs and developments whose failure could have been predicted in advance. The procedures illustrated should be helpful in deciding on the probable merits of a new design as compared to probable competitive practice at the completion of the development.

EXPERIMENTAL DEVELOPMENT

All new engine designs, no matter how conventional, require a period of experimental development. This period will be longer to the degree in which the new design departs from successful practice, and shorter the greater the first-hand experience of the design group.

For all except the most conventional designs, development work on a single cylinder is indispensable. If the design is at all radical in nature, and if cylinders of more than about 10-in. bore (250 mm) are to be used, serious consideration should be given to doing the early development work on a small-scale, single-cylinder model. (The author has found a CFR crankcase very satisfactory—see Chapter 14.) The scale effects discussed in Volume I, Chapter 11, can be used to translate the results into terms of the larger cylinders. Chapter 14 discusses the question of engine test methods appropriate to development work.

ENGINE TABLES

The following tables of engine characteristics are intended to illustrate typical practice. Undated tables were compiled in 1965 and apply to categories where little change has occurred up to 1984. Where significant changes have occurred, 1983–84 data have been used or added. The following symbols apply to all tables: U, or no designation: unsupercharged; S: supercharged; SA: supercharged and aftercooled. Unless otherwise noted, all supercharging is by exhaust-driven units (that is, "turbosuperchargers"). Particular attention is invited to the quantities that are comparable between engines of different cylinder size, namely brake mean effective pressure (bmep), mean piston speed (s), specific output (bhp/in.^{2} piston area), specific weight (lbm/in.^{3} displacement), and specific fuel consumption (sfc, lbm/hphour) or brake thermal efficiency (η).

Table 10–5. Small Engines (Unsupercharged)

Use	Make Model	Cyl. No. Cycle	Bore in.	Stroke in.	Displacement cu in.	Displacement liters	Comp. Ratio	Rated Maximum bhp	Rated Maximum rpm	Rated Maximum bmep psi	Rated Maximum Piston Speed ft/min	Rated Maximum Valve-Gear c.s. ft/min	Stroke/bore	Weight lbm	Weight lbm/cu in.	Weight lbm/bhp	Specific Output bhp/sq in.
m	Tee-Dee 0.10	1–2	0.237	0.226	0.01	1.63×10^{-4}	—	0.028	32,000	34.7	1200	—	0.95	0.031	3.1	1.1	0.63
m	Arden	1–2	0.495	0.516	0.10	1.63×10^{-3}	—	0.136	11,400	47.0	980	—	1.04	0.26	2.6	1.9	0.70
m	Dooling 61	1–2	1.015	0.75	0.607	0.01	—	1.5	14,500	67.5	1810	—	0.74	0.88	1.45	0.6	1.85
O	Evinrude	1–2	1.56	1.38	5.28	0.09	—	3.0	4,000	56.0	920	—	0.88	—	—	—	0.78
K	McCulloch 90	1–2	2.165	1.635	6.0	0.10	9.0	12.0	9,000	66.0	2450	—	0.75	11.0	1.83	0.92	3.26
O	McCulloch 75	3–2	3.125	2.75	63.0	1.03	6.5	75	5,600	84.0	2560	—	0.88	196	3.1	2.6	3.26
MC	Vincent Black Lightning	2–4	3.31	3.54	60.9	1.0	—	70	6,400	137.0	3780	3520	1.07	—	—	—	3.93
O	Chrysler 820	1–2	2.53	1.62	8.20	0.13	10.5	10	8,000	61.0	2160	—	0.64	13.5	1.65	1.35	2.00
O	Chrysler 610	1–2	2.19	1.62	6.10	0.10	10.5	8	8,000	65.0	2160	—	0.78	14.5	2.4	1.80	2.13
MC	Fichtel & Sachs	1–2	1.5	1.73	3.07	0.05	10.0	8	10,000	103.0	2980	—	1.15	—	—	—	4.50
MC	Fichtel & Sachs	1–2	1.5	1.73	3.07	0.05	14.0	12	13,000	180.0	3740	—	1.15	—	—	—	6.70
h	Power Products	1–2	2.09	1.25	4.3	0.07	9.25	3.5	7,000	46.0	1458	—	0.6	8	1.86	2.3	0.51
O	Johnson	4–2	3.63	2.5	89.5	1.46	—	100	5,000	88.5	2090	—	—	—	—	—	2.80
MC	Yamaha YM-1 305	2–2	2.36	2.13	18.6	0.30	7.1	26	7,000	79.0	2460	—	0.71	—	—	—	2.95
MC	Yamaha YDS-3C	2–2	2.21	1.97	15.1	0.24	7.8	21	7,500	73.0	2460	—	—	—	—	—	2.72
MC	Dukati MK 11	1–4	2.98	2.28	15.2	0.25	10.0	30	8,400	186.0	3190	4160	0.90	—	—	—	4.5
MC	Harley-Davidson Super-Sportster	2–4	3.0	3.813	53.9	0.89	10.6	82	6,800	177.0	4310	3400	1.27	—	—	—	5.8
MC	Honda 305 CL-77	2–4	2.36	2.12	18.6	0.31	9.5	27.4	9,000	130.0	3150	3500	0.90	—	—	—	3.10
MC	MZ ES 150	1–2	2.40	2.36	10.7	0.18	12.0	18	7,500	89.0	2950	3000	0.98	—	—	—	3.98
ht	Ohlsson & Rice	1–2	1.25	1.09	1.34	0.02	10.0	0.85	12,600	20.0	2300	—	0.87	3.88	2.9	4.6	0.70

m = model; O = outboard marine; MC = motorcycle; h = home use (lawn mowers, etc.); K = motor "Kart"; ht = hand tools.

Table 10-6A. U.S. Passenger-Automobile Engines

Make Model	Cyl. Arr. No.	Bore in.	Stroke in.	Displacement		Comp. Ratio	Rated Maximum					Stroke/Bore	Specific Output bhp/sq in.
				cu in.	liters		bhp	rpm	bmep psi	Piston Speed ft/min	Valve-Gear c.s. ft/min		
1965 models													
Cadillac	V-8	4.13	4.0	429	7.0	10.5	340	4600	136	3070	3160	0.97	3.16
Chevrolet Corvair	OP-6	3.44	2.94	164	2.68	9.25	140	5200	130	2550	3000	0.85	2.51
Chevrolet Six	I-6	3.88	3.53	250	4.08	8.5	155	4200	117	2470	2710	0.91	2.19
Chevrolet V-8	V-8	4.25	3.76	427	7.0	10.25	390	5200	139	3260	3690	0.89	3.43
Plymouth Satellite	V-8	4.25	3.75	426	6.95	10.25	425	5000	158	3120	3540	0.88	3.73
Ford Mustang 65A	I-6	3.68	3.13	200	3.26	9.2	120	4400	108	2300	2700	0.85	1.88
Ford 239 CID PF	V-8	4.0	2.87	289	4.72	10.5	271	6000	124	2870	3980	0.72	2.70
Lincoln 462-4V	V-8	4.38	3.83	462	7.52	10.25	340	4600	127	2940	3380	0.87	2.83
1983 models													
Chevrolet 2.5L	I-4	4.0	3.0	151	2.5	8.2	92	4000	121	2000	2667	0.75	1.83
Chevrolet 2.8L	V-6	3.5	3.0	172	2.8	8.5	112	4800	107	2392	3111	0.86	1.95
Chevrolet 2.8 LHO	V-6	3.5	3.0	172	2.8	8.9	135	5400	115	2679	3147	0.86	2.35
Chrysler 2.2 L U	I-4	3.44	3.62	135	2.2	9.0	98	5200	111	3137	2988	1.05	2.63
Chrysler 2.2 L S						8.2	142	5600	149	3378	3217	1.05	3.81
Ford 1.6L	I-4	3.15	3.13	97.6	1.6	9.0	80	5400	120	2817	2835	1.0	2.6
Ford Mustang U	I-4	3.78	3.13	140	2.3	9.0	90	4600	111	2392	2869	0.83	2.00
Ford Mustang S						8.0	142	5000	161	2600	3120	0.83	3.16
Ford 5L	V-8	4.0	3.0	302	5.0	8.4	134	3400	103	1700	2667	0.75	1.33

U = unsupercharged; S = supercharged.
Valve gear c.s. (characteristic speed) = piston speed × bore/stroke
1966 ratings are for bare engines, 1983 are with SAE standard equipment. 1983 engines are emission-limited. 1965 engines had few emission controls.

Table 10-6B. Foreign (Non-U.S.) Passenger-Automobile Engines, Typical, 1965 and 1983–84

Make Model	Cyl. Arr. No.	Bore in.	Stroke in.	Displacement cu in.	Displacement liters	Comp. Ratio	Rated Maximum bhp	rpm	bmep psi	Piston Speed ft/min	Valve-Gear c.s. ft/min	Stroke / Bore	Specific Output bhp/sq in.
1965 models													
Austin Mini	I-4	2.48	2.69	51.7	0.84	8.3	34	5500	95	2450	2280	1.08	1.77
Austin A-40 II	I-4	2.54	3.29	67.0	1.09	8.5	48	5100	111	2800	2150	1.30	2.36
Ford Zodiac	I-6	3.25	3.13	156	2.54	8.3	114	4800	120	2500	2600	0.96	2.27
Hillman Minx	I-4	3.21	3.00	97.2	—	8.3	57	4100	113	2050	2180	0.94	1.76
Jaguar E-4.2(s)	I-6	3.63	4.17	258.0	—	9.0	265	5400	150	3760	3270	1.15	4.27
Mercedes 220 SE	I-6	3.16	2.86	133.9	2.18	8.7	134	5000	158	2380	2620	0.91	2.85
Volkswagen 1500-s	OP-4	3.27	2.72	91.1	1.49	8.5	66	4800	119	2180	2620	0.83	1.96
1983–84 models													
Honda C.T.	I-4	2.6	3.54	73.2	1.2	10u / 7.5s	67 / 100	5500	131 / 197	3245	2386	1.36	3.2u / 4.8s
Mazda F8	I-4	3.39	3.03	109	1.8	9.0	100	5700	127	2879	3220	0.89	2.8
Mitsubishi Minica	I-2	2.76	2.8	33	0.55	9.0	31u / 39s	5500	134 / 169	2567	2541	1.01	2.6u / 3.3s
Mercedes 380 SE	V-8	3.46	3.10	234	3.84	8.3	155	4750	110	2454	2478	0.90	2.05
Nissan MA 10	I-4	2.68	2.68	60.2	0.99	9.5	57	6000	125	2680	2680	1.0	2.5
Nissan VG30E	V-6	3.43	3.27	181	2.96	9.0u / 7.8s	160 / 200	5200	135 / 168	2834	2983	0.75	2.9u / 3.6s
Saab	I-4	3.54	3.07	121	1.98	9.3u / 8.5s	110 / 135	5250 / 4800	137 / 184	2686 / 2456	3087 / 2823	0.87	2.8u / 3.4s
Toyota 4A-C	I-4	3.19	3.03	126	2.1	9.0	68.4	4800	90	2424	2551	0.95	1.6
Toyota Celica Supra	V-6	3.27	3.35	168	2.8	8.8	150	5200	136	2903	2846	1.02	2.99
Volkswagen Jetta	I-4	3.13	3.40	105	1.7	8.2	74	5000	112	2833	2599	1.09	2.4
Volvo B231F B21FT	I-4	3.78	3.15	141	2.3	9.5u / 8.7s	111u / 157	5400	115 / 163	2783	3353	0.83	2.6u / 3.7s

1965 data from *Automotive Industries*, March 15, 1965. 1983–84 data from manufacturers' specifications.
u = unsupercharged; s = supercharged. Valve gear c.s. (characteristic speed) = piston speed × bore/stroke.
1983–84 models meet U.S. emission standards.

Table 10-7. Diesel Engines for Passenger Automobiles, 1983–84

Make		Cyl. Arr. No.	Bore in.	Stroke in.	Displacement cu in.	Displacement liters	Comp. Ratio	Rated Maximum bhp	rpm	bmep psi	Piston Speed ft/min	Stroke/Bore	Specific Output bhp/sq in.	Weight lb	lb/hp	lb/in³
General Motors 1983		I-4	3.30	3.23	111	1.8	22	51	5000	73	2691	0.98	1.48	—	—	—
		V-6	4.05	3.38	262	4.3	23	85	3600	71	2028	0.83	1.10	—	—	—
		V-8	4.06	3.38	350	5.7	22	105	3200	74	1802	0.83	1.01	—	—	—
Mercedes 1984	U	I-4	3.43	3.64	134	2.2	22	72	4200	101	2548	1.06	1.95	—	—	—
	S	I-5	3.58	3.64	183	3.0	22	123	4350	122	2639	1.02	2.45	—	—	—
British Ford	U	I-4	3.15	3.15	97	1.61	23	54	4800	92	2520	1.0	1.75	302	5.6	3.1
Daihatsu CL	U	I-3	2.99	2.87	60	1.0	—	38	4800	105	2295	0.96	1.81	245	6.4	4.1
Peugeot 505	U	I-4	3.7	3.26	141	2.3	23	71	4500	89	2445	0.88	1.64	—	—	—
	S						21	80	4150	108	2255		2.5	—	—	—
Volvo 78 GLE	SA	I-6	3.01	3.40	145	2.4	23	103	4800	117	2720	1.13	2.4	—	—	—
VW-Audi	U	I-4	3.01	3.4	97	1.6	23	40	4500	73	2550	1.13	1.4	—	—	—
	S							70		127			2.5			
VW-Audi	S	I-5	3.01	3.4	121	2.0	23	87	4500	127	2550	1.13	2.5	325	3.7	2.7

U = unsupercharged; S = supercharged; SA = supercharged and aftercooled. Diesel engines for automobiles are more widely used abroad than in the U.S. All the above engines use divided combustion chambers (IDI).

Table 10-8. Diesel Engines for Trucks and Buses, 1983-84.

| Make Model | Cyl. Arr. No. | U or S | Displacement | | | | Comp. Ratio | Rated | | | | Specific Output $\frac{bhp}{sq\ in.}$ | $\frac{Stroke}{Bore}$ | Weight | | | Specific Fuel Cons. lb/hph | Source |
			Bore in.	Stroke in.	cu in.	liters		bhp	rpm	bmep psi	Piston Speed ft/min			lmb	$\frac{lbm}{cu\ in.}$	$\frac{lbm}{bhp}$		
Caterpillar 3405B	I-6	SA	4.75	6.0	638	10.5	—	250*	1800	172	1800	2.35	1.26	2040	3.2	8.16	0.340	M
3208	V-8	U	4.5	5.0	636	10.4	—	210	2800	93	2333	1.65	1.11	1340	2.10	6.38	—	M
3408	V-8	SA	5.4	6.0	1099	18	—	450	2100	154	2100	2.46	1.11	3250	—	—	—	M
Cummins																		
L-10-270	I-6	SA	4.92	5.35	611	10	16.3	270*	1900	184	1694	2.36	1.09	1950	3.19	7.22	0.340	M
Twin Turbo 475	I-6	SA	5.5	6.0	855	14	—	475	2100	210	2100	3.33	1.09	2690	3.15	5.66	—	M
GM Detroit Diesel 2-cycle																		
6-71TT	I-6	SA	4.25	5.0	426	7	17	230*	1950	110	1625	2.70	1.18	2195	5.15	9.54	0.370	M
8V-92TA Silver	V-8	SA	4.84	5.0	736	12	17	445	2100	114	1750	3.02	1.03	2415	3.28	5.43	0.362	M
8V-71 TA	V-8	U S	4.25	5.00	568	9.5	17	304 370	2100	101 123	1750	2.67 3.25	1.18	2415	4.25	7.94 6.52	0.38	M
International																		
Harvester DT-466	I-6	S	4.3	5.35	466	7.6	16.3	210	2600	137	2318	2.41	1.24	1400	3.0	6.66	—	M
Mack E6-325R	I-6	SA	4.9	6.0	672	11	15	316*	1800	207	1800	2.8	1.22	2164	3.22	6.35	—	AI

*Economy rating, reduced piston speed and increased supercharge.

All these engines use open combustion chambers (DI). Recent tendency is toward reduced piston speed and increased supercharger pressure to improve fuel economy. Specific fuel consumption is at rating. Values at reduced ratings are lower. Sources: M = manufacturer's bulletin; AI = Automotive Industries, April, 1983 or 1984.

Table 10-9. Locomotive Diesel Engines (Supercharged and Aftercooled)

Make Model	Cyl. Arr. No.	Bore in.	Stroke in.	Displacement cu in.	liters	Comp. Ratio	Rated Maximum bhp	rpm	bmep psi	Piston Speed ft/min	Valve-Gear c.s. ft/min	Stroke/Bore	Weight lbm	lbm/cu in.	lbm/bhp	Specific Output bhp/sq in.	sfc lbm/bhph
1965																	
Alco-251	V-16	9.0	10.5	10,688	174	13.0	3050	1050	215	1840	1570	1.17	42,000	3.9	13.8	3.00	—
Cooper-Bessemer FVBS-16T	V-16	9.0	10.5	10,688	174	10.0	2750	1000	204	1750	1500	1.17	39,000	3.7	14.2	2.70	—
Dorman*	V-12	6.25	6.5	2,400	39	—	608	1800	124	1940	1870	1.04	7,500	3.1	12.3	1.82	—
Electromotive 16-645-E-3	V-16	9.06	10.0	10,320	172	14.5	3300	900	143	1500	1360	1.1	35,050	3.5	10.6	3.26	—
English Electric	V-16 (2)	10.0	12.0	15,000	244	—	2700	850	167	1700	1410	1.2	43,800	2.9	16.2	2.15	—
Fairbanks-Morse	OP-8 (2)	8.13	10.0	8,300	135	16.1	2580	850	145	1415	—	1.23	44,910	5.4	17.4	3.11	—
Gen. Motors 6-110*	I-6 (2)	5.0	5.6	660	10.8	18.0	309	2000	96	1870	1670	1.12	2,260	3.4	7.3	2.72	—
M.A.N. V8V-24/30	V-16	9.45	11.81	13,300	216	12.3	4100	900	271	1790	1430	1.25	46,200	3.5	11.3	3.67	—
Mirrless	V-16	9.75	10.5	12,500	204	—	2300	900	162	1575	1460	1.08	41,000	3.3	17.8	1.93	—
Sulzer 16-LVA-24	V-16	9.45	11.0	12,400	202	—	4000	1100	230	2020	1730	1.16	40,700	3.3	10.2	3.10	—
Maybach 839 Bb	V-16	7.5	9.1	6,364	104	—	2100	1500	174	2280	1090	1.21	16,000	2.5	7.6	3.00	—
1983																	
Electromotive 16-F3B	V-16 (2)	9.06	10.0	10,320	172	16	3800	950	154	1583	1435	1.10	36,800	3.6	9.7	3.68	0.336
Sulzer 16AT25	V-16	9.84	11.8	14,357	239	12.2	4506	1000	249	1967	1154	1.2	55,125	3.8	12.2	3.70	0.346
Sulzer 6ASL25/30*	I-6	9.84	11.81	5,392	90	12.7	1660	900	271	1771	1039	1.2	—	—	—	3.64	0.346

*Railcar engines—all others for mainline locomotives.
(2) Two-cycle.

Note high bmep of the 1983-84 Sulzer engines, used in order to compete with two-cycle engines in specific output and also to promote high mechanical efficiency. Light weight is not important for locomotive engines.

Table 10-10. Medium-Size Industrial and Marine Diesel Engines (Except Locomotive)

Make Model	Cyl. Arr. No.	Bore in.	Stroke in.	Displacement cu in.	Displacement liters	Comp. Ratio	Rated Maximum bhp	rpm	bmep psi	Piston Speed ft/min	Valve-Gear c.s. ft/min	Stroke/Bore	Weight lbm	lbm/cu in.	lbm/bhp	Specific Output bhp/sq in.	sfc lbm/bhph
1965																	
Caterpillar D-399	V-12	6.25	8.00	2,946	48	15.5	1170*	1300	240	1740	1360	1.28	11,650	3.9	10.7	3.16	—
Cooper-Bessemer FVBL12T	V-12	9.0	10.5	8,000	130	—	1815*	1000	180	1750	1500	1.17	32,000	4.0	17.6	2.39	—
Maybach 820-DB	V-12	6.9	8.1	3,612	59.2	—	1200*	1500	175	2020	1730	1.17	8,350	2.36	7.0	2.68	—
Fairbanks-Morse (2)	OP-6	8.125	10.0	6,420	105	—	1500*	850	109	1420	(2)	1.23	—	—	—	2.34	—
MAN VV 30.45	V-12	11.81	17.71	23,300	380	—	1880*	428	149	1263	855	1.5	—	—	—	1.43	—
GMC Detroit Diesel 16-V-149 (2)	V-12	5.75	5.75	2,380	39	18	1325*	1900	116 (2)	1820	(2)	1.0	9,940	4.2	7.4	4.42	—
1983-84																	
Caterpillar D-399	V-16	6.25	8.8	3,927	64.4	—	1125	1225	185	1633	898	1.28	17,240	4.4	15.3	2.29	0.37
Cummins KTA-3067-P	V-16	6.25	6.25	3,067	50.3	—	1600	2100	197	2187	1540	1.0	—	—	—	3.26	—
GM Detroit (2) 16V-92TA	V-16	4.84	5.0	1,472	24	17	880	2100	113 (2)	1750	850	1.03	4,840	3.3	5.5	3.0	0.378 / 0.370
GM-Electromotive 20-645-FB (2)	V-20	9.06	10	12,900	215	16	4000	900	136 (2)	1500	682	1.10	—	—	—	3.10	0.346 / 0.325
Waukesha L6670 DS 1	V-12	9.13	8.5	6,670	111	—	2021	1200	200	1700	1287	0.93	21,000	3.14	10.4	2.6	0.38 / 0.37

All are supercharged unless marked U. (2) = two-cycle. Most are aftercooled.
bsfc upper figure at rating, lower figure minimum (at lower speed).

Table 10-11. Large Diesel Engines, Marine and Stationary. All are Supercharged and Aftercooled.

Make Model	Cycle	Cyl. Arr. No.	Bore (in.)	Stroke (in.)	Displacement per Cylinder		r	Rated Maximum					Weight			Specific Output
					cu in.	liters		bhp	rpm	bmep (psi)	Piston speed (ft/min)	bsfc	1000 lbm	lbm/cu in.	lbm/bhp	bhp/Ap
Medium-speed engines with propeller reduction gears																
MAN-B&W L40/45 (1983)	4	V-12	15.8	17.7	3450	56.5	11.8	9000	600	283	1772	0.311	202	4.9	22.5	3.8
MAN-B&W L32/36 (1983)	4	V-12	12.6	14.17	1833	30.4	12.1	6600	750	317	1771	0.324	111	5.2	16.8	4.4
Pielstick 8PC4-2L	4	V-8	22.4	24.4	9650	161	—	12,000	400	307	1626	0.286	—	—	—	3.79
Low-speed engines with direct propeller drive																
Hanshin 6EL44 (1982)	4	I-6	17.3	34.7	8160	134	12.5	4000	220	294	1270	0.297	157	3.2	39.2	2.82
Nordberg TS-2112-C (1965)	2	I-12	21.5	31.0	11,250	184	9.5	9200	257	105	1320	0.38	372	2.75	40	2.1
Sulzer RD90 (1965)	2	I-12	35.4	61.0	72,000	1200	—	27,600	119	127	1210	0.34	2145	3.0	84	2.33
Sulzer RTA84 (1982)	2	I-12	33.7	94.5	73,600	1206	—	48,360	87	223	1370	0.28	3562	4.9	74	4.69
Burmeister & Wain 84VT (1965)	2	I-12	33.1	71	60,800	997	—	27,800	110	137	1300	0.34	2090	2.9	76	2.7
Mitsui-B&W L90GBE (1982)	2	I-12	35.4	85.8	84,400	1383	—	47,520	97	189	1388	0.289	2900	2.9	61	4.01

In these giant marine engines designed for direct drive of the propeller, the rpm is limited by propeller efficiency. Thus, the stroke is determined by the choice of an appropriate piston speed (usually 1200–1400 ft/min). Recent emphasis on high fuel economy has required supercharging to high indicated mean effective pressures, thus limiting the bore and resulting in usually large stroke-bore ratios. Thermal efficiencies np up to 0.50 are now achieved with this type of engine, a gain of 15% over 1967 models.

Table 10-12. Large Gas Engines (All Turbosupercharged)

Name	Cycle	Cyl. Arr. No.	Bore in.	Bore mm	Stroke in.	Stroke mm	Displ. cu in.	Displ. liters	r	Max Rating* bhp	rpm	bmep psi	Piston Speed ft/min	bsfc Btu/hph	Weight 1000 lbm	lbm/cu in.	lbm/hp	Specific Output bhp/sq in.
Cooper Bess. KSV-16-SG	4	V-16	13.5	343	16.5	419	37,800	615	11.6	4,300	514	175	1410	6400	160	4.25	37	1.87
Cooper Bess. V-250,† 1966	2 ls	V-16	18	457	20	508	81,700	1330	10	7,250	330	106	1100	6700	360	4.4	50	1.77
Ing.-Rand KVR† 1983	4	V-16	17	432	22	559	80,000	1310	8.0 / 10.0	6,000	350	187	1283	6400	318	4.0	48	1.81
Ing.-Rand P-KVH	4	V-16	16	406	22	559	71,000	1160	8.5	5,500	360	156	1320	6900	240	3.4	48	1.56
Worthington "Mainliner"† LE-C	2 pv	V-18	16	406	18	457	65,700	1060	6.8 / 9.8	5,500	310	108	930	6750	292	4.5	53	1.52
Clark Bros. TVM†	2 ls	V-16	8.63	219	9	238	6,780	110	9.7	1,000	600	98	900	7500	—	—	—	1.33
GMC Radial 16-358	2 ls	R-16	12.5	318	14.5	368	28,500	464	—	3,300								
Nordberg Radial	2 ls	R-12	14	345	16	407	29,500	480	11	3,000	400	101	1070	6900	115	3.9	—	1.63
White-Superior	4	V-16	10	254	10.5	267	13,200	215	—	2,000	900	134	1570	—	—	—	—	1.59

ls = loop scavenged

pv = poppet exhaust valves

r = compression ratio—lower number is expansion ratio if different

Compression ratio is based on total stroke for 4-cycle, volume above ports for 2-cycle engines.

* Most gas engines are designed to operate for limited periods at 10 to 25 per cent above rated power.

† Designed for integral compressors. Weight does not include compressors.

Table 10–13. Special Racing Engines of Limited Displacement, and Super-Sports Engines 4-Cycle, Unsupercharged Except as Noted

Make and Model 1965	Cyl. Arr. No.	Bore		Stroke		Displacement		Comp. Ratio	Rated						Weight			Specific Output	
		in.	mm	in.	mm	cu in.	liters		hp	rpm	bmep psi	kg/cm²	Piston Speed ft/min	Valve-Gear c.s. ft/min	lbm	lbm/hp	lbm/cu in.	bhp/cu in.	bhp/sq in.
Coventry-Climax 4-91	I-4	3.20	81	2.80	71	90.7	1.48	10.0	151	7500	176	12.4	3500	4000	275	1.82	3.0	1.65	4.67
Coventry-Climax Flat-16 (1965)	V-16 (180°)	2.13	54	1.6	40.7	90.7	1.48	—	220	12000	160	11.3	3200	4250	—	—	—	2.42	3.89
Coventry-Climax 8-92	V-8	2.48	63	2.36	60	91	1.50	—	191	8400	198	14.0	3300	3470	298	1.56	3.3	2.10	4.95
Ferrari 6-90	V-6	2.87	73	2.32	59	90	1.47	—	190	9500	176	12.4	3660	4520	265	1.40	3.0	2.11	4.87
Ferrari 500	I-4	3.54	90	3.07	78	122	1.99	8.5	180	7000	167	11.8	3580	4130	—	—	—	1.47	4.53
Ferrari 3500	V-12	2.87	73	2.74	69.5	214	3.49	—	350	7200	179	12.6	3280	3430	—	—	—	1.63	4.44
Maserati 150S	I-4	3.19	81	2.83	72	90.7	1.48	8.8	140	7500	163	11.5	3540	4000	—	—	—	1.54	4.37
Ford (1964)*	V-8	3.76	96	2.87	73	255	4.15	12.5	477	7800	190	13.4	3750	3440	344	0.72	1.35	1.87	5.37
Ford-Cosworth DFV 1983†	V-8	3.37	86	2.55	65	182.6	2.99	12.0	510	11200	198	13.9	4760	4430	340	0.67	1.86	2.8	7.12
Ford-Cosworth DFX-"Indy" 1981(S)‡	V-8	3.37	86	2.26	57	161	2.64	10.5	720	11000	321	22.6	4135	4394	—	—	—	4.5	10.1
Drake-Offy Indy (S)**	I-4	4.125	105	3.125	79	168	2.74	—	608	8000	358	25.9	4180	3900	425	0.70	2.52	3.62	11.6
Drake Offenhauser	I-4	4.281	109	4.375	110	255	4.11	15.0	402	6000	210	14.8	4375	3000	458	1.14	1.85	1.60	6.98
Mercedes Benz 300 SL	I-6	3.34	85	3.46	88	184	3.0	9.5	250	6200	173	12.2	3580	3460	—	—	—	1.36	4.69
Porsche 1500RS	I-4	3.34	85	2.7	66	92	1.5	9.8	145	7200	173	12.2	3240	4000	—	—	—	1.57	4.25

(S)=supercharged.

*Converted for Indianapolis. Rating is with ethanol fuel mixture (see p. 399).

†Developed for European Formula 1. Used ethanol at Indianapolis.

‡Supercharged for Indianapolis. Ethanol fuel. Has been consistent winner there since 1979, when advantages of 8 cylinders vs. 4 with limited displacement finally prevailed.

**Consistent winner at Indianapolis through 1978.

Valve-gear speed = piston speed $\times$ (bore/stroke) $\div \sqrt{\text{number of inlet valves per cylinder}}$.

Table 10-14. Military (and Naval) Diesel Engines

Make Model	Cycle	Cyl. Arr. No.	No. & Type Valves	Comp. Ratio	Bore in.	Stroke in.	Displ. cu in.	Maximum Rating		bmep		Piston Speed ft/min	Valve-Gear c.s. ft/min	Weight			Specific Output bhp/sq in.
								bhp	rpm	psi	kg/cm²			lbm	lb/cu in.	lb/bhp	
Continental* (Tank) 1957	4	V-12 ac	2-ov	—	5.75	5.75	1790	750	2400	139	9.8	2300	2300	3800	2.12	5.1	2.42
Continental LDS-427	4	I-6	2-ov	22	4.31	4.88	427	140	2600	102	7.2	2120	1870	—	—	—	1.64
Lycoming S B H	2	V-4 ac	ls	—	4.25	5.5	312	120	2200	69	4.9	1930	—	600	1.92	5.0	2.02
Caterpillar LVMS-1050	4	V-12	4-ov	20.5	4.5	5.5	1050	1000	2800	270	19.0	2570	1480	2480	2.36	2.5	5.26
Mitsubishi 24WZ	2	W-24	4-ov	15.3	5.91	7.87	5208	3000	1600	143	10.0	2100	1110	13200	2.54	4.4	4.66
Rolls-Royce K-60	2	OP 6	OP	—	3.44	3.6	401	240	2400	99	7.0	1440	—	1570	3.9	6.5	2.16
Leyland L-60	2	OP 6	OP	16.8	4.63	5.75	1.160	700	2100	114	8.0	2005	—	4250	3.66	6.1	3.47

All engines are turbosupercharged except Leyland, Lycoming, and Rolls-Royce.

OP = opposed-piston; ov = overhead poppet valves; ls = loop scavenged; ac = air-cooled.

* Continental makes a line of tank engines for the U.S. Army, but details are not generally published. Table is for 1966. Later data are not available.

Table 10–15. Typical American Aircraft Engines (All are spark-ignition.)

Make Model	Cyl. Arr. No.	Bore Stroke in.	Displ. cu in.	Comp. Ratio	Take-Off Rating					Supercharged	Prop. Gear Ratio	Weight			Specific Output bhp/sq in.
					bhp	rpm	bmep	Piston Speed ft/min	Value-Gear c.s. ft/min			lbm	lbm/cu in.	lbm/bhp	
1965															
Continental A-65-8	HO-4	3.875 3.625	171	6.3	65	2300	131	1400	1500	No	1.0	167	0.98	2.6	1.39
10-360-D	HO-6	4.4 3.875	360	8.5	210	2800	165	1760	2050	No	1.0	294	0.82	1.4	2.8
GTS10-520 (geared)	HO-6	5.25 4.0	520	7.5	340	3200	162	2140	2800	No	0.75	552	1.06	1.6	2.62
Lycoming 0-235-CIB	HO-4	4.75 3.88	233	6.75	115	2800	140	1900	2330	No	1.0	212	0.90	1.8	2.02
IGSO-480	HO-6	5.125 3.875	480	7.3	340	3400	165	2200	2920	GS	0.58	514	1.07	1.5	2.75
Wright R-1820-82A	R-9	6.125 6.875	1823	7.21	1525	2800	236	3220	2870	GS	0.563	1479	0.81	0.97	5.75
1983															
Avco Lycoming AE 10-320 BD	HO-4	5.125 3.875	320	8.5	160	2700	147	1743	2305	No	1.0	255	0.80	1.6	1.94
AE 10-540-D	HO-6	5.125 4.375	542	8.5	260	2700	141	1743	2305	No	1.0	381	0.70	1.5	2.10
T10-540-V2AD	HO-6	5.125 4.375	542	7.3	350	2600	197	1679	2220	TA	1.0	547	1.0	1.6	2.83
T1GO-541-DG	HO-6	5.125 4.375	542	7.3	450	3200	205	2333	2733	TA	0.58	706	1.3	1.6	3.63

GS = gear-driven supercharger
TA = turbosupercharged and aftercooled

HO = horizontal opposed
R = radial

Table 10–16. Performance of Engine of Example 10–6 at Various Expansion Ratios

ξ	η_o	$\dfrac{\eta_i}{\eta_o}$	η_i	imep psi	pmep ideal psi	pmep actual psi	fmep psi	bmep psi	bmep kg/cm²	isfc $\dfrac{\text{Btu}}{\text{hph}}$	bsfc Btu/hph	bsfc $\dfrac{\text{kgcal}}{\text{hph}}$	Piston Area ×10⁻³ sq in.	Piston Area ×10⁻³ cm²
8.3	0.465	0.85	0.395	241	−9.2	−4	9	232	16.3	6450	6700	1690	2.37	15.3
10	0.492	0.845	0.416	206	−8.4	−3	10	196	13.8	6120	6430	1620	2.80	18.1
12	0.518	0.84	0.435	176	−6.8	−2.5	10.5	165.5	11.7	5950	6330	1590	3.32	21.4
14	0.54	0.83	0.448	153	−5.4	0	13	140	9.85	5670	6200	1560	3.92	25.3
16	0.558	0.82	0.460	137	−3.0	+1	14	123	8.65	5530	6150	1550	4.47	28.8
18	0.575	0.81	0.465	122	−0.4	+2	15	107	7.54	5470	6250	1570	5.14	33.2
Col. No.	2	3	4	5	6	7	8	9	10	11	12	13	14	15

Col. 2 from Fig. 0–1
Col. 3 from Fig. 10–7
Col. 4 = Col. 2 × Col. 3

Col. 5 from Eq. 10–5
Cols. 6–13 as explained in text
Cols. 14–15 from Eq. 10–1

11 | Engine Design II: *Detail Design Procedure, Power-Section Design*

The subject of detail design of a machine so complex as an engine would require much more space than is available here and could easily fill several volumes if covered in a thorough manner. As far as your author knows, no complete and authoritative book on this topic is now in existence. This condition is attributable in part to the fact that the subject depends greatly on practical experience in a field subject to constant growth and change. In general, people active in this field have neither the time nor the inclination to write. Hence, most existing literature in this field is written from an academic standpoint and is often irrelevant and even misleading. The bibliography supplementing this chapter is selected to include only material of historical or practical value. The present author's defense for writing in this field is that he has had nearly fifty years of responsible experience in engine design, either in industry or as a consultant to industry, and that the discussion is limited as far as possible to basic principles which are unlikely to change with time. *In this chapter it will be possible to emphasize only these major principles and to point out common mistakes and their corrections.* The illustrations representing good practice are subject to improvement as knowledge, experience, and the development of new materials and techniques continue.

In this chapter it will be assumed that the major design decisions have been made, as outlined in Chapter 10, and that the cylinder dimensions, number and arrangement of cylinders, and performance characteristics have been decided upon. At this stage of the design process, preliminary assembly drawings, usually called "layouts," will have been made, showing the relative position of all the major parts of the engine, and in particular the *power section* which is taken to include cylinders, pistons, connecting rods, crankshaft, and their supporting structure. Valves and valve gear will be considered in Chapter 12. Minor or even major changes in the preliminary layout may be necessary as the detail design proceeds.

The detail design of reciprocating internal-combustion engines has developed over a period of some hundred years, during which time nearly all conceivable mechanical arrangements have been tried and the less practical and convenient ones discarded. To date the conventional crank-

and-connecting-rod arrangement has invariably proved best. The question of unconventional arrangements will be dealt with more fully in Chapter 13. The discussion in the present chapter will be confined to conventional systems.

Before proceeding to more detailed considerations, it may be said that all successful modern engine designs are, in principle at least, copies of previous successful designs with, hopefully, as many improvements in detail as the experience and ingenuity of the designer can contribute. A thorough knowledge of current good practice is therefore essential to the creation of a successful new design. In addition to such knowledge, considerable first-hand experience in engine development work appears to be essential. The ablest groups of engineers and designers, if they lack first-hand experience in the actual development of internal-combustion engines, seldom produce a successful engine design until they have accumulated such experience.

Another way of stating the problem of engine design is to say that it requires decisions to be made which cannot be based wholly on mathematical calculations. Such calculations can be very helpful, as will appear, but they seldom lead to answers that do not call for a large amount of interpretation and modification based on experience. These considerations apply, to a greater or lesser extent, to all engineering design, so that the process of design is in most cases more of an art than an exact mathematical procedure.

Within the limits of conventional practice, the success of a design in respect to weight, bulk, and cost per unit of power depends largely, and the all-important factors of reliability and durability almost wholly, on the quality of the detail design.

GENERAL PROBLEMS IN DETAIL DESIGN

Important considerations in the design of any part always include:

1. Structural integrity
2. Method of manufacture
3. Method of assembly and disassembly

In addition to the above fundamental factors, the design may involve problems of:

4. Thermal performance:
    Efficiency
    Operating temperature
    Heat conductivity
    Thermal stresses

5. Fluid-flow performance:
   Flow-passage shape
   Prevention of leakage
6. Durability:
   Wear
   Corrosion
   Accidental breakage
7. Electrical and magnetic characteristics

Structural Integrity

Of all the problems involved in engine design, that of securing a reliable structure to withstand the heavy loads imposed by gas pressure, inertia forces, and thermal expansion, without excessive weight, bulk, or cost is the most difficult and basically the most important. Without a sound, reliable, and durable structure no amount of refinement of other aspects of the design is of any value.

In general, and within the limits imposed by a particular service, that design is the most successful which fulfills its requirements with the smallest amount of material.

The material in the perfect structure would be distributed so that all of it would be stressed to an equal fraction of its endurance limit. The degree to which this ideal can be approached depends on the skill of the designer, the amount of experimental stress analysis available, and the necessary compromises with manufacturing costs, assembly and accessibility problems, and nonstructural functions to be performed by the same parts.

In the case of the reciprocating internal-combustion engine, this ideal was most nearly approached in the late models of large aircraft engines, where the need for a high specific output was so extreme as to justify high manufacturing costs for the individual parts. While this resulted in high cost per unit weight, these engines had a relatively low cost per horsepower because of their very high power-weight ratios. Next to high-production automotive engines, their cost per horsepower was the lowest of any prime mover during the time when they were manufactured in large numbers.

The need to minimize bulk, weight, and cost for a given power output is present to some degree for all engines; from this fact it follows that the basic engine structure will always be relatively highly stressed and that control of stresses will be a principal consideration in the structural design.

Stress Computations

From the discussion in Chapter 9 we have seen that, as far as structural parts are concerned:

1. Failure is characteristically due to fatigue.

2. Failure occurs at points of stress concentration, that is, at points where stress is higher than average.

3. Accurate calculation of maximum or *critical* stress is not usually possible, even where loads are accurately known.

In the case of engines, stress computations are complicated by uncertainties as to loading and load distribution. It will be recalled from Chapter 8 that inertia loads can be computed on a rigid-body basis, and it is obvious that gas loads on the piston and cylinder can be computed from indicator diagrams. But because the applied loads vary with time and the engine structure is flexible, loads due to flexure and vibration are added (algebraically) to the calculated loads. Furthermore, the distribution of these loads through the engine parts depends on elastic factors that are too complex to lend themselves to accurate computation. Thus, the actual loading of the engine structure, as well as the distribution of stress in its various parts, is not accurately predictable.

For these reasons the detail design of structural engine parts has been largely an empirical development, but with important assistance from rational calculation and, especially in recent years, from experimental stress analysis.

Conventional Methods of Stress Analysis. It is customary in most design offices to make stress computations based on nominal loads calculated from gas pressure and inertia forces, using simple assumptions as to load and stress distribution. Examples of such procedure will be found in refs. 11.201–11.2031. The size of the stressed section under design is then chosen so as to have a value of calculated, or so-called *working*, stress found acceptable in practice. This procedure is basically a method of transferring previous experience to new parts of similar shape and function. It is valuable only under the following circumstances:

1. Reasonable similitude between the new design and the designs from which the value of working stress was obtained. This similarity must extend not only to the particular part in question but also to the supporting structure and to the adjacent parts through which the loads are applied.

2. The method of calculation must be the same for the new part as that used to determine the value of working stress from previous practice.

As we have seen from Chapter 9, the critical stresses in engine parts are usually much higher than those calculated by simple tension, compression, and beam theories. For a given load, the critical stress in any part may be defined as the calculated stress multiplied by the stress-concentration factor (see p. 322). Thus, the working stresses found acceptable by the method described above must be in reality a function of the endurance limit of the material divided by the stress-concentration factor. Thus it is not surprising to find that working stresses considered satisfactory have different values for different parts of the same material. For example, the following values of working stress have been used in the past by several manufacturers:

Crankshafts	20,000 psi	1400 kg/cm²
Piston pins	40,000 psi	2800 kg/cm²
Connecting rods	60,000 psi	4200 kg/cm²
Valve springs	80,000 psi	5600 kg/cm²

Since the material for these parts is similar and the loading cycle is also roughly similar, these values reflect an inverse relationship to the stress-concentration factors characteristic of the parts in question, as well as to the degree of approximation in the stress computations and load estimates.

For relatively simple shapes, as mentioned in Chapter 9, the theory of elasticity affords a means of calculating stress distribution but gives only an approximation of the critical stresses in real materials under fatigue conditions. However, the calculated stresses in fillets, holes, etc., as given in ref. 11.200 for example, are of great value to the designer in a qualitative sense. More will be said of this method of approach in connection with the design of particular parts.

Thermal stresses can also be computed for simple shapes when the temperature distribution is known (11.020–11.023). Unfortunately, in engines the temperatures vary with time and are never accurately known; also, most of the shapes are too complex for meaningful computations even with known temperature distribution.

The Goodman Diagram. In cases where the range of stress is known with reasonable accuracy, the "Goodman" diagram, a version of which is shown in Fig. 11–1, is often used as an aid in design (11.207).

Examples where the approximate stresses are known include parts where the stresses have been measured in an appropriate experimental engine or model, or parts that are simple enough in shape so that calculated stresses are reasonably reliable. Examples of the latter category include helical valve springs, long circular shafts where the torque variation is known, and valve-gear push rods. What these parts have in common is a long, uniform section for which the maximum and minimum loads are predictable. Even with such parts, the stresses at the ends or end fastenings are hard to predict.

The Goodman diagram is plotted for a given material on the basis of curves such as those of Figs. 9–10 and 9–12. Figure 11–1 consists of two diagrams, plotted from Figs. 9–10 and 9–12, respectively. Both abscissa and ordinate scales are given in terms of the dimensionless ratio stress divided by ultimate stress. As demonstrated in Chapter 9, this is a useful ratio for expressing endurance-limit values.

Figure 11–1a refers to tension and compression stresses, either direct or in bending. In this figure, tension is taken as positive and compression

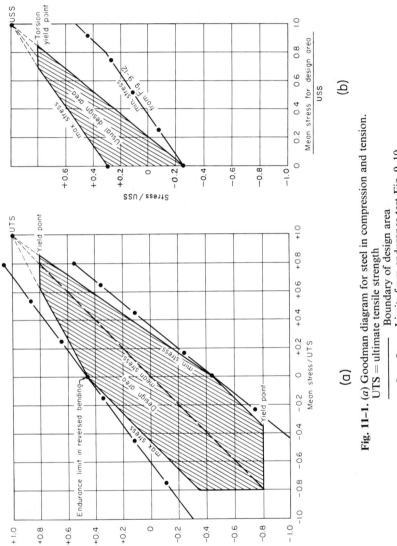

Fig. 11-1. (*a*) Goodman diagram for steel in compression and tension.
UTS = ultimate tensile strength

—— Boundary of design area
— · — Limits from endurance test Fig. 9–10
+ tension, − compression.

(*b*) Goodman diagram for steel in torsion. USS = ultimate shear strength.

as negative. The abscissa scale refers to the mean value of a fluctuating stress, mean being defined as half the algebraic sum of the maximum and minimum stresses. The ultimate tensile stress is plotted at 1.0 on both scales. The plus and minus values of the alternating endurance-limit stress are plotted at zero average stress. On the positive side of mean stress, the alternating-stress–endurance-limit points are connected to the ultimate tensile stress by straight lines, and on the negative side lines parallel to the mean stress are drawn through the alternating-stress–endurance-limit points.

For engine parts for which a very long endurance life is expected, the design is arranged to fall within the shaded area of this diagram. Provided the stress limits are accurately predictable, the result should be a safe design, since actual endurance tests usually give results falling outside the plotted boundaries, as shown by the dot-dash lines.

Figure 11–1b shows a similar diagram for shear stresses, usually the result of torsional loads. Here the positive and negative signs indicate opposite directions only. In this case the shaded area tends to be conservative compared to the endurance-test results indicated by the maximum-stress line and the dot-dash line.

As an example of the use of Fig. 11–1a let it be assumed that photoelastic tests on a pinion tooth show a stress-concentration factor of 1.2 at the tooth fillet. The load on the fillet varies from zero to a maximum in tension. On the basis of a steel with ultimate strength 200,000 psi (14,000 kg/cm^2), what calculated stress in the fillet is allowable?

From Fig. 11–1a when the minimum stress is zero, the Goodman diagram indicates an allowable maximum stress of 0.64 UTS. The maximum calculated stress at the fillet can thus be set at 0.64(200,000)/1.2 = 107,000 psi (7550 kg/cm^2), less whatever allowance for material defects, measurement error, etc. may be considered wise by the designer. Since high-grade gears are very carefully made and inspected, a calculated working maximum stress of 100,000 psi (7000 kg/cm^2) might be chosen.

As an example of the use of Fig. 11–1b, suppose the maximum shear stress in a helical spring is computed at 80,000 psi (5630 kg/cm^2) for steel with an ultimate shear strength of 160,000 psi (11,300 kg/cm^2). The dimensionless stress is 8/16 = 0.5. With reference to Fig. 11–1b, the allowable minimum stress is 0.10 (160,000) = 16,000 psi (1130 kg/cm^2). The allowable maximum and minimum loads on the spring will be in the ratio 80/16 = 5.0. As will be explained in Chapter 12, the endurance properties of valve springs depend heavily on surface conditions and the integrity of the wire used. Thus, for ordinary wire the designer might choose a lower stress range and maximum stress, say from 0.45 to 0.15 on the diagram, with maximum stress 72,000 and minimum 24,000 psi (5080 and 1690 kg/cm^2). On the other hand for especially selected shot-peened wire, limits outside the Goodman diagram can be used, as indicated by Fig. 9–12, which shows the allowable range of stress to be nearly independent of the maximum stress.

It is evident from these examples that even when the Goodman diagram can be used, considerable uncertainty remains which must be resolved on the basis of experience with similar parts and materials.

Stress versus Life. In the case of parts, such as the transmission shafts and transmission gears of automotive vehicles, which may be subject to a limited number of very high stress cycles, the design may be based on S–N curves (Fig. 9–10) for the material in question and on a schedule of stress cycles based on experience (11.203, 11.207, 11.209, 11.210). Engine parts, however, are usually required to last for many millions of cycles, and therefore working stresses should never exceed endurance-limit values.

In view of all the uncertainties involved the dangers inherent in depending exclusively on stress calculations are obvious. At best, such methods give a rough check on the major dimensions of the parts in question. Theory is no substitute for minimization of stress concentrations by all possible means, including experimental stress analysis. By reducing stress concentrations, higher "working stresses" can be used; that is, heavier loads can be carried by a given part, or less material can be used to carry a given load. A review of the development of engine design from the early models at the beginning of this century to the present time indicates that progress in structural strength has been made not so much by changing the relative sizes of the various parts as by securing a more even distribution of stress, that is, *by the reduction of stress concentrations*. Until the advent of modern methods of experimental stress analysis, however, the less obvious points of stress concentration were usually recognized only by failures which occurred in testing or in service. Such testing is still necessary, of course, but the proper use of experimental stress analysis can greatly reduce both the time and expense involved in this phase of engine development.

Experimental Stress Analysis

Since the mid 1930's, experimental stress analysis has been used with great success in improving the structural design of engine parts. The brittle-lacquer method (11.101, 11.102) has been found especially useful. In this procedure a sample of the finished part is coated with lacquer and loaded in a manner similar to that expected in the engine. As the load is increased, a crack pattern develops in the lacquer in such a manner as to show

1. The points of especially high stress concentration
2. The location and direction of the surface stresses
3. An approximate ratio of stress to load over the surface of the sample.

By suitable changes in design the critical stresses may often be drastically reduced, as indicated in ref. 11.101. Results of this technique will be shown in the discussion of important structural parts of the engine.

For more accurate measurement of stress-load ratios, strain gauges of the electrical-resistance (11.110, 11.111) or mechanical type (11.112) may be used. Before applying such gauges, the location and direction of the critical stresses should be determined by means of the brittle-lacquer technique. While strain gauges are quantitatively more accurate than the lacquer, they suffer from the limits of a finite gauge length and from the difficulties of application in a limited space.

One great advantage of the electric-resistance strain gauge is that by means of suitable wiring and slip rings it can be made to function on moving parts while the engine is in operation. Thus, actual stresses under operating conditions can be measured, where the need for such information justifies the time and expense involved (11.110, 11.111).

Another valuable method for experimental stress analysis consists of endurance testing of the complete part under conditions of simulated cyclic loading (11.119). This method is obviously expensive but has been used successfully by organizations able to afford it.

Finally, the supreme check on engine stresses is the long endurance test of the complete engine. Such testing is mandatory for all new designs and nearly always reveals important points of weakness overlooked or not understood by the designer. Until the advent of other methods, in the 1930's, this was the only reliable form of experimental stress analysis available. Reports on engine endurance tests make very instructive reading if properly interpreted.

Method of Manufacture

In nearly every case the geometry of a manufactured part is a compromise between a shape that is ideal from the functional point of view and a shape that facilitates manufacture. The designer therefore must have a basic knowledge of manufacturing methods and, where necessary, should consult with experts in production methods, especially when large-quantity production is planned.

It is beyond the scope of this volume to discuss this question in detail, except as it is mentioned in connection with the discussion of individual parts. In general, it may be said that there is no satisfactory substitute for first-hand experience in this area.

Assembly and Disassembly

For a design satisfactory with respect to assembly, the ideal shape from the functional point of view may again have to be compromised.

Assembly and disassembly problems vary according to whether these operations are to be performed in a well-equipped shop or with the engine in place within the vehicle or building where it operates. In general, provision must be made for a certain amount of in-place servicing. In most cases, the removal and replacement of valves, cylinder sleeves, pistons, connecting rods, and main bearings in the field must be possible. Where cylinders and crankcase are integral, as in the majority of automotive engines, there is the special problem of accessibility to the main and connecting-rod bearings, and of the removal of the piston-and-rod assembly. The latter operation is done either up through the cylinder bore (or the water-jacket space after the cylinder sleeve has been removed), or down past the crankshaft and out through the side cover or bottom cover of the crankcase. The appropriate method will depend on the nature of the installation for which the engine is designed.

In addition to the parts already mentioned auxiliaries, including ignition systems, spark plugs, injectors and injection pumps, oil and water pumps, filters, strainers, etc., must always be easily accessible in the field. This requirement often involves compromise with other considerations.

Thermal Performance

The design factors controlling thermal efficiency have been covered exhaustively in Volume I of this series. The basic problems of cooling and heat flow were dealt with in Volume I, Chapter 8.

As regards detail design, the chief problems concern the avoidance of excessive temperatures and thermal stresses in cylinders, pistons, valves, and ports, and in the exhaust system including the exhaust turbine, if any. In general the solutions involve

1. Use of the correct material
2. Minimizing length of paths for heat flow from hot-side surface to coolant (see Volume I, Fig. 8–7, p. 282)
3. Minimizing or circumventing barriers to heat flow, such as bolted joints and gaskets.

The discussion in Volume I, Chapters 8 and 11, should be helpful to the designer in respect to problems in this area. Questions of detail design to secure satisfactory thermal performance will be discussed further in connection with individual engine parts.

Fluid-Flow Performance

Problems of fluid flow are involved in the inlet system, the valves and ports, the exhaust system, the cooling system, and the fuel system. Volume I, Chapters 6 and 7, discusses these problems in a general way, and they will be covered in more detail in connection with the individual parts involved.

Durability

The question of durability involves chiefly problems of wear and corrosion. The relationship of wear damage to *size* is pointed out in Volume I, pp. 355–356. There is also the problem of provision against damage in handling and shipment.

Questions of wear and corrosion are too complex for treatment in the space here available, except as they are mentioned in connection with individual parts. See Volume I, refs. 9.80–9.84, for an introduction to this subject.

Electrical and Magnetic Characteristics

These are involved with the detail design of ignition systems, starters, generators, etc., and are not included in the scope of this volume.

SCREW FASTENINGS

Since screw fastenings are such an essential element in all engines, it seems appropriate to consider them in some detail at this point. Engine parts are held together almost entirely by screw fastenings, that is, bolts, cap screws, or studs. These three categories are defined in Fig. 11–2.

The screw fastenings in an engine may also be classified as those whose function is primarily structural and those used primarily to secure a tight joint between parts under light loading. In the former category are the fastenings in the power train, while the latter includes the fastenings for covers, gear housings, accessories, strainers, oil pans, etc.

In the following discussion, the word " bolt " will be used to refer to any of the three types shown in Fig. 11–2, unless otherwise specified.

Thread Form. The form of the thread, of course, has an important effect on critical stresses in the threads. Before 1948 the standard U.S. thread forms specified sharp corners at the thread roots—obviously undesirable from the structural point of view. In 1948–1949 the so called " international standards " were adopted for both metric and English-unit threads, and these standards quite properly provide for rounded grooves at the bottom of all threads (11.231).

The thread standards now in general use are given in detail in ref. 11.230. The strongest threads are made by rolling, which prestresses the grooves in compression and also reduces the possibility of inadvertent notches which may be caused by a cutting process (11.242).

Thread Size. Theoretically the finer the thread, the larger the minimum diameter of the bolt, and the stronger the bolt. However, in practice the fact that manufacturing tolerances are the same regardless of thread size leads to an optimum range of thread pitches for each bolt diameter. Reference 11.230 gives standards for many types of screw fastenings. The thread systems include "coarse," "fine," and "extra-fine" series. The "fine" series is most often used in internal-combustion engines. Departures from standard may be desirable for some highly stressed fastenings in the power train, as will appear in the subsequent discussion.

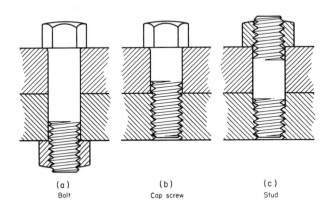

(a)   Bolt

(b)   Cap screw

(c)   Stud

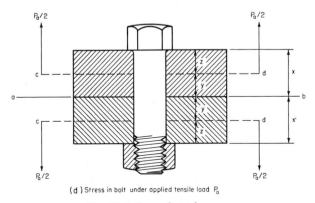

(d) Stress in bolt under applied tensile load P_a

Fig. 11–2. Screw fastenings.

Stresses in Bolts Under Tension. Consider a bolt, as in Fig. 11–2d, holding two plates together. The bolt has been tightened to a tensile load P_x. A load P_a, less than P_x, is applied at the planes c–d. What is the new load on the bolt?

Obviously the new load exceeds the tightening load only if the grip distance, $x + x'$, increases under the applied load.

Let a–b be the axis of symmetry and let the lengths x, y, and z be as represented in Fig. 11–2d.

Dimensions

Let P_x = initial tightening load $|F|$
P_a = applied load $|F|$
K = axial stiffness of each element $|FL^{-1}|$

Let dx, dy, and dz be the changes in length of x, y, and z, respectively, due to the application of the load P_a. Let ΔP_x, ΔP_y, and ΔP_z be the changes in load on the grip elements x, y, and z due to the application of the load P_a. Assuming all stresses to be within the elastic range, the following relations hold for this system:

$$dx = dy - dz;$$

$$dx = \Delta P_x/K_x, \; dy = \Delta P_y/K_y, \; dz = \Delta P_z/K_z;$$

$$\Delta P_z = \Delta P_x.$$

Therefore:

$$\frac{\Delta P_x}{K_x} = \frac{\Delta P_y}{K_y} - \frac{\Delta P_z}{K_z} = \frac{\Delta P_y}{K_y} - \frac{\Delta P_x}{K_z};$$

$$\Delta P_x\left(\frac{1}{K_x} + \frac{1}{K_z}\right) = \frac{\Delta P_y}{K_y};$$

$$\Delta P_y = P_a - \Delta P_x.$$

Therefore:

$$K_y\Delta P_x\left(\frac{1}{K_x} + \frac{1}{K_z}\right) = P_a - \Delta P_x$$

and

$$\Delta P_x = \frac{P_a}{1 + K_y\left(\dfrac{1}{K_x} + \dfrac{1}{K_z}\right)}. \tag{11–1}$$

In the usual situation, K_y and K_z will be nearly equal, and each can be designated as $2K_g$, where K_g is the stiffness of each part of the gripped material. In this case,

$$\Delta P_x = \frac{P_a}{2 + K_g/K_x}. \tag{11–2}$$

Equation 11–2 indicates that, as long as the tightening load is greater than the applied load, so that the two elements of the grip do not separate, the increment of load, and therefore of stress, in the bolt is smaller as the axial stiffness of the grip, K_g, exceeds the axial stiffness of the bolt, K_x.

The case of particular interest here is that of the bolts in the power train, which are subject to applied loads that alternate from zero to a maximum. The load increment ΔP_x then represents the alternating part of the load on the bolt. This can evidently be minimized by minimizing the axial stiffness of the bolt in relation to the axial stiffness of the gripped material. Since the gripped material is usually determined by other considerations, the design objective should be to make the bolt as flexible as possible in the axial direction without allowing the stress at any point to exceed the endurance limit of the material. This end can be accomplished by the modifications to the conventional bolt design indicated by Fig. 11–3. Making the bolt as long as feasible also reduces its axial stiffness. Since the critical stress is ordinarily in the threads, the shank of the bolt can be reduced in diameter until its stress level is only slightly lower than the critical stress.

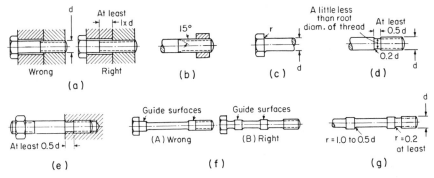

Fig. 11–3. Recommendations for bolt design when high fatigue strength is required. (*a*) Unengaged thread length between the thread end and head should be at least 1 × bolt diam; (*b*) runout angle of the thread should not exceed 15 degrees; (*c*) radius under the head should not be less than 0.08d for small bolts and 0.1d for ¾-in. bolts and larger; (*d*) bolt-thread relief proportions for high fatigue strength, especially when there are bending stresses on the bolt; (*e*) this design of thread relief on studs recommended over (*b*) for greater fatigue strength, especially with incidental bending or repeated tension; (*f*) design of bolts used for centering and locating members where shear and bending forces are present. Minimum body diameter determined by greatest static tension or torsion force; (*g*) proportions for relieved body bolts shown in (*e*) (Staedel, 11.246, translated in 9.101, p. 125).

Bolt Tightening. If P_a exceeds P_x, the bolted surfaces will separate, usually with disastrous results. Equations 11–1 and 11–2 are not valid under such circumstances. It is evident that, in order to utilize the full ability of a bolt, it should be tightened to the maximum allowable value of P_x. From the Goodman diagram, Fig. 11–1*a*, it can be concluded that a bolt holding a joint in tension and tightened to its elastic limit has a small range of additional varying tensile stress in which it can safely operate. Therefore, if the ratio K_g/K_x is large so that very little stress is added by the applied load, the bolt may be tightened to its elastic limit. Tests have shown that under most circumstances tightening to the elastic limit has no deleterious effects on bolt strength (11.235). Experience has also shown that in the case of small bolts the tendency of mechanics is to tighten them until they sense the "give" at the elastic limit, and that in the usual case this procedure is allowable.

Where proper bolt tightening is considered critical, the operation can be controlled by the use of torque wrenches. The relation between torque on the nut or screw and the axial load is quite variable on account of variations in friction. Therefore this practice has its limitations and uncertainties (11.235, 11.239, 11.241). Except in very critical cases, the torque-wrench system appears satisfactory when properly controlled. A more accurate method of control, where possible, is to measure the bolt elongation (11.250).

Bending in Bolts. The load on a real bolt is of course never purely axial, so that some bending stress is always present. In many cases the bending stress is critical. The best method of reducing bending stress is by accurate machining of bolt, nut, and grip and by designing the gripped parts for minimum nonaxial deflection. Spherical seats, as illustrated by Fig. 11–4*f*, can also be used when bending stress is likely to be critical. Long threading or necking down of bolts, as shown in Fig. 11–3, obviously reduces bending stress with a given departure from rectilinear geometry.

Critical Thread Stresses. In the conventional bolted joint, the critical stress is nearly always due to combined tension and bending. It is usually

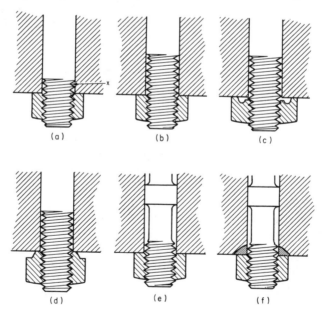

Fig. 11–4. Reduction of critical stress at thread roots. (*a*) Conventional, stress concentration at *x*; (*b*) longer threads (see Fig. 11–3*a*); (*c*) modified nut; (*d*) modified nut; (*e*) stress-relieving neck (see Fig. 11–3*d*); (*f*) spherical seat and washer to reduce bending.

found at the base of the threads, where the threads first pick up the load. With a given thread form, this stress can be reduced by the devices shown in Figs. 11–3 and 11–4.

Obviously, the design of the structure immediately surrounding a bolt may have a powerful influence on bolt stresses. This fact is illustrated by Fig. 11–5, which shows good and bad design for a bolted flange under tension.

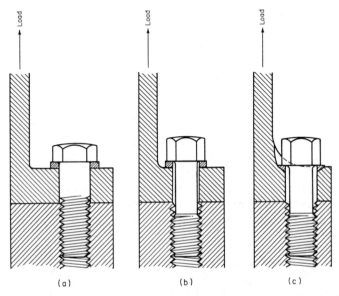

Fig. 11–5. Bolted flange in tension. (*a*) Poor design. Standard cap screw far from load. Small fillets on flange and screw; (*b*) better design. Narrow-hex screw close to load. Screw properly necked. Larger fillets on flange and screw; (*c*) best design. Maximum flange fillet. Spherical washer and seat. Screw same as (*b*).

Threads in Aluminum and Magnesium. Studs permanently threaded into aluminum and magnesium have been found satisfactory in aircraft engines.* However, when a screw set in aluminum or magnesium must be removable, the threads must be of harder material. A helical wire insert (11.234) has been found satisfactory for this purpose.

Locking of Nuts and Screws. Of the thousands of arrangements proposed and available for locking nuts and screws, the cotter pin, lock-wire, or the bent-metal "tab washer" have proved the most satisfactory where the fastening must be released for servicing. Various kinds of lock washers may be used for fastenings that are seldom or never disturbed during the life of the engine (11.251).

---

* In aluminum, studs should have an effective threaded length of at least $2\frac{1}{2}$ diameters (11.235).

Table 11-1. Specifications for Engines of Figs. 11-6 through 11-24

Fig. and Use	Arr.	Bore in.	Bore mm	Stroke in.	Stroke mm	Displacement cu in.	Displacement liters	r	Rated bhp	Rated rpm	Rated bmep	Rated s	Rated s.o.	Weight lb/hp	Weight lb/cu in.
McCulloch M210 (2) 1966	11-6 C	1	1.75	44.5	1.375	35	3.31	0.054	7	6.5	11,000	70	2500	2.65	1.66	3.25
Honda 450 1966	11-7 MC	I-2	2.8	70	2.28	58	28.1	0.468	8.5	43	8,500	147	3230	3.6	3.27	3.26
Nissan MA10 1984	11-8 A	I-4	2.68	68	2.68	68	60.2	0.99	9.5	57	6,000	125	2680	2.54	2.75	2.6
Volvo B23FT 1984	11-9 A	I-4S	3.78	96	3.15	80	141.4	2.31	8.7	157	5,300	166	2783	3.5	2.36	2.62
Chevrolet 2.8L	11-10 A	V-6	3.88	99	3.0	76	172	2.8	8.5	112	4,800	107	2392	1.95	—	—
Ford 5L	11-11 A	V-8	4.0	102	3.0	76	302	5.0	8.4	134	3,400	103	1700	1.33	—	—
Caterpillar 3406B	11-12 T	I-6S	4.75	120	6.0	152	638	10.5	—	250	1,800	172	1800	2.35	8.16	3.2
Detroit Diesel 8V 71(2)	11-13 T	V-8S	4.25	108	5.0	127	568	9.5	17	370	2,100	123(2)	1750	3.25	6.52	4.25
Sulzer 16-LVA-24	11-14 L	V-16S	9.45	250	11.0	280	12,400	202	—	4,000	1,100	230	2020	3.5	10.2	3.3
Electromotive 16-F3B(2)	11-15 L	V-16S	9.06	230	10.0	254	10,320	172	16	3,800	950	154(2)	1583	3.68	—	—
Diahatsu 1 liter	11-16 A	I-3	3.0	76	2.87	73	60	1.0	—	38	4,800	105	2295	1.81	6.4	4.1
Cooper Bessemer (2)	11-17 G	V-16S	18.0	457	20	508	5,089*	85*	10	7,250	330	106(2)	1100	1.77	50	4.4
Nordberg radial (2)	11-18 G	R-12S	14.0	345	16	407	2,458*	41*	11	3,000	400	101(2)	1070	1.63	38	3.9
Fairbanks Morse (2)	11-19 I	V-12 OPS	20	508	21.5	546	7,583*	126*	11.6	12,000	400	130(2)	1433	2.65	22	3.5
Hanshin GEL 44	11-20 M	I-6S	17.3	439	34.7	879	8,160*	134*	12.5	4,000	220	294	1270	2.82	39	3.2
Sulzer RTA-84 (2)	11-21 M	I-12S	33.1	840	94.5	2400	73,600*	1,226*	—	48,360	87	223(2)	1370	5.2	74	4.9
Offenhauser 4.11L	11-22 R	I-4	4.281	109	4.375	110	225	4.11	15	402	6,000	210	4375	6.9	1.14	1.8
Rolls-Royce K-60 (2)	11-24 MT	OP-6	3.44	87	3.60	91.5	401	6.7	—	240	2,400	99(2)	1440	2.16	6.5	3.9
Avco AEIO 540-D	11-25 AC	O-6	5⅛	130	4⅜	111	542	9.0	8.5	260	2,700	141	1969	2.10	1.47	0.70

Use: A, automobile; AC, aircraft; C, chain saw; G, gas pipeline; GI, gas industrial; I, Industrial; L, locomotive; M, marine; MC, motorcycle; MT, military; R, racing; T, truck and bus. r = compression ratio. S = piston speed (ft./min). s.o. = hp per in.^{2} piston area. *For one cylinder. (2), two-cycle.

ENGINE ILLUSTRATIONS

In order to illustrate the discussion of detail design, cross sections of typical engines of various categories are shown in Figs. 11–6 through 11–25. Specifications for these engines are given in Table 11–1 and supplemented by Tables 10–5 through 10–15 and by Tables 11–2 through 11–9, which give important design ratios for the engines illustrated and for a number of others.

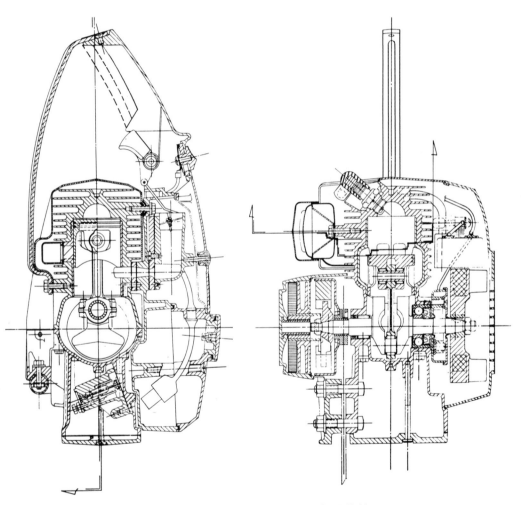

Fig. 11–6. McCulloch chain-saw engine M2-10.
Special features: Small size, light weight (10.75 lbm), 2-cycle, air-cooled, cast-aluminum cylinder and crankcase, ball and roller bearings. *(Courtesy McCulloch Corporation)*

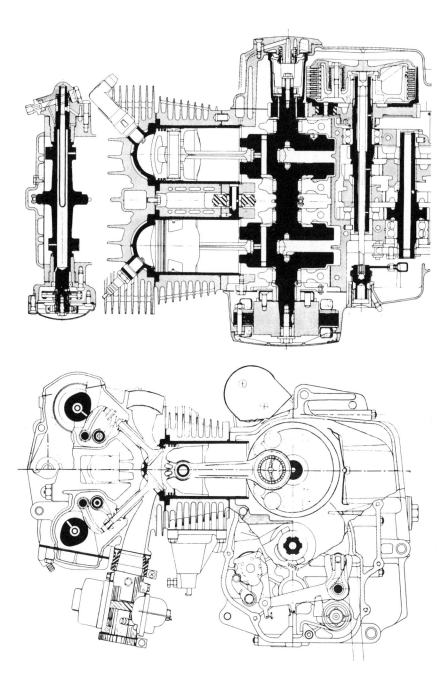

Fig. 11–7. Honda 450 4-cycle motorcycle engine, designed for high specific output, unsupercharged.

Special features: Domed cylinder head with very large valves, excellent valve and port design, two overhead camshafts, valve seats cast into head, torsion-bar valve springs, built-up crankshaft with roller bearings, chain camshaft drive. See also Tables 10–5 and 11–2. (*Courtesy Honda Motor Co., Ltd.*)

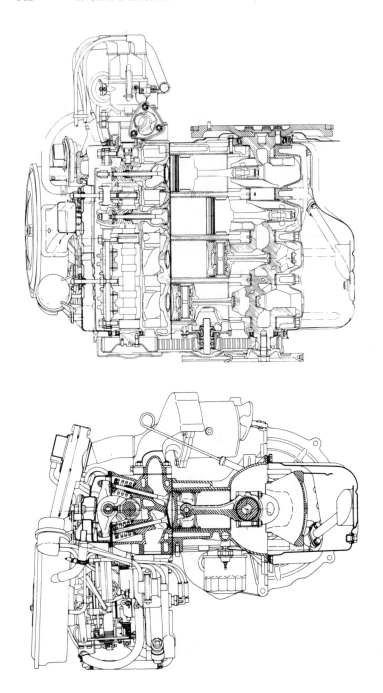

Fig. 11–8. Nissan MA10 four-cylinder 60 cu in. (0.987 liter) automobile engine, 1980. Bore 2.68 in., stroke 2.68 in., $r = 9.5$. Typical of the engines used in small foreign automobiles. Overhead camshaft, aluminum cylinder block with thin cast-iron liners, cast-iron crankshaft, hemispherical combustion chambers without squish areas. (*Courtesy Nissan Motor Company, Ltd.*)

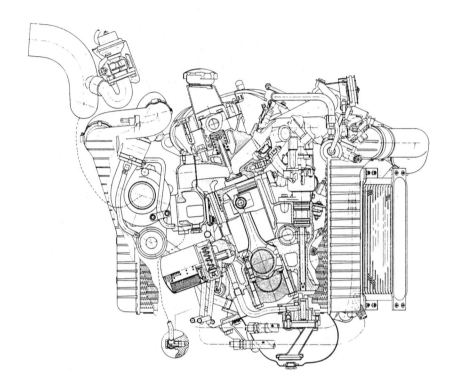

Fig. 11–9. Volvo B23FT 141 cu in. (2.32 liter) supercharged and aftercooled four-cylinder automobile engine, 1984. Bore 3.78 in., stroke 3.15 in., $r = 8.7$. Direct acting overhead camshaft with parallel valves. Combustion chamber with large squish areas. Fuel injection with mechanical plus electronic controls. Air-cooled aftercooler located in front of the engine radiator.
(*Courtesy Volvo Car Corporation.*)

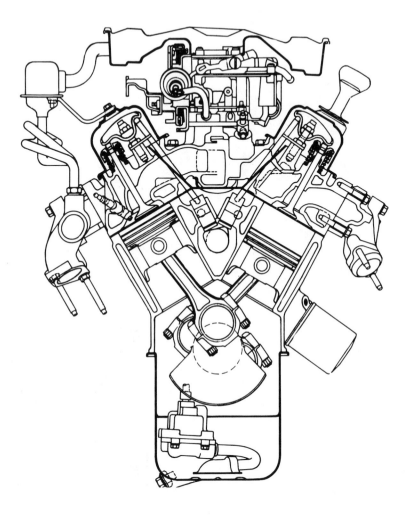

Fig. 11–10. 1983 Chevrolet Citation 60° V-6 engine, 151 cu in. (2.8 liters). Bore 3.5 in., stroke 2.99 in., $r = 8.5$. Replaces earlier I-6 engines so that it can be placed transversely in front-drive cars. Typical U.S. pushrod valve gear and parallel-valve combustion chamber with squish areas. (*Courtesy General Motors Corp., Chevrolet Division.*)

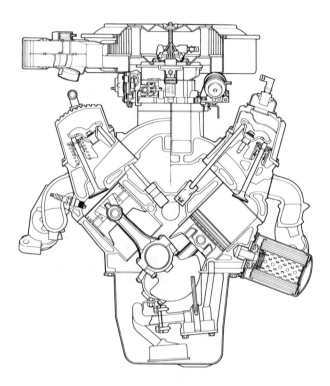

Fig. 11–11. 1984 Ford 90° V-8 automobile engine, 302 cu in. (5 liters). Bore 4 in., stroke 3 in., $r = 8.4$. With only minor changes, this has been Ford's standard engine for large cars since the mid 1960s. It is also offered with fuel injection and supercharging. Typical U.S. style, with push-rods and squish-type combustion chamber. (*Courtesy Ford Motor Company.*)

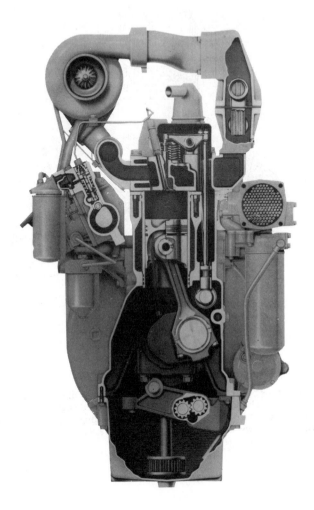

Fig. 11–12. 1983 Caterpillar 3306 six-cylinder truck engine, 638 cu in. (10.5 liters). Bore 4.75 in., stroke 6.0 in. Typical heavy-duty supercharged and aftercooled Diesel engine. Open (DI) combustion chamber, water-cooled aftercooler. Sturdy, simple design. See also Tables 10–8 and 11–1. (*Courtesy Caterpillar Tractor Company.*)

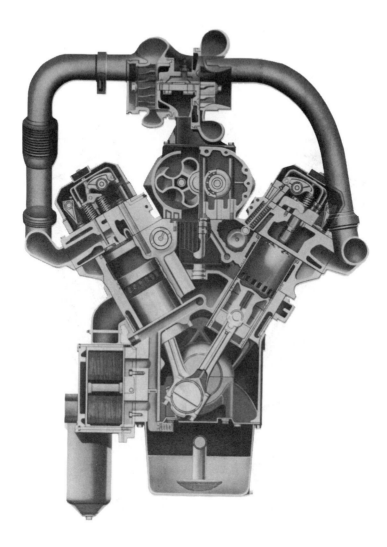

Fig. 11–13. General Motors Detroit Diesel series 71 60° V engine, 1983.
Special features: Only U.S. 2-cycle automotive Diesel. Usually turbosupercharged, with gear-driven Roots-type scavenging pump. Unit injector in each cylinder operating with very high (>20,000 psi) injection pressure. Primary rocking moment balanced by counterweights on camshafts. Four poppet exhaust valves. Made with 6, 8, 12, and 16 cylinders. See also Tables 10–8 and 11–1. (*Courtesy General Motors Corporation, Detroit Diesel Allison Division.*)

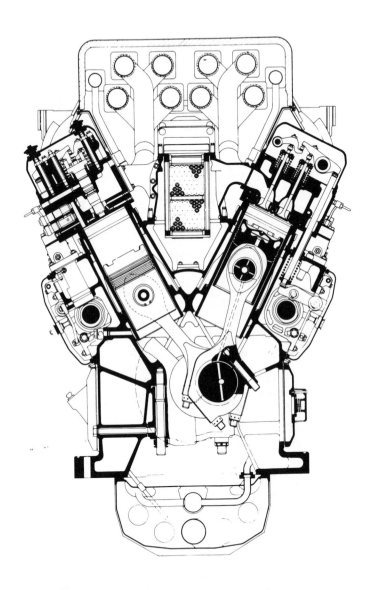

Fig. 11–14. Sulzer 4-cycle locomotive Diesel engine, model ATV 25, 1983.
Special features: Supercharged and aftercooled. Extremely rugged construction for high specific output. Welded steel framework, massive connecting rods split at angle to allow withdrawal through cylinders with maximum crankpin diameter, serrated rod junctions. Four valves per cylinder. Very high bmep. See also Tables 10–9 and 11–1. Multiple exhaust pipes indicate "blowdown"-type turbosupercharger. (*Courtesy Sulzer Brothers, Ltd.*)

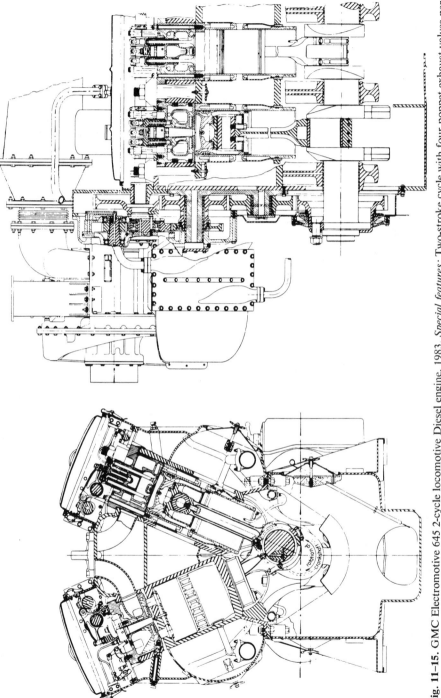

Fig. 11–15. GMC Electromotive 645 2-cycle locomotive Diesel engine, 1983. *Special features:* Two-stroke cycle with four poppet exhaust valves per cylinder. Usually turbosupercharged and aftercooled, with Roots blower in series. Fork-and-blade-type connecting rods with lightweight caps (permissible with low tensile load on rod). Large scavenging air box. Welded steel crankcase, unit injectors. See also Tables 10–9 and 11–1. (*Courtesy General Motors Corporation, Electromotive Division.*)

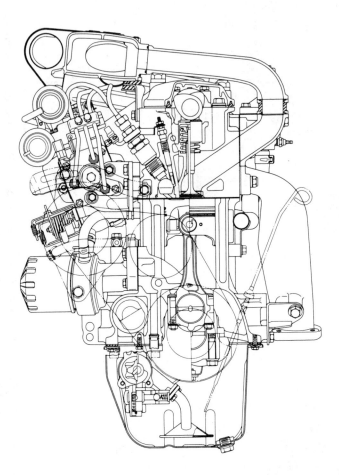

Fig. 11-16 Diahatsu Diesel engine for small automobiles. Three cylinders; 3 in. × 2.9 in.; 1 liter (60.03 cu in.) displacement. Smallest automotive Diesel, 1983. See Tables 10–7 and 11–1. (From Yamaguchi, *Automotive Engineering*, March 1983. Reproduced by permission. © 1983, Society of Automotive Engineers, Inc.)

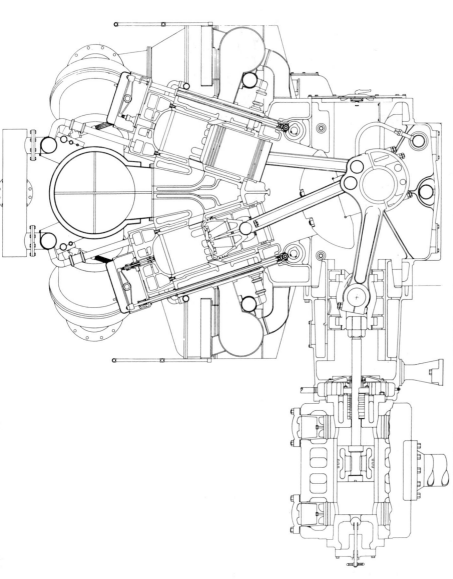

Fig. 11-17. Cooper-Bessemer V-250 2-cycle loop-scavenged gas-engine-compressor unit.

Special features: Scavenged and supercharged by turbosuperchargers. Aftercooled. Relatively light structure adapted to low piston speed and low maximum cylinder pressure. Exhaust ports open at 69 per cent of stroke. Crankshaft demountable through side opening in crankcase. Cast-iron frame work depending in part on foundation for alignment. See also Tables 10–12 and 11–7. (*Courtesy The Cooper-Bessemer Corporation.*)

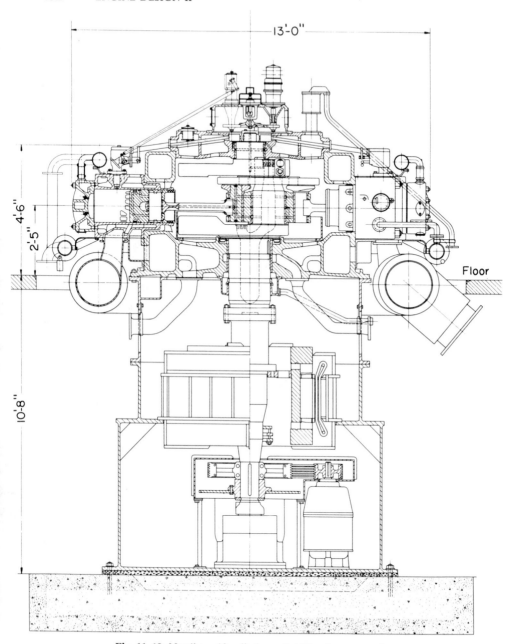

Fig. 11–18. Nordberg 12-cylinder radial Diesel or gas engine.
Special features: Loop-scavenged, 2-cycle radial, with vertical crankshaft. Turbosupercharged and aftercooled. Special connecting-rod system with cluster bearing held from rotation by two link bars and cranks, see ref. 10.76. See also Tables 10–12 and 11–7. (*Courtesy The Nordberg Company.*) Nordberg engines are no longer built; this figure is retained as an example of an unusual design that was successful in its time.

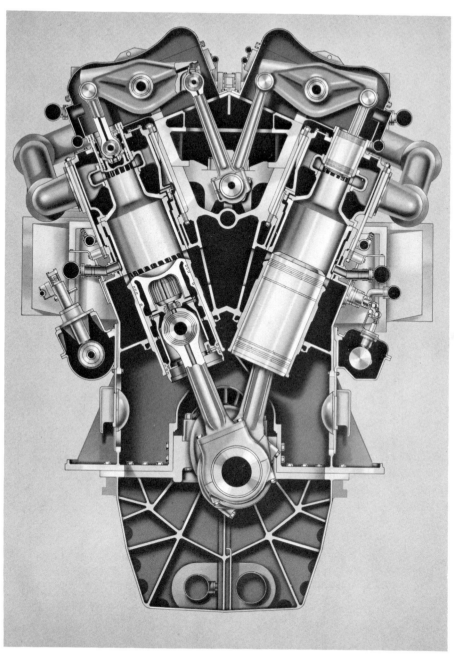

Fig. 11–19. Fairbanks-Morse opposed-piston 2-cycle Diesel engine, 1000 bhp per cylinder. *Special features:* Unusual arrangement for opposed-piston engines reducing exhaust pistons to the category of piston-type valves. Turbosupercharged and aftercooled. Small ratio exhaust/inlet port area. (See Volume I, Chapter 7, for favorable effects.) Exhaust crankshaft is connected to main crankshaft by a train of heavy spur gears. Nodular-iron crankshaft, welded steel frame. Introduced 1965 in I-6, V-8, and V-12 versions. See also Tables 10–11 and 11–8. (*Courtesy Fairbanks-Morse, Inc.*)

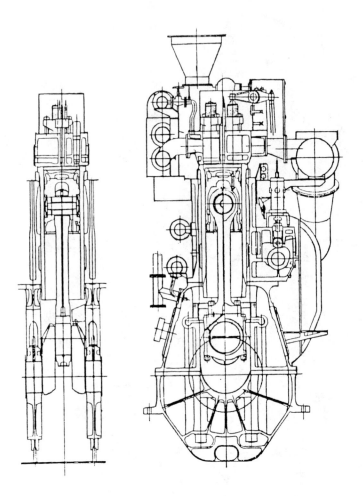

Fig. 11–20. Hanshin GEL44 marine Diesel engine, 1983. Bore 17.3 in., stroke 34.7 in.; 667 hp per cylinder at 220 rpm. Trunk piston engines of this size are unusual. This one is designed for direct propeller drive at 220 rpm. Best-economy piston speed (1270 ft/min) determines stroke. Bore is small to allow very high bmep (294 psi) with consequent high mechanical efficiency, resulting in remarkable fuel economy at rating of 0.297 lbm/bhph or 48 percent brake thermal efficiency. (*Courtesy Hanshin Diesel Works, Ltd.*)

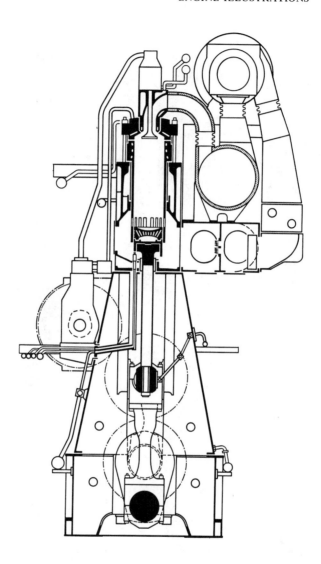

Fig. 11–21. Sulzer RTA84 two-cycle marine Diesel engine, 1983. Cross-head type, 4030 hp per cylinder. Bore 31.5 in., stroke 94.5 in. Available with 4–12 cylinders in line, 16,120–48,360 bhp at 87 rpm. These giant engines (34 ft high) are designed for maximum fuel economy in long-distance supertankers. Stroke is based on most economical propeller rpm and piston speed (1370 ft/min); bore is limited in order to allow supercharging to 250 psi mep (equivalent to 500 psi in a four-cycle engine). Fuel economy at rating 0.280 lbm/bhp, 0.472 brake thermal efficiency.
(Courtesy Sulzer Bros. Ltd.)

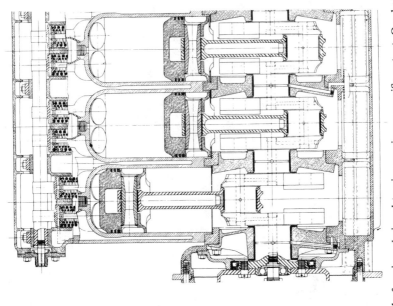

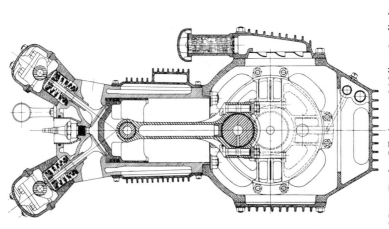

Fig. 11–22. Meyer-Drake Offenhauser 4.16-liter 4-cylinder racing engine. Typical of engines designed to give maximum specific output. Overhead direct-acting camshafts with extra-large valves (four per cylinder); hemispheric heads. Does not take advantage of large number of small cylinders or short stroke-bore ratio useful when displacement is limited. See example 10–5, p. 398, and Tables 10–13 and 11–9. This engine and its 2.74-liter supercharged version were steady winners at Indianapolis for many years. In the 1980s, engines with more cylinders and the same displacement, notably the Ford-Cosworth V-8, displaced this type. The specific output of the 2.74-liter version, 11.6 bhp/in.², remains a near world record (Table 10–13).

(Courtesy Drake Engineering & Sales Corp.)

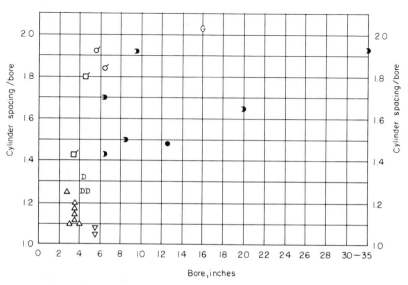

Fig. 11–23. Cylinder spacing measured parallel to crankshaft.

Integral cylinder heads, *liquid-cooled*	*Separate cylinder heads,* *liquid-cooled*	*Separate cylinders,* *air-cooled*
△ automotive, spark ignition	◗ Diesel engines, in-line and V	◁ in-line or V engines
D automotive, Diesel	● Diesel engines, radial	♂ radial engines
▽ aircraft	G gas engine with integral compressors	

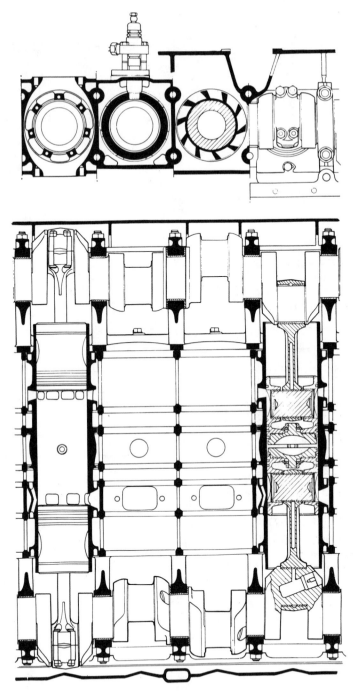

Fig. 11–24. Rolls-Royce K-60 opposed-piston 2-stroke Diesel engine developed especially for military use.

Special features: Opposed piston design with through bolts from bearing cap to bearing cap. Extra-large piston pin. Composite pistons with stainless-steel crowns designed to operate at high temperature. Spur-gear train between crankshafts with geared-up output shaft. See also Tables 10–14 and 11–4. (*Courtesy Rolls-Royce, Ltd.*)

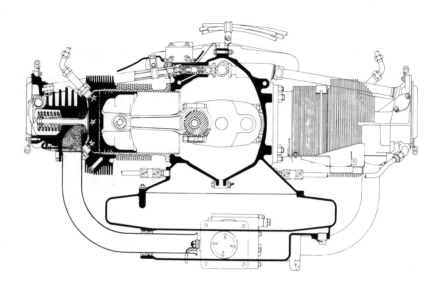

Fig. 11–25. Avco Lycoming airplane engine, 1983. Opposed cylinders, air cooled. Made with 4, 6, and 8 cylinders and with cylinder bores of 4⅜ and 5⅛ in. Compression ratios of 7.0–8.7 depending on fuel to be used. Turbosupercharged and unsupercharged, 105–400 hp. Typical of spark-ignition engines for light aircraft. (*Courtesy Avco, Lycoming Division.*)

DESIGN-RATIO TABLES, 11–2 THROUGH 11–9.

These tables were prepared for typical engines of the late 1960s. Engines designed since then have not shown significant changes in these ratios except in the case of a few heavy-duty Diesel engines where there has been a slight increase in crankshaft diameter to bore. With computers, more attention has been paid to reducing stress concentrations without a change in the tabulated ratios.

Symbols used in the tables are defined as follows:

Piston

l = length, p = pin diam, h = length above pin, N_c = number of compression rings, b = bore, S = stroke.

Connecting Rod

L = length center to center, B = bolt diam, w_r/P = big-end bearing width-diameter.

Crankshaft

P = crankpin diam, M = journal diam, R = crank radius, N = number of journals, t = axial thickness of crankcheek, w_j = length of journal.

Valves

No. = number in one cylinder, D_i = o.d. of inlet valve, D_e = o.d. of exhaust valve.

Weight

W/displacement = engine weight, lbm divided by displacement in cubic inches. Multiply by 2.77 to obtain kg per liter.

Cylinder Spacing

x = distance between cylinders, measured parallel to crankshaft.

Table 11–2. Design Ratios of Small Engines

Class	Make Model Year	Cylinders			Piston				Conn. Rod			Crankshaft					Valves			$\frac{W}{\text{displ.}}$
		No. Arr. Cycle	Bore in.	S/b	l/b	p/b	h/b	N_c	L/R	B/b	w_r/P	P/b	M/b	N	$\frac{w_j}{M}$	t/b	No.	D_i/b	D_e/b	
K	McCulloch 90 1966	1 vert 2	2.165	0.75	0.77	0.23	0.49	2	3.1	—	0.26	0.29	0.32	2	—	0.23	—	—	—	1.3
CS	McCulloch M2-10 1966	1 vert 2	1.75	0.79	0.99	0.21	0.44	2	3.7	—	0.23	0.29	0.34	2	—	0.25	—	—	—	3.27
O	McCulloch 75 1966	3 vert 2	3.125	0.88	—	0.26	—	2	3.2	0.08	0.24	0.38	0.58	4	—	0.18	—	—	—	2.6
O	Chrysler 610	1 vert 2	2.187	0.73	0.83	0.23	0.40	2	3.9	0.085	0.83	0.34	0.36	2	—	—	—	—	—	2.4
O	Chrysler 820	1 vert 2	2.531*	0.64	0.72	0.20	0.35	2	3.9	0.074	0.83	0.30	0.31	2	—	—	—	—	—	1.6*
Mc	Honda 450	1 vert 4	2.80	0.82	0.98	0.28	0.38	2	4.8	—	—	0.47	0.63	2	—	—	2	0.58	0.53	—

* Aluminum cylinder with plated bore. K = motor "Kart," CS = chain saw, O = outboard marine, Mc = motorcycle.

Table 11–3. Design Ratios of Passenger-Automobile Engines

Make Model Year	Cylinders			Piston				Conn. Rod				Crankshaft				Valves			$\frac{x}{b}$
	Arr. and No.	Bore Stroke in.	S/b	l/b	p/b	h/b	N_c	L/R	B/b	w_r/P	P/b	M/b	N	w_j/M	t/b	No.	D_i/b	D_e/b	
Chevrolet 250 cu in. 1965	I-6	3.88 3.53	0.91	1.02	0.24	0.43	2	3.2	0.09	0.4	0.51	0.59	7	0.44	0.39	2	0.44	0.39	1.13
Ford Falcon 1964	I-6	3.5 2.94	0.715	0.92	0.26	0.46	2	3.2	—	0.38	0.61	0.65	4	—	—	2	0.47	0.40	1.11
Plymouth 1964	I-6	3.41 3.13	0.915	0.96	0.26	0.47	2	3.7	—	0.45	0.64	0.81	4	—	—	2	0.48	0.40	—
Cadillac 1964	V-8	4.13 4.00	0.97	0.85	0.24	0.40	2	3.3	—	0.36	0.55	0.73	5	—	—	2	0.45	0.36	—
Ford 1966	V-8	4.00 2.88	0.74	0.88	0.23	0.36	2	3.6	—	0.34	0.53	0.56	5	—	—	2	0.45	0.36	1.10
Oldsmobile 1964	V-8	3.94 3.39	0.86	0.92	0.25	0.46	2	3.5	—	0.39	0.54	0.64	5	—	—	2	0.48	0.40	—
Rambler 1960	V-8	3.13 4.25	1.36	1.0	0.27	0.57	2	3.1	—	0.46	0.67	0.8	7	—	—	2	0.47	0.41	1.17
SAAB* 1965	I-3	2.76 2.87	1.04	1.31	0.25	0.63	2	4.15	—	0.18*	0.39	0.5	4	0.96	0.33	—	—	—	1.25
Chevrolet Corvair‡	OC-6	3.44 2.94	0.86	0.93	0.23	0.46	2	3.2	0.09	0.35	0.51	0.59	4	0.48	0.18	2	0.39	0.36	—
Ford Mustang 65A 1966	I-6	3.68 3.13	0.85	—	0.25	—	2	3.0	—	—	0.58	0.61	4	0.45	—	2	0.45	0.38	—
Rambler Typhoon 290	V-8	3.75 3.25	0.87	0.89	0.25	0.41	2	3.7	—	—	0.54	—	5	—	—	2	0.48	0.37	—
Pontiac 1966†	I-6	3.88 3.25	0.84	0.91	0.24	0.48	2	3.5	—	0.5	0.50	0.50	7	0.50	0.25	2	0.496	0.413	1.15

* Roller bearings, 2-cycle.   † Overhead camshaft, driven by toothed rubber belt.   ‡ Air-cooled.   OC = opposed cylinder

Table 11–4. Design Ratios of Automotive Diesel Engines

Make Model	Cylinders					Piston			Conn. Rod			Crankshaft					Valves			$\frac{W}{\text{displ.}}$	$\frac{x}{b}$
	Arr. and No.	Cycle	Bore Stroke in.	S/b	l/b	p/b	h/b	N_c	L/R	B/b	w_r/P	P/b	M/b	N	w_j/M	t/b	$N.$	D_i/b	D_e/b		
Cummins V-265	V-8	4	5.50 4.13	0.75	0.78	0.32	—	2	4.0	—	—	0.57	0.64	5	—	—	4	0.34	0.34	2.65	—
Caterpillar 61	I-6	4	4.5 5.5	1.22	1.25	0.34	0.85	2	3.4	—	0.43	0.69	0.80	7	0.31	0.24	2	0.41	0.38	4.0	1.31
GMC Detroit Diesel* 12-V-71	V-12	2	4.25 5.00	1.18	1.4	0.35	0.9	4	4.1	0.10	0.36	0.65	0.82	7	0.37	0.19	4	ports	0.28	3.9	1.26
GMC Toro-flow DH-478	V-6	4	5.125 3.86	0.75	0.88	0.31	0.5	2	3.7	0.08	—	0.55	0.74	5	—	—	2	0.49	0.31	2.0	—
Mack End-864	V-8	4	5.0 5.5	1.1	1.1	0.36	0.67	3	3.8	0.11	—	0.70	—	5	—	—	2	0.47	0.39	2.4	—
International	V-8	4	4.50 4.31	0.98	1.05	0.34	0.63	2	4.0	—	0.4	0.58	0.68	5	0.36	0.17	2	0.45	0.35	2.4	1.25
Rolls-Royce K-60	OP-6	2	3.44 3.60	1.05	1.12	0.42	0.73	3	3.4	0.08†	0.7	0.70	0.80	7	0.45	0.31	—	—	—	3.9	1.40
GMC Detroit Diesel 16-V-149	V-16	2	5.75 5.75	1.0	1.17	0.43	0.47	4	4.2	0.10	—	0.68	0.85	9	0.67	0.28	4	—	0.28	4.2	—

* Two-cycle with 4-poppet exhaust valves in cylinder head.   OP = opposed piston.
† Four bolts per rod—others have two.

Table 11–5. Design Ratios of Locomotive Diesel Engines

Make Model	Cylinders					Piston			Conn. Rod			Crankshaft					Valves			$\frac{W}{\text{displ.}}$	$\frac{x}{b}$
	Arr. and No.	Cycle	Bore Stroke in.	S/b	l/b	p/b	h/b	N_c	L/R	B/b	w_r/P	P/b	M/b	N	w_j/M	t/b	No.	D_i/b	D_e/b		
Alco 251-T	V-16	4	9.0 10.5	1.17	1.23	0.33	0.75	4	3.75	—	—	0.67	0.95	9	—	—	4	0.32	—	3.9	—
Electro-Motive 16-645-E-3	V-16	2	9.06 10.0	1.10	1.3	0.41	0.72	4	4.6	—	0.4	0.72	0.83	9	0.57	0.29	4	—	0.28 (4)	3.5	—
Cooper-Bessemer FVBS-16-DT	V-16	4	9.0 10.5	1.17	1.1	0.35	0.61	4	4.4	0.13 (4)	0.38	0.86	0.92	9	0.55	0.33	4	0.32	0.32	3.7	—
M.A.N. V8V 24/30	V-16	4	9.45 11.81	1.25	1.37	0.44	0.78	3	4.1	0.11 (2)	0.48	0.79	0.79	9	0.52	0.40	4	0.34	0.31	3.5	1.95
Sulzer 16LVA24	V-16	4	9.45 11.0	1.16	1.4	0.48	0.92	2	4.1	—	—	0.85	—	9	—	—	4	0.35	0.35	3.3	—

Table 11-6. Design Ratios of Medium-Size (6–13.5-in. bore) Industrial and Marine Diesel Engines. Locomotive Engines, Table 11–5, Are Also Used in This Category

Make Model	Cylinders			Piston					Conn. Rod			Crankshaft					Valves			$\frac{W}{displ.}$	$\frac{x}{b}$
	Arr. and No.	Cycle	Bore Stroke in.	S/b	l/b	p/b	h/b	N_c	L/R	B/b	w_r/P	P/b	M/b	N	w_j/M	t/b	No.	D_i/b	D_e/b		
Maybach M. B. 820 Db	V-12	4	6.9 8.1	1.17	1.07	0.33	0.42	3	3.7	—	—	0.72	—	7	—	—	4	0.30	0.30	2.3	—
Poyaud A8150 SR	V-8	4	5.92 7.09	1.2	1.42	0.38	0.85	4	4.3	—	—	0.65	0.79	5	—	—	4	0.32	—	4.0	—
White Superior	I-6	4	8.5 10.5	1.24	1.44	0.41	0.90	4	3.7	—	0.73	0.65	0.70	7	0.69	0.26	2	0.39	0.39	—	1.52
Waukesha L5788DS1M	V-12	4	8.5 8.5	1.0	1.30	0.35	0.70	4	4.2	—	0.53	0.67	0.74	7	0.58	—	2	0.42	0.37	2.8	—
Caterpillar D-379	V-12	4	6.25 8.00	1.28	1.32	0.38	0.85		4.2	—	0.45	0.80	0.95	7	0.40	0.24	2	—	—	3.9	—
Nordberg VIB-115	V-16	4	13.5 16.5	1.22	1.40	0.41	—	4	4.0	0.10	0.49	0.74	0.96	9	0.40	0.31	2	0.40	0.40	3.4	—

Table 11-7. Design Ratios of Large Gas Engines (13.5–18-in. bore)

Make Model	Cycle	Cylinders				Piston			Conn. Rod			Crankshaft					Valves†			$\frac{W}{\text{displ.}}$	$\frac{x}{b}$
		Arr. and No.	Bore stroke in.	S/b	l/b	p/b	h/b	N_c	L/R	B/b	w_r/P	P/b	M/b	N	w_j/M	t/b	No.	D_l/b	D_e/b		
Cooper-Bessemer KSV-16-SG	4	V-16	13.5 16.5	1.22	1.31	0.38	0.74	4	4.9	0.11 (4)	0.74	0.85	0.85	10	0.50	0.29	4	0.31	0.31	4.2	—
*Cooper-Bessemer V-250	2	V-16 l.s.	18.0 20.0	1.11	1.51	0.29	0.97	4	6.0	0.083 (6)	0.97	0.78	0.78	10	1.07	0.31	0	—	—	4.4*	—
Worthington Mainliner*	2	V-18 pv	16.0 18.0	1.12	1.90	0.41	1.04	4	5.0	0.086 (4)	0.40	0.88	0.88	10	0.63	0.34	2	—	0.40	4.5*	—
Ingersoll-Rand PKVT	4	V-16	16.0 22.0	1.37	1.48	0.44	0.80	4	4.0	0.08 (4)	0.38	0.88	0.88	10	0.38	0.38	4	0.30	0.30	2.7	2.02
Nordberg Radial	2	radial 12 l.s.	14.0 16.0	1.14	1.40	0.41	0.79	4	4.4	0.07	—	1.0	0.85 1.3	2	0.58 0.80	0.36	0	—	—	3.9	—

* Engine-compressor units; weight does not include compressors but includes compressor mountings.

† For data on porting of 2-cycle engines, see Fig. 12–4.

l.s. = loop scavenged; pv = poppet exhaust valves.

Table 11-8. Design Ratios of Large Diesel Engines (17–36.6-in. bore)

Make Model	Cycle	Cylinders					Piston		Conn. Rod				Crankshaft				Valves				
		Arr. and No.	Bore stroke in.	S/b	l/b	p/b	h/b	N_c	L/R	B/b	w_r/P	P/b	M/b	N	w_j/M	t/b	No.	D_i/b	D_e/b	$\dfrac{W}{displ.}$	$\dfrac{x}{b}$
Enterprise RV-16-3	4	V-16	17.0 21.0	1.23	1.77	0.38	0.83	5	4.6	—	—	0.70	—	9	—	—	4	—	0.38	3.4	—
Fairbanks-Morse* 38A20	2	V-12 O.P.	20.0* 21.5	1.07* 1.08	1.75	0.39	1.00	3	4.7	—	—	0.81	—	—	—	—	–	—	—	2.9	1.65
Fiat 900-12S	2	I-12 l.s.	35.4 63.0	1.78	1.90	0.50	—	6	3.9	—	—	0.82	0.82	13	—	—	–	—	—	—	—
M.A.N. K293/170	2	I-12 l.s.	36.6 66.9	1.83	1.90	0.59	—	6	4.0	0.135	0.49	0.78	0.78	14	0.59	0.43	—	—	—	3.4	—
Nordberg TS-2112-SC	2	I-12 l.s.	21.5 31.0	1.44	2.36	0.56	1.47	5	5.0	0.087 (4)	0.40	0.69	0.77	13	0.45	0.32	—	—	—	2.75	—
Sulzer RD/RF90	2	I-12 l.s.	35.4 61.0	1.72	1.10	0.74	—	5	4.6	—	0.51	0.71	0.71	13	0.72	0.41	—	—	—	3.0	1.91
Stork Werkspoor 170-8	2	I-12 p.v.	33.5 67.0	1.74	0.50	0.50	—	–	4.0	—	—	0.77	0.77	13	—	—	4	—	0.3	3.0	—
Burmeister & Wain	2	I-12 p.v.	33.06 70.88	2.12	0.75	0.61	—	6	4.0	—	—	0.66	0.88	13	—	—	1	—	0.60	3.0	—

* Upper figures are for intake piston and lower for exhaust piston (see Fig. 11–19).
l.s. = loop scavenged; p.v. = poppet exhaust valves; O.P. = opposed piston.
All engines are crosshead type except Enterprise, Fairbanks-Morse, and Nordberg.

Table 11-9. Design Ratios of Automobile-Racing and Super-Sports-Car Engines

Make Model	Cylinders				Piston			Conn. Rod				Crankshaft				Valves					
	Arr. and No.	Bore in.	S/b	l/b	p/b	h/b	N_c	L/R	B/b	w_r/P	P/b	M/b	N	w_j/M	t/b	N°.	D_i/b	D_e/b	A_i/A_p*	$\dfrac{W}{\text{displ.}}$	$\dfrac{x}{b}$
Ferrari 6-90	I-6	2.87 2.32	0.81	—	—	—	—	—	—	—	—	—	4	—	—	2	0.57	0.57	0.32	2.9	—
Ferrari-Lancia	V-8	2.99 2.69	0.90	0.78	0.37	0.51	—	3.9	0.11	—	0.72	—	5	—	—	2	0.625	0.625	0.39	—	—
Ford 1964	V-8	3.76 2.87	0.76	0.86	0.27	0.45	2	3.6	0.12	0.35	0.56	0.61	5	0.53	0.18	4	0.39	0.36	0.30	1.57	1.20
Coventry-Climax Flat-16	Flat 16	2.13 1.60	0.75	—	0.22	—	2	—	—	0.37	0.60	0.60	10	0.50	0.22	2	0.57	0.55	0.32	—	—
Coventry-Climax FPE-2.5	V-8	3.00 2.68	0.89	0.83	0.29	0.48	2	3.9	0.12	0.35	0.71	0.84	5	0.35	0.22	2	0.57	0.55	0.32	2.25	1.40
Coventry-Climax FPF	I-4	3.2 2.8	0.88	0.75	0.3	0.45	2	3.4	0.10	0.52	0.75	0.88	5	0.50	0.28	2	0.53	0.45	0.28	2.8	1.18
Drake-Offenhauser	I-4	4.28 4.38	1.02	0.67	0.25	0.34	2	3.5	0.11	0.60	0.50	0.55	5	0.54	0.18	4	0.40	0.40	0.33	1.85	1.17
Drake-Offenhauser INDY	I-4	4.125 3.125	0.76	—	—	—	—	—	—	—	—	—	—	—	—	4	0.37	0.35	0.27	1.8	—

* A_i = inlet valve area based on D_i and number of valves.

The engine illustrations and tabular data have been chosen as examples of good contemporary design as of 1966. Many others of equal quality have been omitted because of space limitations.

THE POWER TRAIN

The power train is here taken to include the assembly of cylinders, pistons, connecting rods, crankshaft, crankcase, supporting structure, and bearings. Valves and valve gear will be considered in Chapter 12.

Gas-Pressure Loads. In most cases the greatest load that must be carried by the power train is that due to the maximum gas pressure in the cylinders. This load is carried as a compressive load from the top of the piston down through the piston pin, connecting rod, and crankshaft to the main-bearing caps. The tensile load from main-bearing caps back to the cylinder heads may be carried in various ways, as will be discussed under the heading of crankcase design.

Inertia Loads. Inertia loads seldom constitute a problem in engines that run at low piston speeds. In high-piston-speed engines the inertia loads require attention chiefly in connecting-rod, crankpin, and bearing design and will be dealt with under those headings.

CYLINDER DESIGN

Good cylinder design is one of the most important features of good engine design. Not only do the cylinders constitute a large fraction of the engine cost, weight (25 per cent is an average figure), and bulk, but they also exert a major influence on engine performance, reliability, durability, and maintenance.

Cylinder Barrel

The barrel functions both as a structural member and a bearing surface for the piston and rings. The most widely used material is cast iron for both the structure and the bearing surface (11.310). For all types of cylinder barrel, proper fillets are necessary at all changes of section, as for all other engine parts. Protection from rust on the outer surface of the barrel, where it comes in contact with cooling water, is essential. Failure of cylinder barrels because of weakening by corrosion is not uncommon where proper rustproofing is not provided.

Water-Cooled Cylinder Barrels. In engines where long operating life is not required, as in most passenger automobiles and small industrial engines, the cylinder barrel is cast integrally with the jacket structure (Figs.

11–9 through 11–11). In this case, worn-out bore surfaces may be refinished to a larger diameter, and over-size pistons inserted.

For long-life industrial and commercial use, separate removable cylinder barrels are used (Figs. 11–12 through 11–21). This method of construction not only facilitates repair and replacement but also allows appropriate materials to be used, regardless of the crankcase requirements. Cast iron is the preferred material, although steel is occasionally used.

Liquid-cooled reciprocating aircraft engines are now obsolete. Descriptions of their various cylinder designs can be found in refs. 10.845, 10.846, and 10.848.

Cylinder-Barrel Loading. Where the cylinder barrel does not carry any axial load, stress due to gas pressure can be computed rather accurately. It will usually be found, however, that the requirements for maintaining circularity, for avoiding damage in handling, and for reboring result in a thickness such that the pressure stress is not the decisive factor.

When the barrel carries the gas load from head to crankcase, the design should be such as to avoid stress concentrations around the bolting flange. The near ultimate in cylinder design from this point of view is illustrated by late practice in aircraft cylinders, Fig. 11–5c. This design incorporates the following special features designed to reduce stress concentrations at the bolting flange:

1. Closely spaced studs—(nine altogether)
2. Flange well filleted into barrel
3. Stud axis as close to bore as possible
4. Spherical seats for the nuts, thus minimizing uneven pressure distribution under the nuts and minimizing bending stresses in the screws
5. Screw designed for minimum stress concentration (see discussion of screw fastenings)
6. Rustproof treatment of all surfaces.

This design was developed through the use of experimental stress analysis and much bench and field testing, and is obviously expensive. Where light weight is not so important, less costly designs are acceptable, as in Fig. 11–5b.

Cylinder-Barrel Cooling. The cylinder barrel is of course a heat-flow path from gases to coolant. The problems associated with this function increase with cylinder size (see Volume I, Chapters 8 and 11). They involve the following considerations:

1. Shortest practicable path from inner surface to coolant, particularly at the head end where heat flow is most rapid
2. Avoidance of distortion of the barrel. Except in the case of loop-scavenged 2-cycle cylinders, the forces for distortion come chiefly from the head and jacket structure. The transmission of distortion

from these elements to the barrel should be minimized in the design. With separate barrels, the clamping and gasket arrangement should be such as to keep the barrel as free as possible from such forces.

Ported Cylinder Barrels in 2-Cycle Engines. Exhaust ports in such cylinders constitute a serious problem, except in small sizes. This problem is especially acute in loop-scavenged cylinders, where the temperature difference between the intake and exhaust sides is large. Except for small cylinders, provision must be made to cool the bridges between the exhaust ports, either by circulation of liquid coolant within the bridges, or by the provision of adequate finning and air circulation (10.820, 10.823, 10.824).

The importance of a high flow coefficient for 2-cycle cylinder ports has been emphasized in Volume I, Chapter 7. This end is achieved by maximum port areas consistent with strength and heat flow, and by rounding the entrance of the inlet ports, where this is practicable. Port design for 2-cycle engines will be considered in detail in Chapter 12.

Cylinder Bores. For most cylinder bores the wearing surface is cast iron. For Diesel engines, a favored method of repairing worn cylinder bores is chromium plating, and some manufacturers even use it for new cylinders (11.315). Steel cylinder barrels for aircraft engines are sometimes nitrided to minimize wear.

Final finish of the bore is usually by honing to a specified degree of surface "roughness." The desirable degree of roughness depends on the materials used for bore, piston, and rings, and to some extent on design details and operating regimes. Surfaces that are too rough cause rapid wear and those too smooth prevent fast "break-in" of piston rings and bore. Specified limits on surface finish vary with different materials and methods of manufacture. Table 11–10 indicates 13 rms* microinches (3.3×10^{-4}mm) as a desirable value for cast-iron bore surfaces.

Cylinder Heads for Water-Cooled Engines

Water-cooled engines use cast iron for cylinder heads except where the need for light weight or other special circumstances justify other materials, usually aluminum. The use of long multiple aluminum head castings on iron cylinder blocks is rare because of differential-expansion problems.

The cylinder head is one of the most critical elements in engine design, since it combines problems of structure, heat flow, and fluid flow in a complex shape. An exception to shape complexity is the cylinder head of loop-scavenged 2-cycle cylinders, for which a simple, symmetrical head design, as shown in Figs. 11–9, 11–17, 11–18, and 11–20, is usually possible.

---

* The abbreviation rms stands for root mean square of the depth of the surface irregularities.

Table 11–10. Typical Values of Surface Roughness for Various Parts.
Root mean square of profilometer reading, microinches

(From 1964 SAE Handbook, ref. 11.230)

$\overset{1}{\sqrt{}}$	Micrometric anvils Mirrors Gauges	$\overset{16}{\sqrt{}}$	Spline shafts Motor-shaft bearings
$\overset{2}{\sqrt{}}$	Shop-gauge faces Comparator anvils	$\overset{32}{\sqrt{}}$	Brake drums Broached holes Bronze bearings Precision parts Gear teeth Ground ball and roller bearings
$\overset{4}{\sqrt{}}$	Vernier-caliper faces Wrist pins Hydraulic piston rods Precision tools Honed roller and ball bearings	$\overset{63}{\sqrt{}}$	Gear-locating faces Gear shafts and bores Cylinder-head faces Cast iron gear-box faces Piston crowns
$\overset{8}{\sqrt{}}$	Crankshaft journals Camshaft journals Connecting-rod journals Valve stems Cam faces Hydraulic cylinder bores Lapped roller and ball bearings	$\overset{125}{\sqrt{}}$	Mating surfaces, no motion
$\overset{13}{\sqrt{}}$	Piston outside diameter Cylinder bores	$\overset{250}{\sqrt{}}$	Clearance surfaces Rough machine parts

The central problem in valved-cylinder-head design is to achieve an arrangement satisfactory from the point of view of valves and ports, which will carry the gas loads and at the same time avoid excessive distortion and stress due to temperature gradients, and also avoid excessively high cost or undue complexity.

As in all parts subject to heat flow and the resultant temperature gradients, the problem becomes more difficult as the bore increases (see Volume I, Chapters 8 and 11). For cylinders up to about 6-in. bore, well-designed iron or aluminum castings which will carry the gas and valve-gear loads generally give no serious troubles from thermal stresses. As the bore increases from this point, the thickness necessary to carry the applied loads becomes large enough to introduce troublesome thermal stresses and distortions. Location of the areas of high thermal stress on a conventional head are shown in Fig. 11–26a. Possible methods of reducing these stresses are illustrated in Figs. 11–26b and 11–26c. In general, head stresses are minimized by:

1. Careful detail design to reduce stress concentrations, including generous fillets

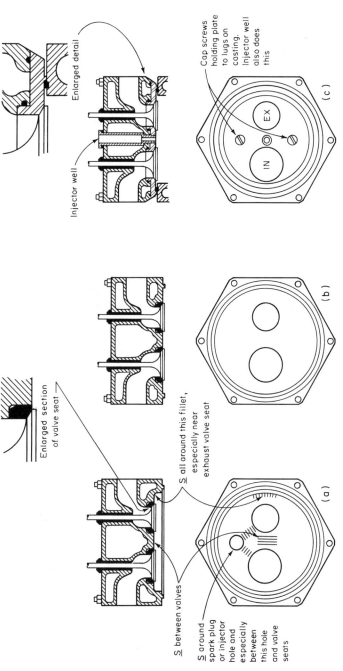

Fig. 11–26. Cylinder-head designs. (*a*) Conventional head showing areas of high stress concentration *S*; (*b*) same as (*a*) except head fillet is eliminated by raising gasket surface to level of head surface; (*c*) "floating" steel-plate head, sealed by O-rings and spigotted at valve ports.

2. Separation of functions of heat flow and load-carrying (Fig. 11–26*c* and Volume I, Fig. 8–7)
3. Use of separate cylinder heads on en-bloc cylinder assemblies
4. Combined with 1 to 3, substitution of aluminum for iron. This device is not generally used for large cylinder heads, but is believed by your author to be worth development. The success of aluminum heads in aircraft engines indicates the possibilities of this material under conditions of heavy loads and rapid heat flow.

A final method of solving the cylinder-head design problem is to eliminate the head altogether by the adoption of the opposed-piston arrangement, Fig. 11–24. However, this arrangement is relatively costly (2 crankshafts) and inappropriate for many applications.

Air-Cooled Cylinder Design

As previously stated, air-cooling is impractical for cylinders larger than about 6-in. (150-mm) bore.

The substitution of cooling fins for water jackets usually involves separate cylinders rather than cylinders grouped together in a common water-jacket structure. Some small, low-output engines use cast-iron cylinders with integral cooling fins. An alternative construction is a one-piece aluminum casting with bore plated with chromium or other wear-resistant material (11.314), or with an iron barrel cast into the aluminum body (Fig. 11–6).

Most air-cooled cylinders designed for high specific output are made with separate heads and barrels. Air-cooled aircraft cylinder barrels are usually steel forgings, machined all over and screwed into the head, as in Fig. 11–27. Steel cylinder bores appear to be satisfactory from the point of view of friction and wear, at least for aircraft requirements. In some cases the bores are nitrided to increase their wear resistance (10.844, 10.847, 10.850–10.854, 11.301). Fins on these barrels may be integrally machined, or applied in the form of sheet-metal rings.

Figures 11–7 and 11–8 show separate barrels, the former having a finned aluminum jacket and the latter a finned iron barrel casting.

Cooling Fins. Much research has been conducted on the subject of cooling-fin design and placement on the cylinder (11.301–11.303, 12.922–12.924). This subject will be mentioned again in Chapter 12 in connection with cooling systems. The basic principles involved are:

1. Cooling fins must be placed as close as possible to the critical sources of heat input, including especially: exhaust-valve seats, exhaust ports, spaces between ports and valves, spark-plug bosses, and exhaust-port bridges of 2-cycle engines.

2. No portion of the cylinder head, or barrel down as far as the position of the piston rings at bottom center, should be unfinned.
3. Fins should be oriented in the direction of air flow.
4. Fins should be as deep and as closely spaced as possible, considering the material and manufacturing process used. The large air-cooled aircraft engines of the 1940's used aluminum fins on their cylinder heads as deep as $2\frac{1}{2}$ in. (60 mm) and spaced as close as 0.20 in. (5 mm) with a thickness of 0.10 in. (2.5 mm). However, such finning had to be machined and would be justified only for engines of very high specific output.

Fig. 11–27. Cylinder head, inlet and exhaust ports of Wright "Cyclone" aircraft engine. Inlet port design (left) gives priority to gas flow considerations. Exhaust port compromises flow for the sake of cooling the valve. Head-fin area is over fifty times piston area, with cast fins. Later versions had machined fins with even larger area.
(*Courtesy Wright Aeronautical Corp.*)

5. For engines of over 5-in. bore, high-conductivity fins applied to the cylinder barrels may be necessary. These usually consist either of a finned cast-aluminum jacket or sheet-aluminum rings pressed or rolled on the barrel (see Figs. 11–7, 11–8, and 11–27).
6. Air flow should be directed at right angles to the cylinder axis, with the exhaust side "upwind."

In the past, various types of applied finning have been used on cylinder heads. Air flow parallel to the cylinder axis has also been used. Experience has shown these methods to be inferior to finning integral with the cylinder head and to flow in the radial direction.

Valve Ports. Valve ports must be designed with due regard for structural strength, gas flow, heat flow, and cooling, especially of the exhaust valve. A general principle in port design is to avoid restrictions to flow smaller than that of the valve at full lift and to secure as uniform a flow passage as is practicable. Good port design is illustrated in Figs. 11–7, 11–8, 11–10, 11–12, 11–22, and 11–27.

Inlet Ports

Since heat flow to and from the inlet valve is relatively small (Volume I, ref. 6.32), flow capacity is the major consideration after structural strength. Good flow capacity calls for the smallest diameter valve stem consistent with structural integrity, and the avoidance of sharp turns or sudden changes of flow area (see especially Figs. 11–22 and 11–27). The effects of valve-seat details and valve flow capacity are discussed further in Chapter 12.

In many engines it is desirable to promote swirl of the charge by means of a tangential inlet flow. Swirl is particularly desirable in open-chamber Diesel engines. The problem of obtaining adequate swirl without unduly reducing the inlet-flow coefficient (see Volume I, Chapter 6) is a difficult one. Its degree of success is generally measured by a swirl coefficient determined from steady-flow tests and defined as follows:

$$C_{sw} = 2g_0\Delta p/\rho v^2, \tag{11–3}$$

where C_{sw} = dimensionless swirl coefficient

Δp = measured pressure drop across inlet port

g_0 = Newton's law coefficient

ρ = air density

v = measured tangential swirl velocity close to the periphery of the combustion chamber, but not in the boundary layer.

The swirl coefficient may be measured at various valve lifts but is most often judged by the result at maximum lift.

A very effective swirl producer is the "shrouded" inlet valve (11.304). This valve has been standard on the CFR fuel-test engine (14.111) since its inception. While it is used on a few commercial Diesel engines, it is not considered desirable by this author because it reduces flow capacity of the valve by about 50 per cent, and it requires the valve stem to be keyed against turning. This latter requirement introduces mechanical complications and prevents desirable rotation of the valve. Thus, most engines which require swirl use inlet ports so designed as to direct the incoming air in a tangential direction through the inlet valve (11.304–11.306).

Exhaust Ports. In order to provide adequate cooling for the exhaust valve, the stem guide must extend as far toward the valve head as possible and have adequate provision for water or air circulation around the seat and stem areas. A large valve-stem diameter assists both in cooling and in providing for adequate valve-stem strength and heat transfer. In general, some sacrifice in flow capacity is allowable in the interest of good cooling and long valve life.

Figure 11–27 shows inlet and exhaust ports of a large aircraft cylinder, where the problems of high flow capacity and adequate strength and cooling have had meticulous attention. Less highly stressed engines could profit by adapting such high-grade port design as far as considerations of cost and space limitations allow.

Valve design will be covered in detail in Chapter 12.

Valve Seats. Valve seats of material harder than that of the cylinder head are mandatory for aluminum heads and are desirable in most cases with cast-iron heads, at least for the exhaust valves. The subject of seat design and material is discussed in ref. 11.012. The most difficult problem is to secure the seat so that it will never come loose and at the same time to provide for good heat flow from seat to coolant. Various devices for securing valve seats include casting into the head, screwing, shrinking, and peening. The most satisfactory method seems to be carefully controlled shrinking by means of inserting cold valve seats into a heated cylinder head. More elaborate systems used in aircraft engines are also described in ref. 11.012.

Valve Guides. Cast iron pressed into place is generally used for valve guides. Many other materials have been used (11.012), but iron appears to be the most popular and usually the most satisfactory.

Cylinder Spacing. In order to keep engines as compact as possible, cylinders in a line should be spaced as closely as design considerations will allow. Data on ratios of cylinder spacing to bore are given in Fig. 11–23 and Tables 11–3 through 11–9.

The closest spacing is possible when the cylinder heads of a given bank of cylinders are cast together, as in most liquid-cooled automotive engines. The ratio of cylinder spacing to bore is from 1.1 to 1.2 for gasoline engines and from 1.2 to 1.35 for Diesel engines in this class. Where there is a special premium on engine compactness, as in aircraft engines, this ratio can be as low as 1.05.

Engines with separate cylinder heads must have wider cylinder spacing in order to accommodate the water-jacket walls and bolts. Figure 11–23 shows spacing ratios from 1.4 to 2.0 for engines in this class.

The closeness of cylinder spacing is sometimes considered to be limited by the required crankpin and main-bearing lengths and the crankcheek widths necessary to limit crankshaft stresses. That this need not be the case for V engines is shown by the fact that the smallest spacing shown in Fig. 11–23 is for the engine with the highest specific output, that is, the very successful Rolls-Royce "Merlin" of 5.4-in. bore with a spacing ratio of 1.05. This engine had a military rating of bmep = 300 psi (21 kg/cm^2) at

3000 ft/min (15 m/sec) piston speed, specific output 6.8 bhp/sq in. Even radial engines, including the Pratt and Whitney 28-cylinder engine with 7 cylinders on each crank, the Wright Turbo-Compound with nine cylinders per crank, and the General Motors gas engine with four cylinders per crank have spacing ratios of less than 2.0.

From these considerations it would seem that spacing ratios greater than 1.2 for integral-head and 1.5 for separate-head engines are unnecessarily large. Special cases where wider spacing may be required include air-cooled engines where the finning between cylinders may control, and integral engine-compressor combinations where compressor spacing may be the governing factor.

When a preliminary engine design emerges with a cylinder spacing greater than that required by the cylinder design itself, serious consideration should be given to the alternative of enlarging the bore and reducing the rated bmep accordingly. Wide gaps between cylinders do not appear to be either necessary or desirable, except under very special circumstances.

PISTON DESIGN

The piston combines problems of structure, heat flow and cooling, friction, lubrication, and wear, and in many cases noise suppression.

Materials

As outlined in Chapter 9, cast aluminum alloy is the most widely used material for pistons. Large aircraft engines and some other high-output engines use aluminum forgings (10.844–10.848, 11.012). Engines of large bore usually use cast iron, cast steel, or sometimes a composite construction involving two materials. In the engine illustrations, the pistons used are as follows:

Cast Aluminum. Figures 11–6 through 11–12, 11–14, and 11–22.

Cast Iron or Cast Steel. Figures 11–13, 11–15 through 11–17, 11–19 through 11–21.

Aluminum Crown, Iron Skirt. Figure 11–18.

Steel Crown, Iron Skirt. Figure 11–24.

For compositions of iron and aluminum for pistons see Tables 9–7 and 9–8.

Piston Geometry

In unsupercharged engines of small bore, the piston may be of fairly light construction, as illustrated by Fig. 11–28a. In passenger cars the problem of noise suppression is critical and is usually solved by flexibility in

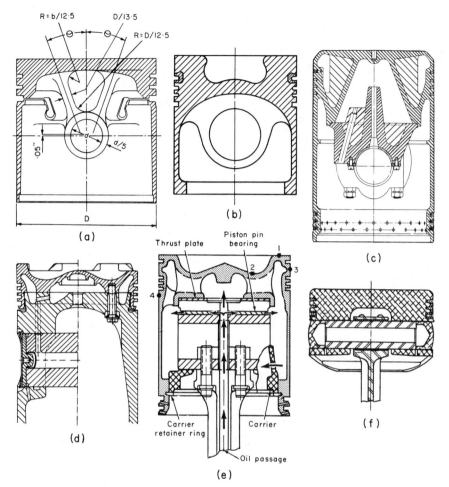

Fig. 11–28. Typical piston designs. (*a*) Passenger-car type, cast-aluminum alloy. See also Figs. 11–8, 11–10, and 11–11. (*b*) Automotive Diesel type, cast-aluminum alloy with cast-in iron ring belt. See also Figs. 11–12 and 11–14. (*c*) Cooper-Bessemer V-250 2-cycle gas engine, Fig. 11–17. Both parts of piston are of cast iron. (*d*) M.A.N. 9.45 × 11.8 in. (240 × 300 mm) V8V24/30 highly supercharged Diesel engine. Piston has steel crown, aluminum-alloy skirt. (M.A.N. Diesel Engine News No. 44, 1965). (*e*) GMC-Detroit Diesel 149 series 5.75 × 5.75 in. (146 × 146 mm) 2-cycle engines. Rotatable piston, cast steel. (*f*) Lycoming 5.375 × 3.875 in. (136 × 98.5 mm) aero engine. Aluminum-alloy piston. Length/diam. = 0.64. (*Courtesy Lycoming Division, Avco Corporation.*)
(For other types of piston see Figs. 11–13 and 11–15, cast steel; Figs. 11–16, and 11–19 cast iron; Figs. 11–20, 11–21 pistons for crosshead engines; Fig. 11–24, cast iron with stainless steel crown.)

the piston-skirt supports, so that the piston runs at all times with near-zero skirt clearance. Another solution is to support the skirts by steel or Inconel plates inserted before pouring. The low expansion coefficient of these plates makes small cold clearance possible. Offsetting the piston pin also helps in noise suppression (11.410). Small-bore industrial engines require less noise suppression and may omit the skirt-flexibility feature in favor of lower cost and more rugged design.

In the case of Diesel engines and supercharged spark-ignition engines, high maximum cylinder pressure requires more rugged design with avoidance of high stress concentrations (Fig. 28b). Experimental stress analysis (11.401) has shown the likely areas of critical stress, principal among which is the upper side of the pin boss. Piston geometry is obviously too complex for theoretical stress analysis.

Piston Cooling

In small-bore (under 4 in.) unsupercharged engines, pistons are usually cooled by conduction of heat from the head to the ring belt and skirt, from which heat then flows to the cylinder bore. For this type of cooling aluminum is desirable, with generous sections for heat conduction from the center of the piston head. Figures 11–9, 11–11, and 11–28a illustrate this method. Heat conduction to the cylinder walls is promoted by small skirt clearances. Rings run cooler as the top "land" (distance from top edge of piston to top ring) is increased.

For high-performance and large-bore spark-ignition engines, and for most Diesel engines, oil cooling of the piston has been found necessary. The oil may be circulated by spraying from a fixed nozzle aimed at the inside of the piston (Figs. 11–13 and 11–15), by supplying oil through the crankshaft and connecting rod to an enclosed space under the piston head (Figs. 11–16, 11–17, 11–19), or through tubular passages in the piston head (Figs. 11–14, 11–18). Large crosshead engines often carry coolant for the piston through jointed (Fig. 11–20) or telescoping pipes (Fig. 11–21). For oil-cooled pistons the rate of oil circulation must be high enough to avoid breakdown of the oil into carbonaceous deposits. The question of piston cooling is discussed in more detail in refs. 11.411 to 11.413.

Piston Pins

Piston pins should be of the *floating* type (loose in both pin and rod) except where noise suppression is very important, in which case the pin is usually clamped in the upper end of the rod. The end play of floating pins must be restricted by suitable snap rings or by end plugs of soft material (aluminum or bronze) finished to a spherical radius slightly less than cylinder radius (11.410).

Piston-pin stress is often computed on the basis of beam theory under maximum gas-pressure load. As a device for extending experience with similar designs, this practice may be useful. However, its validity as a measure of actual stress is obviously bad in view of the unknown distribution of the load from the rod and pin bosses to the pin surfaces. Since the pin is a minor fraction of the piston weight, stiffness, and strength should never be marginal in this element. A rugged pin of large diameter contributes to the reduction of critical stress in all three elements, piston, pin, and connecting rod.

Piston Proportions

Referring to Tables 11–2 through 11–9 and Fig. 11–28, the following is the range for major piston proportions:

	Symbol	Max	Min
Length/bore	l/b	2.36	0.64
Height from pin center to top of bearing surface	h/b	0.97	0.25
Pin diam/bore	p/b	0.48	0.20
Number of compression rings	N_c	6	2

Airplane engines have achieved excellent reliability with piston lengths as short as $0.6 \times$ bore. There would seem to be little reason for pistons to be longer than $1 \times$ bore except where necessary to cover ports at top center in 2-cycle engines. Two-cycle engines with trunk pistons require the piston to cover the ports at piston top center. In such cases, minimum piston length is the stroke plus enough overlap to prevent leakage (see Figs. 11–6, 11–9, 11–13, 11–15, 11–17, 11–18, 11–19). Crosshead engines may not have this requirement (Figs. 11–20, 11–21).

The ratio of pin-diameter to bore (see the Table on this page) runs from 0.20 to 0.48 in trunk-piston engines. Since no space and but little weight is saved by small pins, this author recommends a ratio of 0.35 or greater.

Height Above Pin. The ratio of this height to bore is given in the design-ratio tables. Where possible, it would seem desirable to locate the pin in the middle of the skirt-bearing surface. However, successful engines have been produced with pin locations near the rings and also near the bottom of the skirt.

Figure 11–29 shows principles of good piston design for engines of automotive size or less. Areas of critical stress are greatly reduced in the preferred design. A common fault in piston design is a pin of too small a diameter; in such a case deflection of the pin in bending creates high stress concentrations around the pin bosses. Another common fault is the use of ribs under the piston head. Such ribs act as stress raisers and have been found ineffective in cooling the piston head (11.411). See also later discussion of ribs in connection with crankcase design.

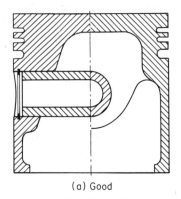 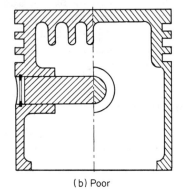

(a) Good   (b) Poor

Fig. 11–29. Principles of good piston design for heavy duty.

(a) Good	(b) Poor
Omit fins	Fins are stress raisers
Narrow-faced compression rings	Rings too wide-faced
Filleted ring grooves	Ring grooves sharp
Large hollow pin	Pin too small
Pin boss well braced to crown	Inadequate strength, pin boss to crown

For more detailed discussions of piston design, see refs. 11.400 to 11.416. See also the pistons illustrated in Figs. 11–6 through 11–22 and 11–24. Useful material on the design of pistons for large-bore engines will be found in refs. 10.401–10.595.

Piston Friction

This subject is discussed in Volume I, Chapter 9. To minimize piston friction, use

1. Small length-to-bore ratio
2. Minimum skirt-bearing area
3. Maximum skirt clearance
4. Small number of rings.

Requirements 1 to 3 are in conflict with minimum noise from piston "slap" which is usually an important factor in engine mechanical noise (8.911). The above features have been used to their maximum effect in aircraft engines, where other sources of noise are so great as to conceal piston noise. On the other hand, pistons for passenger automobiles must be very quiet at the expense of greater piston friction (see refs. 8.900 *et seq.*; also Volume I, pp. 325–337 and Fig. 9–27, p. 350).

Piston Seizure

As the specific output of engines is increased, piston seizure is a common problem. It is usually due to one or more of the following causes:

1. Inadequate clearance in the bore
2. Inadequate cooling
3. Distortion due to heating or loading
4. Heavy detonation, in spark-ignition engines
5. Ring failure, followed by gas blowby and piston burning.

The question of piston cooling is discussed in refs. 11.411–11.413. Tinplating of piston skirts seems to reduce the tendency toward seizure, especially with aluminum pistons.

Piston Rings

Piston rings have constituted a difficult design problem ever since the advent of the internal-combustion engine, and it is only in the period since World War II that they have achieved life and reliability commensurate with those of other wearing parts of the engine.

Piston-Ring Materials. Most piston rings are made of gray cast iron because of its excellent wearing properties in all kinds of cylinder bores. The compositions and the preferred grain type for this purpose are discussed in refs. 11.420–11.425. Where ring leakage is a problem, nodular iron or even steel is used, usually with coated bearing surfaces. For greatest resistance to wear of both ring and bore, rings are faced either with chromium plate or "metallized molybdenum," a porous structure of molybdenum oxides (11.420). Oil-control rings may be of iron or steel.

Piston-Ring Design. The design of piston rings has developed almost entirely on an empirical basis. The literature on the subject is extensive, and only a few important items are listed in the bibliography (11.407–11.409, 11.420–11.425).

Piston rings have two functions: to prevent gas leakage and to hold the flow of oil into the combustion chamber down to the minimum necessary for adequate ring and piston lubrication. In modern engines the necessary oil-flow rate past the rings is extremely small and approaches zero for passenger-automobile engines. All rings take part in the oil-flow control process, but there is usually at least one ring designed for this purpose alone. Such rings are called *oil-control* rings, while the others are *compression* rings.

Compression-Ring Design. The following items have been found to be desirable (Fig. 11–30):

1. Face width of rings should be small. A ratio of face width to cylinder bore of 1/50 is typical of good modern practice (11.420, 11.424). Small face width reduces ring mass and facilitates wear-in.

2. Radial depth of ring must be sufficient to give necessary spring pressure but not so large as to cause rings to break on assembly.

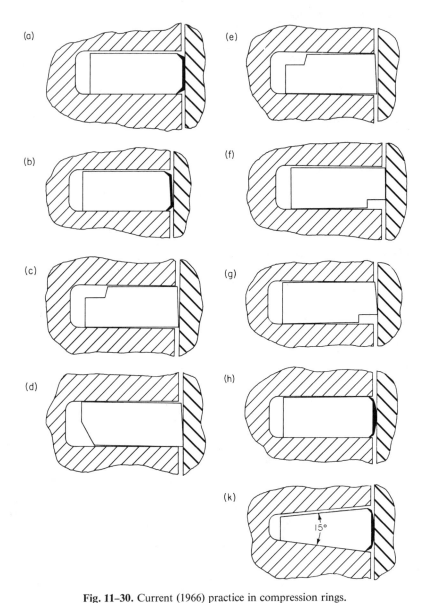

Fig. 11–30. Current (1966) practice in compression rings.
(*a*) Plain compression ring, chrome plated; (*b*) taper-face compression ring, chrome plated; (*c*) torsional-twist compression ring; (*d*) reverse-twist taper-face compression ring; (*e*) torsional-twist taper-face compression ring; (*f*) scraper type 70 compression ring; (*g*) taper-face compression ring; (*h*) barrel-face compression ring, chrome plated. (*k*) wedge-type compression ring, chrome plated (Binford, 11.4201).

3. Spring-pressure of ring against cylinder wall should be controlled to definite limits by the ring manufacturer (11.420, 11.423).

4. Bearing surface of new rings should be slightly conical with larger diameter toward the crankcase, or lower, side (Fig. 11–30*b–g*). This "taper" is often created by machining off the upper inside corner of the ring, after the face has been cylindrically finished (Fig. 11–30*c, e*). The purpose of this taper is both for oil control and for fast "break-in" of new rings. As an alternative to face tapering, the ring may be "barrel"-faced (Fig. 11–31*h, k*), either with or without chromium plating. Barrel-faced rings have been found to give no higher oil consumption than tapered rings, thus invalidating the theory of one-way "oil scraping" due to the taper. A close-fitting worn-in contact seems to be the essential requirement.

5. For heavy-duty service, including most Diesel engines, the face of the top one or two rings is chromium plated. This plating has been found to reduce both ring and bore wear significantly. Other coatings used to resist wear or to assist in the "break-in" process include oxide of molybdenum, iron oxides, and tin plating.

6. Where ring "sticking," that is, seizure in the groove, is a problem, as in many Diesel engines, its incidence can be reduced by wedge-shaped grooves and rings, Fig. 11–30*k*. This shape apparently makes it more difficult for carbonaceous material to hold the ring tight in the groove.

7. Research (Volume I, refs. 9.46, 9.47) shows that gas pressure in the ring grooves is about 50 per cent of cylinder pressure behind the top ring, 20 per cent behind the second ring, and zero behind the third ring. This result leads to the conclusion that two good compression rings are all that is necessary. The use of a greater number can be defended only on the ground of insurance against leakage in the event of failure of the upper ring or rings.

8. Clearance behind the rings must always be so large that the rings can never "bottom" in their grooves. Axial clearance in the grooves must be controlled to reasonably close limits. (See manufacturer's specifications for typical clearance values.) Giving the ring an initial twist (Fig. 11–30*c, d, e*) reduces axial play of the ring in the groove.

9. In aluminum pistons, service life of both piston and ring can be extended by mounting the rings in a belt of harder material (steel or bronze, Fig. 11–28*b*). This practice is increasing, especially for heavy-duty engines (11.404).

Compression-Ring Failures. Such failures occur through sticking, excessive wear, loss of spring tension, or breakage. Clean combustion and adequate piston cooling are the best insurance against all of these failures, except possibly breakage. Breakage has often been attributed to ring "flutter," a type of ring vibration never fully defined or observed. It may exist as a periodic motion of the ring out of contact with the cylinder surface. Small ring-face width tends to reduce the incidence of ring breakage, while breakage seems to be promoted by large axial clearance and very high cylinder pressures. As a last resort to prevent ring breakage, chromium-plated steel rings may be used.

Oil-Control Rings. Most engines require one or more rings specifically

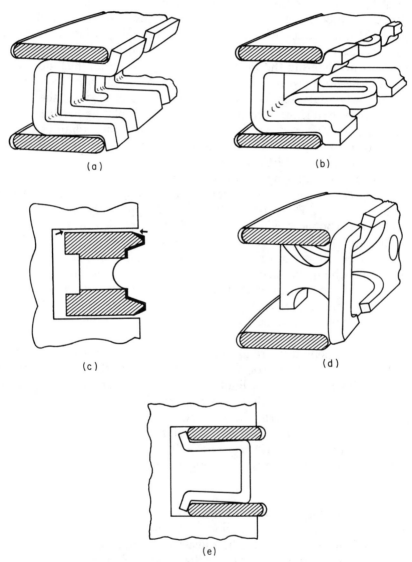

(a)

(b)

(c)

(d)

(e)

Fig. 11–31. Oil-control rings. (*a*) Steel side-sealing oil ring; (*b*) another type of steel oil ring; (*c*) cast iron oil ring (transient position); (*d*) special type of oil ring; (*e*) section of type *a* oil ring, showing side-sealing feature (11.4201).

designed to minimize oil flow to the combustion chamber. Various successful designs for this purpose are shown in Fig. 11–31. These are usually
made up of thin steel stampings which may or may not require a plated
bearing surface against the cylinder bore. This type of ring is not loaded
by gas pressure and must depend on its own spring action to ensure
contact with the bore surface.

For more detailed discussions of ring design, see refs. 11.420–11.425.

Ring Friction. This question is discussed in Volume I, Chapter 9. Ring
friction cannot be separated from piston friction, as the two are inter-
dependent. As far as the rings are concerned, friction is minimized by use
of a minimum number of well-designed rings.

CONNECTING RODS

Because of space restrictions and the usual necessity for minimum mass,
the connecting rod is one of the more heavily stressed engine parts.
Most connecting rods are made from steel forgings. Some small engines
use forged aluminum or cast malleable iron.

Connecting-Rod Loads and Stresses. The theoretical loading on the
connecting rod is easily calculated from the indicator diagram and the
inertia forces discussed in Chapter 9. The heaviest load comes at top center
from peak cylinder pressure. The resultant stress in the column of the rod
is then calculable on the basis of the column theory. (The usual dimensional
ratios are such that the second-order term is negligible.) The actual stress
in the rod column may be far greater than such a calculated value on
account of unsymmetrical loading, vibratory stresses (including the piston
in rotation), and stress concentrations. Stresses due to lateral inertia, or
"whip" of connecting rods, are usually negligibly small.

Heavy unsymmetrical loading comes from the piston axis being out of
line with the rod axis, more or less common to all engines. It has been found
that such loading may cause greater bending moments in the rod than
"whip," and it is one argument for orienting the H section of the rod
column 90 degrees from the usual direction. This design also simplifies forging and machining in many cases (Fig. 11–32). As indicated in Chapter 8,
near top center inertia loads are opposite to gas loads. Inertia therefore
reduces the maximum gas load on the rod. In some cases pistons and connecting rods are made heavier than necessary in order to reduce crankpin-
bearing maximum loading (see Fig. 11–14, for example).

In 4-cycle engines the outward inertia load during the exhaust stroke is
not offset by gas pressure and may be large at high piston speeds. In radial
engines, because of the large mass of the master-rod assembly, the inertia
load on the crankpin and rods can be very large, especially when the engine
is run at high speeds with low gas pressure in the cylinders. This problem

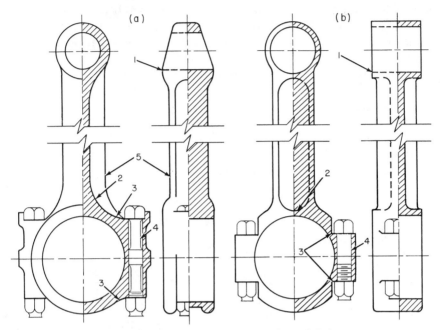

Fig. 11–32. Principles of good connecting-rod design.

(a) Good	(b) Poor
1. Thin edge avoided	1. Stress concentrates at thin edge
2. Plenty of metal over bearing	2. This too thin
3. No sharp grooves	3. Stress concentrates in corners
4. Good bolt design	4. Bolt not properly designed
5. H section turned 90°. Recommended but not mandatory	

became acute in the case of airplane engines in combat during World War II and was responsible for the development of the silver-lead-indium bearing (see later discussion of bearing materials).

In most 2-cycle engines the calculated outward inertia load never exceeds the gas load, and it has been assumed that there are no tension loads in the connecting rod or reversals of load on the piston pin. This assumption is dangerous because it does not recognize the fact that vibrating forces may be added to the inertia forces in such a way that load reversal in the connecting rod actually does take place. In any case, if compression pressure is lost, the full inertia load will fall on the connecting rod. It is therefore necessary to allow for some reversal of load in the case of 2-cycle connecting rods and their bearings.

One vibratory rod load not usually considered is that due to rotary vibration of the piston against the torsional stiffness of the rod. This type of vibration has occasionally proved critical and provides a good reason, among others, for minimizing piston mass and making the rod column torsionally stiff. Crankshaft torsional vibration obviously influences rod loading but is usually more critical to the crankshaft than to the rod.

Connecting rods can be classified as either single or multiple, the latter term applying to the case where more than one rod is combined to act on a single crank.

Single Connecting Rods. Figure 11–32 illustrates good and bad practice in single connecting-rod design for heavy-duty service. The usual points of critical stress are indicated (see also refs. 11.600 and 11.602). Figure 11–33 shows the rod of the high-output Diesel engine of Fig. 11–14. The

Fig. 11–33. Connecting rod of Sulzer locomotive engine, Fig. 11–14. Tubular column with serrated parting surface turned at steep angle to make removal of rod through cylinder block possible with large crankpin diameter. Massive lower end of rod for bearing support. (*Courtesy Sulzer Brothers, Ltd.*)

column is tubular, perhaps to achieve torsional stiffness. The design emphasizes stiff support for the crankpin bearing, with a serrated parting line, which is slanted to permit removal of the disassembled rod through the cylinder opening.

V-Engine Rods. Except where there is a premium on short engine length, as in aircraft, single rods side by side on a long crankpin seem the simplest and most reliable system for V engines (Figs. 11–11 through 11–14, 11–16, and 11–19). Where the crankpin must be shorter, either the forked and plain system shown in Fig. 11–15 or an articulated system as shown in Fig. 11–17 has proved successful. In the case of articulated rods, the geometry of piston motion is different from that of the master rod and varies with the angle chosen between the planes intersecting the crankpin and the link pin and the center line of the master rod, that is, the *link-pin angle*. With this angle the same as the V angle, the stroke will be the same as that of the master rod, but the top center will come at a crank angle not quite at crank alignment with the cylinder axis. With the link-pin angle such that the top center of the link-rod piston will coincide with the top center of the crank, the stroke will be longer than in the master-rod cylinder. Either system gives satisfactory engine performance, provided the compression ratio is held constant by suitable positioning of cylinders with respect to the crankshaft axis.

The design of first-class forked or articulated connecting rods requires great attention to the minimizing of stress concentrations. It is best carried out with the help of experimental stress analyses, including the brittle-lacquer technique. Good designs of forked rods will be found in ref. 10.848. For good articulated-rod design see refs. 10.41 and 10.46.

Radial-Engine Rods. In the master-rod system generally used in radial aircraft engines (Fig. 11–34), link-pin angles are equal to cylinder angles, and schedules of piston position versus crank angle for the link rods differ from each other and from that of the master rod. These differences are not great enough to affect cylinder performance measurably, provided the compression ratio in all cylinders is held the same by proper positioning of the link pins in the master rod. In spite of this fact, some designers have gone to more complex systems for the sake of symmetrical piston motion (Fig. 11–18 and refs. 10.75, 10.76). In view of the great success of the master-rod system in high-performance aircraft engines, the symmetrical designs seem both unnecessary and undesirable.

In any master-and-link-rod system, the gas and inertia forces on the link-rod elements introduce bending moments on the master rod which can be quite large (11.601). The bending strength of the master rod must be considered accordingly.

Figure 11–34b shows an improvement in the design of the connecting-rod system for a large radial aircraft engine, arrived at largely through stress-coat analysis (ref. 11.102). The improved design is stronger, simpler, and lower in cost than the earlier one.

Connecting-Rod Proportions. Tables 11–1 through 11–8 show successful ratios of rod length to crank radius, L/R, ranging from 3.0 to 5.0. Within these limits, this ratio seems to be dictated by space considerations, including the relative importance of low engine height and the question of interference with the cylinder bore. The latter factor accounts for the fact that L/R tends to vary in the same sense as stroke-bore ratio. Bolt diameters for the bearing caps, B/b, are in the range of $\frac{1}{10}$ of the bore for two bolts and somewhat less with four bolts. The crankpin bearing width/diam ratio, w_r/P, can be as small as 0.23. An average figure seems to be about 0.4.

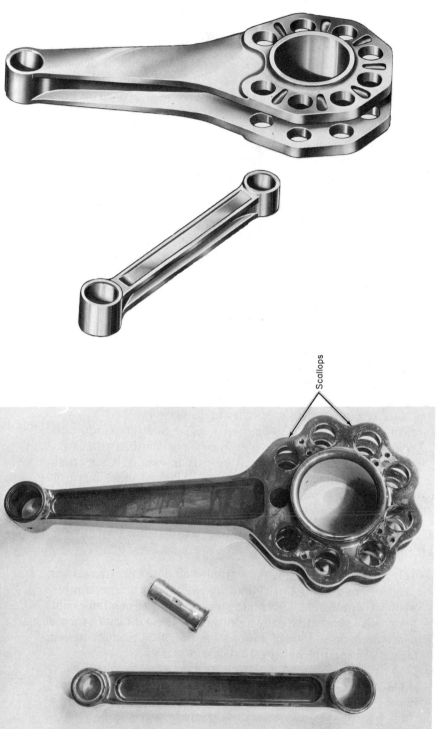

Scallops

Fig. 11–34. Evolution of Wright radial-engine connecting-rod system, largely by means of stresscoat analysis; (*left*) 1936 rods: Complicated geometry, with stress concentrations at holes and scallops, and at transitions from column to ends; (*right*) 1948 rods: H sections turned 90 degrees, simplifying manufacture and reducing stress concentrations at transitions from pins to body-holes. Scallops and threaded holes eliminated. (*Courtesy Wright Aeronautical Corp.*)

CRANKSHAFT DESIGN

As in the case of other engine parts, the crankshaft should be designed on the basis of previous successful practice plus modifications imposed by each individual case, and with hoped-for improvements based on experience and research.

Crankshaft Loads. Most designers plot diagrams of crankpin and main-journal loads based on indicator diagrams and calculated inertia forces, using rigid-body assumptions. The results are translated into nominal working stresses, which are held to limits dictated by experience with similar designs. As previously indicated, such procedures are basically an application of the theory of similitude, since accurate computation of loads and stresses is not possible in such a complex system.

Material. Crankshafts are made from steel forgings, or are cast in steel, nodular iron, malleable iron, or gray iron (see Chapter 9 for basic characteristics of these materials). With a given design the strength of the shaft lies approximately in that order. Where high specific output is required, forged steel is the preferred material (11.706). Gray iron is used only for small, low-cost engines. Modern techniques in metallurgy and foundry practice have made it possible to produce shafts of the best cast materials which are very near to steel shafts in their structural qualities.

Forged steel shafts are generally of 0.3 to 0.4 per cent carbon or alloy steel, treated to 250–275 Brinell hardness, giving tensile strengths in the range 125–150 ksi, 88–106 kg/mm². Compositions for cast materials are given in ref. 11.230 and physical properties in Table 9–4.

The object of using cast shafts in place of steel forgings is to reduce the expense of manufacture, or to achieve geometry not feasible with forged shafts. For example, counterweights can be cast integrally. Cast crankshafts are found most often in passenger-automobile and other relatively small engines. However, some very large engines use cast shafts, as in Fig. 11–19.

Crankshaft Stresses

Qualitatively speaking, the loads on a crankshaft result in stresses due to bending, torsion, and shear throughout its entire length. The complex geometry involved would make accurate stress computations impossible even if the loads were accurately known. In spite of these difficulties, however, much has been done toward rationalization of crankshaft design, largely by means of experimental stress analysis.*

---

* As an introduction to the principles of crankshaft design, a careful reading of ref. 11.700 is highly recommended. Both German and English versions are available.

Results of Experimental Stress Analysis. The most useful information in regard to crankshaft structural design comes from experimental studies of stress distribution in typical crankshaft samples or models subjected to arbitrary loads under laboratory conditions. Figure 11–35*a* shows the enormous effect of fillet radius on the maximum stress in a straight shaft and a simple crank. Given reasonably well-chosen ratios of major dimensions, fillet radius is the most important factor in crankshaft strength. The larger the fillet radius the better, provided adequate space is available for the required bearing lengths. Where bearing length is at a premium, undercut fillets, Fig. 11–35*b*3, are far better than inadequate conventional ones. Fillets with noncircular contours are slightly better than fillets with

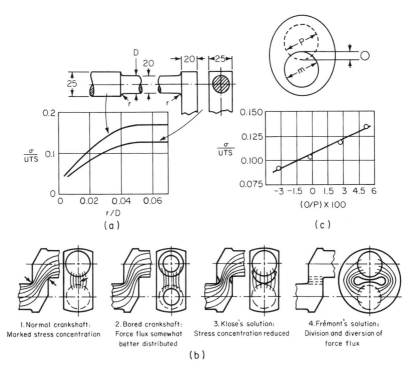

Fig. 11–35. Crank stress patterns (Matinaglia, 11.700). (*a*) Endurance-limit nominal stress in torsion (σ) versus fillet radius. Numbers are dimensions of test pieces in cm; (*b*) "flux" of force from pin to journal. Stress is smaller as flux lines are farther apart. (*c*) Effect of overlap on crank endurance strength in bending. The dimensions held constant, in millimeters are as follows: Crank throw 73; pin length 52; pin bore 40; journal length 37; journal bore 50; cheek thickness 22; cheek width 126. The dimensions varied, in millimeters, were as follows:

O/P	Pin diam	Journal diam
0.055	68	85
0.028	66.4	83.4
0.000	64.4	81.4
−0.032	62.4	79.4

σ = nominal endurance stress   O = overlap   P = pin diam

a fixed radius but are not usually practicable except for large crankshafts (11.700).

The fact that, with the same over-all dimensions and fillet radius, the crank of Fig. 11–35a is weaker than the stepped shaft, shows that the stress between pin and cheek is not uniformly distributed around the end of the pin, but tends to concentrate near the inside corner, as would be expected. Figure 11–35b illustrates this point and shows how better distribution of the forces between pin and journal can be obtained, with consequent lowering of the stress concentrations at the fillets.

Crankshaft Symbols. In subsequent discussions the following symbols will be used to designate crankshaft dimensions. All of these have the dimension of length.

B = diameter of bore in pin or journal
M = diameter of main journals
O = overlap between pin and journal
P = diameter of crankpin
R = crank radius (stroke/2)
r = fillet radius
t = axial thickness of crank cheek
w_c = transverse width of crank cheek
w_j = axial length of journal
w_p = axial length of crankpin

Crankshaft Overlap. When journal and crankpin diameters are sufficiently large in relation to the crank radius, circles representing the journal and pin will overlap when viewed along the crankshaft axis. The amount of overlap will be defined as follows:

$$O = \frac{P + M}{2} - R.$$

When this number is positive, it gives the radial distance by which the two circles overlap. A negative value indicates the minimum distance between the two nonoverlapping circles.

Figure 11–35c shows the effect of overlap on the strength of a crank in bending, as reported in ref. 11.700. As overlap increases, the strength of the shaft, as measured by endurance tests in bending, increases. This trend is due to the fact that with increasing overlap more and more of the load is carried directly from pin to main journal through the overlap area, thereby reducing stress in the fillets.

Figure 11–36 shows results of endurance tests on single cranks, published by Lurenbaum (11.706) in 1937, which have been given wide circulation since that time. The length of the vertical lines represents the reversed-torsional loading which these cranks could withstand for several million cycles without failure. The design ratios held constant for this series of tests were as follows:

$R/M = 1.10$, $P/M = 1.0$, $t/M = 0.45$, $O/M = -0.12$, $r/M = 0.10$ (approximately), $w_p/P = 1.5$ (approximately). For all the bored shafts $B/M = B/P = 0.76$.

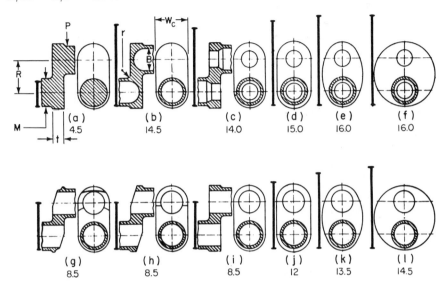

Fig. 11–36. Influence of form on the relative strength of crankshafts made of the same material. The strength was determined by reversed-torsion tests on models. Numbers and vertical lines indicate relative strength of the cranks (Lurenbaum, ref. 11.706).

The following relations are indicated by Fig. 11–36:

1. The solid crank *a* of Fig. 11–36 is by far the weakest of the group, with large improvement in strength attained by boring the pin and main bearings with straight holes *g* and still further gains by more complicated boring *b*, *c*. These trends can be explained on the basis of Fig. 11–35*b*.

2. "Beveling" crank *i* to shape *h* had no effect, but this bevel was used with a relatively thick cheek, $t/M = 0.45$. With thinner cheeks the effect of beveling can be detrimental, as will be explained subsequently.

3. Increasing the cheek contour as viewed along the journal axis had favorable results with the straight-bore crank (*i* to *j*) but was less effective with the 2-diameter bore (*c* to *d*). Further increases in contour, *d* to *f* and *k* to *l*, had little effect.

The results illustrated in Figs. 11–35 and 11–36 simply show the relative loads which the various crank designs can carry without fatigue failure. The actual stresses involved were not measured.

Figure 11–37 shows results of work by Gadd *et al.* (11.707), where the stresses at various points were measured by means of short gauge-length extensometers on crank models subjected to arbitrary moments of bending and torsion. The following design ratios were held constant:

$$P/M = 0.8 \qquad t/M = 0.27$$

$$w_p/P = 0.79 \qquad r/M = 0.05$$

$$w_m/M = 0.57 \qquad w_c/M = 1.18$$

(cheeks were beveled as shown in Fig. 11–37b).

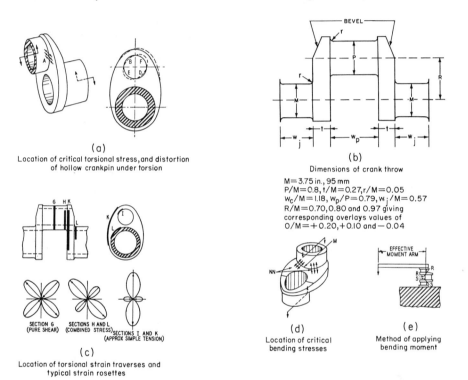

(a)
Location of critical torsional stress, and distortion of hollow crankpin under torsion

(b)
Dimensions of crank throw

M=3.75 in., 95 mm
P/M=0.8, t/M=0.27, r/M=0.05
w$_c$/M=1.18, w$_p$/P=0.79, w$_j$/M=0.57
R/M=0.70,0.80 and 0.97 giving
corresponding overlays values of
O/M=+0.20,+0.10 and −0.04

SECTION G (PURE SHEAR)   SECTIONS H AND L (COMBINED STRESS)   SECTIONS I AND K (APPROX SIMPLE TENSION)

(c)
Location of torsional strain traverses and typical strain rosettes

(d)
Location of critical bending stresses

(e)
Method of applying bending moment

Fig. 11–37. Stresses in model crankshafts, as measured by a short gauge-length extensometer. (Gadd and van Degrift, 11.707).

Bending moment was applied as shown at e in the figure. Torsion was applied to a complete six-throw shaft supported by rollers on the main journals. The results given in Fig. 11–37 apply to No. 2 crank, which was found to have higher stresses than the others. The variables investigated were the diameters of the bore in crankpin and main journal, and the length of the crank throw ($\frac{1}{2}$ engine stroke). Changing the throw varied the overlap between journal and crankpin from $O/M = +0.20$ to $O/M = -0.04$. The short-throw crank with $O/M = +0.20$ can be taken as typical of 1967 automotive practice.

Conclusions which can be drawn from Gadd's work for cranks with his particular design ratios are as follows (letters refer to the various parts of Fig. 11–37):

1. The maximum stress in a crank can be several times higher than the

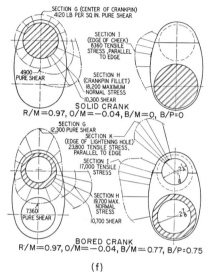

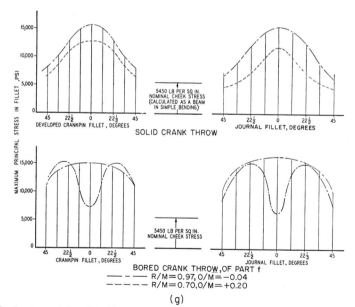

(g)

Bending stresses in the solid and bored cranks of part f, with bending moment applied as shown in part e.

Fig. 11–37 (*continued*)

stress calculated by the simple beam and shaft assumptions (k, l). Fillet stresses up to five times the nominal torsional stress and three times the nominal bending stress in the pin are shown.

2. Reducing the crank throw from $R/M = 0.97$ to $R/M = 0.70$, which increased the overlap from $O/M = -0.04$ to $O/M = +0.20$, reduced

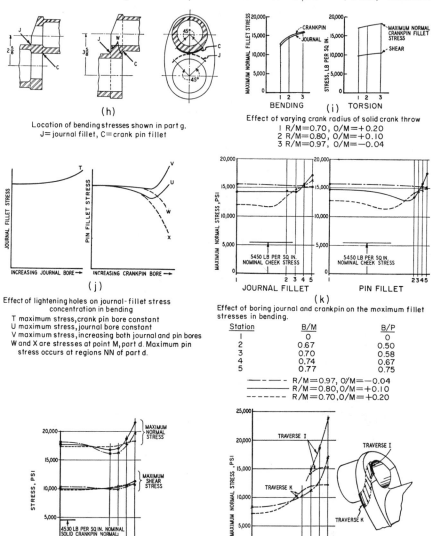

(h)

Location of bending stresses shown in part g.
J=journal fillet, C=crank pin fillet

Effect of varying crank radius of solid crank throw
1 R/M=0.70, O/M=+0.20
2 R/M=0.80, O/M=+0.10
3 R/M=0.97, O/M=−0.04

(j)

Effect of lightening holes on journal-fillet stress concentration in bending

T maximum stress, crank pin bore constant
U maximum stress, journal bore constant
V maximum stress, increasing both journal and pin bores
W and X are stresses at point M, part d. Maximum pin stress occurs at regions NN of part d.

(k)

Effect of boring journal and crankpin on the maximum fillet stresses in bending.

Station	B/M	B/P
1	0	0
2	0.67	0.50
3	0.70	0.58
4	0.74	0.67
5	0.77	0.75

—— —— R/M=0.97, O/M=−0.04
———— R/M=0.80, O/M=+0.10
– – – – R/M=0.70, O/M=+0.20

(l)

Effect of boring journal and crankpin on the maximum fillet stress in torsion. Crank dimensions are the same as in part k.

(m)

Effect of boring journal and crankpin on stresses at edges K and I in torsion. Same crank dimensions as in parts k and l.

Fig. 11–37 (*continued*)

stresses except with large values of the bores in pin and journal (see parts *g* through *m*).

3. Boring of pins and journals to 0.50 times their outside diameters gives an appreciable reduction in fillet stresses. Larger boring tended to increase fillet stresses (*j*, *k*, *l*) and to introduce large stresses at the cheek edges (*M*).

4. The "bevel" shown in Fig. *b* and used in all the tests of Fig. 11–37 was probably detrimental to strength, at least in the case of the bored shafts. Without this bevel, the distortion shown in *a* would have been reduced, and the stresses at section *K* (Fig. *f*) might have been greatly reduced. The same may be said for traverse *I* in *m*.

The stresses shown at the edge of the bored hole could have been reduced by rounding these edges to a radius similar to the fillet radius. Sharp corners invite stress concentration and should be avoided in all engine parts.

Boring of Pins and Journals. In comparing the results of Figs. 11–36 and 11–37 it is well to consider the long-throw shaft of Fig. 11–37 ($R/M =$ 0.97, $O/M = -0.04$) which had a negative overlap near that of the Fig. 11–36 shafts ($R/M = 1.10$, $O/M = -0.12$). In the Lurenbaum work, the effect of boring can be demonstrated by comparing crank *a* with cranks *h* and *i*, Fig. 11–36. In this case boring to 76 per cent outside diameter increased torsional strength by 90 per cent. Gadd's work, shows that with his long-throw crank the strength in bending was little affected (Fig. 11–37*g*, *k*) while the strength in torsion was actually decreased by a similar boring operation (*f*, *l*, *m*).

One reason for the smaller effect of boring in Fig. 11–37 may be the fact that the cheek thickness t/M was only 0.27, while in Fig. 11–36 it was 0.45. This much heavier cheek used by Lurenbaum probably gave high fillet-stress concentration in the solid shaft, which was greatly relieved by boring. Another important difference may also be in the fact that Gadd used a finite gauge length and therefore did not measure actual maximum stress, while the endurance tests of Lurenbaum were responsive to the absolute maximum stress in the shafts.

Other considerations in this connection include the fact that all large aircraft engines used bored shafts which were subjected to very heavy loading. From this fact alone it seems safe to conclude that boring of pins and main journals need not be detrimental to strength, and is probably beneficial when judiciously applied. Figures 11–37*k*, *l*, *m* suggest that a uniform bore should not greatly exceed half the outside diameter. On the other hand, boring of the type shown in Fig. 11–36*c* is probably an effective method of increasing crankshaft strength, where the cost of such an operation can be justified (see also ref. 11.718).

Crank-Cheek Design. In most engines the cylinder size and spacing determine the length of the crankshaft, and whatever length is allotted to the crank cheek is at the expense of length for crankpins and main journals.

Where crankpin and journal lengths tend to be short, the longitudinal thickness, t in Fig. 11–38, must be held to a minimum.

The short-stroke engine ($S/b < 1.0$) always has a crankshaft with considerable overlap between crankpins and main journals as shown at a in

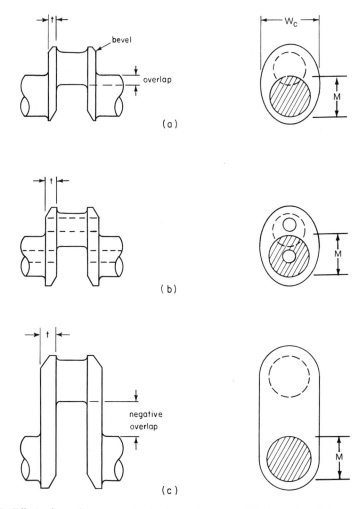

Fig. 11–38. Effect of overlap on crank-cheek requirements. (*a*) Normal, solid, overlapped crank; (*b*) normal crank, with bored pin and journal; (*c*) long-stroke crank, or two cranks with no bearing between.

Fig. 11–38. Since test results (Fig. 11–35c) show a beneficial effect of overlap, it would appear that much of the torsional and bending load is transmitted through the overlap area. This observation, together with good design practice indicates that in well-overlapped solid shafts the ratio t/M

can be quite small, as indicated in Fig. 11–38a. Figures 11–35c and 11–37i and k tend to justify this practice. Examples of this type of shaft appear in Figs. 11–15, 11–16, 11–18, 11–22, and 11–24. The transverse width of the cheek w_c is probably of reduced importance in overlapped shafts, since the overlap area carries so much of the loading.

Figures 11–35c through 11–37 indicate, as would be expected rationally, that cheek design becomes more important when pins and/or journals are bored out. With hollow pins or journals, the cheeks should be designed to distribute the loads around the end periphery of these parts. In such cases, Fig. 11–38a should be modified in the direction of b. In this case t/M should not be too small, and beveling should be minimum. In Fig. 11–37, the use of a small t/M ratio (0.27) together with considerable beveling probably accounts for the weakening effect of boring, whereas with $t/M = 0.45$, boring increased the strength of the shaft (cf. Fig. 11–36).

For engines with large stroke-bore ratios, there is usually a negative overlap. This condition also applies to cheeks connecting cranks which have no journal between them. The cheek dimensions should be still larger in this case, as shown in Fig. 11–38c.

Figures 11–39a, b, and c show interesting methods of reducing fillet stresses in cast crankshafts, especially where the cheek connects two crank-pins without a main journal between them. These designs endeavor to transmit the loads from pin to journal or from pin to pin in such a way as to spread the load around the pin or journal end circumference rather than to allow it to concentrate at the fillet in the region of closest approach.

Counterweights. Counterweights may be necessary to improve engine balance or to relieve main-journal loading (see Chapter 8). An advantage of cast shafts is that counterweights can be included as an integral part of the casting. It is difficult to form adequate counterweights on forged shafts, and welding or bolting is used where such weights are considered essential.

Built-up Crankshafts

Shafts built up from separate pieces are used for the following reasons:

1. To accommodate assembly with one-piece connecting rods. Many small single-crank engines use this arrangement to accommodate anti-friction bearings (Fig. 11–7). Built-up shafts have also been used for radial aircraft engines to avoid splitting the master rod (10.847).

2. To facilitate manufacture of very large crankshafts. For large engines of the types shown in Figs. 11–20 and 11–21, crankshafts are usually built up, as shown in Fig. 11–39d, by shrinking heavy crank cheeks over individual crankpins and/or journals. The ratio t/M runs as high as 0.58 for this type. The width w_c must also be large, as indicated in Fig. 11–39d (see also 11.700, 11.712).

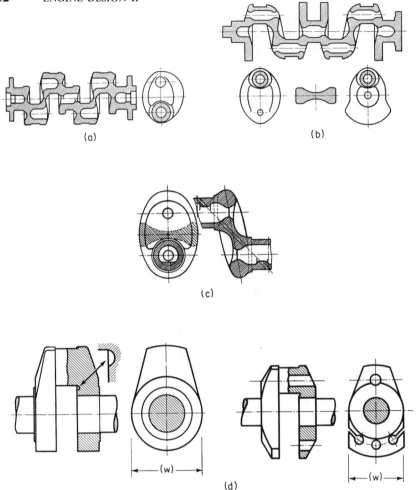

Fig. 11–39. (*a*), (*b*), (*c*), Sulzer designs for cast crankshafts. (*d*) Built-up shaft for large engines (Matinaglia, 11.700).

Mistakes in Crankshaft Design

The most common mistakes involve unnecessary stress concentrations due to oil holes passing too near the fillets, screw threads in stressed areas, sharp corners (either concave or convex), or inadequate fillets. Drastic beveling, especially with bored pins and journals, must also be considered an error.

Crankshaft Surface Finish and Hardening

Bearings and journals must always be finished to a relatively smooth surface, as indicated by the table on p. 472. Crankshafts for very high specific output, including aircraft engines, are generally finished all over to a very smooth surface (see Fig. 9–18 for effect on endurance). Where stresses are not considered to be high, crank cheeks are allowed to remain in the as-forged or as-cast condition.

In order to reduce wear, crankpins and journals are often hardened by local heat-treating of the surfaces, either by flame or by electric induction. When this process includes the fillets, some strengthening of the shaft results. Aircraft and other high-cost engines have used nitrided pins and journals. Another method of improving fillet strength is by rolling (11.714).

Crankshaft Design Ratios

The principal dimensions of a crankshaft in relation to cylinder bore and stroke are limited by practical considerations as well as by considerations of strength and structure. The obvious need for strength and stiffness in the crankshaft has led toward increasing size of crankshafts as the mean pressure and piston speed of engines has increased with time. Tables 11–2 through 11–9 show ratios of crankpin diameter and main-bearing diameter to bore for different classes of recently designed engines. Except for some in-line 6-cylinder engines of Table 11–3, all these shafts have main bearings on both sides of each crank throw.

When it is required that the connecting rod be removable through the cylinder bore, (P/b) is limited to a maximum value of about 0.6. Where this requirement is not present, or where separate cylinder barrels are used, ratios as high as 1.0 are found. Main-journal diameter-to-bore ratios (M/b) run from just under 0.6 to as high as 1.0. It is remarkable that the range of these ratios is about the same for all classes of engines listed and that it seems to be nearly independent of the number and arrangement of the cylinders. It is also remarkable that these design ratios tend to be rather independent of engine specific output. As an example, the crankpin and main-bearing diameters for the engines shown in Fig. 10–2 were not changed during the periods indicated, although loads on the crankshafts increased almost in proportion to the power ratings. Improvement in detail design of the shafts, however, was considerable.

As a general rule, of course, pin and journal diameters should be chosen on the high side for engines of high specific output, especially where cast crankshafts are to be used.

Recommendations for Crankshaft Design

It must by now be apparent that designing a crankshaft is far from an exact science and must depend on an intelligent application of test results, such as those of Figs. 11–35 to 11–37, together with judgment based on previous successful practice. From the foregoing discussion, plus the author's general experience, the following recommendations with regard to crankshaft design are submitted:

1. *Crankpin* diameter should be at least 0.60 times the bore. Crankpin length should be enough to accommodate connecting-rod bearings of not less than 0.30 of the crankpin diameter.

2. *Main-journal* diameter should be somewhat larger than crankpin diameter. Main-journal length can be as short as 0.30 times the journal diameter when centrifugal crank loads are counterbalanced.

3. Overlap of pins and journals improves crankshaft strength, provided fillets are adequate.

4. Judicious boring of pins and journals tends to reduce stresses.

5. With bored shafts, the cheeks should be designed to spread the load over the ends of pins and journals. For this reason, excessive beveling should be avoided.

6. Fillet radii should be as large as possible (at least 0.05 times the journal diameter), and all sharp corners should be well rounded off. Fillets should have very smooth finish, and be rolled where possible (11.714).

7. The shaft should be as free as possible from holes and keyways, and particularly from threaded holes near areas of high stress. All holes should be well chamfered at their openings.

8. Where dimensions are not unduly restricted, cast crankshafts appear to be a satisfactory alternative to forged-steel shafts.

CRANKCASE DESIGN

Design of the crankcase, or engine "frame" as it is called in the case of large engines, has already been briefly discussed in connection with cylinder-barrel design.

Materials

As noted in Chapter 9, crankcase materials include cast iron, cast aluminum, forged aluminum, and forged steel, usually welded in the latter case. Which type is appropriate depends principally on considerations of engine type and manufacturing costs versus the importance of weight saving.

Most automotive engines, both gasoline and Diesel, use en-bloc iron castings, as illustrated in Figs. 11–9 through 11–13. Methods of molding and casting are reviewed in refs. 11.80 and 11.81.

Use of aluminum in place of iron results in weight saving, but at the expense of greater cost.* There was some use of cast aluminum for U.S. passenger cars in the early 1960's, but it was generally abandoned in favor of a return to cast iron, because in this case the saving in weight (about 50 lbm per engine) was not considered sufficiently important to justify the cost increase. However, this development did lead to useful investigations of stress distribution in the automotive-type cylinder block (11.82–11.86).

Engines where weight saving appears to justify cast-aluminum crank-cases, and often cylinders too, include:

1. Small, portable engines, including those for lawn mowers, chain saws, outboard motorboats, etc. For this class of engine, crankcases are small enough to be diecast, which is a cheap process in large-quantity production (11.81). Small engines with aluminum crankcases are shown in Figs. 11–6 and 11–7.

2. Sports-car and racing-type automotive engines, where low cost is not so important as in ordinary automotive engines (Fig. 11–22 and refs. 10.801 *et seq.*).

3. Aircraft engines and military engines, Fig. 11–24. Cast magnesium alloy has also been used to a limited extent in these classes of engine (see Fig. 11–41*b* and ref. 11.83).

Some large radial aircraft engines (now obsolescent) use forged-aluminum or forged-steel crankcases, as indicated in refs. 10.844 and 10.847.

For large Diesel and gas engines, the alternate materials for crankcase (frame) structures are cast iron and welded steel. For stationary purposes where weight is not a paramount consideration, the tendency is to use iron (Figs. 11–16, 11–17, and 11–18).

Welded steel crankcases are generally used in high-output locomotive engines, Figs. 11–14, 11–15, and for the very large marine engines of Figs. 11–19, 11–20, and 11–21. The engine of Fig. 11–21 is available with a cast-iron frame at a weight increase of 8 per cent.

Cast Crankcases

General principles to be followed in the design of cast crankcases include the following:

1. Thick sections cool more slowly than thin sections, and abrupt changes in section thickness should therefore be scrupulously avoided in order to minimize shrinkage strains and cracks (11.82–11.86).

2. Avoid large plane sections. Curve or "dish" all large thin sections.

3. *Ribs.* The custom of adding "ribs" or thin "webs" to improve

---

* One of the earliest cast-aluminum crankcases was that of the Wright Brothers' first air-plane engine (10.840).

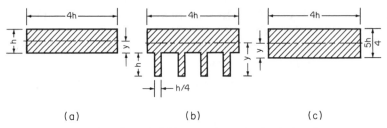

	Section		Maximum
Section	*Modulus*	*y*	*Stress*
(*a*) Plain	0.333 h^4	0.5 h	1.52 M/h^3
(*b*) Ribs added	0.555 h^4	1.3 h	2.44 M/h^3
(*c*) Rib material added	0.644 h^4	0.625 h	0.98 M/h^3

Fig. 11–40. Effect of ribbing on a beam section in bending.

M = bending moment around neutral axis — — – –

strength in bending usually has the opposite effect. With a given moment, the stress in a beam in bending is proportional to y/J, where y is the maximum distance from neutral axis to remotest point and J the section modulus. Adding widely spaced ribs increases y by nearly the depth of the rib but has little effect on J. Even with fairly closely spaced ribs, as shown in Fig. 11–40, the ribbing increases maximum stress enormously. Adding the rib material to the section thickness is a much better use of the additional material, as indicated by Fig. 11–40*c*.

Photoelastic analysis has shown the advantages of avoiding ribbing on crankcases, as indicated in Fig. 11–41. The conclusion is inescapable that ribbing acts as a stress raiser and should not be used where the object

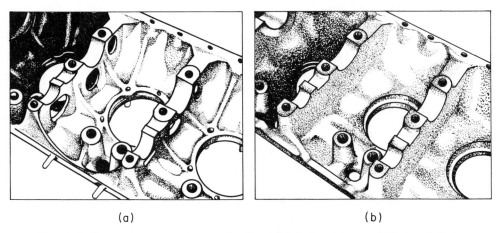

(*a*) (*b*)

Fig. 11–41. Improvement in crankcase design through experimental stress analysis and elimination of ribs and lightening holes. (*a*) Aluminum crankcase as originally designed with its network of stiffening ribs and lightening holes in the diaphragms. (*b*) Same crankcase after stress analysis. Note absence of ribs and development of well-blended sections. Following redesign, it was found possible to use magnesium instead of aluminum, with a consequent weight reduction of 27 per cent (Geschelin, 11.83).

is to increase bending strength. Its use for other purposes such as cooling or reducing vibration is permissible, provided the added stress is not critical. In the case of air-cooled cylinder heads, which must be finned for cooling, it may be necessary to interrupt the finning in order to avoid high stresses at the rib edges.

4. Avoid lightening holes, since stresses concentrate around such openings (see Fig. 11–41).

Welded Steel Crankcases and Frames

The advantages of this method of design include a reduction of 7 to 10 per cent of the total engine weight by the use of thinner sections than would be practical with a casting. Other advantages are reduction of thermal stresses associated with thick sections and avoidance of the uncertainties associated with hidden defects in castings. In general, a welded crankcase structure is more expensive than a casting, except possibly where engine size is as large as in the engines of Figs. 11–19 through 11–21.

Welded steel crankcases are used mainly in Diesel engines of locomotive size or larger (see Figs. 11–14, 11–15, 11–19, 11–20, and 11–21). The high specific output required of locomotive engines is an important factor which makes the steel crankcase attractive for this service. For the enormous marine engines of Figs. 11–20 and 11–21, saving in weight and the difficulties associated with making huge iron castings are important considerations.

Radial engines lend themselves well to the use of forged-steel crankcase sections (10.847).

Steel crankcases require great ingenuity on the part of the designer in order to keep down the cost of manufacture without too much sacrifice in the desired geometry. Flat-plate sections may have to be used, where a cast section would be curved for increased stiffness. The higher modulus of elasticity of steel (30 million psi as compared with about 15 million for iron) is helpful.

Crankcase Structural Design

The heaviest structural load on the crankcase is nearly always the tensile load from cylinder head to crankshaft main bearings caused by maximum cylinder pressure. In small en-bloc engines this load is taken through the casting itself. Examples are Figs. 11–6 and 11–9 through 11–12.

In the case of large or highly stressed in-line engines, the tensile load is often taken by long bolts extending from near the top of the cylinder barrels to the bottom of the frame, as in Figs. 11–20 and 11–21. Alternatively, this load may be taken by long studs from cylinder head to crankcase, as in Figs. 11–8 and 11–25. (The engines of Figs. 11–7, 11–16, 11–17, and 11–18 also use such studs, but these do not show in the illustrations.)

In V engines, in addition to the tensile loads there are heavy side loads on the main bearings at the times of high cylinder pressure. In small gasoline engines, such as the one of Fig. 11–11, the main block casting absorbs this load. However, when cylinder pressures are very high, as in Diesel engines, the crankcase may need reinforcement by horizontal bolts or studs at the level of the main bearings, as in Figs. 11–12 and 11–25. All the high-output V-type aircraft engines use such reinforcement (10.845, 10.846, 10.848).

An interesting case of reinforcement by bolts is shown in Fig. 11–24, where the tensile load between the two crankshafts of an opposed-piston engine is taken by long bolts from bearing cap to bearing cap. Similarly, long studs from cylinder head to cylinder head are possible with opposed-cylinder engines, such as those of Fig. 11–8. However, in Fig. 11–8 this system is replaced by two studs and a bolt, as shown.

Figure 11–25 shows an interesting case of heavy reinforcement of the frame structure by well-anchored bolts. This engine operates with exceptionally high cylinder pressures (271 psi bmep, Table 10–9).

Accessibility

An important factor in crankcase design is the question of accessibility of the moving parts for inspection, repair, and replacement. Solutions of this problem depend on many factors, including cylinder arrangement, engine mounting and surroundings, and methods of maintenance and repair.

En-bloc engines generally require that pistons and connecting rods be removable through the cylinder bores. Accessibility to crankshaft and main bearings in automotive engines is generally provided for by a removable cover at the bottom of the engine and by main-bearing caps that are removable from below, as in Figs. 11–5 through 11–15 and Fig. 11–22.

Stationary and marine engines are not usually accessible from below, and with such engines the usual practice is to provide main-bearing caps bolted on top of the crankshaft and accessible through side plates. Examples are shown in Figs. 11–16, 11–19, 11–20, and 11–21. In the engine of Fig. 11–17, the crankshaft is removable from the side. In all these engines the main-bearing shells are usually replaceable by turning the lower bearing shell around to the top or side, without removing the shaft.

Aircraft engines and military engines are usually removed as a whole from the vehicle for repairs that involve major parts. This system is also generally used for locomotive engines and for fleets of trucks and buses.

Summary of Principles in Crankcase Design

1. Adequate provision for the tensile load from cylinder heads to main bearings and the side load on the main bearings of V engines.

2. Adequate provision for the required inspection and repair of the engine as installed.

3. Avoidance of stress-raising ribs and abrupt changes of section thickness.

4. As in all engine parts, reduction of stress concentrations to a minimum through the use of adequate fillets.

5. Use of experimental stress analysis, including the brittle-lacquer technique (refs. 11.101 *et seq.*), as far as practicable in the design development.

ENGINE BEARINGS

There is space here for only a brief discussion of general principles in the selection of materials and in bearing-detail design. For a more complete discussion, see refs. 11.900 *et seq.*

Journal Bearings

The bearings in the power train are critical elements in the design of any engine. Most engines use plain journal bearings. The use of "antifriction," that is, ball, roller, or needle, bearings, will be discussed under a separate heading. The general subject of bearing lubrication and friction has been dealt with in Volume I, Chapter 9. Here the emphasis will be on detail design of bearings in the power train.

Bearing Materials. Power-train journal bearings consist of an iron or steel journal supported in a bearing of some other material, usually much softer than the journal. Table 11–11 shows materials generally used. Desirable characteristics of the bearing material include:

1. Adequate *endurance strength*

2. *Deformability*, that is, the ability to deform plastically at points of very high pressure and, generally, to conform to the shape of the journal

3. *Embeddability*, that is, the ability to allow particles of foreign matter to embed in the bearing material rather than to circulate in the oil film where they would score or wear the journal and bearing surfaces

4. *Resistance to scoring and seizure*, or welding of the bearing material to the shaft

5. *Resistance to corrosion*, usually from elements in the lubricant.

The relative importance of these characteristics varies with the type of bearing and conditions of operation, as will appear in subsequent discussions. Table 11–12 shows relative characteristics of the most common journal-bearing materials.

Oscillating Journal Bearings. (11.921) In the power train, these are the piston-pin bearings and the link-pin bearings of articulated connecting

Table 11–11. Classification and Uses of Bearing and Bushing Alloys

SAE No.	Nominal Composition, %	Method of Manufacture	Characteristics	Applications
Tin-base alloys	Sn / Sb / Cu	Cast on steel, bronze or brass backs, or directly in the bearing housing.	Soft, corrosion resistant, moderately fatigue resistant.	Main and connecting-rod bearings; motor bushings. Operates with either hard or soft journal.
11	87.5 / 6.75 / 5.75			
12	89 / 7.5 / 3.5			
Lead-base alloys	Pb / Sn / Sb / As	SAE 15 is cast on steel; others are cast on steel, bronze, or brass, or in the bearing housing.	Soft, moderately fatigue resistant, corrosion resistant.	Main and connecting-rod bearings. Operates with hard or soft journal. Requires good journal finish.
13	rem / 6 / 10 / —			
14	rem / 10 / 15 / —			
15	rem / 1 / 15 / 1			
Lead-tin overlays	Pb / Sn	Electrodeposited as a thin layer on copper-lead or silver bearing faces.	Soft, corrosion resistant. Bearings so coated run satisfactorily against soft shafts throughout the life of the coating.	Heavy-duty, high-speed main and connecting-rod bearings.
19	90 / 10			
190	93 / 7			
Copper-lead alloys	Cu / Pb		Moderately hard. Subject to acid attack; modern oils have minimized this trouble. Fatigue resistance good to fairly good. Listed in order of decreasing hardness and fatigue resistance.	Main and connecting-rod bearings. The higher lead alloys can be used unplated against a soft shaft, although an overlay is helpful. The lower lead alloys may be used against a hard shaft, or with an overlay against a soft one.
49	76 / 24	Cast or sintered on steel back.		
48	70 / 30			
480	65 / 35			
481	60 / 40	Cast on steel back.		

Table 11-11. *Continued*

Aluminum-base alloys

Alloy	Sn	Cu	Ni	Al	Si	Cd	Method	Properties	Application
770	6.25	1	1	rem	—	—	Cast in permanent molds; work-hardened to improve physical properties.	Hard. Extremely fatigue resistant. Resistant to oil corrosion.	Main and connecting-rod bearings. Generally used with suitable overlay.
780	6	1	0.5	rem	1.5	—	Bonded to steel back. Also procurable in strip form without steel backing.	Same as 770.	Same as 770.
781	—	—	—	rem	4	1	Bonded to steel back. Also produced in castings or wrought strip, without steel back.	Same as 770.	Main and connecting-rod bearings generally used with overlay. Bushings and thrust bearings with or without overlay.

Other copper base alloys

Alloy	Cu	Sn	Pb	Zn	Method	Properties	Application
795	90	0.5	—	rem	Wrought solid bronze.	Hard, strong, good fatigue resistance.	Intermediate-load oscillating motion such as tie rods, brake shafts, and so forth.
791	rem	4	4	4	Wrought solid bronze.	General-purpose bearing material, good shock and load capacity. High-temperature resistant. Hard shaft desirable. Less score-resistant than higher lead alloys.	Moderate to high loads. Transmission bushings and thrust washers, piston pin bushings, and so forth.
793	rem	4	8	4 max	Cast on steel back.	General-purpose bearing material, good shock and load capacity. High-temperature resistant. Hard shaft desirable. Less score resistant than higher lead alloys.	Medium to high loading. Typical applications: transmission and chassis bushings and thrust washers.
798					Sintered on steel back.		
792	80	10	10	—	Cast on steel back.	Has maximum shock and load-carrying capacity of conventional cast bearing alloys; hard, both fatigue and corrosion resistant. Hard shaft desirable.	Heavy loads with oscillating or rotating motion. Typical applications: piston pin, steering knuckle, differential axles, thrust washers, wear plates, and so forth.
797					Sintered on steel back.		
794	73.5	3.5	23	—	Cast on steel back.	Higher lead content gives improved surface action for higher speeds but results in somewhat less corrosion resistance.	Intermediate-load application for both oscillating and rotating shafts, that is rocker-arm bushings, transmissions, and farm implements.
799					Sintered on steel back.		

(From 1964 SAE Handbook reference 11.230.)

rods. The "slipper"-type bearings sometimes used in radial engines (11.920) also fall into this class.

Since oscillating bearings operate at relatively low average surface speeds, the problem of wear due to abrasive material is less serious than in rotating bearings. Also, the work done in friction, and therefore the heat generated, is relatively small. Under these circumstances, high endurance strength takes precedence over embeddability. A good bearing bronze, such as SAE 791, or a bearing-grade aluminum is satisfactory for such bearings. With aluminum pistons, the piston pin can bear directly in the aluminum, and some piston pins bear directly in cast-iron pistons. Hard-surface pins are desirable.

With proper lubrication, oscillating bearings can stand a very high maximum unit pressure, as evidenced by the fact that their projected area is usually less than half that of the crankpin bearing which carries nearly the same maximum load.

Lubrication of Journal Bearings. Crankpin and main-journal bearings require "force-feed" lubrication, that is, a supply of oil under pressure into the bearing. Older methods of crankpin oil supply, as by "splash," or "dip" feed, are now considered obsolete except for some very small engines. For a discussion of oil holes and grooves in journal bearings, see Volume I, Chapter 9. Articulated-rod link pins should also be pressure lubricated.

In small engines, pistons and piston-pin bearings usually receive adequate lubrication from the oil thrown off by the crankpins. This method is employed in most automotive and smaller engines. It is successfully used in nearly all airplane engines.

In engines having oil-cooled pistons, the upper-end rod bearing, and often the pin bearings in the piston, are usually pressure lubricated from the oil passages that supply the piston. In large engines, especially those that run at low piston speeds, positive lubrication of the pistons and pins is necessary. The reason can be seen in the fact that the length of trajectory of a body in a gravity field depends on its initial velocity. As engines get larger, the required trajectory of the oil from crankpin to piston and pin becomes longer, while the initial velocity depends only on the piston speed, which is usually less than that in smaller engines.

Crankshaft Bearings. These are among the most critical elements in engine design. Early engines, and even now some low-specific-output engines, use tin-base or lead-base "Babbitt" material such as SAE 11 through 15 of Table 11–11. Such material has all the desirable characteristics listed at the beginning of this section, except the first, high endurance strength. It was formerly cast directly into the connecting rod and finished after casting. Later practice has been to cast it onto removable steel or bronze bushings. Babbitt has been superseded in most modern engines for reasons of strength alone, as it is otherwise a nearly ideal bearing material. Typical structural failures of Babbitt bearings are illustrated in refs. 11.900 and 11.906.

Materials similar to Babbitt, called lead-tin "overlays," SAE 19 and 190 of Table 11–10, are widely used as a very thin coating on stronger bearing materials, such as copper-lead, silver, or aluminum. This overlay acts as a protection against seizure and as an absorber of foreign material smaller than the thickness of the overlay.

The poor structural strength of Babbitt has led to the development of the other materials listed in Tables 11–11 and 11–12. The most widely used of these are copper-lead and aluminum.

Copper-Lead Bearings. (11.900–11.908) Copper-lead consists of a spongy matrix of copper with lead particles distributed in the interstices of the spongy structure. This combination was developed in order to combine the structural strength of copper with the desirable bearing properties of lead. As of 1966 it was the most widely used material for crankpin and crankshaft bearings.

Table 11–12 indicates that this material ranks next to Babbitt in resistance to scoring and seizure. It has good endurance strength. Its principal weakness is the susceptibility of lead to corrosion by acid constituents in the lubricant. This trouble has been largely overcome by the use of oils with suitable anticorrosion additives (11.900) and by the addition of indium to the lead.

Copper-lead bearings are given added resistance to corrosion and seizure by a thin overlay of lead-tin Babbitt (see Tables 11–11 and 11–12).

Aluminum Bearings. (11.910–11.913) The use of aluminum as a bearing material has increased rapidly since about 1960. As shown in Table 11–12, aluminum ranks high in strength and resistance to seizure, but low in deformability and embeddability. Modern methods of high-precision manufacture and effective oil filtration, have reduced the necessity for a high degree of those two characteristics. Many engine designers (and, of course, the aluminum manufacturers) consider this the best available material for crankshaft bearings. For heavy duty, the bearing should consist of a thin layer of aluminum on a steel shell, preferably with a very thin lead-tin overlay.

Bearing Loads. The load on power-section bearings can be plotted against crank angle from the indicator diagram and the inertia forces for any assumed condition of engine speed and load (see Volume I, Fig. 9–7, p. 324). Such calculations give results well within the accuracy of estimates of what a given bearing will carry.

Bearing Load Capacity. As discussed in Volume I, Chapter 9, the load that may be carried by a journal bearing depends on many factors, including characteristics of the lubricant as well as of the bearing. Theoretically, the oil film will not fail if $\mu N/p$ (viscosity $\times$ rpm/unit pressure) does not fall below its critical value (see Volume I, Figs. 9–3 and 9–4, pp. 320, 321).

Table 11–12. Relative Characteristics of Bearing Materials for
Internal-Combustion Engines

A **Fatigue Strength and Deformability**

	Order of Fatigue Strength	Approximate Maximum Load, psi	Order of Deform-ability
Bronzes	1	10,000	7
Copper-lead with tin or silver	2	3000–4000	6
Thin Babbitt overlays (0.003 in. or less)	3	2000–4000	5
Aluminum alloys	4	2000–3000	4
Copper-lead	5	1500–2500	3
Cadmium alloys	6	1200–1500	2
Lead and tin-base Babbitts	7	800–1500	1

B **Embeddability**

No.	Materials	Rating
1	Thick Babbitt (0.032 in. thick)	124
2	Durex 100A	100
3	Copper (oxygen-free)	0.5
4	Gridded copper (48 pitch)	32
5	Silver (oxygen-free)	0.7
6	Copper + 0.0008 in. Pb-Sn alloy	50
7	Copper-lead (60 Cu, 40 Pb)	7.5
8	Copper-lead (65 Cu, 35 Pb)	5.5
9	Copper-lead (70 Cu, 30 Pb)	5.0
10	Copper-lead-tin (75 Cu, 24.5 Pb, 0.25 Sn)	4.0
11	Copper-lead-tin + 0.001-in. Pb-Sn alloy	69
12	Aluminum (95 Al, 4 Si, 1 Cd)	10
13	Aluminum + 0.0008-in. Pb-Sn alloy	68

Table 11–12 (*continued*)

C **Score Resistance and Application**

	Copper	Zinc	Tin	Lead	Order of Score Resistance	Order of Inherent Hardness or Strength	Suggested Application
Brass — Bronze	85–90	Fe–2	Al–4–10	—	8	1	Very poor antiscore quality
	(T) 86	Fe–2	Al–4	—			Used in minor locations
	60–90	10–40	—	0–2	7	2	Same as above
	(T) 90	9	—	$\frac{1}{2}$			Used with hypoid-gear oils
	85–90	$\frac{1}{2}$–3	7–10	0–5	6	3	For heavily loaded* piston pins and similar motions where surface speed* is very low
	(T) 88	2	10	—			
	80–90	0–5	5–10	5–10	5	4	Average bushing materials for low-speed, small-diameter shafts
	(T) 88	4	4	4			
	(T) 80	—	10	10			
	70–75	—	4–10	20–25	4	5	For moderate-speed bushings and where lubrication is not well maintained
	(T) 72	—	4	20			

	Order of Score Resistance	Order of Inherent Hardness or Strength	Suggested Application
Aluminum alloys	3	6	For heavily loaded moderate-speed bearings
Copper-leads	2	7	For heavily loaded high-speed bearings
White-metal alloys (Cadmium and Babbitts)	1	8	Low to moderately loaded, low to high-speed bearings where dirt and deflections are present

(T) Typical analysis

* Conditions of load and speed	Load — psi	Speed — ft/sec
Low	1000 and under	up to 5–10
Moderate	up to 2000	up to 15–20
High	over 3000	over 30

D **Steady-Load Capacity in Test Machine for 100 hrs**

		lbf
F–1	Tin-Base 0.022 Babbitt	1800
F–23	Lead-Base 0.022 Babbitt	2100
F–1	Micro Tin Base 0.004 Babbitt	2700
F–23	Micro Lead Base 0.004 Babbitt	2700
F–17	Trimetal 0.004 F–1 on F–19 Leaded Bronze	2700
F–12	Copper Lead (35%) 0.022 Thickness	3600
F–81	Aluminum Alloy Solid Construction	5400
Super Trimetal	Trimetal 0.001 F–38 on F–27 Leaded Bronze (Circular Shapes only)	7500
F–21 Plus Overlay	F–21 Silver Plate with 0.001 Pb-In Overlay	7500
F–77	Trimetal 0.001 Pb-Sn-Cu on S–77 Copper Lead	7500

A,B,C from Etchells, 11.902
D from Schager, 11.908

In practice, few bearing failures are attributable directly to overloading of the oil film.

Empirical formulas for the limiting load capacity of bearings have been used by many engine designers. Such formulas usually include a maximum unit pressure for each type of bearing (piston pin, crankpin, main, etc.) and a maximum value of pV, the unit pressure multiplied by the surface velocity. The maximum pressure involves strength, and the pV factor influences the rate of heat generation of the bearing. The values of these "limiting" factors have risen with time as the design of the surrounding structure and of bearing details has been improved. A spectacular example was the increase in load and pV capacity of the crankpin bearings of radial aircraft engines forced by combat conditions in World War II. By improving the master connecting-rod design and by using new bearing materials (in this case a thin silver layer on a steel shell, with lead-indium overlay), the allowable values of p and pV were raised to several times the values previously thought to be maximum. Table 11–12D gives bearing-pressure test ratings for various materials. These data are probably valid in a relative sense for the conditions prevailing in modern automotive engines. (See also refs. 11.922–11.924.)

Bearing-Length–Diameter Ratio. From considerations of unit pressure it would seem that the longer a bearing, the more load it can carry. On the other hand, the longer a bearing, the greater the chance of distortion or axial misalignment. Also, with a given clearance, the longer the bearing, the more dangerous is a given angle of axial misalignment. For these reasons, it is well to keep length-diameter ratios for crankpin and crankshaft bearings (w_r/P and w_j/M in Tables 11–2 through 11–9) to less than 1.0, even where space considerations might allow a longer bearing. The ratios used in practice range mostly between 0.3 and 0.5. For main bearings with noncounterweighted cranks at near the same angle on each side, ratios up to 1.0 may be necessary.

Journal-Bearing Clearance. The usual recommendation for crankshaft bearing clearance has been in the neighborhood of 0.001 times the journal diameter. The author's experience, confirmed by most bearing experts, is that this is a *minimum* value and that values up to twice this amount are acceptable and often desirable. Small clearances are dangerous both because they do not allow enough for distortion and because they have less safety against wear and scoring by foreign particles. Figures 9–3 and 9–4 of Volume I show that bearing coefficient of friction decreases as clearance/diameter increases.

The objections to large clearance include a lowering of the value of the Sommerfeld variable (Volume I, p. 320) and hence a lower factor of safety against oil-film breakdown. However, most bearings fail for reasons other than direct oil-film failure. The present author cannot recall a bearing failure due directly to excessive clearance, whereas he has seen a great many bearing failures attributable to inadequate clearance.

Because manufacturing tolerances are more or less independent of size, small bearings may have to have larger clearance/diameter ratios than would otherwise be used. No journal bearing in an engine, no matter how small, should have a minimum clearance less than 0.001 in. (0.025 mm).

Bearing Structural Failure. (11.900, 11.906) Failure not due to corrosion or wear is here defined as structural failure. For bearings operating within what experience indicates should be safe nominal loading, so that the bearing material does not fail by fatigue, structural failure is usually due to one or both of the following causes:

1. Failure of oil supply to the loaded area
2. Distortion of the intended bearing geometry

Such failures are usually attributable to poor detail design of the oil-supply system or to poor design of the bearing or shaft-support structure. As discussed in Volume I, Chapter 9, the principles of good oil supply are:

1. Introduction of the oil at points of minimum load
2. Avoidance of oil grooves or holes in heavily loaded areas

Distortion of the bearing geometry may act to cut down the oil supply to the loaded areas, or simply to concentrate the bearing load over a small area of the bearing. A particularly harmful type of distortion is misalignment of the axes of bearing and journal, usually due to bending of the crankshaft or engine frame, or both. Bearings with length-diameter ratios greater than 1.0 are especially prone to this trouble. Maintenance of the cylindrical shape of the bearings, together with good axial alignment, should be a paramount consideration in the design of connecting rods, crankshafts, and engine frames.

Bearing Detail Design. For many years the manufacture of bearing shells for crankpins and crankshafts of internal-combustion engines has been done by separate organizations specializing in such work. The advice of such organizations should be obtained in making decisions as to materials and dimensions of bearing shells. Many bearing manufacturers make a line of standardized shells which include sizes and designs appropriate for many types of engine. In all but the smallest and cheapest engines, crankpin and crankshaft bearings are composed of thin steel or bronze shells lined with a thin layer of the bearing material. These parts are made with sufficient precision so that they are interchangeable and require little or no "running in."

Antifriction Bearings

The use of antifriction bearings in the power train, in place of journal bearings, may be for one or more of the following reasons:

1. *Lubrication.* Ball and roller bearings do not require flooding with oil. They are adequately lubricated by a thin mist of oil in the crankcase or by

oil mixed with the fuel in the case of two-cycle engines with crankcase compression.

2. *Space.* For a given load capacity, ball and roller bearings can be axially shorter than journal bearings but are considerably larger in outside diameter.

3. *Friction.* Use of antifriction bearings significantly reduces starting friction and may have a small favorable effect on over-all running friction. (See Volume I, Chapter 9.)

Disadvantages of ball and roller bearings include the problem of assembly on a crankshaft and the fact that bearing failure is likely to be more sudden, more disastrous, and with less warning than the in case of journal bearings. Corrosion may also be a serious problem, since heavy-duty ball and roller bearings cannot be made of stainless material. Once a bearing starts to corrode, its life is very short.

Ball Bearings. (11.951–11.953, 11.955–11.957) Ball bearings are made in standard sizes by specialists (11.950). Since it is not usually practical to allow the balls to bear directly on anything but separate "races," they are applied in engines only where they can be used and assembled as complete units, as in Fig. 11–9. They are especially useful as *thrust* bearings and are so used in aircraft engines (10.840–10.848).

Roller Bearings. (11.954–11.957) Roller bearings lend themselves to direct use against hardened cylindrical shafts and are therefore used on crankshafts in preference to ball bearings. The problem of assembling roller bearings on a crankshaft is solved by one of the following methods:

1. For crankpin bearings, since there is little side load, the rod may be split and the rollers allowed to bear directly in the rod material (Fig. 11–6).

2. The crankshaft may be separable to allow assembly as in Figs. 11–7 and 11–9.

3. Rollers may be assembled piece by piece through a suitable slot, with the cage added as a final operation (10.46, 10.47, 10.846).

For single-crank engines, of course, main bearings may be assembled over the ends of the crankshaft (Figs. 10–6 and 10–7).

Needle Bearings. (11.960–11.964). These are roller bearings where the diameter-length ratio of the rollers is small. They are usually used without separation of the rollers by means of cages, and require more lubrication than conventional ball or roller bearings. They are used in the piston-pin ends of connecting rods, principally for small 2-cycle engines. They also serve as crankpin bearings on some small engines (Figs. 11–6 and 11–9).

The 2-cycle piston-pin journal bearing is a special problem because of the fact that the load is almost entirely, and often entirely, in one direction, and the bearing does not rotate to maintain an oil film under the load. Theoretically no oil film can be maintained under such circumstances

(11.921). In practice, however, many 2-cycle engines do operate satisfactorily with plain bearings at the piston-pin end, as in Fig. 11–28e (see also Figs. 11–13, 11–15, and 11–17 – 11–21 and ref. 11.920).

Load-Carrying Ability of Antifriction Bearings. The typical failure of an antifriction bearing is by fatigue. Therefore, bearing life is a function of loading on a given bearing size, as shown in Fig. 11–42.

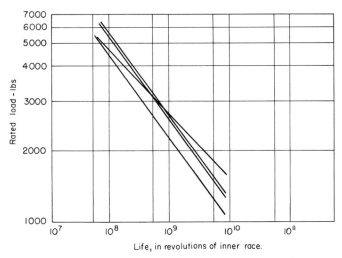

Fig. 11–42. Catalog-rated load versus catalog-rated life of four makes of ball bearings, all of same size, to show differences in load ratings of the different makers (Almen, 9.205).

Antifriction Bearing Standards. The external dimensions of antifriction bearings have been standardized (11.950). Manufacturers of such bearings also publish load ratings and tables of expected life. Except in the case of small, low-cost engines, standard bearings are seldom used for crankshafts and crankpins, because space and weight considerations usually call for special arrangements.

Detail Design of Antifriction Bearings. As in the case of journal bearings, antifriction bearings are made by specialists. When it is desired to use such bearings in an engine, detail design should be worked out with the advice of competent bearing manufacturers.

Journal Versus Antifriction Bearings

On account of the possibility of sudden failure and problems of space, assembly, and corrosion, most engines use plain journal bearings. However, the performance of antifriction bearings in small engines appears to be satisfactory within the limited requirements of such engines. For high-output engines of larger size, experience seems to indicate that journal bearings are to be preferred except under very special circumstances.

Nonmetallic Bearings

Nonmetallic bearings, principally of Teflon and related materials (11.970), are finding use in many applications where loads are light and temperatures relatively low. They have not yet found application in the power train of internal-combustion engines.

12 | Engine Design III: Valves and Valve Gear. Gears and Auxiliary Systems

Nomenclature

Because in the English language usage differs in respect to the naming of valve-gear parts, Fig. 12–1 is presented to show the nomenclature to be used here.

Types of Valves

The poppet valve is now universally used in 4-cycle engines. Sleeve and piston valves have been used in the past but are now obsolete because of their high cost, considerable friction, and their adverse effect on oil consumption. The same disadvantages apply in an even larger degree to the great number of rotary valves, slide valves, etc., which have been proposed from time to time but have never proved commercially successful (12.00).

In 2-cycle engines, as we have seen (Volume I, Chapter 7 and Volume II, Chapter 11), ports controlled by piston motion are used, either exclusively or in combination with poppet valves. Port design has been discussed in the chapters cited, and a supplementary discussion is included in this present chapter. As indicated in Volume I, Chapter 7, some large 2-cycle engines use auxiliary automatic or rotary valves in connection with the ports of loop-scavenged cylinders in order to achieve unsymmetrical timing. Some crankcase-compression 2-cycle engines use rotary or automatic inlet valves for the crankcase inlet.

POPPET VALVES

The advantages of the poppet valve over other types include the following:

1. It can give larger values of valve-flow area to piston area than most other types.
2. It has excellent flow coefficients if properly designed.
3. The manufacturing cost of the poppet-valve system is lower than that of any other type.

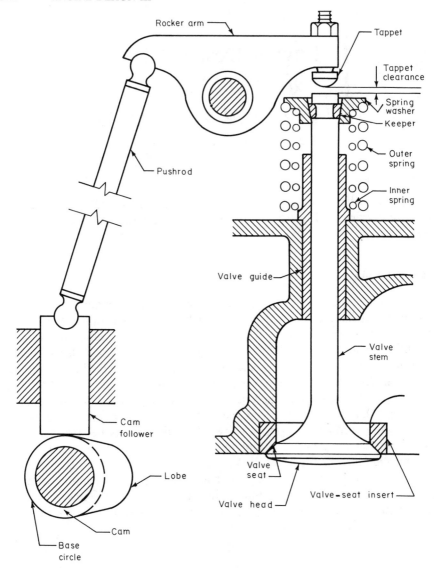

Fig. 12–1. Poppet-valve nomenclature.

4. It involves very little friction and requires less lubrication than any other type (no bearing surfaces are under cylinder pressure while the valve is in motion).

Its only disadvantage is concerned with the problem of cooling exhaust valves. This problem will be explained in a later section.

Poppet-valve design must achieve satisfactory results in respect to

1. Gas flow
2. Cooling and heat flow

3. Structural strength
4. Lubrication and wear
5. Provision for repair and replacement.

Inlet-Valve Flow Capacity

The importance of adequate valve flow capacity is explained in Chapters 6 and 7 of Volume I. For 4-cycle engines it was shown that the inlet-valve Mach index is critical. This index is defined as follows:

$$Z = \left(\frac{A_p}{A_i}\right) \frac{s}{C_i a}, \tag{12–1}$$

where A_p = piston area
$\quad A_i$ = inlet valve nominal area, $\pi D^2/4$, times number of valves
$\quad D$ = valve outside diameter
$\quad s$ = mean piston speed
$\quad a$ = sound velocity in inlet gases
$\quad C_i$ = mean valve flow coefficient based on area $\pi D^2/4$ and determined as explained in Volume I, Chapter 6.

It was shown in Volume I, Chapter 6, that to maintain good volumetric efficiency the value of Z should not exceed 0.6, with a preferred value in the neighborhood of 0.4 and a minimum value of about 0.25 for rated engine conditions.

Exhaust-Valve Flow Capacity

It was also shown in the same chapter that the exhaust-valve capacity can be less than the inlet-valve capacity (see Table 12–1). In order to secure the largest feasible inlet-valve capacity, a good design compromise is to make the ratio of exhaust-valve to inlet-valve flow capacity 0.70 to 0.75. Assuming equal flow coefficients and an equal number of exhaust and inlet valves, this means an exhaust-valve diameter 0.83 to 0.87 of the inlet-valve diameter.

Valve Capacity and Cylinder-Head Design

Figure 12–2 shows maximum feasible valve-diameter-to-bore ratios for flat cylinder heads with an exhaust-to-inlet area ratio of 0.87. The limiting piston speeds for each arrangement can be computed from eq. 12–1. Values of this limit for $Z = 0.6$, sound velocity 1200 ft/sec (366 m/sec) and a value of C_i of 0.35 are given below the figure. The assumed flow coefficient can be achieved by good inlet-valve and port design with a lift-diameter ratio of 0.25.

Table 12–1. Valve Capacity of 4-Cycle Engines

Data from Tables 11–2 through 11–9

Class	S-I or Diesel	Inlet No.	D_i/b max	D_i/b min	A_i/A_p max	A_i/A_p min	Exhaust No.	D_e/b max	D_e/b min	A_e/A_p max	A_e/A_p min	A_e/A_i max	A_e/A_i min
Small	S-I	1	0.58		0.34		1	0.53		0.28		0.82	0.82
Passenger auto	S-I	1	0.49	0.39	0.24	0.15	1	0.41	0.36	0.17	0.13	0.85	0.83
Automotive Diesel	D	1	0.49	0.45	0.24	0.20	1	0.39	0.31	0.15	0.10	0.75	0.70
	D	2	0.34	0.34	0.23	0.23	2	0.34	0.34	0.23	0.23	1.0	1.0
Locomotive Diesel	D	2	0.35	0.32	0.24	0.20	2	0.35	0.31	0.24	0.19	1.0	0.83
Large Gas	S-I	2	0.31	0.30	0.19	0.18	2	0.31	0.30	0.19	0.18	1.0	1.0
Medium-size Diesel	D	2	0.32	0.30	0.20	0.18	2	0.30	0.30	0.18	0.18	1.0	1.0
	D	1	0.42	0.40	0.18	0.16	1	0.40	0.37	0.16	0.14	1.0	0.78
Auto racing	S-I	1	0.63	0.53	0.40	0.28	1	0.63	0.55	0.4	0.30	1.0	0.72
	S-I	2	0.39	0.37	0.30	0.27	2	0.36	0.35	0.26	0.25	0.92	0.85

D_i = outside diameter, inlet valve; D_e = outside diameter, exhaust valve; b = cylinder bore; $A_i = \pi D_i^2/4$; $A_p = \pi b^2/4$; $A_e = \pi D_e^2/4$.

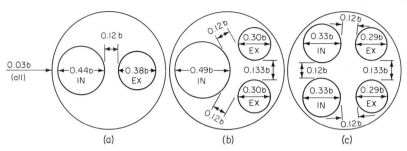

Fig. 12–2. Valve area ratios for a flat cylinder head.

Arrangement	$\dfrac{D_i}{b}$	$\dfrac{D_e}{b}$	$\dfrac{A_i}{A_p}$	Piston Speed for $Z = 0.6$ ft/min	m/sec
a Two valves	0.44	0.38	0.193	2900	14.7
b One inlet Two exhaust	0.49	0.30	0.24	3620	18.4
c Four valves	0.33	0.29	0.218	3300	16.7

For all cases clearance to bore $= 0.03b$, space between valves $= 0.12b$, except $0.133b$ between two exhaust valves. $A_e/A_i = 0.75$, $a_i = 1200$ ft/sec, $C_i = 0.35$.

Nearly all 4-cycle Diesel and gas engines use flat cylinder heads (see Figs. 11–12 through 11–16). This choice appears rational on the basis of combustion-chamber geometry and simplicity of the valve gear. It is evident that the valve flow capacity with a flat cylinder head can be adequate for all but exceptionally high piston speeds. An appreciable gain in valve size can be made by making the head inside diameter slightly larger than the bore. Surprisingly, the proximity of a valve to the side of the head or bore seems not to reduce its flow capacity (Volume I, ref. 6.40 *et seq.*).

Wedge-Shaped Cylinder Heads. The heads of Figs. 11–10 and 11–11 have about the same valve-area limitations as those of 2-valve flat heads, provided "squish-area" requirements do not encroach upon the area available for valves. Where Z would otherwise be too high at the required piston speed, the head space can be slightly larger than the bore, as in the case of flat cylinder heads.

Domed Cylinder Heads. Where maximum valve capacity is required, as in racing engines and aircraft engines, the cylinder head can be "domed," as illustrated in Figs. 11–7 and 11–22. Figure 12–3 shows that the valve area/piston area can be drastically increased by "doming" the head of a 2-valve cylinder. The disadvantages of doming include more complex valve-gear geometry and higher costs of manufacture. Domed heads used in some U.S. passenger automobiles around 1960 proved to have excellent performance characteristics when clean, but appeared especially prone to octane-number depreciation and to "wild ping" and "rumble" (see Chapter 2),

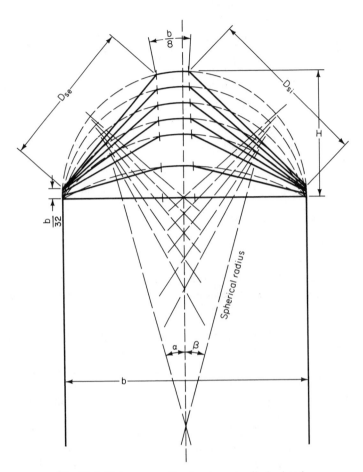

Fig. 12–3. Valve-area ratios for domed cylinder heads.

$\dfrac{H}{b}$	$\dfrac{D_i}{b}$	$\dfrac{A_i}{A_p}$	$\alpha°$	$\beta°$	s at $Z = 0.5$ ft/min	m/sec
0.50	0.65	0.42	49.3	43.9	4980	25.1
0.44	0.61	0.37	45.2	40.6	4400	22.2
0.38	0.56	0.31	40.4	36.6	3680	18.5
0.31	0.53	0.28	35.1	31.9	3320	16.7
0.25	0.48	0.23	29.1	26.6	2730	13.8
0.13	0.44	0.19	15.7	14.1	2260	11.4
0	0.42	0.18	0	0	2140	10.8

$A = \pi D^2/4$
D = valve diam
D_{si} = inlet-valve seat diam
D_{se} = exhaust-valve seat diam

Assumptions
$A_e/A_i = 0.81$

Values of s are at
$D_i = 0.975\ D_{si}$, $D_e = 0.975\ D_{se}$
$a = 1200$ ft/sec, 366 m/sec
$C_i = 0.33$

$$Z = \left(\frac{b}{D_i}\right)^2 \frac{s}{aC_i}$$

as deposits were accumulated. A number of passenger-car engines made outside the U.S.A. use this type of head, as do some U.S. sports-car conversions.

L-Head, F-Head, and T-Head. L-head, F-head, and T-head engines obviously have no theoretical limits on valve-area piston-area ratio but, as stated in Chapter 2, the disadvantages of such designs, from the combustion-chamber point of view, have rendered them virtually obsolete, at least for consideration in new designs.

Valve Area. Having chosen the appropriate cylinder-head design, it would seem advisable to design for the largest feasible valve area in order to minimize inlet-pressure loss, exhaust blowdown loss, and pumping losses (see Volume I, Chapters 6 and 9). If this procedure results in too low values of Z at the designed piston speed, valve lift can be adjusted accordingly. Future increases in piston speed are then taken care of by the simple expedient of increasing valve lift.

Table 12–1 and Fig. 12–4 give valve-area ratios for a number of contemporary (1966) 4-cycle engines.

Valve Flow Coefficients

In Volume I, Chapter 6, a method of determining the steady-flow coefficient of poppet valves at various values of the lift-diameter ratio is described. The *average* flow coefficient used in computing Z is obtained by integrating the flow coefficients at various L/D ratios over the schedule of valve lift versus crank angle.

Figure 12–5 shows modifications that can be made on conventional inlet valves in order to improve their flow coefficient. To be effective, the geometry of the modifications must be carefully controlled and maintained in spite of wear and deposits. Since proper controls of this kind are difficult, such modifications are seldom used except where the resulting improvement in flow coefficient is very important, as in racing and aircraft engines. In those cases, machining of the ports to the required shape may be justified, as well as frequent removal of deposits and replacement of worn parts.

Similar work on exhaust valves (12.03, 12.04) has shown the possibility of some increase in flow coefficient, but the modifications are generally not justified because of the insensitivity of most engines to small changes in exhaust-valve capacity (see Volume I, Fig. 6–24, p. 194).

Valve Lift

Figure 12–5 shows that lifting the valve more than one quarter of its diameter gives diminishing returns in flow capacity. Since valve-gear loads increase as the lift squared, it is desirable to make a reasonable compromise between maximum flow capacity and valve-gear loading. In many cases

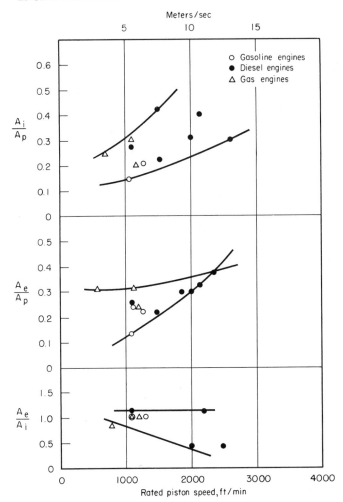

Fig. 12–4. Valve-area ratios of 4-cycle poppet-valve engines. A_i = inlet-valve area; A_e = exhaust-valve area; A_p = piston area. Areas are taken as $\pi D^2/4$, where D is outside diameter of valve.

there is also a problem of interference between valves and piston at top center. The usual compromise is to use lift-diameter ratios of about 0.25. Values higher than this one would be justified only where the inlet Mach index Z would otherwise be greater than 0.6.

Multiple Valves

Figure 12–2 shows that valve-area/piston-area ratios for multiple valves are somewhat larger than those for two valves in a flat cylinder head. In addition to improved flow capacity, compelling reasons for using multiple

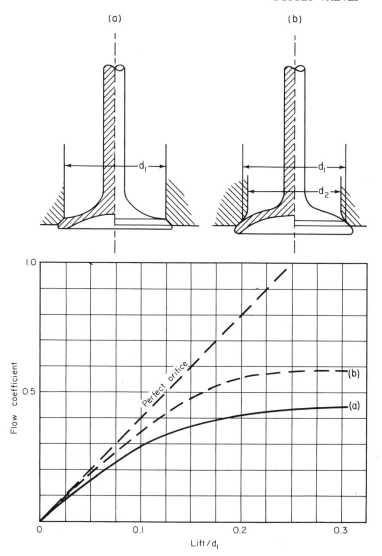

Fig. 12–5. Flow coefficient versus lift d_1 for inlet valves. (*a*) Conventional design; (*b*) modified valve and port, $d_2/d_1 = 0.80$.

valves include reduction of temperatures and thermal stresses and reduced valve-gear stresses at a given piston speed. As indicated in Volume I, Chapter 11, with similar geometry at a given engine rpm, inertia stresses in the valve gear will be inversely proportional to the valve diameter squared. It was also shown in Volume I, Chapter 8, that temperatures will be lower and that, even at the same temperatures, thermal stresses will decrease with decreasing valve size.

In most cases where it is advisable to use multiple valves, the 4-valve design is the best compromise. This design has the further advantage of allowing for a centrally located injector or spark plug. Only with large cylinders would more than two valves of each kind be justified (see ref. 10.46 for an engine using three inlet and three exhaust valves).

Valve Cooling

While the principal valve-cooling problems concern the exhaust valve, inlet valves operate at surprisingly high temperatures, as indicated by Fig. 12–6a. However, inlet valves seem to require no special cooling arrangements that are not taken care of by good structural and flow design.

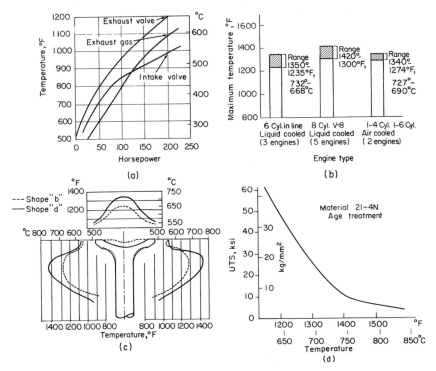

Fig. 12–6. Typical poppet-valve temperatures. (a) Measurements from a Diesel engine (12.20); (b) exhaust-valve temperatures of passenger-automobile engines (12.21); (c) effect of valve-head design on exhaust-valve temperatures (12.21); (d) effect of temperature on strength of an exhaust-valve steel (12.21).

Exhaust valves have presented a difficult cooling problem ever since the advent of the internal-combustion engine. The heat-receiving surface is much larger than the heat-rejecting surface, which consists only of the stem and seat. Furthermore, both the mean gas temperature and the gas velocity relative to the surface are higher than at any other point in the engine.

The difficulty of cooling poppet exhaust valves has been the principal stimulant for the development of other types of valve, and the solution of this problem has been the principal factor in making other valve types obsolete. This solution has been due both to improved design and to the development of valve materials that can withstand high temperatures without failure due to breakage, warping, burning, or excessive wear. Valve materials have been mentioned in Chapter 9 and are listed in Table 12–2. Most engines use the austenitic steels, EV 3 to EV 11. However, even with the best materials the question of cooling and heat flow must receive the designer's careful attention. While the valve is open, the only path for heat to flow to the engine coolant is through the stem and valve guide, but this path is obviously a poor one. The most important path for heat flow (75 per cent according to ref. 12.21) is through the valve seat while the valve is closed.

Valve Temperatures. Figure 12–6 shows valve temperatures in several successful engines. It is interesting to note that the temperatures are about the same in both water-cooled and air-cooled types. The highest temperatures shown in this figure may be considered the maximum allowable in commercial practice. Reference 12.21 indicates how even a small reduction in valve temperature, as from 1400° to 1350°F (777° to 750°C), results in a marked improvement in valve life and reliability. Note also the improvement in strength of the material of Fig. 12–6d. The designer should strive for even lower temperatures as far as possible.

Valve Temperature and Design. Since gas temperature, coolant temperature, and gas flow are established by other engine requirements, for a valve of given size its temperature may be minimized by

1. Minimizing heat-receiving area by extending valve guide boss as near to the valve head as possible
2. Maximizing heat-rejecting areas, that is, valve-seat and valve-stem areas
3. Minimizing length of path from valve guide and valve seat to coolant
4. Maximizing valve heat conductivity by generous valve-head thicknesses.

Figures 12–7b and c show exhaust valves designed so as to afford the maximum cooling consistent with good flow characteristics. The provisions for good cooling include

Relatively large stem diameter
Plenty of material in the valve head
Minimum exposure of stem to hot gas
Coolant passages all around seat and stem
Minimum length of heat paths to coolant.

Internal Cooling. As pointed out in Volume I, Chapter 8, cooling problems become more difficult as size increases. For valves of more than about 2-in. (50-mm) diam, or even for valves of smaller size subjected to severe conditions (as in aircraft), *internal cooling* may be necessary.

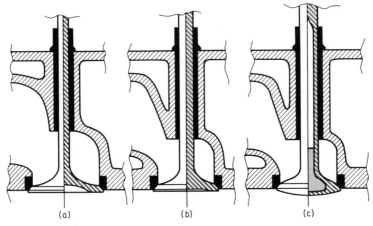

Fig. 12–7. Well-designed poppet valves.

(a) *Inlet valve*
1. Emphasis is on high flow capacity
2. Diam = 1.15 × exhaust-valve diam
3. Stem diam small (0.15 × D_i)
4. Free length of stem long
5. Port has generous flow area

(b) *Exhaust valve*
1. Emphasis is on cooling
2. Stem diameter large (0.25 × D_e)
3. Free length of stem short
4. Coolant passages all around seat and guide

(c) *Sodium-cooled exhaust valve*

Figures 12–7c and 12–8 show internally cooled exhaust valves. The hollow space in the valve is half filled with sodium, which liquefies at valve temperature and " sloshes " back and forth in the valve as the valve moves. Thus, the heat conductivity from the exposed surface of the valve to the seat and stem is greatly increased.

The development of internally cooled valves is largely the result of the work of one man, S. D. Heron (10.843). Valves with internal cooling are now made by a number of manufacturers in the U.S.A. and elsewhere. Internally cooled valves are expensive and are used only under circumstances where valve performance would otherwise be unsatisfactory. Their use is most frequent in the case of engines of high specific output, such as aircraft, locomotive, racing, and military engines.

Effect of Valve Size on Cooling. As explained in Volume I, Chapters 8 and 11, reduced unit size is a powerful means of improving thermal performance. Thus, valve cooling can be greatly improved by the replacement of a single valve by two similar valves having a total flow capacity equal to the single valve they replace, that is, with dimensions $1/\sqrt{2} = 0.71$ times those

Fig. 12–8. Sodium-cooled exhaust valve. Hollow space should be about half filled with sodium.

of the single valve. This change reduces the length of heat-flow paths in the valve by that factor and reduces valve temperature accordingly.

Valve Failure

The question of valve failures and how they may be prevented is discussed in refs. 12.10, 12.23. The most typical failure of exhaust valves starts with leakage due to poor seating. This may be due to warping of the valve, valve seat, or valve-guide support structure, or to accumulation of deposits on the seat. Severe leakage results in overheating the valve and ultimately in complete failure. Failure may be by burning and/or breakage of the stem just above the head. Some attempts have been made to introduce flexibility in seat inserts or valves to prevent leakage due to warping (12.14), but such designs have not proved necessary or even desirable in most cases. Failure of valves may also be due to high seating velocity because of excessive engine speed, improperly designed valve gear, or excessive valve clearance. These questions will be discussed later in this chapter.

In addition to failures of the valve head or at the junction of head and stem, a possible point of failure is at the grooves for the *keeper*, that is, the anchorage of the spring support, Fig. 12–9. These grooves should be of minimum depth, carefully filleted and smoothly finished. As in the case of threads, rolling of grooves is to be preferred over machining. Figure 12–9 shows good and poor valve-stem and spring-anchorage designs.

The best insurance against valve failure is the use of appropriate materials together with good detail design, as illustrated by Figs. 12–7 through 12–9. As previously suggested, the development of improved materials (austenitic steels) has probably been the most important factor contributing to the present reliability of well-designed poppet exhaust valves. Another factor is the control of fuel and oil composition so as to avoid excessive build-up of deposits on valve seats.

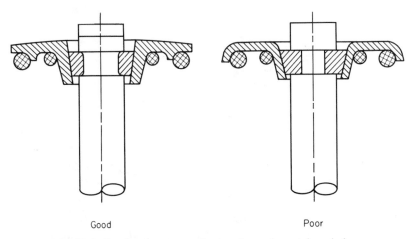

Good   Poor

Fig. 12–9. Good and poor practice in valve-spring retainer design.

Good	*Poor*
Maximum neck diameter	Neck diameter too small
Well-filleted neck	No fillets in neck
Groove rolled	Groove machine finish
Sturdy washer	Washer too thin
Springs centered on inside	Spring centered on outside
Hard button on top	Soft top

Valve Lubrication and Wear

It is now universal practice to enclose the valves and valve gear and to circulate oil over the valve-spring ends of all valves. A difficult problem here is to provide enough lubrication for the valve stems without involving excessive oil consumption. Loss of oil through the inlet-valve guides is a special problem for spark-ignition engines that run throttled most of the time, as in the case of passenger automobiles. When an engine is throttled, there is a large pressure difference between valve housing and inlet port. Control of this type of oil consumption is achieved by avoiding any accumulation of oil which will submerge the top of the valve guide. Lubrication of valve stems is then achieved by splash from the springs, cams, and rocker arms. Many engines use an oil-seal ring or a metal *umbrella* at the top of the valve guide in order to minimize loss of oil through the guides.

Valve Seats and Valve-Stem Guides

Proper choice of material for these parts is important. The best material for guides seems to be a grade of cast iron selected especially for this service (10.847). Exhaust-valve-seat inserts are generally made of steel similar to that used in the valves themselves (12.32, 12.33). In order to minimize valve-guide wear, it is well to minimize side forces on the valve stem by suitable design of valve-gear geometry. Roller tappets are especially effective in this respect.

Valve Performance

Gradual improvement in design and materials, plus adequate lubrication, plus internal cooling where required, has brought poppet valves to the point where they no longer constitute a critical problem in most services. However, valve failures still do occur, and exhaust valves must be considered an element in engine design still subject to improvement.

VALVES AND PORTS FOR 2-CYCLE ENGINES

The subject of porting of 2-cycle engines has already been considered in some detail in Volume I, Chapter 7. Further references to this subject are 12.05–12.092.

For loop-scavenged engines of the cross-flow type, the Schnurle system (often called the Curtiss system in the U.S.A.) is in general use, except in the case of small engines. Details of Schnurle port design will be found in Volume I, refs. 7.13–7.15, 7.19, and 7.20. Engines using this system are described in refs. 10.272, 10.392, 10.3991, 10.56, 10.594, 10.595, 10.596, 10.71, 10.820, and some of the small 2-cycle engines in refs. 10.640–10.696. The engines of Figs. 11–9, 11–17, 11–18, and 11–20 also use this system.

Many small 2-cycle engines use *opposite* porting, that is, two sets of inlet ports facing each other 180 degrees apart on the cylinder circumference, with two sets of exhaust ports in a plane at 90 degrees from the inlet ports. Your author has found no scientific measurements regarding the effectiveness of such ports.

Where a baffle on the piston is allowable, effective scavenging can be secured with radial porting, as reported in Volume I, refs. 7.19 and 7.20.

2-Cycle Exhaust Ports

These ports require careful attention to cooling, particularly for the bridges between adjacent ports. In water-cooled engines these bridges should be drilled for coolant circulation, except in the case of very small cylinders. The bridges of air-cooled cylinders should be provided with adequate finning and air circulation as close to the bridges as possible.

Ports and Piston Rings

The circumferential width of ports must be limited in order to avoid damage to piston rings as they slide over the ports. The allowable width varies with the relative dimensions of piston rings. Widths in use can be estimated from some of the engine cross sections illustrating Chapter 11 and from refs. 10.272, 10.37, 10.392, 10.396, 10.42, 10.51–10.593 and from Volume I, refs. 7.14, 7.19, 7.20, 7.22, 7.23, 7.37–7.392.

Most, but not all, 2-cycle engines prevent rotation of the piston rings by means of pins fastened in the piston and projecting into the ring gap. The purpose of this design is to ensure that the ring gap never passes over a port. With this arrangement, exhaust blowdown through the gap of the top ring is avoided, and larger port widths can be used without danger of ring damage.

Port Areas

Figure 12–10 shows port and valve area ratios for a number of 2-cycle engines (see also Volume I, pp. 246 and 248). Reference 10.3991 gives an empirical formula for port height as a function of piston speed. Actually, the design of 2-cycle porting has never been adequately rationalized, and each new design seems to require the development of optimum porting by means of cut-and-try methods. In the author's experience, the performance of 2-cycle engines is surprisingly insensitive to small changes in port design in the region of optimum porting. This probably accounts in part for the wide variations in port area shown in Fig. 12–10.

Poppet Exhaust Valves for 2-Cycle Engines

Such valves should be as large as feasible, because a rapid blowdown is an essential characteristic for this type of engine. Since these valves must open and close in about half the crank angle available in 4-cycle engines, there is an especially strong case for using multiple valves in order to reduce loads on the valve gear, as well as to reduce over-all engine height and to facilitate valve cooling. For example, with equal flow areas, four valves will have half the dimensions of a single valve and their total mass will be half that of the single valve they replace (see Figs. 11–13 and 11–15). In Fig. 11–21 a single exhaust valve is used in an engine designed for low piston speed.

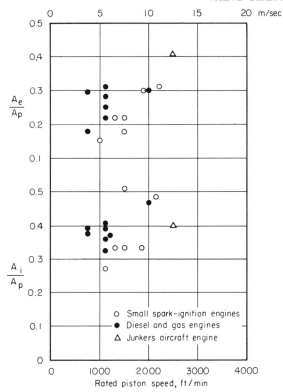

Fig. 12–10. Port-area to piston-area ratios of commercial 2-cycle engines. A_i = inlet area; A_e = exhaust area; A_p = piston area.

VALVE-GEAR DESIGN

Since the cam-operated valve-gear system using helical valve springs is now almost universal, this type will be called the "conventional" valve gear and will be considered here. The question of less conventional gears is briefly discussed at the end of this section. For definitions of the terms used, see Fig. 12–1.

Cam Design

Because of stress limitations, poppet valves cannot be opened and closed suddenly but must follow a motion pattern of the general character shown in Fig. 12–11. The actual motion of the valve will be the designed motion, as shown in this figure, modified by the elastic characteristics of the valve-gear mechanism.

In order that the valve always seats firmly, and to allow for adjustments, wear, and expansion and contraction due to temperature changes, some

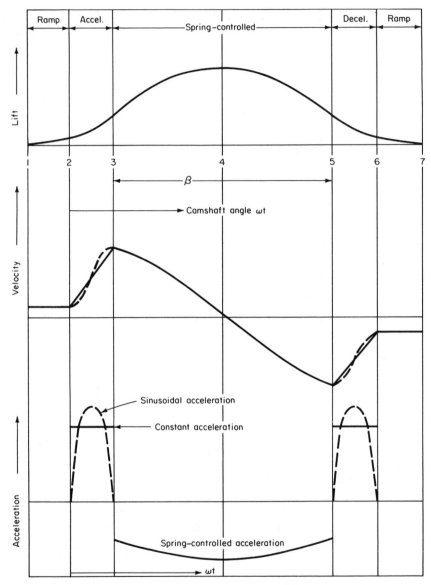

Fig. 12–11. Theoretical valve lift, velocity, and acceleration versus camshaft angle. ω = crankshaft angular velocity, t = time.

clearance or *lash* is always required (see Fig. 12–1). This clearance should be the minimum necessary to ensure that the valve seats firmly under all normal conditions, including a reasonable allowance for errors in adjustment.

As the cam meets the follower at the beginning of the opening cycle, the clearance should be "picked up" by a *ramp* of constant velocity (velocity is here referred to as the lift per unit time at a given crank speed, as shown by the middle curve of Fig. 12–11).

When the cam follower strikes the ramp, the impact stresses with a given design will be a function of the ramp velocity. These stresses are difficult to predict but are usually controlled by using a ramp velocity which experience has shown to be tolerable. In large aircraft engines the usual ramp velocity is 2 ft/sec (0.63 m/sec) at rated speed. Since the point of impact varies with the clearance, which in turn is a variable, the ramp velocity should be the same at any expected clearance as shown in Fig. 12–11.

Lift. Because of the uncertainty regarding the point at which the clearance is taken up, it is customary to define the valve lift as the lift above the ramp and to define the valve-opening angle as the crank angle between stations 2 and 6 of Fig. 12–11.

The Use of Hydraulic Followers. Use of hydraulic cam followers (Fig. 12–12) is now quite general for passenger-car engines, where a low noise level is considered essential (12.37–12.39). Since the hydraulic system cushions the clearance pick-up, the ramps for use with hydraulic followers

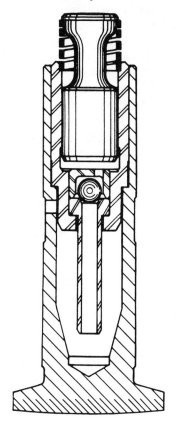

Fig. 12–12. Hydraulic valve lifter with mushroom face and removable hydraulic unit. (*Courtesy Eaton Manufacturing Company.*)

can have a higher velocity than those with mechanical systems. Another practical advantage of hydraulic followers is that they adjust automatically for valve-gear wear therefore eliminating the need for periodic clearance adjustment.

Disadvantages in the use of hydraulic followers include greater variation in the valve opening and closing angles, and the danger of the hydraulic system becoming inoperative due to clogging of the passages or to sticking of the bearing surfaces. The latter dangers are greatly reduced through the use of detergent lubricating oils. In the early development of these devices, they would sometimes " pump up " (that is, fill with oil to the upper limit of the plunger travel) and hold the valve open. This trouble is now infrequent. Hydraulic followers are not generally used for very high-performance engines, or where control of engine noise is not of great importance.

Theoretical Valve Motion. For convenience in introducing the subject of cam design, it will first be assumed that all the moving parts of the valve gear are rigid bodies. The subject of flexibility in the valve mechanism and its effects will be introduced later. Valve motion under this rigid-body assumption will be called *theoretical* valve motion.

Theoretical Acceleration. At the end of the ramp, the valve is accelerated by a curve of increasing slope (period 2–3 in Fig. 12–11). Many successful cams have been designed on the theory of constant acceleration during this period. However, even with a cam designed for constant acceleration, the valve will not follow such a curve, because it calls for the instantaneous transmission of a sudden change in load. Resilience in the valve gear will modify the valve motion, hopefully to a curve such as that shown by the dashed lines of Fig. 12–11. In view of these facts it would seem logical to design for an acceleration curve such as that indicated by the dashed line. Here it must be remembered that the work of acceleration, i.e., the area under the acceleration-time curve, must be the same for any curve which gives the same velocity at station 3.

Theoretical Motion under Spring Control. At the end of the acceleration curve, station 3, the valve spring must decelerate the valve to zero velocity at station 4, and then accelerate it downward to station 5, where the load is again transferred to the cam for deceleration to ramp velocity. Ideally, the motion during the spring-controlled period, 3–5, will be such that the deceleration and acceleration load will be a constant fraction of the available spring pressure. Referring to Fig. 12–13, to achieve this result, period 3–5 must be part of a sine curve such that

$$L = L_s \sin \alpha - (L_s - L_{max}), \qquad (12\text{-}2)$$

where L = valve lift (above ramp)
  L_{max} = maximum valve lift
    L_s = spring deflection from zero load to maximum lift
    α = angle generating the sinusoid. This angle is so chosen that the

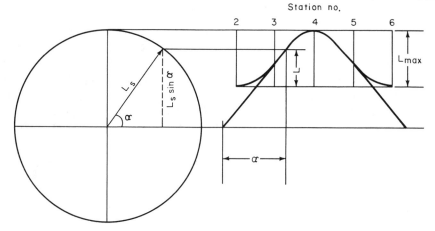

Fig. 12–13. Fitting of the spring-controlled curve to the acceleration curve.
L_s = deflection of valve spring from maximum lift to zero load
L = valve lift above clearance ramps
α = angle generating sine curve

sinusoid is tangent to the acceleration and deceleration curves and to the maximum lift at station 4. Its value can be determined graphically on a large-scale drawing, or by mathematical means.

Let $\alpha = \omega_s t_s$, where t_s = time and ω_s is the angular velocity of α. By differentiating twice with respect to time, the acceleration between points 3 and 5 is:

$$\ddot{L} = -L_s \omega_s^2 \sin \omega_s t_s. \tag{12–3}$$

This equation can be written in terms of camshaft angle as follows:

$$\ddot{L} = -L_s (\beta \omega)^2 \sin \beta \omega (t - t_0), \tag{12–4}$$

where ωt = camshaft angle
$\beta = \omega_s / \omega$
$t_0 = t$ when $t_s = 0$.

The force on the cam is

$$F_c = \left[K - \frac{M}{g_0} (\beta \omega)^2 \right] L_s \sin \beta \omega (t - t_0), \tag{12–5}$$

where F_c is the force on cam, K the spring rate, and M the mass to be accelerated, that is, the inertia of the moving parts of the valve and gear, reduced to an equivalent mass at the spring. From Eq. 12–5 the crankshaft speed for a 4-cycle engine at which the force F_c becomes zero is

$$\Omega_J = 2\omega_J = \frac{2}{\beta} \sqrt{K g_0 / M}, \tag{12–6}$$

where Ω_J is the crankshaft angular velocity at which the theoretical force on the cam is zero between stations 3 and 5 of Fig. 12–11 and may be called the *theoretical jumping speed* for the valve gear. It would be the actual jumping speed if the valve gear were perfectly rigid. With the assumption of rigidity and with the curve 3–5 chosen in accordance with Fig. 12–13, the pressure on the cam will be proportional to L_s at any speed lower than the jumping speed. Since the spring force is also proportional to L_s, the curve specified between stations 3 and 5 gives the highest possible theoretical jumping speed for given values of K and M. Any other curve with the same tangencies will give a higher acceleration at some point, and therefore will have a lower theoretical jumping speed or will require a heavier spring.

Theoretical Deceleration. With rigid-body assumptions and the same limit on valve-gear load, the deceleration curve 5–6 will be the same as the acceleration curve 2–3 and the whole motion curve will be symmetrical with respect to station 4.

Limits on the Valve-Motion Curve. Practical limits on the curves of Figs. 12–11 and 12–13 are imposed by the following factors:

Points 2, 4, and 6 are determined by valve timing and valve lift.

Points 3 and 5 are determined by chosen limits on valve-gear load and spring load.

It is obvious that for a given lift, timing, and speed, the shorter the periods 2–3 and 5–6, the greater the acceleration loads on the valve gear, and the less the spring load required. In high-speed valve gears, the ratio of periods 2–3 plus 5–6 to period 2–6 is usually in the range 0.25 to 0.35.

The foregoing method of designing the cam so as to conform to the available spring force affords maximum theoretical jumping speed with a given lift and a given spring. It is therefore a desirable base design for valve gears that must run at high characteristic speeds.

In practice, the period 3–4–5 is not always designed to conform with the available spring force. A common compromise is shown in Fig. 12–14, which assumes fixed values of maximum lift and timing, and fixed acceleration and deceleration curves.

Curve *a* corresponds to the spring-conforming design of Fig. 12–13. Curve *b* is a type used in many cases in order to increase valve opening in the cross-hatched areas. This design increases acceleration in these regions and requires a heavier spring for the same theoretical jumping speed. Curve *c* shows the theoretical valve motion at jumping speed. The cam could be designed for this curve with the same spring, provided the increased lift is permissible. The possible gain in valve-flow coefficient due to the change from curve *a* to curve *b* increases as the engine piston speed decreases, but the importance of the gain (due to decrease in Z) decreases. Except for engines designed for low piston speeds, the basic design method suggested by Fig. 12–13 would seem appropriate.

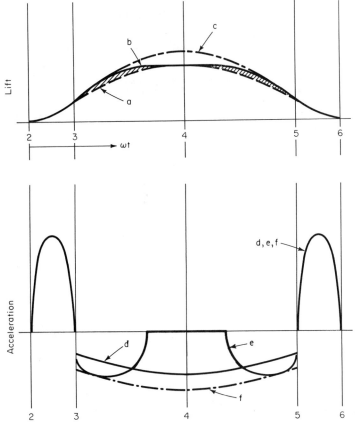

Fig. 12–14. Various types of spring-controlled motion. *Upper diagram, lift versus camshaft angle:* (*a*) Valve lift for design of Fig. 12–13; (*b*) flattened lift curve to give increased valve lift in cross-hatched areas; (*c*) trajectory of valve at jumping-speed with curve (*b*). *Lower diagram, acceleration versus ωt at jumping speed:* (*d*) For lift curve (*a*), spring force divided by mass coincides with acceleration 3–5; (*e*) acceleration for lift curve (*b*) at same speed; (*f*) spring force divided by mass for curve (*c*) with same jumping speed as curve (*a*).

Actual Valve Motion

The foregoing analysis based on rigid-body assumptions provides a good framework on which to base cam design. However, some modifications may be necessary because of valve-gear elasticity. Since the valve-operating mechanism, including its supporting structure, is composed of elastic bodies, the actual motion of the valve will be the designed motion plus motion due to the elasticity of the gear. A basic problem in cam and valve-gear design is to achieve an actual valve motion as close to the theoretical rigid-body motion as possible. In the subsequent discussion, the theoretical rigid-body motion is called the *designed motion*, and the departure from the designed motion is called *false motion*.

Figure 12–15 shows typical examples of false motion of a valve. At low speed the valve follows the designed motion very closely because the forces of cam acceleration and deceleration are small and are applied over

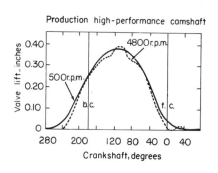

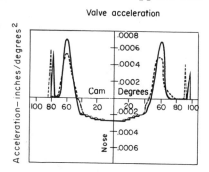

(a)   Exhaust valve–high speed motion study

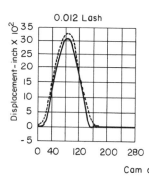

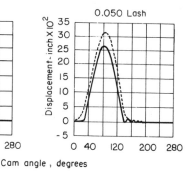

(b)   Influence of lash on valve performance. GM Diesel series 53 engine; cam speed: 2700 rpm.

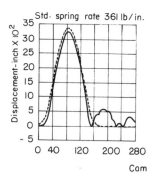

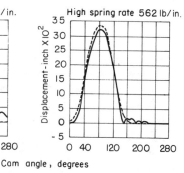

(c)   Valve performance with two different cam follower springs. GM series 53 engine; cam speed: 4000 rpm

Fig. 12–15. Actual versus theoretical valve motion. (*a*) Valve motion and acceleration for 1963 Chevrolet "Corvair" engine (Pechenik, 12.45); (*b*) valve motion at two clearances (Johnson, 12.44); (*c*) valve motion with two spring rates (Johnson, 12.44). In *b* and *c*, dashed lines are the designed valve motion.

relatively long time periods (see Fig. 12–15a). These forces increase as the speed squared; therefore, as the speed increases a point is reached at which false motion becomes apparent. As the speed increases further, the amplitude of this motion may reach a point where the follower leaves the cam at one or more points. This is the *actual jumping speed* of the valve gear, and is always lower than the theoretical jumping speed.

Curve c of Fig. 12–15 shows how increased spring force may reduce false motion, in particular the seating velocity and the degree of *bounce*. Thus high spring force would seem desirable, within the limits imposed by space for the spring and wear on the valve-gear parts (see later discussion of spring design).

Some degree of false motion is always present at high engine speeds. It can be tolerated only up to the point where there is danger of failure or excessive wear of the mechanism. The usual limit is set by the seating velocity of the valves. It will be noted in Fig. 12–15 that this velocity increases to values well above the theoretical seating velocity, whenever false motion is considerable. In every case above 500 rpm shown in Fig. 12–15, the valve seats so fast that it rebounds, or "bounces," sometimes several times. The danger of failure, particularly of the exhaust valve, increases with increasing seating velocity, which increases tensile stress in the valve. It is obvious that one principal objective of valve-gear design should be to minimize false motion.

Analysis of Actual Valve-Gear Motion. Analysis of actual valve-gear motion is difficult because of the fact that both flexibility and mass are distributed in a complex manner, and that vibratory motions are restrained in such a way that deflections are not necessarily proportional to loads.

Figure 12–16a shows a simplified system which helps to explain some of the major factors involved. In this figure the valve-gear system is represented by a stiff spring of rate K (dimensions F/L) and mass M. The valve spring rate is represented by K_s, which is very small compared with K. The cam is represented by the function $y(\Theta)$, where Θ is the cam angle and is a function of time; x is the resultant position of mass M, and its motion represents the valve motion. By proper choice of the values M, K, and K_s, the actual motion of a valve can be approximated (12.42).

As long as the follower is in contact with the cam, neglecting the relatively small spring force, the motion of M is described by the relation:

$$\frac{dx}{d\Theta} = \left(\frac{\omega_n}{\omega_c}\right)^2 \int_0^\Theta (y - x)\, d\Theta, \qquad (12\text{–}7)$$

where $\omega_n = (Kg_0/M)^{\frac{1}{2}}$ and $\omega_c = \Theta/t$. Figure 12–16b shows a graphical interpretation of this equation for an assumed curve of y vs. Θ. Part c of the same figure shows resultant values of x for low and high values of ω_n/ω_c. With a given y curve at a given speed, it is evident that the higher the natural frequency ω_n, the smaller will be the amount of false motion, $y - x$.

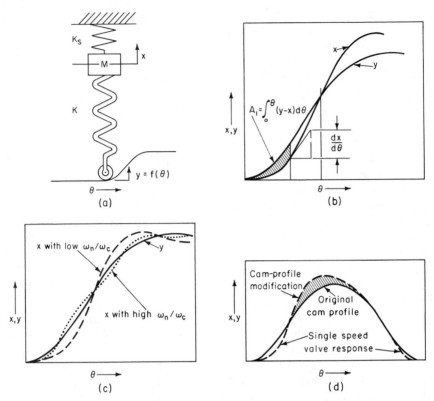

Fig. 12–16. Simplified representation of valve-gear dynamics. (*a*) Simplified system; (*b*) graphic interpretation of Eq. 12–7; (*c*) x versus θ at two values of ω_n/ω_c; (*d*) possible cam design to give smooth valve action at a single speed. $y =$ cam contour $= f(\theta)$, $x =$ position of mass M representing valve, $K =$ valve-gear stiffness, $K_s =$ spring stiffness, $\omega_n = \sqrt{k g_0/M}$, $\omega_c =$ camshaft angular velocity, θ/t (Taylor and Olmstead, 12.42).

After the x curve crosses the y curve, Fig. 12–16*b*, the cam follower may jump off the cam. Jumping will occur when the distance $y - x$ exceeds the expansion of the valve-gear parts (push rod, etc.) as they are relieved of the acceleration loads. While the gear is free from the cam, valve motion will be controlled by the valve-spring rate K_s and the mass M, and will oscillate at the frequency $(K_s g_0/M)^{\frac{1}{2}}$.

Part d of Fig. 12–16 shows how a cam contour might be modified so as to avoid jumping at one particular speed. The use of such a contour may be limited by undesirable effects at other speeds.

Valve-Gear Motion by Computer. By the use of computers, a more realistic model of an engine valve gear than that shown in Fig. 12–16 can be employed. Figure 12–17 shows false-motion amplitude curves from computer analysis. These curves reinforce the conclusions reached from Fig. 12–16 and show in addition the effect of different acceleration schedules

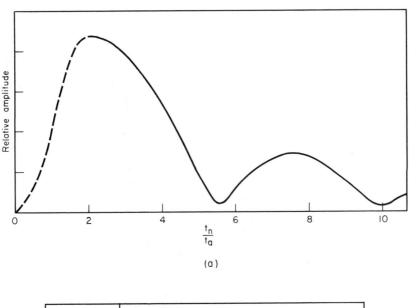

(a)

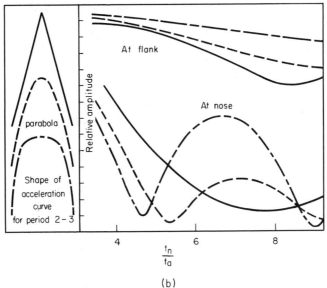

(b)

Fig. 12–17. Valve false motion. (*a*) Relative amplitude of false motion at nose of cam versus t_n/t_a; (*b*) effect of acceleration curve on false motion; $t_a =$ time of acceleration, period 2–3, Fig. 12–11; $t_n =$ natural period of valve gear $= 2\pi/\omega_n$ (computer results by Erisman, 12.43).

for the period 2–3. In actual practice, the acceleration curve will never contain sharp inflections, both because of local deflections at or near the cam-and-follower contact and because the physical shape of the cam surface cannot be held to limits close enough to make such detail differences in acceleration possible.

The important variable, as might be expected, is not the shape of the acceleration curve but rather the ratio of the natural period of the valve gear, $2\pi/\omega_n$, to the time of the acceleration period 2–3 of Fig. 12–11. Figure 12–17 shows maximum values of false motion at $t_n/t_a = 2$, with a lesser peak between 6 and 8. It is clear, then, that the basic problem of avoiding false motion of the valve gear is to avoid the condition where the acceleration time is close to one half of the natural period of the elastic system consisting of the valve and its operating mechanism.

Control of False Motion. As suggested by Fig. 12–16d, false motion at a given speed may be reduced by modifying the cam contour. By means of computer techniques, such modifications can be explored over the entire speed range, and a cam contour can be selected which will minimize false motion over the desired speed range.

In attacking this problem, the valve lift can be expressed either as a power function of time and crank angle, the so-called *exponential* cam, or as a continuous sine and cosine function, the *sinusoidal* cam. The result usually calls for a closing shape (4–5–6 of Fig. 12–11) somewhat different from the opening shape, 2–3–4. The reason for this difference is that vibratory motion is assumed to be zero at point 2 but may be considerable at point 4. In practice, cams so designed usually show an improvement over those designed on a rigid-body basis, at identical values of t_n/t_a. However, the importance of avoiding critical ratios of t_n/t_a is paramount and transcends any other factor in the control of false motion.

With engine speed and valve timing determined by other considerations, the designer has only limited control over t_a, the time of load application. The best he can do is to make the angles of cam acceleration and deceleration (periods 2–3 and 5–6 of Fig. 12–11) as large as possible, consistent with acceptable valve-spring loads.

With t_a determined, the designer must concentrate on reducing t_n, that is, on securing the highest possible natural frequency of the valve gear. Obviously, this end is accomplished by minimum mass and maximum stiffness.

Valve-Gear Stiffness. Important factors in promoting valve-gear stiffness include

1. Camshafts of generous diameter, with supporting bearings close to all cams
2. Stiff rocker-arm bearing supports
3. Push rods and rocker arms of generous dimensions, or
4. Elimination of push rods by means of overhead camshafts.

Valve-Gear Mass. The mass of the moving valve-gear parts can be minimized by the use of overhead camshafts and of multiple valves. In

Volume I, Chapter 11 we have seen that the natural frequency of similar systems is inversely proportional to the characteristic length. Assuming similar design, this means that t_n will be inversely proportional to $\sqrt{N}$, where N is the number of valves with flow area equal to one valve.

With a given cam contour, when stiffness is a maximum, mass a minimum, and spring force as high as practicable, the speed at which prohibitive false motion occurs will be as high as practicable for a given engine.

Literature on Valve-Gear Design. There exists a good deal of published material concerning cam design and valve-gear motion (12.40–12.49). The work includes mathematical models of typical valve-gear mechanisms, computer programs, and various suggestions for reducing false motion by special cam design. Much of this literature obscures the basic importance of the t_n/t_a relation, and in some cases calls for modifications of normal cam contours outside the limits of manufacturing accuracy, or impossible to maintain under conditions of normal wear.

Valve-Gear Details

Cam Contours. The curve of valve lift versus crank angle (Fig. 12–11) must be translated into the required cam contour, which depends on the design of the valve mechanism, including, of course, the shape of the cam follower. Flat followers (see Figs. 11–10 and 11–11) acting directly on the cam surface are the simplest and cheapest kind and have the advantage of being free to rotate in their guides. Actually the "flat" surfaces are often slightly curved so as to allow for any small misalignment with the cam surface. A disadvantage of flat followers is that pressures between cam and follower are limited to those at which the sliding surfaces show a satisfactory freedom from wear. This consideration places limits on spring loads lower than for roller followers. As mentioned in Chapter 9, a good material for these parts is a special chilled cast iron (12.30, 12.34, 12.35).

Where cam loading must be high, as in engines that run for long periods at high piston speeds, roller followers made of surface-hardened steel and operating on surface-hardened cams are called for (see Figs. 11–12, 11–15, and 11–16). Roller followers are also advisable where very long engine life is required, as in large marine and stationary engines.

Push Rods. These must be stiff in compression and also as a column. Steel tubing seems the most appropriate material. Ball ends and ball cups must be well hardened. See Figs. 11–10 and 11–11 for good proportions.

Rocker Arms. Rocker arms should be designed for stiffness of the arm itself, the rocker bearing, and the supporting structure. Good proportions are shown in Figs. 11–10 to 11–13. The use of stamped rocker arms (12.36) is now quite common in low-cost engines.

Piston Speed and Valve-Gear Design. In 4-stroke engines the maximum allowable piston speed is often limited by the valve-gear rather than by the crank-rod-piston system. This limitation will be lower as the stroke-bore ratio is reduced, since the valve and valve gear mass depend on the cube of the bore but is independent of the stroke. Thus, with a given cylinder-head and valve-gear design, at a given piston speed the stresses in the valve gear increase as the square of the bore-stroke ratio. The *characteristic valve-gear speed* shown in Tables 10–6 through 10–15 is $s(b/S)/\sqrt{N_v}$ where s is piston speed, b is bore, S is stroke, and N_v is the smallest number of valves of a given kind in one cylinder, either inlet or exhaust. The square of this speed is a relative measure of valve-gear stresses for the largest valve, in systems of similar valve-gear design.

Poppet Valve-Gear for 2-Cycle Engines. Poppet exhaust valves for 2-cycle engines must open and close in a much shorter crank angle (120° to 160°) than in the case of 4-cycle exhaust valves (220° to 280°, see Volume I, Fig. 7–20, p. 246). For engines designed to run at high piston speeds, this limitation makes multiple valves (usually four) advisable. Even with best possible design, the valve-gear limit on piston speed will be lower than that of a corresponding 4-cycle engine. However, 2-cycle piston speeds are also limited by scavenging problems, as explained in Volume I, Chapter 7, and therefore the valve-gear limitation is not usually critical. Some 2-cycle engines use only one exhaust valve per cylinder as in Fig. 11–21. This design puts a low ceiling on piston speed.

Valve-Spring Design

Materials. Steel with a carbon content of 0.50 per cent is the generally accepted material for valve springs. Both carbon and alloy steels are used, the latter being preferred where cost is not a primary consideration. (See Chapter 9, Fig. 9–24, and Table 9–6.)

Literature. Selected references on the subject of spring materials and spring design are given in refs. 12.50 *et seq.* The following discussion is a summary of present good practice.

Spring Loads. The minimum load, i.e., the load with valve closed, should be high enough to hold the valve firmly on its seat during the period when the valve is closed. In engines that are throttled, the exhaust valve must stay closed at highest manifold vacuum, and in supercharged engines the inlet valve must not be opened by the highest expected manifold pressure. The dynamic requirements are set by the required engine operating speeds, and may be calculated to the first approximation from the cam-lift curve, using rigid-body assumptions. The true jumping speed, i.e., the speed at which the follower leaves the cam at some point in the lift curve, will always be lower than the theoretical jumping speed, as previously explained.

Of all the various types of valve spring proposed and used over the past seventy years, the simple cylindrical helix or " coil " spring has proved the most reliable and economical. The coil spring using cylindrical wire is one of the few parts in an engine where stresses can be computed with a good degree of accuracy. The formulas are well known, but are repeated here for convenience:

$$\sigma = \frac{8PC^3}{\pi D^2}(Y) = \frac{\Delta}{N}\frac{G}{\pi dC^2}(Y); \tag{12-8}$$

$$\partial = \frac{\Delta}{N} = \frac{8PC^4}{GD} = \frac{\pi dC^2}{GY}\sigma; \tag{12-9}$$

$$k = \frac{P}{\Delta} = \frac{GD}{8C^4N}. \tag{12-10}$$

Symbols	Dimensions
σ = maximum shear stress in spring	FL^{-2}
∂ = axial deflection of spring, per coil	L
k = spring stiffness	FL^{-1}
P = load	F
$Y = \dfrac{4C-1}{4C-4} + \dfrac{0.615}{C}$	1
C = "spring index" = D/d	1
N = number of working coils	1
G = shear modulus of material*	FL^{-2}
D = pitch diameter of the coil	L
d = wire diameter	L
Δ = deflection of spring	L

*11.5 × 10⁶ psi (8.1 × 10⁵ kg/cm²) for usual spring steel.

Charts in the quick solution of these equations can be obtained from most spring manufacturers. The formulas assume cylindrical wire in combined torsion and bending, with no points of stress concentration.

Spring Surge. The equations given assume that the deflection of each coil is the same. This assumption does not take into account intercoil vibration, called *surge*, which is always present in more or less degree. With intercoil vibration, the maximum stress will be higher than the calculated stress in the ratio of actual to assumed deflection of the coils. It is obviously desirable to reduce the amplitude of intercoil vibration to a minimum.

The first-mode natural frequencies of intercoil vibration as a function of stress range and lift are plotted in Fig. 12–18. Increasing stress, increasing stress range, and decreasing lift will increase the spring frequency. Figure 12–19 shows the relation of the number of active coils to natural frequency with a given lift and maximum stress. Harmonic coefficients for a typical cam contour are shown in Fig. 12–20. The latter plot indicates that if the

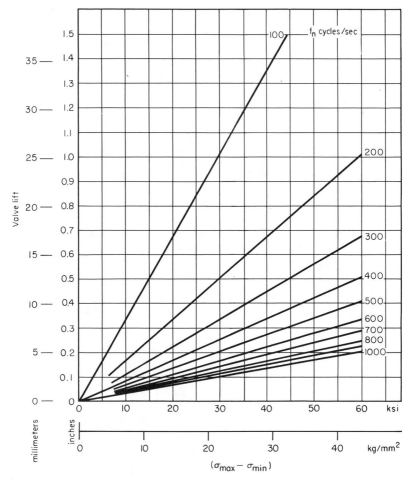

Fig. 12–18. First-mode natural frequency of helical springs, round wire; $Y = 1.15$; $\sigma =$ stress at surface of wire.

spring natural frequency is higher than 5 times the camshaft speed, spring vibration should be small.

Methods of securing a high ratio of spring frequency to camshaft frequency include minimizing valve lift, as by the use of multiple valves, as well as a high maximum stress and a low number of active spring coils, to give a high stress ratio. In aircraft-engine practice, spring-stress ratios as high as 2 have been successfully used. This method, together with close coiling, has substantially eliminated trouble from intercoil vibration. Other engine types have tended to use much lower stress ratios; this results in springs of many coils and can produce intercoil vibrations of objectionable amplitude. Such vibrations may be reduced by means of friction dampers of various kinds, or by nonuniform helix angles (12.56, 12.57). Close coiling,

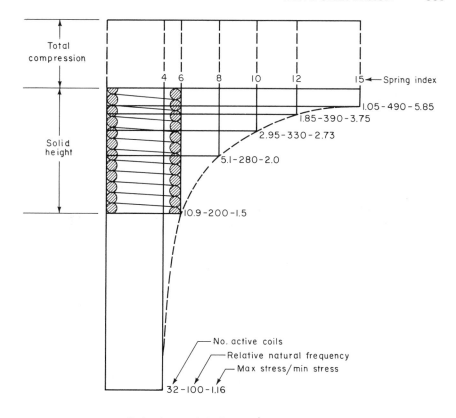

Spring index = pitch diameter / wire diameter

Fig. 12–19. Helical springs of equal stiffness and equal maximum stress under the same maximum load.

so that vibration amplitude is restricted when the valve is open, is an effective method of reducing spring vibration, but may not be feasible in many cases on account of noise or because of the accurate control of dimensions required for successful application.

It appears that the use of high stress ratios, i.e., springs of few coils, is advisable wherever possible. This method should be feasible except where the maximum spring load is limited by considerations of cam and cam-follower wear. With roller followers, high stress ratios are usually possible and desirable.

Spring Stress Limits. The valve spring is obviously subject to fatigue loading, and its stresses must be kept within the fatigue limit of the material (12.51–12.55). Figure 12–21 shows a Goodman type diagram for valve-spring steel, together with safe limits for the stress range. These limits assume high-quality springs, preferably shotblasted. By high quality is meant wire free from surface defects and excessive inclusions (see

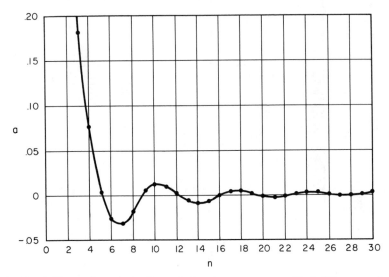

Fig. 12–20. Harmonic analysis of a typical engine cam. a = amplitude/lift; n = order of harmonic based on camshaft speed; $a = 0.48$ (lift) at $n = 1$.

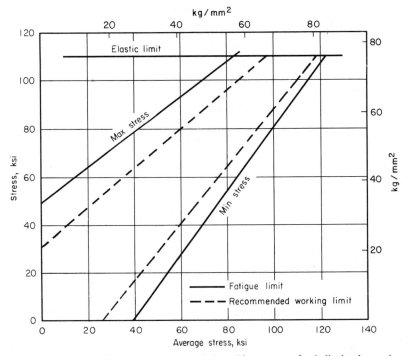

Fig. 12–21. Fatigue limits and recommended working stresses for helical valve springs.

Chapter 9). Many engines are using nominal stresses in the neighborhood of 80,000 psi maximum, 40,000 minimum (5600 and 2800 kg/cm², respectively).

Concentric Springs. Where space is at a premium, it may be advisable to use two or even three springs, one inside the other. In order to have the same stress and stress range in the inner springs as in the outer spring, the spring index D/d should be the same for inner and outer springs and the number of active coils in the springs should be inversely proportional to the diameter of the coil. Also, the free length of both springs should be the same. Under these conditions the solid height* of the two springs will be the same. Proper clearance must be allowed between concentric springs to prevent interference. Assuming a radial clearance between coils of one half the diameter of the inner spring wire, the load-carrying capacity of the combination, $P_1 + P_2$, divided by the load-carrying capacity of the outside spring, P_1, is a function of the spring index, D/d. This relation is plotted in Fig. 12–22.

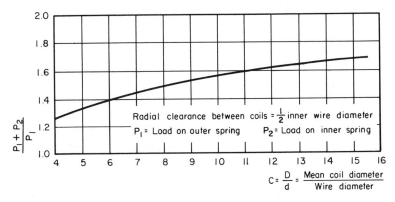

Fig. 12–22. Carrying capacity of two concentric coil springs compared to the capacity of a single spring. Springs have equal maximum stress.

Similitude of Springs. Springs are geometrically similar when the number of active coils and the value of D/d are the same. Geometrically similar springs have vibration frequencies inversely proportional to their linear dimensions.

High-Temperature Effects. Ordinary spring steels are likely to lose tension when operated at temperatures above 350°F (177°C). For operation at higher temperatures, an appropriate austenitic steel should be used (see Chapter 9).

Finish. Chronic failure of well-designed springs has usually been traced to surface defects, such as seams created during the drawing process.

---

* Solid height is the length of the spring when compressed to contact between all coils.

Shotblasting helps to reduce stress concentrations due to surface defects and should be specified for all high-grade engines. However, shotblasting will not cover up serious defects in the original wire, which must be free from surface imperfections or seams developed in the drawing process. It is very important to protect valve springs from corrosion by proper lubrication and ventilation of the valve-gear enclosure.

Valve-Spring Retainers. Since the diameter of helical springs increases slightly as the load increases, springs should be centered on their inside diameter, as indicated in Fig. 12–9.

Valve Gears Without Springs

There have been many proposals involving the use of mechanical arrangements which supply the forces ordinarily provided by the spring in the conventional valve mechanism. The usual objective of such mechanisms is to reduce false motion and thereby to attain a higher limit on operating speed, as compared with the conventional arrangement. Such devices always involve a considerable degree of mechanical complication, including arrangements for pulling as well as pushing on the valve stem, and a cam mechanism which can exert forces in two directions.

Actually it is not possible to eliminate spring action entirely from a poppet-valve mechanism, because an allowance must always be made for clearance and clearance variation, which can only be compensated for by some sort of spring. However, since the clearance is relatively small compared with the lift, such a spring device operates only over a relatively small travel and can be designed for a small stress range. In most designs this spring action is incorporated in some part of the mechanism, such as a shaft in torsion.

In practice it is almost always found that a conventional spring mechanism can be designed to function at the required speed, provided the basic design principles, already discussed, are understood and applied. There are very few, if any, successful engines that use any but the conventional valve-gear mechanism. Note that the engines with the highest rated piston speeds listed in Table 10–13 all use helical valve springs.*

GEARING

Under normal circumstances the only heavily loaded gears in internal-combustion engines are those involved in the power train, that is, gears, if any, between crankshaft and output shaft, or gears connecting the crankshafts of opposed-piston engines. There is a vast literature on gear design.

---

* The engine of Fig. 11–7 uses shafts in torsion instead of helical springs.

Selected items are included in the bibliography, refs. 12.601 *et seq.* The question of tooth geometry is well covered in this literature and will not be discussed here except in a general way. Only those gears that are built into the engine structure will be considered.

Plain spur gearing has been found adequate in most engine applications. Some aircraft engines have used a planetary arrangement of spur or bevel gears (10.844, 10.847), and some large engines have used herringbone gearing. A few opposed-piston engines have used bevel gearing or chains to connect the crankshafts, but most use plain spur gearing.

Tooth Form

The involute system is almost invariably used (12.608). The minimum number of teeth should be so chosen as to avoid *undercutting*, that is, reduction of tooth thickness below the pitch line. For the usual involute system, this minimum is 15 to 20 teeth. To reduce bending stresses, the *stub* form of tooth, that is, a relatively short tooth height-to-thickness ratio, is generally used (12.609). Ample fillet radius at the base of the tooth is essential, as indicated by Fig. 12–23 (see also ref. 12.612).

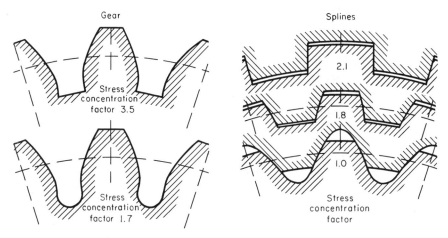

Fig. 12–23. Effect of fillet radius on stress concentration of gear teeth and splines in torsion.

Materials

For the type of gearing in question the tooth material (12.601–12.604) must be a high-grade casehardening alloy steel such as SAE 4320 or 5115, and the tooth surfaces must be well casehardened and smoothly finished. The smallest, or *pinion*, pitch diameter is determined by the chosen pitch and the minimum number of teeth considered appropriate for the pinion.

The tooth pitch should be large enough so that bending stress at the root of the pinion teeth is not critical, assuming proper fillets and finish at this point.

Tooth Loading

Except under conditions of unacceptable torsional vibration, the load on a gear tooth varies from zero to a maximum in one direction only. The maximum tooth load can be regarded as consisting of the theoretical average load, based on the mean torque, plus loads due to

1. Engine torque variation (Chapter 8)
2. Drive line torsional vibration
3. Gear-tooth geometric error.

With gears made by the best manufacturing processes, tooth geometrical error is so small as to be a negligible factor. Theoretical engine-torque variation is calculated by the methods discussed in Chapter 8. This leaves the question of loads due to torsional vibration of the system as the chief unknown factor in the loading cycle. For a new design these loads should be estimated and later measured on an experimental engine. Reference 8.400 discusses methods of estimating and measurement.

Tooth Stresses

Tooth stresses include stresses due to bending of the tooth as a cantilever beam and the stresses associated with the load on the contact surfaces. Such stresses will increase as the number of teeth decreases, and therefore the smallest gear (pinion) is critical. These stresses have in the past been computed by the well-known *Lewis* formulas for tooth bending and the *Hertz* formulas for compression stress at the junction of curved surfaces (12.611). Various modifications of these formulas, based on experimental stress analysis and endurance testing, are now generally used (12.613, 12.617).

Face Width

The wider a gear tooth, the more difficult it becomes to make sure that the tooth is loaded uniformly along its width. In general, face width of gears in the engine power train should not exceed half the pitch diameter. Wider gears may actually be weaker because of resultant poor load distribution along the contact surface. However, when this rule results in unduly large gear diameters, the face width of the pinion can be wider, provided all possible means are used to hold the axes in line and to avoid tooth distortion. In gas-turbine gears, where ratios as high as 10 are used in a single pair, pinions of a face width about equal to their diameter are generally used.

Pinion Size

For heavily loaded gears, the minimum number of teeth on a pinion will be in the range of 18 to 20. Having chosen this number and the face width to be employed, the diametral pitch, and therefore the diameter, will be determined by the assumed limitations on computed compressive and bending stress. Increasing the pinion diameter reduces tooth maximum stress and increases the probable life of the gearing. The best design will be an appropriate compromise between size and stress limitations.

Gear Support Structure

Once the pinion size is chosen, the size of the other gear or gears in the train is determined by the desired speed ratio, and the problem of design becomes one of the gear-wheel details and mounting structure. The mounting structure should be such as to ensure *accurate axial alignment of the gears under all loading conditions.*

Examples of good practice in this respect can be found in the large reciprocating aircraft engines, where minimizing of bulk and weight was a paramount consideration.

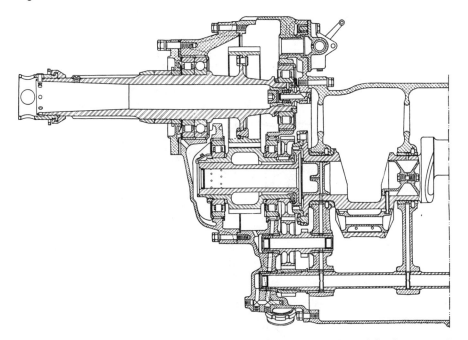

Fig. 12–24. Spur propeller-reduction gear of an aircraft engine. Note how pinion is supported independently of the crankshaft. (*Courtesy Ranger Engineering Corporation.*)

Figure 12–24 shows the spur reduction gear of a V-type aircraft engine, the notable features being as follows:

1. The pinion is not mounted directly on the crankshaft but is held in its own bearings and splined to the crankshaft in such a way that crankshaft bending motion is not transmitted to the pinion.

2. The gear-case structure, including the supports for all bearings, is such as to hold gear and pinion in strict alignment. Both pinion and gear are supported by bearings on both sides.

Figure 12–25 shows a planetary spur gear used on the radial engine of ref. 10.847. The pinion carrier is mounted in such a way as to allow it some freedom to move in response to the tooth loads and thus adopt a position of equal load distribution between the pinions.

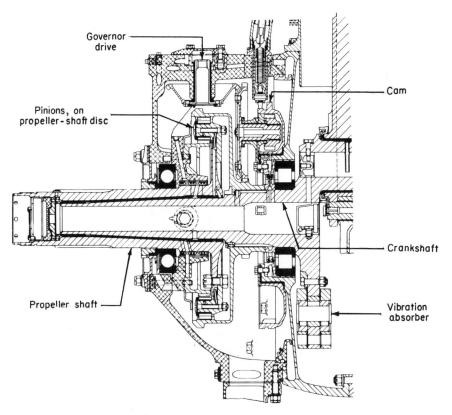

Fig. 12–25. Planetary propeller-reduction gear of the Wright R-1820 radial aircraft engine. Internal gear is mounted on crankshaft. External gear is stationary, splined to crankcase. Pinions are mounted on disc which drives propeller shaft.
(*Courtesy Wright Aeronautical Corp.*)

Gear Wheels

Gear wheels should have a stiff rim, at least twice the tooth depth in thickness. The whole wheel should be designed to minimize possible deflections that might lead to tooth misalignment.

As an example of extremely high-grade gear design, the following data for the propeller reduction gear of the Rolls-Royce *Merlin* airplane engine (10.845) should be of interest:

	Pinion	Gear
Diametral pitch	4	4
Number of teeth	21	44
Pitch diameter, in.	5.25	11
mm	133	280
Face width, in.	2.572	2.572
mm	66	66
Nominal tooth load,* lbf	16,000	16,000
kg	7280	7280
Hertz stress at tooth surface,* psi	230,000	—
kg/cm²	16,200	—

* Based on average torque at 2000 bhp, 3000 rpm, assuming all load on one tooth.

Figure 8–29, which applies to this same engine, shows torque peaks in 3000 rpm up to 25 per cent in excess of the mean torque. On the other hand, with a pinion of 21 teeth not all the load falls on one tooth. The calculated stress is therefore probably quite near to the actual stress, at least within the limits of the Hertz formulas. Experience with the Rolls-Royce *Merlin* showed that performance of the gear was extremely reliable and tooth wear very small during the life of the engine. For nonaeronautical applications, however, a more conservative design is recommended.

Camshaft Drives

The torque required for camshafts is relatively small, and gearing will in general be lightly loaded. The designer's problem here is one of convenient and economical arrangement rather than of load and stress analysis. Although spur gearing is the usual system, chain drives are common (12.701, 12.704) and even toothed belts have been used (12.703). Where noise suppression is important, as in passenger cars, the tendency is to use chains. Nonmetallic gears (not pinions) have also been used on the camshafts of engines of automotive size or smaller.

Auxiliaries that require a fixed relation to the crankshaft, that is, ignition systems and injection systems, may be driven by gears, chains, or toothed belts. Convenience of location, accessibility, and simplicity of the system, rather than load and stress considerations, are the basic design problems.

SUPERCHARGERS AND SCAVENGING PUMPS

Nearly all supercharging is now accomplished by exhaust-driven *turbo-superchargers*. These are manufactured for engines of all sizes above about 100 hp. Most supercharged engines are Diesel engines, for reasons already given. The only spark-ignition engines generally supercharged are large gas engines (Figs. 11–16, 11–17, 11–18) and those used for aircraft and for unlimited racing. One U.S. manufacturer offers a supercharged version of a sports car (10.293).

The question of supercharger performance and its matching to engine requirements is discussed in Volume I, Chapters 10 and 13. Scavenging pumps are discussed in Volume I, Chapters 7 and 10. Additional material in this area is presented in refs. 12.800 *et seq*. Space available here permits only a very limited discussion of the detail design of these elements.

Compressors

As explained in Volume I, Chapters 10 and 13, displacement-type compressors are suitable only for engine-driven installations with low pressure-ratio requirements. This type is most frequently used for scavenging 2-cycle engines.

When centrifugal superchargers are used for scavenging rather than for supercharging 2-cycle engines, they are driven through gearing from the engine crankshaft. This method of drive is also in general use for the now obsolescent large reciprocating aircraft engines. One of the problems in this type of drive is to avoid torsional vibration in the supercharger drive system, since this usually involves two elements with large inertia, coupled with a relatively flexible drive system. The effective inertia of the super-charger rotor is large because, relative to the crankshaft, its moment of inertia is equal to the moment of inertia of the wheel multiplied by the gear-ratio squared. With normal gear ratios in the range of 10, the result is a large number. For this reason, some kind of very flexible coupling is usually necessary to keep the natural frequency of the system below the operating range. Examples will be found in refs. 10.845–10.848.

Turbo-superchargers always employ compressors of the dynamic type on account of the possibility of direct connection between compressor and turbine. Nearly all are single-stage centrifugal, which permits pressure ratios up to 3.5 with acceptable efficiency (see Volume I, Fig. 10–9, p. 379). A very few engines have used multistage axial compressors, but these are much more expensive and have a narrower range of efficient operation. The detail design of centrifugal compressors is included in refs. 12.800 *et seq.* In choosing a size and design for a given engine, operation over the entire speed and load range must be considered. It is often necessary to sacrifice peak efficiency in favor of a wide operating range, as for example by using a vaneless diffuser. When the pressure ratio exceeds about 3.5, two-stage compressors will be required.

Exhaust Turbines

These can be of either the radial or axial-flow type. Detail design is covered in refs. 12.800 *et seq.*

The radial type is favored for engines of less than about 1000 bhp because of its low cost. Both types are in use for larger engines.

Blowdown versus Steady-Flow Turbines. Most exhaust turbines that use multiple exhaust pipes in an attempt to capture exhaust-blowdown energy should be classed as *mixed-flow* turbines, because during the period between blowdown pulses they operate with inlet pressure much higher than atmospheric pressure (see Volume I, Chapter 10, pp. 475–480).

For automotive applications that must operate over a wide speed-load range, and for small engines in general, it is seldom worth while to use a mixed-flow system, with its requirement for many individual exhaust pipes and nozzle boxes. Two-cycle automotive engines usually require an engine-driven scavenging pump to cover the required wide range of speed and load, whatever type of turbine is used. In such cases, the scavenging-pump capacity can be so chosen that a steady-flow turbine is adequate.

The mixed-flow type of turbine finds its widest use in constant-speed stationary engines and in marine engines where the speed-load range is limited by propeller characteristics. As indicated by Volume I, Fig. 16–11 (p. 479), a well-designed mixed-flow turbine will give an appreciably higher output than a steady-flow turbine at a given supercharger pressure ratio. In the case of 2-cycle engines, a mixed-flow system is usually required if a gear-driven scavenging pump is to be eliminated.

The design of mixed-flow turbines and exhaust systems still involves many unknown factors and should be the subject of further research and analysis (see Volume I, Chapter 10, and refs. 12.814–12.820). However, some very effective high-pressure-ratio systems now employ mixed-flow turbines, including most of the locomotive engines of refs. 10.401 *et seq.*

Many engine builders purchase their turbo-supercharger equipment from specialists in this area, who will furnish valuable advice on problems involving their use.

Compound Engines

Compound engines are defined as engines whose exhaust-driven turbine is geared to the load system rather than to the compressor alone. The best-known current example is that of the Wright Turbo-Compound Airplane engine (10.847). In Volume I, Fig. 13–10 (p. 477) shows the computed difference in performance between a 4-cycle engine with turbo-super-charger (curve no. 1) and the same engine with the turbine geared to the shaft (curve no. 2). These curves were based on a set of representative assumptions for Diesel-engine operation, and while they will not apply to all practical conditions, they serve to illustrate the point that the possible gain in performance due to gearing the exhaust turbine to the crankshaft may be quite appreciable. Whether or not the considerable added compli-cation and cost of the geared system are justified depends on particular cir-cumstances. To date, the compound engine is not used except in very special aeronautical or military applications (10.829, 10.847).

MANIFOLDING

The design of manifolding may have a large effect on engine perform-ance, as illustrated in Volume I, Figs. 6–27 (pp. 198–199), 7–22 (p. 252), and 7–23 (p. 254).

Inlet Manifolding

In carburated engines, inlet manifolding must be designed to distribute fuel and air to the various cylinders in as uniform a manner as is feasible. At the same time the manifolding must offer the minimum resistance to flow. General principles in securing these objectives are

1. Symmetry of piping from carburetor to cylinders, that is, branches of nearly equal length and area
2. Symmetry in time, that is, evenly spaced suction events on each car-buretor barrel
3. Highest feasible velocity, that is, smallest flow areas consistent with acceptable pressure loss through the system.

The result must be a compromise between good distribution and maximum air capacity, the latter requirement being in conflict with the high velocity required under 3.

In the case of engines using separate inlet pipes, or engines that inject the fuel at the inlet ports or in the cylinders, the requirement for high air veloc-ity is reduced and the manifold can be large enough so that pressure loss is negligible. Symmetry, or else very large inlet-manifold volume, is required to secure even air distribution. Many Diesel and gas engines sacrifice good air distribution in favor of simple or compact inlet manifolding. One extreme example of this is the long uniform pipe from one end of the cylinder block to the other, with air entrance at one end. The air distribution in such a manifold improves as the diameter of the manifold increases, so that velocities are low throughout its length. Two-cycle engines, being very sensitive to inlet-system pressure fluctuations, are usually provided with inlet manifolds or reservoirs of large capacity (Figs. 11–13, 11–15, and 11–17 to 11–21). Often the space between cylinders and in the V is employed as the inlet reservoir (Figs. 11–13, 11–15).

"Ram" Manifolds

Figures 6–26 and 6–27 (pp. 196–199) in Volume I show that considerable gains in volumetric efficiency at certain speeds can be obtained by proper choice of inlet-pipe length and diameter. Such pipes must either have separate air inlets for each cylinder or else be connected to a relatively large common pipe or reservoir. The large gains occur at certain ranges of speed and are often accompanied by losses at other speeds. Figures 6–26 and 6–27 of Volume I can be used as a design base for this type of inlet system. References 12.05 and 12.08 show effects of inlet-pipe geometry on crankcase-scavenged 2-cycle engines.

Exhaust Manifolding

Four-cycle engines are relatively insensitive to moderate variations in exhaust-system design, as long as high back-pressure is avoided. Reference 10.801 shows that power gains as high as 7 per cent are possible at certain speeds through the use of "tuned" exhaust pipes. Such pipes are relatively long and must not be obstructed by mufflers. Their use is largely confined to racing engines (10.801 *et seq*).

Two-cycle engines, on the other hand, are very sensitive to exhaust-system effects, as shown by Volume I, Figs. 7–22 and 7–23 (pp. 252 and 254). For most 2-cycle engines, exhaust ports should discharge into large reservoirs through the shortest possible passages. References 12.07 and 12.08 report some gains to be made by means of exhaust piping for 2-stroke engines with crankcase compression. Two-cycle engines with displacement-type scavenging pumps will accept more impediments to flow in the exhaust system than will engines using centrifugal scavenging pumps. This result comes from the fact that the delivery rate of a positive pump is much less affected by delivery pressure than is the case with a centrifugal blower.

This relation can be seen by comparing the slope of the curves of air flow-versus-pressure ratio at a given speed (Volume I, Fig. 10–5, p. 370).

As already mentioned, exhaust systems for use with mixed-flow turbo-superchargers require multiple piping of very special design.

IGNITION SYSTEMS

Ignition systems cannot be considered here in detail. Only some important general observations will be offered.

Spark-plug location and number are important design decisions. Chapters 1 and 2 of this volume discuss the effects of these variables on combustion and detonation. In the case of stratified-charge engines or large gas engines with gas injection, spark-plug location is critical and must be worked out experimentally.

Spark-plug design is also critical because it controls the delicate balance between preignition and fouling and may therefore have an important influence on reliability and maintenance. Some degree of experiment, under guidance of a reliable manufacturer, is usually necessary to achieve good results. For example, different types of service may require different spark plugs in the same engine model.

Small engines, including automotive gasoline engines, usually find the conventional breaker-coil-distributor system with storage battery quite adequate. Beginning in the 1960's, transistor-type systems have been developed, and their use is increasing. Whether these offer sufficient advantages to justify their probable higher cost is not yet clear. Many large engines use low-tension distribution systems with high-tension transformers (coils) located on or near the individual cylinders. This system is also used on some large airplane engines.

Since ignition systems are supplied by specialists, they present no serious problems to the engine designer except for the location and proper choice of spark plugs.

INJECTION SYSTEMS

Injection systems have been discussed in Chapter 7 of this volume. Detail design is usually handled by specialists. A few Diesel-engine builders design and manufacture their own injection systems.

Most large gas engines employ injection systems of the *common-rail* type, i.e., timed valves that open into each cylinder, supplied with gas from a manifold held at a controlled pressure. The result is usually a partly mixed cylinder charge at ignition. Experience shows that detail design, location, and timing of such valves may have an important effect on combustion, including regularity of firing. Achieving optimum results usually requires a period of experimentation with injection-valve location, design, and timing.

COOLING SYSTEMS, LIQUID

The principles of engine cooling are discussed in Volume I, Chapter 8. Water passages and jackets have been discussed by implication under cylinder and cylinder-head design. The approximate over-all rate of flow required can be computed from Volume I, Fig. 8–11, with allowances for the effects of combustion-chamber type and of cooling by air blown through the combustion chamber during the overlap period. Some typical rates of water-flow volume to piston displacement per unit time are: passenger-automobile engines 0.0005 to 0.002; supercharged Diesel engines 0.005 to 0.010. Here, displacement per unit time is taken as piston displacement multiplied by crankshaft rpm.

The references on liquid-cooling system design, 12.910 *et seq.* should be found helpful. Important general principles to be observed are as follows:

1. Cylinder-head water passages should be so designed as to achieve rapid circulation over the exhaust ports and exhaust-valve seat areas. Less rapid circulation can be tolerated around the inlet ports and cylinder barrels.
2. Pockets where vapor can collect must be studiously avoided.
3. All water passages should be protected against corrosion, unless anti-corrosive coolant additives are to be used at all times.
4. Heat rejection to the cooling system is decreased by:
   a. Highest practicable inlet and outlet coolant temperatures.
   b. As little jacketed surface as possible exposed to the exhaust gases.
   c. Compact, open combustion chambers (12.910).
   d. Large valve overlap for supercharged engines, allowing blow-through of inlet air during the overlap period. Overlaps up to 150 degrees crank travel are now common (12.910).

Water pumps are usually of the centrifugal type, although many marine engines use gear-type pumps. The shaft seal must be designed so as to avoid undue friction and wear. Commercial seals for this particular purpose are usually satisfactory, if chosen with the advice of the manufacturer.

COOLING SYSTEMS, AIR

Basic principles of cooling are discussed in Volume I, Chapter 8, and those of cylinder design in Chapter 11 of this volume. References 12.920

et seq. cover research in cooling-fin and cooling-system design and perform-ance. With a given engine design, principles to be observed in the design of the air-flow system are as follows:

1. Air-flow system should be designed such that all air from the fan is confined so as to pass between the fins. Air that does not pass between the fins is wasted as far as cooling is concerned.
2. The air-duct system should be designed for minimum pressure drop.
3. Fans should be of high efficiency.

A study of the cylinder and cooling-system design of the large air-cooled radial aircraft engines should be helpful to the designer of any air-cooled engine. See also Figs. 11–6, 11–7, and 11–8.

LUBRICATION SYSTEMS

Chapter 9 in Volume I discusses theory and practice of engine lubrica-tion. Supplementary discussion of the lubrication of pistons and bearings has been offered in Chapter 11 of this volume, in connection with power-train design. Full pressure feed to all bearings in the power train is the accepted practice for all except small 2-cycle engines lubricated by oil in the gasoline. Most other plain bearings, including auxiliary drives, are best pressure lubricated.

Oil Pumps

The gear-type pump or some similar type is usually the best choice because, like the displacement compressor, its flow rate tends to be inde-pendent of the outlet pressure. With a given flow rate, this pressure will vary with wear in the engine and condition of the oil filters. The flow rate required depends on detail engine design as well as on size and speed of the engine. One important factor is the question whether oil cooling of the pistons is used. Typical oil-flow rates for a Diesel engine with cooled pistons and with new bearings are 0.0025 to 0.0035 ratio volume of oil to piston dis-placement $\times$ rpm. To allow for wear and emergencies, pump capacity should be double these values. Pressures used at bearing feed points are usually between 50 and 100 psi (3.5 and 7 kg/cm^2).

Oil Filters

Effective oil filtration is of course essential in all engines. For large, long-life engines, very elaborate filtration or even centrifuge systems may be justified. In automotive engines, full-flow filtration is now preferred over the by-pass filtration formerly used. Oil-filtration equipment is usually supplied by specialists in this field, whose advice should always be solicited.

AUXILIARIES

Auxiliaries are taken to include starters, electric generators, hydraulic pumps, and other equipment mounted on the engine but not essential to its operation. These items are usually purchased from specialized manufacturers, and the engine designer's problem is to choose appropriate types and sizes and to mount and drive them so that they are properly accessible and so that the drive-train mechanisms are simple, reliable, and of minimum cost. In choosing types and sizes, consultation with the manufacturers is essential.

Auxiliary Drives

Standardized shaft ends and mountings for the auxiliaries used in U.S. automotive practice are published in the SAE Handbook. In designing the drive systems, attention must be given to the question of torsional vibration, especially with units of considerable inertia, such as electric generators. Electric generators driven by the engine are required for all automotive engines, and for many others. The size and type depend on particular requirements.

Starters

Most small- and medium-size engines are started electrically. Starting motors are usually supplied by specialists who will furnish information on types, sizes, and mountings (see ref. 9.00 for standards for starting motors, starter drives, and flywheel ring gears for automotive engines).

Large Diesel engines are usually started by means of compressed air supplied to each cylinder through timed valves. It is outside the scope of this volume to discuss detail design of these systems. Reference to the literature concerning large marine and stationary Diesel engines is advised.

Air Filtration

Some degree of air filtration is required for all types of engines. The type and quality of the equipment required depend on the concentrations of foreign matter expected in service. This problem is now handled by the specialists who manufacture such equipment, in consultation with the engine designer. As shown by Fig. 11-11, the air filter, which is also an intake silencer, may be an item of considerable size.

GASKETS AND SEALS

Gaskets and seals are an essential part of all engines and must be properly designed and installed.

Cylinder-Head Gaskets

For small and automotive-size spark-ignition engines, these gaskets are usually of the copper-asbestos type. Such gaskets seal both the cylinder head proper and the water passages from head to block (12.714). Some Diesel engines of less than 6-in. (152-mm) bore also use this type of gasket. The success of such gaskets depends heavily on arranging the cylinder-head bolting system so as to ensure that all areas of the gasket are equally compressed.

For cylinders above about 6-in. bore, and for some even smaller Diesel engines, a metal ring gasket is used to seal each combustion chamber. Separate soft gaskets of *neoprene*, or some similar synthetic rubber are used to seal the water passages between head and block structure (12.711). Metal gaskets may be of copper, aluminum, or soft steel (12.710, 12.715, 12.717). They should be accurately located by suitable grooves or pilots. The success of such a gasket depends on securing an adequate, uniform bolt pressure on its entire circumference, on good accuracy of dimensions of the gasket and of the surfaces on which it fits, and on proper control of surface smoothness.

Water Seals

For sealing the water spaces of wet cylinder lines and for other water-passage seals, the O-ring of artificial rubber is widely used (12.711). In designing for O-rings it should be remembered that rubber does not compress easily and that, when assembled, the cross-sectional area of the groove should be a little larger than that of the ring. Sealing is achieved by distorting rather than by compressing the rubber. Properly chosen and installed, O-rings are very effective *static seals*, that is, seals between parts that have no relative motion.

Seals for Moving Parts

Seals for moving parts, such as water pumps, shafts, piston rods, and supercharger rotors, present special problems for which rubber O-rings are not suitable. Effective water-pump seals are now available from specialists. Supercharger-rotor sealing is discussed in some of the references on supercharger design and in ref. 12.810. Piston rods are used only in double-acting engines, which are now obsolete.

Crankcase Sealing

Where two major structural elements of a crankcase are bolted together, good practice calls for smoothness and accuracy of the mating surfaces such that no gasket will be required. Where a shaft projects through the crankcase, an oil-slinger disc, plus a soft gasket or a helical return groove, are called for (Figs. 11–20, 11–22). Alternatively, commercial seal assemblies (Fig. 11–6) or metallic split rings (Fig. 11–9) may be used. Oil seals for projecting shafts are discussed in ref. 12.716.

Soft Gaskets

Gaskets made of cork, asbestos, felt, paper, etc., may be used for oil pans and for preventing oil leakage around the mounting flanges of water pumps, oil pumps, ignition apparatus, starters, etc. (12.712). Such gaskets are not suitable for sealing against high pressure or in locations exposed to high temperature.

Space limitation prevents more detailed discussion of sealing problems. It should be realized, however, that seal design is an important element in successful engine design.

OVER-ALL DESIGN CRITERIA

A complete engine design can be evaluated and compared with competing designs by means of a number of important criteria.

Engineering Criteria

To be of most value, these criteria should be basically independent of cylinder size. Valuable parameters in this category are

1. Thermal efficiency over useful load and speed range
2. Mean effective pressure
3. Piston speed
4. Specific output (power per unit of piston area)
5. Weight per unit of displacement (Fig. 12–26)
6. Over-all or "box" volume per unit of displacement (Fig. 12–26)

Items 1 through 4 have been the subject of intense discussion throughout both volumes of this work and need no further comment here. Figure 12–26 and Tables 10–5 through 10–16 as well as Volume I, Fig. 11–5 (p. 407), show that weight per cubic unit of displacement is substantially independent of cylinder bore and need not exceed 4 lbm/cu in. (112 kg/liter) for any type of engine. Diesel engines average about 3 lbm/cu in. (84 kg/liter) and spark-ignition engines, not including gas engines, average about 2.5

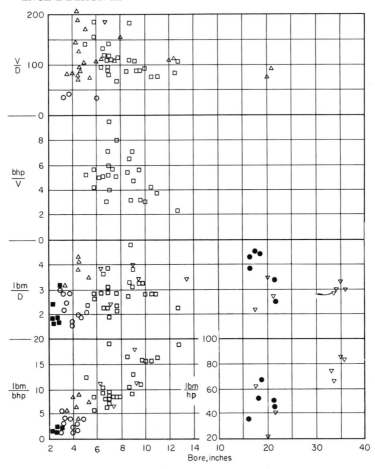

Fig. 12-26. Over-all engine characteristics.

bhp = maximum rated horsepower	V/D is dimensionless
D = displacement volume, cu ft	■ small engines
lbm = weight of engine	○ automotive spark-ignition engines
V = box volume, length × width × height, cu ft	△ automotive Diesel engines
	□ locomotive Diesel engines
bhp/V = bhp per cu ft box volume	▽ industrial and marine Diesel
lbm/D = pounds weight per cu in. displacement	engines
lbm/bhp = pounds weight per bhp	● gas engines

lbm/cu in. (70 kg/liter). The large aircraft reciprocating engines weighed about 1 lbm/cu in. (28 kg/liter) of displacement.

A criterion very valuable to the designer is the ratio of over-all, or "box," volume to the displaced volume of an engine. The data given in Fig. 12–26 are based on published figures of over-all length, width, and height, and box volume is taken as the product of these three numbers. A serious difficulty here is that there is no uniformity in defining these dimensions, so that in some cases projecting auxiliaries are included and in some cases not. Indications are that ratios of less than 100 are possible in most cases.

User's Criteria

The foregoing engineering criteria do not directly concern the user. For example, an engine with a very high specific output may have comparatively large weight and bulk and may involve excessive cost in proportion to its output. An engine of high efficiency may require a very expensive fuel.

From a practical point of view, the user of an engine should be concerned with the following parameters:

7. Over-all weight per rated horsepower (Fig. 12–26)
8. Rated horsepower per unit box volume (Fig. 12–26)
9. Installed cost per rated horsepower
10. Cost of fuel and oil per operating hour
11. Cost of maintenance and supervision

As we have seen, criteria 7 and 8 increase rapidly with cylinder size and vary both with type of service and with rating policy. They may be usefully compared only in connection with a particular type of service.

Installed cost depends not only on the cost of the engine but also on costs of such foundations, housing, auxiliary equipment, etc., as are required for a particular service. Experience is the chief guide in making estimates of these factors.

As to cost of fuel and oil, the unit cost of these items as such is predictable for the near future on the basis of current trends. Over-all fuel costs, however, depend both on engine efficiency throughout the operating range and on the time of operation at each load and speed. With a given fuel, of course, the higher the efficiency, the lower the fuel cost.

Oil consumption per unit of power output tends to be higher with 2-cycle than with 4-cycle engines because of the inevitable loss of oil through piston-controlled ports. Usually, however, oil cost is a minor item.

Cost of maintenance depends on engine reliability and wear rate under service conditions. Maintenance and reliability considerations tend to set limits on specific output ratings for each particular service. As we have seen, for a given set of service conditions, the larger the cylinder bore, the slower the effective rates of wear of the various engine parts. Actual costs can be estimated only on the basis of previous experience in a given type of service.

Cost of supervision depends on the confidence of the user in the reliability of the installation, as well as on labor costs. Most small engines can go unsupervised for long periods, while large engines (more than about 12-in. bore) are usually considered to require rather continuous surveillance. Figure 13–9 gives some interesting data on estimated 1966 costs of Diesel engines and gas turbines.

Table 12–2. Engine Poppet-Valve Materials

SAE No.	C	Mn	Si	Cr	Ni	Co	W	Mo	Fe, max	Other	Similar Commercial Designations
Intake Valves											
Constructional											
NV 1 (SAE 1041)	0.41	1.50	0.25	—	—	—	—	—	—	—	—
NV 2 (SAE 1047)	0.47	1.50	0.25	—	—	—	—	—	—	—	—
NV 3	0.50	0.80	0.30	0.40	0.30	—	—	0.15	—	—	NE 8150
NV 4 (SAE 3140)	0.40	0.80	0.30	0.65	1.25	—	—	—	—	—	—
NV 5 (SAE 8645)	0.45	0.90	0.30	0.50	0.55	—	—	0.20	—	—	—
NV 6 (SAE 5150)	0.50	0.80	0.30	0.80	—	—	—	—	—	—	—
High alloy											
HNV 1	0.55	0.40	1.50	8.0	—	—	—	0.75	—	—	Sil 2
HNV 2	0.40	0.30	3.9	2.2	—	—	—	—	—	—	Sil F
HNV 3	0.45	0.40	3.3	8.5	—	—	—	—	—	—	Sil 1
HNV 4	0.45	0.40	3.3	7.0	1.0	—	—	0.50	—	—	731
HNV 5	0.35	0.40	2.5	13.0	8.0	—	—	—	—	—	CNS
HNV 6	0.80	0.40	2.3	20.0	1.3	—	—	—	—	—	XB
HNV 7 (SAE 71360)	0.55	0.20	0.20	3.5	—	—	14.00	—	—	—	—
Exhaust Valves											
Martensitic											
HNV 3	0.45	0.40	3.3	8.5	—	—	—	—	—	—	Sil 1
HNV 6	0.80	0.40	2.3	20.0	1.3	—	—	—	—	—	XB
Sigma phase											
EV 1	0.45	0.50	0.50	23.5	4.8	—	—	2.8	—	—	XCR
EV 2	0.40	4.3	0.80	24.0	3.8	—	—	1.4	—	—	TXCR
Austenitic											
EV 3	0.20	1.3	1.0	21.0	11.5	—	—	—	—	—	21-12
EV 4	0.20	1.3	1.0	21.0	11.5	—	—	—	—	N-0.15-0.25	21-12N
EV 5	0.38	1.0	3.0	19.0	8.0	—	—	—	—	—	Sil 10

Table 12-2 (*continued*)

SAE No.	C	Mn	Si	Cr	Ni	Co	W	Mo	Fe, max	Other	Similar Commercial Designations
Exhaust Valves, *continued*											
EV 6	0.38	1.0	3.0	19.00	8.0	—	—	—	—	N-0.15-0.25	Sil 10N
EV 7	0.20	5.00	0.50	21.00	4.50	—	—	—	—	N-0.30	2155N
EV 8	0.53	9.00	0.15	21.00	3.75	—	—	—	—	N-0.42, S-0.07	21-4N
EV 9	0.45	0.50	0.60	14.00	14.00	—	2.40	0.35	—	—	TPA
EV 10	1.00	0.8	3.0	14.5	14.5	—	—	—	—	—	CAST 14-14
EV 11	0.70	6.3	0.55	21.0	1.9max	—	—	—	—	N-0.23	B-312
Super alloys											
HEV 1	0.10	1.50	0.50	21.30	20.00	20.00	2.5	3.00	—	Ni-0.15 Cb and Ta-1.00	N-155
HEV 2	0.05	2.30	0.05	16.00	Base	0.50	—	—	—	Al-0.05, Ti-3.10	TPM
HEV 3	0.05	0.60	0.30	15.00	Base	0.70	—	—	—	Al-0.70, Ti-2.50	Inconel X
HEV 4	0.10	1.50	0.8	20.00	10.00	Base	15.00	—	—	Cb and Ta-1.00	HS 25
HEV 5	0.05	1.0	0.6	20.00	Base	2.0max	—	—	5.0	Al-1.2, Ti-2.5	Nimonic 80A
HEV 6	0.05	1.0	1.5max	20.00	Base	18.00	—	—	—	Al-1.2, Ti-2.5	Nimonic 90
Facings											
Alloy overlays											
VF 1	0.20	0.80	0.20	20.00	Base	—	—	0.50	1.0	—	80-20 Ni-Cr
VF 2	1.20	0.5	1.20	28.00	3.00	Base	4.50	—	3.0	—	Stellite 6
VF 3	2.40	—	0.70	29.00	39.00	10.00	15.00	—	6.5	—	Eatonite
VF 4	2.00	0.30	0.30	26.00	Base	0.50	8.75	—	4.0	—	X-782
VF 5	1.75	0.30	1.00	25.00	22.00	Base	12.00	—	2.0	—	Stellite F

This table is intended only to supply information on typical valves and is not a complete list of all materials used for valve steels.

Recommendations

Martensitic steels for exhaust valves below 1200°F.

Sigma phase steels for exhaust valves to 1400°F. Stems are hardenable for good wear resistance.

Austenitic for exhaust valves up to 1600°F. Often used with facing materials on seat surfaces.

Super alloys. For gas-turbine blades and very "heavy-duty" exhaust valves.

From SAE handbook, ref. 9.00.

13 | Future of the Internal-Combustion Engine. *Comparison with Other Prime Movers*

The reciprocating internal-combustion engine is at the present time (1967) without serious competitors in road and railroad transportation and for small power plants for home use, including outboard motorboat engines, lawnmower engines, etc. It is also without serious competition in the marine field for conventional boats and ships under about 1000 tons,* and even for larger ships it is now much more widely used than steam (Fig. 13–1). Ships use Diesel engines of over 30,000 hp per unit, as shown by Table 10–11. For this type of service, the advantage of Diesel over steam plants in regard to fuel economy is illustrated in Fig. 13–2.

Except for very large installations (where gas turbines are sometimes used), the reciprocating engine is also dominant for all classes of portable commercial power, including contractors' machinery, roadbuilding machinery, mining and oil-well machinery, and agricultural machinery.

Electric-power stations of less than about 10,000 kW are almost universally Diesel powered, and there are many much larger Diesel-electric installations. However, gas turbines are beginning to be used for peak and standby power as the capacity of existing stations is increased or new stations are established.

As mentioned in Volume I, Chapter 1, the total power of reciprocating internal-combustion engines now in use exceeds that of all other prime movers combined by at least one order of magnitude (ten times). Most of this installed power is in the form of spark-ignition engines of less than 250 hp for motor vehicles. Road transportation depends entirely on reciprocating engines and accounts for about one third of the total energy consumption of the world. The question to be discussed in this chapter is the probability of changes in the foreseeable future.

The reasons for the predominance of reciprocating engines in the services mentioned are centered around their major characteristics as compared with those of other types. These are

---

* Many unconventional vessels, such as hydrofoil and air-cushion boats, and small high-speed naval vessels, are now using gas turbines where reciprocating engines would formerly have been used.

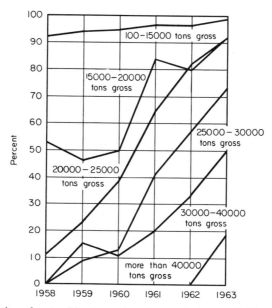

Fig. 13–1. Proportion of motorships in number of annually launched ships (Sorenson, 13.04).

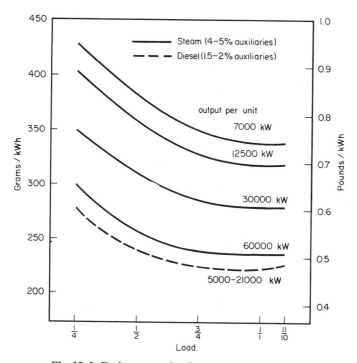

Fig. 13–2. Fuel consumption for power stations (13.04).

Lighter weight (except for gas turbines)

Smaller size (except for gas turbines)

Lower cost (except in large units)

Better fuel economy (see Fig. 13-2, for example)

Simplicity of operation and control

Flexibility (ability to operate over a wide range of speed-load combinations).

Possible future competitors of the conventional reciprocating internal-combustion engine may be taken to include

Unconventional displacement engines

Gas turbines

External-combustion engines (steam or other working fluids)

Electric motors powered by batteries, fuel cells, thermoelectric devices, magnetohydrodynamic devices, etc.

Nuclear reactors acting through external-combustion engines or by electrical energy conversion, with electric motors as the final power.

MODIFICATIONS TO THE CONVENTIONAL TYPES

The question of engines with unconventional valves was discussed in Chapter 12. There seems little likelihood of general adoption of any of these arrangements.

Stratified-charge engines, discussed in Chapters 1 and 2, may be classed as unconventional. As stated there, the difficulties of securing good charge stratification and the marginal nature of the results so far obtained give little ground for predicting any widespread acceptance of this idea.

Variable compression-ratio devices, especially for Diesel engines, have also been mentioned (11.414, 11.415). While the advantages of variable compression ratios are definite, the question is whether they justify the necessary increase in mechanical complexity and chance of failure. A military Diesel engine with a variable compression ratio is now being produced (11.414). Your author believes that the use of such devices will be restricted to special applications, of which the foregoing is an example.

There has been considerable interest since 1980 in the so-called "adiabatic" Diesel engine. The object is to reduce the rate of cylinder cooling so that the combustion chamber can operate in the range of 1200°F (649°C) in order to reduce heat loss and to increase exhaust temperature for use in supercharging. Ceramic materials would be used for combustion chamber walls (including pistons and valves). Even if such an engine should prove mechanically possible, there would still be a serious question whether the gain in efficiency would be sufficient to justify the added expense, danger, and inconvenience of a "red-hot" engine.

UNCONVENTIONAL DISPLACEMENT ENGINES

The patent offices of the world are loaded with ideas for replacing the conventional connecting-rod–crankshaft system with devices thought to be superior from one point of view or another. To date none has been wholly successful, as will appear in the following discussion.

Cams, Levers, Scotch Yokes, etc.

Engines with various types of cams or levers introduced between the piston and output shaft have been proposed in many forms (13.21 *et seq.*). The only type that has achieved even a limited commercial application is the single-crankshaft opposed-piston type with rocker arms, a recent example of which is shown in Fig. 13–3*a*. This arrangement changes the shape of an opposed-piston engine to a form which in certain applications may be more convenient than the more usual two-crankshaft arrangement shown in Fig. 11–24. It allows scavenging pistons to be incorporated in a compact manner, but increases the number of heavily loaded bearings. The engine shown in Fig. 13–3*a* is designed for military use and is experimental as of 1966 (13.01). A few similar arrangements have been used in farm tractors. The type does not appear to have an important place in future applications.

Another type in this class, which came near to success, was the 4-cylinder Fairchild-Caminez radial engine illustrated in Fig. 13–3*b*. It is evidently somewhat more compact than a conventional radial engine with the same number and size of cylinders. Built and successfully endurance-tested in 1927, it received U.S. Dept. of Commerce type certificate No. 1, the first engine so certified (13.24). Flight tests revealed what should have been predicted: a very large fourth-order torque variation due to the unusually heavy piston assemblies with their large ball-bearing rollers. Since the cam had two lobes, the second-order inertia torque of the conventional 4-cylinder engine (see Chapter 8) becomes fourth order in the arrangement in question. The engine was abandoned on this account.

Barrel or Revolver Engines

These are piston engines with cylinders arranged parallel to the main shaft and acting on it by means of some kind of cam or "wobble-plate" mechanism. Examples are described in refs. 13.25 to 13.292, and a typical mechanism is shown in Fig. 13–3*c*. The objective of such designs is to secure a more compact arrangement than is possible with a conventional crank-and-rod mechanism. Many such engines have been built and tested, but few have been found sufficiently rugged to withstand normal piston speeds

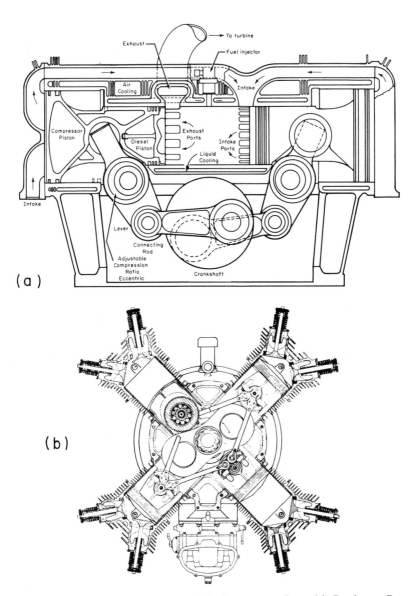

Fig. 13–3. Three types of unconventional displacement engines: (*a*) Southwest Research Institute opposed-piston engine-compressor unit designed for use with an exhaust turbine geared to output shaft (13.01); (*b*) Fairchild-Caminez cam engine for airplanes (13.24); (*c*) moving parts of Alfaro barrel-type engine (13.26).

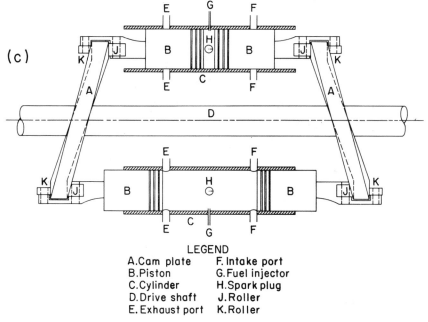

(c)

LEGEND
A. Cam plate   F. Intake port
B. Piston   G. Fuel injector
C. Cylinder   H. Spark plug
D. Drive shaft   J. Roller
E. Exhaust port   K. Roller

Fig. 13–3 (*continued*)

and mean effective pressures for any but short periods. A marine Diesel engine of this type was put on the market in 1945 but was not a commercial success (13.27). Such cylinder arrangements have serious disadvantages with regard to accessibility and mounting structure, which would make them undesirable for most services even if a reliable mechanism could be developed. There is no likelihood of such engines becoming important competitors to the conventional types.

Toroidal Engines

A great many ideas for engines in which toroidal pistons rotate or reciprocate within toroidal cylinders have been advanced (13.30, 13.31). The difficulties of connecting such pistons to the output shaft by a simple and reliable mechanism, together with the problem of sealing the sliding surfaces involved, make such ideas little more than amusing adventures in ingenuity.

Rotary Displacement Engines

These are engines in which a rotating member is arranged to vary the working volume in a manner similar to that of the vane-type compressor (13.46) or by some kind of eccentric motion of a rotor within a cylindrical

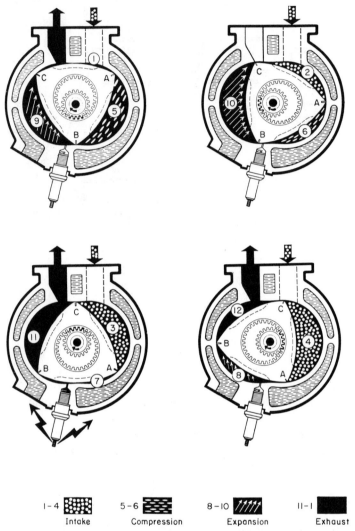

I-4 ▦ Intake   5-6 ▧ Compression   8-10 ▨ Expansion   11-1 ■ Exhaust

Fig. 13–4. The Wankel rotary engine (13.40–13.44).

(usually noncircular) space (13.40–13.44). Many engines of this category have been built, tested, and finally abandoned, often after the expenditure of vast sums of money.

The most difficult problem in such engines is that of sealing the combustion chamber against leakage without excessive friction and wear. This problem is far more difficult than that with conventional piston rings, for the following reasons:

1. "Line contact" rather than surface contact is usually involved.
2. The surfaces to be sealed are discontinuous, with sharp corners.

3. The velocity of the seal is high during the high-pressure portion of the cycle, in contrast to piston rings whose velocity is near zero at maximum cylinder pressure.

The only engine of this type that has been developed to the point of quantity production is the Wankel, shown diagrammatically in Fig. 13–4. It is now used in a line of sports-type cars by Mazda of Japan, whose main production is conventional engines. The Wankel engine is indeed smaller and lighter and has less vibration than conventional engines of the same output. There is no evidence that it is cheaper to produce. The sealing problem seems to have been solved as far as reasonable durability is concerned, but there is evidence of considerable seal leakage. This defect and the attenuated shape of the combustion chamber are responsible for poorer fuel economy as compared with the equivalent conventional engine.

The Wankel engine has been considered, but not yet adopted, by other car manufacturers, and also for some other applications. A few fairly large gas engines of this type have been put into service on an experimental basis. Although development of this type is still under way, the Wankel's future is still uncertain. It does not appear to be likely to become a serious competitor to conventional engines.

Free-Piston Engines

Free-piston engines supplying hot gas to a power turbine are discussed at some length in Volume I, Chapter 13. After an initial impetus led by Sigma of France (13.504–13.506) and the General Motors Corporation under Sigma license (13.507–13.510), there was a period of intense discussion and development of this type of power plant in the late 1950's, but experience since that time has greatly reduced expectations for its commercial success. The U.S. Shipping Board installed a free-piston plant in a merchant ship which was tested in competition with similar ships powered by steam, gas-turbine, and Diesel engines. The installation was not considered successful and has been withdrawn from service (13.510).* Trials of free-piston power plants in automobiles (13.507), tractors (13.513), and railroad locomotives (13.505) have also been unsuccessful. The French Navy has a number of small vessels with this type of power supply (13.504), but other navies are not following suit.

---

* One of the serious difficulties was gas-column surging in the hot-gas piping system, with resultant noise and pipe breakage.

The free-piston power plant can be regarded as a Diesel engine whose connecting-rod–crankshaft system has been replaced by a hot-gas transmission system involving hot-gas receiver and piping, and a gas turbine driving the output shaft. Its advantages over a conventional Diesel engine appear to be

1. Elimination of many highly stressed parts and heavily loaded bearings.
2. Possibility of connecting a number of free-piston units to one power turbine, thus taking advantage of the use of a single, standardized unit for a large range of power.
3. Greater tolerance to fuels of low cetane rating, because the compression ratio automatically increases to the point of ignition (13.508).
4. Full mechanical balance.
5. High torque of the turbine at starting (of interest only for rail and automotive purposes).

As yet these advantages have not been found sufficient to offset the disadvantages, chief among which are

6. Lower efficiency than the Diesel engine, especially at part load (see Volume I, Fig. 13–10, curves No. 5).
7. High cost of the gas turbine.
8. Starting and idling problems.
9. Excessive noise and other difficulties due to unsteady gas flow (13.510).

As to the advantages, No. 2 can be obtained more easily by gearing or by Diesel-electric drive, as in railroad-locomotive practice; No. 3 has been largely nullified by the development of "multifuel" ability in conventional Diesel engines; No. 4 is offset by disadvantage No. 9; advantage No. 5 is more effectively and economically attained with conventional engines by means of a hydraulic torque converter. There appears to be little likelihood that free-piston power plants will offer serious competition to the more conventional engines.

Where the output desired is compressed air rather than power from a shaft, the free-piston compressor unit looks quite attractive. Small units of this kind are marketed in Europe but have not been used to any great extent in the U.S.A. The principal reason for this situation is probably the fact that conventional engine-compressor units are well developed and are available at low cost because the engines are built in large quantities for other purposes. The free-piston compressor is a single-purpose machine which has not yet and may never be highly developed, because present conventional types are so inexpensive and satisfactory.

The free-piston principle has also been used to a very limited extent for pile-drivers and portable hammers (13.515). These usually have starting difficulties and have not succeeded in displacing the more conventional pneumatic machines to any important extent.

Crankshaft Engines as Gas Generators

Several versions of this type have been tried experimentally (13.60–13.62). This arrangement has all the disadvantages of the free-piston arrangement (except No. 8), without the advantage of fewer moving parts. The transmission efficiency is such that the shaft output of the Diesel engine to the connected compressor is usually greater than the output of the turbine. It would seem poor engineering judgment to spend money on the development of this type.

GAS TURBINES

The concept of the gas turbine is old, but it did not materialize as a practical power source until after World War II. Its commercial development was stimulated by the successful introduction of turbojet engines for aircraft in England and Germany near the end of that war. These developments in turn were based on improvements in the efficiency of centrifugal air compressors, made in the course of improving the turbosupercharger for military use. Later development of highly efficient multistage axial compressors has also been an important factor, especially for large gas-turbine installations and for aircraft.

Gas-Turbine Classification

A convenient system of designating the various possible arrangements of gas-turbine components is the following (referring to Fig. 13–5):

$$C = \text{compressor}$$
$$B = \text{burner}$$
$$T = \text{turbine}$$
$$X = \text{heat exchanger}$$
$$I = \text{intercooler}$$

The letters are given in order in the flow system. A dash indicates that the rotor is mounted on a separate shaft, and the subscript p signifies a separate power-output element. For example:

CBT means a simple single-shaft compressor-burner-turbine unit

CXBTX indicates the above with heat exchanger

CXBT–T_pX indicates a compressor-burner-turbine unit supplying gas to a separate or "free" power turbine, T_p, with a heat exchanger from T_p to the usual location between C and B.

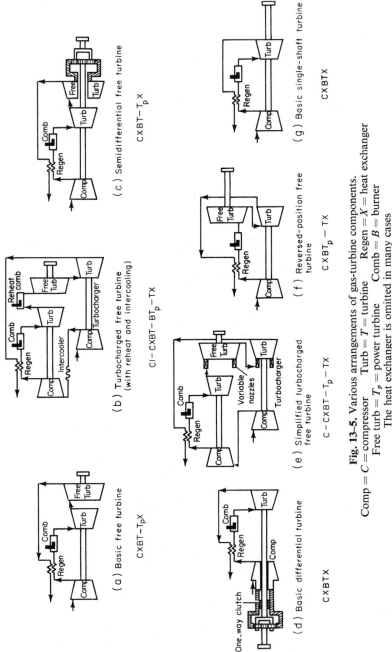

Fig. 13-5. Various arrangements of gas-turbine components.

Comp = C = compressor Turb = T = turbine Regen = X = heat exchanger
Free turb = T_p = power turbine Comb = B = burner
The heat exchanger is omitted in many cases

(a) Basic free turbine

$CXBT-T_pX$

(b) Turbocharged free turbine (with reheat and intercooling)

$CI-CXBT-BT_p-TX$

(c) Semidifferential free turbine

$CXBT-T_pX$

(d) Basic differential turbine

$CXBTX$

(e) Simplified turbocharged free turbine

$C-CXBT-T_p-TX$

(f) Reversed-position free turbine

$CXBT_p-TX$

(g) Basic single-shaft turbine

$CXBTX$

The Ford 704 compound turbine (13.705) is classed as C_1I–$C_2XB_1T_2$– B_2T_p–T_1X. This turbine has three shafts, one located between T_1 and C_1, one between T_2 and C_2, plus the power shaft of T_p.

Potential Gas-Turbine Performance

Figures 13–6 and 13–7 show the potential performance of CBT and CXBTX arrangements as a function of the ratio of turbine-inlet stagnation temperature T_{03} to compressor-inlet stagnation temperature T_{01}. For these figures the assumed compressor and turbine efficiencies represent best probable values for large, well-developed machines. The heat-exchanger effectiveness assumed for Fig. 13–7 (0.75) is representative of a good compromise between size and effectiveness (see Volume I, p. 392).

Examples of the use of Figs. 13–6 and 13–7 are given in Table 13–1. It is evident from these data that the gas turbine has a potential fuel economy at its rating point comparable with that of the Diesel engine, provided turbine-inlet temperatures above 2000°R (1110°K) are used. Even better potential economies are predictable with higher temperatures or more complicated turbine arrangements (13.700, 13.701).

Table 13–1. Potential Performance of Gas Turbines as a Function of T_{03} (from Figs. 13–6 and 13–7)

T_{03} (°R)	1500	2000	2500
(°K)	834	1110	1390
T_{03}/T_{01}*	2.8	3.7	4.63
CBT turbine			
best efficiency	0.22	0.32	0.38
sfc† (lbm/bhph)	0.625	0.43	0.36
(gm/bhph)	284	196	164
CXBTX turbine			
best efficiency	0.28	0.35	0.41
sfc† (lbm/bhph)	0.49	0.39	0.336
(gm/bhph)	222	177	153

* $T_{01} = 540°R = 300°K$
† $Q_c = 18,500$ Btu/lbm

The allowable turbine-inlet temperature at the rating point depends on materials used, on detail design including provisions for blade cooling, and on the type of service. As in the case of reciprocating engines, services where long life is required or where extended operation at rating is expected call for conservative ratings. In the case of gas turbines, conservative rating means relatively low turbine-inlet temperature at the rating point.

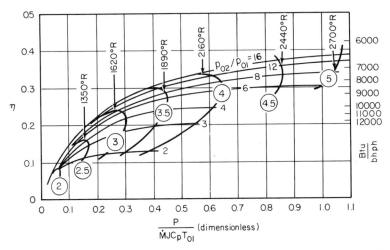

Fig. 13–6. Computed performance at rating point of CBT gas turbines. Turbine efficiency = 0.85; compressor efficiency = 0.90; T_{03} = stagnation temp. at nozzle; T_{01} = stagnation temp. at compressor entrance. Numbers in circles are T_{03}/T_{01}, temperatures shown are T_{03} when T_{01} is 540°R (300°K).

η = thermal efficiency $\qquad$ J = Joule's constant
P = power in basic units $\qquad$ C_p = spec. heat of gas, const. press.
$\dot{M}$ = mass air per unit time $\qquad$ p_{01} = comp. inlet stagnation press.
$\qquad\qquad\qquad\qquad\qquad\qquad$ p_{02} = comp. outlet stagnation press.

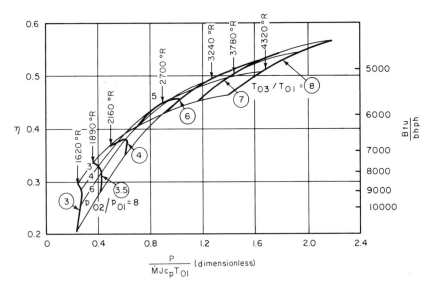

Fig. 13–7. Computed performance at rating point of CXBTX gas turbines. Symbols and conditions same as for Fig. 13–6. Heat-exchanger effectiveness = 0.75.

Part-Load Economy

Figure 13–8 shows potential part-load performance of various gas-turbine arrangements and also typical performance maps for automotive Diesel engines. These curves (from ref. 13.700) are based on estimates of future potential performance with $T_{01} = 540°R$ ($300°K$) and $T_{03} = 2540°R$ ($1412°K$) for equipment as installed in motor vehicles with the necessary auxiliaries.

Significant relations shown in Fig. 13–8 and in ref. 13.700 are as follows:

1. Gas turbines without heat exchanger have relatively poor part-load and light-load fuel economy.
2. The economy of gas turbines with heat exchangers is potentially competitive with that of Diesel engines.
3. Fuel economy can be improved by increased complexity, including such items as variable nozzles, intercoolers, reheat stages, etc.

Weight and Bulk

Simple gas turbines without heat exchangers are much lighter and smaller than competitive reciprocating engines of equal power. For example, large reciprocating aircraft engine installations weigh about 1.5 lbm per take-off hp, while weights of 0.5 lbm per bhp or less are characteristic for turbopropeller installations. The volume-to-horsepower ratio is also considerably less for turbines without heat exchangers than for competitive reciprocating engines. On the other hand, when a heat exchanger of high effectiveness is used, the bulk of a gas-turbine installation increases to values comparable to that for a competitive reciprocating engine, and the weight tends to be equal or only slightly less. For example, the Chrysler automotive turbine (13.704) of 130 hp has about the same bulk and weight as the gasoline engine it replaces.

Size Effects

Figure 13–9 indicates a trend of efficiencies of gas turbines to increase with size. This relation is both because of favorable Reynolds number effects and because the necessary refinements to achieve high efficiency are more easily made as size becomes larger. Furthermore, as in most types of power machinery, the importance of high efficiency relative to first cost is greater in applications calling for large unit size. The relative effects of size on specific weight and volume, as pointed out in Volume I, Chapter 11, are the same for gas turbines as for other types of machinery.

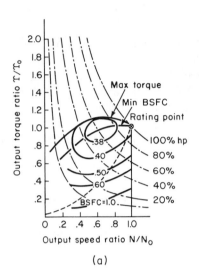

(a)

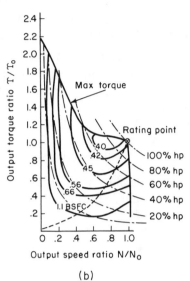

(b)

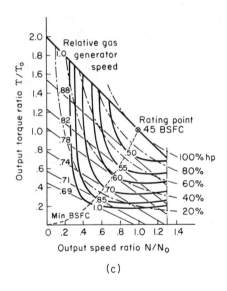

(c)

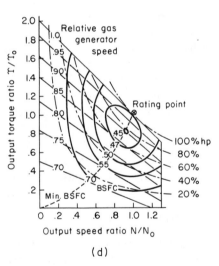

(d)

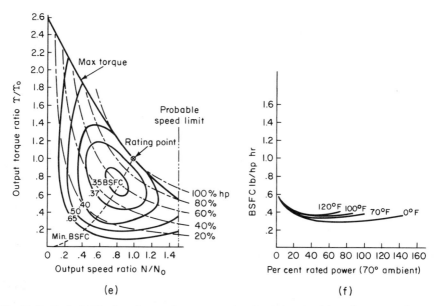

(e)   (f)

Fig. 13–8. Typical Diesel-engine performance and estimated potential performance of gas-turbine power plants for road vehicles with turbine–inlet temperature = 2460°R (1370°K). – – – – – indicates level-road regime. (*a*) Typical Diesel engine; (*b*) Diesel engine with "low stall" torque converter; (*c*) advanced nonrecuperative, free turbine CBT-T_p; (*d*) typical regenerative gas turbine CXBT-T_pX; (*e*) recuperative gas turbine with multistage free turbine, CXBT-T_pX; (*f*) effect of inlet air temperature on gas-turbine performance, optimum speed-load schedule. Rating is at 70°F (21°C) (Wood, 13.700).

Fuel-economy comparison, road load

		Rated *Point*	*0.75 Load*	*0.40 Load*
a	Diesel engine, geared	0.39 - 177	0.50 - 227	0.8 - 364
b	Diesel engine, torque conv.	0.41 - 186	0.54 - 246	1.1 - 500
c	CBT-T_p turbine	0.45 - 204	0.60 - 272	1.5 - 680
d	CXBTX-T_pX turbine	0.45 - 204	0.52 - 236	1.0 - 500
e	CXBTX-T_pX multistage	0.40 - 181	0.36 - 164	0.8 - 364

First number is lbm/bhp hr, second number is grams/bhp hr.

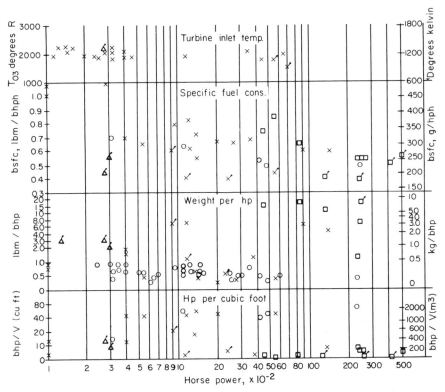

Fig. 13–9. Characteristics of gas turbines in production, 1966. (From *Gas Turbine*, Jan.–Feb. 1966, and *Automotive Industries*, March 15, 1965).

☐   central-station power
○   aeronautical turbo-propeller
×   industrial
△   automotive (experimental — not in production)
kg or lbm = weight given by manufacturer
          V = box volume, length × width × height
flag ✗ indicates heat exchanger

Cost of Gas Turbines

No background of experience in the manufacture of gas turbines in the very large quantities characteristic of automobile production exists so far. However, even in mass production it appears that gas turbines with heat exchangers would be more expensive than the gasoline engines now used in passenger automobiles. The heat exchanger is basically an expensive item, as are the rotating parts which require special materials and methods of fabrication. It seems unlikely that costs of suitable gas turbines with heat exchangers can be brought down to compete with the present astonishingly low cost of passenger-car gasoline engines.

At the present time Diesel engines are made in smaller quantities than automobile engines and, as we have seen, their cost is inherently higher. Here the cost of gas turbines, even with heat exchangers, may be competitive in the near future, particularly in the case of large units. Without heat exchangers, gas turbines of 1000 hp or more are not more expensive, and may even be cheaper, than Diesel engines of the same power rating.

Present Status of the Gas Turbine

Reference 13.703 lists 541 models and 94 manufacturers of gas turbines in 9 countries as of 1966. This includes only one Russian model out of what is probably a large number. Figure 13–9* shows reported fuel economy and turbine inlet temperature for a number of existing turbines. This figure shows values of T_{03} in the vicinity of 2000°R (1111°K) and fuel economy over a wide range. The low boundary of fuel consumption is from heat-exchanger installations. Rated powers in a single unit range from 44 to 25,000 hp, fuel consumption from 0.40 to 1.6 lbm/bhph (182 to 728 grams) and pressure ratios from 2.5 to 16. Many types are included, with and without heat exchangers.

Figure 13–9 shows that the efficiency of commercial gas turbines is generally lower than the potentials shown in Figs. 13–6 to 13–8. However, present trends in design and materials indicate that performances such as those shown in Figs. 13–6 to 13–8 will be possible in the near future. For example, usable values of turbine inlet temperature (T_{03}) have been increasing year by year since gas turbines were introduced.

Nearly all successful gas-turbine applications to date (1984) are for services where fuel economy, and especially part-load fuel economy, can be sacrificed for light weight, small bulk, and sometimes reduced first cost —in other words where simple CBT or CBT-T_p arrangements can be used. These applications include

Airplanes and helicopters
High-speed air-cushion and hydrofoil boats
Short-range high-performance military vehicles and watercraft
Gas pipe-line pumping stations
Peak-load electric power units
Peak-load power for naval vessels
Standby or emergency power
Fire-fighting equipment
Starters for large jet engines
Some railroad locomotives

---

* Compare hp/V in Fig. 13–9, which runs as high as 60 hp/cu ft. with hp/V in Fig. 12–26 where the maximum value is 9.5.

Gas turbines are not yet (1984) used for:

Road vehicles
Small low-cost power units
Nonmilitary ships and motorboats

Aircraft Gas Turbines. To date the only service where the reciprocating internal-combustion engine has been completely displaced is for new transport and military airplanes, where the jet or the turbopropeller engine is now the accepted form. The basic reason for this change is that gas turbines, as compared with reciprocating engines for the same service, are lighter in weight and enable the aircraft to attain much greater speeds and to fly at much higher altitudes. The poorer fuel economy of the CBT gas turbine on a horsepower basis has been offset by higher aircraft speed and higher-altitude operation, so that the fuel consumption per ton mile traveled is actually less than it was at the lower speeds and lower altitudes at which reciprocating-engine power was used.

Note that aircraft do not operate, except for short periods, at engine loads less than about half maximum power, so that the poor economy of CBT gas turbines at light loads is not a serious drawback. In this service, high initial cost of the power plant is also not a serious factor, considering the other gains made possible by its use.

After some 25 years of experience with aircraft gas-turbine engines, it has been found that they are much more reliable and require less frequent overhaul than the very high-strung reciprocating engines which they replaced. The future of the gas turbine seems secure in the large-aircraft field of application, and it will gradually displace the reciprocating engine for smaller airplanes as the cost of manufacture is reduced and the performance of small units improved.

A similar situation applies to helicopters, where the larger sizes already use gas turbines and the smaller sizes will probably follow this trend.

Military Applications Other than Aircraft. Gas turbines are in use or under development for military (or naval) services where high specific output rather than low fuel consumption is a primary consideration. Examples include short-range assault vehicles, both land- and water-borne, and starters for large jet engines. The use of gas turbines for emergency or combat power of naval vessels which cruise at lower speeds on steam or Diesel power is an interesting possibility which is being developed (13.751, 13.752).

Gas Turbines for Peak-Load or Standby Service. In this type of service fuel economy may be sacrificed for other advantages, such as reliability and quick start-up, freedom from noise and vibration, low installation and foundation expense, all of which favor the gas turbine provided it can be obtained at competitive cost per horsepower.

For large peak-load or standby installations, the CBT or CBT–T_p type of gas turbine may be lower in cost than either steam or Diesel power, and it is being used to an increasing extent. Small gas turbines are also being used for standby power for hospitals, office buildings, etc., in competition with gasoline and Diesel engines.

It seems probable that the present trend toward replacing reciprocating engines by gas turbines for purposes of standby, emergency, and peak-load power will increase as the art of gas-turbine design and manufacture improves.

Total-Energy Installations. When the waste heat is useful, the low efficiency of the CBT turbine can be accepted even for continuous operation. Installations where waste heat is used are called *total-energy* installations. Similar installations are also possible with reciprocating engines, but in this case less waste heat is available in proportion to shaft power (13.755). On the whole, the gas turbine appears to be quite attractive for this type of service.

Gas Turbines in Gas-Pipeline Service. An area where the gas turbine has made inroads on the use of reciprocating engines is in the field of gas-pipeline pumping stations (13.732). To date only the CBT–T_b type of turbine has been used. The costs of housing, attendance, and maintenance are less than for the alternative large reciprocating engines, and for very large units even the initial cost may be less. Since the fuel (gas being pumped) is low in cost, the foregoing savings may offset the lower efficiency of the gas-turbine installations compared with reciprocating engines. Two special factors also favor the gas turbine in this field: a favorable tax policy and the fact that aircraft jet engines are used in which enormous development costs have already been paid for by government and which can therefore be purchased at a reasonable price (13.731, 13.732). Increased use of gas turbines in this service seems very likely.

Railroad Power. Gas-turbine-powered locomotives using fuel oil are in use on one section of a U.S. railroad for the purpose of hauling freight up a very long, steep grade (13.720, 13.721). In this application the turbines are simple CBT units with electric transmission. They are used at full rated power for the whole uphill run (they coast back), and the poor part-load performance of this type of power is not a drawback. Many railroads have experimented with gas-turbine locomotives, but they are in general use only in the one instance cited. It is still an open question whether gas-turbine locomotives have a significant future.

Coal-Burning Gas Turbines. Much effort has been expended by the Association of American Railroads on the development of a coal-burning gas turbine for locomotives (13.722–13.725). The most difficult problem is to supply the turbine with an ash-free and noncorrosive gas from coal. As this development proceeds, the size, cost, and heat losses involved in ash separation have grown to such an extent that it seems questionable whether a competitive power plant will ever result, even if a reliable and durable one should be finally developed.

Marine Gas Turbines. An experimental gas-turbine power plant was operated by the U.S. Shipping Board in competition with steam, Diesel, and free-piston plants in similar vessels. The results have not started a trend in the direction of gas turbines. Where turbine drive is desired, the steam plant appears to be much better adapted, especially with regard to fuel economy and ability to use low-grade fuels. Actually, direct-drive Diesel engines dominate in the merchant ships of the world and are likely to continue to do so in the foreseeable future (see Fig. 13–1). As already noted, there is a trend toward the application of gas turbines for short-duration high-power craft, such as small naval attack vessels, hydrofoil boats, and air-cushion craft.

Automotive Gas Turbines. Since passenger-automobile and light commercial-vehicle service involves a great deal of operation at light loads, automotive gas turbines require heat exchangers to achieve acceptable fuel economy. Since about 1950, large sums have been spent on gas-turbine development by a number of automotive companies (13.704–13.7091). The Chrysler Corporation has even arrived at the point of putting a few turbine-powered automobiles in the hands of the public but has not as yet publicized the results. Most of these developments have been with the basic arrangement CXBT–T_pX (13.06 *et seq.*). The weight and bulk of these installations is about the same as those of the reciprocating engines they replace. The cost of these turbines in the small quantities so far produced has been impractically high. As previously stated, it is unlikely that, even in similar quantities, the automotive gas turbine can be manufactured as cheaply as the present type of automobile engine.

Looked at from another point of view, the present type of gasoline engine used in passenger cars and light commercial vehicles is an eminently satisfactory power plant. As installed, it is free from objectionable noise and vibration, is manufactured at almost ridiculously low cost, is very reliable, very flexible, requires little maintenance, and has fuel economy at least equal to that predicted for the alternative gas turbine. It is difficult to see how a gas-turbine plant could improve on these qualities in any significant way. Similar considerations hold for gas turbines as a replacement for the Diesel engines now used in light road vehicles.

On the other hand, a place where gas turbines may eventually replace Diesel engines for road service is that of heavy long-distance trucks. The gas-turbine developments by Ford (13.705, 13.708) and by General Motors (13.707, 13.709) have been directed toward this type of service. Figure 13–10 shows progress in performance of this type of turbine as reported by General Motors. Table 13–2 gives estimates of over-all initial and operating costs of 400-hp units for this type of service, based on the 1966 state of the art. It is evident that only a small improvement in gas-turbine technology, including especially a reduction in first cost, is necessary to make the regenerative turbine competitive with the Diesel engine in this category. Similar considerations should apply to heavy off-the-road equipment, such as that used in mining and construction work. It should be remembered that in both the services here in question, the power plant operates a large part of the time at or near maximum load.

Table 13–2. Estimated Operating Costs of 400-hp Power Plants for Long-Distance Trucking, 1966

ASSUMPTIONS			
	Diesel	Nonregenerative Turbine	Regenerative Turbine
Horsepower	400	400	400
First Cost ($/hp)	20	40	50
Engine Life (hrs)	7500	15,000	15,000
Maintenance Cost ($/hr)	1.00	0.35	0.50
Lubrication Cost ($/hr)	0.35	0.10	0.10
Specific Fuel Consumption at rating	0.42	0.60	0.45

Engine Operating Costs, $/hr			
	Diesel	Nonregenerative Turbine	Regenerative Turbine
Fuel	$3.50	$5.40	$4.10
Maintenance	1.00	0.35	0.50
Lubrication	0.35	0.10	0.10
Depreciation	0.91	0.91	1.13
Interest, insurance, and taxes	0.27	0.61	0.76
Totals	$6.03	$7.37	$6.59
Percentage of Diesel cost	100	122	109

(From Kahle and Hung, 13.761)

The foregoing statements about gas turbines for automotive use must be modified for 1984. The Diesel engine, usually turbosupercharged, is now firmly established as the preferred power plant for heavy on-road and off-road vehicles. Higher cost and poorer fuel economy are probably the major reasons for the failure of gas turbines in this application.

The gas turbine has been effective in special racing automobiles. The near victory of a turbine-powered car in the 1967 Indianapolis 500 (13.7091) led to a change in the rules to ban such engines. In this application heat exchangers are not necessary because fuel economy is unimportant,* and the high output per unit weight and bulk of the simple CBT–T$_p$ turbine is a significant advantage. Also, the cost of the power plant is not a major consideration.

Gas Turbines Versus Small, Low-Cost Engines

It is extremely doubtful that the cost of alternative gas turbines of any type can be as low as that for the small, mostly single-cylinder, engines for personal and industrial use, as in lawn mowers, chain saws, home lighting plants, and outboard motorboat engines. In these applications it is very unlikely that the reciprocating engine will be replaced by any other type in the foreseeable future.

Smog and the Gas Turbine

The exhaust emissions of gas turbines include appreciable fractions of oxides of nitrogen and particulates, but these are in relatively low concentrations because of the very lean mixtures used. In this respect turbines have an advantage over conventional spark-ignition and Diesel engines.

Reciprocating Engines in Parallel with Gas Turbines

The use of an exhaust-driven turbine geared to the output shaft of a reciprocating engine has been discussed in Volume I, Chapter 13 and at various points in both Volume I and Volume II. To date it has been used to a limited extent for aircraft and military power (10.829, 10.847, 13.740 *et seq.*). In the usual case the performance of such a combination is only marginally better than that of the corresponding turbosupercharged engine (compare Volume I, Fig. 13-10, curves 1 and 2, at the same imep), and except in unusual circumstances the requisite additional mechanical complications, including high-ratio gearing with its associated shafts, bearings, and vibration-isolating couplings, do not appear justified. Only very limited use of this system is foreseen.

Gas turbines in parallel with reciprocating engines are being developed for military and naval purposes (13.751, 13.752). Such arrangements are a form of standby power and seem appropriate only where the reciprocating engine is to be used most of the time, with only short-time requirements for the high-power potential of the gas turbine, as under combat conditions.

---

* Note from example 10–5, page 398, that reciprocating racing engines generally have high fuel consumption.

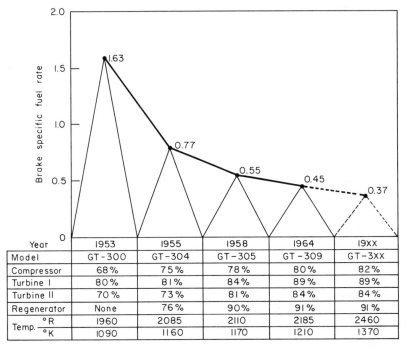

Year	1953	1955	1958	1964	19XX
Model	GT-300	GT-304	GT-305	GT-309	GT-3XX
Compressor	68%	75%	78%	80%	82%
Turbine I	80%	81%	84%	89%	89%
Turbine II	70%	73%	81%	84%	84%
Regenerator	None	76%	90%	91%	91%
Temp. °R	1960	2085	2110	2185	2460
Temp. °K	1090	1160	1170	1210	1370

Fig. 13–10. Progress in automotive gas-turbine performance as reported by General Motors Corp. Fuel economy, components efficiencies, and turbine-inlet temperatures at rating. Fuel rate is in lbm/hph. 19XX indicates estimate for near future (13.709).

EXTERNAL-COMBUSTION POWER

Power plants that depend on the transfer of heat from the combustion of fuel and air through a heat exchanger to another working fluid will be called *external-combustion* power plants. In order to conserve the working fluid and to achieve high efficiency, such plants must also have a heat exchanger to cool the working fluid. In the familiar steam plant, these heat exchangers are called the *boiler* and the *condenser*, respectively.

Present Status of Steam Power. Steam is of course well established as a source of power for the production of electrical energy, especially in large units where its efficiency can compete with that of the Diesel engine. In large units its advantages over Diesel engines include the ability to burn almost any type of fuel, including coal, its generally lower maintenance requirements, and the possibility of locating the elements of the plant in various ways to suit particular installations. Since all modern steam plants use turbines, the steam plant is comparatively free from the balance and vibration problems discussed in Chapter 8. For these reasons steam is the most generally used heat engine for the generation of electrical energy in units above about 10,000 hp. On the other hand, in the last twenty or thirty years steam has been giving way to Diesel-engine power in the smaller installations of stationary power (10.51), and there are numerous large installations of Diesel power and gas-engine power where the fuel situation is favorable. A basic requirement of the efficient steam plant is a large supply of cool water for condensing. Where this is not available, Diesel engines or gas turbines are a logical alternative source of power.

The other application where steam is widely used is for large ships, especially naval and passenger vessels. Merchants ships and smaller passenger vessels are moving toward Diesel power, as indicated by Fig. 13–1. This trend is due to the development of systems for burning the cheap grades of fuel oil formerly only usable in steam plants (10.51). Apparently the superior fuel economy of large marine Diesel engines (Fig. 13–2) more than offsets their greater maintenance requirements in most merchant-ship services.

Diesel locomotives have replaced steam completely in the U.S.A. and are in the process of doing so all over the world. The basic reason is the great reduction in out-of-service time, which was very large in the non-condensing steam locomotive because of boiler maintenance. The lower fuel consumption and absence of high water consumption are also important factors in the railroad Diesel's favor. There is no likelihood of the return of steam in this service (see previous discussion of gas turbines for locomotive service).

Steam Road Vehicles. Very early road vehicles were steam powered because no other portable power was available. Steam passenger automobiles were produced in the U.S.A. in important numbers from 1900 to 1916, but declined rapidly after that time and completely disappeared in 1924 (13.802). All later attempts to revive steam automobiles have failed (13.803). Feeble endeavors to apply steam to aircraft have been completely unsuccessful (13.804). Coal-fired steam motor trucks ("lorries") were popular in England until the mid-1930's. The White Motor Car Company built both gasoline and steam automobiles until about 1910, at which time they abandoned steam permanently.

The reasons for the abandonment of steam in favor of internal-combustion engines for all but very large units are easy to see. The steam plant requires not only an engine (or turbine) but also furnace, boiler, feed-pump, condenser, and rather elaborate controls. The internal-combustion engine requires only small and inexpensive auxiliaries, namely, the ignition system, carburetor or injection system, and starting system. In addition, it is more efficient. Its disadvantage of being less tolerant of fuel quality has been overcome by the discoveries of enormous petroleum resources and the development of cheap refining techniques. The disadvantage of requiring special starting equipment and of having zero stall torque have been overcome by electric starters and hydraulic torque converters.

THE STIRLING ENGINE

Another external-combustion engine that has received some recent attention is the Stirling type of hot-gas engine (13.850 to 13.857). By using hydrogen as the working fluid, reasonably good efficiencies and modest outputs per unit piston area have been achieved. However, both hot and cold heat exchangers are required in addition to a burner-furnace system. These auxiliaries are large, heavy, and expensive compared with the auxiliaries required for an internal-combustion engine of equal power. It is difficult to see anything but highly specialized applications of this type.

Many different working fluids have been proposed for external-combustion engines (or turbines) including air, mercury, Freon, etc. (13.801). None of these eliminates the requirement for large hot and cold heat exchangers and for burner-furnace and control systems, which are the basic disadvantage of all such power plants. There is very little likelihood of the external-combustion engine becoming a serious competitor to the internal-combustion engine in its present applications.

ELECTRIC POWER

The electric motor itself is a most convenient, simple, flexible, and adaptable device for converting electric energy into mechanical power. Where electric energy is plentiful and cheap, it is the most logical source of mechanical power for most purposes. The difficulty comes when its use is considered for moving vehicles (road, rail, air, or marine) or for use in locations where centrally produced electric energy is not available. Possible sources of electric power for vehicles include storage batteries, thermo-electric generators, and fuel cells.

Storage-Battery Vehicles

At present, electric power with storage batteries is used in some fleets of small vehicles operated within the limited radius of a single industrial or commercial enterprise. However, except under circumstances where the exhaust gases are particularly objectionable, most such fleets use internal-combustion engines.

Prior to about 1925, battery-powered electric passenger automobiles were widely used within city limits. Since that time the demand for longer range, higher speed, and lower initial cost of the vehicle has made this type obsolete. No developments in storage batteries are at present foreseen that are likely to bring the battery-powered vehicle into competition with the gasoline automobile, except possibly for comparatively low-speed, short-range use. The important part which conventional automobiles play in producing smog and other forms of air pollution has stimulated the study of the possibilities of the electric vehicle for use in cities (13.900–13.902). Whether the traveling public can be induced to leave their gasoline automobiles outside city limits and use electric vehicles inside these limits is a question fraught with such difficult political and economic problems that the future of such schemes appears highly uncertain.

Nonbattery Portable Electric Power

For some years there has been active development work with thermoelectric devices and fuel cells (13.903, 13.910).

Thermoelectric devices produce electricity by the heating of dissimilar elements as in thermocouples, or by heating so as to produce an electric potential between two closely spaced plates. Such devices are similar to external-combustion engines in requiring a furnace and at least one heat exchanger. This requirement, together with the electric motors and controls also required, make it appear unlikely that this form of power plant will ever become attractive for vehicle propulsion.

A fuel cell is a device which produces electricity through the chemical reaction of fuel and oxygen by means of electrolytic action at relatively low temperatures. At the present state of development, fuel cells must have clean gaseous fuels (usually oxygen for one) and are large, heavy, and expensive in proportion to their output. Assuming that fuel cells will some day be developed to the point where they can use gasoline and air economically and reliably, the total power plant including fuel cell, electric motors, controls, and auxiliaries is almost sure to be larger, heavier, and more expensive than the present forms of internal-combustion engine.

Plasma-Jet Power

The production of electricity by means of high-temperature, high-velocity plasma jets, which has been proposed for large stationary power units, would seem singularly inappropriate for vehicle power even if developed for large thermal power plants.

NON-AIR-BREATHING POWER PLANTS

In this category may be included nuclear reactors, solar power plants, and various forms of chemical power not involving atmospheric oxygen.

Nuclear Power

Nuclear power is an alternative source of heat for use with external-combustion engines or thermoelectric devices. In the former category it is already in use for submarines, where it is displacing the Diesel engine, and for large ships, where it displaces the fuel-burning steam boiler and furnace. It seems entirely out of the question for road or rail vehicles or for small power plants in general because of the large amount of shielding required and the very high costs and potential hazards of atomic reactors.

Solar power has a very limited application except in outer space, because of the enormous heat-collector areas required and because of the intermittent nature of sunlight. It is obviously inappropriate for most purposes where internal-combustion engines are now used.

Reciprocating Engines in Space

Because of its requirement for air, the conventional internal-combustion engine is not suitable for use in space. Such use is possible only when the engine is modified to use bipropellants, that is, chemicals that react together to supply an expanding gas (13.910–13.913). The least modification would be required for the use of stored oxygen plus a petroleum or similar fuel. Since temperatures of reaction with combustible mixtures of oxygen and fuel would be dangerously high, they should be reduced as required, by recirculating exhaust products. This method has been successfully tried on an experimental basis.

An example of successful use of a bipropellant in a reciprocating engine was in some naval torpedoes during World War II. The propellants used were hydrogen peroxide and liquid petroleum. The peroxide was delivered in a liquid form to a catalytic reaction chamber where it turned to gas in the reaction $2H_2O_2 \rightarrow 2H_2O + O_2$. This gas flowed under pressure to the engine cylinder. Fuel was injected into the cylinder in sufficient quantity to consume the oxygen during expansion. Burned gases were expelled on the return stroke.

In this case, as in most cases where bipropellant operation is contemplated, the bipropellants will be delivered to the engine as gases under high pressure. If this pressure energy is not to be wasted, the engine inlet system must be operated at high pressure, with the unburned mixture admitted during the early part of the expansion stroke, as in a steam engine. The gases are expelled during the return stroke, again as in a steam engine. The essential difference, as compared with a steam engine, is combustion or other chemical reaction during the expansion process. To date, engines of this type are not generally used, nor have they been used in space so far as is known to this author. Possible bipropellant combinations are discussed in refs. 13.901–13.903.

SUMMARY

The present type of reciprocating internal-combustion engine has the following characteristics, which have made it the dominant source of heat power all over the world for most uses except large electric power plants and aircraft:

1. Relative mechanical simplicity and low cost.

2. Flexibility, that is, the ability to run over a wide range of speeds and loads with reasonable efficiency.

3. Adaptability to a wide range of applications over an extraordinary range of sizes (from less than 1 hp to 50,000 hp in one unit).

4. The highest efficiency of any fuel-burning engine, especially in combination with exhaust-driven turbines.

5. Freedom from external high-temperature elements such as furnaces and boilers, and from large high-energy elements such as large turbine rotors and high-pressure storage tanks.

6. Adaptability to existing supplies of fuel and to most suggested liquid and gaseous fuel alternatives.

With one exception, the hundreds of suggested alternative engines (refs. 13.00, 13.01) have not proved successful. Somewhat surprisingly, there is now limited use of the Wankel rotary engine (see pages 581–583) in light road vehicles. That this application will expand seems doubtful, but not impossible.

Air-pollution problems are serious with all fuel-burning power plants, and there is little evidence that alternative types would be much better in this respect. For example, steady-flow burners produce less nitrous oxides and CO but more particulates.

It has proved dangerous to attempt to predict technological developments beyond the "foreseeable future." For this period, the author would predict continued use of conventional reciprocating spark-ignition and Diesel engines in the fields where they are now dominant.

14 | Engine Research and Testing
Equipment—Measurements—Safety

It is not within the scope of this volume to present an exhaustive treatment of the subject of research and testing. It is, rather, the object of this chapter to record the author's observations and opinions based on many years of experience in engine research, particularly with regard to those aspects that are important in promoting accuracy of measurement, in holding down costs for educational and other institutions whose funds for such equipment may be limited, and in providing safety precautions. For additional information in this field see the references for this chapter.

Emphasis in this chapter will be on engine research, that is, work designed to discover basic relationships applicable either to all internal-combustion engines or to some category of engine, such as spark-ignition or Diesel engines. Only brief attention will be given to routine testing, that is, performance or endurance testing of particular engines and/or their auxiliaries, or to routine testing of fuels and oils.

Descriptions of complete engine laboratories will be found in references 14.021 *et seq.*

BASIC EQUIPMENT

Some research applicable to internal-combustion engines can be carried out on special equipment such as combustion bombs (1.20–1.29, 3.31–3.36) and rapid-compression machines (2.20–2.22, 3.37–3.401). This is highly specialized work, requiring a wide background in physical and chemical research. It is best studied by consulting the references just cited.

Basic engine research should be carried out on single-cylinder test engines, with necessary modifications to fit the particular problem in hand. Engines of more than one cylinder are required for problems that involve the relationship between cylinders and the equipment that connects them. Examples of problems of this kind are studies of manifolding, fuel and air distribution, vibration, and friction.

Test Engines

Many types of single-cylinder engines designed for research purposes have been built and used, and several types are available on the open market (14.111–14.121).

The engine chiefly used by your author, and commonly used by many others, is the CFR (Cooperative Fuel Research) single-cylinder engine, 3.25-in. (82.5-mm) bore by 4.5-in. (114-mm) stroke, manufactured by the Waukeska Engine Co. of Waukeska, Wis. (14.111). This engine was originally designed for fuel octane-number determination, and is still the standard engine for this purpose when equipped with the necessary auxiliary equipment (2.06–2.093).

For research or instructional purposes, these engines are purchased without carburetor or other auxiliary equipment. The so-called "high-speed" camshaft should be specified, although for speeds in excess of about 3000 rpm special cams must be designed in accordance with the recommendations in Chapter 12. There is a version of this engine with two first-order balancing pistons (see Chapter 8), but this model is not recommended, because of its abnormally high friction. Use of a spring-mounted installation, to be described later, makes balancing pistons unnecessary.

The basic structure of this engine is extremely rugged, having withstood piston speeds of over 3000 ft/min, and in one instance an indicated mean effective pressure of over 1000 psia (77.5 kg/cm^2). Special cylinder hold-down equipment was added for the very-high mep tests (14.112).

Dynamometers

The d-c electric cradle-type dynamometer is the most versatile and convenient type, since it can "motor" the engine for friction tests or for research work with the engine not firing. In its commercial forms this type is by far the most expensive, and it may require very costly auxiliary and control equipment, such as motor-generator equipment, *amplidyne* units, and power-absorbing resistance units. For a given power rating, the maximum allowable speed of the electric type is usually lower than that of the other types (14.201).

For use with the CFR and other single-cylinder engines of small size (less than 40 hp) the Sloan Laboratories at MIT have constructed low-cost electric dynamometers by purchasing a suitable commercial direct-current generator and mounting it on trunnion supports and bearings made in the MIT shops (see Fig. 14–1). If a suitable supply of d-c electric current is available, this equipment can be constructed and installed at relatively low cost. By means of a field rheostat, a wide range of speed and load can be obtained with the generator armature connected directly across the d-c supply lines. The power is thus absorbed in the lines. An armature rheostat will extend the range to satisfy the entire speed-load requirement within the speed limits of the generator.

Next in order of versatility to the d-c dynamometer is the combination of a constant-speed a-c motor and a water-cooled eddy-current brake. This type of unit can motor up to the speed of the a-c unit only (3600 rpm at 60 cps, 3000 rpm at 50 cps). It can usually be operated directly from the a-c power lines with a minimum of auxiliary equipment. The power is absorbed in the a-c lines and by the cooling water used for the eddy-current brake. This type of dynamometer is much lower in cost than the commercial d-c dynamometer (14.202). Eddy-current brakes alone have not been very successful, because of their limited torque-speed characteristics.

The commercial water brake is the cheapest form of cradle dynamometer and is satisfactory where no motoring is required. However, without any means for motoring, indicated performance can be measured only by means of an accurate cylinder-pressure indicator. Since single-cylinder-engine friction is not usually representative of multicylinder friction, means for motoring are a great convenience in engine research work (see Volume I, Chapter 9 for a discussion of friction measurement by motoring).

To make accurate torque measurements possible, a cradle dynamometer should have minimum friction in the trunnion bearings, be carefully balanced, and have adequate flexibility in all connections, such as pipes or wires. The best type of trunnion bearing is one floated on an oil film by means of oil under pressure from a pump (14.315, 14.318). Next best are ball trunnion bearings, but these must be replaced before they become worn or pitted. The author has experienced much inconvenience from ball trunnion bearings that have become inaccurate because of corrosion and/or wear.

Noncradle dynamometers, where power is measured by means other than direct torque measurements, are not recommended for any but routine or endurance testing in which accuracy of torque measurement is not required.

MEASUREMENT EQUIPMENT AND TECHNIQUES

Torque Meters

After long experience with beam scales and dial scales, your author has found the hydraulic scale illustrated in Fig. 14–1 to be the most satisfactory type. This scale has no knife edges or similar parts to wear or to be distorted by vibration. It has been found to retain its calibration indefinitely. The motion of the piston between no load and maximum load is so small that errors due to imbalance of the dynamometer are minimized. The scale

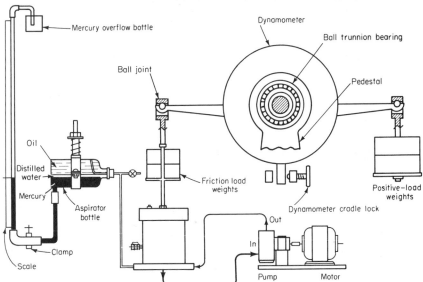

Fig. 14–1. MIT hydraulic torque scale.

requires no adjustment and little or no maintenance beyond seeing that the mercury manometer is in good order. Its sensitivity can be varied by changing the piston diameter or the length of the torque arm, and the damping can be controlled by a pinchcock on a rubber manometer line. The unit is easily constructed in any good machine shop, except for the oil pump and electric motor, which are available commercially at reasonable cost. This unit, complete, costs less than a high-grade dial scale and requires far less maintenance and calibration (14.314, 14.315).

There is a pneumatic torquemeter operating on the same principle, available commercially (14.316), but its cost is relatively high.

Torque Linkage

The knife edges in reversible torque linkages are subject to wear and distortion. Your author much prefers the simple torque arm with provision for torque reversal by means of the addition of calibrated weights at a carefully measured distance from the torque axis.

Dynamometer Speed Controls

Many commercial dynamometers are supplied with automatic speed-control equipment. Such devices can be very troublesome unless their rate and damping can be adjusted to fit the requirements of the engine under test. The control rate must be much slower than the rate of torque variation in the engine. For single-cylinder engines at low rpm, the ordinary speed control has a relatively high rate, which may cause serious fluctuations in the brake load. In such cases hand control of speed is necessary.

In any event, there must be some way of indicating when the engine and dynamometer reach a constant speed, that is, a point without torque due to acceleration or deceleration. The stroboscopic lamp described below is an excellent indicator of this kind.

Speed Measurement

Most commercial tachometers depend on centrifugal force, magnetic drag, or the voltage of a small electric generator. Few such instruments can be relied on to give speed measurements of the accuracy required without frequent calibration and adjustment. In the Sloan Automotive Laboratories at MIT an inexpensive but accurate method of using a tachometer in connection with speed measurements is used. The equipment consists of a commercial tachometer to indicate the approximate speed, plus a neon lamp arranged to flash with the 60-cycle a-c current supply (Fig. 14-2). Thirty-six evenly spaced white stripes on the flywheel or coupling will appear stationary under the flashing lamp at even hundreds of rpm. Small

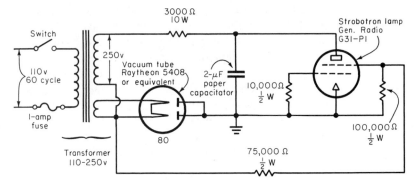

Fig. 14–2. Sixty-cycle stroboscope circuit diagram.

amounts of acceleration or deceleration of the engine are shown immediately by an apparent slow rotation of the stripes. The tachometer need only be accurate enough to identify increments of 100 rpm. Use of this system, with hand control of speed, has been found both convenient and extremely accurate. For 50-cycle systems, 30 stripes on the flywheel or coupling will indicate even hundreds of rpm.

Speed Counters

Speed counters give measurements of the number of revolutions in a given time, the accuracy depending on that of the time measurement. When this is done by means of a hand-controlled stopwatch, the period of measurement should be at least 30 seconds. The difficulty here is that there is no indication of speed deviations during that period.

Automatic electric counters that operate over very short periods of time and display the result every few seconds are available in convenient, though expensive, form. Because of the frequency of the count there is good indication of whether or not speed is being held constant. Accuracy here depends chiefly on the accuracy of the time measurement, which is usually based on the frequency of the a-c supply. In the U.S.A. the standard 60-cps frequency is held to extreme accuracy by the large power networks, and it is the most reliable and convenient time base available in most laboratories (14.321–14.323).

Surge Tanks

The inlet and exhaust systems for laboratory engines must be closed both for reasons of safety and to make flow measurements possible. In order to facilitate such measurements and also to protect the engine from the influence of pressure waves in long systems of piping, it is necessary to install large reservoirs, or *surge tanks*, close to the engine, for both inlet and exhaust. For single-cylinder engines, two inlet surge tanks are required. Four-cycle cylinders should have inlet surge tanks of at least 50-cylinder volumes each. Because of the sensitivity of 2-cycle cylinders to unsteady flow, 100-cylinder volumes are a desirable size for the final inlet surge tank. One surge tank of 50 times the volume of one cylinder is sufficient for multicylinder engines. The exhaust surge tank should be about the same size as the final inlet surge tank. Very short connections between cylinder and tank should be used, to avoid the influence of long inlet and exhaust pipes on engine performance (see Volume I, Fig. 1–1 and Chapters 6 and 7).

Airmeters

Of the many forms of airmeters available, your author has found the orifice meter to be the most satisfactory, provided it is used in a steady-flow system. This requirement means that the flow through the orifice must be protected from the pulsations of the engine inlet system by locating it on the upstream side of the two inlet surge tanks. Of the various forms of orifice meter, the standard ASME orifices (14.401–14.404) have been found both cheap to make and accurate on the basis of the ASME calibration, provided the design is *exactly* as recommended. With such orifices, the accuracy of measurement depends chiefly on having a steady flow and on measuring the pressure difference by means of an accurate manometer. Corrections for air temperature and humidity are contained in the ASME instructions.

Fuel Measurement

In most engine laboratories, fuel flow is measured by weight or by volume, with timing by a stopwatch which may be either manually or automatically controlled. In order to obtain an accurate measurement, the time period must be relatively long, but during such a period the engine speed or load may fluctuate.

A more satisfactory system of fuel-flow measurement is through the use of a reliable flowmeter (14.412–14.414), carefully calibrated for the particular fuel and fuel temperature to be used in the test. The calibration can be made in unlimited time increments and without the distractions associated with an engine in operation. The calibrated flowmeter gives instantaneous indications of fuel-flow rate, and neither an extended period of steady operation nor a hand-operated stopwatch is required for accurate results.

Pressure Measurements

For accurate measurements of small steady pressures, manometers are the most reliable means. Commercial or carefully handmade manometers are acceptable. In the engine laboratories at MIT, slant gauges are used for pressures up to 13 in. (33 cm) of water and vertical mercury manometers for pressures up to 40 in. (100 cm) of mercury (14.421–14.423). For still higher pressures, first-class commercial pressure gauges are satisfactory, provided they are carefully calibrated. Inlet and exhaust pressures can be measured in the surge tanks next to the engine. In passages where the velocity is high, care must be exercised in locating the pressure pickup holes so as to be sure that either static or total pressure is being measured, as desired.

Fluctuating Pressures. The reading of any pressure gauge or manometer in a system with rapidly fluctuating pressure is dependent on the ratio of pulsation frequency to gauge natural frequency (14.407–14.4091). For rapidly fluctuating pressures, ordinary pressure gauges or manometers are quite unsatisfactory. In the case of engines, the most important fluctuating pressures are those in the cylinder. An apparatus designed especially for measurement of cylinder pressures is generally called an engine *indicator*. Such apparatus is also suitable for measuring fluctuating pressures in inlet and exhaust systems.

Engine Indicators

This type of apparatus is discussed in Volume I, Chapter 5. For displaying individual or successive pressure cycles on an oscillograph screen, some form of electric indicator is necessary (14.434–14.436). The best electric indicators have now reached the stage where reasonably accurate measurements of pressure differences are possible, provided great care is used in calibration, operation, and maintenance. As yet, such indicators involve some uncertainty as to the absolute value of the pressure. However, if the pressure at some point in the cycle is known from other considerations, absolute pressures can be measured. More or less successful pressure transducers have been developed on the principles of changing electrical resistance, capacity, pressure on a crystal, or change in a magnetic field (see Fig. 2–1). The most difficult problem has been to eliminate the influence of changing temperature and to make the transducers rugged enough to withstand the high pressures and high rates of pressure change encountered in engine cylinders.

In addition to their ability to record a single cycle, an advantage of electric indicators over others is that, by means of a suitable transducer signaling cylinder volume, pressure-volume diagrams can be directly displayed or photographed. The inconveniences associated with photographic recording have been greatly relieved by the Polaroid and other makes of self-developing camera.

The most accurate indicators for engines are of the balanced-pressure type, where the crank angle at which cylinder pressure equals a known

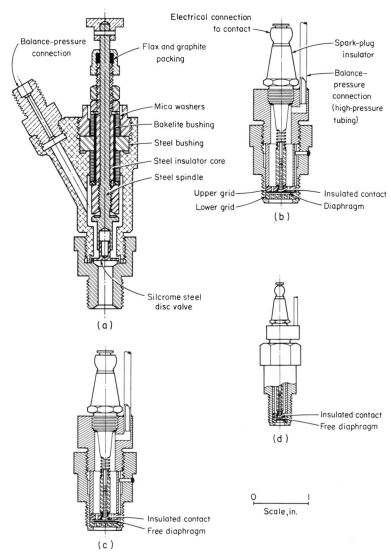

Fig. 14–3. Balanced-pressure indicator units. (*a*) Dobbie-McInnes Farnborough type; (*b*) MIT diaphragm type; ((*c*) MIT free-diaphragm type; (*d*) suggested miniaturized free-diaphragm type.

steady pressure is recorded over the range of cylinder pressures (14.430–14.433). The success of such indicators depends chiefly on the design and maintenance of the pressure-balancing, or *pickup*, unit. Cross sections of well-designed units are shown in Fig. 14–3. These units contain either a thin diaphragm clamped at its circumference, or a disc free to move between two restraining elements. The advantages and disadvantages of these two types are as follows:

Free units tend to clog with oil or deposits and are therefore troublesome in engines with high oil consumption and in Diesel engines operating at any but low fuel-air ratios.

Clamped-diaphragm units involve the force necessary to deflect the diaphragm. This force can be measured by applying known static pressures to the unit, but it may be quite different under operating conditions because of distortion of the diaphragm at high temperatures. Also, since the diaphragm must be as thin as possible, it may fail at high pressures. Because there is no gas flow past the diaphragm, as is the case with free units, the diaphragm unit is less sensitive to oil and deposits. The miniaturized free unit shown at *d* of Fig. 14–3 is suggested on the basis that the smaller the moving unit, the stronger it is at a given ratio of area to mass. Minimum current should be passed through the contacts; this means that an amplifying circuit, such as that shown in Fig. 14–4 is most desirable.

The MIT (Volume I, Figs. 5–3 and 5–4, pp. 115–116) and the Farnboro (Volume I, ref. 5.02) recording apparatus for balanced-pressure indicators are both very expensive. A cheap form of recording mechanism is shown in Fig. 14–5. All that is needed besides the necessary tubing and wiring is a good pickup unit, a telephone head set, one or two dry batteries, a tank of compressed nitrogen, and an electric contactor or *timer* on the crankshaft, rotatable through 360 degrees. To operate, the timer is set for a given crank angle, and the nitrogen pressure is increased until a signal is heard in the receivers. The pressure and crank angle are recorded. By resetting the timer at various crank angles, a complete record of pressure versus crank angle can be made. The only disadvantage of this method is the time and labor required to record and plot an indicator diagram by hand. The system is particularly useful when only maximum pressure, or pressure over a limited range of crank angle, is required.

Balanced-pressure indicators in good condition give very accurate values of average, maximum, and mean cylinder pressure, but cannot, of course, display a single cycle. When such indicators are not in working order, they usually signal this condition by records that have no relation to an indicator diagram. Automatic recording is limited to plotting pressure–crank·angle diagrams, which may then be translated to a pressure-volume basis by hand or by a special device, such as that shown in Fig. 14–6.

A long, narrow passage between cylinder and sensitive element may lead to errors in measurement with any type of indicator. For a discussion of the accuracy of various types of indicators, see Volume I, refs. 5.04, 5.09, 5.092.

In the author's experience, first-class maintenance of balanced-pressure units is necessary. They must be frequently cleaned, kept free from corrosion, and should be stored in air-tight containers with silica gel. They

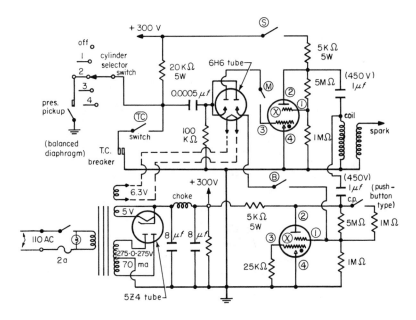

	ⓜ	Ⓑ	Ⓢ	T.C.	Note
Off	Open	O	O	O	Calibrate, const press.
Make and break	Closed	C	C	O	Indicator card
Top center	O	C	O	C	Top center line
Make only	C	O	C	O	Rising pressure
Break only	O	C	O	O	Falling pressure

Fig. 14–4. Electronic circuit for MIT balanced-pressure indicator. T.C. switch is fixed to engine crankshaft so as to close circuit at top dead center. χ's are General Radio 631-P1 strobotrons (see also Volume I, pp. 114–119).

should be calibrated at each assembly to make sure that the pressure difference required to make or break the electric circuit is small.

Temperature Measurement

One of the most difficult measurements to make with accuracy is that of the temperature of a flowing fluid when the containing walls are at a temperature different from that of the fluid itself. In such cases it is hard to protect the thermometric element from the influence of the containing

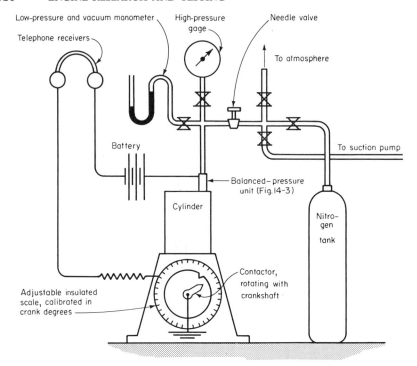

Fig. 14–5. Earphone system of using balanced-pressure indicator units. Pressure gauge should be accurate and carefully calibrated. *X*s indicate hand-operated valves.

walls. There is considerable literature on this subject, which should be consulted in cases where great accuracy is required (14.440–14.452).

For engine inlet-air temperature measurements a glass thermometer or thermocouple in the inlet surge tank will usually suffice. Water and oil temperatures are usually measured with satisfactory accuracy by glass thermometers or thermocouples projecting into the pipes. The use of metal thermometer wells is not conducive to accuracy.

Accurate measurement of true exhaust temperature is nearly impossible because of the highly fluctuating temperatures and velocities of the exhaust gases and the great differences in temperature between the gases and their containing walls. Reference 14.452 discusses several possible methods and their relative accuracy. The reading of a thermometer in the exhaust surge tank has little value. Many Diesel engines are equipped with thermocouples installed near the exhaust port of each cylinder. While these are valuable as indicators of the degree of uniformity with which the cylinders are performing, the temperatures indicated have only comparative significance.

Cylinder-Gas Temperature. Measurement of cylinder-gas temperatures is not only technically extremely difficult but is complicated by velocities of uncertain magnitude and direction, by nonuniformity of temperature within the body of gas, and by extremely rapid temperature fluctuations during the cycle.

Direct measurement by means of a thermocouple or resistance wire is virtually hopeless, both because of the lag of wire temperature behind gas temperature and of the probability of burning up any wire small enough

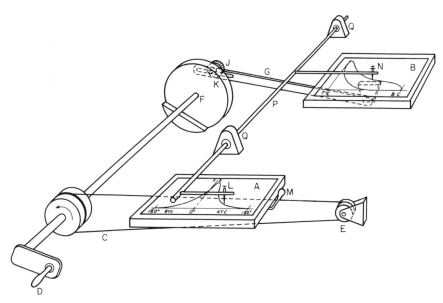

Fig. 14–6. Transfer table for conversion of p-θ diagrams to p-V diagrams.

A	sliding table	G	connecting rod
C	steel tape	B	reciprocating table
D	hand crank	P	sliding shaft
E	fixed pulley	Q	bearings for sliding shaft
F	crank disc	L	p-θ follower
J	adjustment for R/L ratio	N	p-V tracer
		M	clamp, table A to tape

Operation

1. p-θ diagram on table A; 2. Set table B to top-center position; 3. Adjust table A along steel tape to top center and clamp to tape; 4. Set hand crank to desired crank angle; 5. Set L on p-θ curve. 6. Press down N to mark point on p-V diagram. (DEMA Bulletin, 14.431).

to be useful. Indirect measurement of cylinder-gas temperature has been attempted, with more or less success, by three methods: spectroscopic, radiation, and sound velocity. For the first two methods the measurement applies to a large portion of the gas and can give only some kind of an average result. The sound-velocity method, on the other hand, measures within a small element of the gas, and gives a localized result (see Volume I refs. 5.10–5.16 and 5.30–5.32).

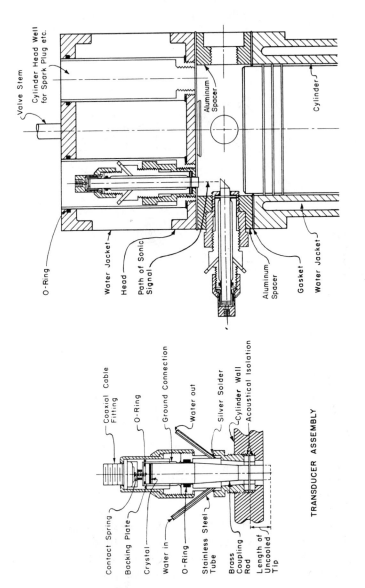

SECTION SHOWING PATH OF SONIC SIGNAL
IN FLAT CYLINDRICAL CHAMBER

TRANSDUCER ASSEMBLY

Fig. 14–7. Sound-velocity transducers and their application to a combustion chamber (Volume I, ref. 5.16).

Measurement of temperature by means of sound-velocity depends on the relation given by Volume I, eq. A–10, which is repeated here for convenience. For a perfect gas:

$$a^2 = \frac{g_0 kRT}{m},$$

$(14-1)$

where a = velocity of sound
  T = absolute temperature
  k = specific-heat ratio
  m = molecular weight
  R = universal gas constant
  g_0 = Newton's law coefficient.

From the specific-heat ratio and molecular weight of the gas, the temperature can be computed for a given value of a.

The sound-velocity method has been quite successful in measuring charge temperatures before combustion, and has been used especially for temperatures of the end gas just before detonation occurs. Figure 14–7 shows the transducers, and Fig. 14–8 is a block diagram of the electronic

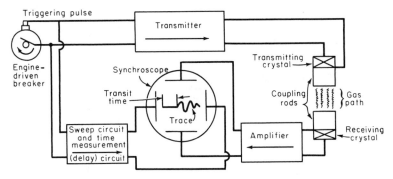

Fig. 14–8. Block diagram of sound-velocity apparatus (Volume I, ref. 5.16).

circuit used in sound-velocity measurements. This method requires two holes suitably located in the cylinder head. However, this would appear to be a less difficult requirement than the transparent windows necessary for radiation or spectroscopic systems.

COMPLETE SINGLE-CYLINDER TEST INSTALLATIONS

Figure 14–9 shows a complete single-cylinder installation, typical of those which have been used for most of the research work in internal-combustion engines done at the Sloan Laboratories for Automotive and Aircraft

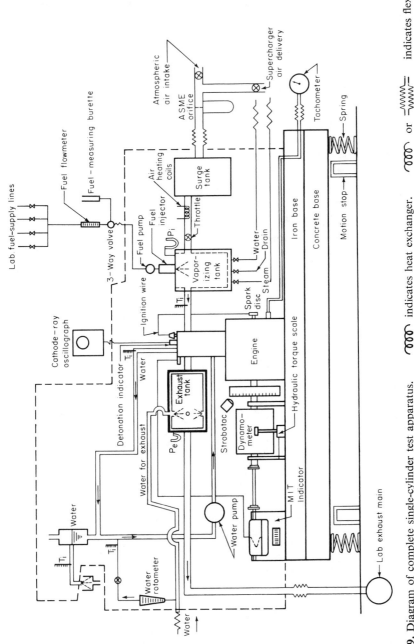

Fig. 14-9. Diagram of complete single-cylinder test apparatus. $\widehat{\text{OOO}}$ indicates heat exchanger. $\widehat{\text{OOO}}$ or $\sim\!\!\!\wedge\!\!\!\wedge\!\!\!\sim$ indicates flexible connection. Everything within the dashed line is mounted on and moves with the mounting base. Control valves may be remotely controlled by flexible electrical or mechanical means. Pressure and temperature readings may also be remote by means of flexible manometer tubes and thermocouple wires.

Engines at MIT. A large part of the material in both volumes of this present work comes from tests made by means of this equipment.

Dynamometer

The dynamometer is made up from a commercial d-c generator, as previously described. The generator should be capable of operating up to at least 3000 rpm and of absorbing 15 hp (30 hp supercharged) at this speed. The hydraulic torque scale (Fig. 14–1) is used.

Spring Mounting of Engine Test Beds

In order to prevent the transmission of engine vibrations to floors or other parts of building, the whole engine-dynamometer unit can be mounted on springs in the manner illustrated by Fig. 14–9. Either steel or rubber springs can be used. At MIT steel coil springs have been found more satisfactory and lower in cost. The springs must be so chosen that the primary natural frequencies in vertical and angular motion are below the lowest steady operating speed of the engine. The reinforced-concrete base is necessary because the usual commercial cast-iron bed plate is not sufficiently rigid by itself to avoid objectionable plate vibrations in torsional and bending modes.

The chief problem involved in the use of such spring mounts is that of making all pipe connections and control connections flexible. In the author's experience, the spring-mounted test bed is very desirable, especially for single-cylinder engines that run at high piston speeds (see refs. 14.250–14.254 for further details).

Fuel System

Where accurate control of fuel-air ratio is required, a carburetor is inconvenient for a single-cylinder engine. In its place, two systems of fuel supply have been used successfully. In one, Fig. 14–10a, fuel is supplied at constant pressure to a needle-valve, from which the fuel is discharged into the inlet surge tank. This tank must be water-jacketed for heating to a temperature higher than that for complete fuel evaporation (for motor gasoline 140°F, 60°C). The interior of the tank is baffled so as to secure good mixing and complete evaporation of the fuel. Complete evaporation is insured if no liquid can be drained from the petcock at the bottom of the tank. Advantages of this system are

1. Engine is supplied with a stabilized gaseous mixture of fuel and air.
2. Inlet temperature can be measured accurately.

Disadvantages are

1. Inlet temperature with normal gasoline cannot be lower than about 140°F (60°C).
2. Backfires in the tank are common until the operator has learned to avoid them by proper control of fuel-air ratio.
3. The tank must be designed for a pressure of 100 psia (7 kg/cm²) and have a good safety valve.
4. Some time is necessary for the fuel-air ratio to stabilize after each adjustment.

A more convenient fuel-supply system involves the use of a timed fuel pump (usually a Diesel type injection pump) by means of which fuel is injected into the inlet port (Fig. 14–10b). Advantages of this system are

1. Any reasonable inlet-air temperature can be used.
2. Explosions in the inlet tank are rare (they occur only if fuel is accidentally allowed to accumulate in the tank).
3. The discharge from the injection pump can be also directed into the inlet tank for use with the first arrangement.
4. It is easier to control fuel-air ratio as engine speed changes.
5. It is readily adapted to Diesel operation, as will be mentioned later.

The only disadvantages of system b for use with spark-ignition engines are the cost of a Diesel fuel pump and the fact that fuel evaporates during the inlet process, thereby influencing volumetric efficiency as a function of fuel-air ratio (see Volume I, Chapter 6 and Volume II, pp. 157 to 159). For tests with a constant fuel-air ratio, the system of Fig. 14–10b is quite satisfactory if it is realized that when the fuel-air ratio is changed, volumetric efficiency is also changed.

Diesel Operation. System b is easily adapted for operation of the engine on the Diesel cycle. The injection pump is merely connected to the Diesel injector unit rather than to the inlet-port injector. A commercial injection pump driven by the half-time shaft (crankshaft in the case of 2-cycle engines), together with a commercial injector, forms the basis of a satisfactory system, provided arrangements are made for varying the injection timing while the engine is running and for easily changing the nozzle tip so as to change the spray pattern as required.

On account of its versatility, the use of a Diesel injection pump for the three systems — inlet-tank, inlet-port, or Diesel — is recommended. When using a Diesel injection pump with gasoline, the manufacturer's recommendations for proper lubrication should be followed.

Ignition System

In all research work with spark-ignition systems it is essential to know the spark timing at all times and to be able to change it while the engine is

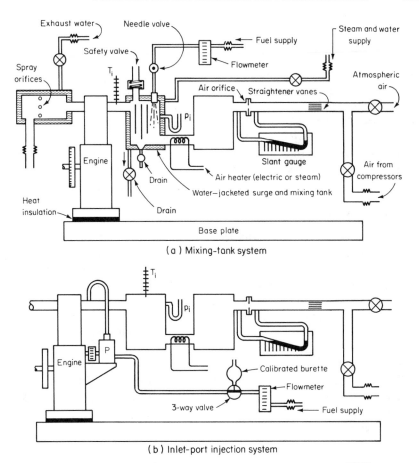

(a) Mixing-tank system

(b) Inlet-port injection system

Fig. 14–10. Fuel and air systems for single-cylinder research engine. ⟨〰⟩ indicates heat exchanger. 〜〰〜 indicates flexible coupling. The exhaust coupling can be rubber, with water in the exhaust.

P = Diesel injection pump with timer coupling, 360° range
T_i = thermometer for inlet temperature
p_i = manometer for inlet pressure

All equipment inside flexible couplings is attached to engine bedplate. Exhaust system is not shown in lower diagram, but is the same as for the upper one. In both cases a throttle valve should be located between surge tanks as in Fig. 14–9.

running. The ignition breaker should be mounted so that it can be set to any desired timing by means of a calibrated disc. For work where more exact measurement is necessary, as in combustion and detonation studies, a high-tension spark indicator, in series with the spark plug, is used. This consists of an insulated arm rotating with the crankshaft inside an insulated disc calibrated in crankshaft degrees. As the spark jumps the small gap between them, its exact angular position is indicated.

Cooling System

The cooling-water system should have a relatively small capacity in order to facilitate rapid attainment of temperature equilibrium. A simple loop circuit with a small storage tank outside the circulation system is recommended (see Fig. 14–9). Temperature is controlled by the make-up-water valve. For measurements of heat rejection to the coolant, the system should be well heat-insulated. Heat flow is measured by measuring the cooling-water flow rate and the temperature rise between cooling-water inlet and water overflow.

Exhaust System

In many engine-test installations, the engine exhausts into a vertical pipe extending through the roof. This system is practical only where the resultant noise is acceptable. When exhaust noise must be suppressed, the most satisfactory method is a spray of water in the exhaust system. In most cases it is well to water-jacket the exhaust surge tank and to deliver the silencing water through the jacket to the interior of the tank, as shown in Fig. 14–9. In this way a minimum of exhaust heat is delivered to the test room. Drainage of water back into the engine must be absolutely prevented by having the exhaust outlet from the surge tank well below the engine exhaust ports and by providing for automatic drainage of water whenever the engine is shut down. Any laboratory exhaust system should be kept under continuous negative pressure (2 to 3 in. of water) by means of an adequate exhaust fan.

Auxiliaries

Water pump and oil pump are best driven separately by small electric motors. This system has two advantages: (1) a variable source of engine friction is eliminated, and (2) the engine may be warmed up by circulating heated oil and water through the system without operating the engine itself. This feature saves wear and tear on the engine. The pumps can be safely operated without attention while the test personnel are otherwise engaged. Of course the final warming up before actual testing must be done with the engine in operation. Heat exchangers with steam or electric heating are necessary, as indicated in Fig. 14–9.

Design Changes

One great advantage of the small single-cylinder engine for instruction or for research work is the relatively low cost and short time required for structural changes. Among the changes made during the course of the author's research work have been

New cylinder heads, with different valve and port arrangements (Volume I, refs. 6.41, 6.44).

Two-cycle cylinders of both the loop-scavenged and poppet-valve type (Volume I, refs. 7.22, 7.23).

A wide variety of pistons. (Experimental pistons are most easily made by machining from a round bar of aluminum.)

Many different camshafts (Volume I, refs. 6.41, 6.44, 6.63).

Cylinder heads for special purposes, such as temperature measurement (Volume I, ref. 5.16) and flame photography (*1.50 et seq.*).

Diesel cylinders of various types.

Nearly all of this equipment has been made in the machine shop connected with Sloan Laboratories for Aircraft and Automotive Engines at MIT.

TEST PROCEDURE

Although no detailed instructions as to test procedure can be given here, a few general remarks based on the author's experience may be found useful.

Temperature Equilibrium. The need to have all temperatures stabilized is an obvious one, since engine temperatures affect nearly all aspects of engine performance. Helpful practice in this regard includes

1. Engine base should be thermally insulated from the bedplate.

2. Separately driven oil and water pumps, together with adequate heaters, will facilitate the warm-up process.

3. With similar installations, the warm-up time required will increase nearly in proportion to the cylinder bore.

Speed Equilibrium. Speed equilibrium is very important, as mentioned in connection with speed measurement.

Adequate manpower is important, so that instrument-reading time can be minimized, with one man always free to devote full time to engine control and observation.

SAFETY

General Safety Measures

In experimental work there is often a tendency to use improper equipment in the interest of saving money and time. Long experience has taught your author that "jerry-built" test equipment is usually dangerous as well as unsatisfactory from the technical point of view. There is no substitute for careful design and installation of all piping, wiring, couplings, controls, and so on to avoid failures in operation. Time spent to secure a thoroughly sound and secure installation not only promotes safety but saves time in the end by avoiding shutdowns for repair.

All laboratories where power machinery and inflammables are present should be properly equipped with fire extinguishers, fire blankets, emergency showers, and a stretcher.

From the safety point of view, test beds should not be confined in small rooms unless all controls are located outside. For college laboratories especially, a single large space is to be preferred over individual test rooms. Not only is the large space less likely to accumulate dangerous concentrations of fumes, but it affords easy escape and easy surveillance.

Another important general rule in regard to safety is *never to leave an operating engine unattended.* Furthermore, no engine should ever be run without at least one person in addition to the operator present or very near at hand. In general, students operating engines should be in teams of not less than two and should have professional supervision at all times.

The chief dangers in engine testing are

1. Explosion and/or fire due to accumulation of fuel or fuel vapor
2. Bursting or breakage of tanks, pipes, or hoses
3. Mechanical failure of flywheels, couplings, etc., due to overspeed
4. Asphyxiation
5. Electric shock
6. Human contact with moving parts of the installation
7. Burns due to contact with hot exhaust-system parts or with escaping hot water or hot oil.

Fuel Fires and Explosions

Experience shows this to be the most serious danger connected with the testing of internal-combustion engines. The worst accidents have been due to overnight accumulation of fuel vapor in the test room. Necessary precautions include

Continuous forced ventilation, 24 hours a day

Use of pumped fuel-supply systems rather than gravity feed. Such systems should display continuous warning lights when pumps are in operation.

Avoidance of trenches, sumps, etc., below floor level, where fuel or oil can accumulate unnoticed

Use of nonsparking light switches and electric motors

Frequent inspection.

Fuel fires also can occur when fuel-supply lines break during a test. Only well-supported tubing, or hose especially designed for fuel, should be allowed.

When a local supply of special fuel must be obtained from a tank or drum, it should be pumped to the engine by means of a closed system. The tank or drum should never be hoisted above the engine for the sake of gravity feed or for any other reason.

Another danger characteristic of engine testing is the possible rupture of piping that contains hot liquids or inflammable material. The vibration associated with engine operation requires special precautions in this regard. All joints should be carefully made, and all piping should be supported so as to minimize vibration. Flexible hose for fuel and oil must be of a type and material especially designed for such liquids.

Another source of possible danger is rupture of surge tanks or piping due to explosions of fuel-air mixture. Such explosions may be caused by backfires in inlet systems or by the accumulation of unburned mixture in the exhaust system. Necessary precautions include the installation of suitable safety valves or blowout plugs at points where they will not impinge on personnel or equipment. In spite of all normal precautions, such explosions do occur. Therefore the equipment should be designed so that they can take place without damage.

Mechanical Failures

This type of failure is most often due to overspeeding of the engine. Excessive engine speed is most likely in the hands of inexperienced operators, including students. Important precautions in this regard are

1. Locating instruments and controls at the ends, rather than at the sides, of the engine and dynamometer
2. Use of steel rather than cast-iron flywheels wherever possible (all single-cylinder engines at MIT have steel flywheels)
3. Use of approved couplings only between engine and dynamometer
4. Careful training of operators
5. Use of overspeed shut-off devices when practicable
6. Great care in the design and installation of all rotating test equipment.

Asphyxiation

Dangerous fumes include engine exhaust, fumes from the crankcase breather, and fuel vapor, particularly when it contains tetraethyl lead. The forced ventilation necessary for fire prevention is the best preventative for this type of accident, together with the obvious requirement for well-designed closed fuel systems and closed exhaust systems. Breathers should be piped into exhaust ducts. Precautions against dangers of electric shock and contact with moving parts are those which should be used in any well-regulated building where electric current and power machinery is in operation. General current supply should never exceed 220 volts. Higher-voltage equipment should be enclosed and provided with danger signs. Belts, couplings, and other rotating elements should be properly guarded.

Accidents

The author has known of, or experienced, the following accidents in connection with engine testing:

Two test-room explosions from overnight accumulation of fuel vapor (both fatal).

One catastrophic failure of an automatic transmission due to engine overspeed.

Many fires of accumulated fuel and oil in exhaust systems.

Numerous fires following fuel-pipe failures or fuel accidentally spilled.

Serious burns due to breakage of hot-water pipes or pipes containing hot oil.

Safety-valve failure on surge tank, caused by improperly installed valve.

Crankcase explosions in very large engines (some fatal).

Numerous instances of crankshaft or connecting-rod failure at normal engine speeds. Such failures have seldom thrown material out at dangerous velocity. However, oil-vapor fires quite often result if the crankcase is ruptured.

Conforming to the safety precautions recommended should make such accidents highly improbable.

Addendum, 1984

Since this chapter was written (1968), the development of electronic control systems has revolutionized instrumentation for the testing of engines and the recording of test results. In the case of road-vehicle engines, elaborate equipment for measuring and recording exhaust emissions has become available. These new systems have improved the accuracy and speed of obtaining test results, especially in the case of routine testing.

Symbols and Their Dimensions

(See also special symbols for Chapter 8, pp. 241–243.)

Symbol	Name	Dimension
A	area	L^2
a	acceleration	Lt^{-2}
a	velocity of sound	Lt^{-1}
amep	accessory-drive mep	FL^{-2}
atc	after top center	
atm	atmosphere	FL^{-2}
B	a coefficient	any
B	barometric pressure	FL^{-2}
BHN	Brinnell hardness number	
Btu	British thermal unit	Q
b	bore	L
bbc	before bottom center	
bc	before center	
bhp	brake horsepower	FLt^{-1}
bmep	brake mean effective pressure	FL^{-2}
bpsa	best-power spark advance	
bsac	brake specific air consumption	$MF^{-1}L^{-1}$
bsfc	brake specific fuel consumption	$MF^{-1}L^{-1}$
btc	before top center	
°C	degrees Centigrade	θ
C	engine flow coefficient	1 *
C	a dimensionless coefficient	1
C	clearance of a bearing	L
C_e	exhaust-valve flow coefficient	1
C_i	inlet-valve flow coefficient	1
C_n	turbine-nozzle flow coefficient	1
C_p	specific heat at constant pressure	$QM^{-1}\theta^{-1}$
C_v	specific heat at constant volume	$QM^{-1}\theta^{-1}$
cg	center of gravity	
cm	centimeter	L
cmep	compressor mean effective pressure	FL^{-2}

* The number 1 indicates a dimensionless quantity, that is, a ratio. A blank space indicates an abbreviation.

629

cps	cycles per second	t^{-1}
D	diameter	L
d	diameter	L
d	differential operator	
d	depth of wear	L
E	internal energy (excluding potential, electric, etc.) per unit mass	QM^{-1}
E^*	internal energy for $(1 + F)$ pounds-mass	QM^{-1}
E°	internal energy for one pound-mole	QM^{-1}
E_{lg}	internal energy of liquid referred to gaseous state	QM^{-1}
E_{sf}	sensible internal energy per unit mass of gaseous fuel	QM^{-1}
e_s	scavenging efficiency	1
e_v	measured or actual volumetric efficiency	1
e_{vi}	volumetric efficiency with the ideal induction process	1
e_{vo}	volumetric efficiency at reference conditions, or *base* volumetric efficiency	1
°F	degrees Fahrenheit	θ
F	force (fundamental unit)	F
F	fuel-air ratio (over-all)	1
F'	combustion fuel-air ratio	1
F_R	F/F_c	1
F_c	stoichiometric, or chemically correct, fuel-air ratio	1
f	followed by a parenthesis indicates a function of the material in parentheses	
f	residual-gas fraction, coefficient of friction	1
fhp	friction horsepower	FLt^{-1}
fmep	friction mean effective pressure	FL^{-2}
fpm	feet per minute	Lt^{-1}
fps	feet per second	Lt^{-1}
ft	foot	L
G	mass flow per unit area, or mass velocity	$ML^{-2}t^{-1}$
g	acceleration of gravity	Lt^{-2}
g_o	force-mass-acceleration coefficient (Newton's law coefficient)	$MLt^{-2}F^{-1}$
H	enthalpy per unit mass	QM^{-1}
H^*	enthalpy of $(1 + F)$ pounds-mass	QM^{-1}
H°	enthalpy per pound mole	QM^{-1}
H_{lg}	enthalpy of liquid referred to gaseous state	QM^{-1}
H_o	stagnation enthalpy per unit mass	QM^{-1}
H_{sf}	sensible enthalpy per unit mass of gaseous fuel	QM^{-1}
h	ratio mass water vapor to mass air	1
h	heat-transfer coefficient	$Q\theta^{-1}L^{-2}t^{-1}$
h	hardness	FL^{-2}
hp	horsepower	FLt^{-1}

hp-hr	horsepower hour	FL
hucr	highest useful compression ratio	1
I	mass moment of inertia	ML^2
ihp	indicated horsepower	FLt^{-1}
imep	indicated mean effective pressure	FL^{-2}
in.	inch	L
isac	indicated specific air consumption	$MF^{-1}L^{-1}$
isfc	indicated specific fuel consumption	$MF^{-1}L^{-1}$
J	Joule's law coefficient	FLQ^{-1}
K	coefficient with dimensions	any
k	specific heat ratio C_p/C_v	1
kg	kilogram	M or F
k_c	thermal-conductivity coefficient of coolant	$QL^{-1}\theta^{-1}t^{-1}$
k_g	thermal-conductivity coefficient of gas	$QL^{-1}\theta^{-1}t^{-1}$
k_s	stiffness coefficient	FL^{-1}
k_w	thermal-conductivity coefficient of a solid wall	$QL^{-1}\theta^{-1}t^{-1}$
klimep	knock-limited indicated mean effective pressure	FL^{-2}
ksi	stress, thousands of pounds per square inch	FL^{-2}
L	length (fundamental unit)	L
l	length	L
lb mole	pound-mole	M
lbf	pound-force	F
lbm	pound-mass	M
M	mass (fundamental unit)	M
$\mathbf{M}$	Mach number	1
$\dot{M}$	mass flow per unit time	Mt^{-1}
$\dot{M}^*$	critical mass flow per unit time	Mt^{-1}
m	molecular weight	1
m	meter	L
mep	mean effective pressure	FL^{-2}
min	minute	t
mmep	mechanical-friction mean effective pressure	FL^{-2}
mpg	miles per gallon	L^{-2}
N	number of moles	1
N	revolutions per unit time	t^{-1}
n	a number, an exponent, a harmonic order	1
O	zero	1
P	power	FLt^{-1}
P	load due to pressure	F
$\mathbf{P}$	Prandtl number	1
p	pressure, absolute	FL^{-2}
p_e	exhaust pressure	FL^{-2}
p_i	inlet pressure	FL^{-2}
p_o	stagnation pressure, absolute	FL^{-2}

pmep	pumping mean effective pressure	FL^{-2}
PN	performance number	1
psi	pounds per square inch	FL^{-2}
psia	pounds per square inch absolute	FL^{-2}
psig	pounds per square inch gauge	FL^{-2}
Q	quantity of heat (fundamental unit)	Q
$\dot{Q}$	heat flow per unit time	Qt^{-1}
Q_c	heat of combustion per unit mass	QM^{-1}
q	internal energy of residual gas per unit mass	QM^{-1}
°R	degrees Rankine (°F + 460)	θ
R	a design ratio $(R_1, R_2, \ldots, R_n)$	1
R	universal gas constant	$FL\theta^{-1}M^{-1}$
R	Reynolds number	1
R_i	ratio imep to reference imep	1
R_s	scavenging ratio	1
r	compression ratio	1
r	pressure ratio, absolute	1
rpm	revolutions per minute	t^{-1}
S	stroke	L
S	entropy per unit mass	$Q\theta^{-1}M^{-1}$
s	mean piston speed	Lt^{-1}
sa	spark advance angle	1
sac	specific air consumption	$MF^{-1}L^{-1}$
sec	second	t
sfc	specific fuel consumption	$MF^{-1}L^{-1}$
T	temperature (fundamental unit)	θ
T_i	inlet temperature	θ
T_o	stagnation temperature	θ
t	time (fundamental unit)	t
t	thickness	L
tmep	turbine mean effective pressure	FL^{-2}
U	internal energy, total from all sources	Q
UTS	ultimate tensile strength	FL^{-2}
u	velocity (usually of a fluid stream)	Lt^{-1}
V	volume	L^3
V^*	volume of $1 + F$ pounds-mass	L^3M^{-1}
V°	volume of one pound mole	L^3M^{-1}
V_c	maximum cylinder volume	L^3
V_d	displacement volume	L^3
v	volume of a unit mass	L^3M^{-1}
W	weight, load	F
w	work	FL
w_s	shaft work	FL
x	an unknown quantity	any

x	a length	L
Y_c	adiabatic compression factor	1
Y_t	adiabatic expansion factor	1
y	a length	L
Z	inlet-valve Mach index	1
Z'	compressor inlet-valve Mach index	1
α	an angle or a ratio, aftercooler effectiveness	1
β	an angle or a ratio	1
Γ	trapping efficiency	1
γ	ratio of exhaust valve to inlet valve, flow capacity = $C_e A_e / C_i A_i$	1
Δ	increment sign (ΔT, Δp, etc.)	
ϵ	base of Naperian logarithms	1
η	efficiency	1
η_a	efficiency of the air cycle	1
η_b	brake efficiency	1
η_c	compressor efficiency	1
η_i	indicated efficiency	1
η_{kb}	blowdown-turbine kinetic efficiency	1
η_m	mechanical efficiency	1
η_o	efficiency of the fuel-air cycle	1
η_{ts}	turbine efficiency, steady-flow	1
η'	efficiency based on fuel retained in cylinder	1
θ	temperature (fundamental unit)	θ
θ	crank angle	1
Λ	reduced port area divided by piston area	1
μ	viscosity	$FL^{-2}t$
π	circumference/diameter	1
ρ	density	ML^{-3}
ρ_a	density of air at inlet valve	ML^{-3}
ρ_i	inlet density	ML^{-3}
ρ_o	stagnation density	ML^{-3}
ρ_s	scavenging density	ML^{-3}
σ	unit stress	FL^{-2}
Φ	a function of what follows in parentheses	
Ω	angular velocity	t^{-1}
ω	angular frequency	t^{-1}

SUBSCRIPTS

a	air or atmosphere, adiabatic
b	brake, base value, burned
c	coolant, combustion, or compression

d	displacement (volume)
e	exhaust or expansion
f	fuel
g	gas
i	inlet, indicated, ideal
m	mixture of fuel and air
n	natural (frequency)
o	reference state or stagnation state
p	refers to horsepower constant, piston, or constant-pressure
r	residual gas, ratio
s	surface, scavenging, sensible
u	unburned
v	volumetric, constant-volume, or vapor
w	wall (material) or water
x	at point of inlet-valve opening
y	at point of inlet-valve closing

SUPERSCRIPTS

m	exponent for Prandtl number
n	exponent for Reynolds number
$'$	indicates special condition

CONVERSION FACTORS, ENGLISH TO METRIC

Multiply	by	to obtain
Btu	0.252	kg calories
psi	0.0703	kg/cm^2
ksi	0.703	kg/mm^2
lbm/cu ft	16.02	kg/m^3
lbm/cu in.	27.68	kg/liter
°R	0.555	°K
(°F − 32)	0.555	°C
inches	2.54	cm
inches	25.4	mm
feet	0.305	m
lbm or lbf	0.454	kg
lbm or lbf	454	g
sq in.	6.45	cm^2
cu in.	16.39	cm^3
cu in.	0.01639	liters
ft/min	0.00508	m/sec
lb sec/sq ft (viscosity)	478	poises

1984 REFERENCES FOR VOL. 2

Since the Bibliography beginning on page 523 was printed, the literature on the subject of internal-combustion engines has grown beyond the possibility of detailed listing. Publications in English for following continuing developments include:

Publications of the Society of Automotive Engineers (SAE), Warrendale, PA, 15096, as follows:

Automotive Engineering, a monthly magazine including reviews of current practice and condensations of important papers.

SAE Reports, classified by subject matter and containing selected *SAE Papers* on various subjects. Lists of each year's reports are available annually on request.

Individual *SAE Papers*, available in limited quantities on application or at SAE meeting.

Automotive Engineer, monthly, Box 24, Bury St. Edmunds, Suffolk, IP326BW, England. Published bi-monthly by Inst. of Mechanical Engineers, Automotive Division.

Bulletin of Marine Engineering Society in Japan, quarterly. The Marine Engineering Society of Japan, 1-2-2 Uchisaiwai-Cho, Chiyoda-Peu, Tokyo, 100, Japan.

Chilton Automotive Industries, monthly. Chilton Way, Radnor, PA, 19089. Statistical issue, April each year.

Diesel and Gas Turbine World Wide Catalog, 13555 Bishop Court, Brookfield, Wisconsin, 53005-0943. Annual.

Bibliography

ABBREVIATIONS

AER	Aeronautical Engineering Review, publ. by IAS, New York
AFA	American Foundryman's Association
AIME	American Institute of Mining and Metallurgical Engineers
API	American Petroleum Institute
ASM	American Society for Metals
ASME	American Society of Mechanical Engineers
SNME	American Society of Naval Architects and Marine Engineers
ASTM	American Society for Testing Materials
ATZ	Automobiltechnische Zeitschrift
Auto. Ind.	Automotive Industries
CFR	Cooperative Fuel Research (now CRC)
CIMAC	Congrès International des Moteurs à Combustion Interne
CIT	California Institute of Technology
CRC	Coordinating Research Council, New York
DEMA	Diesel Engine Manufacturers Association, Chicago, Illinois
EES	Engineering Experiment Station
IAE	Institute of Automobile Engineers, London
IAS	Institute of the Aeronautical Sciences, New York*
ICE	Internal Combustion Engine(s)
IME	Institute of Mechanical Engineers, London
IPT	Institute of Petroleum Technologists, London
ISC	International Symposium on Combustion
JSME	Japanese Society of Mechanical Engineers, Tokyo
M.A.N.	Maschinenfabrik Augsburg-Nürnberg
MIT	Massachusetts Institute of Technology, Cambridge, Mass.
MTZ	Motortechnische Zeitschrift
NACA	United States National Advisory Committee for Aeronautics. Since 1958 renamed the National Aeronautics and Space Agency. Published material is listed as follows:

ARR	Advance Research Reports
MR	Memorandum Reports
TR	Technical Reports
TN	Technical Notes
TM	Technical Memoranda
WR	Wartime Reports

* Since 1963, American Institute of Aeronautics and Astronautics.

NASA	National Aeronautics and Space Agency
NPL	National Physics Laboratory (England)
ONI	Office of Naval Intelligence
RAE	Royal Aeronautical Establishment (England)
RAS	Royal Aeronautical Society (London)
SAE	Society of Automotive Engineers
SAL	Sloan Laboratories for Aircraft and Automotive Engines, MIT
SESA	Society for Experimental Stress Analysis
S-I	Spark-Ignition
SP	Special Publication (each consists of a group of papers on various subjects, see ref. 11.002)
Trans.	Transactions
WADC	Wright Air Development Center, Dayton, Ohio
ZVDI	Zeitschrift des Vereines deutscher Ingenieure, Berlin

INTRODUCTION

The number of publications in this area since Volume I was published in 1960 has been so huge that only a sampling can be given here. The items listed have all been reviewed by the author and are limited to those considered to be important contributions.

General and Historical

0 010 See Vol. 1, p. 524, references 1.01–1.16.

0 011 Ricardo, *Memories and Machines*, Anchor Press Ltd., Tiptree, Essex, England.

0.012 Taylor, *Aircraft Propulsion*, Smithsonian Institution, Washington, D.C.

0.020 Proceedings of CIMAC, Congrès International des Moteurs à Combustion Interne. Published every two years by that organization. (Papers from this international meeting held every two years in Europe contain many important contributions in the field of engines, with major emphasis usually on the larger sizes of Diesel engines. Copies are obtainable through Library Service Association, 11 rue Lavoisier, Paris 8, France.)

Thermodynamic Data

(See also Volume I, Chapter 3 and refs. 3.20–3.54.)

0.030 Newhall and Starkman, "Thermodynamic Properties of Octane and Air for Engine Performance Calculations," in *Digital Calculations of Engine Cycles*, Vol. 7, SAE Special Publication *Progress in Technology* (TP7), 1964, p. 38. See also bibliography in that publication. (These charts, computer based, cover wider ranges than charts *C*-1 through *C*-4 in the back cover of Volume I.)

0.032 Bell and Starkman, "A Consideration of the High Temperature Thermodynamics of Internal Combustion Engines," in *Digital Calculations of Engine Cycles*, Vol. 7, SAE Special Publication *Progress in Technology* (TP7), 1964, p. 1. (Computer calculations of equilibrium compositions versus temperature and pressure.)

0.033 Kopa *et al.*, "Combustion, Temperature, Pressure, and Products at Chemical Equilibrium," in *Digital Calculations of Engine Cycles*, Vol. 7, SAE Special Publication *Progress in Technology* (TP7), 1964, p. 10.

0.034 Starkman *et al.*, "Comparative Performance of Alcohol and Hydrocarbon Fuels," *SAE Paper* SP-254, June 1964. (Contains thermodynamic charts for ethyl and methyl alcohol.)

0.035 Starkman, "Thermodynamics of Propane-Air and Methane-Air Mixture for Engine Combustion Calculations," *SAE Paper* 670466, 1967.

Fuel-Air Cycles

(See also Volume I, Chapter 4, and references.)

0.040 Edson and Taylor, "The Limits of Engine Performance," in *Digital Calculations of Engine Cycles*, Vol. 7, SAE Special Publication *Progress in Technology* (TP7), 1964, p. 65. (Characteristics of constant-volume fuel-air cycles from F_R 0.4 to 1.4, r from 6 to 24, p_1 from 1 to 6 atm, f from 0 to 0.10, h from 0 to 0.06. Cycles are based on conditions at point 1. Basis of Figs. 0–1 through 0–6. Also basis of Fig. 4–5 in revised edition of Volume I, 1966.)

0.041 Edson, "The Influence of Compression Ratio and Dissociation on Ideal Otto Cycle Engine Thermal Efficiency," in *Digital Calculations of Engine Cycles*, Vol. 7, SAE Special Publication *Progress in Technology* (TP7), 1964, p. 49. (Fuel-air cycles up to compression ratio 300. Uses ideal four-stroke inlet and exhaust processes. Outlines computer program.)

0.042 Bonamy, "An Analysis of the Otto Cycle Taking into Consideration the Effect of Varying Operating Parameters," *J. Eng. Power* (*Trans. ASME*) 85, 165 (July 1963). (Fuel-air cycles by computer, with ideal four-stroke inlet-exhaust process.)

0.043 Van der Werf, "The Idealized Limited-Pressure Compression-Ignition Cycle: An Empirical Treatment of Its Response to Varying Parameters," *ASME Paper* 63–OGP–S, May 1963. (Uses ideal four-stroke inlet-exhaust process. Empirical relations for computing over a wide range of conditions.)

0.044 Bonamy and Van der Werf, "Constant-Volume and Limited-Pressure Engine Cycles: A Theoretical Study Applicable to Both Two-Stroke and Four-Stroke Engines," *ASME Paper* 65–OGP–1, 1965. (Digital-computer results, based on conditions at point 1. Nomographs for quick solutions for efficiency, mep, and work. Range: p_1 from 6 to 50 psia, T_1 600° to 1000°R, F_R 1.0 to 1.5 for constant-volume cycles, and 0.3 to 0.8 for LP cycles. Residual fraction 0 to 0.10. Important reference, superseding in value all work to date on the limited-pressure cycle.)

0.045 Sabat and Ahmed, "An Analysis of the Limited-Pressure Fuel-Air Cycle," S.M. Thesis, Department of Naval Architecture and Marine Engineering, Massachusetts Institute of Technology, August 1962. (Curves for LP cycle with $T_1 = 520°R$ (289°K), F_R 0.25 to 0.90, p_3/p_1 60, 75, 90, and also at $p_3 = 1500$ psia (106 kg/cm²). (Student work, based on Hottel charts, Volume I, ref. 3.20. Included here for comparison with previous reference.)

Actual Cycle Analysis
(See also Volume I, Chapter 5.)

0.050 Kerley and Thurston, "The Indicated Performance of Otto-Cycle Engines," *Trans. SAE 70*, 5 (1962). (Engine tests with compression ratios 7–14 and relative fuel-air ratios 0.52–1.5. Indicator diagrams included. Gasoline and methanol fuels.)

0.051 Tsao and Wu, "On the Mathematical Model of Motored Compression Temperature," *SAE Paper* 650453, May 1965.

Four-Stroke Air Capacity
(See also Volume I, Chapter 6.)

0.060 Huber and Brown, "Computation of Instantaneous Air Flow and Volumetric Efficiency," *Trans. SAE 73*, 698 (1965). (Effects of valve flow characteristics and piston speed, with constant pressures at inlet and exhaust. Fair correlation with measurements on actual engine.)

0.061 Wolgemuth and Olson, "A Study of Engine Breathing Characteristics," *SAE Paper* 650448, May 1965. (Computer results, based on simplified assumptions. Useful for trends as a function of design and operating parameters.)

0.062 Tsao and Wu (see ref. 0.051).

Two-Stroke Engine Research
(Volume I, Chapter 7; see also refs. 12.05–12.092.)

0.070 Miyabe and Shimomura, "Measurement of Trapping Efficiencies of a Two-Stroke Engine," *Bull. JSME 6*, No. 22, 297 (1963). (Exhaust-gas analysis with a crankcase-scavenged engine using a known fuel-air mixture. Considerations of accuracy included. Engine shows equal or slightly better than perfect-mixing curve for scavenging ratios between 0.3 and 0.65.)

Heat Losses and Heat Transfer Research
(Volume I, Chapter 8; see also refs. 10.880–10.884.)

0.080 Annand, "Heat Transfer in the Cylinders of Reciprocating Internal Combustion Engines," *Proc. IME* (London) *177*, 973 (1963). (Review of the literature and re-evaluation of existing data. Thirty-five items in bibliography.)

0.081 Moore *et al.*, "Heat Transfer and Engine Cooling, Aluminum versus Cast Iron," *Trans. SAE 71*, 152 (1963). (Measurements of temperatures and rates in an "all aluminum" and "all cast-iron" engine. Octane requirements found the same.)

0.082 Overbye *et al.*, "Unsteady Heat Transfer in Engines," *Trans. SAE 69*, 461 (1961). (Exhaustive data based on the temperature of an insulated thermocouple at the cylinder-head surface. Analysis based on the doubtful assumption that thermocouple reading represents surface temperature.)

0.083 Ebersole *et al.*, "The Radiant and Convective Components of Diesel-Engine Heat Transfer," *SAE Paper* 701C, June 1963. (Indicated radiation found to be 5 to 10 per cent of *local* heat transfer at light load and 35 to 45 per cent at rated load for open chamber GM-71 engine at one point in cylinder head.)

0.084 Woschni, "Computer Programs to Determine the Relationship Between Pressure Flow, Heat Release, and Thermal Load in Diesel Engines," *SAE Paper* 650450, May 1965. (Analytical work at M.A.N., Augsburg. Theoretical trends. No comparison with tests.)

0.085 Alcock, "Heat Transfer in Diesel Engines," *Proc. Int'l Heat-Transfer Conference*, Boulder, Colo., 1961, publ. by IME, London, 1961, p. 174. (Measurements of local rates by means of wall thermocouples in series. Bibliography.)

0.086 Oguri, "On the Coefficient of Heat Transfer Between Gases and Cylinder Walls of the Spark-Ignition Engine," *Bull. JSME 3*, 363 (1960).

0.087 Gross-Gronomski, "Piston Temperature in Compression-Ignition Engines," *SAE Paper* 670469, May 1967. (Measurements on several engines. Speed effects both at constant power and constant bmep. Bibliography.)

Engine Friction and Lubrication

(See also Volume I, Chapter 9.)

0.090 Whitehouse and Metcalf, "The Power Losses of Reciprocating I.C. Engines: Some Preliminary Motoring Tests on Two Small Modern Engines," Motor Industry Research Assn. (Lindley, England), *Report* No. 1957/5. (Gives friction breakdown data in terms of fhp and rpm.)

0.091 Whitehouse and Metcalf, "The Mechanical Efficiency of I.C. Engines," Motor Industry Research Assn. (Lindley, England), June 1958. (A review of published information, unfortunately based on hp and rpm instead of fmep and piston speed.)

0.092 Bishop, "Effect of Design Variables on Friction and Economy, *SAE Paper* 812A, Jan. 1964. (An attempt, not entirely successful, to establish friction contributions of various engine parts, including number of rings, ring tension, etc. Based on motoring results. Discussion important.)

0.093 Starkman, "A Radioactive Tracer Study of Lubricating Oil Consumption," *Trans. SAE 69*, 86 (1961). (A valuable study of oil consumption in an operating engine.)

0.094 Various authors, "Symposium on Lubricants," *SAE Special Publication SP-140*, 1955.

0.095 Various authors, "Symposium on Wear in Engines," *SAE Special Publication SP-116*, 1958.

0.096 Hersey, *Theory and Research in Lubrication*, John Wiley & Sons, Inc., New York, 1966. (Latest by a well-known authority in this field.)

0.097 Rabinowicz, *Friction and Wear of Materials*, John Wiley & Sons, Inc., New York, 1965. (Results of extensive research at MIT, theory and experiment.)

Size Effects

(See also Volume I, Chapter 11.)

0.110 Taylor, "Effect of Cylinder Size on Detonation and Octane Requirements," *Trans. SAE 60*, 175 (1962). (Report on work with the MIT similar engines.)

0.112 Gray, *How Animals Move*, Penguin Books, London, 1965. (This and the following reference show size effects in nature similar to those applying to engines and other machinery. Further proof of the relations is outlined in Volume I, Chapter 11.)

0.113 Greenewalt, *Humming Birds*, Doubleday, Garden City, N.Y., 1960. (Includes data on weights and wing-beat frequencies as a function of length.)

0.114 Collier, " Multi-Engine Marine Drive Meets Varying Road Demands," *Diesel and Gas Turbine Progress*, 31 (Feb. 1967). (Description of Great Lakes vessels using from two to six Fairbanks-Morse o.p. engines geared to a single propeller shaft. Typical gear ratio 750 to 200 rpm.)

Engine Performance, Unsupercharged
(See also Volume I, Chapter 12.)

0.120 Dicksee, "Influence of Atmospheric Pressure and Temperature upon the Performance of a Naturally Aspirated Four-Stroke C.I. Engine," IME (London), *Proc. Automobile Div.* 1959–1960, p. 83. (Experimental determination and correlation of results at fixed rates of fuel injection. Important contribution.)

0.121 Patterson and Van Wylen, "A Digital Computer Simulation for Spark-Ignited Engine Cycles," in *Digital Calculations of Engine Cycles*, Vol. 7, SAE Special Publication *Progress in Technology* (TP7), 1964, p. 82. (Allowances for combustion time and heat losses give fair simulation of actual cycle in a test engine.)

0.122 Strange, "An Analysis of the Ideal Otto Cycle, Including the Effects of Heat Transfer, Finite Combustion Rates, Chemical Dissociation, and Mechanical Losses," in *Digital Calculations of Engine Cycles*, Vol. 7, SAE Special Publication *Progress in Technology* (TP7), 1964, p. 92. (Similar work to ref. 0.120. Correlation with actual cycles fair, using empirical coefficients from the test engines.)

0.123 Caris and Nelson, "A New Look at High Compression Engines," *Trans. SAE 67*, 112 (1959). (Performance of V-8 automotive engines with various combustion chambers and with compression ratios up to 25. Maximum output at $r = 17$. Octane requirements, engine noise, friction, pumping mep, heat balance, combustion time, indicator diagrams.)

Engine Performance, Supercharged
(Volume I, Chapter 12; see also refs. 10.860 *et. seq.*)

CHAPTER 1

Combustion, General

1.10 Lewis and Von Elbe, *Combustion, Flames and Explosions of Gases*, Cambridge University Press, New York, 1938.

1.12 Burk and Grummit, *The Chemical Background of Engine Research*, Interscience Publishers, New York, 1943.

1.131 Toong, *Combustion Dynamics*, McGraw-Hill Book Company, 1983.

1.14 International Symposia on Combustion, from the Third (1949) to the Tenth (1963), published in the U.S.A. by various book publishers. For reviews of the First Symposium see *Ind. Eng. Chem.* 20, 998–1057 (1928); for the Second Symposium see *Chem. Rev.* 21, 209–460 (1937); 22, 1–310 (1938). (Most of the work covered by these symposia is of a general nature. The most important articles directly concerned with reciprocating engines are listed separately in the bibliographies for Chaps. 1–3.)

1.15 Proc. Joint ASME and IME Conference on Combustion, Boston, June 1955 and London, Oct. 1955, published by IME (London) and ASME, New York.

1.16 Starkman, "Reciprocating Engine Combustion Research — A Status Report," *Ninth* (*International*) *Symposium on Combustion*, Academic Press, 1963, p. 1005. (Brief review of Combustion Research to that date with bibliography of 147 important references.)

Combustion in Bombs; Premixed Charge

1.20 Hopkinson, "Explosions in Gaseous Mixtures and the Specific Heat of the Products," *Engineering* (London), 81, 1906.

1.21 Rosencranz, "An Investigation of the Mechanism of Explosive Reactions," Univ. of Illinois, Exp. Eng. Station, *Bull.* 157, July 1926.

1.22 Stevens, "The Gaseous Explosive Reaction — The Effect of Pressure on the Rate of Propagation of the Reaction Zone," *NACA TR* 372, 1930.

1.23 Fiock and King, "The Effect of Water Vapor on the Flame Velocity in Equivalent $CO-O_2$ Mixtures," *NACA TR* 531, 1935.

1.24 Fiock and Roeder, "The Soap-Bubble Method of Studying the Combustion of Mixtures of CO and O_2," *NACA TR* 532, 1935.

1.25 Fiock and Roeder, "Some Effects of Argon and Helium upon Explosions of Carbon Monoxide and Oxygen," *NACA TR* 553, 1936.

1.26 Lewis and Von Elbe, "Determination of the Speed of Flames and the Temperature Distribution in a Spherical Bomb from Time-Pressure Explosion Records," *J. Chem. Phys. 2*, 283 (1934).

1.27 Fiock *et al.*, "Flame Speeds and Energy Considerations in a Spherical Bomb," *NACA TR* 682, 1940.

1.28 Damkohler, "The Effect of Turbulence on Flame Velocity in Gas Mixtures," *NACA TM* 1112, Apr., 1947.

1.29 Miao *et al.*, "Transient Gas-Flame Temperatures in a Spherical Bomb," *Trans. ASME 77*, 89 (Jan. 1955).

Gas Temperatures in Spark-Ignition Engines

1.30 Hersey and Paton, "Flame Temperatures in an Internal Combustion Engine Measured by Spectral Line Reversal," Univ. of Illinois, Exp. Eng. Station *Bull.* 262, 1933.

1.31 Rassweiler and Withrow, "Flame Temperatures Vary with Knock and Combustion-Chamber Position," *Trans. SAE 30*, 125 (1935).

1.32 Brevoort, "Combustion-Engine Temperatures by the Sodium-Line Reversal Method," *NACA TN* 559, 1936.

1.40 Livengood *et al.*, "Ultrasonic Temperature Measurement in Internal Combustion Engine Chamber," *J. Acoust. Soc. Am. 26*, 824 (1954).

1.41 Livengood *et al.*, "Measurement of Gas Temperature in an Engine by the Velocity of Sound Method," *Trans. SAE 66*, 683 (1958).

1.42 Chen *et al.*, "Compression and End-Gas Temperature from Iodine Absorption Spectra," *Trans. SAE 62*, 503 (1954).

Flame Motion and Pressure Development in Engines

1.50 Marvin, "Combustion Time in the Engine Cylinder," *NACA TR* 276, 1927.

1.51 Marvin and Best, "Flame Movement and Pressure Development in an Engine Cylinder," *NACA TR* 399, 1931.

1.52 Withrow and Boyd, "Photographic Flame Studies in a Gasoline Engine," *Ind. Eng. Chem. 23*, 539 (1931).

1.53 Schnauffer, "Engine-Cylinder Flame-Propagation Studied by New Methods," *Jour. SAE 34*, 17 (1934).

1.54 Rabezzana and Kalmar, "Factors Controlling Engine Combustion," *Auto. Ind.*, 324, 354, 394 (March 2, 9, and 16, 1935).

1.55 Minter, "Flame Movement and Pressure Development in Gasoline Engines," *Jour. SAE 36*, 89 (1935).

1.551 Taylor, "The Thermodynamics of Combustion in the Otto-Cycle Engine," *NACA TN* 533, 1935.

1.56 Marvin *et al.*, "Further Studies of Flame Movement and Pressure Development in an Engine Cylinder," *NACA TR* 556, 1936.

1.57 Withrow and Rassweiler, "Slow Motion Shows Knocking and Nonknocking Explosions," *Jour. SAE 39*, 297 (1936).

1.58 Bouchard *et al.*, "Variables Affecting Flame Speed in the Otto-Cycle Engine," *Jour. SAE 41*, 514 (1937).

1.59 Rothrock and Spencer, "A Preliminary Study of Flame Propagation in Spark-Ignition Engine," *NACA TN* 603, June 1937.

1.60 Rassweiler and Withrow, "Motion Pictures of Engine Flames Correlated with Pressure Cards," *Jour. SAE 42*, 185 (1938).

1.61 Rassweiler *et al.*, "Engine Combustion and Pressure Development," *Trans. SAE 46*, 25 (1940). (Extensive flame photographs correlated with pressure measurements; spark timing fixed.)

Combustion in Engines, Miscellaneous Research

1.71 Midgeley and Gilkey, "Spectroscopic Investigations of Internal Combustion," *Trans. SAE 17*, Pt. 1, 114 (1922).

1.72 Withrow, "Following Combustion in the Gasoline Engine by Chemical Means," *Ind. Eng. Chem. 22*, 945 (1930).

1.73 Withrow and Boyd, "Spectroscopic Studies of Engine Combustion," *Ind. Eng. Chem. 23*, 769 (1931).

1.74 Spanogle and Buckley, "The NACA Gas-Sampling Valve and Some Preliminary Test Results," *NACA TN* 454, 1933.

1.75 Marvin *et al.*, "Infrared Radiation from Explosions in a Spark-Ignition Engine," *NACA TR* 486, 1934.

1.76 Rothrock and Cohn, "Some Factors Affecting Combustion in an Internal Combustion Engine," *NACA TR* 512, 1934.

1.77 Clarke, "Initiation and Some Controlling Parameters of Combustion in the Automobile Engine," *Trans. SAE 70*, 240 (1962). (Valuable record of experience of Joseph Lucas, Ltd., including studies of ignition limits, mixing process, flame photographs, etc., for both spark-ignition and Diesel engines.)

1.78 Boyd, "Engine Flame Researches," *Trans. SAE 45*, 421 (1939). (Historical review of spark-ignition engine combustion research with special emphasis on flame observation, pressure waves, relation of pressure to flame travel, etc. Good bibliography.)

1.79 Bolt and Holkeboer, "Lean Fuel-Air Mixtures for High-Compression Spark-Ignited Engines," *Trans. SAE 70*, 195 (1962). (Tests on CFR engine show that lean firing limit is at reduced fuel-air ratio as compression ratio and inlet temperature increase. Pressure-crank angle curves included.)

1.791 Johnson, "Effect of Swirl on Flame Propagation in a Spark-Ignition Engine," *SAE Paper* 565C, Sept. 1962. (CFR engine with ionization gaps. Shrouded valve more than doubles flame speed. Flame time is practically independent of spark timing.)

1.792 Wentworth and Daniel, "Flame Photographs of Light-Load Combustion Point the Way to Reduction of Hydrocarbons in Exhaust Gas," *SAE Paper* 425, Jan. 1955. (Good photographs under various operating conditions, including extremes of fuel-air ratio and residual content.)

1.793 Kumagai and Kudo, "Flame Studies by Means of Ionization Gap in a High-Speed Spark-Ignition Engine," *Ninth (International) Symposium on Combustion*, Academic Press, New York, 1963, p. 1083. (Work with a 1-cylinder 4-cycle Honda small air-cooled engine, $r = 10.2$ shows crank-angle for flame travel increasing from 51° at 9000 rpm to 53.2° at 17,000 rpm. Octane requirement 68 at 12,000, and 50 at 16,000 rpm.)

1.794 Brewster and Kerley, "Automotive Fuels and Combustion Problems," *SAE Paper* 725C, 1963. (Excellent summary of effects of engine and fuel variables on performance, exhaust emissions, starting, preignition, etc., for spark ignition and Diesel engines. Includes data on spark-plug and exhaust temperatures. Ethyl Corporation work.)

1.795 Bolt and Harrington, "The Effects of Mixture Motion Upon the Lean Limit and Combustion of Spark-Ignited Mixtures," *SAE Paper* 670467, 1967.

Cyclic Variation in Flame Speed and Pressure

1.80 Averette, "A Study of Peak Pressure Variations in the Otto-Cycle Engine," report on file in Sloan Automotive Laboratories, Massachusetts Institute of Technology, Cambridge, Mass., September to June, 1964. (Measures peak-pressure dispersion versus engine operating parameters with and without shrouded inlet valve in various positions and with various positions of the spark plug in CFR engine.)

1.81 Vichnievsky, "Combustion in Petrol Engines," *Proc. Joint Conference on Combustion*, in Boston 15th–17th June, in London 25th–27th Oct. 1955, IME (London), 1956, Section IV, p. 288. (Excellent study by French engineer of cyclic variation as a function of fuel and engine parameters. Covers normal combustion and also detonation.)

1.82 Daniel, "Flame Quenching on Walls of an Internal Combustion Engine," *Sixth* (*International*) *Symposium on Combustion*, Reinhold Publishing Corporation, New York, 1956, p. 886. (Flame photographs show dead zone near walls.)

1.83 Arnold and Sherburne, "Observations of the Ignition and Incipient Flame Growth in Hydrocarbon-Air Mixtures," *Fourth* (*International*) *Symposium on Combustion*, Williams & Wilkins, Baltimore, 1953, p. 139.

1.831 Olsen, Gayhart, and Edmonson, "Propagation of Incipient Spark-Ignited Flames in Hydrocarbon-Air and Propane-Air Mixtures," *Fourth* (*International*) *Symposium on Combustion*, Williams & Wilkins, Baltimore, 1953, p. 144.

1.832 Soltan, "Cylinder Pressure Variations in Petrol Engines," IME (London), *Proc. Automobile Div.* 1960–1961, No. 2, p. 99. (Shows variations increase on either side of max power F_R, also with increasing residuals. Variation mostly in early stage of combustion.)

1.833 Patterson, "Cylinder Pressure Variations, a Fundamental Combustion Problem," *SAE Paper* 660129, Jan. 1966. (Gen. Motors research finds that the major cause is "mixture velocity variations that exist within the cylinder near the spark plug at ignition. Ignition energy and residual gas fraction show no effect. Lean mixtures give greater variation than "standard" mixture ratio.)

1.834 Roberts, "An Investigation into the Cycle to Cycle Variation in the Combustion of a Premixed Charge," S.M. Thesis, Department of Mechanical Engineering, Massachusetts Institute of Technology, Cambridge, Mass., 1966.

Effects of Cylinder Size on Combustion

1.840 See refs. 2.671–2.679 in this volume and 11.10–11.44 in Volume I.

1.841 Taylor, *Dimensional Analysis for Engineers*, Oxford University Press.

Exhaust Composition versus Fuel-Air Ratio

1.85 Gerrish and Meem, "Relation of Constituents in Normal Exhaust Gas to Fuel-Air Ratio," *NACA TR* 757, 1943.

1.86 D'Alleva and Lovell, "Relation of Exhaust-Gas Composition to Air-Fuel Ratio," *Trans. SAE 31*, 90 (1936).

Exhaust Emissions and their Control

1.870 *Vehicle Emissions*, SAE Technical Progress Series TP6, 1964. (Collection of papers on chemistry of smog, methods of emission measurement, engine emissions and their control, 1955–1963. Bibliography and abstracts.)

1.871 Way and Fagley, "Field Survey of Exhaust Gas Composition," *SAE Paper* 11A, Jan. 1958.

1.872 Brownson and Stebar, "Factors Influencing the Effectiveness of Air Injection in Reducing Exhaust Emissions," *SAE Paper* 650526, 1965.

1.873 Cantwell and Pahnke, "Design Factors Affecting the Performance of Exhaust Manifold Reactors," *SAE Paper* 650527, 1965.

1.874 See refs. 1.792 and 1.794.

1.875 Wimmer and McReynolds, "Nitrogen Oxides and Engine Combustion," *Trans. SAE 70*, 733 (1961). (Primarily devoted to the effects of nitrogen oxides, generated during combustion, on the combustion process itself.)

1.876 Larson *et al.*, "A Study of Distribution and Effects of Automotive Exhaust Gases in Los Angeles," *Trans. SAE 63*, 567 (1955). (Measurements of quality and quantity of air pollution and its relation to motor-car traffic.)

1.877 Daniel, "Flame Quenching at the Walls of an Internal Combustion Engine," *Sixth (International) Symposium on Combustion*, Reinhold Publishing Corporation, New York, 1957, p. 886.

1.878 Anon, "The General Motors Air Injection Reactor System for the Control of Exhaust Emissions," *Gen. Motors Eng. J. 13*, Third Quarter, 20 (1966). (Development, design, and test results.)

1.879 Van Derveer and Chandler, "The Development of a Catalytic Converter for the Oxidation of Exhaust Hydrocarbons," *SAE Paper* 295, Jan. 1959.

1.8791 See also the following earlier articles published in the *SAE Transactions:* Hutchinson and Holden, *Trans. SAE 63*, 581 (1955); Haagen-Smit and Fox, *ibid.*, p. 575; Larson *et al.*, *ibid.*, p. 567; Wentworth and Daniel, *ibid.*, p. 602; Rounds *et al.*, *ibid.*, p. 591; AMA Task Group, *ibid.*, 66, 383 (1958).

1.8792 "Hydrocarbon Reactivity—A New Look at the Smog Problem," *Search 1*, No. 3, May/June (1966), published by General Motors Research Labs., Detroit, Michigan. (Excellent summary of the chemistry of smog formation from automobile exhaust. Shows rate of reaction toward smog varies enormously with the type of hydrocarbon, olefines being from 10 to 50 times more reactive than paraffins and benzines. Includes a typical exhaust-hydrocarbon analysis.)

Combustion-Chamber Design and Engine Roughness

Most work on combustion-chamber design for spark-ignition engines is centered around prevention of detonation and is covered in refs. 2.60–2.67. The following references are concerned with reducing the rate of pressure rise by attenuating the combustion chamber. See also refs. 8.929–8.940 and 11.002, Nos. 119 and 137.

1.90   Ricardo, "Cylinder-Head Design," *Auto. Eng.* (London), July (1929).

1.91   Janeway, "Combustion Control by Cylinder-Head Design," *Trans. SAE 24*, 139 (1929).

1.92   Whatmough, "Combustion-Chamber Design in Theory and Practice," *Jour. SAE 25*, 375 (1929).

1.93   Taub, "Combustion-Chamber Progress Correlated," *Jour. SAE 27*, 413 (1930).

1.931 Manty, "The Effect of Combustion-Chamber Shape on Smoothness of Power," *SAE Special Publication SP-137*, 1955. (Valuable discussion of Cadillac experience with many shapes of overhead valve chambers, from point of view of rate-of-pressure rise and performance.)

1.94   Sampietro and Matthews, "The New Overhead Camshaft Willis Engine," *SAE Paper* 532A, June 1962. (Interesting domed-head 2-valve design which showed marked gain in performance over F-head and L-head chambers.)

CHAPTER 2

History of Detonation Research

2.00 Midgeley, "The Combustion of Fuels in Internal-Combustion Engines," *Trans. SAE 15*, Pt. 2, 659 (1920). (Early work by the discoverer of tetraethyl lead.)

2.01 Tizard and Pye, "Ignition of Gases by Sudden Compression," *Phil. Mag. 44* [6], 79 (1922). (Work with early rapid-compression machine.)

2.02 Ricardo, "Recent Research Work on the Internal-Combustion Engine," *Trans. SAE 17*, Pt. 1, 1 (1922). (Opinions of a pioneer in internal-combustion engine research.)

2.03 Boyd, "Pathfinding in Fuels and Engines," *SAE Quart. Trans. 4*, 182 (1950). (History of the work of Midgeley, Boyd, *et al.* and the discovery of tetraethyl lead. Includes recent work by General Motors. Good bibliography.)

Pressure Waves

2.04 Draper, "The Physical Effects of Detonation in a Closed Cylindrical Chamber," *NACA TR* 493, 1935.

2.05 Draper, "Pressure Waves Accompanying Detonation in the Internal Combustion Engine," *J. Aeron. Sci. 5*, 219 (1938). (Shows relation between pressure-wave frequency and natural frequency of gases in combustion chamber and audible frequency.)

2.051 Freiwald and Koch, "Spherical Detonations of Acetylene-Oxygen-Nitrogen Mixtures as a Function of Nature and Strength of Initiation," *Ninth (International) Symposium on Combustion*, Academic Press, New York, 1963, p. 275. (Detonation waves observed in mixtures contained in spherical rubber balloons at atmospheric pressure require large energy in the ignition spark to produce waves with air. Waves increasingly prevalent as nitrogen content is reduced. These "detonations" are sudden accelerations of a continuous flame front and have never been convincingly observed in engines.)

Apparatus for Detecting and Measuring Detonation

2.06 Horning, "The Cooperative Fuel-Research Committee Engine," *Trans. SAE 26*, 436 (1931). (The standard engine used for detonation measurement and generally for detonation research.)

2.061 Taylor *et al.*, "New Instrument Devised for the Study of Combustion," *Jour. SAE 34*, 59 (1934).

2.07 Beale and Stansfield, "The Standard Sunbury Engine Indicator," *Engineer* (London), Dec. 13, 20, 27, 1935.

2.08 Bogen and Faust, "Aircraft Detonation Indicators," *SAE Quart. Trans. 1*, 450 (1947).

2.081 Draper and Li, "A New High Performance Engine Indicator of the Strain Gage Type," *J. Aeron. Sci. 16*, 593 (1949).

2.09   Li, "High Frequency Pressure Indicators for Aerodynamic Problems," *NACA TN* 3042, Nov. 1953. (Review of various electric pick-up methods.)

2.091   General Motors Corp., "SLM Piezo-Electric Pressure Indicator," *Tech. Memo. Rep.* M-174; *Eng. Staff Power Development*, GM Technical Center, Detroit, Jan. 21, 1954. See also *DEMA Bull.* 14, Jan. 1956.

2.092   *CRC Handbook*, annual publication by CRC, Rockefeller Center, New York.

2.093   *ASTM Standards, Part IIIa for Non-metallic Materials.* (Standard method of test for knock characteristics of motor fuels. Revised periodically by ASTM, Philadelphia, Pa.

Detonation Damage
(See also ref. 2.85.)

2.100   Bierman and Covington, "Relation of Preignition and Knock to Allowable Engine Temperatures," *NACA ARR* 3G14 F-134, July 1934. (Cylinder temperatures are not increased by detonation until the power is increased 12% above that of first audible knock. Preignition rather than knock may be the cause of failure.)

2.101   MacCoull, "Power Loss Accompanying Detonation," *Trans. SAE 34*, 154 (1939). (Study includes effects of spark timing and compression ratio, preignition spark-plug design, etc.)

2.102   Gerber and Schindler, "A Study of Knock Induced Stress in the Connecting Rod of a CFR Engine," S.B. Thesis, Mechanical Engineering Department, Massachusetts Institute of Technology, Cambridge, Mass., 1956.

2.103   del Valle and Cruz, "A Study of the Knock-Induced Stress in the Connecting Rod of a CFR Engine," S.B. Thesis, Mechanical Engineering Department, Massachusetts Institute of Technology, Cambridge, Mass., 1959. (This and the preceding reference show that detonation does increase maximum rod stress by a small amount.)

Chemistry of Detonation
(See also refs. 1.10–1.13.)

2.11   Lewis and von Elbe, "Hydrocarbon Reactions and Knock in Internal Combustion Engine," *Ind. Eng. Chem. 29*, 551 (1937).

2.12   Sokolik, "Detonation and Peroxides in the Internal Combustion Engine," *Acta Physicochemica* (USSR) *9*, Nos. 3–4, 593 (1938). RTP translation 2429, Durand Publishing Co., in care of CIT, Pasadena, Calif.

2.13   Reman *et al.*, "Pre-flame Reactions in Hydrocarbon Air Mixtures," Delft Publication 75, Royal Dutch Shell Labs., Delft, Netherlands, April 1947. (Research in a steady-flow system and with a special CFR engine with windowed cylinder head. Spectroscopic and chemical sampling. Tetraethyl lead reduces peroxides.)

2.131   Downs *et al.*, "A Study of the Reactions That Lead to Knock in the Spark-Ignition Engine," *Phil. Trans. Roy. Soc.* (*London*) *A243*, 463 (1951).

2.14   Downs and Wheeler, "Recent Developments in 'Knock' Research," IME (London), *Proc. Automobile Div.* 1951–1952, p. 89. (Correlation between peroxides and detonation.)

2.15 Davis *et al.*, "A Comparison of the Intermediate Combustion Products Formed in an Engine with and without Ignition," *Trans. SAE 63*, 386 (1955).

2.16 Gaydon *et al.*, "Chemical and Spectroscopic Studies of Blue Flames in the Auto-ignition of Methane," *Proc. Roy. Soc.* (*London*) *A230*, 12 (1955).

2.17 Sturgis, "Some Concepts of Knock and Anti-knock Action," *Trans. SAE 63*, 253 (1955). (Especially concerned with action of TEL.)

2.18 Pipenberg and Pahnke, "Spectrometric Investigations of *n*-Heptane Preflame Reactions in a Motored Engine," *Ind. Eng. Chem. 49*, 2067 (1957). (Spectroscopy of reactions in a motored engine; effects of TEL.)

2.19 Lovell, "Some Chemistry of Future High-Compression Engines," IME (London), *Proc. Automobile Div.* 1959–1960, p. 101. (Also published by Ethyl Corporation, Detroit, Mich.)

2.191 Welling *et al.*, "Concurrent Pyrolytic and Oxidative Reaction Mechanisms in Precombustion of Hydrocarbons," *Trans. SAE 69*, 448 (1961). (Results of gas sampling from a motored CFR engine.)

2.192 Walsh, "The Knock Ratings of Fuels," *Ninth* (*International*) *Symposium on Combustion*, Academic Press, New York, 1963, p. 1046. (An attempt at a chemical explanation of end-gas knocking reaction.)

Rapid-Compression Machines
(See also ref. 2.01.)

2.20 Leary *et al.*, "A Rapid-Compression Machine Suitable for Studying Short Ignition Delays," *NACA TN* 1332, Feb. 1948. (Contains bibliography on previous machines and their results.)

2.21 Taylor *et al.*, "Ignition of Fuels by Rapid Compression," *SAE Quart. Trans. 4*, 232 (1950). (Autoignition studies of pressure-time and flame development on several pure compounds, including benzene, heptane, and isooctane.)

2.211 Jost, "Knock Reaction," *Ninth* (*International*) *Symposium on Combustion*, Academic Press, New York, 1963, p. 1013. (Work with a rapid-compression machine with analysis of delay or "induction time." See bibliography in ref. 2.20 for previous work by Jost.)

2.22 Livengood and Wu, "Correlation of Autoignition Phenomena in Internal Combustion Engines and Rapid Compression Machines," *Fifth* (*International*) *Symposium on Combustion*, Reinhold Publishing Corp., New York, 1955, p. 347. (Describes method of predicting detonation in engines on basis of compression-machine data.)

Detonation in Engines

2.26 Miller *et al.*, "Analysis of Spark-Ignition Engine Knock as Seen in Photographs Taken at 200,000 Frames per Second," *NACA TR* 857, 1946. (See also *NACA TR* 727, 761, 785, by Miller *et al.*)

2.27 Osterstrom, "Knocking Combustion Observed in a Spark-Ignition Engine with Simultaneous Direct and Schlieren High-Speed Motion Pictures and Pressure Records," *NACA TR* 987, 1948.

2.28 Lawrence and Hall, "Study of the Effect of Rate of Compression on the Auto-ignition Pressure of Several Fuels," S.B. Thesis, Mechanical Engineering

Department, Massachusetts Institute of Technology, Cambridge, Mass., June 1950. (Effect of rate of compression on autoignition pressure-time characteristics of di-isobutylene, triptane, ethyl benzene, *n*-heptane, and isooctane at stoichiometric ratio.)

2.29   Kumagai and Kimura, "On Engine Knock with Special Reference to Its Intensity," paper presented at 29th Annual Lecture Meeting, JSME, April 3, 1952.

2.30   "New Data on Combustion in Automotive Engines," *Auto. Ind.*, Sept. 15 (1953), 34. (Work done by Levedahl; pressure, radiation, and chemical samples in "one-shot" engine; emphasis on two-stage reaction and cool flames.)

2.31   MacPherson *et al.*, "The Relationship Between Time, End-Gas Pressure, Temperature, and Knock," *SAE Paper* 3A, Jan. 1958.

2.32   Gluckstein *et al.*, "End-Gas Temperature-Pressure Histories and Their Relation to Knock," *SAE Paper*, June 1960. (Also published by Ethyl Corporation, Detroit, Mich.)

2.321   Male, "Photographs at 500,000 Frames per Second of Combustion and Detonation in Reciprocating Engine," *Third* (*International*) *Symposium on Combustion*, Williams & Wilkins Co., Baltimore, 1949, p. 721. (Interesting Schlieren photographs but over-interpreted by the author.)

2.322   Anzillotti and Tomsic, "Combustion of Hydrogen and Carbon Monoxide as Related to Knock," *Fifth* (*International*) *Symposium on Combustion*, Reinhold Publishing Corp., New York, 1955, p. 356. (CFR engine tests, including effects of tetraethyl lead, burning rates, etc.)

2.323   Firey, "A Detonation Wave Theory of Engine Knock," *Sixth* (*International*) *Symposium on Combustion*, Reinhold Publishing Corp., New York, 1956, p. 878. (Tests using metal strips in cylinder. Not convincing.)

2.324   Curry, S., "A Three-Dimensional Study of Flame Propagation in a Spark Ignition Engine," *SAE Paper* 452B, 1962, and *Trans. SAE 71*, 628 (1963). (Measurements with ionized gaps. Effects of compression ratio. Cyclic variation, influence of detonation on flame velocity. Interpretation questionable.)

2.325   Haskell and Bame, "Engine Knock and End-Gas Explosion," *SAE Paper* 650506, May 1965. (Single-cylinder study with defense of the autoignition theory.)

End-Gas Temperatures

(See also refs. 1.40–1.42.)

2.33   Gluckstein and Walcutt, "End-Gas Temperature-Pressure Histories and Their Relation to Knock," *Trans. SAE 69*, 529 (1961). (Measurements with many fuels under many operating conditions show surprising constancy of end-gas temperature at knock for a given fuel-air mixture. An important contribution to detonation research.)

2.34   Agnew, "End-Gas Temperature by a Two-Wavelength Infrared Radiation Method," *Trans. SAE 69*, 495 (1961). (An alternative to the sound-velocity method, developed at General Motors.)

2.341   Johnson and Uyehara, "End-Gas Temperatures, Pressures, Reaction Rates and Knock," *SAE Paper* 650505, May 1965. (Temperature by infrared method.)

Preflame Reactions in Motored Engines
(See also refs. 2.15, 2.18, 2.30, 2.32.)

2.35 Pastell, "Precombustion Reactions in a Motored Engine," *SAE Quart. Trans. 4*, 571 (1950). (Effects of pressure, temperature, fuel, etc., on precombustion reactions and cool flames.)

2.36 Retailliau *et al.*, "Precombustion Reaction in the Spark-Ignition Engine," *SAE Quart. Trans. 4*, 438 (1950). (Prereactions in motored engines measured by response of a thermal plug.)

2.37 Rifkin *et al.*, "Early Combustion Reactions in Engine Operation," *SAE Quart. Trans. 6*, 472 (1952). (Motored-engine work. Criterion is amount of "heat" evolved; cool flames observed.)

2.38 Corzilius *et al.*, "Some Factors Affecting Precombustion Reactions in Engines," *Trans. SAE 61*, 386 (1953). (Autoignition limits, cool-flame limits, lead effects, speed and temperature effects.)

2.39 Mason and Hesselberg, "Engine Knock as Influenced by Precombustion Reactions," *Trans. SAE 62*, 141 (1954). (A firing engine is supplied with partially reacted gases from a motored engine.)

2.391 Calder, "Auto-ignition and Combustion of a Pre-mixed Charge of Normal Heptane in a CFR Engine," S.B. Thesis, Mechanical Engineering Department, Massachusetts Institute of Technology, Cambridge, Mass., Feb. 1956.

2.392 Ball, "Photographic Studies of Cool Flames and Knock in an Engine," *Fifth (International) Symposium on Combustion*, Reinhold Publishing Corp., New York, 1955, p. 366. (CFR L-head engine with transparent head. High-compression motoring tests with *n*-heptane show cool flames preceding normal flame spread.)

2.393 Levedahl, "Multistage Autoignition of Engine Fuels," *Fifth (International) Symposium on Combustion*, Reinhold Publishing Corp., 1955, p. 372. [Combined radiation and pressure measurements in a motored CFR engine with autoignition compared with autoignition in Schmidt's rapid-compression machine at Aachen; *n*-heptane, isooctane, and 40 and 80 octane-number blends all showed nearly the same temperature for autoignition (1550° to 1600°R), but more resistant blends required more drastic initial conditions to achieve this temperature, because the cool flames occurred at higher temperatures as the octane number increased. See Fig. 2–11*a*.]

2.394 Melmberg *et al.*, "A Study of Cool Flames and Associated Reactions in an Engine," *Fifth (International) Symposium on Combustion*, Reinhold Publishing Corp., p. 385. (Motored CFR engine with autoignition. Pressure and photomultiplier readings plus chemical samples for several different fuels. Effects of fuel composition, fuel-air ratio and compression ratio on intensity and duration of cool flames.)

Effect of Operating Variables on Detonation
(See also refs. 2.26–2.325.)

2.40 Imming, "The Effect of Piston Head Temperature on Knock-Limited Power," *NACA ARR* E4G13 E-35, 1944. (Cooling piston head from 340 to 260°F increased klimep up to 18 per cent.)

2.41   Sanders *et al.*, "Relative Effects of Cylinder-Head and Inlet-Mixture Temperatures upon Knock Limits of Fuels," *NACA RB* E4J13 E-37, 1944.

2.42   Sanders and Peters, "The Effect of High Temperature of the Cylinder Head on the Knocking Tendency of an Air-cooled Engine Cylinder," *NACA ARR* E5A29a E-39, 1945.

2.43   Raviolo, "Engine Design and Its Relation to Fuels and Lubricants," *Auto. Ind.*, Sept. (1950), 44. (Effect of exhaust-valve temperature on performance and octane requirements found negligible.)

2.44   Popinski and Reedy, "The Effect of Inlet Temperature, Jacket Temperature, and Engine Speed on the Detonation Characteristics of Di-isobutylene and Normal Heptane," S.B. Thesis, Department of Mechanical Engineering, Massachusetts Institute of Technology, Cambridge, Mass., May 1951. (At constant mass of air per stroke, DIB shows reduced hucr with increasing speed. Normal heptane shows opposite slope for one temperature.)

2.45   Diggs, "The Effect of Combustion Time on Knock in a Spark-Ignition Engine," *SAE Trans. 61*, 402 (1953).

2.46   Tower and Alquist, "Correlation of Effects of Fuel-Air Ratio, Compression Ratio, and Inlet-Air Temperature on Knock Limits of Aviation Fuels," *NACA TN* 2066, Apr. 1950.

2.461   Taylor *et al.*, "Effect of Fuel-Air Ratio, Inlet Temperature and Exhaust Pressure on Detonation," *NACA TR* 699 (1940).

2.47   Porter *et al.*, "Test Results with Dual-Fuel Carburetion at High Compression," *Auto. Ind.*, Jan. 15, 1952.

2.48   Hurley and Cook, "The Influence of Turbulence on the Highest Useful Compression Ratio," *Engineering* (London) *130*, Sept. 5, 1930.

2.49   Bolz and Breitweiser, "Effect of Compression Ratio, Cooled Exhaust Gas and Inlet-Air Temperature on Knock-Limited Performance," *NACA ARR* E-124 (1943).

2.50   Cook *et al.*, "Effect of Several Methods of Increasing Knock-limited Power on Cylinder Temperatures," *NACA ARR* E4I15 E-36, 1944.

2.51   Barber, "Knock-limited Performance of Several Automobile Engines," *SAE Quart. Trans. 2*, 401 (1948). (Effect of spark advance, rpm, etc., three-dimensional graphs.)

2.52   Gibson, "Factors Affecting Octane-Number Requirement," *SAE Quart. Trans. 3*, 557 (1949). (Atmospheric conditions, deposits, adjustments all have strong effects.)

2.53   Heron and Felt, "Cylinder Performance Compression Ratio and Mechanical Octane-Number Effects," *SAE Quart. Trans. 4*, 455 (1950). (Magnum opus on effect of operating conditions, valve shrouding, and piston shape on knock-limited performance.)

2.54   Hesselberg and Lovell, "What Fuel Antiknock Quality Means in Engine Performance," *Jour. SAE 59*, April 1951, 32. (Performance number versus knock-limited speed, squish, sensitivity effects. Correlates octane number with performance number.)

2.55   Roensch and Hughes, "Evaluation of Motor Fuels for Higher Compression Engines," *SAE Quart. Trans. 5*, 71 (1951). (Speed and compression-ratio effects, octane requirements.)

2.551   Felt *et al.*, "The Effect of Fuel Injection on Knocking Behavior," *SAE Preprint* 3B, Jan. 1958.

2.56 Bigley *et al.*, "Effect of Engine Variables on Octane Number of Passenger Cars," *API Refining Section*, 32, Pt. 3, 1952, p. 174. (Comprehensive study of many variables, including cylinder deposits.)

2.57 Davis, "Factors Affecting the Utilization of Anti-knock Quality in Automobile Engines," IME (London), *Proc. Automobile Div.* 1952–1953, Pt. III. (Comprehensive work with engine operating variables.)

2.58 Kerley and Thurston, "The Knocking Behavior of Fuels and Engines," *Trans. SAE 64*, 554 (1956).

2.59 Potter *et al.*, "Weather or Knock," *Trans. SAE 62*, 346 (1954). (Atmospheric temperature, pressure, and humidity effects.)

Effects of Combustion-Chamber Geometry on Detonation
(See also refs. 2.53, 2.55, 11.002, Nos. 119 and 137.)

2.60 King and Moss, "Detonation, Spark-plug Position and Engine Speed," *Aeron. Res. Comm.* (London), R and M 1525, 1933.

2.61 Kerley, "Octane-Aiding Engines for Gold-Plated Fuels," Ethyl Corp. Res. Labs., 1950, Collected Technical Papers. Summary of Heron's work on effect of squish and shroud and Roensch's work on octane requirements.

2.62 Matthews and Turlay, "New Buick V-8 Engine Offers Many Advantages," *Trans. SAE 61*, 478 (1953). (Shows a number of combustion-chamber forms and their octane requirements.)

2.63 Geschelin, "Advanced Combustion Chamber Design," *Auto. Ind.*, Mar. (1955). (Review of changes in compression ratio, head design, and octane requirement since new V-8's were introduced.)

2.631 Caris *et al.*, "Mechanical Octanes for Higher Efficiency," *Trans. SAE 64*, 76 (1956). (Combustion-chamber shape versus octane requirements. Important.)

2.64 Geschelin, "New Combustion Chamber Designs in 1958 Passenger Car Engines," *Auto. Ind.*, Dec. 15 (1957). (Excellent review.)

2.65 Hughes, "Combustion Chamber Geometry and Fuel Utilization," *SAE Paper* 53, 1957. (Compression ratio, quench area, and fuel consumption varied in a careful consistent manner. Valuable reference.)

2.651 Rifkin, "The Role of Physical Factors in Knock," *API Paper*, 1958.

2.652 Grant and Newman, "New Tools for Understanding Two-Cycle Engine and Its Octane Requirements," *SAE Paper* 442B, Jan. 1962. (Outboard engine—one of very few measurements on two-cycle engines.)

2.653 Weil *et al.*, "Development of the New Family of Cummins Natural Gas Engines," *ASME Paper* 65–OGP–13, April 1965. (Shows significant gain in detonation limit by changing from a flat piston to cupped piston with edge "squish.")

Number and Position of Spark Plugs
(See also refs. 2.53 and 2.60.)

2.66 Ricardo, *The High-Speed Internal-Combustion Engine*, Blackie & Son, London, 1933, pp. 45–46.

2.67 Diggs, "Effect of Combustion Time on Knock in a Spark-Ignition Engine," *Trans. SAE 61*, 403 (1953). (One to seventeen plugs in one cylinder.)

Effects of Cylinder Size on Detonation

2.671 Kamm, "Ergebnisse von Versuchen mit geometrisch ähnlich gebauten Zylindern verschiedener Grösse und Folgerungen für die Flugmotorenentwicklung," *Schriften der deutschen Akademie der Luftfahrtforschung*, Heft 12, March 1939.

2.672 Gaboury *et al.*, "A Study of Friction and Detonation in Geometrically Similar Engines," S.M. Thesis, Aeronautical Engineering Department, Massachusetts Institute of Technology, Cambridge, Mass., June 1950. (Includes data on friction mep, flame speeds, and knock-limited imep. Flat (open) combustion chambers, $r = 5.53$.)

2.673 Gall, "A Study of Detonation Characteristics of Geometrically Similar Engines," S.M. Thesis, Department of Mechanical Engineering, Massachusetts Institute of Technology, Cambridge, Mass., June 1958. (Comparative flame speeds and knock-limited imep with open combustion chambers, $r = 5.53$.)

2.674 Talbot, "A Study of Octane Requirement as a Function of Size in Geometrically Similar Engines," S.B. Thesis, Mechanical Engineering Department, Massachusetts Institute of Technology, Cambridge, Mass., 1959.

2.675 Hamzeh and Nunes, Jr., "Highest Useful Compression Ratios for Geometrically Similar Engines," S.M. Thesis, Mechanical Engineering Department, Massachusetts Institute of Technology, Cambridge, Mass., 1959.

2.676 Cline and Raymond, "A Study of Detonation Characteristics Utilizing End-Gas Temperature Measurement for Geometrically Similar Engines," S.M. Thesis, Mechanical Engineering Department, Massachusetts Institute of Technology, Cambridge, Mass., Sept. 1960.

2.677 Taylor, "The Effects of Cylinder Size on Detonation and Octane Requirement," *Trans. SAE 70*, 175 (1962). (Summary of the work of refs. 2.672–2.676 covering work with MIT similar engines to that date.)

2.678 Swenson, Jr., and Wilcox, "Detonation Characteristics and Related Combustion Phenomena in Highly Turbulent Combustion Chambers," S.B. Thesis, Mechanical Engineering Department, Massachusetts Institute of Technology, Cambridge, Mass., January 1963. (Squish-type combustion chambers, $r = 8.62$ to 8.88, and flat chamber for 4-in. bore only, $r = 8.74$. Performance, heat rejection, friction, flame speeds, pressure-rise rates, octane requirements.)

2.679 Swenson, Jr., and Wilcox, "Detonation Characteristics and Related Combustion Phenomena in Highly Turbulent Combustion Chambers," *Supplement A* to S.B. Thesis (ref. 2.678), March 1, 1963, on file at Sloan Automotive Laboratories, Massachusetts Institute of Technology, Cambridge, Mass. (Performance and octane requirements of MIT 4-in. bore and 6-in. bore similar engines with fuel blends of sensitivity 10, squish pistons, $r = 8.62$ to 8.88.)

Effect of Stroke-Bore Ratio

2.6791 Swenson, Jr., and Wilcox, "Detonation Characteristics and Related Combustion Phenomena in Highly Turbulent Combustion Chambers," *Supplement B* to S.B. Thesis (ref. 2.678), May 20, 1963, on file at Sloan Automotive Laboratories. (Performance, friction, flame speeds, and octane requirements of 4-in. bore engine with 3.0-in. stroke compared with 4.8-in. stroke in the same engine, squish-type combustion chambers. 3.0-in. stroke shows lower octane requirement below 3000 rpm, above which the requirement is the same for both strokes. See also Volume I, ref. 6.45.)

Detonation Control
(See also Chapter 4 and refs. 4.210–4.219.)

2.70   Cook *et al.*, "Effect of Several Methods of Increasing Knock-limited Power," *NACA WR* E-36 (1944).

2.71   Rowell and Williams, "Spark Governor Controlled by Engine Load Proposed," *Auto. Ind.*, Aug. 10, 1929.

2.72   Heldt, "Spark Timing Control Actuated by Pressure in Engine Cylinder," *Auto. Ind.*, Feb. 14, 1931.

2.73   " Mallory Combines Vacuum and Speed Control of Spark," *Auto. Ind.*, June 18, 1932.

2.74   Taylor *et al.*, "Spark Control of Supercharged Engines," *J. Aeron. Sci. 3*, 326 (1936).

2.75   Rothrock *et al.*, " The Induction of Water to the Inlet Air as a Means of Internal Cooling in Aircraft-Engine Cylinders," *NACA TR* 756, 1943.

2.76   Bellman and Evvard, " Knock-limited Performance of Several Internal Coolants," *NACA TR* 812 (*ARR*), 1945.

2.77   Obert, " Detonation and Internal Coolants," *SAE Quart. Trans. 2*, 52 (1948). (Indicator diagrams with and without water-alcohol injection.)

2.78   Miller and Suliss, " Recycling Exhaust Gas for Suppression of Knock in Internal Combustion Engines," presented at API meeting, May 14, 1953, preprint. (Shows that cool exhaust gas has strong antiknock effect. Exhaust back pressure has small effect on octane requirement.)

2.781 Brun *et al.*, " End-Zone Water Injection as a Means of Suppressing Knock in a Spark-Ignition Engine," *NACA WR* E4127, E-72, Sept. 1944. (Water injected into end zone of the one-cycle window test engine. Photos show water spray into cylinder. Water at the rate of 30 per cent fuel found effective. Tests very limited in scope.)

Stratified-Charge Engines

2.790 Ricardo, "The High-Speed Internal-Combustion Engine," 3rd rev. ed., edited by Clyde, Blackie & Son, London, 1941.

2.7901 Mock, "Engine Types and Requirements for Preparation of Fuels," *Trans. SAE 38*, 257 (1936). (Includes description of prechamber spark-ignition engine.)

2.791 Dillstrom, " A High Power Spark-Ignition Fuel Injection Engine," *Trans. SAE 35*, 431 (1934). See also Pope, *ibid.*, p. 385.

2.80   Barber *et al.*, " The Elimination of Combustion Knock," *J. Franklin Inst. 241*, 275 (1946).

2.81   Barber *et al.*, " The Elimination of Combustion Knock — Texaco Combustion Process," *SAE Quart. Trans. 5*, 26 (1951).

2.811 Conta and Durbetaki, " How to Step Up Fuel Economy — Charge Stratification

2.8111 Turkish, "3-Valve Stratified Charge Engines," SAE Paper 741163, 1974.

See also SAE Publications ME 132 (1980) and MEP-1 (1977).

Shows a 20% Boost to Part-Throttle Efficiency and Good Detonation Performance at High Compression Ratio," *Jour. SAE 67*, March (1959), 56. (Tests with a divided combustion chamber.)

2.812 Hussman *et al.*, "Charge Stratification by Fuel Injection into Swirling Air," *Trans. SAE 71*, 421 (1963). (Elaborate theoretical treatment plus tests in a bomb. Value in real applications doubtful.)

2.813 Schweitzer and Grunder, "Hybrid Engines," *Trans. SAE 71*, 541 (1963). (Historical review of attempts to develop stratified-charge and other unconventional combustion systems.)

2.814 Davis *et al.*, "Fuel Injection and Positive Ignition," *Trans. SAE 69*, 120 (1961). (Development to 1961 on the Texaco Combustion Process.)

2.815 Conta *et al.*, "Stratified Charge Operation of Spark Ignition Engines," *SAE Paper* 375B, 1961. (Work with divided combustion chambers on CFR engine. Bibliography.)

2.816 Heywood and Tabaczyniski, "Current Developments in Spark-Ignition Engines," *SAE Paper* 760606, 1976.

Octane Requirements of Engines in Service

(See also refs. 4.515–4.524.)

2.820 "Octane Number Requirement Surveys" and "Standard Equipment Surveys," published annually by Coordinating Research Council, 30 Rockefeller Plaza, New York.

2.821 Gerard and DiPerna, "Multifaceted Octane Numbers for Diverse Engine Requirements," *SAE Paper* 976A, Jan. 1965. (Effect of operating variables, fuel sensitivity, throttle setting, compression ratio, etc., on "road" octane numbers of typical U.S. passenger automobiles.)

2.822 Uchinuma and Nakajime, "Road Knock Occurrence in Japan," *SAE Paper* 967B, Jan. 1965.

2.83 "Antiknock Performance Clocked on Turbocharged Engines of Corvair and Oldsmobile," *Jour. SAE 72*, Aug. 1963, 92.

2.831 Kerley and Thurston, "Knocking Behavior of Fuels and Engines," *Trans. SAE 64*, 554 (1956). (Contains valuable basic information including engine and operating variables, fuel sensitivity, performance numbers, etc.)

2.840 Toulmin and Lenane, "Part-Throttle Knock," *SAE Paper* 128T, Oct. 1959. See also *Jour. SAE 69*, March 1960, 42.

2.841 Freeman and Forrest, "Part-Throttle Economy and Octane Quality Demand of Current V-8 Engines," *SAE Paper* 128U, Oct. 1959.

2.842 Sanders and Scott, "Seasonal Octane Control," *SAE Preprint* 127T, Oct. 1959.

2.843 Best *et al.*, "Anti-knock Requirements of Passenger Cars," *SAE Quart. Trans. 5*, 335 (1951).

2.844 For data on octane requirements prior to 1960 see *SAE Quart. Trans. 4*, 93 (1950); *ibid.*, *5*, 335 (1951); *Trans. SAE 62*, 288 (1954); *ibid.*, *64*, 554 (1956); *ibid.*, *65*, 175 (1957); *Jour. SAE 51*, June 1943, 198; *ibid.*, *57*, Dec. 1949, 24; *ibid.*, *63*, Dec. 1955, 52; *ibid.*, *68*, Jan. 1959, 109; *68*, Feb. 1959, 60; *ibid.*, *69*, Jan. 1960, 56; *ibid.*, *72*, Aug. 1963, 58.

Preignition
(See also refs. 4.530–4.533.)

2.85 Hundere and Best, " Preignition and Its Deleterious Effects in Aircraft Engines," *SAE Quart. Trans. 2*, 546 (1948). (Preignition rather than detonation causes failures.)

2.86 Winch and Mayes, "A Method for Identifying Preignition," *Trans. SAE 61*, 453 (1953). (Method depends on variation in resistance of the spark gaps; discussion describes other methods.)

2.87 Warren and Hinkamp, " New Instrumentation for Engine Combustion Studies," *Trans. SAE 64*, 665 (1956). (Use of ionization gaps.)

2.871 Winch, " The Occurrence of Preignition in Present Day Cars in Normal Service," *Trans. SAE 62*, 50 (1954). (Road tests show incidence of preignition in a number of 1952 and 1953 models. Clean engines showed no preignition. Octane number increase decreases preignition.)

2.88 Downs and Pigneguy, "An Experimental Investigation into Preignition in the Spark-Ignition Engine," IME (London), *Proc. Automobile Div.* 1950–1951, Pt. IV. (Exhaustive data on experiments in an engine using different forms of hot bodies.)

2.89 Melby *et al.*, "An Investigation of Preignition in Engines," *Trans. SAE 62*, 32 (1954). (Multiple spark plugs used to simulate preignition.)

2.90 Heron, " Uncontrolled Combustion in Spark-Ignition Engines," *ibid.*, *62*, 24 (1954). (Good general discussion.)

2.91 Black, " Some Aspects of Surface Ignition in an Internal Combustion Engine," S.M. Thesis, Department of Mechanical Engineering, Massachusetts Institute of Technology, Cambridge, Mass., May 1954. (Measurement of hot-surface temperatures required to ignite various fuels in a motored engine.)

2.92 Meagher *et al.*, " Correlation of Engine Noises with Combustion Phenomena," *Trans. SAE 63*, 481 (1955). (dp/dt records used to measure and interpret experimental results.)

2.93 Anderson, " Pre-ignition in Aircraft Reciprocating Engines," *Trans. SAE 65*, 739 (1957).

2.94 Bowditch and Stebar, " Autoignition Associated with Hot Starting," *Trans. SAE 66*, 179 (1958). (Flame photographs and pressure measurements.)

2.95 Robison *et al.*, " Investigating Rumble in Single-Cylinder Engines," *Trans. SAE 66*, 549 (1958).

2.951 Livengood *et al.*, " Surface Ignition in a Motored Engine," *Proc. Joint Conference on Combustion*, in Boston 15th–17th June, in London 25th–27th Oct., 1955, IME (London), 1956, Section IV, p. 235. (Results of tests at MIT with hot surface flush with combustion-chamber walls. Effect of operating parameters and fuels.)

2.952 Tomsie, " Surface Ignition Behavior of Fuels," *Trans. SAE 70*, 74 (1962). (Study of peak-pressure variations with and without surface ignition. Increasing compression ratio increases early peak-pressure angles. Effects of phosphorus additives, fuel and oil composition included. Important contribution.)

2.953 Palinke, " Surface Ignition — Factors Affecting Its Occurrence in Engines," *Trans. SAE 71*, 663 (1963). (Tests of various pellet materials and fuels for their surface-ignition tendency. 30 items in bibliography.)

Deposits, Effect on Detonation and Preignition
(See also refs. 2.85–2.953.)

2.96   Dumont, "Possible Mechanisms by Which Combustion-Chamber Deposits Accumulate and Influence Knock," *SAE Quart. Trans. 5*, 565 (1951). (Attempts to separate volume, heat-transfer, and chemical effects.)

2.97   Withrow and Bowditch, "Flame Photographs of Autoignition Induced by Combustion-Chamber Deposits," *SAE Quart. Trans. 6*, 724 (1952). (First good pictures of preignition caused by deposits.)

2.98   Gibson *et al.*, "Combustion-Chamber Deposits and Knock," *Trans. SAE 61*, 361 (1953).

2.99   McNab *et al.*, "The Effect of Lubricant Composition on Combustion Chamber Deposits," *Trans. SAE 62*, 228 (1954).

2.991   Hirschler *et al.*, "Deposit-Induced Ignition," *Trans. SAE 62*, 40 (1954). (Single-cylinder work with ionization gaps.)

2.992   Warren, "Combustion-Chamber Deposits and Octane-Number Requirement," *Trans. SAE 62*, 582 (1954). (Separation of various effects: thermal, chamber volume, air consumption, etc.)

2.993   Wimmer, "Some Factors Involved in Surface Ignition in Spark Ignition Engines," *SAE Paper* 872, 1956. (Effect of deposit composition on preignition.)

2.994   Felt *et al.*, "Rumble — A Deposit Effect at High Compression Ratios," *SAE Paper* 58–9, June 1958. (Single-cylinder investigation of causes.)

2.995   Bowditch and Yu, "A Consideration of Deposit Ignition Mechanism," *Trans. SAE 69*, 435 (1961). (Includes flame photographs.)

2.996   Dumont, "Possible Mechanisms by Which Combustion Chamber Deposits Accumulate and Influence Knock," *SAE Paper* 623, June 1951. (Significant exploration of the nature of deposits, effects of lead and sulphur, and the increase in octane requirement.)

2.997   Robinson *et al.*, "Investigating Rumble in Single-Cylinder Engines," *SAE Paper* 61H, 1958. (Work by Ford locates source of deposit ignition. Shows that small quench height increases dp/dt.)

2.998   "Engine Noise Symposium," *SAE Special Publication SP-157*, 1958. (A collection of 1958 papers on preignition by representatives of fuel and engine companies. Fuel composition effects, as well as operating variables and combustion-chamber geometry. See also earlier symposium, *SAE Special Publication SP-113*, 1952.)

CHAPTER 3

General

3.01   Ricardo, "The High-Speed Compression-Ignition Engine," *Aircraft Eng.* (London), May 1929.

3.02   Ricardo, "Combustion in Diesel Engines," *Proc. IAE* (London), Mar. 1930.

3.03   Rothrock, "Combustion in a High-Speed Compression-Ignition Engine," *NACA TR* 401, 1931.

3.04   Ricardo, "Diesel Engines," *J. Roy. Soc. Arts* (London), Jan., Feb. 1932.

3.05 Dicksee, "Some Problems Connected with High-Speed Compression-Ignition Engine Development," *Proc. IAE* (London), Mar. 1932.

3.06 Ricardo and Pitchford, "Design Developments in European Automotive Diesel Engines," *Trans. SAE 32*, 405 (1937).

3.07 Whitney and Foster, "The Diesel as a High-Output Engine for Aircraft," *Trans. SAE 33*, 161 (1938).

3.08 Dicksee, "The High-Speed Compression-Ignition Engine," Blackie & Son, London, 1940.

3.09 Elliott, "Combustion of Diesel Fuel," *SAE Quart. Trans. 3*, 490 (1949). (Comprehensive review of fundamentals and their relation to engine combustion.)

Flame and Spray Photographs
(See also refs. 7.10–7.50.)

3.11 Underwood, "The Combustion of an Oil Jet in an Engine Cylinder," *Proc. IME* (London) *121*, Dec., 379 (1931).

3.12 Rothrock, "The NACA Apparatus for Studying the Formation and Combustion of Fuel Sprays and the Results from Preliminary Tests," *NACA TR* 429, 1932.

3.13 Rothrock and Waldron, "Fuel Vaporization and its Effect on Combustion in a High-Speed Compression-Ignition Engine," *NACA TR* 435, 1932.

3.14 Rothrock and Waldron, "Some Effects of Injection Advance Angle, Engine-Jacket Temperature, and Speed on Combustion in a Compression-Ignition Engine," *NACA TR* 525, 1934.

3.15 Rothrock, "Photographic Study of Combustion in Compression-Ignition Engine," *Jour. SAE 34*, June (1934), 203.

3.16 Rothrock and Waldron, "Effects of Air-Fuel Ratio on Fuel Spray and Flame Formation in a Compression-Ignition Engine," *NACA TR* 545, 1935.

3.17 Rothrock and Waldron, "Effect of Nozzle Design on Fuel Spray and Flame Formation in a High-Speed Compression-Ignition Engine," *NACA TR* 561, 1936.

3.18 Rothrock, "A Visual Study in a Displacer-Piston Compression-Ignition Engine," *Trans. SAE 32*, 22 (1937).

3.19 Ricardo and Co., "Combustion Photographs on C.I. Engines," report by Ricardo and Co., Shoreham (Sussex), England, Mar. 10, 1955.

3.191 Alcock and Scott, "Some More Light on Diesel Combustion," *SAE Paper* 872A, June 1964. (Excellent presentation of color photos and pressure curves showing combustion in various forms of combustion chamber. Good comparison of open and "Comet" type chambers. Comparison with gasoline and Diesel oil. Bibliography of 10 important items.)

3.192 Nagao and Kakimoto, "Swirl and Combustion in Divided-Combustion-Chamber Type Diesel Engines," *Trans. SAE 70*, 680 (1962). (Flame photos and pressure measurements. Similar to ref. 3.191 but without color. Good contribution.)

Combustion of Fuel Droplets
(See also refs. 3.09, 3.56.)

3.21 El-Wakil *et al.*, "A Theoretical Investigation of the Heating-up Period of Injected Fuel Droplets Vaporizing in Air," *NACA TN* 3179, May 1954.

3.22   Nishiwaki and Hirata, "Some Phenomena on the Combustion of Fuel Droplets," special report on file at Sloan Laboratories, MIT, 1958. (Photographic observation of burning drops; see bibliography included.)

3.23   El-Wakil and Abdou, "Ignition Delay Analyzed from Self-Ignition of Fuel Drops," *Jour. SAE 71*, May (1963), 42, and *SAE Paper* 598B, 1963. (Unconvincing attempt to separate "physical" and "chemical" delays.)

Bomb and Compression-Machine Experiments

3.31   Cohn and Spencer, "Combustion in a Bomb with a Fuel-Injection System," *NACA TR* 544, 1935.

3.32   Michailova and Neumann, "The Cetene Scale and the Induction Period Preceding the Spontaneous Ignition of Diesel Fuels in Bombs," *NACA TM* 813, Dec. 1936.

3.33   Selden, "Auto-Ignition and Combustion of Diesel Fuel in a Constant-Volume Bomb," *NACA TR* 617, 1938.

3.34   Hurn and Hughes, "Combustion Characteristics of Diesel Fuels as Measured in a Constant-Volume Bomb," *SAE Quart. Trans. 6*, 24 (1952).

3.35   Hurn *et al.*, "Fuel Heat Gain and Release in Bomb Autoignition," *Trans. SAE 64*, 703 (1956).

3.36   Battelle Memorial Inst., "Injection and Combustion of Liquid Fuels," *WADC Report* 623.74703 B33, Mar. 1957.

3.37   Rogowski, "A Machine for Studying Combustion of Fuel Sprays with Controlled Air Motion," *SAE Paper* 436F, Nov. 1961. (MIT 4-in. bore rapid-compression apparatus.)

3.38   Brinkmann and Singh, "Study of Compression Ignition of Fuel Sprays in a Controlled Environment," S.M. Thesis, Department of Mechanical Engineering, Massachusetts Institute of Technology, Cambridge, Mass., Jan. 1962. (Describes MIT Diesel rapid-compression machine and work with screens to create turbulence. Preliminary results.)

3.39   Landis, "The Effect of Turbulence on the Combustion of Fuel Sprays," S.M. Thesis, Department of Mechanical Engineering, Massachusetts Institute of Technology, Cambridge, Mass., Jan. 1963. (Describes the MIT rapid-compression spray-combustion machine, methods of measuring temperature, pressure, and turbulence by means of hot-wire device. Tests with and without wire-mesh screens at top of piston travel. Heat capacity of screens seems to mask their effects on combustion.)

3.40   Fehrenbach and Nayak, "An Investigation into the Effects of Swirl Combustion," report on file at Sloan Automotive Laboratories, Massachusetts Institute of Technology, Cambridge, Mass., Oct. 29, 1964. (Work with swirl induced by an air jet before compression, using the MIT Diesel rapid-compression machine. Results inconclusive.)

3.401   Fehrenbach and Ghosh, "The Effect of Air Swirl on Spray Combustion in the MIT Rapid-Compression Machine," S.M. Thesis, Department of Mechanical Engineering, Massachusetts Institute of Technology, Cambridge, Mass., May 1965. (Swirl induced by air jet prior to compression and injection increases maximum pressure and reduces combustion time. Rates of swirl used were small compared with those in Diesel engines employing swirl.)

Ignition Delay and Rate of Pressure Rise
(See also refs. 3.09 and 3.31–3.36.)

3.41 Matthews and Gardiner, "The Influence of Inlet-Air Temperature and Jacket-Water Temperature on Initiating Combustion in a High-Speed Compression-Ignition Engine," *NACA TN* 185, May 1924.

3.42 Sass, "Ignition and Combustion Phenomena in Diesel Engines," *NACA TM* 482 and *Diesel Power*, Mar. and June 1930.

3.43 LeMesurier and Stansfield, "Fuel Testing in High and Slow Speed Diesel Engines," *Trans. IPT* (London) *17*, July 1931.

3.44 Davies and Giffen, "Injection Ignition and Combustion in High-Speed Heavy-Oil Engines," *Proc. IAE* (London), Mar. 1931.

3.45 Boerlage and Broeze, "Ignition Quality of Diesel-Engine Fuels as Expressed in Cetene Numbers," *Jour. SAE 31*, July 1932, 283.

3.46 Gerrish and Voss, "Influence of Several Factors on Ignition Lag in a Compression-Ignition Engine," *NACA TN* 434, Nov. 1932.

3.47 Pope and Murdock, "Compression-Ignition Characteristics of Injection-Engine Fuels," *Trans. SAE 30*, 136 (1932).

3.48 Schweitzer, "Combustion Knock in Diesel Engines," Paper presented at ASME Meeting, 1932.

3.49 Boerlage and Van Dyck, "Causes of Detonation in Petrol and Diesel Engines," *Jour. IAE* (London) *1*, April 1934, 11; reprinted in *Jour. RAE* (London), Dec. 1934, 953.

3.50 Spanogle, "A Comparison of Several Methods of Measuring Ignition Lag in a Compression-Ignition Engine," *NACA TR* 485, 1934.

3.51 Gerrish and Ayer, "Influence of Fuel-Oil Temperature on the Combustion in a Pre-Chamber Compression-Ignition Engine," *NACA TN* 565, Apr. 1936.

3.52 Wentzel, "Ignition Process in Diesel Engines," *NACA TM* 797, June 1936.

3.53 Holfelder, "Ignition and Flame Development in the Case of Diesel Fuel Injection," *NACA TM* 790, Mar. 1936.

3.54 Davies, "Combustion in Compression-Ignition Oil Engines," *The Chartered Mechanical Engineer* (London), Jan. 1954. (Relation of vaporization to ignition and combustion.)

3.55 Yu *et al.*, "Physical and Chemical Ignition Delay in an Operating Diesel Engine Using the Hot Motored Technique," *Trans. SAE 64*, 690 (1956).

3.551 Chian *et al.*, "Physical and Chemical Delay in a Diesel Engine Using Hot-Motoring Technique, Pt. II," *Trans. SAE 68*, 562 (1960). (Authors recognize a "physical" delay not convincingly demonstrated to exist in normal engine operation.)

3.56 El-Wakil *et al.*, "Fuel Vaporization and Ignition Lag in Diesel Combustion," *Trans. SAE 64*, 712 (1956).

3.57 Nagao *et al.*, "Idling Knock in a Pre-Combustion Type Diesel Engine," *Bull. JSME* (Tokyo), *2*, 175 (1959).

3.58 Lyn, "Study of Burning Rate and Nature of Combustion in Diesel Engines," *Ninth (International) Symposium on Combustion*, Academic Press, New York, 1963, p. 1069. (Excellent study of the three phases of combustion for three types of combustion chambers. Well supported by experimental data.)

Effects of Spray Characteristics and Air Motion on Diesel Combustion
(See also refs. 3.01–3.192.)

3.60 Work by the U.S. National Advisory Committee for Aeronautics (now NASA):

Author	Subject	NACA Publication
Spanogle	Nozzles for quiescent chambers	TN 382, June 1931
Spanogle	High-velocity turbulence	TN 396, Oct. 1931
Spanogle	Injection rate	TN 402, Dec. 1931
Rothrock	Injection timing	TR 525, 1935
Rothrock	Injection quantity	TR 545, 1935
Rothrock	Spray characteristics	TR 561, 1936
Rothrock	Air motion	TR 588, 1937
Joachim	Combustion-chamber type	TR 282, 1928
Rothrock	Combustion photos	TR 401, 1931
Rothrock	Fuel vaporization effects	TR 435, 1932
Buckley	Flame photographs	TN 496, 1934
Moore	Combustion-chamber shape	TN 514, Dec. 1934
Rothrock	Spray geometry	TR 561, 1936
Rothrock	Flame photographs	TR 588, 1937

3.61 Bird, "Some Characteristics of Nozzles and Sprays for Oil Engines," *Proc. Third Int. Oil Power Conf.*, Berlin, 1930.

3.62 Loschge, "Vergleichende Druckindizierversuche mit Luftspeicherung an einem Vorkammerdieselmotor," *ATZ* (Berlin) *35*, Dec. 10, 1932, 562.

3.63 Taylor and Huckle, "Effect of Turbulence on the Performance of a Sleeve-Valve Compression-Ignition Engine," *Diesel Power*, Sept. 1933.

3.631 Alcock, "Air Swirl in Oil Engines," *Proc. IME* (London) Apr. 1934.

3.64 Cook, "Diesel Engine Cycle Analysis of Relationships of Fuel Injection to Fuel Compression Ignition Characteristics and Best Fuel Utilization," *SAE Paper* 650449, May 1965. (Computer results showing trends due to changing injection rates and timing with fuels having assumed evaporation and delay characteristics. Results not compared with practical measurements except for one condition.)

Effects of Combustion Chamber Design on Diesel-Engine Performance

3.650 Work of the U.S. National Advisory Committee for Aeronautics (now NASA):

Author	Subject	NACA Publication
Spanogle	Airflow in combustion chambers	TN 414, Apr. 1932
Moore	Geometry of divided chamber	TN 435, Nov. 1932
Moore	Connecting-passage diameter	TN 436, Nov. 1932
Moore	Combustion-chamber shape	TN 514, Dec. 1934
Moore and Collins	Displacer piston	{ TN 518, Feb. 1935 { TR 577, 1937
Foster	Open chamber, no swirl	TR 568, 1936
Moore	Displacer piston	TN 569, May 1936
Earle	Air entry angle	TN 610, Aug. 1937
Spanogle	Divided chamber	TN 396, Oct. 1931

3.660 Whitney, "High-Speed Compression-Ignition Performance — Three Types of Combustion Chamber," *Trans. SAE 30*, 328 (1935).

3.670 Dicksee, "Open Combustion-Chamber Diesel Engines in Britain," *SAE Quart. Trans. 3*, 89 (1949); also *ibid.*, *5*, 151 (1951).

3.680 Gosling, "The Small Automotive Diesel Engine in Great Britain and Europe," *SAE Paper* 120, Aug. 1953.

3.682 Meurer, "The M Combustion of MAN," *Trans. SAE 64*, 250 (1956).

3.683 Hockel, "The MWM Balanced-Pressure Precombustion Chamber System for High-Speed Diesel Engines," *Trans. SAE 66*, 38 (1958). (Report of excellent fuel economy with a divided combustion chamber.)

3.690 Dawson, "Diesel Development in Medium and Small Commercial Applications," *SAE Paper* 215A, Sept. 1960. (Gives good comparison between open and divided combustion-chamber performance based on Perkins experience.)

3.691 Jackson, "Combustion in Large Diesel Engines," *Proc. Joint Conference on Combustion*, in Boston, 15th–17th June, in London, 25th–27th Oct. 1955, (IME London), 1956, Section IV, p. 253. (General experience with large Sun-Doxford marine engines. Empirical.)

3.692 Clawson, "A Study of Combustion in Small High-Speed Diesel Engines," S.M. Thesis, Department of Mechanical Engineering, Massachusetts Institute of Technology, Cambridge, Mass., Aug. 1963. (Results of development work and tests of 2.5-in. × 1.875-in. single-cylinder 4-stroke 4-valve Diesel engine show possibilities for good performance up to 4500 rpm, 1400 ft/min [7.13 m/sec] piston speed, provided high dp/dt and high max pressures are accepted.)

3.693 Davies *et al.*, "Combustion in Diesel Engines with Divided Combustion Chambers," *IME Joint Conference on Combustion*, Section 4, 1955, pp. 51–67. (Descriptions, performance, pressure measurements in pre and main chambers, loss analysis of several commerical types. Especial emphasis on heat-loss rates and their effects.)

3.694 Gallois, "Direct Injection and Precombustion Chamber Injection of High Supercharged Four-Stroke Engines," *Seventh CIMAC Congress*, London, 1965. (Comparison of performance, economy, heat rejection, firing pressures, and piston temperatures of locomotive-size engines of these two types.)

3.695 Loeffler, "Development of an Improved Automotive Diesel Combustion System," *Trans. SAE 62*, 243 (1954). (Mack Motor Truck Co., switches from divided to open chamber, with performance improvement.)

3.696 Boll, "The New Cummins V-Type Diesel Engines," *SAE Preprint* 553B, Aug. 1952. (Details and performance of short-stroke, $5\frac{1}{8}$-in. × $4\frac{1}{8}$-in., engines with very thin combustion space, giving good performance at 2600 rpm.)

3.697 Henny and Herrmann, "Development and Performance of the Hispano-Suiza Turbulence Chamber," *SAE Paper* 650732, 1965.

3.698 Fleming and Bayer, "Diesel Combustion Phenomena as Studied in Free Piston Gasifiers," *Trans. SAE 71*, 74 (1963). (Interesting data on delay, rate of pressure rise, indicator diagrams, as a function of the variable compression ratio. Includes flame photographs.)

3.699 Linnenkohl, "Conversion of High-Speed Air-Cooled Diesels from Pre-Combustion Chamber Process to Direct Injection," *SAE Paper* 660010, Jan. 1966. (MWM, Germany, shows marked improvement. Swirl required.)

Gas Temperatures in Diesel Engines

3.71 Meyers and Uyehara, " Flame Temperature Measurements," *SAE Quart. Trans. 1*, 592 (1947).

3.72 Uyehara and Meyers, " Diesel Combustion Temperatures," *SAE Quart. Trans. 3*, 178 (1949).

3.73 Tsao *et al.*, " Gas Temperatures During Compression in Motored and Fired Diesel Engines," *SAE Paper* 272B, Jan. 1961.

Induction of Auxiliary Fuel
(See also refs. 3.907–3.909.)

3.80 Lyn, " Experimental Investigation of Effect of Fuel Addition to Intake Air on Performance of Compression-Ignition Engine," *Proc. IME* (London) *168*, 265 (1954).

3.81 Derry *et al.*, " Effect of Auxiliary Fuels on Smoke-limited Power Output of Diesel Engines," *Proc. IME* (London) *168*, 280 (1954).

3.82 Arnold *et al.*, " Bifuel Approach to Burning Residual Fuels in Diesel Engines," *Trans. SAE 66*, 54 (1958).

3.83 Alperstein *et al.*, " Fumigation Kills Smoke — Improves Diesel Performance," *Trans. SAE 66*, 574 (1958). (Introduction of premixed fuel into the inlet.)

Smoke and Other Exhaust Products of Diesel Engines
(See also refs. 4.600–4.608 and 14.701–14.704.)

3.851 Manville *et al.*, " Control of Smoke in Automotive Diesel," *Trans. SAE 47*, 397 (1940).

3.852 Schweitzer, " Must Diesel Engines Smoke?" *SAE Quart. Trans. 1*, 476 (1947).

3.853 Elliot and Davis, " Composition of Diesel Exhaust Gas," *SAE Quart. Trans. 3*, 330 (1950).

3.854 Various authors, " Diesel Engine Smoke Symposium," *SAE Special Publication* SP-101, 1948. (Collection of preprints 121 to 126, Jan. 1948.)

3.855 Millington and French, " Diesel Exhaust — A European Viewpoint," *SAE Paper* 660549, Aug. 1966. (Chiefly concerned with air-pollution problems, especially smoke. Useful data on Diesel smoke and smoke measurements. Ricardo & Co. experience.)

Starting Diesel Engines

3.891 Shea and Ayton, " Cool Weather Operation of Diesel Engines — a Bibliography," Library of Congress, Ref. Department, Sept. 1952. (Complete list of references to that date.)

3.892 Blose, " Factors Affecting Diesel-Engine Starting," *Jour. SAE 60*, April 1952, 40.

3.893 Kilty *et al.*, " Starting Cold Diesels," *SAE Paper* 795, Sept. 1952.

3.894 Blose, " Effect of Low Ambient Temperatures on Cranking and Starting of Diesel Engines," *SAE Paper* 399, Oct. 1954.

3.895 Meyer *et al.*, "Engine Cranking at Arctic Temperatures," *Trans. SAE 63*, 515 (1955).

3.896 Bowditch and Stebar, "Why Hot Compression-Ignition Engines Start Harder than Cold Ones," *Jour. SAE 65*, Sept. 1957, 72.

Size Effects in Diesel Engines
(See also Volume I, Chapter 11.)

3.897 Taylor, "Effect of Size on the Design and Performance of Internal Combustion Engines," *Trans. ASME 72*, 633 (1950).

3.898 Taylor, "Correlation and Presentation of Diesel-Engine Performance Data," *SAE Quart. Trans. 5*, 194 (1951).

Multifuel Engines

3.900 Woomert and Pelizzoni, "Compression Ratio and Multi-fuel Capability," *SAE Paper* 929C, 1964. [Indicated power almost exactly proportional to fuel density at given pump setting, at either high (20.5) or low (15.8) compression. Higher ratio required for starting with 18 cetane-number gasoline.]

3.901 Isley, "Development of Multi-fuel Features of the LD-465 and LDS-465 Military Engines," *SAE Paper* 929D, 1964. (In-line engines with M.A.N. spherical combustion chamber, inlet swirl, and fuel-pump compensator to deliver equal weight of gasoline or Diesel oil at the same throttle setting. Inlet-air heater for starting.)

3.902 Meurer, "Multifuel Engine Practice," *Trans. SAE 70*, 712 (1962). (Detailed record of M.A.N. developments with M combustion system, ref. 3.682, complete and well documented.)

3.903 Anon, "The Development of a Complete Family of Compression-Ignition Multi-fuel Engines," *Gen. Motors Eng. J. 11*, First Quarter, 2 (1964). (Conversion of General Motors line of 2-cycle Diesel engines to multifuel operation, chiefly by raising the compression ratio from 17.5 to 23.0. Includes useful performance data.)

3.904 Schweitzer and Hussman, "Lycoming S & H, The Compact Multi-fuel Engine," *SAE Paper* 790C, Jan. 1964. See also *Auto. Ind.*, April 1 and 15, 1963. (History of development and present performance of a 2-cycle air-cooled automotive Diesel engine. Shows performance with several combustion-chamber shapes.)

3.905 Schweitzer and Grunder, "Hybrid Engines," *Trans. SAE 71*, 541 (1963). (Review of developments to date in stratified-charge and multi-fuel engines. Largely descriptive. Good bibliography.)

3.906 Various authors, "Multi-fuel Symposium," *SAE Special Publication* SP-158, 1958.

Dual-Fuel Engines

3.907 Moore and Mitchell, "Combustion in Dual-fuel Engines," *Proc. Joint Conference on Combustion*, in Boston 15th–17th June, in London 25th–27th Oct. 1955, IME, London, 1956, Section IV, p. 300. (Engine performance versus operating parameters including pilot-fuel quantity and ignition accelerators. Properties of producer, town, blast-furnace, sewage, coke-oven, and natural gas in England.)

3.908 Boyer, "Status of Dual-Fuel Engine Development," *SAE Paper* 296, Jan. 1949. (General development history. Performance of Cooper-Bessemer dual-fuel engines.)

3.909 Harnsberger and Martin, "Dual-Fuel Engine Performance," S.B. Thesis, Department of Mechanical Engineering, Massachusetts Institute of Technology, Cambridge, Mass., 1950. (Performance of a CFR cylinder operating as gas engine, Diesel engine, and gas-Diesel engine over a wide range of mixture ratios and compression ratios.)

CHAPTER 4

Important Books and Articles on the Subject of Fuels
(See also refs. 1.10–1.16 and 11.002, No. 137.)

4.00 "Development and Application of Automotive Fuels, Gasoline and Diesel," *SAE Special Publication* SP-137, Section I, Diesel Fuels, by Wilson; Section II, Gasoline, by Felt. (Discussion of resources, production and refining, engine requirements, etc. Excellent practical review of the fuel situation from Ethyl Corp. Laboratories. Extensive bibliography.)

4.01 Nash and Howes, *The Principles of Motor Fuel Preparation and Application*, Vols. I and II, John Wiley & Sons, Inc., New York, 1936.

4.02 *ASTM Manual of Engine Test Methods for Rating Fuels*, ASTM, Baltimore. (Standard methods for rating motor, aviation, and Diesel-engine fuels. There is a wealth of supplementary information concerning not only operation and maintenance details but also building and utility requirements, installation and assembly of the test engine, and reference fuels.)

4.03 Smith, "Composition and Properties of Diesel Fuels," *SAE Quart. Trans. 3*, 164 (1949). (Comprehensive survey.)

4.04 Barnard, "Role of Gasoline in Engine Development," *SAE Quart. Trans. 5*, 272 (1951). (General article, stressing trends and economics.)

4.05 ASTM, *Standards on Petroleum Products and Lubricants*, ASTM, Baltimore, 1957. (Basic test methods, specifications, and standards in United States.)

4.051 Popovich and Hering, *Fuels and Lubricants*, John Wiley & Sons, Inc., New York, 1959. (Wide coverage, including specifications, tests, bibliography.)

4.06 Brewster and Kerley, "Automotive Fuels and Combustion Problems," *SAE Paper* 725C, Aug. 1963. (Valuable summary of fuel and combustion problems in both S-I and Diesel engines. Includes many data on gas and spark-plug temperatures.)

4.07 Phillips Petroleum Co., "Phillips 66 Hydrocarbons," published every few years by Phillips Petroleum Co., Bartlesville, Okla. (Valuable reference on composition and properties of available pure hydrocarbon products.)

4.08 Campbell, "Looking Ahead in Fuels for Automotive Transportation," *Jour. SAE 65*, Dec. 1957, 41. (Excellent historical review, data on trends in fuel production, use and characteristics, together with discussion of present problems and probable future trends.)

4.09 Various authors, "Symposium on Fuels," *SAE Special Publication* SP-139, 1955.

Heat of Combustion of Fuels

(See also ref. 4.101.)

4.091 Kharasch, "Heats of Combustion of Organic Compounds," *J. Res. Natl. Bur. Standards 2*, 359 (1929).

4.092 Cragoe, "Thermal Properties of Petroleum Products," U.S. Natl. Bur. Standards, *Misc. Publ.* 97, Nov. 1929.

4.093 Rossini, "Calorimetric Determination of the Heats of Combustion of Ethane, Propane, Normal Butane, and Normal Pentane," *J. Res. Natl. Bur. Standards 12*, 735 (1934).

4.094 Fiock *et al.*, "Definition of Heats of Combustion of a Fuel and Current Methods for Their Determination," *Trans. ASME 70*, 811 (1948).

4.095 Christie, "Uses of High and Low Heat Values," *Trans. ASME 70*, 819 (1948).

4.096 Fein *et al.*, "Net Heat of Combustion of Petroleum Hydrocarbons," *Ind. Eng. Chem. 46*, 610 (1953).

Physical and Chemical Characteristics of Fuels

(See also Table 4–1 and refs. 0.030–0.035.)

4.100 Egloff, "Physical Constants of Hydrocarbons," Reinhold Publishing Corp., New York, 1939.

4.101 Maxwell, "Data Book on Hydrocarbons," D. Van Nostrand Company, Princeton, N.J., 1950. (Properties, physical and thermodynamic, of petroleum and nonpetroleum fuel constituents.)

4.102 Brooks *et al.*, "The Chemistry of Petroleum Hydrocarbons," Reinhold Publishing Corp., New York, Vol. I, 1954; Vol. II, 1955.

4.103 Phillips Petroleum Co., "Characteristics of Liquid Petroleum Products," *Bull.* 01.0–31.3, 1956–1957. (Data on density, volatility, vapor pressure, Reynolds number, friction factor, bibliography.)

4.104 Moore and Mitchell, "Combustion in Dual-Fuel Engines," *Proc. Joint Conference on Combustion*, IME, London, 1956, Section IV, pp. 80–89. (Includes properties of producer, town, blast-furnace, sewage, coke-oven, and natural gas in England.)

4.105 U.S. Bureau of Mines, Mineral Industry Surveys, "Motor Gasolines," published twice a year (for "winter" and "summer" fuels) by U.S. Department of the Interior, Washington, D.C.

4.106 Schnidman *et al.*, "Gaseous Fuels," published by American Gas Association, 1954.

4.107 Finley, "Handbook of Butane-Propane Gases," Western Gas, 2nd ed., 1935.

4.108 Alden and Selim, "LP Gas for Motor Fuel," *SAE Paper* 548, Jan. 1951. (Economics and supply.)

4.109 Heat of Combustion, see refs. 4.091–4.096.

4.1091 Octane ratings of, see refs. 4.200–4.293.

4.1092 For price of fuels, see ref. 4.119.

Fuel Resources, Supply, Economics

(See also refs. 4.00, 4.04, 4.08, 4.09, 4.358.)

4.110 Mather, "Enough and to Spare," Harper, New York, 1944. (Broad treatment of world resources, including fuels.)

4.111 Merkel, "U.S. Motor Fuel Trends," published by E. I. du Pont de Nemours & Co., Inc., May 11, 1964, A.P.I. Div., St. Louis, Mo. (Survey of present and future petroleum resources.)

4.112 Withrow, "U.S. Resources Assure Endless Fuel Supply," *Jour. SAE 58*, Sept. 1950, 77. (Data on production and reserves of various fuels.)

4.113 Barnard, "Role of Gasoline in Engine Development," *SAE Quart. Trans. 5*, 273 (1951). (General article, stressing trends and economics.)

4.114 Davis, "Petroleum Products: There Will Be Enough If We Expand All Facilities in Proportion," *Jour. SAE 60*, Aug. 1952, 17. (Data on world resources.)

4.115 Van Sant and Orr, "Trends in Residual Fuels," *SAE Preprint* 27, Jan. 1957. (Excellent summary of the problem of petroleum resources and supply in the United States.)

4.116 Blade, "Mineral Industry Survey, Diesel Fuel Oils, 1957," U.S. Bureau of Mines, Petroleum Products Survey, No. 2, 1957.

4.117 Fieldner, "Production of Gasoline Substitutes from Coal," *Trans. SAE 22*, 78 (1927).

4.118 "Symposium on Coal Hydrogenation," *Trans. ASME 72*, 349 (1950). (Descriptions of apparatus and processes.)

4.119 Clark and Bunch, "Dual Fuel Combustion in a Railroad Diesel Engine," *Trans. SAE 71*, 91 (1963). (Data on fuel prices in the U.S., 1940–1960.)

Fuels, Safety of

4.120 Taylor and Taylor, "Crash Fire Tests with Diesel Oil," *Aviation*, Nov. 1930. (Ignitability and subsequent fires with gasoline and Diesel oil compared.)

4.121 Scull, "Relation Between Inflammables and Ignition Sources in Aircraft Environments," *NACA TN* 2227, Dec. 1950. (Summary of the literature and compilation of curves from many sources. Directed specifically toward fire hazard.)

4.122 Klinkenberg and van der Mijn, *Electrostatics in the Petroleum Industry*, Elsevier Pub. Co., Amsterdam, London, New York, and Princeton, N.J., 1958.

Unconventional Fuels

(See also refs. 4.201–4.205, 13.910–13.913.)

4.123 Pawlikowski, "Rupa Pulverized Coal Engine," *Engineering* (London) *126*, 409 (1928).

4.124 "Le Moteur à Charbon Pulvérisé," *La Technique Moderne* (Paris), July 15, 1935, 490.

4.125 Masi *et al.*, "Oxygen Boost of Engine Power at Altitude," *SAE Quart. Trans. 1*, 76 (1947).

4.126 Starkman, "Nitromethane as a Piston Engine Fuel," *SAE Paper* 410, Nov. 1954.

4.127 Starkman, "Ammonia as a Fuel for the Internal-Combustion Engine," *SAE Paper* 660155, Jan. 1966. (Fuel-air-cycle calculations and engine tests—power 70 per cent and bsfc 200—250 per cent of gasoline operation. Ammonia must be gaseous and contain 4 per cent hydrogen. Exhaust contains more nitric oxide than with gasoline.)

4.128 Gerrish and Foster, "Hydrogen as an Auxiliary Fuel in Compression-Ignition Engines," *NACA TR* 535, 1935.

FUELS FOR SPARK-IGNITION ENGINES

Alcohol

(See also ref. 0.034.)

4.201 "Comparative Fuel Economy with Gasoline and with Alcohol-Gasoline Blends," U.S. Natl. Bur. Standards, *Tech. News Bull.* 197, Sept. 1933.

4.202 Rogowski and Taylor, "Comparative Performance of Alcohol-Gasoline Blends," *J. Aeron. Sci. 8*, 384 (1941).

4.203 Retel, "Utilization of Alcohol as Motor Fuel by Direct Injection," *Engineers' Digest*, Dec. 1945. From *La Vie Automobile*, no. 1279/1280, July 10–25, 1944; no. 1281/1282, Aug. 10–25, 1944; no. 1283/1284, Sept. 10–25, 1944. (Discussion of mixture control and comparative performance on alcohol.)

4.204 Wiebe and Nowakowska, "The Technical Literature of Agricultural Motor Fuel," U.S. Department of Agriculture, *Bibliographical Bull.* 10, Feb. 1949. 4.2041 Kilayko, "Ethyl Alcohol as Fuel for Internal Combustion Engines," S.M. Thesis, Mechanical Engineering Department, Massachusetts Institute of Technology, Cambridge, Mass., 1960.

4.205 Various authors, "Alcohols and Hydrocarbons as Motor Fuels," *SAE Special Publication* SP-254, June 1964. (Comparative performance of alcohol and gasoline, including exhaust emissions.)

Antiknock Materials

4.210 Boyd, "Pathfinding in Fuels and Engines," *SAE Quart. Trans. 4*, 182 (1950). (History of the work of Midgeley, Boyd, and others, and the discovery of tetraethyl lead. GM work. Bibliography.)

4.211 Boyd, "Quantitative Effects of Some Compounds upon Detonation in Internal Combustion Engines," *International Critical Tables*, Vol. II. Published by McGraw-Hill Book Company for the National Research Council, 1927, p. 162.

4.212 Campbell *et al.*, "Anti-Knock Effect of Tetraethyl Lead in Increasing the Critical Compression Ratio of Individual Hydrocarbons," *Ind. Eng. Chem. 27*, 593 (1935).

4.213 Zang *et al.*, "Tetraethyl Lead in Various Pure Hydrocarbons; Antiknock Effectiveness," *Ind. Eng. Chem. 43*, 2826 (1951).

4.214 Rifkin and Walcutt, "A Basis for Understanding Anti-Knock Action," *Trans. SAE 65*, 552 (1957). (Theory of antiknock action, especially with tetraethyl lead.)

4.215 Broun and Lovell, "A New Manganese Anti-Knock," paper presented at the Meeting of American Chemical Society, San Francisco, April 1958. (Compound called AK-33X.)

4.216 Gibson *et al.*, "Anti-Knock Compounds, Research Development and Refinery Application," paper presented at the Fifth World Petroleum Congress, New York, May 1959. (Details of AK-33X and its use.)

4.217 See also the following NACA publications on lead susceptibility of various fuels: *TM* 940, Chemical Composition, April 1940; *ARR* 3E26, Paraffins and Olefins, May 1943; *ARR* E4J02, Paraffinic Fuels, Oct. 1944.

4.218 Hesselberg and Howard, "The Antiknock Behavior of the Alkyl Lead Compounds," *Trans. SAE 69*, 5 (1961). (Includes laboratory and road ratings.)

4.219 Richardson *et al.*, "Organolead Antiknock Agents—Their Performance and Mode of Action," *Ninth (International) Symposium on Combustion*, Academic Press, New York, 1963, p. 1023. (Test results and theory of behavior.)

Detonation Characteristics versus Fuel Composition

(See also refs. 2.00–2.03, 2.11–2.192, 4.353.)

4.220 Midgeley, Jr., and Boyd, "Detonation Characteristics of Aromatic and Paraffin Hydrocarbons," *Ind. Eng. Chem. 14*, 589 (1922).

4.221 Lovell, "Engine Knock and Molecular Structure of Hydrocarbons," *SAE Quart. Trans. 2*, 532 (1948).

4.23 Livingston, "Knock Resistance of Pure Hydrocarbons: Correlation with Chemical Structure," *Ind. Eng. Chem. 43*, 2834 (1951).

4.24 Rifkin *et al.*, "Early Combustion Reactions in Engine Operation," *SAE Preprint* 689, 1951. (Effects of fuel octane number and lead.)

4.25 Cornelius and Caplan, "Some Effects of Fuel Structure, Tetraethyl Lead, and Engine Deposits on Precombustion Reactions in a Firing Engine," *SAE Quart. Trans. 6*, 488 (1952). (Effect on p-θ diagram, cool-flame radiation, and on imep.)

4.26 Fleming *et al.*, "Control of Engine Knock Through Gasoline and Oil Composition," *SAE Preprint* 428, Jan. 10, 1955.

4.27 NACA index of technical publications 1915–1947, pp. 229–233, under "Fuels—Relation to Engine Performance," contains numerous references to reports on knock-limited engine performance as affected by fuel structure.

4.28 Meyer and Goldthwait, "Effect of Hydrocarbon Type and Distribution in the Boiling Range on Road Antiknock Performance," *SAE Preprint* 258, 1957.

4.29 Scott, "Knock Characteristics of Hydrocarbon Mixtures," paper presented at American Petroleum Institute meeting, 1958.

4.291 Armor *et al.*, "LP-Gas Octane Numbers and Their Relationship to Engine Performance," *SAE Preprint* 6U, Jan. 1959.

4.292 Frey *et al.*, "Knock Characteristics of Sixteen Pure Gaseous Hydrocarbons," *Bull.* 252, Phillips Petroleum Company, Bartlesville, Oklahoma, and also *Oil and Gas Journal*, Sept. 1948.

4.293 Scheller, "A General Study of Knock-Limited Mean Effective Pressures in an Engine Using a Gaseous Fuel," B.S. Thesis, Mechanical Engineering Department, Massachusetts Institute of Technology, Cambridge, Mass., May 1956. (Knock-limited mep versus fuel-air ratio, compression ratio, etc.)

Deposits and Wear versus Fuel Composition in S-I Engines
(See also refs. 2.96–2.998.)

4.320 Hunn, "Gum in Gasoline," *Jour. SAE 26*, Jan. 1930, 31.

4.321 Aldrich and Robie, "Gum Stability of Gasoline," *Trans. SAE 26*, 476 (1930); *ibid.*, *26*, 484 (1930).

4.322 Rogers *et al.*, "Gum Formation in Gasoline," *Ind. Eng. Chem. 25*, 397 (1933); Winning and Thomas, *ibid.*, *25*, 511 (1933).

4.323 Egloff *et al.*, "Inhibitors in Cracked Gasoline, I. Relation of Structure to Inhibiting Effectiveness," *Ind. Eng. Chem. 24*, 1375 (1932); Lowry, Jr., *et al.*, "II. Correlation of Inhibiting Action and Oxidation-Reduction Potential," *ibid.*, *25*, 804 (1933).

4.330 Bowhay and Koenig, "Factors Affecting Formation of Low Temperature Engine Deposits," *SAE Quart. Trans. 2*, 132 (1948). (Laboratory tests correlated with tests of delivery trucks and passenger cars.)

4.331 Backoff, "Study of Varnish and Sludge Tendencies of Fuels," *SAE Quart. Trans. 2*, 88 (1948).

4.332 Jeffrey *et al.*, "Effects of Sulfur in Motor Gasoline on Engine Operation," published by CRC, Inc., New York, 1950. (Results of fleet tests. Shows increase of wear due to sulfur only for intermittent type of service.)

4.333 Gibson *et al.*, "Combustion-Chamber Deposits and Power Loss," *SAE Quart. Trans. 6*, 595 (1952). (Includes data on the influence of TEL and fuel composition.)

4.334 Street, "Mode of Formation of Lead Deposits in Gasoline Engines," *Trans. SAE 61*, 442 (1953). (Analysis of solid deposits. Theory of lead effect suggested.)

4.335 Greenshields, "Spark-Plug Fouling Studies," *Trans. SAE 61*, 3 (1953). (Lead fouling and effects of tricresyl phosphate.)

4.336 "Symposium on Engine Deposits," *SAE Special Publication* SP-113, 1952. (A collection of SAE papers dated 1939–1952 concerning deposit problems in S-I and Diesel engines.)

Effect of Fuel on Performance of S-I Engines
(See also refs. 4.552–4.558.)

4.350 Brown, "The Relation of Motor Fuel Characteristics to Engine Performance," *Univ. of Michigan, Eng. Res. Bull.* 7, May, 1927. (Early comprehensive study, principally with regard to volatility.)

4.351 Wilson, "Significance of Tests for Motor Fuels," *Trans. SAE 26*, 151 (1930).

4.352 Wilson and Barnard, "Chemical Hay for Mechanical Horses," *Trans. SAE 35*, 359 (1934).

4.353 Heron, "Engine Performance with High Octane Number Fuels," Ethyl Corp., Detroit, Feb., 1949. (Extensive work with 17.6-cu in. supercharged Ethyl Corp. engine. Knock-limited imep and isfc with various fuels at various operating conditions.)

4.354 Barnett, "NACA Investigation of Fuel Performance in Piston-Type Engines," *NACA TR* 1026, 1951. (A compilation of many of the pertinent research data acquired by the NACA on fuel performance in piston engines.)

4.355 Alquist *et al.*, "Aviation Fuel Economy and Quality," *Trans. SAE 61*, 717 (1953). (Effects of performance number and volatility on performance in the air.)

4.356 Holaday, "What Can We Get from Higher Octane Number Fuels?" *SAE Preprint* 225, Jan. 11, 1954. (Predicts important future increases in octane number by improved methods of refining. Estimates car mileage versus octane number.)

4.357 ASTM, *Significance of ASTM Tests for Petroleum Products*, ASTM, Baltimore, 1955.

4.358 MacGregor *et al.*, "The Economics of High-Octane Gasolines," *Trans. SAE 67*, 343 (1959).

4.359 Kerley and Thurston, "The Indicated Performance of Otto-Cycle Engines," *Trans. SAE 70*, 5 (1962).

Gaseous Fuels versus Engine Performance

4.400 "French Fuel Demonstration Ends," *Auto. Ind.*, Sept. 1 (1928). (Producer gas for automobiles.)

4.410 Benz and Beckwith, "Liquefied Petroleum Gases Boost Power, Cut Fuel Consumption," *Auto. Ind.*, May 26 (1934).

4.420 Vogt, "Tests Show Merits of Butane as Internal-Combustion Engine Fuel," *Auto. Ind.*, Sept. 22 (1934).

4.440 Barnard, "Butane as an Automotive Fuel," *Trans. SAE 29*, 399 (1934).

4.442 Glidewell, "Engines to Digest the Vitamin-Enriched Fuel — Elpeegee," *Trans. ASTM 61*, 131 (1953). (Properties of gaseous fuels. Carburetion and carburetor design.)

4.443 Samuelson, "Comparative Operating Data of Tractors Using Gasoline or Liquefied Petroleum Gas Fuel," *SAE Preprint* 550, Jan. 1951. (Comparison of engine deposits. Favorable to gas.)

4.445 Raymond, "Liquefied Petroleum Gas as Fuel for Automotive Vehicles," *SAE Preprint* 535, Nov. 1950. (General discussion, details of a bus application.)

4.446 Ebuiger *et al.*, "The Evaluation of the Effect of Propylene in Propane Used as a Motor Fuel," *SAE Preprint* 409, Nov. 1954. (15 per cent propylene does no harm.)

4.447 Schnidman *et al.*, "Gaseous Fuels," published by American Gas Association, 1954.

4.448 Adams and Bolt, "What Engines Say About Propane Fuel Mixtures," *Trans. SAE 73*, 718 (1965). (Comparison of engine performance, knock limit, etc., between gas and gasoline. Important.)

4.449 Holvenstot, "Performance of Spark-Ignition Four-Cycle Engines on Various Fuels," *ASME Paper* 61-OGP-14, 1961. (Performance of large-cylinder spark-ignition engine on hydrogen, methane, propane, and butane and mixtures of hydrogen and methane. Includes effect of nitrogen additions.)

4.4491 See refs. 10.70–10.76.

4.4492 Various authors, "Symposium on LP-Gas for Engines," *SAE Special Publications* SP-114, 1951, and 285, 1966.

Knock Rating, Methods

(See refs. 2.092, 2.093, 4.01, 4.02, 4.05.)

Knock Rating Historical
(See also refs. 2.00–2.03, 4.210.)

4.500 Dickinson, "The Co-operative Fuel Research and Its Results," *Trans. SAE 24*, 262 (1929).

4.501 Campbell, "Detonation Characteristics of Some of the Fuels Suggested as Standards of Anti-Knock Quality," *Trans. SAE 25*, 126 (1930).

4.502 Horning, "The Co-operative Fuel Research Committee Engine," *Trans. SAE 26*, 436 (1931).

Knock Rating, Effect of Operating Variables

4.503 Brooks *et al.* (Atmospheric conditions), *Jour. SAE 25*, 136 (1930).

4.504 Edgar (Jacket and head temperature), *Jour. SAE 29*, July 1931, 52.

4.505 Stacey (Oil temperature and consumption), *Jour. SAE 29*, July 1931, 57.

4.506 Campbell *et al.* (Carburetor setting and spark timing), *Jour. SAE 29*, Aug. 1931, 129.

4.507 Huf *et al.* (Sound intensity), *Jour. SAE 29*, Aug. 1931, 134.

4.508 MacGregor (Humidity), *Jour. SAE 40*, 243 (1937).

4.509 Becker (Test conditions), *Jour. SAE 42*, Feb. 1938, 63.

4.510 Hebl and Rendel (Spark timing), *Jour. SAE 44*, May 1939, 210.

Aviation Knock Rating
(See also refs. 2.092, 2.093, 4.01, 4.02, 4.05.)

4.511 Veal, "Rating Aviation-Fuels in Full-Scale Aircraft Engines," *Trans. SAE 31*, 161 (1936).

4.512 Heron and Gillig, "Supercharged Knock Testing," paper read at the Second World Petroleum Congress, Paris, June 1937.

4.513 Heron, "Fuel Sensitivity and Engine Severity in Aircraft Engines," *Trans. SAE 54*, 394 (1946).

4.514 "A Method for Expressing Antiknock Ratings of Motor Fuels above 100," Cooperative Research Council Inc., New York, 1956.

Knock Rating, Correlation with Road Tests
(See also refs. 4.351 and 4.357.)

4.515 The following are surveys of the octane requirements of cars on the road: CRC, *Jour. SAE 57*, Sept. 1949; CRC, Inc., New York, Aug. 4, 1949; Edgar *et al.*, *SAE Quart. Trans. 4*, 93 (1950) (commercial vehicles); CRC, Inc., New York, May 1950; Best, *SAE Quart. Trans. 5*, 335 (1951); Best *et al.*, *Trans. SAE 62*, 288 (1954).

4.516 The following SAE articles report progress in correlating road and laboratory ratings (see also ref. 13.741): Veal *et al.*, *Trans. SAE 28*, 105 (1933); Veal, *Jour. SAE 36*, May, 1935, 165; Brooks, *ibid.*, *36*, June, 1935, 215; Campbell *et al.*, *Jour. SAE 40*, Apr. 1937, 144; Campbell *et al.*, *Trans. SAE 50*, 458 (1942), Barber,

SAE Quart. Trans. *2*, 401 (1948); Fell and Hollister, *SAE Preprint* 263, 1957; Morris, *ibid.*, *2*, 260, 1957; Caputo, *ibid.*, *2*, 260, 1957; Marshall and Gilbert, *ibid.*, *2*, 259, 1957; Hebl and Rendel, *Jour. SAE 44*, May 1939, 210.

4.517 "Accuracy of Road Rating Techniques Established by 1952, 1953 and 1954 Rating Programs," Co-operative Research Council, Inc., New York, Apr. 1958.

4.518 Bartholomew *et al.*, "The Effect of Carburetion and Manifolding on the Relative Antiknock Value of Fuels in Multi-Cylinder Engines," *Jour. SAE 42*, April, 141 (1938).

4.519 Gay and Meller, "Fuel Antiknock Quality — A Basis for Selection of Compression Ratio," *SAE Quart. Trans. 1*, 514 (1947). (Relates fuel sensitivity to knock-limited engine operation.)

4.520 Thaler, "Knock-rating of Fuels and Octane Requirements of French Cars" (in French), Publication 183, Institut Français du Petrole, Jan. 19, 1949.

4.521 Wintringham, "Improving Road Antiknock Rating of Fuels," Ethyl Corp. Research Laboratories, 1950, Collected Technical Papers. (Chassis dynamometer work on multicylinder knock rating.)

4.5211 Mizelle *et al.*, "Road Rating with a Chassis Dynamometer," *SAE Preprint* 731, 1955.

4.522 Kerley and Thurston, "Knocking Behavior of Fuels and Engines," *Trans. SAE 64*, 554 (1956). (Discussion of road versus laboratory ratings as affected by operating conditions, sensitivity, and engine severity.)

4.523 Barber *et al.*, "An Approach to Obtaining Road Octane Ratings in a Single-Cylinder Engine," *Trans. SAE 65*, 175 (1957). (Use of a more representative cylinder head than the CFR with altered operating conditions gives good agreement.)

4.524 Buerstetta *et al.*, "Road Antiknock Performance and the Boiling Range of Hydrocarbon Types," *SAE Preprint*, Nov. 1958. (Effect of fuel composition on deviation of road octane rating from laboratory ratings.)

Preignition of Fuels

(See also refs. 1.794, 2.85–2.953.)

4.530 Spencer, "Preignition Characteristics of Several Fuels Under Simulated Engine Conditions," *NACA TN* 710, 1941. See also *NACA Reports* ESJ11, Nov. 1945, and *ARR* E5 A11, Jan. 1945.

4.531 Male, D. W., "Effect of Engine Variables on Preignition-Limited Performance of Three Fuels," *NACA TN* 113, Sept. 1946.

4.532 Williams and Landis, "Some Effects of Fuels and Lubricants on Autoignition in Cars on the Road," *Trans. SAE 62*, 57 (1954).

4.533 Livengood *et al.*, "Surface Ignition in a Motored Engine," *Proc. Joint Conference on Combustion*, in Boston, 15th–17th June, in London, 25th–27th October 1955, IME, London, 1956, Section IV, p. 235.

Fuel Surveys

4.542 *National Motor-Gasoline Survey*, published semiannually by Bureau of Mines. Gives principal characteristics of retail motor-fuel samples taken throughout the United States, summer and winter each year.

Vapor Pressure

4.5421 For data on vapor pressure versus temperature of various organic liquids see: Cox, *Ind. Eng. Chem. 15*, 592 (1923); Wilson and Bahlke, *ibid., 16*, 115 (1924); Calingaert and Davis, *ibid., 17*, 1287 (1925); Marks, *Mechanical Engineers' Handbook*, 4th ed., McGraw-Hill Book Co., Inc., New York, p. 304.

Acceleration and Warm-up

4.543 Eisinger, "Engine-Acceleration Tests," *Trans. SAE 22*, Pt. 2 (1927).

4.544 Brooks, "The Influence of Fuel Characteristics on Engine Acceleration," *Trans. SAE 23*, 337 (1928).

4.545 Brooks, "Economic Fuel—Volatility and Engine Acceleration," *Trans. SAE 24*, 229 (1929).

4.546 Brooks, "Operating Factors and Engine Acceleration," *Trans. SAE, 24*, 1929.

4.547 Brooks and Bruce, "Effect of Design on Engine Acceleration," *Trans. SAE 26*, 63 (1930).

4.548 Rendel, "Warming-up Quality Proposed as Good Single Index of Gasoline Volatility," *Auto. Ind.* 1935, July 13.

4.549 Taylor and Gibson, "New Approach to Evaluation of Fuel Volatility," *SAE Quart. Trans. 3*, Oct. 1949, 306. (Selected fuels tested for warm-up performance from 20 mph. ASTM 50 and 90 per cent points studied with 10 per cent point constant.)

4.550 Moore *et al.*, "Effect of Fuel Volatility on Starting and Warm-up of New Automobiles," *Trans. SAE 65*, 692 (1957). (Starting below 20°F improves with increasing front-end volatility. Warm-up depends on 50 to 70 per cent temperatures of ASTM curves.)

4.551 Bridgeman and Aldrich, "How Fuel Makeup Affects Car Pickup," *Oil and Gas Jour.*, June 2, 1958.

Effects of Volatility on Engine Performance
(See also refs. 4.543–4.551, 4.580–4.584.)

4.552 Edgar *et al.*, "The Meaning of the Gasoline Distillation Curve," *API Paper*, Nov. 12, 1930.

4.553 MacCoull, "Effect of Gasoline Volatility on Engine Economy," *Jour. SAE 33*, Nov. 1933, 363.

4.554 Bridgeman, "Equilibrium Volatility of Motor Fuels from the Standpoint of Their Use in Internal Combustion Engines," *J. Res. Natl. Bur. Standards 13*, 53 (1934).

4.555 Eisinger and Barnard, "A Forgotten Property of Gasoline," *Trans. SAE 30*, 293 (1935).

4.556 *Review of Existing Pertinent Data on the Influence of Upper Boiling Range on the Performance of Motor Fuels*, CRC, Inc., New York, revised to Mar. 21, 1949. (Concludes that 50 and 90 per cent points of present government specifications can be revised without detriment to engine performance.)

4.557 Barber and MacPherson, "Application of Equilibrium Air Distillation to Gasoline Performance Calculations," *SAE Quart. Trans. 4*, 15 (1950). (Calculation of EAD from ASTM data, based on experimental correlations.)

4.558 Holaday and Heath, "Motor Fuel Volatility Trends," *SAE Quart. Trans. 5*, 429 (1951). (Survey of volatility trends since 1920. Summary of current information on effect of volatility on starting, warm-up, vapor lock, evaporation loss.)

Measurement of Volatility
(See also ASTM publications, refs. 4.02, 4.05.)

4.560 Sligh, "Volatility Tests for Automobile Fuels," *Trans. SAE 21*, Pt. II, 182 (1926).

4.561 Bridgeman, "Volatility Data from Gasoline Distillation Curves," *Trans. SAE 23*, 42 (1928).

4.562 Bridgeman, "Volatility Data on Natural Gasoline and Blended Fuels," *Trans. SAE 24*, 236 (1929).

4.563 Bridgeman, "Present Status of Equilibrium Volatility Work at the Bureau of Standards," *Trans. SAE 24*, 240 (1929).

4.564 Brown, "The Volatility of Motor Fuels," Univ. of Michigan, *Eng. Res. Bull.* 14, 1930.

Volatility; Oil Dilution

4.570 *Effect of Fuel Volatility on Crankcase Oil Dilution*, Report of Panel on Dilution, Volatility Study Group, Motors Fuels Division, CRC, New York, 1954. [Tests on relation of fuel volatility (end point) to dilution.]

Volatility; Starting
(See also refs. 4.350, 4.351, 4.552–4.558.)

4.580 Cragoe and Eisinger, "Fuel Requirements for Engine-Starting," *Trans. SAE 22*, Pt. I, 1 (1927).

4.581 Ritchie, "Starting Ability of Fuels Compared," *Trans. SAE 22*, Pt. II, 11 (1927).

4.582 Egbert, "Starting Efficiency of Gasoline Varies with Atmospheric Changes," *Auto. Ind.*, July 13, 1929.

4.583 Wawrziniok, "Das Anspringen der Motoren und die Eignung verschiedener Kraftstoffe hierzu," *ATZ* (Berlin), Sept. 25 and Oct. 10, 1933.

4.584 Nelson and Ulzheimer, "Ethyl Ether Best for Starting Cold Diesels," *Jour. SAE 58*, March 1950, 42. (Effects of fuel and additives on cold-weather starting.)

Volatility; Vapor Lock

4.585 "CRC Issues Critique on Vapor Lock Tests," *Jour. SAE 58*, Jan. 1950, 73. (Discussion of the CRC-223 report on vapor-lock road tests.)

4.586 Barnard *et al.*, "Weather or Lock — Vapor Lock Study, Road and Laboratory," *SAE Preprint* 866, 1956.

4.587 Bridgeman and Aldrich, "Cold Weather Vapor Lock Control," *Oil and Gas Journal* 1958, April 28.

4.588 Caplan and Brady, "Vapor Locking Tendencies of Fuels," *Trans. SAE 66*, 327 (1958). Suggests a new criterion for vapor lock.)

4.589 Percy *et al.*, "CRC Vapor Lock Technique—Its Development and Application," *Trans. SAE 71*, 122 (1963).

4.590 Symposium, "Fuels and Fuel Systems," publ. by Natural Gasoline Association of America, Tulsa, 1955, 108 pages. (Largely devoted to the vapor-lock problem.)

FUELS FOR DIESEL ENGINES

Deposits, Smoke, and Wear versus Fuel Composition in Diesel Engines
(See also refs. 3.851–3.855.)

4.600 Moore and Kent, "Effect of Sulphur and Nitrogen Content on Fuels on Diesel-Engine Wear," *SAE Quart. Trans. 1*, 687 (1947).

4.601 Blanc, "Effect of Diesel Fuel on Deposits and Wear," *SAE Quart. Trans. 2*, 306 (1948). (Sulfur found to be the principal offender.)

4.602 Lander, "Combustion Characteristics of Diesel Fuels," *SAE Quart. Trans. 3*, 200 (1949). (Deposits and smoke.)

4.603 Seniff and Robbins, "Full Scale Field Service Tests of Railroad Diesel Fuels," *Trans. SAE 65*, 46 (1957). (Deposits, corrosion, etc., as affected by fuel.)

4.604 Neely *et al.*, "Dark Diets for Diesels," *Trans. SAE 65*, 516 (1957). (Light fuels necessary for light loads; residual fuels acceptable for heavy loads. Proposes a fuel-blending system.)

4.605 Lyn, "Combustion Products and Wear in High-speed Compression-ignition Engines, with Particular Reference to Use of Lower Grade Fuels," *Proc. Joint Conference on Combustion*, Oct., 1955, IME, London, 1956, Section IV, p. 242. (Study of wear and deposits as a function of fuel and oil.)

4.606 McConnell, "Diesel Fuel Properties and Exhaust Gas—Distant Relations?" *SAE Preprint* 670091, Jan. 1967.

4.607 Golothan, "Diesel Engine Exhaust Smoke: The Influence of Fuel Properties and the Effects of Using a Barium-containing Fuel Additive," *SAE Paper* 670092, Jan. 1967.

4.608 Miller, "Diesel Smoke Suppression by Fuel Additive Treatment," *SAE Paper* 670093, Jan. 1967.

Effect of Fuel on Engine Performance, Diesel Engines
(See also refs. 4.700–4.705.)

4.610 Withers, "The Performance of High-Speed Diesel Engines for Road Transport as Affected by Nature of the Fuel," *J. Inst. Petrol. Technologists* (London), Aug. (1933).

4.611 McGregor, "Diesel Fuels—Significance of Ignition Characteristics," *Jour. SAE 38*, 217 (1936).

4.612 Good, "Cetane Numbers, Life Size," *Trans. SAE 41*, 343 (1937).

4.613 Burk *et al.*, "Fuel Requirements of Automotive Diesel Engines," *Trans. SAE 53*, 166 (1945). (Comprehensive report on nearly all aspects of Diesel fuels as related to engine performance.)

4.614 ASME, *Diesel Fuel Oils: Production, Characteristics and Combustion*, ASME, New York, 1948.

4.615 Broeze and Stillbroer, "Fuels for Automotive and Railroad Diesel Engines," *SAE Reprint*, Sept. 7–9, 1948, and *Jour. SAE 57*, March 1949, 64. (Excellent data on volatility effects, as well as other fuel properties. Shows large differences due to engine design.)

4.616 Crampton *et al.*, "Wide-Cut Diesel Fuels Help Extend Refinery Output, Perform Well in Railway Tests," *Jour. SAE 57*, Sept., 35 (1949). (Shows small differences in engine performance between normal and wide-range volatility fuel. Thermal efficiency independent of fuel volatility within range of tests.)

4.617 Vaile, "Research on the Compression-Ignition Engine and Its Fuels," *Proc. IME* (London) *162*, 13 (1950).

4.618 Horiak, "The New Hercules Polydiesel Engine," *SAE Paper* 118T, Nov., 1959. (Performance on gasoline and Diesel oil.)

4.620 Spanogle, "Compression-Ignition Engine Tests of Several Fuels," *NACA TN* 418, May 1932.

Nonpetroleum Liquid Fuels in Diesel Engines

4.621 Gautier, "Use of Vegetable Oils in Diesel Engines," *J. Inst. Petrol. Technologists* (London), May (1933).

4.622 Tu and Ku, "Cottonseed Oil as a Diesel Oil," *J. Chem. Eng.* (China), Vol. III, Sept. (1936).

4.623 "Indian Vegetable Oils for Diesel Engines," *Gas and Oil Power*, May (1942).

4.624 Loh *et al.*, "Compression-Ignition Engine Performance with Soybean Oil, Cottonseed Oil and Their Blends," S.M. Thesis, Department of Mechanical Engineering, Massachusetts Institute of Technology, Cambridge, Mass., 1944.

Measurement of Ignition Quality

(See refs. 3.32, 3.34, 4.02, and 4.05.)

Ignition-Quality, Additives

4.700 Robbins *et al.*, "Performance and Stability of Some Diesel Fuel Ignition Quality Improvers," *SAE Quart. Trans. 5*, 404 (1951). (Effect of additives on cetane number, dp/dt, maximum pressure, maintenance problems.)

4.701 Anderson and Wilson, "Effectiveness of Amyl Nitrate in a Full-Scale Diesel Engine," *SAE Quart. Trans. 6*, 230 (1952). (Test results with a GM 3-71 engine. Good measurements of delay as function of fuel, additive, and operative variables.)

Ignition Quality of Diesel Fuels
(See also refs. 2.20, 2.22, 3.41, 3.43, 3.45, 3.47.)

4.702 Michailova and Neumann, "The Cetene Scale and the Induction Period Preceding the Spontaneous Ignition of Diesel Fuels in Bombs," *NACA TM* 813, 1936.

4.703 Puckett and Candle, "Ignition Qualities of Hydrocarbons in the Diesel-fuel Boiling Range," *U.S. Bur. of Mines Information Circ.* 7474, 1948.

4.704 Jackson, "Spontaneous Ignition Temperatures of Pure Hydrocarbons and Commercial Fluids," *NACA RM* E50J10, Dec., 1950. (Fuel drops falling into a heated cup.)

4.705 Hurn and Hughes, "Combustion Characteristics of Diesel Fuels as Measured in a Constant-Volume Bomb," *SAE Quart. Trans. 6*, 24 (1952). (Measurements of delay as a function of pressure, temperature, and fuel composition.)

Fuels for Dual- and Multi-Fuel Engines
(See refs. 3.900–3.909.)

CHAPTER 5

Mixture Requirements

5.10 Tice, "Carbureting Conditions Characteristic of Aircraft Engines," *NACA TR* 48, 1919.

5.20 Berry and Keggerreis, "The Carburetion of Gasoline," Purdue Univ., *EES Bull.* 5, 1920.

5.30 Berry and Chenowith, "Car Carburetion Requirements," Purdue Univ., *EES Bull.* 17, 1924.

5.40 Sparrow, "Relation of Fuel-Air Ratio to Engine Performance," *NACA TR* 189, 1924.

5.50 Keggerreis *et al.*, "Carburetion of Kerosene," Purdue Univ., *EES Bull.* 27, 1927.

5.51 Mock, "Engine Types and Requirements for Preparation of Fuels," *Trans. SAE 39*, 257 (1936).

5.60 Fawkes *et al.*, "The Mixture Requirements of an Internal-Combustion Engine at Various Speeds and Loads," S.M. Thesis, Department of Aeronautical Engineering, Massachusetts Institute of Technology, Cambridge, Mass., 1941.

5.61 Collmus and Friberger, "Gasoline Consumption of a Highly Throttled Multicylinder Engine," S.B. Thesis, Department of Mechanical Engineering, Massachusetts Institute of Technology, Cambridge, Mass., 1945. (Mixture requirements at idling.)

5.62 MacClain, "Factors Producing Erratic Engine Operation," *SAE Quart. Trans. 1*, 28 (1947). (Includes discussion of mixture requirements of aircraft engines.)

5.63 Johnston and Lane, "A Study of the Best-Economy Performance of a Typical Automotive Engine," S.B. Thesis, Department of Mechanical Engineering,

Massachusetts Institute of Technology, Cambridge, Mass., 1950. (Includes curves of fuel-air ratio, bsfc, manifold pressure, and spark advance for best-economy operation of Plymouth and CFR engines.)

5.64 Chamberlain and Mayer, "An Investigation of the Best-Economy Performance of an Automotive Engine Using Gaseous Fuel," S.B. Thesis, Department of Mechanical Engineering, Massachusetts Institute of Technology, Cambridge, Mass., 1951. (Same engine as ref. 5.63. A large inlet manifold was used to elim-inate inlet dynamics. Adequate precautions were taken to ensure good mixing of air and gaseous fuel. Maps of fuel-air ratio, spark advance, manifold pressure, and bsfc are included.)

5.65 Hoyer, "A Study of the Acceleration of a Multicylinder Spark Ignition Engine," S.B. Thesis, Department of Mechanical Engineering, Massachusetts Institute of Technology, Cambridge, Mass., 1956. (Mixture requirements for good acceler-ation.)

Distribution

5.70 Sparrow, "The Arithmetic of Distribution in a Multicylinder Engine," *NACA TN* 162, 1923.

5.71 Mock, "Dual Carburetion and Manifold Design," *Trans. SAE 24*, 168 (1929).

5.72 Taylor, "Effect of Centrifugal Supercharger on Fuel Vaporization," *Trans. SAE 24*, 294 (1929).

5.73 Taub, "Mixture Distribution," *Trans. SAE 25*, 54 (1930).

5.74 Rabezza and Kalmar, "Mixture Distribution in Cylinder Studied by Measuring Spark-Plug Temperature," *Auto. Ind. 66*, Mar. 19 and 26, 1932.

5.75 Stokes and Holland, "Comparative Tests with Petrol and Butane on Air and Water Cooled Aircraft Engines," *Aeron. Res. Comm.* (London), R and M 1570 (1934).

5.76 Gerrish and Voss, "Mixture Distribution in a Single-Row Radial Engine," *NACA TN* 583, Oct. 1936.

5.77 Turner *et al.*, "Investigation of the Liquid Layer in a Transparent Manifold," Report for Course 2.802, 1942, Sloan Automotive Laboratories, MIT.

5.78 Marble *et al.*, "Study of the Mixture Distribution of a Double-Row Radial Aircraft Engine," *NACA ARR* E-5105 E-47, Oct. 1945.

5.79 Quentert and Ferkan, "Charge-Air Distribution among the Cylinders of a Double-Row Radial Aircraft Engine," *NACA WTR* E-282, July 1946.

5.80 Donahue and Kent, "A Study of Mixture Distribution," *SAE Quart. Trans. 4*, 546 (1950). (Exhaustive measurements with both liquid and gaseous fuels over wide ranges of operation. Distribution of fuel components and octane quality in addition to over-all distribution.)

5.81 Cooper *et al.*, "Radioactive Tracers Cast New Light on Fuel Distribution," *Trans. SAE 67*, 619 (1959). (New technique found very effective. Results of work on V-8 automobile engines as a function of speed, etc. Fuel-component distribu-tion and octane number also studied.)

5.82 Yu, "Fuel Distribution Studies — A New Look at an Old Problem," *Trans. SAE 71*, 596 (1963). (Excellent study of both time and geometric fuel-air-ratio varia-tion. Cycle-to-cycle gas sampling. Bibliography.)

CHAPTER 6

Carburetor Research

6.10 "Properties of Submerged Carburetor Jets," *Auto. Ind.*, Oct. 6 (1928).

6.11 Santer, "Investigation of Atomization in Carburetors," *NACA TM* 518, 1929.

6.12 Schebel, "On Atomization in Carburetors," *NACA TM* 644, 1931.

6.13 Dodson, Booth, and Metsger, "An Investigation of the Air Metering Characteristics of Carburator Elements," S.M. Thesis, Department of Aeronautical Engineering, Massachusetts Institute of Technology, Cambridge, Mass., 1940. (Important basic work. Source material for Fig. 6–10.)

6.14 Hwa, "A Study of Double Venturis for Aircraft Carburators," S.M. Thesis, Department of Mechanical Engineering, Massachusetts Institute of Technology, Cambridge, Mass., 1941.

6.15 Ting, "Analysis of Air Bleeds and Several Typical Idling Systems for Carburators," S.M. Thesis, Aeronautical Engineering Department, Massachusetts Institute of Technology, Cambridge, Mass., 1946.

6.16 Prien, "A Study of the Effect of Air Pulsations on the Operation of a Simple Carburator," S.B., General Science Thesis, Massachusetts Institute of Technology, Cambridge, Mass., 1954.

6.17 Fisher, "Some Notes on Carburation and Other Fuel System Problems," IME (London), *Proc. Automobile Div.* 1955–1956.

6.18 Wentworth, "Carburator Evaporation Losses," *SAE Preprint* 12B, Jan. 1958.

Carburetor Design

6.20 O'Neill, "The Techniques Employed in the Application of Carburettors to Vehicle Engines," IME (London), *Proc. Automobile Div.* 1962–1963, p. 113. (Valuable account of practical methods used.)

6.21 Bolton, "Carburetor Design and Calibration," *SAE Special Publication* SP-119, 1953. (Brief outline of carburetion principles and Marvel-Schebler practice.)

Proprietary Carburetors
(See also manuals and descriptive matter published by the various carburetor manufacturers.)

6.30 Lucke and Willhofft, "Carburetor Design — A Preliminary Study of the State of the Art," *NACA TR* 11, 1916. (Description of many early types.)

6.31 Tice, "Metering Characteristics of Carburetors," *NACA TR* 49, 1919, p. 5.

6.32 Keggerreis *et al.*, "Commercial Carburetor Characteristics," Purdue Univ., *EES Bull.* 21, Aug. 1925. (Steady-flow tests.)

6.33 "New Carburetor Designed for all Automotive Needs Except Aircraft," *Auto. Ind.*, Nov. 1 (1930). (Zenith open-passage type widely used 1930 to 1940.)

6.34 "Original Design Features of Rochester Carburetor on '49 Oldsmobile Engine," *Auto Ind.*, Dec. 15 (1948). (Includes cross section and photos. Compact open-passage design.)

6.35 Braun, "The Buick Air Power Carburetor," *Gen. Motors Eng. J. 1*, Feb. 26, 1954. (Open-passage type with compound venturis, four barrels.)

6.36 Knight, "The S.E. Carburetor," *Gen. Motors Eng. J. 1*, Feb., 128 (1954).

6.37 Kehoe, "The Quadrajet: A New Approach to Four-Barrel Two-Stage Carburetion," *Gen. Motors Eng. J. 13*, Third Quarter, 11 (1966). (Two small barrels with spring-loaded air valve system for idling and part-throttle operation. Two larger open-passage [about 2 × diam] barrels in parallel for heavy load operation.)

6.38 Heldt "Schebler Offering New Carburetor of Air-Valve Type," *Auto. Ind.*, Feb. 25, 1928, 40. (Venturi idle system by-passed by spring loaded air valve with dash pot. Typical air-valve design.)

Aircraft Carburetors

6.40 "Holley Carburetor Model CG for Aircraft Engines," *Auto. Ind.*, July 15, 1939.

6.41 Wiegand, "Carburetion for the Aircraft Engine," *Trans. SAE 51*, 294 (1943). (Outline of problem with descriptions of Wright Aeronautical Corp. equipment.)

6.42 Thorner, *Aircraft Carburation*, John Wiley & Sons, Inc., New York, 1956. (Qualitative description of carburetor behavior and design and operation of existing aircraft carburetors.)

6.43 Reid, "Aero-Engine Injection Carburettors," *Aircraft Eng.* (London), Feb., 1947. (General discussion of design requirements of aircraft carburetors. Description of leading British injection carburetors responsive to variations in engine speed and manifold density.)

6.44 "Speed-Density Carburetion System," publ. by Bendix Products Division, South Bend, Ind., 1947.

Gas Carburetors

6.50 Glidewell, "Engines to Digest the Vitamin-Enriched Fuel — Elpeegee," *Trans. SAE 61*, 131 (1953). (Gas composition; carburetor design and performance.)

6.51 St. George, "Carburation Equipment for LPG," *Jour. SAE 59*, Aug. 1951, 66.

Carburetor Icing

6.71 Allen *et al.*, *Jour. SAE 35*, Dec. 1934, 417.

6.72 Johnson, *Aviation*, May 1936.

6.73 Sanders, *Jour. SAE 45*, Jan. 1939, 14.

6.74 Chapman, *NACA MR* E6G11, July 1956.

6.75 Kunc *et al.*, *SAE Paper* 557, Jan. 1951.

6.76 Brown, *Jour. SAE 61*, June 1953, 44.

6.77 Dugan, *SAE Paper* 331, June 1954.

6.78 Dugan, *Jour. SAE 63*, March 1955, 39.

6.79 Barnard, *SAE Paper* 6R, Jan. 1955.

Emissions Control

See references 1.870–1.8792, 2.8111, 2.816, and references listed on pages 213 and 636.

CHAPTER 7

Spray Mechanics

7.00 DeJuhasz and Meyer, *Bibliography on Sprays*, 2nd ed., The Texas Co., New York, Apr. 1953. (Comprehensive, with both author and subject index. Short abstract included for each. U.S.A. and foreign.)

7.10 Research in Spray Mechanics by the U.S. National Advisory Committee for Aeronautics (now the NASA):

	Author	*Subject*	*NACA Publication*
7.11	Rothrock	Air flow effects	*TN* 329, Dec. 1939
7.12	Gelalles	Air and fuel temperature	*TN* 338, May 1930
7.13	Gelalles	Orifice length/diameter	*TN* 352, Oct. 1930
7.14	Rothrock	Open nozzles	*TN* 356, Nov. 1930
7.15	Gelalles	Discharge coefficients	*TR* 373, 1931
7.16	Gelalles	Orifice length/diameter	*TR* 402, 1931
7.17	Gelalles	Orifice length/diameter	*TN* 369, Mar. 1931
7.18	Rothrock	Air velocity effects	*TN* 376, May 1931
7.19	Rothrock	Low-pressure sprays	*TN* 399, Nov. 1931
7.20	Lee	Nozzle design	*TR* 425, 1932
7.21	Lee	Distribution in sprays	*TR* 438, 1932
7.22	Rothrock	Spray vaporization	*TN* 408, Feb. 1932
7.23	Lee and Spencer	Photomicrographs of sprays	*TR* 454, 1933
7.24	Lee	Nozzle type and design	*TR* 520, 1935
7.25	Lee	Distribution in sprays	*TR* 565, 1936

7.30 Bird, "Some Characteristics of Nozzles and Sprays for Oil Engines," *Proc. Third Int. Oil Power Conf.*, Berlin, 1930.

7.31 Castlemen, "The Mechanism of Atomization Accompanying Solid Injection," *NACA TR* 440, 1932.

7.32 DeJuhasz *et al.*, "On the Formation and Dispersion of Oil Sprays," Penn. State College, *EES Bull.* 40, 1932.

7.40 Selden and Spencer, "Heat Transfer to Fuel Sprays Injected into a Heated Gas," *NACA TR* 580, 1937.

7.41 "Investigations into Fuel Injection in Diesel Engines," *Sulzer Tech. Rev.*, No. 3 (1949). (Theory and experimental results; excellent photographs of spray development.)

7.42 York and Stubbs, "Photographic Analysis of Sprays," *Trans. ASME 74*, Oct. 1952, 1157. (A reasonably simple method for photographing drop size and velocities.)

7.50 Griffin and Muraszew, *The Atomization of Liquid Fuels*, John Wiley & Sons, Inc., New York, 1953. (Well-documented review of theory and experiment to date of publication.)

Mechanics of Fuel Delivery

7.60 Gelalles and Marsh, "Rates of Fuel Discharge as Affected by Design of Fuel Injection Systems," *NACA TR* 433, 1932.

7.601 Rothrock and Marsh, "Distribution and Regularity of Injection from a Multi-Cylinder Fuel-Injection Pump," *NACA TR* 533, 1935.

7.61 Lee, "Discharge Characteristics of a Double Injection-Valve Single-Pump Injection System," *NACA TN* 600, May 1937.

7.62 Nagao and Kakuda, "The Volumetric Efficiency of a Fuel Injection Pump," *JSME* (Tokyo) *paper*, Mar. 16, 1939.

7.64 Wehrman *et al.*, "Measuring Rate of Fuel Injection in an Operating Engine," *Trans. SAE 61*, 542 (1953). (Delivery pressure and rate curves for GM injectors.)

Fuel Injection for Diesel Engines
(See also ref. 11.002, No. 143.)

7.70 "Injection Pumps," *Auto. Ind.*, June 12, 881 (1937).

7.71 Alden and Weldy, "A New Fuel Injection Nozzle for High Speed Diesel and Gasoline Injection Engines," *SAE Paper* 70.4, Jan. 1942. (Small inward-opening nozzle.)

7.72 Austen and Goodridge, "Wear of Fuel Injection Equipment and Filtration of Fuel for Compression-Ignition Engines," IME (London), *Proc. Automobile Div.* 1950–1951, Pt. III.

7.73 Ricardo, "Pintaux Nozzle," *Diesel Power 29*, June 1951, 73. (Nozzle injects fuel into center of chamber at starting speeds to improve starting characteristics of comet head.)

7.74 Roosa, "Simplifying Fuel Injection," *SAE Preprint* 28, Jan. 1953. (Plunger-distributor type of pump.)

7.75 Miller, "How to Design Fuel Injection Equipment," *Jour. SAE 60*, June 1952, 31. (Bosch equipment.)

7.76 Rowell and Boll, "The New Cummins PT Diesel Fuel System," *SAE Paper* 334, Dec. 1954.

7.77 "Many Parts Eliminated in New Diesel Fuel System," *Auto. Ind.*, Sept. 1954, 1.

7.78 Miller and Hess, "A Diesel Injection System with New Features," *SAE Paper* 417, Jan., 1955. (Bosch single-plunger-distributor pump.)

7.79 Dolza, Kehoe, and Arkus-Duntov, "The General Motors Fuel Injection System," *SAE Preprint* 16, Jan. 1957. (Largely descriptive.)

7.791 "Worthington's Dual-Fuel Developments," *Diesel Power and Diesel Transportation*, Sept. (1949). (Fuel pump contains a pilot fuel plunger and a main plunger.)

7.792 Bowers *et al.*, "Development of a Single-Plunger Injection Pump for Four-Cylinder Diesel Engines," *Trans. SAE 52*, 296 (1944). (Single-plunger pump with piston-valve distributor.)

7.793 Mansfield, "A New Servo-Operated Fuel Injection System for Diesel Engines," *SAE Paper* 650432, 1965. (System powered by fuel pressure. Performance favorable.)

Effects of Spray Characteristics on Performance
(See also refs. 3.60–3.64.)

7.80 Ricardo, "Combustion in Diesel Engines," *Proc. IME* (London) *162*, 145 (1950). (Ricardo's theories regarding delay, effect of drop size, cetane number, air swirl, etc.)

7.81 Parker, "Fuel Injection, Engineering Know-How in Engine Design, Part 4," *SAE Special Publication* SP-143, 1956.

7.82 Miller, "Design and Application of Diesel Fuel Injection Equipment," *SAE Paper* 675, Oct., 1951. (Review of American Bosch experience.)

7.83 "Diesel Engine Improvements Through Experimentation with Fuel Injection and Combustion Chamber Relationships," *Gen. Motors Eng. J. 1*, Jan.–Feb. 1954, 32. (GM experience with sprays and chamber shapes.)

Fuel Injection with Spark Ignition
(See also refs. 2.790–2.817 and 11.002, Nos. 143 and 152.)

7.900 Taylor *et al.*, "Fuel Injection with Spark Ignition in an Otto-Cycle Engine," *Trans. SAE 26*, 346 (1931).

7.901 Taylor and Williams, "Further Investigation of Fuel Injection in an Engine Having Spark Ignition," *Trans. SAE 27*, 24 (1932).

7.902 Schey and Young, "Performance of a Fuel-Injection Spark-Ignition Engine Using a Hydrogenated Safety Fuel," *NACA TR* 471, 1933.

7.903 Schey and Clark, "Comparative Performance of Engines Using Carburetor, Manifold Injection and Cylinder Injection," *NACA TN* 688, Feb. 1939.

7.904 Wiegand and Meador, "Fuel Injection for the Aircraft Engine," *Trans. SAE 53*, 208 (1945). (Shows gains in performance with cylinder injection.)

7.905 Lange, "Fuel Injection for Light Aircraft," *SAE Quart. Trans. 2*, Apr., 210 (1948). (Shows performance improvement with port injection.)

7.906 Barrington and Downing, "An Application of Petrol Injection to Automobile Engines," IME (London), *Proc. Automobile Div.* 1949–1950, Pt. III, p. 133. (Extensive tests on 4-cylinder auto engines—most successful with port injection.)

7.907 Schmidt, "Benzinmotoren mit Kraftstoffeinspritzung und Funkenzündung" (gasoline engines with fuel injection and spark ignition), Jahrgang 11, Nov.–Dec. 1950, MTZ (Berlin), 6. (Based on experimental work in Germany.)

7.908 Nagao and Ohigashi, "On the Vaporization and Combustion Properties of Gasoline Spray," *Mem. Fac. Eng., Kyoto Univ. 14*, 38 (1952). (Test in a bomb with spark ignition.)

7.909 Nagao and Ohigashi, "Flame Development and Combustion Pressure with a Gasoline Injection Pump," *Mem. Fac. Eng.* Kyoto Univ. *14*, 112 (1952). (Flame-travel pictures and pressure measurements.)

7.910 Nagao and Ohigashi, "Performance and Combustion in Gasoline-Injection Engine," *Mem. Fac. Eng.* Kyoto Univ. *14*, 168 (1952). (Comparative performance with manifold, carburetor, and cylinder injection.)

7.911 Hirigoyen, "Study of End-Gas Temperature in a Fuel-Injection Spark-Ignition Engine," S.M. Thesis, MIT, Department of Mechanical Engineering, Massachusetts Institute of Technology, Cambridge, Mass., 1958. (Differences in charge temperature before ignition with premixed charge and with same fuel injected into the cylinder during the inlet stroke.)

7.912 Felt *et al.*, "The Effect of Fuel Injection on Knocking Behavior," *SAE Preprint* 3B, Jan. 1958. (Shows measurable improvement with inlet-port injection.)

7.920 Campbell, "Fuel Injection as Applied to Aircraft Engines," *Trans. SAE 37*, 77 (1935).

7.921 "Fuel Injection System Used on German Planes," *Auto. Ind.*, April 1940, 15.

7.922 "Three German Engine Fuel Systems," *Aircraft Eng.* (London) *15*, 253, 293 (1943).

7.923 Bolt, "Fuel Metering by Engine Speed and Manifold Density," *Trans. SAE 55*, 498 (1947).

7.924 Brook and Nicolls, "Some Aspects of Petrol Injection Equipment Development," IME (London), *Proc. Automobile Div.* 1947–1948, Pt. I. (Equipment requirements and design.)

7.925 Lange and van Overbeke, "Fuel Injection for Spark-Ignited Automotive Engines," *SAE Paper* 166, Jan., 1948. (Classification of systems; several examples of improved engine performance.)

7.926 Marshall, "Direct Injection for Wright 18-Cylinder Engine," *SAE Paper* 127, Jan., 1948. (Details of system using air-flow control.)

7.927 Mock and Suttle, "Problems of Fuel Injection for Gasoline Automobile Engines," *SAE Paper* 646, Nov., 1955. (Review of various types of equipment.)

7.928 Miller and Hossak, "Fuel Injection for Gasoline Automobiles," *Jour. SAE 66*, Jan. 1956, 61; Feb. 1956, 27. (Description of various commercial types.) See also ref. 11.002, No. 143.

7.929 "Simmonds Fuel Injection System Features Simplified Design," *Auto. Ind.*, Nov. 1957, 15. (Distributor-type pump with speed-density control.)

7.930 Dolza *et al.*, "The General Motors Fuel-Injection System," *Trans. SAE 65*, 739 (1957). (Steady-flow inlet-port injection with air-flow control system.)

7.931 Winkler and Sutton, "Electrojector-Bendix Electronic Fuel Injection System," *Trans. SAE 65*, 758 (1957). (Electronic control from manifold pressure, atmospheric pressure, and engine temperature.)

7.932 Wiseman, "Continental's Fuel Injection for Business Aircraft," *Auto. Ind.*, Apr. 15, 1958, 58. (Simple throttle-controlled system for light airplanes.)

7.933 "Improved Fuel Injection for Mercedes-Benz Engine," *Auto. Ind.*, Oct. 15, 1958, 63. (Distributor pump with speed-density control.)

7.934 Nystrom, "Automotive Gasoline Injection," *Trans. SAE 66*, 65 (1958). (Latest Bosch system with single-plunger distribution pump and control by speed-density system. See also article on the same system by Miller, *Trans. SAE 64*, 458 (1956).

7.935 Allen and Scibbe, "Low Pressure Timed Injection and Control System for the Otto-Cycle Engine," *Trans. SAE 71*, 293 (1963). (Includes analysis of speed-density mechanical control system.)

7.936 "Fuel Injection System," pamphlet form B-10-90, issued by Continental Motors Corporation, Aircraft Engine Division, 1962. (Describes system similar in principle to that shown in Fig. 7–14.)

7.937 Meyer, "Application of Fuel Injection to Spark Ignition Engines," *SAE Special Publication* SP-152, 1958. (Comprehensive discussion including description of the Marvel-Schebler system of injection and control.)

Fuel Injection for Gas Engines
(See refs. 10.70–10.77.)

CHAPTER 8

Engine Balance, General

8.00 Lanchester, "Engine Balancing," *Proc. IAE* (London) (1914). (Describes Lanchester secondary balancer.)
8.01 Root, *Dynamics of Engine and Shaft*, John Wiley & Sons, Inc., New York, 1932. (Fundamentals in brief.)
8.02 Schroen, *Die Dynamik der Verbrennungskraftmaschine*, Springer-Verlag, Vienna, 1942.
8.03 Wilson, "The Fundamentals of Engine Balancing," *Marine Engineer and Naval Architect* (London), five parts, Nov. 1954 to March 1955. (Last part gives balance characteristics of a large number of cylinder arrangements for 2- and 4-cycle engines.)
8.04 Wilson, "The Fundamentals of Engine Balancing," *Gas and Oil Power*, July, August, and September (1955). (Condensation of preceding reference.)
8.05 Wilson, *The Balancing of Oil Engines*, C. Griffin & Co., Ltd., London, 1929.
8.06 Wilson, *Vibration Engineering, A Practical Treatise on the Balancing of Engines, Mechanical Vibration, and Vibration Isolation*, C. Griffin & Co., Ltd., London, 1959. (Includes theory of engine balance, data on unbalanced forces, torsional vibration theory and practice, vibration isolation and engine mounting theory and practice.)
8.07 Kimball, "Optimum Ratio Between Crank Arm and Connecting Rod for Smooth Operation of Reciprocating Engines," *J. Franklin Inst. 246*, 409 (1948).

Engine Balance, Particular Types

8.10 Tanaka, "The Inertia Forces and Couples and Their Balancing of the Star Type Engine," report of The Aeronautical Research Inst., Tokyo Imperial University, Vol. 1, No. 10, March 1925. (Authoritative analytical treatment including influence of the master-rod system for radial and rotary engines.)
8.11 Heldt, "Balance Conditions in a 90-Degree 10-Cylinder Engine," *Auto. Ind.*, April 23, 616 (1927).
8.12 Taylor, E. S., "Balancing the Four-Cylinder Aircraft Engine," paper delivered at the annual meeting of the SAE, Jan., 1932. See also *Aircraft Eng.* (London) *4*, May, 129 (1932); Western Flying *12*, July, 43 (1932).
8.13 Nakanishi, "On the Balancing of Two-Stroke Twelve-Cylinder Engines," report of The Aeronautical Research Inst., Tokyo Imperial University, Vol. 7, No. 5, Sept. 1932. (Shows that, with even firing, twelve cylinders or more are required for complete balance of 2-cycle engines.)
8.14 Heldt, "Narrow-Angle V-8 Engine in Dynamic Balance," *Auto. Ind.*, Dec. 2, 1933, 664.

8.15 Kalb, "Engine Types Adapted to Car Trends," *Jour. SAE 35*, Oct., 13 (1934). (Contains table of unbalanced forces and moments for various types of 4-stroke engines. Source of Table 8–2.)

8.16 Coppens, "Improved Formula for Computing Counterweights of ... Radial Engines," *Trans. SAE 29*, 101 (1934).

8.17 Coppens, "Etude de l'équilibrage des moteurs en étoile," 2me Congrès National des Sciences, Section des Sciences Appliquées, Bruxelles, published by M. Hayez, Rue de Louvain 112, Brussels, 1935. (Briefer than previous reference but includes master-rod effects. Torques not included.)

8.18 Heldt, "Balance Conditions in Five-Cylinder In-Line Engines," *Auto. and Aviation Ind.*, Jan. 1947, 15. (Includes various crank arrangements.)

8.19 Heldt, "An Analysis of Vee-Type and Flat Six-Cylinder Engines," *Auto. Ind.*, Aug. 15, 1948, 32; *ibid.*, July 15, 1953, 42.

8.191 Heldt, "An Evaluation of the 90-Deg. V-Six Engine," *Auto. Ind.*, July 15 (1953). (Balance and torque analysis with 60° and 90° V angles.)

8.192 Cormac, "The Design of Dynamically Balanced Crankshafts for Two-Stroke-Cycle Engines," *Engineering* (London), Oct. 11, 1929, 458. (Effects of crank arrangement, cylinder spacing, firing order. Formulas for balancing weights for engines of 1 to 16 cylinders.)

Torque Harmonic Analysis

8.20 Taylor and Morris, "Harmonic Analysis of Engine Torque ...," *J. Aeron. Sci. 3*, 129 (1936). (Coefficients for spark-ignition engines measured from indicator diagrams.)

8.21 Bentley and Taylor, "Gas Pressure Torque of Radial Engines," *J. Aeron. Sci. 6*, 1 (1938).

8.22 Porter, "Harmonic Coefficients of Engine Torque Curves," *J. Appl. Mech.* (ASME), *10*, Mar. 1943, A-33. (Harmonic coefficients for various types of 2-stroke and 4-stroke engines from theoretical indicator diagrams.)

8.23 Evans, "Harmonic Gas Torques in Petrol Engines," Royal Aircraft Establishment (HM Stationery Office, London) R + M 1964, July 1943. (Measured from aircraft engines.)

8.24 *M.A.N. Diesel Engine News* No. 34, April 1937. (Harmonic torque components for a Diesel cylinder.)

Vibration, General

8.30 Kimball, *Vibration Prevention in Engineering*, John Wiley & Sons, Inc., New York, 1932.

8.31 Manley, *Fundamentals of Vibration Study*, John Wiley & Sons, Inc., New York, 1943. (Brief introductory treatment.)

8.32 Freberg and Kemler, *Elements of Mechanical Vibration*, John Wiley & Sons, Inc., New York, 1943.

8.33 Myklestad, *Vibration Analysis*, McGraw-Hill Book Company, New York, 1944.

8.34 Church, *Elementary Mechanical Vibrations*, Pitman Publishing Company, New York, 1948.

8.35 Oldenburger, ed., *Frequency Response*, The Macmillan Company, New York, 1956. (Symposium containing 20 articles related to vibration and control problems, with extensive bibliography.)

8.36 Den Hartog, *Mechanical Vibrations*, 4th ed., McGraw-Hill Book Company, New York, 1956.

8.361 Hansen-Chenea, *Mechanics of Vibration*, John Wiley & Sons, Inc., New York, 1952.

8.37 MacDuff and Felgar, "Vibration Design Charts," *Trans. ASME 79*, 1459 (1957). (Useful data on natural frequencies of beams, plates etc.)

8.38 Safford, "Vibration Damage Evaluation," *Jour. SAE 65*, Nov. 1957, 48.

8.39 Thomson, *Vibration Theory and Application*, Prentice-Hall, New York, 1965. (Includes introduction to random vibrations.)

Crankshaft Vibration

8.400 Wilson, "Practical Solution of Torsional Vibration Problems," John Wiley & Sons, Inc., New York, three volumes — Vols. I and II, 2nd ed., 1940, and Vol. III, "Strength Calculations," Chapman and Hall, London, 1965. (The most complete treatment of the subject available in the English language. Contains numerous examples and an enormous amount of data. Vol. III contains an exhaustive treatment of crankshaft design.)

8.401 Lewis, "Torsional Vibration in the Diesel Engine," *Trans. Soc. Naval Architects and Marine Engineers 33*, 109 (1925). (Basic discussion with especial emphasis on marine installations with long propeller shafts.)

8.410 "Graphic Determination of Natural Frequency of Torsional Oscillations," *Sulzer Tech. Rev.*, No. 3, 14 (1936).

8.420 Hazen and Montieth, "Torsional Vibration of In-line Aircraft Engines," *Trans. SAE 33*, 335 (1938). (With especial reference to the Allison V-12 engine.)

8.430 Malychevitch, "Torsional Vibration of Diesel-Engine Crankshafts," *Auto. Ind.*, Pt. 1, Mar. 18, 1939; Pt. 2, Mar. 25, 1939. (Brief but clear treatment of theory and Holtzer method.)

8.440 Masi, "Permissible Amplitudes of Torsional Vibration in Aircraft Engines," *Trans. SAE 34*, 311 (1939). (Chiefly valuable for review of difficulties of predicting crankshaft stresses as brought out in the discussion.)

8.450 Biot, "Equations of Finite Differences Applied to Torsional Oscillations of Crankshafts," *J. Appl. Phys. 11*, 530 (1940). See also, same author in *J. Aeron. Sci. 7*, 107 (1940); *7*, 376 (1940).

8.460 Meyer, "The Coupling of Flexural Propeller Vibrations with Torsional Crankshaft Vibrations," *NACA TM* 1051, Dec. 1943.

8.470 Morris and Evans, "Engine Crankshaft Frequency Curves," *Aircraft Eng.* (London), June, 156 (1943); see also *Aeron. Eng. Rev.*, Sept. 1943, 58.

8.481 Fisher, "Torsional Vibration of Multi-Disk Systems," *J. Appl. Phys. 15*, 676 (1944).

8.482 Geiger, "Determination of Crankshaft Stresses in Critical Regions with Damping," *Auto. Aviation Ind. 92*, April 1, 1945, 28.

8.483 Stewart, "Electrical Model for Investigation of Crankshaft Torsional Vibrations," *Trans. SAE 54*, 238 (1946).

8.484 Den Hartog, "Forced Torsional Vibrations with Damping: an Extension of Holzer's Method," *J. Appl. Mech.* (ASME) *13*, Dec. 1946, A-276.

8.485 Kleiner, "Angular Amplitude and Speed Fluctuations in Reciprocating Engines," *Sulzer Tech. Rev.*, No. 3/4, 23 (1947). (Method of computing torsional characteristics of multicylinder engine-generator sets.)

8.486 Bogdanoff, "A Method for Simplifying the Calculations of the Natural Frequencies for a System Consisting of *n* Rigid Rotating Discs Mounted on an Elastic Shaft, *J. Aeron. Sci. 14*, 5 (1947).

8.487 Hussman, "Calculation of Torsional Vibrations in Engines," Proc. 20th National Oil and Gas Power Conference (1948), *Trans. ASME 71*, Feb. 1949, 21.

8.488 Spaetgens, "Holzer Method for Force-Damped Torsional Vibrations," *Trans. ASME 72*, March 1950, 59.

8.489 Goohs, "Torsional Vibrations in Diesel Engines," *Gen. Motors Eng. J.*, Nov.–Dec. 1953. (Several examples of Holzer method.)

8.490 Nestorides, *A Handbook of Torsional Vibrations*, Cambridge University Press, New York, 1958.

8.491 "Torsional Vibrations and Fracture of Crankshafts," *Sulzer Tech. Rev.*, No. 4 (1943). (Effects of number of cylinders, firing order, etc., in in-line engines up to 10 cylinders. Examples of crankshaft failure in torsion and bending.)

8.492 Carter and Forshaw, "Torsiograph Observations on a Merlin II Engine . . . with Five Pitch Settings of the Propeller Blades," *ARC Tech. Rep.* R + M, No. 1983, HM Stationery Office, London. (Torsiograph curves showing propeller effects.)

8.493 Den Hartog and Butterfield, "The Torsional Critical Speeds of Geared Airplane Engines," *J. Aeron. Sci. 4*, 487 (1937).

8.494 Critchlow and Bean, "Crankshaft Bending Vibration," *SAE Quart. Trans. 1*, July 1947, 380. (Discussion and examples of failures in aircraft engines.)

8.495 Colby, "Torsional Vibration — A Case History," *Trans. SAE 72*, 186 (1964). (Engine versus transmission connected by long shaft. Measurements and cure with flexible couplings.)

8.496 Hoffmeister, "Torsional Vibration Control in Diesel-Engine Drive," *Trans. SAE 72*, 544 (1964). (Similar in content to previous reference.)

8.497 Hamilton, "The Use of an Electrical Model to Analyze Torsional Vibration in a Diesel Engine," *Gen. Motors Eng. J.*, Oct.–Nov.–Dec. 1958.

8.498 Wilson, "Understanding Torsional Vibrations," *Gen. Motors Eng. J. 9*, No. 1 (1962). (Includes drive line of motor cars.)

Damping in the Crankshaft System
(See also ref. 8.84.)

8.50 Hunsaker, "Damping in an Internal-Combustion Engine," S.M. Thesis, Department of Mechanical Engineering, Massachusetts Institute of Technology, Cambridge, Mass., 1940. (Torsional damping of Ford V-8 engine found to be about 0.10 critical damping.)

8.51 Draminsky, "Crankshaft Damping," *Proc. IME* (London), *159*, No. 46, 416 (1948). (Measurements of damping on a number of engines.)

8.52 Schaefer and Winter, "Hydraulic Torque Converter—Its Effect on the Power Train," *Trans. SAE 61*, 142 (1953). (Strain-gauge measurements in truck operation show reduced stress when torque converter is used.)

Crankshaft Dampers and Absorbers
(See also ref. 11.002, No. 163.)

8.60 Taylor, "Eliminating Crankshaft Torsional Vibration in Radial Aircraft Engines," *Trans. SAE 38*, 81 (1936). (First U.S. presentation of pendulous absorber. See ref. 8.63 for applications to aircraft engines.)

8.61 "A Dynamic Damper for Torsional Vibrations," *Sulzer Tech. Rev.*, No. 1 (1938). (Example of Sulzer design of pendulous absorber.)

8.62 Stieglitz, "Control of Torsional Vibrations by Pendulum Masses," *NACA TM* 1035, Nov. 1942.

8.63 Moore, "The Control of Torsional Vibration in Radial Aircraft Engines by Means of Tuned Pendulums," *J. Aeron. Sci. 9*, 229 (1942). (Theory and practice at Wright Aeronautical Corp.)

8.64 Shieh, "Optimum Performance of Pendulum-Type Torsional Vibration Absorber," *J. Aeron. Sci. 9*, 337 (1942). (Equations for the two natural frequencies when applied to a one-degree-of-freedom system.)

8.65 Manley, "Pendulum-Type Vibration Absorbers," *J. Aeron. Sci. 10*, 38 (1943). (Alternative theoretical solutions.)

8.66 Carter, "Torsional Vibration of Crankshafts: Analytical Note on Dynamic Absorbers," *ARC Tech. Report* R + M 1973. (HM Stationery Office, London.)

8.67 Kleiner, "Dynamic Damping of Vibrations," *Sulzer Tech. Rev.*, No. 1 (1945). (Pendulous absorbers.)

8.68 Bailey *et al.*, "The Dynamic Absorber and Its Application to Multi-Throw Crankshafts," *Proc. IME* (London) *145*, 73 (1941). (Theory and application.)

8.69 Peirce, "Torsional Vibration Dampers," *Trans. SAE 53*, 480 (1945). (Rubber-mounted dampers, examples of application.)

8.691 Zdanowich and Moyal, "Some Practical Applications of Rubber Dampers for the Suppression of Torsional Vibrations in Engine Systems," *Proc. IME* (London) *153*, 61 (1945). (Definitive treatment of theory and experiment. Measurements of amplitude with and without dampers.)

8.692 Brock, "A Note on the Damped Vibration Absorber," *J. Appl. Mech.* (ASME) *13*, No. 4, A-284 (1946).

8.693 O'Connor, "The Viscous Torsional Vibration Damper," *SAE Quart. Trans. 1*, 87 (1947). (Mathematics and one example.)

8.694 Rumsey, "Viscous Torsional Vibration Damper," *SAE Special Publication* SP-163, Pt. 7, 1959. (Design practice and methods of testing.)

Engine Flexible Mountings
(See also ref. 8.06.)

8.70 Rubin, "Design Procedure for Vibration Isolation on Nonrigid Supporting Structures," *Trans. SAE 68*, 318 (1960). (Important contribution, taking into account that most mounting structures, such as automobile and aircraft frames, are nonrigid. Includes useful curves and equations.)

8.701 Haushalter, "Rubber as a Load-Carrying Material," *Trans. SAE 45*, 15 (1939). (Mechanical properties of rubber and design of rubber springs.)

8.71 Roche, "Engine Smoothness and Protection through Shear Rubber Mountings," *Trans. SAE 50*, 314 (1942). (Characteristics of rubber mounts.)

8.73 Müller, "Progressive Federung durch anschlagfreie Kombination," *ATZ* (Berlin) *56*, Oct. 1954, 272. (Discussion of the stiffness versus deflection of various shapes of rubber shock mountings.)

Automotive-Engine Mounting

(See also ref. 11.002, No. 163.)

8.80 Riesing, "Resilient Mountings for Passenger-Car Powerplants," *SAE Quart. Trans. 4*, 38 (1950). (Description of existing rubber engine mounts for U.S. passenger cars. Data on physical characteristics of various rubber compositions. See also *Jour. SAE 57*, Dec., 41 (1949).)

8.81 Denham, "Engine in New Plymouth Is Cradle Mounted," *Auto. Ind. 65*, July 4, 12 (1931). (Describes new decoupled engine mount called "floating power.")

8.82 Heldt, "Up from 1925 Flexible Engine Mountings Progress to Damping of Torsional as Well as Reciprocatory Vibration," *Auto. Ind. 66*, Feb., 184 (1932). (History of flexible automobile-engine mounts to date.)

8.83 Taub, "Resilient Mountings as Applied to Automotive Engines," *Trans. SAE 34*, 136 (1934). (Theory and practice.)

8.84 Peirce and Robinson, "Engine Mounting and Torsional Vibration Dampers for Trucks and Buses," *SAE Quart. Trans. 6*, 36 (1952). (Design detail of rubber engine mounting and torsional vibration dampers — data on torsional amplitude of various engines with and without torsional dampers — correlation with piston speeds.)

Aircraft-Engine Mounting

8.85 Draper and Bentley, "Measurement of Aircraft Vibration During Flight," *J. Aeron. Sci. 3*, 116 (1936). (Measurements on radial engine.)

8.86 Air Corps, "Vibration Isolating Radial Engine Mounts," Air Corps Inf. Circular, Vol. VIII, No. 707, Feb. 1, 1937. (Theory; measurements on one engine.)

8.87 Browne, "Dynamic Suspension — A Method of Aircraft-Engine Mounting," *Trans. SAE 45*, 185 (1939). (Theory and practice of a decoupled mounting for radial engines.)

8.88 Bentley, "Vibration of Radial Aircraft Engines, Part I," *J. Aeron. Sci. 6*, 278 (1939); Pt. II, *ibid.*, *6*, 333 (1939). (Measured motion of engine in very "soft" spring mounting.)

8.89 Taylor and Browne, "Vibration Isolation of Aircraft Power Plants," *J. Aeron. Sci. 6*, 43 (1938). (Same subject matter as ref. 8.87.)

Stationary and Marine-Engine Mounting

8.891 Newcomb, "Principles of Foundation Design for Engines and Compressors," *ASME Preprint* No. 50–OG–5, March 1950.

8.892 Mohr, "Vibration and Noise Control of Outboard Motors and Other Products," *SAE Paper* 183B, June 1960. (Flexible mountings, mufflers, etc.)
8.893 Wilson, ref. 8.05, Chap. XIII, and ref. 8.06, Chaps. XV and XVI.

Vibration Measurements
(See refs. 8.400, 8.490, 14.301–14.304, 14.801, 14.802.)

Acoustic Theory of Engine Noise

8.900 Beranek, ed., *Noise Reduction*, McGraw-Hill Book Company, New York, 1960.
8.901 Harris, *Handbook of Noise Control*, McGraw-Hill Book Company, New York, 1957.
8.902 Beranek, *Acoustics*, McGraw-Hill Book Company, New York, 1954.

Engine Noise General

8.909 Various authors, "Engine Noise Symposium," *SAE Special Publication* SP-157, 1958.
8.910 Priede, "Noise of Internal Combustion Engines," *Paper C-2*, NPL Symposium on the Control of Noise, June 1961. (Comprehensive study of S-I and Diesel engines at C.A.V. Ltd., Acton, England.)
8.911 Ross and Ungar, "On Piston Slap as a Source of Engine Noise," ASME Paper 65–OGP–10, 1965. (Correlation of noise measurements from many engines shows piston slap to be a very important factor. Offset piston pins recommended. Bibliography nineteen items.)

Small-Engine Noise

8.912 Mohr, "Vibration and Noise Control of Outboard Motors and Other Products," *SAE Paper* 183A, June 1960.
8.913 "Muting of Power Lawnmowers," *Noise Abatement Digest 1*, No. 5 (1960).

Exhaust and Intake Noise and Mufflers

8.920 "Exhaust and Intake Silencers," Part 6 of "Engineering Know-How in Engine Design," *SAE Special Publication* SP-163, Society of Automotive Engineers, New York, 1959.
8.921 Watters *et al.*, "Designing a Muffler for Small Engines," *Noise Control 5*, March 1959, 18.
8.922 Rothfuss, "Exhaust Noise and Its Suppression in 2-Stroke (Diesel) Engine Trucks," *Tech. Mitt. Krupp 17*, 172 (1959) (in German).
8.923 Martin, "Acoustic Filter and Exhaust-System Design for Small and Large Flow Fluctuations" (in German), *ATZ* (Berlin) *61*, 253 (1959). (Mathematical and practical approach.)
8.924 Dyer, "Noise Attenuation of Dissipative Mufflers," *Noise Control 2*, May 1956, 51.

8.925 Staff Report, "Programs in Exhaust Silencing," *Noise Control 2*, May 1956, 18.

8.926 Nelson, "Truck Muffler Design," *Noise Control 2*, May 1956, 24.

8.927 Spahr, "The Development of an Engine Intake Silencer," *Noise Control 2*, May 1956, 34.

8.928 Davis *et al.*, "Theoretical and Experimental Investigation of Mufflers with Comments on Engine-Exhaust Muffler Design," *NACA Report* 1192, 1954.

Combustion Noise and Engine Roughness in Spark-Ignition Engines

8.929 Anon, "Engine Combustion Noises." Technical note published by Ethyl Corporation, Detroit. (Excellent introduction to and summary of "knock," "surface ignition," "wild ping," "rumble," etc.)

8.930 Various authors. "Engine Noise Symposium," *SAE Special Publication* SP-157 1958. (11 papers on subject of combustion noise in passenger car engines. See also articles on this subject in SP-163, 1959, and SP-271, 1965.)

8.931 Andon and Marks, "Engine Roughness — The Key to Lower Octane Requirement," *Trans. SAE 72*, 636 (1964).

8.932 Craig, "Theoretical Relationship Between Combustion Pressure and Engine Vibration," *SAE Preprint* 647C, Jan. 1963. (Using CFR engine, finds that at higher frequencies vibration intensities are proportional to rate of change of slope of pressure-time curve, at mid-frequencies proportional to slope of pressure-time curve, at low frequencies proportional to combustion pressure.)

8.933 Starkman and Sytz, "The Investigation and Characterization of Rumble and Thud," *Trans. SAE 68*, 93 (1960). (Attributes "rumble" with sound spectrum between 400 and 2000 cps to bending vibrations of the crankshaft induced by high rates of combustion-pressure rise.)

8.934 Stebar *et al.*, "New Studies Provide More Information on Engine Rumble-Phenomenon of High Compression Ratio Engines," *Gen. Motors Eng. J. 7*, 22 (1960).

8.935 Meagher *et al.*, "Correlation of Engine Noises with Combustion Phenomena," *Trans. SAE 63*, 481 (1955).

8.936 Ball, "Passenger Car Noise — Engine Harshness," *SAE Preprint*, March 1952. (Measurements of frequencies of crankshaft bending vibration. Suggests combustion-chamber modifications. Abridgement in *Jour. SAE 60*, Oct. 1952, 61.)

8.937 Hinze, "Effect of Cylinder Pressure Rise on Engine Vibrations," *Proc. 21st National Oil and Gas Power Conf. Trans. ASME 71*, 603 (1949). (Short review of the subject with good list of references.)

8.938 Fry *et al.*, "Shock Excited Transient Vibrations Associated with Combustion Roughness," *SAE Quart. Trans. 1*, Jan. 1947, 164. (Report of extensive work, with measurements, on engines.)

8.939 Withrow and Fry, "Physical Characteristics of Roughness in Internal-Combustion Engines," *Trans. SAE 52*, 100 (1944). (Engine vibration measurements show crankshaft bending to be principal factor.)

8.940 Janeway, "Fast Burn Makes Engines Rough," *Jour. SAE 72*, May 1963, 67. (Elementary explanation with diagrams.)

Diesel-Engine Noise

(See also refs. 8.910 and 8.911.)

8.950 Skorecki, "Vibration and Noise of Diesel Engines," *ASME Paper* 63–OGP–2, August 1962. (Compares sound emitted by an engine with its vibratory acceleration. Cites combustion pressure and mechanical impact as the two main sources of vibrations. Comparison of power and motoring runs showed combustion pressures to be the major high-frequency noise source.)

8.951 Zinchenko, "Noise of Marine Diesel Engines," ONI Translation No. 401 (from Russian of 1957) by Donald Ross, publ. by Bolt, Beranek, and Newman, Cambridge, Mass., July 1962. (Condensation from a larger and very comprehensive work in Russian. Bibliography of Russian publications.)

8.952 Priede, "Relation Between Form of Cylinder-Pressure Diagram and Noise in Diesel Engines," *IME* (London), paper, Nov. 1960, 8. (Thorough and authoritative.)

8.953 Austen and Priede, "Origins of Diesel Engine Noise," *SAE Paper* 125T, October 1959; also Symposium on Engine Noise and Suppression, IME (London), 1959.

8.954 Mercy, "Analysis of the Basic Noise Sources in the Diesel Engine," *ASME Paper* 55–OGP–4, March 1955.

8.955 Mercy, "The Diesel Engine Noise Problem," paper presented at the Ship Noise Symposium, 1954, U.S. Bureau of Ships, *Report* No. 371–N–22.

8.956 Ancell, "Practical Noise Reduction Treatment for Diesel Engines," *J. Acoust. Soc. Am. 25*, 1163 (1953).

8.957 Bradbury, "The Measurement and Interpretation of Mechanical Noise with Special Reference to Oil Engines," IME (London), *Proc. Automobile Div.* 1952–53, IB, p. 1.

8.958 Davies, "Injection Characteristics and Diesel Knock," IME (London), *Proc. Automobile Div.* 1951–52, p. 214.

8.959 Austen and Priede, "Diesel Engines Do Not Have to Be So Noisy," *Jour. SAE 74*, May 1965, 71. (Frequency analysis and discussion, including results from special "Structure Research Engine.")

CHAPTER 9

The literature on the subject of engineering materials is so vast that only the most pertinent references familiar to the author will be included here. Most of these references themselves contain extensive bibliographies. For selection of materials for particular engine parts, see also Chapters 10 to 12.

General

9.00 *SAE Handbook*, published annually by SAE, New York. (Specifications, heat treatments, properties of, uses of, inspection and test methods for, and much useful advisory information on all materials used in automotive engines and vehicles. Bibliographies included in all sections. An essential reference for engine-design purposes.)

9.01   *SAE Aeronautical Material Specifications Index*, published by SAE several times a year, gives cross references to material specifications of ASM, American Society for Metals, AISI, American Iron and Steel Institute, and all pertinent U.S. Government specifications.

9.02   *ASTM Standards*, published periodically by ASTM, Philadelphia. (Accepted U.S. standards for materials and materials testing).

9.03   ASTM Materials Handbook, published periodically by ASTM, Philadelphia. (Important basic reference book.)

9.04   *Handbook of Metals*, American Society for Metals (ASM), Cleveland, Ohio. Published periodically (complete reference material in two volumes).

9.041   *Trans. American Society for Metals* (ASM), Proceedings and Special Publications. (Continuous publication of important developments in the field of engineering materials.)

9.042   Gillett, *The Behavior of Engineering Metals*, John Wiley & Sons, Inc., New York, 1957. (Authoritative brief review from the engineer's viewpoint. Excellent bibliography for each chapter.)

Corrosion and Its Effects on Fatigue
(See also refs. 9.200–9.208.)

9.050   Uhlig, "Corrosion and Corrosion Control," John Wiley & Sons, Inc., New York, 1963.

9.051   McAdam and Clyne, "Influence of Chemically and Mechanically Formed Notches on Fatigue of Metals," *J. Res. Natl. Bur. Standards 13*, 527 (1934).

Choice of Engine Materials, General
(See also ref. 9.410.)

For materials used in individual engine parts, see discussion and bibliography in Chapters 11 and 12 under particular parts, such as cylinders, pistons, crankshafts, etc. See also Fig. 9–24 and Tables 9–1, 9–6, 9–7, and 9–8. References referring to materials used in complete engines include the following:

9.06   McVey *et al.*, "Materials in Engine Design," *SAE Special Publication* 137, 1955. (Materials used in International Harvester gasoline and Diesel engines.) See also articles describing proprietary engines, refs. 10.20–10.854, especially refs. 10.802, 10.803, and 10.804.

9.07   Sigwart, "Material Aspects in Mercedes Benz Diesel Engines," *SAE Paper* 972A, Jan. 1965. (Experience with various materials and hardening processes from the point of view of resistance to fatigue, thermal stresses, wear, etc.)

9.08   Schilling, "Operational Stresses in Automotive Parts," *SAE Quart. Trans. 5*, 292 (1951). (Source of Fig. 9–24. General Motors practice in steel selection. Stress cycles in service. Full-scale testing.)

Strength of Materials

9.100   Timoshenko, *Strength of Materials*, Parts I and II, D. Van Nostrand, New York, 1947. (A highly regarded work on the subject, mainly from the theoretical point of view.)

9.101 Lessells, *Strength and Resistance of Metals*, John Wiley & Sons, Inc., New York, 1954. (Discussion of tensile, impact, creep, and fatigue properties and their relation to physical and chemical properties of metals and full-scale machine parts.)

9.102 Polakowski and Ripling, *Strength and Structure of Engineering Materials*, Prentice-Hall, Inc., Englewood Cliffs, N.J., 1966. (Parts I–III on theory of stress and strain, rheology, etc.; Part IV on mechanical behavior of all types, including creep and fatigue. Bibliography.)

9.103 McClintock and Argon, *Mechanical Behavior of Materials*, Addison-Wesley Publishing Co., Reading, Mass., 1966. (Late book by recognized authorities.)

9.104 Clark, *Metals at High Temperatures*, Reinhold Publishing Corporation, New York, 1950.

Fatigue of Materials

Note: The literature on fatigue is enormous. The following references have been selected as excellent summaries of existing knowledge. They are well documented as to source material.

9.190 Moore and Kommers, *Fatigue of Metals*, McGraw-Hill Book Company, New York, 1927. (Early book on the subject. Still useful.)

9.200 Forrest, *Fatigue of Metals*, Pergamon Press, London, 1962. (Comprehensive, complete, authoritative. Contains extensive bibliography.)

9.201 Grover *et al.*, *Fatigue of Metals and Structures*, Thames and Hudson, London, 1956. (Comprehensive review of what is known and unknown about fatigue failure in theory and practice. Oriented toward practical design. Extensive bibliography.)

9.202 Matthaes, "Fatigue Strength of Airplane and Engine Materials," *NACA Tech. Memo* 743, April 1934. (Translated from *Zeitschrift für Flugtechnik und Motorluftschiffahrt* (Berlin), *24*, Nov. 4 and Nov. 28 (1933). (Excellent short review of metal fatigue and its influence on design.)

9.203 Battelle Memorial Institute, *Prevention of Failure of Metals Under Repeated Stress*, John Wiley & Sons, Inc., New York, 1941. (Examples of fatigue failure of many machine parts. Summary of fatigue research to that date. Discussion of practical methods of preventing fatigue failure.)

9.204 Murray, *Fatigue and Fracture of Metals*, The Technology Press, Cambridge, Mass., 1950. (Symposium on fatigue and fracture in airplanes, ships, machinery, etc.; brittle behavior in metals, influence of transition temperature, metallographic structure, etc.)

9.205 Almen and Black, *Residual Stress and Fatigue of Metals*, McGraw-Hill Book Company, New York, 1963. (Summary of many years' work by Almen on the study of and improvement in fatigue properties. Practical. Important in design.) See also refs. 9.2051–9.2055 as follows by Almen:

9.2051 *Trans. SAE 50*, 52 (1942).

9.2052 *Trans. SAE 51*, 248 (1943).

9.2053 *Auto. Ind.*, Feb. 1 and Feb. 15, 1943.

9.2054 *Prod. Eng. 21*, 118 (1950).

9.2055 *Metal Progr. 46*, Dec. 1944, 1263.

9.206   Cazaud, *Fatigue of Metals*, Philosophical Library, New York, 1953. (Translation of " La Fatigue des Métaux " by a distinguished French authority.)

9.207   Gohn, " Fatigue and Its Relation to the Mechanical and Metallurgical Properties of Metals," *Trans. SAE 64*, 31 (1956). (Excellent elementary summary. Based in part on Bell Telephone Laboratories research.)

9.208   *Proceedings International Conference on Fatigue of Metals*, IME (London), 1956. (Large number of important papers on almost all aspects of fatigue including size effects, prestressing, combined stresses, component testing of bolts and welds, temperature effects, corrosion, shot peening, cumulative damage, screw fastenings, rollers, steel inclusions, service failures, journal bearings.)

9.2081 "Symposium on Fatigue with Emphasis on Statistical Approach," *Special Publication* 121, Pt. I and Pt. II, ASTM, Philadelphia, 1951 and 1952.

9.2082 "Symposium on Basic Mechanisms of Fatigue," ASTM, Philadelphia, June 1958.

9.2083 ASTM, Committee E-9 on Fatigue, "References on Fatigue," ASTM, Philadelphia, 1950. (Includes brief abstracts.)

9.2084 Valluri, "Effect of Frequency and Temperature on Fatigue of Metals," *NACA Tech. Note* 3972, Feb. 1957.

9.2085 Gadd *et al.*, "Some Factors Affecting Fatigue Strength of Steel Members," *Trans. SAE 63*, 362 (1955). (Testing of full-scale parts, size effects, surface treatments, etc.)

9.2086 Sines, "Failure of Materials under Combined Repeated Stresses," *NACA TN* 3945, 1955.

9.2087 Kuhn and Hardrath, "An Engineering Method for Estimating Notch-Size Effect on Steel," *NACA TN* 2805, Oct. 1952. (Calculation of notch effects and comparison with tests. Bibliography.)

9.2088 Smith, "Effect of Range of Stress on Fatigue Strength," Univ. of Illinois, Eng. Station, Exp. *Bull.* 334, 1942.

9.2089 Pomp and Hempel, "The Fatigue Behavior of Cast Iron and Malleable Iron" (in German), *Mitt. Kaiser-Wilhelm Inst. für Eisenforschung 22*, 169 (1940).

9.209   Noll and Lipsin, "Allowable Working Stresses," *Proc. Soc. Exp. Stress Anal.* III, 89 (1946).

9.2091 Woodward *et al.*, " Effect of Mean Stress on the Fatigue of Aluminum Alloys," in *Proceedings International Conference on Fatigue of Metals*, IME (London), 158 (1956).

9.2092 O'Connor and Morrison, "Effect of Mean Stress on the Push-Pull Fatigue Properties of an Alloy Steel," *Proceedings International Conference on Fatigue of Metals*, IME (London), 102 (1956).

Fatigue Testing

(See also refs. 9.08, 9.190, 9.203.)

9.211   ASTM, Committee E-9 on Fatigue, "A Guide for Fatigue Testing and the Statistical Analysis of Fatigue Data," 2nd ed. ASTM, Philadelphia, 1964.

9.211   ASTM, Committee E-9 on Fatigue, "Manual on Fatigue Testing," ASTM, Philadelphia, Special Publication 91, 1950.

9.212 ASTM, "Symposium on Testing of Parts and Assemblies," *ASTM Tech. Publication* 72, June 1946.

9.213 "Non-Destructive Testing," *Proc. Fourth International Conference on Non-Destructive Testing*, Butterworth, Inc., Washington, D.C., 1960. (All types of materials by many different methods.)

9.214 Hogarth and Blitz, "Techniques of Non-Destructive Testing," Butterworth, Inc., Washington, D.C., 1960. (216 pages ill.) (Ultrasonic, X-ray, magnetic methods, etc.)

Fatigue — Improvement by Surface Treatment

9.220 "Shot Peening," *SAE Special Publication* SP-84, 1952. (Collection of useful articles on the subject.)

9.221 Almen and Black, ref. 9205 (important!).

9.222 Anon, "Shot Peening Applications." Published by Metal Improvement Company, October 1958. (Contains useful information on the uses and effectiveness of shot-peening on various machine parts.)

9.223 Liss *et al.*, "The Development of Heat-Treat Stresses and Their Effect on Fatigue Strength of Hardened Steels," *SAE Paper* 650517, May 1965. (Control of strengthening effects by means of severe quenching so as to introduce high compressive stresses at the surface of carbon steels.)

9.225 "What's Been Written About Shot Peening," *SAE Special Publication* SP-126a, 1954. (Abstracts of thirty-six papers and references to catalog literature.)

9.226 "Bibliography on Residual Stress," *SAE Special Publication* SP-231, Aug. 1962. (702 abstracts, 1957–1960. See also earlier edition, *Special Publication* 125.)

9.227 Almen, "Fatigue Weakness of Surfaces," *Prod. Eng. 21*, 118 (1950).

9.229 Tarasov and Grover, "Effects of Grinding and Other Finishing Processes on the Fatigue Strength of Hardened Steel," *Proc. ASTM 50*, 668 (1950).

Stress-Concentration Factors

9.300 Timoshenko, *Theory of Elasticity*, 2nd ed., McGraw-Hill Book Company, New York, 1951. (Standard work by a recognized authority.)

9.301 Peterson, "Stress Concentration Design Factors," John Wiley & Sons, Inc., New York, 1953. (Well organized book giving stress-concentration factors for large numbers of shapes and typical machine configurations.)

9.302 Neugebauer, "Stress Concentration Factors and Their Effects on Design — I," *Prod. Eng. 14*, Feb. (1943); Part II, *ibid.*, *14*, March (1943). (Correlation of theoretical stress concentration factors with results of endurance tests on plates and on shafts in bending and torsion with fillets, keyways, holes, etc. Contains twenty-item bibliography.)

9.303 Spotts, "Impact Stress in Elastic Bodies Calculated by the Energy Method," *Prod. Eng. 17*, March (1946). (Calculation of stresses caused by falling bodies, etc.)

Materials — Creep

(See also refs. 9.101–9.104.)

9.35 ASME-ASTM, "Compilation of Available High-Temperature Creep Characteristics of Metals and Alloys," ASTM, Philadelphia, 1938.

9.36 Kennedy, *Processes of Creep and Fatigue in Metals*, Oliver and Boyd, London, 1962, and John Wiley & Sons, New York, 1963.

9.37 Hult, "Creep in Engineering Structures," Blaisdell Publishing Co., Waltham, Mass., 1966. (115 pages. Theory and practice. Bibliography.)

Materials Inspection and Quality Control
(See also SAE Handbook, ref. 9.00, and refs. 9.213, 9.214.)

9.401 McQuaid, "The New Technique of Steel Selection for Automotive Applications," *SAE Paper* 650518, May 1965. (Effects of inclusions on endurance limit and suggestions for reducing inclusions by metallurgical and chemical controls.)

9.402 Cummings *et al.*, "Relation of Inclusions to the Fatigue Properties of SAE 4340 Steel," *Trans. ASM 49*, 482 (1957).

9.403 Clements, "Magnaflux Indications Interpreted," *Trans. SAE 45*, 686 (1939). (Results of work at Wright Aeronautical Corp.)

9.404 Johnson, "Magnaflux — What Does It Show?" *Trans. SAE 45*, 59 (1939). (Theory and practice with photographic examples.)

9.405 Doane, "Principles of Magnaflux Inspection," 1941–43, Magnaflux Corp., 6800 E. Washington Boulevard, Los Angeles, Calif.

9.405 Koved and Rospond, "Detection of Potential Fatigue Nuclei in Rolling Contact Bearings," *SAE Paper* 640585 (893B), 1965. (Ultrasonic methods.)

Materials, Processing

9.409 Clark, "Engineering Materials and Processes," 3rd ed., International Textbook Co., Scranton, Pa., 1960. (Good elementary and introductory work from the engineer's point of view. Manufacturing processes well described.)

Steel, Wrought
(See also refs. 9.00–9.04.)

9.410 "Selection of Steel for Automobile Parts — What Engineers Should Know Today About Hardenability-Band Steels." Serial publication in Journal SAE:

Part		Issue	Page
I	Introduction to Hardenability	*57*, Aug. 1949	17
II	Significance of Hardness	*57*, Sept. 1949	25
III	Steel Composition Related to Hardenability	*57*, Oct. 1949	33
IV	Hardenability Selection Method	*57*, Nov. 1949	39
V	Another Method of Hardenability Correlation	*57*, Dec. 1949	29
VI	Where H-Bands do not Apply	*58*, Jan. 1950	47

(Important discussion of the concept of hardenability and its relation to the selection and use of various steels.)

9.4101 Bullens, *Steel and Its Heat Treatment*, in three volumes, John Wiley & Sons, Inc., New York, 1948–1949.

Steel, Cast
(See also refs. 9.00–9.04.)

9.411 Evans *et al.*, "Fatigue Properties of Comparable Cast and Wrought Steels," *Proc. ASTM 56*, 979 (1956).

Iron, Cast
(See also refs. 9.00–9.04.)

9.415 Angus, "Physical and Engineering Properties of Cast Iron," Cast Iron Research Association, Birmingham, England, 1960.
9.416 Vennerholm, "Nodular Cast Iron," *SAE Quart. Trans. 4*, 422 (1950). (Physical properties, costs, and discussion of applications.)
9.417 Reese *et al.*, "Effects of Chemistry and Section Size on Properties of Ductile Iron," *SAE Quart. Trans. 6*, 385 (1952).
9.418 Morrogh, *Fatigue of Metals*, Chapman Hall, London, 1959. See especially chapter on Fatigue of Cast Iron.
9.419 "Malleable Iron Castings," published by Malleable Founders Society, Cleveland, Ohio, 1960. (Definitive work on this subject.)
9.420 Smith, "Effect of Range of Stress on Fatigue Strength," Univ. of Illinois, Exp. Eng. Station *Bull.* 334, 1942.
9.421 Pomp and Hempel, "The Fatigue Behavior of Cast Iron and Malleable Iron" (in German), *Mitt. Kaiser-Wilhelm Inst. für Eisenforschung, 22*, 169 (1940).

Aluminum and Magnesium
(See also refs. 9.00–9.04.)

9.51 See publications of Aluminum Co. of America, Pittsburgh, Pa., available on request, and publications on magnesium by Dow Chemical Co., Midland, Michigan, available on request.
9.52 Voorhees and Freeman, "Report on the Elevated Temperature Properties of Aluminum and Magnesium Alloys," *Special Tech. Publ.* 291, ASTM, Philadelphia, 1960.
9.53 Harvey, "Manufacture, Characteristics, and Uses of Magnesium Castings," *Trans. SAE 33*, 43 (1938). (Typical applications, physical properties, pattern and casting details.)

Titanium
(See also refs. 9.00–9.04.)

9.60 "Metals Handbook Supplement," ASTM, Philadelphia, 1954.
9.61 "Handbook on Titanium Metal," Titanium Metals Corp. of America, New York.
9.62 "Titanium," *SAE Special Publication* SP-117, 1954.
9.63 Parcel, "A Realistic Look at Titanium," *Jour. SAE 60*, Dec. 1952, 29. (Practical aspects of the use of titanium — tables of physical properties.)

Sintered Materials

9.70   Talmadge, "Potential of Powder Metallurgy Is Tied to New Strength and Versatility," *Jour. SAE 73*, Nov. 1965, 30. (Applications to automobiles and table of properties of many sintered materials.)

9.71   Koehring, "Powder Metallurgy Advances," *SAE Paper* 628A, 1963. (Includes examples of valve seats, gears, etc.)

Rubber and Plastics

9.80   "Plastics for Electrical Insulation," *ASTM Standards, ASTM Committee D-9*, 28th ed., Philadelphia, Dec., 1961.

9.81   Carlotte and Hobein, "Limitations of Synthetic Rubber Packings for High Temperature Application," *SAE Paper* 72, April, 1953. (Physical properties and limits of application.)

CHAPTER 10

This bibliography is limited to publications familiar to the author and considered by him to be helpful in engine design. Lack of space and time to review prevents inclusion of many others of value, especially those published in languages other than English. References to proprietary engines are intended to show examples of good current design, with the understanding that many examples not included may be equally good.

For references on detail design see Bibliography for Chapters 11 and 12.

Periodicals That Emphasize Current Engine Research and/or Design Practice

10.01   *Journal of the Society of Automotive Engineers*, publ. by SAE, New York. (Monthly. Contains brief articles and abstracts of papers relating to engine design and performance.)

10.02   *Automotive Industries*, Chilton Co., Philadelphia. Published semimonthly. (Covers current practice in automotive, locomotive, industrial, and aircraft engines. Includes annual statistical issue.)

10.03   *Motor* (London). (Monthly magazine in popular style, but containing useful data, especially on European passenger cars and engines.)

10.031   *Cycle World*, Parkhurst Publishing Co., Long Beach, Calif. (Devoted to motorcycles and motorcycling. Includes technical data on motorcycle engines.)

10.04   *Motortechnische Zeitschrift* (MTZ) (Stuttgart). (High-grade technical material with emphasis on German developments. In German.)

10.05   *Automobiltechnische Zeitschrift* (ATZ) (Berlin). (High-grade technical magazine. In German.)

10.06   *Automobile Engineer* (London). (High-grade technical magazine.)

10.07   *Automobile Facts and Figures* and *Motor Truck Facts and Figures*, Automobile Manufacturers Association, Detroit, Mich. (Useful statistical material.)

10.08 *General Motors Engineering Journal.* Quarterly, General Motors Corp., Detroit, Mich. (G. M. technical progress, well reported.)

10.09 *Diesel and Gas Turbine Progress*, Box 7406, Milwaukee, Wis. (Descriptions of new engine designs and new installations. Occasional technical articles. Annual statistical issue.)

10.091 *Gas and Oil Power*, Whitehall Technical Press, Seven Gales, Kent, England. (Trade journal, with useful descriptive material.)

10.092 *Oil Engine and Gas Turbine*, Temple Press, London. (Useful descriptive material on current practice.)

10.093 *Sulzer Technical Review* (Quarterly), Sulzer Bros., Winterthur, Switzerland. (Valuable technical material on Sulzer practice.)

10.094 *M.A.N. Diesel Engine News* (Quarterly), Maschinenfabrik Augsburg-Nürnberg A.G., Augsburg, Germany. (Valuable technical material on M.A.N. practice.)

10.095 *Motorship* (London). (Includes large marine engines.)

10.096 *Automotive Design Engineering* (Monthly). Rowse Muir Publications, Ltd., 77–79 Charlotte St., London W1. (Useful design information, Diesel and spark-ignition.)

10.097 *Lubrication* (Monthly), published by the Texas Company, 135 E. 42nd St., New York, N.Y.

Proceedings of Technical Societies Which Emphasize Internal-Combustion Engines

10.10 *Trans. SAE.* Annual. Bound volumes of technical papers, 1917–1965. Since 1965 an index of SAE papers, published separately. (Best reference material on internal-combustion engines available in the U.S.A.)

10.11 *Trans. ASME, Journal of Engineering for Power*. Formerly Oil and Gas Power Division (OGP). ASME, New York. (Includes valuable material on stationary and marine-engine research and practice.)

10.12 *Proc. Congrès International de Moteurs à Combustion*, published every two years, beginning 1951, by CIMAC, 10 Avenue Hoche, Paris 8. Available from Library Service Association, 11 rue Lavoisier, Paris 8, France. (Semi-annual congresses held in various cities of Europe. Papers mostly describe proprietary developments in large reciprocating engines and gas turbines. Occasional papers of more fundamental interest.)

10.13 *Proceedings Institution of Mechanical Engineers*. Automobile Division, London. (High-grade technical material.)

10.131 *J. Société des Ingénieurs de l'Automobile*, Paris (Professional Society publication, in French.)

10.14 *Jour. Am. Soc. Naval Architects and Marine Engineers*. (Quarterly), Washington, D.C. (See especially bound volume of papers entitled "Where Are Marine Power Plants Headed?" presented at May 1966 meeting.)

10.15 *Trans. Inst. Marine Engineers*, London.

10.16 *N.E. Coast Inst. Engineers and Shipbuilders*, Newcastle upon Tyne, England.

10.17 *Naval Engineers Journal*. Published by American Society of Naval Engineers, Washington, D.C. (Includes occasional articles on internal-combustion engines for ships.)

10.18   *European Shipbuilding*, Oslo, Norway. (Journal of the Ship-Technical Society of Norway.)

10.181   *Reports of the Motor Industry Research Association*, Lindley, England. (Work sponsored by the British Motor Industry, published at irregular intervals.)

Special Engine Publications
(See also ref. 11.002.)

10.19   *Diesel and Gas Engine Catalog*, publ. annually by Diesel and Gas Turbine Progress, Box 7406, Milwaukee, Wis. (Data on U.S. and many foreign Diesel engines, gas engines, gas turbines, compressors, and equipment.)

10.191   *British Diesel-Engine Catalog*, publ. periodically by George Newnes, Ltd., London, for the British Internal Combustion Engine Manufacturers Association, 6 Grafton St., London. (Data on British and many other Diesel engines.)

10.192   *Engine Design Portfolio, SAE Special Publication*, 1964. (Gives cross-sectional views and major statistics for representative spark-ignition and Diesel engines up to locomotive size.)

Passenger-Automobile Engines
(See also refs. 10.01–10.08 and 10.181.)

10.20   *Auto. Ind.*, Oct. 15, 61 (1963). (Development curves for passenger-car engines. See also annual statistical issues.)

10.21   Walder, "Some Problems in the Design and Development of High-Speed Diesel Engines," *SAE Paper* 978A, Jan. 1965. (Comparison of Diesel versus spark-ignition engines for passenger automobiles. Discusses problems of noise, specific output, cost, etc.)

10.22   Weinert, "Volkswagen Design and Production," *Auto. Ind.*, Aug. 15, 65 (1961). (Engine cross section, data on materials and production methods for revised 1961 engine. For earlier design see Boehner, *SAE Paper* 13, Jan. 1953.)

10.23   Campbell, "Looking Ahead in Fuels for Automotive Transportation," *Jour. SAE 66*, Dec., 41 (1957). (Excellent summary of past developments in passenger-car engines.)

10.24   McPherson and Keinath, "The Chevy II Engines," *SAE Paper* 484B, Jan. 1962. (Design and performance details of in-line four- and six-cylinder engines, new as of 1961–1962.)

10.25   Smith, "Ford's New 240 I-6 Engine," *SAE Paper* 650260 (966C) and also *Jour. SAE* 74, June 1965, 86. (1965 model vertical 6 engine.)

10.26   Hansen *et al.*, "The Chevrolet Corvair," *SAE Paper* 140C, 1960. See also Thoreson "The Corvair Turbosupercharged Engine," *SAE Paper* 531A, June 1962. (Development problems and design details of the only air-cooled passenger-car engine built in the U.S.A. Six-cylinder, horizontal opposed type. Cross sections included.)

10.27   Potter *et al.*, "The New Rambler Six Engine — Torque Command 232," *SAE Paper* 884B, June 1964. (Six-in-line engine with malleable cast iron connecting rods and crankshaft. Useful development and design data including cross sections.)

10.271 Weertman and Beckman, "Chrysler Corp. 273 cu. in. V-8 Engine," *SAE Paper* 640178, 1965. (Design of conventional V-8 passenger-car engine. Forged steel rods and shaft.)

10.272 Mellde, "SAAB Reveals Conclusions from 15 Years with 2-Stroke Engines in Passenger Cars," *SAE Preprint* 650008 (954A). Jan. 1965. Summarized in *Jour. SAE 74*, July 1965, 78. (One of the few successful 2-cycle passenger-car engines — vertical, 3-cylinder, water-cooled.)

10.28 Tanaka, "An Example of the Development in Automotive High-Speed Diesel Engine," *SAE Paper* 978C, 1965. (High-grade small Diesel engines suitable for light vehicles.)

10.29 Hoffman, "Present and Future Developments of Small High-Speed Diesel Engines and Discussion of the New Daimler Benz Engines" (in German), *ATZ* (Berlin) *61*, June 1959, 151. (Widely-used passenger-car and taxicab Diesel engine.)

10.291 Brown, "Diesels for Taxicabs," *Diesel Progress*, May 1959, 22.

10.292 Sarnpiefro and Matthews, "The New Overhead Camshaft Willys Engine," *SAE Preprint* 532A, June 1962. (Details of design, development, and performance of 6-cylinder in-line engine. Hemispherical head gives 25 per cent higher output than previous "wedge" combustion chamber at 2160 ft/min piston speed.)

10.294 McKellar, "Design Features of the Pontiac Six-Cylinder Overhead Camshaft Engine," *Gen. Motors Eng. J. 13*, Third Quarter, 3 (1966). (New six-in-line engine, 3.88 × 3.25-in., 230-cu in. engine with seven-bearing crankshaft, overhead camshaft with levers incorporating hydraulic tappet-adjusters. Toothed rubber belt drives camshaft. Useful data on cam contours, timing, performance, detail design.)

Truck, Bus, Light Marine, and Industrial Engines Not Over 6-in. Bore

10.30 Wright and Tignor, "Relationship Between Gross Weights and Horsepower of Commercial Vehicles Operating on Public Highways," U.S. Dept. of Commerce, Bureau of Public Roads, Oct. 1966.

10.31 Pitchford, "The Development of the Small Automotive Diesel in Western Europe and Its Likely Role in the U.S.A.," *SAE Paper* 215B, 1960. (Excellent technical and critical review from Ricardo & Co. Includes valuable design and development data and thoughtful prediction of future trends.)

10.32 "Inboard Marine Engines," *Lubrication* 45, The Texas Co., N.Y., June 1959. (Excellent summary of applications, types, market in U.S.A., and lubrication problems.)

10.33 Hull, "High Output Diesel Engines," *Trans. SAE 72*, 68 (1964). (Tests on single-cylinder, open-chamber engine at high supercharge pressures give useful data on friction, volumetric efficiency, and performance versus inlet pressure.)

10.331 "Looking at 1962," *Diesel Equipment Superintendent*, Jan. 1962, p. 16. (Record of use and sales of Diesel engines, 1948–1961.)

10.34 Alford and Paterson, "Caterpillar's New 5.4-in. Bore Engines," *SAE Paper* 478A, Jan. 1962. (See also *ASME Paper* 62–OGP–8, 1962, and *SAE Paper* 254A, Oct. 1960.)

10.35 Hull, "Diesels Can Be Built to Give One Bhp per Cubic Inch," *SAE Paper* 631B, 1963. (Report on high-supercharged development work with $4\frac{1}{2} \times 5\frac{1}{2}$-in. Caterpillar engine, 6-atm inlet pressure, bmep 302 psi (21.3 kg/cm²), piston speed 2200 ft/min (11 m/sec), max. cyl. pressure 2700 psi (190 kg/cm²). Performance versus inlet pressure and compression ratio.)

10.351 Wittek, "Development of New Allis-Chalmers Diesel Engines," *Trans. SAE 68*, 169 (1960). (Design and performance details of 4-cycle Diesel engines.)

10.36 Boll, "The New Cummins V-Type Diesel Engines," *SAE Paper* 523B, Aug. 1962. (One of the first short-stroke Diesel engines. See also Schmidt, "Cummins Diesels for Stop-and-Go Service," *SAE Paper* 626A, Jan. 1963.)

10.37 Hulsing and Ervin, "G. M. Diesels Additional Engines," *SAE Paper* 1R, Jan. 1959. (Comprehensive description of improved line of 2-cycle poppet-valve engines. Cross sections, design details, performance data.)

10.38 Geschelin, "New GMC V-6 Diesels Feature Four-Stroke Design," *Auto. Ind.*, Feb. 15, 1964, 63. (New short-stroke 4-cycle design, 5.6 lbm/bhp. See also *Diesel and Gas Engine Progress*, Feb. 1964. General Motors entry into the 4-cycle Diesel-engine field.)

10.39 Pelizzoni *et al.*, "Mack Designs V-8 Diesel for Modern Turnpike Service," *Jour. SAE 72*, May 1964, 72. (Four-cycle, open chamber type.)

10.391 Malcolm, "International's New Motor Truck V-8 Diesel Engines," *SAE Paper* 660076, Jan. 1966. (Performance and design details of 4-stroke, near-square engine using M.A.N. "M"-type combustion system. See Mueller and Lacy, *SAE Preprint* 993A, Jan. 1965, and Dewsberry, *Trans. SAE 68*, 501 (1960) for development work on earlier models.)

10.392 Witteck *et al.*, "A Family of Lightweight Aircooled Industrial Diesel Engines," *SAE Paper* 761A, 1963. (Two-cycle loop-scavenged engines of Hans List design. Air-cooled, stroke/bore ratio = 1.0, 3000 rpm. See also *ATZ* (Berlin) *61*, 11 (1959).

10.393 Whiteside, "Rolls-Royce Diesel Engines," *Trans. SAE 68*, 101 (1960). (Performance and useful design details of 4-cycle in-line open-chamber line. Rod bolt heads secured by *outside* bosses to allow proper fillets inside. Some have gear-driven Roots' superchargers.)

10.394 Schafer, "General Motors 6-110 Diesel Engine for Rail-Car Motive Power," *SAE Preprint* 529, 1950. (Similar to General Motors automotive 2-cycle engines, ref. 10.37, but with larger cylinder dimensions, 5.5 × 5.6 in.)

10.395 Wadman, "Waukesha's New Vee Engine Series," *Diesel and Gas Turbine Progress*, May 1966, 33. (New line of 8- and 12-cylinder $5\frac{3}{4} \times 5\frac{3}{16}$-in. 4-cycle Diesel engines, toroidal open chamber, 4-cycle, 4-valve, of very rugged design.)

10.396 Sass and Schweitzer, "Almost 300-Ton Miles per Gallon with Two-Stroke Diesel," *Auto. Ind.*, Feb. 15, 1952. (Performance data on Krauss-Maffei, Schnürle design, open chamber, loop-scavenged, 4-cylinder 90° V.)

10.397 Paquette, "The P and H Uniflow-Scavenged Two-Stroke Diesel Engine," *SAE Paper* 246, Nov. 1957. (Description, cross sections, and performance curves. Single poppet valve in head. Bmep 130 psi unsupercharged.)

10.398 Fodor, "Engineering Considerations in the Development of Air-Cooled M.W.M. Diesel Engines," *SAE Paper* 735, Aug. 1957. (Characteristic European design practice for air-cooled Diesel engines.)

10.399 Schleicher and Schük, "M.A.N. Motor Truck Diesel Engines in Rail Vehicles,"

M.A.N. Diesel Engine News, No. 36, 97 (1958). (Short article showing applications.)

10.3991 List, "High-Speed, High-Output Loop-Scavenged Two-Cycle Diesel Engines," *Trans. SAE 65*, 780 (1957). (Valuable data on porting and port design and author's wide experience on other problems and developments of this engine type. Review of current practice with engine cross sections.)

10.3992 Reddy *et al.*, "Detroit Diesel Series 149 Engines," *SAE Paper* 660604 Sept. 1966. (New series of 5.75 × 5.75-in. (227-mm) 2-cycle poppet-valve engines similar to the seventy-one series. Eight, twelve, and sixteen cylinders, 450–1325 bhp at 1900 rpm. Cross sections and other design details.)

10.3993 Bachle, "Air Cooled Diesel Engine Appraisal," *SAE Paper* 154, Aug. 1957. (Summary of the more important air-cooled Diesels of that time, including the Continental line of military engines. Arguments for air cooling. Short bibliography.)

Medium-Size Diesel Engines for Locomotive, Marine, and Stationary Use

10.401 "American Locomotive 12½ × 13-in. Diesel Engine," *Diesel Power and Diesel Transportation*, May 1944, 491. See also Vaughn, ref. 10.494, and bulletin "Alco 251 Diesels," published by Alco Products, Inc., 530 Fifth Ave., New York. (Four-cycle locomotive engines up to 3000 bhp.)

10.41 Schultz, "New 2500 hp Mainline G. E. Diesel Locomotive," *Diesel and Gas-Engine Progress*, Aug. 1960, 36. (Four-cycle, sixteen-cylinder Cooper-Bessemer engine 9 × 10.5 in. rated at bmep 204 psi piston speed 1760 ft/min.)

10.42 Kettering, "History and Development of the 567 Series General Motors Locomotive Engine," paper before DEMA Conference, publ. by Cleveland Diesel Engine Div., G.M.C., Oct. 13, 1953. See also *SAE Paper* 846, Oct. 1952. (Valuable data on developments, difficulties, and performance of present and past models of 2-cycle poppet-valve engines up to 3000 hp in 1965.)

10.43 Lassberg, "Stationary and Marine Diesel Engines in the Medium-Power Bracket," and Seitz, "The Diesel Engine in Rail Traction," *M.A.N. Diesel Engine News*, No. 36, 77, 88 (1958) (M.A.N. practice in this field.)

10.44 Schläpfer, "A New Range of Locomotive V Engines," *Sulzer Tech. Rev.*, No. 2 (1965). (Design and performance details of 9.43 × 11.0 in. (240 × 280 mm) 4-cycle engines rated at 230 psi, 16.2 kg/cm², mep, 2000 ft/min, 10.3 m/sec piston speed.)

10.45 French and Lilly, "The Locomotive Diesel Engine," report published by Ricardo & Co., Ltd., Shoreham, Sussex, England, 1966. (Excellent review of current practice. Typical engine cross sections, plus data for main-line engines of many nations. Discussion of heat loads, reliability, open versus divided chambers, etc. Statistics on performance, weight, and volume of current engines. Emphasis on thermal limitations. Gives 3.55 hp per sq in. piston area as the present practical limit rating. See Table 10–9.)

10.46 Garin, "4000 hp Diesel-Hydraulic Locomotive Units," *Trans. SAE 71*, 5 (1963). (Describes Maybach sixteen-cylinder 7.3 × 7.9-in. engine, turbo-supercharged and aftercooled, 2000 bhp at 1580 rpm, 14,500 lbm, bmep = 198 psi, piston speed 1970 ft/min, 6 valves per cylinder.)

10.47 Kleinlin and Maybach, "High-Speed High-Output Diesel Engines—35 Years of Railroad and Marine Applications," *SAE Paper* 367A 1961 and *Trans. SAE*

70, 212 (1962). (Comprehensive review of Maybach history and design philosophy. Size effects based on bore/stroke assumption. Detail design including piston, crankshaft, rods, valves. Use of roller main bearings. Experimental stress analysis, including stress coat and full-scale crankshaft fatigue testing. Instrumentation techniques and results. Data on 1966 locomotive engine, 6 valves per cylinder, roller-bearing crankshaft, forked and plain rods, welded-steel crankcase, bmep 200-psi piston speed 2000 ft/min.)

10.48   Antonsen, "The Development of a Supercharged Medium Speed Two-Cycle Opposed Piston Engine," *ASME Paper* 56–A–99, Nov. 1958. (See also Paper 56–A–99.) (Fairbanks-Morse opposed-piston locomotive engine.)

10.49   Davids, "Design Features of an Opposed Piston Diesel Engine, 300 to 750 hp," *SAE Special Publication* SP-137, 1955. (Description of $8\frac{1}{8} \times 10$ in. Fairbanks-Morse engines, 4 to 12 cylinders.)

10.491 Guglielmotti, "A 4000 bhp Lightweight Engine," *ASME Paper* 64–WA/OGP–2, Dec. 1964. (Four-cycle, V-12 Fiat Diesel 11.8×14.2 in., 300×360 mm, operating at 830 rpm, bmep $= 182$ psi, 12.8 kg/cm², and piston speed 1965 ft/min, 10 m/sec.)

10.492 Radford *et al.*, "Experimental Techniques in the Development of Highly-Rated Four Stroke Diesel Engines," *Seventh CIMAC Congress*, London, 1965. (Useful data on operating temperatures and measured stresses in piston, rod, and cylinder head as a function of design. Valve flow characteristics and performance data versus bmep for Mirrlees locomotive-size engine.)

10.493 Brock *et al.*, "Some Research and Development Investigations on Medium Speed Oil Engines at 200 psi (14.1 kg/cm²) bmep," *Seventh CIMAC Congress*, London, 1965, p. 393. (Temperature and stress measurements for piston, cylinder head, and bolts, crankcase. Valve seating velocities. Ruston and Hornsby locomotive-engine performance data. Exhaust pulse pressures.)

10.494 Other valuable design data on highly supercharged engines of locomotive size are available in *Seventh CIMAC Congress*, London, 1965. (Most of these articles include temperature and stress measurements. See especially articles by Dingle and Stent on Paxman engines, p. 507, and by Vaughn on Alco engines, p. 545.)

10.495 Herschmann, "Daimler-Benz High Output Engines — A Study in Compact Design," *SAE Paper* 67519, 1967. (Four-cycle 6.5×6.9-in., 165×175-mm V engines. Rating is at bmep $= 157$ psi (11 kg/cm²) and 2480 ft/min (12.6 m/sec) piston speed, specific output 2.95 bhp/in². Divided combustion chamber, 4 valves per cylinder. Stress patterns in crankcase and connecting rod are illustrated, as are valve-gear dynamics.)

Large Trunk-Piston Diesel Engines for Marine and Stationary Service

10.51   Kirkwood, "Tomorrow's Stationary Power Engines," *Diesel and Gas Turbine Progress*, Pt. I, Jan. 1966, Pt. II, Feb. 1966. (Data on present status and probable future growth of municipal Diesel-electric power units. Includes a plea for the development of new units of over 6000 hp.)

10.52   Moriarty and Schowalter, "Application of Medium-Speed Diesels to Marine Propulsion," paper before ASNAME, May 1966. (Discussion of geared "medium-speed" versus direct-drive "low-speed" engines.)

10.53 "European Medium-Speed Marine Diesels," *Lubrication 51*, 9 (1965). (Excellent discussion of the problem of using heavy residual fuel in trunk-piston engines.)

10.54 "From the M.A.N. Four-Stroke Engine Development Program," *M.A.N. Diesel Engine News*, No. 44, June (1965). (Excellent review of M.A.N. practice to date. Reasons for using 4 cycles for engines when bore is less than 20.5 in. (520 mm). Details of combustion chambers, pistons, cylinder-head gaskets, engine cross sections, and performance maps. For earlier developments of M.A.N. 4-cycle engines, see *ibid.*, No. 25, April 1952. The latter includes data on rotating auxiliary exhaust valves.)

10.55 Lassberg, "A New Highly-Turbocharged Four-Cycle M.A.N. Marine Engine," *Diesel and Gas Turbine Progress*, July 1966, 31. (Line of 15.75 × 21.26 in., 400 × 540 mm, 6 to 18 cylinders, 3250 to 9800 hp, at 400 rpm 1420 ft/min, bmep = 256 psi, 18 kg/cm². Designed for multiengined geared propeller drive, trunk piston, articulated rods, composite pistons, heavy fuel.)

10.56 "New 6000 kw Nordberg Diesel Engines," *Diesel Power and Diesel Transportation*, June, 585 (1944). (For later models of Nordberg engines, see manufacturer's bulletins.)

10.57 Fairbanks-Morse Model 38A20 Diesel Engine, 1000 bhp per cylinder. Manufacturer's *Bull.* 3800A20, 1965. (Two-cycle opposed-piston type with small exhaust pistons. Vertical and V, 6000 to 12,000 hp. See also *Diesel and Gas Turbine Progress*, June 1965.)

10.58 The trunk-piston radial gas engines of reference 10.76 are also used as Diesel engines in some stationary applications.

10.59 Louzexky, "Design and Development of a Two-Cycle Turbocharged Diesel Engine," *ASME Paper* 56–A–100, Nov. 1956. (General Motors' experience with 8 to 9½ in. bore marine engines.)

Large "Low-Speed" Cross-Head Type Marine Engines
(See also ref. 10.883.)

10.591 Powell, "Estimation of Machinery Weights," *Trans. SNAME*, March, 721 (1958). (Weights of large marine installations and their components, engine, gears, auxiliaries, etc.)

10.592 Sorensen, "Large-Bore Diesels for Modern Power Stations and Sea-Going Ships," *ASME Paper* 65–OGP–11, 1965. (Review of current practice, and use of Diesel versus steam in ships and power stations.)

10.593 Andresen, "Slow-Running Marine Diesel Engine Plants," Meeting of *SNAME*, New York, *Paper* No. 17, May 1966. (General review of state of the art with details of Götaverken engines.)

10.594 Aue, "Aspects in the Designing of a Homogeneous Series of Large Marine Diesel Engines," *Sulzer Tech. Rev.*, No. 4, 205 (1963); also available from Sulzer Bros., Winterthur, Switzerland. (Review of the current status and some development problems of Sulzer crosshead-type marine engines, 2000 to 27,500 bhp. All are in-line, loop-scavenged, 2-cycle with auxiliary rotating exhaust valves, welded steel crankcases, turbo-supercharged. See also the earlier articles on Sulzer practice and developments: Kinchelmann, *Sulzer Tech. Rev.*, No. 2 (1953); Zwicky, *Sulzer Tech. Rev.*, No. 1, 13 (1958).

10.595 Schuler, "Trends in Diesel-Engine Design at the Augsburg Works of M.A.N. After 1945," and Schmidt, "High-Powered Diesel Engines for Marine Propulsion," *M.A.N. Diesel Engine News*, No. 36, 23, 29 (1958). (History and present status of large M.A.N. 2-cycle loop-scavenged crosshead-type marine engines. See also *Diesel and Gas Engine Progress*, April (1965).)

10.596 Hellström, "Some Results from Götaverken Experimental Engines," *Seventh CIMAC Congress*, London, 1965, p. 393. (Temperatures, pressures, stress measurements, and performance of 33.5-in. (850 mm) bore by 67-in. (1700 mm) stroke crosshead type, poppet valve marine engine at bmep up to 170 psi 12 kg/cm². Highest specific output ever documented for this type of engine. New rating is 145 psi at piston speed 1380 ft/min, 7 m/sec, specific output 3.03 bhp per in² piston area. See also Broeze on Stork engines, p. 349 in the same volume.)

10.597 See also marine-engineering publications such as refs. 10.12, 10.14–10.18, and *Naval Architect and Engineer*, Seven Oaks, Kent, England, and *International Shipbuilding Progress*, Rotterdam, Holland.

Miniature Engines (*for Model Airplanes, etc.*)
(See also Volume I, Fig. 11–7 and Table 11–2.)

10.60 Bowden, *Model Glow Plug Engines*, Percival Marshall & Co., London. (History and description of model engines using glow-plug ignition. Notes on installation and operation.)

10.61 Foot, *Model Airplane Engines*, A. S. Barnes and Co., New York.

10.62 "Performance of Miniature Engines," *WADC Tech. Report* 53–180, ASTIA number AD 130 940, Part V. (By Battelle Mem. Inst. for U.S. Govt. Test data on a large number of miniature engines. Never published.)

10.63 See also catalog literature of the following: Pal Engineering, 53 16th Ave., S.W., Cedar Rapids, Iowa; Dooling Bros., 5452 West Adams Blvd., Los Angeles, Calif.

10.631 Rice, "Transition from Toy to Tool," *SAE Paper* 943B, 1964. (Details of a 1.25 × 1.05 in. engine said to give 0.85 bhp and weigh 3.75 lbm. Used for portable tools, mfd. by Ohlsson Rice, Inc., Los Angeles, Calif.)

Small Engines
(For motorcycles, home uses, portable industrial, outboard marine, etc., see also Tables 10–5 and 11–1.)

10.640 Meyer, "Europe's Small Air-Cooled Engines Today," *SAE Paper* 394A, 1961. (Review of small gasoline and Diesel types. Several cross sections shown.)

10.650 Meyer, "European Solutions of the Design of Small Air-Cooled Engines," *SAE Paper*, 1957. See also *Jour. SAE 66*, Sept. 1957, 81. (Statistics on European and U.S. gasoline, and some Diesel, including plots of bmep, piston speed, weights, etc.)

10.651 Heidner and Pike, "Small Engine Development," *SAE Special Publication* SP-143, 1956. (Practical discussion of West Bend approach to the design of an outboard marine engine.)

10.66 Perlewitz, "3.5 hp in an 8-lb Package," *SAE Paper* 660007, Jan. 1966. (Power Products Co. 2.94 × 1.25 in. 2-cycle engine.)

10.67 Smith, "The Development of a Lightweight Air-Cooled Industrial Engine," *ASME Paper* 64–OGP–14, May 1964. (Details of the Petter 3-hp 2.375 × 3.0 in. air-cooled Diesel engine weighing 85 lbm.)

10.68 Paquette and Brooks, "Small High-Speed Two-Stroke Diesels Can Match Their Big Brothers," *SAE Paper* 646A, Jan. 1963. (McColluch "Scott 22D" Diesel, 2.225 × 2.75 in. air-cooled, 15 bhp, 120 lbm. See also *SAE Paper* 650719, Oct. 1965, on McColluch engines.)

10.69 "Power in a Small Package," *Lubrication 45*, May (1959). (Useful summary of applications, types, production figures, and lubrication problems.)

10.691 Onan, "Development of J Engine Family," *SAE Paper* 537A, June 1962. (Onan line of small air-cooled industrial engines, 7 to 40 hp. Good cross sections.)

10.692 Waker, "Present Day Efficiency and Factors Governing Performance of Small Two-Stroke Engines," *SAE Paper* 660009, Jan. 1966. (Design and performance of Fichtel and Sachs motorcycle engines of 50 cc (3 cu in.) displacement. Shows large effects of exhaust-system geometry, 6.7 hp sq in., see Table 10–5.)

10.693 Nakamura, "Small High-Speed, High-Performance Gasoline Engine," *SAE Preprint* 640664, May 1965. See also *Jour. SAE 74*, Oct., 44 (1965); *SAE Paper* 888A, Aug. 1964. (Experience of Honda Co. including combustion, detonation, air capacity, valve gear, inlet dynamics, and friction of small engines running up to 22,000 rpm, 4000 ft/min.)

10.694 Kruckenberg, "McCulloch Three-Cylinder, Two-Cycle Outboard Engine," *Trans. SAE 71*, 561 (1963). (Included in discussion of paper by Schweitzer. Gives difference in fuel economy between carburetion and cylinder injection, 25 to 30 per cent.)

10.695 Grant, "Outboard Engine Fuel Economy," *SAE Paper* 707B, 1963. (Effects of improved inlet reed-valve, combustion chamber and port design on performance of small air-cooled 2-cycle engine.)

10.696 Naito and Taguchi, "Some Development Aspects of Two-Stroke Motor-Cycle Engines," *SAE Paper* 660394, Jan. 1966. (Comprehensive story of the development of Yamaha engines including effects of porting, crankcase intake-valve design, exhaust-system design, piston design, engine cross sections. Includes normal and racing types.)

Large Gas Engines

(Many small Diesel engines also offer alternative gas-burning models.)

10.70 Helmich and Ulrey, "The Design and Development of KSV Engine," *ASME Paper* 62–OGP–1, 1962. (Cooper-Bessemer 4-cycle engine with cooling by expansion in a over-supercharged inlet system.)

10.71 Dorton and Schaub, "Effects of Operational Parameters on a Large 'V' Angle Compressor Engine," *ASME Paper* 64–OGP–9, May 1964. (Cooper-Bessemer 5500-hp loop-scavenged, 2-cycle engine-compressor unit. Contains useful information on pressures, temperatures, and development changes. See also *Diesel and Gas Turbine Progress*, Oct., 23 (1966), showing changes made to rate this engine up to 7250 bhp.)

10.72 Land and Caramos, "Turbocharged Two-Stroke-Cycle Gas Engines," *ASME Paper* 65–OGP–6, May 1965. See also Land, "Gas Engine Compressor — the Worthington Mainliner," *Engine Design and Applications*, June (1965). (Worthington 2-cycle engine-compressor units. Valuable development information including comparison of blow-down and steady-flow supercharger systems. Uses more-complete-expansion cycle.)

10.73 Ingersoll-Rand Company, *KVR, KVT and KVH Gas-Engine Compressors*, Bulletins 1966 and 1961, Ingersoll-Rand Co., Inc., 11 Broadway, New York. See also other bulletins obtainable on request. (Four-cycle gas engines, some with the more-complete-expansion cycle.)

10.74 Weber and Boyd, "Design and Development of a High Specific Output Packaged Gas Engine-Driven Compressor Unit," *ASME Paper* 62–OGP–9, Feb. 1963. (Clark 1000-hp 2-cycle, loop-scavenged engine-compressor unit.)

10.75 Brater, *Engineering Features of a Lightweight Two-Cycle Engine*, Cleveland Diesel Engine Division, General Motors Corp., Oct. 13, 1953. (Detailed description of General Motors 16-cylinder radial engines, well illustrated. See also *Diesel Progress*, Oct. 1954; *SAE Paper* 778, June 1952.)

10.76 Bohn and Grieshaber, "Design Features of the Nordberg Radial Engine," *ASME Preprint* No. 50–OGP–1, April 7 (1950). See also *Trans. ASME 73*, Aug. 1951, 795; and Nordberg Bulletin 200, 1952, showing 40 of these engines installed in Alcoa Plant in Texas.

10.77 Wadman, "Nordberg Centrifugal Gas Compression Unit," *Diesel Progress* (now *Diesel and Gas Turbine Progress*), April (1955). (V-16, 4-cycle "Superair Thermal," or more-complete-expansion-cycle engine geared to centrifugal gas compressor.) (See also ref. 10.857.)

Racing and Sports-Car Engines

(See also Table 10–13.)

10.801 Campbell, *The Sports Car — Its Design and Performance*, Robt. Bentley, Inc., Cambridge, Mass., 1959. (Good summary of the subject, with valuable illustrations, statistics, and discussion. See also, by the same author and publisher, *The Sports-Car Engine — Its Tuning and Modification*, 1964. Includes differences between conventional passenger-automobile engines and sports-car types. Useful to the designer.)

10.802 Gay, "A Ford Engine for Indianapolis Competition," *SAE Paper* 818A, Jan. 1964. (History and details of development from stock 260 cu in. engine. Push-rod, 2-valve engine compared with Offenhauser 4-cylinder engine, ref. 10.807. See also ref. 10.815.)

10.803 Scussel, "Ford's DOHC Competition Engine," *Jour. SAE 72*, Sept. 1964, 68. See also *SAE Paper* 640602 (S397), 640166 (818A), and Geschelin, "Ford's New Overhead Camshaft Race Engine," *Auto. Ind.*, April 15, 1964, 74. (Contains much useful engineering material regarding changes from standard passenger-car engine, development problems, materials of construction, important dimensions, etc.)

10.804 Faustyn and Eastman, "The Ford 427 Cubic Inch, Single-Overhead Camshaft Engine," *SAE Paper* 650497, May 1965. (Further development of engines of previous reference. Useful design information.)

10.805 Scott, "Flat-16 Racing Engine Has Two End-On Crankshafts," *Auto. Ind.*, April, 40 (1965). (1.5-liter Coventry Climax racing engine.)

10.806 *British and European Passenger Cars and Sports Cars*, Bull. P-952, Quaker State Oil Refining Corp., Oil City, Pa., 1958. (Includes some racing types, cross sections, and principal data up to 1958.)

10.807 Meyer and Goossen, "New Offenhauser Engine for Indianapolis Race," *Auto. Ind.*, April 1950, 46–47, 82–84. (Unusually complete dimensional and performance data on this 4-cylinder 4.125 × 3.125 in. supercharged racing engine. See also *Machine Design*, May 12, 1966.)

10.808 Reiners and Schmidt, "The Experimental High-Speed Cummins Diesel Engine," (4000 rpm, 345 hp, 401 cu in.), *SAE Preprint* 626, June 1951. (Conversion from a truck Diesel engine, modified for racing and tried out with good results at Indianapolis.)

10.809 Scherenberg, "Aus der Konstruktion und Berechnung des 2,5-1 Mercedes-Benz Formel-Rennwagens," *ATZ* (Berlin) *60*, June (1955). (Details of this super-racing car from 1934 to date. Valve data, crankshaft and its vibration, performance with carburetor and fuel injection.)

10.810 Braunschweig, "Design Features of the Lancia V-8 Competition Car," *Auto. Ind.*, June 15 (1955). (Eight-cyl., 2.99 × 2.70 in., 152 cu in., 2.5 liters, 250 hp at 8250 rpm, 158 bmep. Section of engine shown.)

10.811 Meyer and Goossen, "The Improved 1949 Offenhauser Midget Racing Engine," *Auto. Ind.*, April (1949). (3 × 3.63 in. 4-cylinder engine for small track racing.)

10.812 "New Ferrari Engine," *Auto. Ind.*, Feb. 15, 1950, 41. (Twelve-cylinder, 91 cu in., 1.5-liter racer.)

10.813 "New British Racer Has Super-Speed Engine," *Auto. Ind.*, Feb. 15, 1950, p. 54. (Sixteen-cylinder, 91 cu in., 1.5-liter engine. Cooperative development by British industry. Rates at 8.6 hp per sq in. piston area.)

10.814 Hassan, "Some Notes on the Coventry Climax Engines," *SAE Paper* 256A, Nov. 1960. (Includes much useful design information on these high-performance engines, including the 2.5 liter, 3 × 2.675 in. V-8 racing engine. Includes engine cross sections and other design details.)

10.815 Macura and Bowers, "Mark II-427 GT Engine," *SAE Paper* 670066, Jan. 1967. (Details of Ford racing engine, winner at Le Mans, 1966. V-8 with push-rod valve gear, four-barrel carburetor. Based on a stock engine with few changes. (See Table 10–13 for performance. See *SAE Paper* 670067, Jan. 1967, for details of induction system.)

10.816 Thoreson and Brafford, "The Corvair Turbosupercharged Engine," *SAE Paper* 531A, June 1962. (Discussion of problems and solutions to adapt the Corvair engine, ref. 10.26, to turbo-supercharging for sports-car use. Compression ratio reduced from 9 to 8. Max. bmep increased from 130 to 220 at 1500 ft/min, 7.5 m/sec, piston speed. Manifold pressure 25 psia, 1.76 kg/cm².)

Military Engines Other than Aircraft

(See also Table 10–14.)

10.820 Schweitzer *et al.*, "Lycoming S and H, The Compact Multifuel Engine," *SAE Paper* 790C, Jan. 1964. (Two-cycle air-cooled V, loop-scavenged Diesel engine, multifuel, unsupercharged.)

10.821 "Continental Hypercycle Six," *Auto. Ind.*, June 1960, 1. (Four-cycle in-line Diesel using M.A.N. spherical combustion system. Multifuel, supercharged.)

10.822 Paluska *et al.*, "Design and Development of a Very High Output (VHO) Multifuel Engine," *SAE Paper* 670520, 1967. See also bull. form 40-20826 DP(1–65) by Caterpillar Tractor Co., Peoria, Ill. (Interesting Caterpillar Co. line of 4.5 × 5.5 in. 4-cycle supercharged Diesel engines with 4, 6, V-8, and V-12 cylinders. Rating is at bmep = 258 psi (18.1 kg/cm^2) and 2570 ft/min (13 m/sec) piston speed, specific output = 5 bhp/in.^{2} (0.78 bhp/cm^2). Crankshaft main journals are the circular crankchecks. Main bearing caps bolted parallel to cylinder axes in V engines. Usual Caterpillar pre-combustion chamber. Unusual connecting rod big ends with straps bolted at right angles to rod axis.)

10.823 *Leyland 700 bhp Engine L60*, Leaflet No. 892, Leyland Motors, Ltd., Leyland, Lancashire, England. (Two-cycle, 6-cylinder, opposed-piston Diesel engine rated at bmep = 114 psi, s = 2020 ft/min. Multifuel, gear-driven Roots blower.)

10.824 "Higher Power for Modern Transport," *The Oil Engine and Gas Turbine*, London, Sept. 1962. (Rolls-Royce opposed-piston Diesel engines up to 300 hp sponsored by British War Dept. See also earlier article, *ibid.*, Dec. 1959.)

10.825 Okamura, "Big Diesel Runs Smoothly," *Jour. SAE 71*, July, 102 (1963). (Twenty-four-cylinder, 4-cycle W-type Diesel engine rated at 3000 hp, bmep = 143, s = 2100 ft/min for naval use.)

10.826 Ware *et al.*, "The New Packard Lightweight Diesel Engines," *SAE Paper* 187, Nov. 3–4, 1953. (Performance data, assembly drawings, bearing-load diagrams, etc., for 12-cylinder, 4-cycle marine engine for small naval craft.)

10.827 Hulsing *et al.*, "The Development of a Complete Family of Compression Ignition Multifuel Engines," *Gen. Motors Eng. J. 11*, First Quarter, 1964. (Conversion of standard G. M. 2-cycle Diesels for military use.)

10.828 Haas and Klinge, "The Continental 750-Horsepower Air-Cooled Diesel Engine," *Trans. SAE 65*, 641 (1957). (Complete story of development of 4-cycle V-12 with much quantitative material.)

10.829 Chatterton, "The Napier Deltic Diesel Engine," *SAE Preprint* 632, November 1955. (Description of unusual triangular, three-crankshaft, opposed-piston, high-output Diesel engine for light high-speed marine applications.)

10.830 Dilworth, "The Sky is the Limit for Two-Cycle Engines," *SAE Paper* 599A, 1962. (Brief description and performance data on 125-hp air-cooled horizontal opposed engines for target airplanes, supercharged to 72 per cent power at 30,000 ft altitude; 50 per cent power at 40,000 ft.)

Reciprocating Aircraft Engines, Historical
(See also Table 10–15.)

10.840 Taylor, "Aircraft Propulsion — A Review of the Evolution of Aircraft Power Plants," Smithsonian Institution, Washington, D.C., Publication 4546, 1963. (Significant trends and examples from early experiments to 1963.)

10.841 Wilkinson, *Aircraft Diesel Engines*, Pitman, 1940. (Since there are no longer any aircraft Diesels, this book is of historical value only.)

10.842 Schlaifer and Heron, "Development of Aircraft Engines and Fuels," Harvard University, Grad. School of Business Administration, Cambridge, Mass., 1950.

10.843 Heron, "History of the Aircraft Engine," publ. by Ethyl Corp., Detroit, 1961.

Typical Military and Commercial Aircraft Engines

10.844 Ryder, "Recent Developments in the R-4360 Engine," *SAE Quart. Trans. 4*, 559 (1950). (Details of the 28-cylinder Pratt and Whitney air-cooled 4-row radial. Development time four to six years. Forged aluminum crankcase and cylinder heads. Temperature surveys of head and exhaust valves. Shielded exhaust port. Steel or forged aluminum pistons said to be satisfactory.)

10.845 Ellor, "The Development of the (Rolls-Royce) Merlin Engine," *Trans. SAE 52*, 385 (1944). See also *Aircraft Eng.* (London), July, 218 (1946); and *Flight* (London), May 7 (1954). (Development of the best-known liquid-cooled engine of World War II.)

10.846 For German aircraft engines of World War II, see *Trans. SAE 49*, 409 (1941); *ibid.*, *50*, 465 (1942).

10.847 Wiegand and Olson, "Postwar Development of the Reciprocating Engine," *SAE Quart. Trans. 4*, 8 (1950). See also Wiegand and Eichberg, *Trans. SAE 62*, 265 (1954). (Improvements in design of Wright Turbo-Compound radial airplane engines, including materials for valve guides, steel crankcase, new piston design, crankshaft "tuning fork" vibration, and its prevention by a lateral damper.)

10.848 Hazen, "The Allison Aircraft Engine Development," *Trans. SAE 49*, 488 (1941). (Design details of cylinders and section of engine. Propaganda for liquid cooling.) See also *Flight* (London), March 26, 1942, 281.

Typical Aircraft Engines for Light Airplanes and Helicopters

10.850 Bachle, "Progress in Light Aircraft Engines," *Trans. SAE 46*, 243 (1940). (Continental opposed-cylinder air-cooled line.)

10.851 "New Continental Aircraft Engines," *Automotive and Aviation Industries 93*, Dec. 15, 1945. (Postwar series of horizontal-opposed air-cooled engines.)

10.852 Wiegman, "Geared Engines for Light Airplanes," *Trans. SAE 47*, 301 (1940). (Lycoming 4-cylinder horizontal-opposed engine with internal spur gear and second-order and fourth-order pendulum vibration absorbers.)

10.853 "New Lycoming Aircraft Engines," *Auto. Ind.*, March 1941, 332. (Postwar series of horizontal-opposed air-cooled engines. See also bulletins of Lycoming Division, AVCO Corp., Williamsport, Pa.)

10.854 Wiseman, "Altitude Performance with Turbosupercharged Light Aircraft Engines," *SAE Paper* 622A, 1963. (Shows critical altitude to 17,000 ft for Continental air-cooled, opposed-cylinder engines.)

More-Complete-Expansion Cycle
(See also refs. 10.72, 10.73, 10.872.)

10.855 Miller, "A Low-Temperature Supercharging System for Compression, Pilot Oil, and Spark-Ignition Engines," *ASME Paper* 57-A-250, Dec. 1957. (More complete-expansion cycle with high supercharge pressure and early inlet-valve closing to limit maximum pressure and temperature.)

10.856 Miller and Leiberherr, "The Miller Supercharging System for Diesel and Gas Engines," CIMAC International Conference on Internal Combustion Engines, Wiesbaden, 1959. (Same subject matter as ref. 0.80.)

10.857 Brimson, "High Performance Gas Burning Engines," *Seventh CIMAC Congress*, London, 1965, p. 603. (Useful design and performance data on Nordberg 13.5 × 16.5 in. 4-cycle engine using more-complete-expansion cycle.)

Supercharging

(See also Volume I, refs. 10.01–10.92, 13.00–13.45; and Volume II, refs. 12.80–12.821.)

Effects of Supercharging on Power and Economy

10.860 Hull, "High-Output Diesel Engines," *Trans. SAE 72*, 68 (1964). (Tests on an open-chamber one-cylinder engine up to bmep $=302$ psi, $p_i = 44$ psia, $T_i = 660°$ R, $p_{max} = 2700$ psia.)

10.861 Nagao and Hirako, "Estimation of the Part-Load Performance of Turbocharged Two-Cycle Diesel Engine," *Bull. JSME 2*, 390 (1959).

10.862 Nagao and Hirako, "Estimation of the Operating Characteristics of a Turbocharged Four-Cycle Diesel Engine," *Bull. JSME*, Vol. 2, 1959.

10.863 Awano, "Performance of Supercharged Four-Cycle Diesel Engines," Report of the Research Institute of Technology, Nihon University, Tokyo, No. 2, Nov. 1954.

10.864 McAulay *et al.*, "Development and Evaluation of the Simulation of Compression Ignition Engines," *SAE Paper* No. 650415, May 1965.
a) Development of the Simulation Program
b) An Engineering Evaluation of the Simulation Program
(Cooperative work develops effective program to indicate trends based on reasonable assumptions. Results agree with tests on one engine.)

10.865 Cook, "Digital Computer Assists Designers in Engine Development," *Jour. SAE 73*, April 1965, 60. See also *SAE Paper* 3S, 1959, by the same author. (Work at Thompson-Ramo-Wooldridge on simulation of effects of valve timing, intake manifold ram, exhaust pipe dimensions, burning effects in terms of combustion rates, accumulated delay for detonation, etc. Curves show significant trends. Performance with turbo-supercharger included.)

10.866 "Symposium on Superchargers and Supercharging," *IME* (London), *Proc. Automobile Div.* 1956–1957, No. 6. (Valuable articles by qualified experts.)

10.867 Johnson, "Supercharged Diesel Performance Versus Intake and Exhaust Conditions," *Trans. SAE 61*, 34 (1953). (Good study on brake basis—needs to be generalized and put on indicated basis. Air-flow figures included.)

10.868 Von der Nuell, "Notes on Turbocharged Two-Stroke Cycle Diesel Engines," *SAE Preprint* 799, Aug. 1956.

10.869 Nagao and Hirako, "Basic Design of Turbocharged Two-Cycle Diesel Engine," *Bull. JSME 2*, 156 (1959).

10.870 Wieberdink and Hootsen, "Supercharging by Means of Turboblowers Applied to Two-Stroke Diesel Engines of Large Output...," *CIMAC Congress*, The Hague, 1955. (Stork experience with self-sustaining supercharger system without auxiliary scavenging pumps for large cross-head type marine engines.)

10.871 Pyles, "Diesel Supercharging—Its Effect on Design and Performance," *Trans. SAE 33*, 215 (1938). (Clark Bros. experience shows effects on performance, bearing loads, durability by the application of Roots gear-driven supercharger to $8 \times 10\frac{1}{2}$ in. engine.)

10.872 Wadman, "Tandem Supercharging Delivers 250 BMEP Rating," *Diesel and Gas Turbine Progress*, July 1967. (Nordberg 4-cycle 13.5×16.5 in., 344×420 mm engines with two turbo-superchargers in series, with inlet pressure ratio 3.0. Rating is at bmep = 250 psi (17.6 kg/cm²) piston speed = 1415 ft/min (7.2 m/sec) maximum cylinder pressure = 1470 psi (103 kg/cm²). Uses more-complete-expansion cycle.)

Effects of Supercharging on Stresses, Temperatures, Heat Flow, Durability
(See also refs. 10.45, 10.47, 10.492–10.495.)

10.880 Robinson and Mitchell, "The Development of a 300 psi (21.1 kg/cm²) Continuous Duty Diesel Engine," *Seventh CIMAC Congress*, London, 1965, p. 269. (Tests on Caterpillar 4.5-in. bore prechamber engine with 2-stage supercharger, inter- and after-cooler. Temperatures at various points, heat rejection, peak pressures, and over-all performance up to bmep = 400 psi, 28 kg/cm². Peak pressure at bmep = 350 about 2500 psi, 175 kg/cm².)

10.881 Mansfield *et al.*, "Development of the Turbocharged Diesel Engine to High Mean Effective Pressures without High Mechanical or Thermal Loading," *Seventh CIMAC Congress*, London, 1965, p. 239. (Use of a hydraulic-controlled, variable-compression-ratio piston allows bmep = 300 psi, 21.1 kg/cm², with only 1250 psi, 87.9 kg/cm², peak pressure. Best economy occurs at bmep = 100–150 psi, 7 to 11 kg/cm². Cylinder-head temperature increases linearly with bmep.)

10.882 Kuechler, "Temperature Distribution in Thermally Loaded Components of High-Speed Prechamber Diesel Engines ...," *Seventh CIMAC Congress*, London, 1965, p. 71. (Temperatures and heat transfer rates as a function of crank angle, speed and bmep up to 200 psi, 14 kg/cm² Daimler-Benz experience.)

10.883 French *et al.*, "Thermal Loading of Highly-Rated Two-Cycle Marine Diesel Engines," *Seventh CIMAC Congress*, London, 1965, p. 309. (Results of tests on a 26.4-in., 670 mm, bore opposed-piston Doxford engine. Local heat flux and metal temperatures, as a function of fuel flow, crank angle, location, etc. Excellent correlation with Volume I, Fig. 8–11, p. 289.)

10.884 Vincent and Henein, "Thermal Loading and Wall Temperature as Functions of Performance in Turbosupercharged Compression-Ignition Engines," *Trans. SAE 67*, 478 (1959). (Theoretical analysis and test results showing effects of charge cooling and operating variables on performance, heat flow, and temperatures, using a 4-cycle divided-chamber engine. Useful discussion.)

10.885 May and Reddy, "Durability Tests of Two-Stroke Diesels Show Exhaust Valves Are Only Part Adversely Affected by Turbocharging," *SAE Paper*, June 1957, abstracted in *Jour. SAE 65*, July 1957, 39. (Measured exhaust-valve temperatures on GM-71 engine with and without supercharging. Discussion shows reduced temperature in another engine due to reduced fuel-air ratio.)

10.886 Boyer, "Design Aspects of Supercharged Diesel Engines," *Mech. Eng.* (ASME) *67*, June 1945, 392. (Data on changes in valve timing, heat rejection, and bearing load diagrams due to supercharging Cooper-Bessemer engines.)

Intercooling and Aftercooling
(See also refs. 10.880–10.886.)

10.900 Mitchell, "An Evaluation of Aftercooling in Supercharged Diesel Engine Performance," *Trans. SAE 67*, 401 (1959). (Review of Caterpillar experience. Valuable discussion by engineers from competing companies. Includes some data on durability as well as performance.)

10.901 Helmich, "Development of Combustion Air Refrigeration System...," *Seventh CIMAC Congress*, London, 1965, p. 637. (Consideration of several means of inlet cooling and data on the Cooper-Bessemer LSV 4-cycle, 15.5×22 in. gas engine using expansion-turbine driving a Freon system with evaporator before the inlet manifold. Inlet temperature 34°F with ambient 85°F resulting in significant improvement in fuel economy at bmep = 165 psi.)

10.902 Syassen, "Influence of Very Low Induction Temperatures in Higher-Ratio Supercharging on Thermal and Mechanical Stress of Diesel Engines," *Seventh CIMAC Congress*, London, 1965, p. 179. (Effects on firing pressure, exhaust temperature power, and economy on a M.A.N. turbo-supercharged loco-motive-size engine. Over-all effects on the complete engine-supercharger com-binations.)

Compound Engines (Exhaust Turbines Geared to Crankshaft)
(See Volume I, refs. 13.11–13.23; Volume II, refs. 10.847, 13.74–13.743.)

CHAPTER 11

Machine Design, General

11.000 *ASME Handbook: Metals Engineering — Design*, 2nd ed., 1965. (Methods, facts, and data to help solve problems of metal machine and product parts design, brought together by more than 60 experts.)

11.001 Heywood, *Designing Against Fatigue*, Chapman Hall, London, 1962.

Engine Design, General
(See also ref. 10.096.)

The author has found no authentic, up-to-date book in the English language on this subject. Existing books are generally obsolete and in many cases merely descriptive of designs current at the time of writing. The following references are considered to have material of permanent value to the designer.

11.002 SAE special publications under the title "Engineering Know-How in Engine Design." (Some of these contain valuable design information. Others are brief outlines. Many contain good bibliographies. The more valuable parts are included in this bibliography under individual subject headings.)

No.	Subjects	Year
SP-23	Torquing of nuts.	1946
SP-29	Spring problems.	1952
SP-119	Combustion chambers, pistons and rings, carburetors, valves and valve gear, engine testing.	1953
SP-122	Single-cylinder, 2- and 4-cycle engines, experimental engines, Ignition systems, Bearings and lubrication, induction systems, radiators and cooling systems.	1954
SP-127	Valve tappet wear.	1954
SP-137	Engine materials, engine dynamics, opposed-piston engine, 2-cycle outboard engines, combustion chamber shape, gasoline and Diesel fuels.	1955
SP-143	Gear design and mfg., supercharging, small-engine development, instrumentation, gas turbines, fuel injection for S-I engines, Diesel fuel injection.	1956
SP-148	Crankshafts and rods, bearings and lubrication, pistons, piston rings.	1957
SP-152	Valve gear and cams, 2-cycle engines, fuel injection for S-I engines, liquid cooling systems, air-cooled cylinder design.	1958
SP-159	Aluminum engine blocks.	1958
SP-163	Engine "roughness," viscous torsional damper, flexible engine mounts, air cleaners and intake silencers, exhaust and intake silencers.	1959
SP-178	Valve-gear dynamics, dynamometer for engine tests, use of radio isotopes in wear measurement and oil contamination, instruments for engine testing, field testing of Diesel engines.	1960
SP-192	Speed governors, disc clutches, hydrostatic transmissions, mechanical transmissions, torque converters.	1961
SP-224	Piston ring and cylinder bore wear, chromium plating, piston-ring materials, and design, etc.	1961
SP-237	Lubrication, wear, lubricants, ignition systems.	1962
SP-256	"Digital" engine, crankshaft testing, lubrication of small two-cycle engines.	1964
SP-270	Power plants for industrial and commerical vehicles, spark ignition, Diesel, gas-turbine. Present status and predicted figure trends.	1965
SP-271	Preignition, rumble, piston rings, gas-engine-valve-seat wear, digital-computer applications.	1965
SP-274	Engine bearings.	1966
SP-280	High-output diesel engines with variable compression ratio.	1966
SP-283	Valves, crankshafts, turbosuperchargers, pistons, bushings, and thrust bearings.	1966

(See also future publications in this series. Information is obtainable from Society of Automotive Engineers, Inc., 2 Pennsylvania Plaza, New York.)

11.011 Taylor, "Radial Engines, Their Power and Frontal Area," *Aviation*, July 1933, 201. (Rational approach to choice of stroke-bore ratio for radial engines. An example of good analysis of a major design problem.)

11.012 Wiegand and Olson, "Postwar Development of the Reciprocating Engine," *SAE Quart. Trans. 4*, Jan. 1950, 8. (Valuable reports on design developments in forged and cast cylinder heads, valve seats and valve guides, pistons and connecting rods, forged steel crankcases, crankshaft with vibration absorbers for large radial aircraft engine.)

11.013 Haas and Klinge, "The Continental 750-Horsepower Air-cooled Diesel Engine," *Trans. SAE 65*, 641 (1957). (Useful discussion of developments in cooling, bearing loads and materials, connecting rod and bolts, pistons, for a very highly stressed engine.)

11.0131 Rosen, "A Half Century of Diesel Progress," *SAE Paper* 660602, Sept. 1966. (Useful review of past and current practice, with good illustrations of combustion chambers and pistons. Some unusual engines.)

Stresses; Thermal, General

11.020 Horvay, "Transient Thermal Stresses in Circular Disks and Cylinders," *ASME Paper* No. 53–SA–51, 1953.

11.021 Palmblad, *Thermal Stresses in a Tube. The Engineers' Digest 3*, February, 1946. (From *Teknisk Tidskrift 29*, July 21, 1945, 821.) (Mathematical expressions for stresses in a hollow cylinder with a temperature gradient across the walls. Example worked out for a Diesel-engine piston rod.)

11.022 Goodier, "Thermal Stresses," *J. Appl. Mech.* (ASME) *3*, March, 1937.

11.023 Kent, "Thermal Stress in Spheres and Cylinders Produced by Temperature Varying with Time," *Trans. ASME 54*, 185 (1932).

Experimental Stress Analysis, Brittle Coatings
(See refs. 11.206, 11.211–11.213.)

11.101 Dietrich and Lehr, "Das Dehnungslinienverfahren," *ZVDI 76*, Oct., 973 (1932). (One of the first publications on the use of brittle lacquer for stress measurements in engines. Examples of improvements in strength of crankshafts, crankcases, connecting rods, and pistons by means of this technique. This article inspired development of "Stresscoat" in the U.S.A. by E. S. Taylor and G. Ellis.)

11.102 Ellis, "Stress Determination by Brittle Coatings," *ASME Paper* No. 47–SA–11, 1947. See also (1) publications of Magnaflux Corp., 5900 N. W. Highway, Chicago; (2) Ellis and Stern, *Experimental Stress Analysis* I, 1943 and VIII, 1945; (3) "How to Organize for Experimental Stress Analysis," *Prod. Eng. 19*, April (1948); (4) De Forest and Ellis, *J. Appl. Mech.* (ASME) *9*, A-184 (1942).

11.103 Racine *et al.*, "The Development of a Non-Flammable Resin System for Engineering Stress-Strain Applications," *Gen. Motors Eng. J. 12*, Fourth Quarter, 8 (1965). (Eliminates both the fire hazard and toxic elements of earlier lacquer systems. New resin is called *Strain tech*. Constituents not given. Presumably available from G.M.C.)

11.104 Bulletins of the Magnaflux Corp., Chicago, Ill., specialists in equipment for inspection and stress analysis.

Experimental Stress Analysis — Photoelasticity
(See also refs. 11.211, 11.247.)

11.105 Coker and Filon, *Treatise on Photoelasticity*, Cambridge University Press, 1931.
11.106 Frocht, "Factors of Stress Concentration Photoelastically Determined," *J. Appl. Mech.* (ASME) 2, A-67 (1935).
11.107 Oppel, "Photoelastic Investigation of Three Dimensional Stresses," *NACA TM* 824.
11.108 Hetenyi, "Photoelastic Studies of Three-Dimensional Stress Problems," *5th International Congress for Applied Mechanics*, Cambridge, Mass., 1938, John Wiley & Sons, New York, 1939.

Experimental Stress Analysis — Strain Gauges
(See also refs. 11.211, 11.713.)

11.110 Gorton and Pratt, "Strain Measurements on Rotating Parts," *SAE Quart. Trans. 3*, 540, Oct. 1949. (Technique with strain gauges, details of commutators, circuits, and some results. Bibliography.)
11.111 Dutee *et al.*, "Operating Stresses in Aircraft-Engine Crankshafts and Connecting Rods. I. Slip-Ring and Brush Combinations for Dynamic-Strain Measurements," *NACA Wartime Report* No. E-187, 1943. (Contains performance data on brush and slip-ring materials.)
11.112 For mechanical strain gauges see refs. 11.211 and 11.214.

Experimental Stress Analysis — Miscellaneous Techniques

11.115 Larkin, "Determination of Crankshaft Stresses by Means of a Rubber Model," S.M. Thesis Department of Aeronautical Engineering, Massachusetts Institute of Technology, Cambridge, Mass., 1934.
11.116 Kirkland, "Rubber Models Insure Accurate Stress Analysis," *Machine Design 7*, Jan. 1935.
11.117 Seely and Dolan, "Stress Concentrations as Found by the Plaster Model Method," Univ. of Illinois, Eng. Exp. Station, *Bull.* 276, June 18, 1935.
11.118 Barrett and Gensamer, "Stress Analysis by X-Ray Diffraction," *Physics 7*, 1 (1936).

Stresses in Engine Parts
(See also references to stress measurements on individual parts listed hereafter.)

11.119 Endurance Testing of Full-Scale Engine Parts. See refs. 9.2085, 9.212–9.214, 11.2131, 11.240, 11.246, 11.248, 11.600, 11.701, 11.706, 11.709, 11.710.
11.200 For data on stress concentration in various standard shapes—fillets, holes, shaft steps, etc., see refs. 9.300–9.303.

11.201 Caminez and Iseler, "Standard Method of Engine Calculations," Air Service Information Circular V, No. 421, April 1, 1923. (Classic arbitrary methods of calculation by tension, torsion, and beam theory.)

11.202 "Stresses in Combustion Engines" (Abstract from *ATZ* (Berlin), Sept. 1930), *J. Roy. Aeron. Soc. 35*, Nov., 1084 (1931).

11.203 Schilling, "Operational Stress Data," *Jour. SAE 59*, May 1951, 36. (Calculated working stresses used by G.M.C. for engine and chassis components.)

11.2031 Noll and Lipson, "Allowable Working Stresses," *Proc. SESA 3*, 89 (1946).

11.204 Taylor, "Critical Stresses in Aircraft Engine Parts," *Trans. SAE 37*, 412 (1935). (Early article calling attention to the importance of stress concentrations and fatigue failure in engine design.)

11.205 Kraemer, "Heat-Stressed Structural Components in Combustion-Engine Design." (From *ZVDI*, *82*, Mar. 1938, 12; English translation: *NACA TM 875*, Sept. 1938.)

11.206 Law, "New Techniques in Stress Analysis," *SAE Paper* No. 255B, 1960.

11.207 Burke, "Designing Shafts for Finite Life," *SAE Preprint* 317A, 1961. (Procedure based on Goodman diagram and data on life versus stress. Useful especially for parts with infrequent cycles of high stress.)

11.208 Holliday, "Use of Allowable Stress Diagrams for Designing Automotive Parts," *SAE Paper* 572E, 1962. (Method of using stress data in design. Example of Chrysler connecting rod.)

11.209 Graham, "Use of Cumulative Damage in Designing to Resist Fatigue," *SAE Paper* 572F, Sept. 1962. (Method of designing for limited life.)

11.210 Hoffmeister, "Application of Cumulative Fatigue Damage Theory to Practical Problems," *Trans. SAE 68*, 274 (1960). (Design of parts subjected to a limited number of high-stress cycles.)

11.211 *Experimental Stress Analysis*, Proc. Soc. for Experimental Stress Analysis, Addison-Wesley Press, Inc., Cambridge, Mass. (Periodical devoted to reporting methods and results. Much work on engine parts during the period 1943–1948.)

11.212 Fuller and Stimson, "Experimental Techniques in the Stress Analysis of Automotive Components," *SAE Paper* 532, 1955. (Useful description of Ford techniques in photoelastic, stress-coat, and strain-gauge measurements.)

11.213 Geschelin, "Continental Finds Stress-Coat Analysis Valuable," *Auto. Ind.*, Aug. 15, 1946. (Examples on crankshaft, gear, connecting-rod fillet, and elimination of ribbing in a crankcase casting.)

11.2131 "Symposium on Testing of Parts and Assemblies," *ASTM Tech. Pub.* 72, June 26, 1946. (Contains examples of crankshaft, rod, crankcase, bearing cap, automobile rear axles.)

11.214 Gadd *et al.*, "Stress Concentration and Fatigue Strength of Engine Components," ibid., p. 76. (Comparison of local stress as measured by $\frac{1}{8}''$ or $\frac{1}{16}''$ extensometers with endurance limit data. Crankshaft, rod, and plain shafts used as examples. Good correlation. Allowance for size and variability is made.)

11.215 Lipson, "Methods of Stress Determination in Engine Parts," *Jour. SAE 51*, April 1943, 105. (Photoelastic and strain-gauge techniques, with examples from Chrysler laboratories.)

11.216 Deardon, "Residual Thermal Stress in Compression Ignition Engines," *J. Brit. Cast Iron Res. Assn. 9*, 540 (1961).

11.217 Anderson, "Improving Engine Parts by Direct Measurement Strain," *Trans.*

SAE 54, 466 (1946). (Measurements on radial-engine crankcase, theory of bolt stresses under varying loads, prestress effects, piston and cylinder-head stresses, Goodman diagrams for various materials.)

11.218 Stark, "Short Range Telemetry System Provides Test Data on Rotating Parts," *Gen. Motors Eng. J. 12*, First Quarter, 23 (1965). (Picking up signals from strain gauges, etc., by radio.)

11.219 Gorton and Pratt, "Strain Measurements on Rotating Parts," *SAE Quart. Trans. 4*, Oct. 1949, 540. (Details of strain gauge attachment, slip rings, etc., on Pratt and Whitney aircraft engines.)

Standards and Specifications for Screw Fastenings

11.230 *SAE Handbook*, ref. 9.00, published annually by SAE. (Contains standards for screw threads, bolts, nuts, studs, screws, rivets, bearings, etc.)

11.231 "International Unified Accord on Screw Thread Standards," *Auto. Ind.*, Jan. 1, 1949, 26. (Announcement and details.)

11.2311 Peterka, "Bolts, Nuts, and Screws," Bulletin of the Lamson & Sessions Company, Cleveland, Ohio, 1944.

11.232 Stewart, "Applications, Materials, and Specifications of Bolts," *SAE Quart. Trans. 2*, July 1948, 413.

11.233 Soled, *Fasteners Handbook*, McGraw-Hill Book Company, New York, 1957. (Covers all kinds of fastenings including screws of all kinds, rivets, clamps, washers, locking devices, etc. Convenient for reference.)

11.234 "Aero-Thread Screw Thread System," Bulletin of Aircraft Screw Products Co., Inc., Long Island City, N.Y. (Details and specifications of the *Heli-coil* system of wire thread inserts for studs and screws in aluminum.)

Design of Screw Fastenings

11.235 *Buick Fastener Manual*, General Motors Corp., Buick Motor Division, Flint, Mich., 1963. (Theory of bolt load versus load on parts joined. Practical data on thread size, torquing, plating, etc.)

11.236 Geschelin, "A Guide to Automotive Fasteners," *Auto. Ind.*, Oct. 15 (1963). (Summary of Buick Fastener Manual noted above.)

11.237 Sorensen *et al.*, "New Fastener Tests Strengthen Joint Reliability," *Jour. SAE 73*, July 1965, 50. (Résumé of current testing methods for strength in fatigue, locking and loosening criteria, effect of hardness of the steel, etc., bibliography.)

11.238 La Belle, "Brittleness in Bolts and Other Threaded Fastenings," *SAE Paper 824C*, 1965. (Relationship of hardness to durability in fatigue.)

11.239 Waltermire, "Torquing for Reliability," *SAE Paper 824F*, 1965.

11.240 Clark, "Static and Dynamic Tests of Self-Locking Bolts," *Trans. SAE 73*, 39 (1965). (Experiments on loosing of standard and slotted-head bolts.)

11.241 Almen, J. O., "On the Strength of Highly Stressed, Dynamically Loaded Bolts and Studs," *Trans. SAE 52*, 151 (1944). (Excellent article from the practical design point of view. Limitations of torquing discussed.)

11.242 Dinner and Felix, "Rolling of Screw Threads," *Sulzer Tech. Rev.*, No. 1, 131

(1945). (Shows superior fatigue strength and desirable grain arrangement with rolled threads.)

11.243 Stewart, "Applications, Materials, and Specifications of Bolts," *SAE Quart. Trans. 2*, July 1948, 412. (Properties of various grades. Distribution of stress in threads, etc.)

11.244 Thurston, "The Fatigue Strength of Threaded Connections," *Trans. ASME 73*, Nov. 1951, 1085. (Good general discussion — some data.)

11.245 Thum and Staedel, "On the Endurance Strength of Screws and the Influence of Their Shape" (Über die Dauerfestigkeit von Schrauben in ihrer Beeinflussung durch Formgebung), *Maschinenbau 11*, June 2, 230 (1932). Abstract of above (in English) in *Metals and Alloys*, Sept. 1933, 286.

11.246 Staedel, "Fatigue Strength of Screws," *Mitt. Deutsch. Metallprüfungsanstalt 4*, Darmstadt, 1933. (In German. Repeated impact tests with varying necking diameters. For results see ref. 9.101, p. 125.)

11.247 Hall, "Determination of Stress Concentration in Screw Threads by the Photo-Elastic Method," Univ. of Illinois, Eng. Exp. Station, *Bull.* 245, 1932.

11.248 Henwood and Moore, "The Strength of Screw Threads Under Repeated Tension," Univ. of Illinois, Eng. Exp. Station, *Bull.* 264, March 13, 1934. (Endurance tests with various thread profiles, hardness, etc. Bibliography.)

11.249 Wiegand, "Fatigue Properties of Bolts and Nuts in Dependence on the Shape of the Nut" (Die Dauerfestigkeit der Schraube in Abhängigkeit von der Mutterform), Schriften der Hessichen Hochschulen, 1933, No. 2, p. 67. Abstract of above (in English) in *Metals and Alloys*, May 1934.

11.250 "Torquing of Nuts in Aircraft Engines," *SAE Special Publication*, Dec. 1945. (Tests and recommendations for determining nut tightness. Due to variations in friction, torque wrench gives a poor measure of nut tightness. Only reliable criterion appears to be stretch of bolt.)

11.252 "Screws and Fastenings," *Prod. Eng. 15*, Dec. 1944. (Describes various types and locking devices.)

Cylinder and Cylinder-Head Design

For effects of combustion-chamber design on the combustion process, see refs. 1.90 *et seq.* and 2.60 *et seq.* for spark-ignition engines and 3.650 *et seq.* for Diesel engines; also ref. 11.002, Nos. 119 and 137.

11.300 For design of cylinder structures incorporated in the crankcase-cylinder structure, see refs. 11.80–11.891.

11.301 Klinge, "Air-Cooled Engine Design Criteria," *SAE Special Publication* SP-152, 1958. (Introduction to the subject, with illustrations of current practice in aircraft and military engines at Continental Aviation and Engineering Corp. Chiefly concerned with cylinder and fin design. See also ref. 11.013.)

11.302 For large air-cooled aircraft cylinder designs, see refs. 10.844, 10.847, and 11.012. For large liquid-cooled aircraft cylinders, see refs. 10.845, 10.846, 10.848.

11.303 Cunningham, "High-Conductivity Cooling Fins for Aircraft Engines," *Trans. SAE 53*, 742 (1945). (Wright Aeronautical Corp., methods of applying fins to cylinder barrels.)

11.3031 For excellent fundamental research on cooling-fin design for air-cooled cylinders, see the following publications of NACA (now NASA) Washington, D.C.:

Technical Reports 488, 511, 555, 612, 676, 726; Technical Notes 331, 429, 602, 621, 649, 779; ARR by Silverstein, Jan. 1943; ACR by Brevoort *et al.*, Feb. 1943; ARR by Maurice *et al.*, Mar. 1943; ARR 3H16, Aug. 1943.

11.3032 Piry, "Cooling Characteristics of Steel and Aluminum Finned Cylinder Barrels," *Trans. SAE 53*, 630 (1945). (Useful data from tests of an aircraft-type cylinder.)

Port Design for Swirl, 4-Cycle Engines

11.304 Meurer, "The Generation, Through Inlet Ports, of Rotating Movements of the Air in the Cylinders of High-Speed Four-Stroke Diesel Engines," M.A.N. Forschungsheft, 1951. (In German. English translation in Texaco Laboratories Library, Beacon, N.Y. Steady flow work on port and valve design for high swirl efficiency.)

11.305 Watts, "Diesel Engine Port Design," *Automotive Design Engineering* (London), March 1964. (Excellent general summary by Ricardo engineer, but suffers from lack of dimensionless parameters.)

11.306 Other references on swirl: Dicksee, *Automobile Engineer* (London), April 1942, 159; Hurley and Cook, *Engineering* (London), Sept. 5, 1930, 290; abstracted in *Auto. Ind.*, Sept. 27, 1930, 446; Chorlton, *Engineering* (London), Feb. 26, 1932, 237; Davis, *Engineering* (London), Jan. 9, 1931, 57.

Two-Cycle Cylinder Design

See Volume I, Chapter 7 and references, and in Volume II refs. 10.37, 10.392, 10.394, 10.396, 10.397, 10.3991, 10.3992, 10.42, 10.48, 10.49, 10.57, 10.59, 10.591–10.597, 10.68, 10.692, 10.694, 10.696, and 11.002, Nos. 137, 143, 152.

Cylinder Materials

(See also refs. 9.06, 9.07, and 11.002, No. 224.)

11.310 Eagan, "Wear Resistance of Gray Iron Diesel Engine Liners," *The Foundry*, 1955. (Excellent study of the influence of the metallurgical structure of cast iron on wear.)

11.311 Moore *et al.*, "Heat Transfer and Engine Cooling, Aluminum versus Cast Iron," *Trans. SAE 71*, 152 (1963). (Study of engine temperature, heat rejection, and octane requirement, comparing "all-aluminum" and "all-cast-iron" engines in cars on the road. Temperatures lower (30° to 100°F) in aluminum engine, but octane requirement the same.)

11.312 Whitfield and Sheshunoff, "Al-Fin Process Bonds Aluminum to Steel," *Automotive and Aviation Industries*, Mar. 15, 1944, 148. (Description of popular method of bonding, much used for finning aircraft cylinders.)

11.313 Whitfield, "Metallurgical Bonding in Cylinder Construction," *SAE Paper* 369A, 1961.

11.314 *SAE Paper* 369C, 1961, p. A11; *Auto. Ind.*, Feb., 32 (1951). (Coating of aluminum cylinder bores.)

11.315 *SAE Paper* 369B, 1961, p. A11; *SAE Paper* 640782, 1965, p. A11. (Chromium-plating of cylinder bores.)

Piston Design
(See also ref. 11.002, Nos. 119, 148, 280, 283.)

11.400 Winship, "Designing an Automotive Piston," *SAE Paper* 670020, Jan., 1967. Also published as part of *Special Publication* SP-283, Pt. 14, 1966 (ref. 11.002). (Design details, materials, clearances, ring grooves, etc., from experience of Bohn Aluminum and Brass Co., large manufacturers of aluminum pistons.)

11.401 Robinson, "Piston-Boss and Wrist Pin Design," *ASME Paper* 65–OGP–4, 1965. (Excellent experimental study of stress distribution in these parts, together with design recommendations. References to previous work in Germany and England.)

11.402 Cavileer, "Piston Design Improvement Through Research Investigations," *SAE Preprint* 636B, 1963. (Work at U.S. Naval Experiment Station on ring wear, piston temperatures, and temperature control by design changes. Rotatable cast-iron pistons in opposed-piston submarine engine.)

11.403 "Basic Production Methods for Aluminum Pistons," *Auto. Ind.* (in three parts) June 1, 15, and July 1, 1954. (Useful information based on Alcoa symposium on aluminum pistons.)

11.404 Geschelin, "Piston Survey," *Auto. Ind.*, June 15, 1961, 80. (Review of some current practice in piston design for Diesel and spark-ignition automotive engines. Shows typical sections and various arrangements for controlling piston expansion and ring-groove wear.)

11.405 Holcombe, "Piston Design," *SAE Special Publication* 119, 1953. (Discussion of basic factors and 1953 practice.)

11.406 Renfrew, "Trends in Piston Design," *SAE Special Publication* 148, 1957. (Brief general discussion for small pistons.)

11.407 Mahle and Rohrle, "Technical Status of Pistons and Piston Rings in German Passenger Automobiles," *ATZ* (Berlin), No. 11, 310 (1958). (In German.)

11.408 Mahle, "Technique of Pistons, Piston Rings, and Pins in German Cars," *ATZ* (Berlin), No. 12, 343 (1958). (In German.)

11.409 Mahle and Rohrle, "Comparison of Pistons and Rings in German and American Cars," *ATZ* (Berlin), 162 (1959). (In German. Includes dimensions and clearances.)

11.410 Cross, "Gudgeon Pin Location," IME (London), *Proc. Automobile Div.* 1956–1957, p. 193. (Various methods of restricting endwise motion of the piston pin, with critical comments.)

11.411 Bush and London, "Design Data for 'Cocktail Shaker' Cooled Pistons and Valves," *SAE Paper* 650727, Oct. 1965. (Mathematical approach to the problem of heat transferred by a liquid partially filling a cylindrical space. Correlation of theory with laboratory experiments. Sample solution for a piston. Ten-item bibliography.)

11.412 Newton, "Stroboscopic Photography of Piston Cooling in Action," *SAE*

Preprint 7B, Jan. 1958. (Photographs of oil flow to the underside of the piston using a motored engine model.)

11.413 Gross-Gronomski, "Piston Heating in a Diesel Engine," *SAE Preprint* 1005A, Jan. 1965. (Temperature measurements as a function of location on piston and operating variables. 7.7 × 10.25 in. cylinders with open chamber. Bibliography.)

11.414 Wallace and Lux, "A Variable Compression Ratio Engine Development," *Trans. SAE 72*, 680 (1964). (Height of piston crown above piston-pins hydraulically controlled. Engine tests show considerable gain in light-load economy. Historical survey of other mechanisms.)

11.415 Paul and Humphreys, "Humphreys Constant Compression Engine," *SAE Quart. Trans. 6*, April, 259 (1952). (Description and some performance curves on engine with spring-loaded cylinder head. See also ref. 13.01.)

11.416 Gross-Gronomski, "Piston Temperature in Compression-Ignition Engines," *SAE Paper* 670469, 1967. (Test results and correlations on several engines. Speed effects at constant power as well as at constant bmep. Bibliography.)

Piston Rings
(See also ref. 11.002, Nos. 224, 271.)

11.420 Prasse *et al.*, "Automotive Piston Rings 1967 State of the Art," *SAE Paper* 670019, Jan. 1967. Also published as part of SP-283, Part 14, 1966. (Materials, design, surface finish, and coatings. Experience of Thompson-Ramo-Wooldridge.)

11.4201 Binford, "Piston Ring Designs — Have They Changed?" *SAE Paper* 650483, 1965. (Latest practice, materials, designs, methods of manufacture, current state of the art for passenger automobile engines.)

11.421 "Piston Rings for Transportation Diesels," *Trans. SAE 69*, 283 (1961). (Summarizes current practice in design and materials.)

11.422 Harris, "Piston Ring Design and Application," *SAE Special Publication* 148, 1957. (Brief discussion, illustrated, by Perfect-Circle engineer. Includes iron structure, chromium plating, steel oil rings.)

11.423 Tales, "Recent Developments in Piston-Ring Materials," *Trans. SAE 45*, 49 (1939). (Effects of iron structure and composition, piston coatings, etc., on corrosion, bore wear.)

11.424 Lane and Nixon, "Engineering Piston Rings for High-Speed Diesels," *Trans. SAE 50*, 528 (1942). (Design data for rings and piston-ring belts. Shows marked advantage of narrow ring.)

11.425 Piston rings for large aircraft engines are discussed in refs. 10.844–10.848.

Connecting-Rod Design
(See also refs. 11.013, 11.101, and 11.002, No. 148.)

11.600 Fetter, "Connecting-Rod Fatigue Testing," *ASME Paper* 65–OGP–3, April, 1965. (Fatigue tests on complete rod with emphasis on effect of bolt tension. This effect is small as long as there is no separation.)

11.601 Taylor, "New Data on Bending Moments in the Master Connecting-Rod," Automotive Research, *Jour. SAE 33*, June 1933, 26. See corrections in July,

1933 issue. Also published in *Automobile Engineer* (London), Sept. (1933). (Computations of bending due to action of articulated rods in radial engines.)

11.602 Kalelkar, "Stress Concentration in the Eye Section of a Connecting Rod," S.M. Thesis, Department of Mechanical Engineering, Massachusetts Institute of Technology, Cambridge, Mass., July 1941. (Photoelastic, stress-coat, and theoretical analysis of a connecting rod upper end. Review of the literature. Bibliography. Exhaustive on one design only, tensile loading only.)

11.603 "Operating Stresses in Aircraft Crankshafts and Connecting Rods. II— Instrumentation and Test Results," *NACA Wartime Report* No. E-191, 1943.

11.604 Lovesey, "Development of Rolls-Royce Merlin from 1939–1945," *Aircraft Eng.* (London), July 1946, 218. (Performance, cross-section drawings, connecting-rod stress analysis and improvement of forked-and-blade rods.)

Crankshaft Design

11.700 Matinaglia, "The Structural Durability of Crankshafts," *Sulzer Tech. Rev.*, No. 2 (1943). (Excellent introduction to the principles of crankshaft design. Results of experimental stress analysis. Detailed design of forged, cast, and built-up shafts. Required reading.)

11.701 Gadd, "Full Scale Testing of Crankshafts," *Proc. SESA*, 150 (1944).

11.702 Williams and Brown, "Fatigue Strength of Crankshafts," *Engineering* (London), *154*, 58 (1942).

11.703 Love, "Cast Crankshafts," *J. Iron Steel Inst.* (London), *159*, 247 (1948). (A survey of published information.)

11.7031 Mills and Love, "Fatigue Strength of Cast Crankshafts," IME (London), *Proc. Automobile Div.* 1948–1949, p. 81.

11.704 Lowell, "Practical Applications of Crankshaft-Geometry Theory," *ASME Paper* 64–WA/OGP–5, Dec. 1964. (Measurements of fillet stress as a function of design ratios confirm the Kano, ref. 11.709, formulae.)

11.705 Harker, "Determination of Engine Crankshaft and Connecting Rod Loading," *SAE Special Publication* SP-148, 1957.

11.706 Lurenbaum, Einfluss von Formgebung und Werkstoff auf die Gestaltfestigkeit geschmiedeter und gegossener Flugmotoren-Kurbelwellen," Jahrbuch 1937 der deutschen Luftfahrtforschung, II/129. (Fatigue strength of cranks as a function of design. Includes picture of torsional testing machine on which this famous work on cranks was done. Graphs give relative strength of different crankshaft materials. Source of Fig. 11–36.)

11.707 Gadd and Van Degrift, "A Short-Gage-Length Extensometer and Its Application to Study of Crankshaft Stresses," *J. Appl. Mech.* (ASME) *9*, March, A-15 (1942). (Source of Fig. 11–37. See also ref. 11.701.)

11.708 Gassner and Schutz, "The Fatigue Strength of Vehicle-Engine Crankshafts," *Engineers Digest 22*, Sept., 79 (1961).

11.709 Kano, "Influence of Geometrical Design Factors on the Bending Fatigue Strength of Crankshafts," *Trans. ASME*, Ser. A, *J. Eng. Power 85*, 177 (1963). (Fatigue tests in bending of full-scale single cranks. Results well presented and correlated.)

11.710 Cornelius, "Versuchsmethoden und Versuchseinrichtungen für Kurbelwellen schnell-laufender Verbrennungs-Maschinen," *ATZ* (Berlin), Vol. 42, April 10,

1939, p. 90. See also, by same author, *ibid.*, July 25, 1939, p. 385, and ZVDI, July 23, 1938, article by Cornelius and Heinrich. (Method and test results on crankshaft strength under torsional vibration. Design and materials varied. Case hardening and nitriding included.)

11.711 Walstrom, "Measurement of Operating Stresses in an Aircraft Engine Crankshaft under Power," *NACA Advance Restricted Report* E5B01, February 1945, File Number E-41. (Released from restriction 1946.)

11.712 Kleiner, "Torsional Vibrations and Fracture of Crankshafts," *Sulzer Tech. Rev.*, No. 4 (1943). (Valuable data on crankshaft failures and how to avoid them by proper design. Effect of grain structure on strength. Unusual designs for cast crankshafts.)

11.713 Goloff "Determination of Operating Loads and Stresses in Crankshafts," *Proc. SESA*, Vol. II, No. 2, 1944, p. 139. (Caterpillar Co. methods of using strain-gauge measurements in an operating engine. Particular attention to stresses in torsional vibration.)

11.714 "The Tangential Rolling of Crankshaft Fillets," *Gen. Motors Eng. J. 11*, Third Quarter, 42 (1964). (Methods and strength improvement.)

11.715 Suppes, *Gen. Motors Eng. 9*, Fourth Quarter, 21 (1962); Anon, *Auto. Ind.*, Feb. 15, 55 (1963); Valentine, *SAE Preprint 15*, Jan. 1957. (Malleable iron crankshafts.)

11.716 Gadd and Ochiltree, "Full Scale Fatigue Testing of Crankshafts," *Proc. SESA*, Vol. II, No. 2, 1944, p. 150. (Equipment and methods, characteristic failures, effects of stress relief, nitriding, shot blasting.)

11.717 Frey, "The Metallurgy and Processing of the Packard-Built Rolls-Royce Merlin Crankshaft," *Proc. SESA*, Vol. II, No. 2, 1944, p. 158. (Material is 0.40 per cent carbon chrome-nickel-molybdenum steel. Describes entire process including forging, machining, heat treatment, nitriding, and polishing.)

11.718 Oldberg and Lipson, "Structural Evolution of a Crankshaft," *Proc. SESA*, Vol. II, No. 2, 1944, p. 118. (Describes design process including calculations, experimental stress analysis and fatigue testing. Shows beneficial influence of barrel boring versus straight boring.)

Crankcase Design
(See also ref. 11.002, No. 159, and refs. 11.012, 11.013.)

11.80 McKellar, "A Study of the Design of Sand-Moulded Engine Castings," *Gen. Motors Eng. J. 3*, March–April 1956, 12. (Review of G. M. methods and techniques.)

11.81 Bauer, "Engine Blocks and Their Components in Aluminum Die Casting," *Trans. SAE 68*, 385 (1960). (Current practice in water-cooled blocks including various types of wear-resistant cylinder bores and useful design examples.)

11.82 Vandeventer and McFarland, "Measurement and Control of Residual Stresses in Cylinder-Block Castings," *Trans. SAE 62*, 68 (1954). (Shows importance of avoiding designs which give large differences in casting cooling rate.)

11.83 Geschelin, "Continental Finds Stresscoat Analysis Valuable in Redesigning Engine Parts," *Aut. Ind.*, Aug. 15, 1946. (Includes dramatic improvement in crankcase design through elimination of ribs. See Fig. 11–41.)

11.84 Stroebel, " Measurement and Interpretation of Stress in an Aluminum Engine,"
       SAE Paper 494B, 1962. (Stresscoat and strain-gauge measurements at critical
       points of the main block casting. Thermal stresses included. Good discussion
       and interpretation.)

11.85 Ravitch, "Some Applications of Stress Analysis Techniques in Improving
       Casting Designs," *Gen. Motors Eng. J. 5*, Fourth Quarter, 22 (1958). (Stress-
       coat and results of application.)

11.86 Van Camp, "Stress Analysis of Aluminum V-8 Cylinder Block," *SAE Paper*
       255A, 1960. (Shows methods and location of critical stresses.)

11.87 Steel crankcase for aircraft engines, see ref. 10.847.

11.88 Young, "Fabrication of Welded Steel Crankcase for Large Two-Cycle Diesel
       or Natural Gas Engine," *Gen. Motors Eng. J. 3*, Third Quarter (1956). (Details
       of crankcase for G. M. 16-cylinder radial engine of ref. 10.75.)

11.89 For fabricated steel crankcases for large marine engines, see refs. 10.56, 10.57,
       10.594, and 10.595.

11.891 For fabricated steel crankcases for locomotive Diesel engines, see refs. 10.401
       et seq.

11.892 Owens, "The Evolution of Diesel Engine Block Weldment Design and Fabri-
       cation," *Welding Journal*, March 1947. (Fairbanks, Morse & Co. practice.)

Bearings, Journal

For theory of lubrication and friction of bearings, see Volume I, Chapter 9,
and refs. 9.01–9.502. The literature on bearings is enormous. The following
references have been selected as being particularly helpful to the engine designer.
(See also ref. 11.002, No. 283.)

11.900 " Plain Bearing Failures," *Lubrication 50*, July (1964). (Review of typical failures
       and their causes. Useful introduction to bearing design.)

11.901 Haugen, "A Review of Design, Material and Performance of Automotive
       Engine Hydrodynamic Journal Bearings," *Gen. Motors Eng. J. 10*, Fourth
       Quarter (1963). (Useful data on materials and design from G. M. experience.
       Short bibliography.)

11.902 Etchells, " How Engineers Select Materials for Oil-Film Bearing Applications,"
       Gen. Motors Eng. J. 1, March–April, 1954. (Chevrolet practice of that date.
       Useful basic and applied discussion of bearing materials and bearing structure.
       Bibliography.)

11.903 Crankshaw and Osters, " Practice Makes the Bearing," *SAE Special Publication*
       SP-148, 1957. (Good practical discussion of materials and detail design of plain
       bearings.)

11.904 Forrester, "How to Choose Materials for Bearing Surfaces," *Eng. Mater.
       Design* (London), *2*, 494 (1959).

11.905 *Automotive Sleeve Type Half Bearing Design Guide*, booklet from Federal-
       Mogul-Bower Bearings, Inc., 1956. (Materials, dimensions, tolerances, stand-
       ards for Babbitt and copper-lead bearings.)

11.906 Duckwater and Walter, " Fatigue of Plain Bearings," *Proceedings International
       Conference on Fatigue of Metals*, IME (London), 1956, p. 585.

11.907 Crankshaw and Savage, "Journal Bearing Design," SAE Special Publication, refs. 11.002, No. 122, abridged in *Jour. SAE 61*, Oct., 26 (1953). (Comparison of materials, advice on detail design.)

11.908 Schager, "Sleeve Bearings — Design, Manufacture, Installation," *SAE Quart. Trans. 6*, 165 (1952). (Excellent treatment of materials and design fundamentals from experience of Cleveland Graphite & Bronze Co., large manufacturers of bearings.)

11.909 "Engine Bearings," *SAE Special Publication* SP-274. (Three 1964 papers and discussions. Aluminum-on-steel bearings, imperfect journal geometry.)

11.910 McNall and Beardsley, "Gas Engine-Driven Compressor Bearing Evaluation," *ASME Paper* 63–OGP–3, 1963. (Clark Bros. experience with solid aluminum bearings in large engine-compressor units.)

11.911 Frank and Lux, "A Billion Engine Hours on Aluminum Bearings," *Trans. SAE 64*, 655 (1956). (Caterpillar experience with solid and steel-backed bearings. Useful design data.)

11.912 Shaw, "Properties and Production of Aluminum Bearings," *Gen. Motors Eng. J. 1*, June–July, 1953. (Experiences of Moraine Products Division of G. M. Co. showing high load-carrying ability of steel-aluminum bearings with very thin overlay of Babbitt.)

11.913 Hunsicker, "Aluminum Alloy Bearings — Metallurgy, Design and Service Characteristics," reprint from *Sleeve Bearing Materials*, ASM, 1949. (Technology of aluminum bearings as of 1949. Extensive bibliography.)

11.914 Earlier Articles on Journal Bearings, Still Useful.

Author	Publication
a. Anon	*Lubrication 39*, May, 1953.
b. Crankshaw	*SAE Paper*, Feb. 10, 1953.
c. Stokely	*SAE Quart. Trans. 3*, 319 (1949).
d. Raymond	*SAE Trans. 50*, 553 (1942).
e. Tichvinsky	*SAE Trans. 51*, 69 (1943).

Oscillating Journal Bearings

(See also ref. 11.924.)

11.920 Underwood and Roach, "Slipper-Type Bearings for Two-Cycle Diesel Connecting Rods," *SAE Paper* 715, Jan. 1952. (Experience with G. M. radial Diesel and gas engines, see ref. 10.75.)

11.921 Burwell, "The Calculated Performance of Dynamically Loaded Sleeve Bearings — II," *J. Appl. Mech.* (ASME) *16*, Dec. 1949, 358. (Theory indicates oil film cannot be maintained in oscillating bearings with load always in one direction.)

Load Capacity of Journal Bearings

11.922 See Volume I, refs. 9.20–9.273, Volume II, refs. 11.900–11.914, and Table 11–11.

11.923 Dayton *et al.*, "Discrepancies Between Theory and Practice of Cyclically Loaded Bearings," *NACA TN* 2545, Nov. 1951.

11.924 Underwood, "Load-Carrying Capacity of Journal Bearings," *SAE Quart. Trans. 1*, 56 (1947). (Tests with rotating and oscillating loads. Measurements of oil-film thickness and load limits. Useful discussion.)

Bearings, Ball and Roller
(See also Volume I, refs. 9.35–9.352.)

11.950 For standards as to exterior dimensions, etc., see *SAE Handbook*, ref. 9.00. See also bulletins by the various manufacturers.

11.951 "Ball and Roller Bearings," *Lubrication 50*, Dec. (1964). (Good introduction to current practice.)

11.952 Barwell and Scott, "Effect of Lubricant on Pitting Failure of Ball Bearings," *Engineering* (London), *182*, 9 (1956).

11.953 Lieblein and Zelen, "Statistical Investigation of Fatigue Life of Deep-Groove Ball Bearings," *J. Res. Natl. Bur. Standards 57*, 273 (1956).

11.954 Hustead, "Consideration of Cylindrical Roller Bearing Load Rating Formula," *Trans. SAE 71*, 202 (1963). (Effects of design, press fit, materials, etc., on bearing life. Useful discussions.)

11.955 Lepeigneux, "Considérations sur les Roulements à Billes et à Rouleaux," *J. Soc. Ing. Automobile* (Paris), May, 2750 (1934). (Permissible loads, friction, choice of correct size, adjustment of antifriction bearings.)

11.956 "Anti-Friction Bearing Seminar," *SAE Special Publication* SP-168. (Several 1959 papers and discussion.)

11.957 Harris, "Rolling Bearing Analysis," John Wiley & Sons, Inc., New York, 1966. (From experience of SKF.)

Needle Bearings

11.960 Publications of Bantam Ball Bearing Co., South Bend, Ind.

11.961 *Auto. Ind.*, May 27, 641 (1933).

11.962 *Auto. Ind.*, Dec. 13, 869 (1930).

11.963 Pitner, "Les Roulements à Aiguilles et leurs Applications," *J. Soc. Ing. Automobile* (Paris), June 1930, p. 1051. (Comprehensive treatise on this type of bearing.)

11.964 Witte, "Das Nadellager," and "Die Nadellagerung von Kolbenbolzen," *Deutsche Motor-Zeitschrift*, 1929, Heft 10, p. 508, and Heft 11, p. 564. (Discussion of various applications to engines and vehicles.)

Nonmetallic Bearings

11.970 O'Rourke *et al.*, "Performance of Teflon Fluorocarbon Resins as Bearing Materials," *ASME Paper* 61–WA–334, 1961. (Data on the possibilities of this material for nonlubricated bearings, piston rings, etc.)

CHAPTER 12

Non-Poppet Valves

12.00 Hunter, "Rotary Valve Engines," John Wiley & Sons, Inc., New York, 1946. (Descriptions of many types—all now obsolete.)

Valve Flow Capacity
(See also Volume I, refs. 6.40–6.46.)

12.01 Tanaka, "Air Flow Through Suction Valve of Conical Seat," Part I, Experimental Research Report of Aeronautical Research Inst., Tokyo Imperial University, No. 50, Oct. 1929. (Steady flow tests. Exhaustive study of effects of valve-seat geometry on flow coefficient.)

12.02 Tanaka, Part II of above report "Analytical Investigation," Experimental Research Report of Aeronautical Research Inst., Tokyo Imperial University, No. 51, Nov. 1929.

12.03 Tanaka, "Air Flow Through Exhaust Valve of Conical Seat," Experimental Research Report of Aeronautical Research Inst., Tokyo Imperial University, No. 67, Jan. 1931. (Study of effects of exhaust-valve geometry.)

12.04 Hu, "Study of Exhaust Valve Design from the Gas Flow Standpoint," *J. Inst. Aeron. Sci. 11*, 13 (1944). (Report on tests using blowdown apparatus.)

Two-Cycle-Engine Porting
(See also Volume I, refs. 7.01 *et seq.*, and refs. 10.392, 10.3991, 10.594–10.596, 10.694–10.696, 10.71, 10.72, 10.75, 10.820.)

12.05 Nagao and Shimamoto, "Effect of the Induction Pipe of a Crankcase-Scavenged Two-Cycle Engine," *Trans. JSME 26*, 1675 (1960).

12.06 Nagao *et al.*, "Effect of Crankcase Volume on the Delivery Ratio of a Crankcase-Scavenged Two-Cycle Engine," *Trans. JSME 25*, 1959, p. 314; *Bull. JSME 2*, 578 (1959).

12.07 Watanabe *et al.*, "Effect of Exhaust-Pipe System upon the Performance Characteristics of a Two-Cycle Diesel Engine," *Trans. SAE 70*, 602 (1962). (Tests on a poppet-valve engine in series with a Roots blower, constant fuel quantity. Hard to interpret in general terms.)

12.08 Nagao, "The Effect of Crankcase Volume and the Inlet System on Delivery Ratio of Two-Stroke Cycle Engines," *SAE Paper* 670030, Jan. 1967. (Excellent presentation of results of varying dimensions of inlet pipe, exhaust pipe, port areas, etc.)

12.09 Tanaka, "Influence of Port Flow Coefficient on Performance of a Two-Stroke Through-Scavenged Engine . . .," S.M. Thesis, Department of Mechanical Engineering, Massachusetts Institute of Technology, Cambridge, Mass., Aug. 20, 1962. (Attempt to correlate 2-cycle scavenging efficiency with the inlet-port Z factor (Mach Index) as with 4-cycle engines. Port timing and inlet/exhaust flow-coefficient ratio held constant with three different port areas. Correlation poor, scavenging efficiency increasing with increasing piston speed at the same values of Z and R_s.)

12.091 Hayashi and Meyer, "An Analytical Method for Optimizing the Scavenging Process of Uniflow Two-Cycle Diesel Engines," *SAE Paper* 650447, May 1965. (Computer program based on simplified assumptions. Results not generalized.)

12.092 Blair, "Direct Evaluation of the Exhaust Lead of a Two-Stroke-Cycle Diesel Engine," *SAE Paper* 650728, 1965. (Computer program for optimizing exhaust-port lead.)

Valve-Gear Design, General
(See also ref. 11.002, No. 152.)

12.10 Tauschek, "Valve Train Engineering — Fundamentals for the Engine Designer," *SAE Special Publication* SP-152, 1958. (Excellent introduction to design problems, materials, and good practice by chief engineer of Thompson Products, Inc.)

12.11 Giles, "Fundamentals of Valve Design and Material Selection," *SAE Special Publication* SP-283, 1966. (Thompson-Ramo-Wooldridge experience.)

12.12 *Wil-Rich Forum*, Publ. by Eaton Mfg. Co., Detroit 13, Mich., especially the following issues:

Vol.	*Date*
II	March 1941; June 1941
III	Jan. 1942
IV	May 1943
V	Dec. 1944
VII	May 1946
VIII	June 1947
X	Nov. 1949; Dec. 1949
XI	March 1950
XV	April 1954

12.13 Colwell, "The Trend in Poppet Valves," *Trans. SAE 45*, 295 (1939). (State of the art in design, internal cooling, materials, etc., for aircraft-engine valves.)

12.14 Horan, "Overhead Valve-Gear Problems," *Trans. SAE 61*, 679 (1953). (Useful article oriented toward design. Discusses dynamics, materials, design details. Alignment chart for checking over-all valve-gear design. Bibliography.)

Exhaust Valves
(See also ref. 11.002, Nos. 119, 127, 152, 283.)

12.20 Johnson and Galen, "Diesel Exhaust Valves," *SAE Paper* 660034, Jan. 1966. (Late experience of Eaton, manufacturer of engine valves. Types of valve failure, temperatures including inlet valves, materials, fuel effects, valve testing, bibliography.)

12.21 Cherrie, "Factors Influencing Valve Temperatures in Passenger Car Engines," *SAE Paper* 650484, May 1965. (Experience of Thompson, manufacturer of valves. Effects of temperature on strength, corrosion, and life. Materials and their improvement with time. Effect of head shape, stem size, and clearance on temperatures. Bibliography.)

12.22 Tauschek, "Metallurgical Considerations for Automotive Exhaust Values," Thompson Products Engineering Bulletin, Vol. 1, No. 1, January 1956. (Data on temperature and other characteristics of valve materials.)

12.23 Colwell, "Modern Aircraft Valves," *Trans. SAE 35*, 147 (1940). (Current practice in materials and design. Internally cooled exhaust valves and their evolution. Valve temperatures and failures.)

Valve Seats, Guides, Followers, etc.
(See also ref. 11.002, Nos. 119, 127, 271.)

12.30   Chase, "Special Hardenable Iron for Tappets," *Auto. Ind.*, May 1 (1955). (Technical and metallurgical discussion of cast cam-follower materials.)

12.31   Valve-guide material and design — see ref. 10.847.

12.32   Valve-seat insert design — see ref. 10.847.

12.33   Newton and Tauschek, "Valve Seat Distortion," *SAE Paper* No. 64, March 1953. (Qualitative discussion. Effects of seat distortion, coolant boiling, etc.)

12.34   Chase, "Special Hardenable Iron for Tappets (cam followers)," *Auto. Ind.*, May 1 (1955). (Technical and metallurgical discussion of cast materials.)

12.35   Other articles on cam and follower design and materials: *Trans. SAE 64*, Etchells *et al.*, p. 161; Garwood *et al.*, p. 138; Laird *et al.*, p. 153.

12.36   Chase, "Cost Reduced with Stamped Rocker-Arms," *Auto. Ind.*, May 1952, 16. (Design and fabrication data.)

12.37   "Hydraulic Valve Lifters for Passenger Car Engines," *Lubrication 40*, May 1954. (Good brief review of current designs, lubrication, and maintenance.)

12.38   "Eaton Zero Lash Hydraulic Lifters," *Eaton Forum*, publ. by Eaton Mfg. Co., Detroit 13, Nov. 1949.

12.39   Young, "Aircraft-Engine Valve Mechanisms," *Trans. SAE 45*, 109 (1939). (Observations of false motion, limits of seating velocity, hydraulic lifters, failures, temperatures.)

Valve Gear Dynamics and Cam Design
(See also ref. 11.002, No. 178.)

12.40   "Application of Computers in Valve Gear Design," *SAE Technical Progress Series* TPS-5. (Digital and analog systems; friction and damping forces; correlation with test results.)

12.41   Hrones, "An Analysis of the Dynamic Forces in a Cam-Driven System," *Trans. ASME 70*, 473 (1948). (Effects of mass, elasticity, and damping on constant acceleration, sinusoidal and cycloidal cams. Theoretical.)

12.42   Taylor and Olmstead, "Poppet-Valve Dynamics," *J. Inst. Aeron. Sci. 6*, July, 370 (1939). (Early work done at MIT on the measurement, causes, and cure of false motion. Good introduction to the subject.)

12.43   Erisman, "Automotive Cam Profile Synthesis and Valve Gear Dynamics from Dimensionless Analysis," *SAE Paper* 660032, Jan. 1966. (Good mathematical analysis of the relations of valve motion and accelerations to cam profile and natural frequency of valve gear. Useful charts of design data from computer studies of typical examples. Spring design chart. Bibliography.)

12.44   Johnson, "Computers Help Design Valve Trains," *SAE Paper* 596C, 1963. (Computation of valve motion and effects of design variables for the exhaust valve of the G. M. Detroit Diesel 71 series 2-cycle engines.

12.45   Pechenik, "Special Problems in Designing Corvair Spyder Camshaft," *SAE Paper* 5358, Jan. 1963. (Measurements of valve motion for Chevrolet Corvair engine. Methods of cam design. Effects of variables on valve motion. See also *Jour. SAE 71*, 82, April 1963.)

12.46 Nourse *et al.*, "Recent Developments in Cam Design," *Trans. SAE 69*, 585 (1961). (Review of the "polydyne" system plus summary of the mathematics involved. No experimental results or comparisons with other types of cam.)

12.47 Roggenbuck, "Designing Cam Profile for Low Vibration at High Speeds," *Trans. SAE 61*, 701 (1953). (Valve gear motion represented by electrical analog. Not generalized. Ford practice.)

12.48 Barkan, "Calculation of High-Speed Valve Motion with a Flexible Overhead Linkage," *Trans. SAE 61*, 687 (1953). (Well-presented theoretical study with experimental comparisons. Computer computations check well with measurements. Friction factor estimations. Bibliography.)

12.49 Thoren *et al.*, "Cam Design as Related to Valve Train Dynamics," *SAE Quart. Trans. 6*, 1 (1952). (Measurements of valve motion. Development of the "polydyne" system of cam design.)

Valve Springs

(See also ref. 11.002, Nos. 29 and 152.)

12.50 Turkish, "Relationship of Valve-Spring Design to Valve Gear Dynamics and Hydraulic Lifter Pump-Up," *Trans. SAE 61*, 706 (1953). (Measurements of valve motion on an engine. Analysis and useful conclusions regarding cam and spring design. Shows differences in behavior favorable to sinusoidal rather than constant-acceleration curves. Good bibliography.)

12.51 Rickett and Mason, "Fatigue Properties of Springs," *Metal Progr. 63*, March, 107 (1953).

12.52 Bardgett and Gartside, "Fatigue of Coiled Springs," *Iron and Steel* (London) *24*, 375, 411, 454 (1951).

12.53 Coates and Pope, "Fatigue Testing of Coil Springs," *Proceedings International Conference on Fatigue in Metals*, IME (London), 1956, p. 604.

12.54 Geschelin, "Designing Valve Springs for High Performance Engines," *Auto. Ind.*, Jan. 15, 1965, 61; Feb. 1, 1965, 41. (Review of current practice in design, materials, and testing.)

12.55 Nixon, "Design of Valve Springs," *Aircraft Eng.*, Sept. 1933.

12.56 Marti, "Vibrations in Valve Springs," *NACA TM* 818, March 1937.

12.57 Donkin and Clark, "Valve Spring Surge," *Trans. SAE 24*, 185 (1929).

Gear Design, Materials

12.601 Hense *et al.*, "Automotive Gear Steels . . . How Industry Selects and Heat-Treats Them." (Excerpts from *SAE Preprint*, "Economics of Automotive Gear Steel and Their Heat-Treatment," June 7, 1950.) *Jour. SAE 58*, Oct., 25 (1950). (Tables of materials and heat treatment.)

12.602 Schwitter, "Two New Gear Materials," *Auto. Ind.*, Jan. 1, 1951, 42. (Ani trided steel and a nodular cast iron.)

12.603 Hense and Buswell, "A Discussion of Economic Factors Affecting the Steel Selection and Heat Treatment for Automotive Gears," General Motors Engineering Journal, Sept.–Oct. 1955.

12.604 Knowlton and Kincaid, "Induction-Hardened Gears," *SAE Quart. Trans. 4*, 116 (1950). (Procedures, results, precautions. See also Redmond, "High-Frequency Treatment of Gears," same publication.)

Gear Design, General

12.605 Dudley, "Gear Handbook," McGraw-Hill Book Company, New York, 1965. (Includes design and manufacturing data.)

12.606 Michaelec, "Precision Gearing, Theory and Practice," John Wiley & Sons, Inc., New York, 1966. (Recent, authoritative. Includes theory, materials, design, manufacturing.)

12.607 Black, "Fundamentals of Gear Design and Manufacture," *SAE Special Publication* SP-143, 1956. (Valuable practical discussion.)

12.608 "Involute Gear Geometry," *Trans. ASME 71*, July (1949). (Presentation by means of dimensional analysis.)

12.609 Heer, "Modifying Standard Proportions of Involute Gears," *Jour. SAE 57*, Nov. 1949, 19. (Objectives of modifications generally used.)

12.610 "Heavy Duty Gears," *Lubrication 40*, June 1954. (Valuable discussion of gear failures and gear lubrication. See also *Lubrication 36*, April 1950.)

12.611 Buckingham, "Analytical Mechanics of Gears," McGraw-Hill Book Company, New York, 1949; and "Operational Stresses in Automotive Gears," *SAE Quart. Trans. 5*, 43 (1951). (Stress distribution in gear teeth. Operation loads, stress computations.)

12.612 Dolan and Broghammer, "A Photoelastic Study of Stresses in Gear-Tooth Fillets," Univ. of Illinois Eng. Exp. Station, *Bull.* 31, Mar. 24, 1942. (Effects of tooth contour and fillet radius.)

12.613 "Stresses Due to the Pressure of One Elastic Solid Upon Another," Univ. of Illinois, Eng. Exp. Station, *Bull.* 212, 1930.

12.614 Love, "Bending Fatigue Strength of Carburized Gears," British Motor Industry Research Association, *Report* 1953/4, September 1953.

12.615 Coleman, "Improved Method for Estimating Fatigue Life of Bevel Gears and Hypoid Gears," *SAE Quart. Trans. 6*, 314 (1952). (Important experimental and theoretical approach.)

12.616 Van Zandt, "Beam Strength of Spur Gears," *SAE Quart. Trans. 6*, 252 (1952). (Tooth deflection and load distribution.)

12.617 Wellauer, "What Do Calculated Gear Stresses Mean?" *SAE Preprint* 806, 1956. (Valuable experimental and analytical approach.)

12.618 Nieman *et al.*, "Some Possibilities in Increasing the Load-Carrying Capacity of Gears," *Trans. SAE 71*, 169 (1963). (Comprehensive and well documented account of German work on many aspects of gear design. Important contribution, including discussions.)

12.619 Kelly, "The Design of Planetary Gear Trains," *Trans. SAE 67*, 495 (1959). (Useful engineering information, including design details, oriented toward vehicle automatic transmissions.)

12.620 "Heavy Duty Gears," *Lubrication 40*, June, 1954. (Types of heavy-duty gearing, types and causes of failure, lubrication and lubricants.)

Chains, Belts

12.700 "Chain Drives," *Lubrication 48*, June, 1962. (Good summary of chain design lubrication, failures, and their correction. Eighteen-item bibliography.)

12.702 Hofmeister, "Application of Cumulative Fatigue Damage Theory to Practical Problems," *SAE Paper* S163, 1959. (Formulas used by Chain Belt Co.)

12.703 See ref. 10.294 for description of a Neoprene rubber fibreglass reinforced toothed belt used to drive the overhead camshaft of the 1966 Pontiac 6-in-line engine.

12.704 Riopelle, "Designing Camshaft Chain Drives," *Auto. Ind.*, Oct. 15, 1948, 37. (Short article giving useful design information.)

Gaskets and Seals

12.710 Czernik *et al.*, "The Relationship of a Gasket's Physical Properties to the Sealing Phenomenon," *SAE Paper* 650431, 1965. (Cylinder-head gaskets for Diesel engines, theory, and practice. Various types compared.)

12.711 "Rubber, 'O' Rings for Automotive Seal and Packing Applications," *SAE Handbook*, J120A. (Standards, testing, and recommended practice.)

12.712 "Nonmetallic Gaskets for General Automotive Purposes," *SAE Handbook*, J90a. (Asbestos, cork, cellulose, etc. Standards and specifications.)

12.713 "Radial Seal Nomenclature," *SAE Handbook*, J111. (Includes useful design data.)

12.714 Kassebaum and Ogden, "Automotive Engine Head and Gasket Problems," *SAE Preprint* 335, August 1954.

12.715 Hepworth, "Sealing Rings for Wet Cylinder Liners," Four papers on engine detail components, IME (London), *Proc. Automobile Div.* 1956–1957, No. 5.

12.716 Thill, "Developments in Oil Seals," *Auto Ind.*, July 15, 1962, 94.

12.717 Smoley, "Flange Loadings in Flange Design and Gasket Selection," *SAE Preprint* 13A, Jan. 1958.

12.718 Webster and Larkin, "How to Determine Compatibility of Seals and Fluids," *Jour. SAE 67*, March 1959, 41.

Ignition Systems

(See also ref. 11.002, No. 122.)

12.720 Sharpe, "Transistorized Ignition for High Speed Gasoline Engines," *SAE Paper* 650498, 1965. (Experience and practice of Lucas, a large English manufacturer.)

12.721 Hardin, "Capacitor Discharge Ignition: The System Approach to Extended Ignition Performance and Life," *SAE Preprint* S414, December 1964.

12.722 A group of SAE papers on conventional and transistorized ignition systems, Jan. 1963.

Author	No.
Burke and Dunne	652A
Larger	652B
Eason	652C
Miller	652D
Henske	652E

12.723 Frederick, "Notes on Electric Ignition for High-Compression Engines," *Diesel Power* III, p. 3. (Cooper-Bessemer practice for large gas engines.)

12.724 Kamo and Cooper, "Modern Ignition Systems for Gas Engines," *ASME Paper* 64–OGP–15, 1964.

12.725 Crankshaw and Arnold, "A Piezoelectric Ignition System for Small Engines," *SAE Preprint* 375A, 1961. (Pressure on a crystal generates ignition spark for a single-cylinder engine.)

12.726 "Engineering Know-How in Engine Design — Part 10 — 1962," *SAE Special Publication* SP-237, 1962. (Ignition systems; lubrication.)

12.727 "Capacitor Discharge, Piezo Electric and Transistorized Spark Ignition Systems," *SAE Technical Progress Series* TP8. (Theoretical and practical considerations; electrical characteristics.)

Supercharger Design

(See also refs. 11.002, No. 283, and 10.860 *et seq.*)

12.800 Hawthorne (editor), "Aerodynamics of Turbines and Compressors," Princeton University Press, Princeton, N.J., 1964. (A symposium volume, with chapters by experts in each field. Includes both axial and radial flow compressors and turbines. Extensive bibliography. Little detail design beyond blade geometry.)

12.801 Smith, "Aircraft Gas Turbines," John Wiley & Sons, Inc., New York, 1956. (Includes some useful practical design data on centrifugal and axial compressors and turbines. Based on engineering point of view, General Electric experience.)

12.802 Toolin and Mochel, "The High Temperature Fatigue Strength of Several Gas-Turbine Alloys," *Proc. ASTM 47*, 677 (1947).

12.803 Betteridge, *The Nimonic Alloys*, Edward Arnold, London, 1959.

12.804 Ault and Deutsch, "Review of NACA Research on Materials for Gas-Turbine Blades," *SAE Paper*, 9–13, 1950.

12.805 Egli, "Basic Design of Superchargers for Diesel Engines," *Trans. SAE 68*, 29 (1960). (Comprehensive article from experience of AiResearch Industrial Division, The Garrett Corp., with radial-inflow turbines for automotive-size engines.)

12.8051 Parker, "Problems in Matching Turbochargers to High Speed Diesel Engines," *SAE Paper* 71C, Aug. 1958.

12.8052 Nagao and Hirako, "Evaluation of the Capacity of the Auxiliary Blower Connected to a Constant-Pressure Turbocharged Two-Cycle Diesel Engine," *Bull. JSME 2*, 1959.

12.8053 "Centrifugal Compressors," *SAE Technology Book* TP-3. (Impeller and diffuser characteristics; combined axial and centrifugal rotors; acceleration limits of automotive gas turbines.)

12.806 King, "Axial versus Centrifugal Superchargers for Aircraft Engines," *Trans. SAE 53*, 736, 1945.

12.807 Concordia and Dowell, "Analytical Design of Centrifugal Air Compressors," *J. Appl. Mech.* (ASME) *13*, A-271, Dec. 1946.

12.808 Penn, "The Design and Operation of Superchargers," *J. Roy. Aeron. Soc.* (London) *48*, 495, Nov. 1944.

12.809 Judson, "The Design and Development of Small Radial-Inflow Turbo-superchargers," Proc. IME, London, *Proc. Automobile Div.*, 1956-7, No. 6.

12.810 Zola, "Seals for Preventing Oil Leakage in High-Speed Superchargers," *Prod. Eng. 17*, Feb. 1946.

12.811 Zola, "Bearing Design and Installation in High-Speed Superchargers," *Prod. Eng. 17*, Jan. 1946.

12.812 Berchtold, "The Comprex Diesel Supercharger," *Trans. SAE 67*, 5 (1959). (See also *SAE Preprint* 118U. Unusual supercharger concept, not used commercially up to 1966.)

12.813 Smith, "Problems in the Mechanical Design of Gas Turbines," *J. Appl. Mech.* (ASME) *14*, June 1947, A-99. (Creep and fatigue failure of metal at temperatures above 1200° F. Discussions on design practice at Elliot Company.)

Computation and Measurement of Exhaust Energy with Particular Reference to Exhaust Blowdown

(See also Volume I, refs. 10.60–10.691.)

12.814 Hussmann and Pullman, "Formation of Pressure Pulses by Exhaust Blowdown," *ASME Paper* 58–A–145, November 1958. (Analysis of blowdown through exhaust valves or ports.)

12.815 Reynolds and Kays, "Blowdown and Charging Processes in a Single Gas Receiver with Heat Transfer," *Trans. ASME 80*, 1160 (1958).

12.816 Wallace and Mitchell, "Wave Action Following The Sudden Release of Air Through an Engine Port System," IME (London), *Proc. Automobile Div.*, 1952–1953, Vol. 1B, p. 343.

12.817 Woods, "Tests to Examine High-Pressure Pulse Charging on a Two-Cycle Diesel Engine," *ASME Paper* 66–DGEP–4, April 1966. (Measurements of blowdown energy with nozzle area/piston area = 0.038 and $F_R = 0.30$.)

12.818 Nagao *et al.*, "A Method of Measuring the Exhaust-Gas Energy Available for Turbocharging of a Diesel Engine," *Trans. JSME 25*, 309 (1959); *Bull. JSME 2*, 573, 1959 (English).

12.819 Nagao and Shimamoto, "On the Transmission of the Blowdown Energy in the Exhaust System of a Diesel Engine," *Trans. JSME 24*, 592 (1958); *Bull. JSME 2*, 170, 1959 (English).

12.820 Büchi, "Exhaust Turbocharging of Internal Combustion Engine—Its Origin, Evolution, Present State of Development and Future Potentialities," a Monograph published by the Journal of Franklin Institute (see *J. Frankl. Inst. 256*, 57, 1953. (Application of Büchi blowdown system with performance results. Nonanalytical.)

Heat Exchangers

12.821 Kays and London, *Compact Heat Exchangers*, McGraw-Hill Book Company, New York, 2nd ed., 1964. (Theory and design of nonrotating exchangers.)

Cooling Systems, Liquid

(See also ref. 11.002, Nos. 122 and 152.)

12.910 French and Hartles, "Engine Temperatures and Heat Flow Under High-Load Conditions," IME (London), Paper No. 8, Birmingham, Oct. 1964. (Measurements of temperatures and local heat flux on a 4.75-inch bore Diesel cylinder

Effects of valve overlap, inlet temperature, exhaust pressure, open and divided combustion chambers. Markedly reduced cylinder-head temperatures were obtained by machining the cast surfaces on the water side.)

12.911 Alcock, "Thermal Loading of Diesel Engines," *Trans. Inst. Marine Engrs.* (London) *77*, Oct. 1965, 279. (Ricardo measurements of local heat flux and recommendations for piston and cylinder design. Emphasis on "strong-back" system of separating heat-flow and strength functions as in Vol. I, Chap. 8, and Fig. 8–7. At high heat flux, wall temperature is controlled by boiling-point of the coolant.)

12.912 Gelnes, "An Analysis of Engine Cooling in Modern Passenger Cars," *SAE Paper* 660c, 1963. (Heat flux versus road speed, coolant velocities on cylinder head, pictures of nucleate boiling, water flow in typical engines, effects of coolant characteristics, car radiator characteristics.)

12.913 Schmitt, "Analysis of Diesel Cooling Systems," *SAE Preprint* 887A, Aug. 1964. (Data on radiator and fan requirements.)

12.914 For work on liquid-cooled aircraft engines by U.S. National Advisory Committee for Aeronautics (now NACA), Washington, D.C. See *NACA TN* 476, 1933, and *NACA TM* 649, 1931.

12.915 Moore *et al.*, "Heat Transfer and Cooling, Aluminum vs. Cast Iron," *SAE Paper* 494A, Mar. 1962. (Rates of heat rejection nearly the same, but hot-side temperatures 30° to 100° F lower with aluminum. Octane requirements equal.)

12.916 Corrosion of cooling systems:

Author	Publication	Date
Blackwood	*SAE Paper* 761	June 1956
Joyner	*SAE Paper* 760	June 1956
Trock	*SAE Paper* 758	June 1956
Bailey	*SAE Paper* 717B	June 1963

Air Cooling

(See also refs. 11.002, No. 152, and 11.301–11.3032.)

12.920 Bachle, "Air-Cooled Diesel Engine Appraisal," *SAE Paper* 154, Sept. 1957. (Review of state of the art with a number of illustrations of cylinder design, fan design, fan power, comparative weights versus liquid cooling. Includes industrial and military engines. See also Haas, *Trans. SAE 65*, 641 (1957), for design of Diesel air-cooled tank engine.)

12.921 Bochner, "The Volkswagen Air-Cooled Engine," *SAE Paper* 13, Jan. 1953. (Study of cylinder temperatures and air-flow patterns through system. No fan power or pressure readings.)

12.922 Piry, "Cooling Characteristics of Steel and Aluminum Finned Cylinder Barrels," *Trans. SAE 53*, 630 (1945). (Useful data from tests of an aircraft type cylinder.)

12.923 Piry, "Single Cylinder Engine High Altitude Cooling Tests," *SAE Paper*, May 1, 1945. (Effects of altitude on cooling of air-cooled aircraft cylinder.)

12.924 For the excellent fundamental research on cooling-fin design for air-cooled cylinders, see the following publications of NACA (now NASA) Washington, D.C.: Technical Reports 488, 511, 555, 612, 676, 726; Technical Notes 331, 429, 602, 621, 649, 779; ARR by Silverstein, Jan. 1943; ACR by Brevoort *et al.*, Feb. 1943; ARR by Maurice *et al.*, Mar. 1943; ARR 3H16, Aug. 1943.

Engine Auxiliaries

Most engine auxiliaries such as starters, generators, radiators, fans, governors, air filters, oil filters, etc., are designed and marketed by specialists. Technical information may be obtained by consulting the manufacturers. For names and addresses, listed by products, see Product Design Catalog File published annually by Sweet's Catalog Service, F. W. Dodge Company, 330 W. 42nd Street, New York 10036 and "Mechanical Catalog" published annually by ASME, 29 W. 39th Street, New York. The following bibliography contains a few references familiar to the author.

Radiators and Cooling Systems

12.980 Young, "Developments in Engine Cooling Systems," *SAE Preprint*, November 1947. (Descriptive. Includes fans, installations, etc.)
12.981 Habel and Gallagher, "Tests to Determine the Effect of Heat on the Pressure Drop Through Radiator Tubes," *NACA TN* 1362, July 1947.
12.982 Gale, "The Soldered Aluminum-Fin Brass-Tube Radiator," *SAE Paper* 767, 1956.
12.983 Dowd *et al.*, "Aluminum for Tomorrow's Radiators," *SAE Paper* 766, 1956.
12.984 Emmons, "All Aluminum Brazed Heat Transfer Equipment," *SAE Paper* 768, 1956.

Fans

12.985 Reynolds *et al.*, "Fans Save Horsepower," *Jour. SAE 64*, Oct. 1956, 67.
12.986 Nedley, A. Lloyd, "Combining Experimental and Computer Techniques for Reducing Fan and Blower Tonal Noise," *SAE Paper* 825D, March 1964.

Electrical

12.987 King, "New Dimensions for Designing and Testing the 1963 'Delcotron' System" (Generator and electrical system for passenger cars), *SAE Preprint* 617A, 1963.

Governors

12.988 "Engine Speed Governors," *SAE Special Publication* SP-166, 1959 (seven papers on the subject) and SP-192 (one paper).

CHAPTER 13

Predictions of Future Trends
(See also ref. 10.31 and Volume I, refs. 13.50–13.52.)

13.00 Jet Propulsion Laboratory, California Institute of Technology, "Should We Have a New Engine?" J.P.L. report SP 43-17, 1975. Available as SAE Publication SP 400 S.

13.01 Jet Propulsion Laboratory, California Institute of Technology, "Compendium of Critiques of JPL report SP 43-17." JPL publication 77-40.

13.011 Richardson, "Automotive Engines for the 1980's," Eaton Corporation, Southfield, Michigan, 1973. Soundest of recent predictions.

13.02 "Where Are Marine Power Plants Headed?" Symposium of twenty-six papers from the May, 1966, Philadelphia meeting of the Society of Naval Architects and Marine Engineers. (Steam versus Diesel, geared versus direct-drive Diesels, fuel cells, and other unconventional possibilities.)

13.03 Kirkwood, "Tomorrow's Stationary Power Engines," *Diesel and Gas Engine Progress*, Jan. and Feb. (1966). (Analysis of past use of large Diesels and attempt to predict future trends in this market.)

13.04 Sorensen, "Large-Bore Diesels for Modern Power Stations and Sea-Going Ships," *ASME Paper* 65-OGP-11, 1965. (Review of current types, statistics on use, comparison with steam plants. Insufficient consideration of the increasing use of smaller, trunk-piston, geared engines.)

13.041 Kyropoulos, "The Potential of Unconventional Power Plants for Vehicle Propulsion," *Trans. SAE 67*, 319 (1959). (Discussion of new energy sources, Diesel versus S-I engines, gas turbine, and free-piston status as of 1959. Predicts no displacement of conventional types in foreseeable future.)

13.05 Taylor, "The Big Reciprocating Engine," *Mech. Eng.* (ASME) *85*, 46 (1963). (Emphasizes the probable increase in use of the gas turbine for large units.)

13.06 Other forecasts:

    a. Rosen, *Trans. SAE 61*, 197 (1953).

    b. Campbell, *Jour. SAE 65*, Dec. 1957, 41.

    c. Marks, *SAE Paper* 71B, Aug. 1958, p. 1.

    d. Rosen, *Jour. SAE 65*, Jan. 1957, 42; *SAE Paper* 212A, Aug. 1960, p. 1.

    e. Pitchford, *Jour. SAE 68*, Dec. 1960, 40.

    f. Caris, *SAE Paper*, Oct. 1956, p. 1.

    g. Broeze, IME (London), *Proc. Automobile Div.* 1953–1954, p. 135.

Modifications to Existing Types
(See also 11.414, 11.415, 13.00, 13.01.)

13.06 "Designing the Adiabatic Diesel Engine," *SAE Report* SP 543. 1983.

13.10 Lundquist, "The Dyna-Star Power Plant Concept for Compact Diesel and Spark Ignition Engines," *SAE Paper* 770A, 1963. (Use of U-type 2-cycle cylinders, Volume I, Fig. 7–1*d*, in a radial arrangement. Not clear why this should be better than a similar arrangement of conventional cylinders with the same total displacement.)

13.11 Schulz, "Compact Powerplant Weighs 4 lbs/hp," *Diesel and Gas Turbine Progress*, Feb. 1967, 50. (Description and 1967 rating of Dynastar U-cylinder radial engine 3 × 3.5 in., bmep = 80 psi, 5.6 kg/cm², piston speed 1450 ft/min, 7.1 m/sec.)

Engines with Levers, Cams, Scotch Yokes, etc.

13.21   Clark and Skinner, "High Output Version of the Model 4A032 Military Standard Engine," *SAE Paper* 650516, 1965. (Details of rocker-arm type opposed-piston Diesel engine by Southwest Research Inst.)

13.22   Heldt, "Two Stroke Diesel Has No Crankshaft," *Auto. Ind.*, June 15, 1955. (Opposed-piston lever engine with eccentrics and half-speed output shaft. Built by Svanewolle Wharf Co., Copenhagen.)

13.23   "The Bourke Two-Stroke," *Hot Rod Magazine*, July 1954. (Account of a high-speed flat-twin crankcase-scavenged engine using *scotch yoke* linkage. Claims about 100 bmep at 1000 rpm or about 4000 ft/min piston speed.)

13.24   "The Fairchild-Caminez Engine," *Aviation*, May 26, 1926, 788; July 4, 1927, 20; Nov. 7, 1927, 1114; *Auto. Ind.*, May 27, 1926, 891; July 30, 1927, 160.

Barrel or "Revolver" Type Engines

13.25   Earl, "A Compact Aeronautic Engine," *Jour. SAE 26*, Mar. 1930, 341. (Axial cam mechanism with rollers in piston assemblies.)

13.26   Cullom, "The Alfaro Engine," Civil Aeron. Authority Tech. Develop. Report No. 4, Jan. 1939. See also *Auto. Ind.*, Dec. 1, 1939, 590. (Opposed-piston cam type shown in Fig. 13–3c.)

13.27   The Sterling Viking Diesel, *Diesel Power and Diesel Trans.*, June 1945, 698. See also *Auto. Ind.*, Feb. 3, 1934, 139 and Dec. 19, 1936, 856. (Opposed-piston two-stroke double-swashplate engines briefly put on the market by Sterling Engine Co., Buffalo, N.Y. Not to be confused with the Stirling hot-gas engine of refs. 13.850–13.857.)

13.28   Hall, "More Power from Less Engine," *Trans. SAE 47*, 504 (1940). (History and descriptive material on *barrel-type* engines, all unsuccessful.)

13.29   Herrmann, "New Light Weight Power Plants for Post-War Airplanes," *SAE Preprint*, Nov. 1944. See also *Auto. Ind.*, Dec. 25, 1937, 902. (Single-cam type, 4-cycle with poppet valves.)

13.291 Sherman, "Slant Mechanism — Its Fundamentals and Application to Diesels for Aircraft and Other Low Weight Fields," *SAE Paper*, June 1947. (Some test data on a 2-stroke opposed-piston crankless engine, originally designed for 1 lbm per bhp, but weighed 2 lbs. per bhp on test. See also *Machine Design 19*, Nov. 1947.)

13.292 Heldt, "Swashplate Replaces Crankshaft in Michell Five-Cylinder Automobile Engine," *Auto. Ind.*, Nov. 24, 1928, 767.

Toroidal or "Doughnut" Engines

13.30   Scott, "Construction and Operation of the Bradshaw Omega Engine," *Auto. Ind.*, January 15, 1956. (Pistons reciprocate in toroidal cylinder by means of complex arrangement of cranks and shafts.)

13.31   Scott, "Rotary Cylinder Engine Has Non-Reciprocating Pistons," *Auto. Ind.*, Aug. 1, 1961, 50. (Toroidal-shaped pistons reciprocate [*sic*] in toroidal cylinders by means of a slant mechanism.)

Rotating-Displacer Engines

13.40 Wankel *et al.*, "Bauart und gegenwärtiger Entwicklungsstand einer Trochoiden-Rotationskolbenmaschine," *MTZ 21*, Feb. 1960, 33. (In German. Theory, geometry, and practice in an authoritative article by the inventor.)

13.41 Huf, "Zur Geschichte der Rotationskolbenmaschinen," *Revue Automobile* (Paris), No. 49, 1961. (In German and French. History of rotary engines from 1588 to the NSU Wankel. R. Cooley of Boston given much credit for his patent for a rotary steam engine.)

13.42 Bentele and Jones, "Curtiss Wright's New Rotating Combustion Units," *Diesel and Gas Turbine Progress*, April 1966, 55. (General review and data on recent models and expected applications. 60 kW a-c generator set weighs 1080 lbm, as compared with 1450 lbm for gasoline engine set, and is somewhat smaller in volume.)

13.43 Jones, "The Curtiss-Wright Rotating Combustion Engines Today," *Trans. SAE 73*, 127 (1965). (Good coverage of development work to date. Optimistic forecast of future applications.)

13.44 Other articles on the Wankel engine: Froede, *Trans. SAE 69*, 179, 194, 641 (1961); Jones, *SAE Paper* 650723, 1965; Moerman, *SAE Paper*, 1961; Wankel, *MTZ 21*, Heft 2, 33 (1960); Huber, *ibid.*, 62, Heft 3, 57 (1960); Bentele, *SAE Paper* 288B, 1961; Bentele, *SAE Paper* S348, 1962; Roepke, *SAE Paper 636A*, 1963; Bensinger, *ATZ 66*, 120 (1964).

13.45 Sisto, "Comparison of Some Rotary Piston Engines," *SAE Paper* 770B, Oct. 1963.

13.46 "Backlash Fires Rotary Engine," *Mech. Eng.* (ASME) *84*, 67 (1962). (Operates on the principle of a vane-type compressor. See Volume I, Fig. 10–3c.)

13.47 Other rotary engines, *Jour. SAE 72*, Jan. 1964, 37.

Free-Piston Power Plants

(See also Volume I, Chapter 13 and refs. 13.30–13.392.)

There is much literature on this subject, most of it dated during the period 1948 to 1960 when there was much interest in this type. Development slowed after 1960.

13.500 Scanlan and Jennings, "Bibliography — Free-Piston Engines and Compressors," *Mech. Eng.* (ASME) *79*, 339 (1957).

13.501 "Free Piston Engines," *Lubrication 44*, Sept. 1958. (Excellent summary of theory and practice up to that date.)

13.502 London, "Free-Piston and Turbine Compound Engine — Status of Development," *Trans. SAE 62*, 426 (1954).

13.503 London and Oppenheim, "The Free-Piston Engine Development — Present Status and Design Aspects," *Trans. ASME 74*, 1349 (1952).

Sigma Free-Piston Work

13.504 Huber, "Present State and Future Outlook of the Free-Piston Engine," *Trans. ASME 80*, 1779 (1958). (Sigma of France, experience with stationary and marine installations.)

13.505 Picard and Huber, "L'emploi des Générateurs à pistons libres en traction ferroviaire," extrait du Bulletin no. 3 (décembre 1951) de la S.F.M. Editions Science et Industrie. (Sigma-powered experimental locomotive, unsuccessful.)

13.506 Barnett *et al.*, "Free Piston Engines Power — 8000 hp in Situ Combustion Compressor Plant," *World Oil*, July 1965, 71. Published by Mobil Oil Co., New York. (Air compressor plant using nine sigma gasifiers to operate large gas-turbine driven air compressor.)

General Motors Free-Piston Developments

13.507 Underwood, "The GMR 4-4 Hyprex Engine," *Trans. SAE 65*, 377 (1957). (Experimental 2-cylinder plant for a passenger automobile. Heavier, more bulky, and much more complicated than a conventional engine. A total failure.)

13.508 Fleming and Bayer, "Diesel Combustion Phenomena as Studied in Free-Piston Gasifiers," *SAE Paper* 472A, 1962. (G. M. work shows tolerance to low-grade fuel.)

13.509 Flynn, "Observations on 25,000 Hours of Free Piston Engine Operation," *SAE Paper* 802, Sept. 1956. (Report on endurance testing of General Motors engine.)

13.510 Spech, "Evaluation of Free-Piston Gas-Turbine Marine Propulsion Machinery in GTS William Patterson," *SAE Paper* 604A, Oct.–Nov. 1962. (Report of rather unsuccessful trials of this experimental General Motors installation, later abandoned.)

Other Free-Piston Developments

13.511 Lasley and Lewis, *ASME Paper* 53–S–34, 1953. (Baldwin-Lima-Hamilton free-piston-gas-turbine experimental work.)

13.512 McMullen and Payne, *Trans. ASME 76*, 1 (1954). (Cooper-Bessemer experimental free-piston-turbine power plant.)

13.513 Noren, *Jour. SAE 65*, July, 44 (1957); and *SAE Paper* 150, June 1957. (Ford builds an experimental free-piston unit for tractors.)

13.515 "Diesel Pile Hammer Speeds Expressway Job," *Diesel and Gas Engine Progress*, May, 32 (1961). (Description of a free-piston Diesel type hammer built by Link Belt Co. Many free-piston Diesel type hammers and rock drills have been designed and built, but none has been commercially successful in competition with the conventional air-driven type.)

Gas-Generator-Turbine Combination with Crankshaft Engines

13.60 Chatterton, "The Diesel Engine in Association with the Gas Turbine," *Proc. IME* (London), *174*, 409 (1960). (Discussion of supercharging, compounding, and using the engine as a gas generator for the turbine. Doubtful conclusions favor the latter. Good discussions included.)

13.61 Tenger, "Crank-Piston Gas-Generator Locomotives," *SAE Paper* 2140, 1960. (Swedish experimental locomotive with opposed-piston Diesel gas generator. Fuel economy not given.)

13.62 Hooker, "A Gas-Generator Compound Engine," *Trans. SAE 65*, 293 (1947). (Two-cycle Diesel engine supplying hot gas to turbine developed by General Electric Co. with funds from U.S. government. Unsuccessful.)

Gas Turbines

The literature on gas turbines is extensive. See also under "Supercharger Design," refs. 12.800 and 12.801 and bibliography contained in ref. 13.702.

13.700 Wood, "Gas Turbines in Future Industrial Vehicles," *Paper* 650480 in *SAE Special Publication* SP-270, p. 27, 1965. (One of the best brief presentations of potential performance of various types. Source of Fig. 13–8. Good bibliography.)

13.701 Marwood and Presser, "Potential Performance of Gas Turbine Powerplants," *SAE Paper* 650715, 1965. (Performance potential of industrial and marine gas turbines, including various open cycles with intercooling, regeneration and/or reheat.)

13.702 Taylor, "The Internal Combustion Engine," 2nd ed., International Textbook Co., Scranton, Pa., 1960, chapter 20 and refs. 20.10–20.78.

13.703 "1966 Gas Turbine Specifications," *Gas Turbine*, Stamford, Conn., Jan.–Feb. 1966, S-1. (Specifications, performance, fuel consumption, weights, and dimensions for U.S. and foreign gas turbines. Exhaust temperatures but no turbine inlet temperatures. Includes rated powers from 44 to 25,000 in a single unit, bsfc from 0.40 to 1.6 lbm/hp hr, pressure ratios from 2.5 to 16.

13.7031 Giroux *et al.*, "A 5000-kW Railway-Mounted Gas-Turbine Power Plant," *ASME Paper* 54–A–191, 1954. (Shows possibility of high-powered transportable units.)

Automotive Gas Turbines

(See ref. 13.700 and also refs. 20.70–20.78 of ref. 13.702.)

13.704 Huebner, "The Chrysler Regenerative Turbine-Powered Passenger Car," *SAE Paper* 777A, Jan. 1964. See also Chapman, "Chrysler's Gas-Turbine Car," *SAE Paper* 777B, Jan. 1964. (The most advanced project for passenger cars. Highly experimental as of 1966.)

13.705 Swatman and Malohn, "An Advanced Automotive Gas-Turbine Concept," *SAE Paper* 187A, 1960. (Brief description and performance curves of Ford three-shaft compound automotive gas turbine, compared with other types. See also *Trans. SAE 70*, 109, 1962; and *SAE Preprint* 991B, 1965.)

13.706 Penney, "Rover Case History of Small Gas Turbines," *Trans. SAE 72*, 131 (1964). (Development for passenger-car use. Very complete data on development work and performance, with and without static heat exchanger.)

13.707 Turunen and Collman, "General Motors Research GT-309 Gas Turbine Engine," *SAE Paper* 650714, 1965. (Latest unit in G. M. automotive turbine development. Gasifier and power shafts are connected through a controlled torque convertor in a "power transfer" system. Test results indicate excellent full- and part-load economy.)

13.708 Wadman, "Ford's New Generation Truck Gas Turbine," *Diesel and Gas Turbine Progress*, Dec., 22 (1966). (New experimental CXBT-T$_p$X model, 375-hp 1800° F turbine inlet, ceramic rotary heat exchangers. Apparently supersedes the compound turbine of ref. 13.705.)

13.709 Anon, "A Design Summary of the General Motors Research G7-309 Gas Turbine Engine," *Gen. Motors Eng. J. 13*, Fourth Quarter, 2 (1966). (Material supplementing ref. 13.707. Efficiency of components, component design, development history.)

13.7091 Schultz, "The Storm after the Calm at Indy," *Diesel and Gas Turbine Progress*, July 1967, 46. (Description of the gas turbine racing car which nearly won the 1967 Indianapolis race. Well illustrated.)

Small Industrial Gas Turbines
(See also ref. 13.703.)

13.710 "Fiat Small Gas Turbine Development," *Diesel and Gas Turbine Progress*, Mar. 1966, 62 (490-hp, 584-lbm, 17-cu ft compact unit with heat exchanger, min bsfc 0.4 lbm/bhph.)

13.712 Clark, "Materials, Costs, and the Small Gas Turbine Engine," *Diesel and Gas Turbine Progress*, Mar. 1966, 72. (Good summary of required special materials, their properties, and applications.)

13.713 Pitt, "What's Ahead for the Small Gas Turbine?" *Trans. SAE 72*, 62 (1964). (Data on present applications and probable future developments. Heat exchangers too heavy and expensive. If they could be made for < 2 lbm/hp and 10 per cent of the overall cost, they would be competitive.)

Gas Turbines in Railroad Service
(See also ref. 13.703.)

13.720 Smith, "8500 hp Locomotive Power Plants," *Railway Locomotives and Cars*, Vol. 131, July 1957, p. 53. (20 per cent thermal efficiency for Union Pacific locomotives.)

13.721 Gowans, "Design of 8500 hp Gas-Turbine Electric Locomotives for High Speed Freight Service," *ASME Paper 57–A–149*, Dec. 1957.

13.722 "Coal-fired Gas Turbine Is Road Tested in Freight Service," *Railway Locomotives and Cars*, Vol. 137, Jan. 1963, p. 28.

13.723 Joseph, "8500 hp Dual-fuel Oil and Coal Turbine First on Rails," *Diesel and Gas Engine Progress 3*, April 1964, 56. (Union Pacific coal-burning gas-turbine electric locomotive.)

13.724 Barlow, "The Development of the First Gas-Turbine Mechanical-Drive Locomotive," *ASME Paper 54–A–183*, 1954.

13.725 Yellot et al., "Development of Pressurizing, Combustion, and Ash Separation Equipment for a Direct-Fired, Coal-Burning, Gas-Turbine Locomotive," *ASME Paper 54–A–201*, 1954. (Progress in the struggle to make gas turbines accept powdered coal as fuel.)

Large Gas Turbines (except aircraft)
(See also ref. 13.703.)

13.731 "Nine Utilities Buy Jets for Peak Power," *Power Plant*, Pratt and Whitney Aircraft, Hartford, Conn., Mar. 21, 1966. (Indicates growing use of aircraft jet engines as the gas-generator units for peak electric power-units with J-75 engine, rated at 15,000 kW each.)

13.732 Heard and Wright, " Gas Turbines for the Pipe Lines," *Gas Turbine*, Stamford, Conn., Jan.–Feb. 1966, p. 59. Types used and costs of installation and operation. Combination with steam cycle. Costs of complete turbine and compressor installations given as follows:

	$ per hp installed	
	no	with
HP	Heat Exch.	Heat Exch.
5,000	200	240
10,000	130	140
15,000	90	100

Gas-Turbines Compounded with Reciprocating Engines (Geared-in Exhaust Turbines)
(See also Volume I, refs. 13.11–13.22, 13.25, and Volume II, refs. 10.847, 13.01.)

13.740 "The Supercharging of Two-Stroke Diesels," *Sulzer Tech. Rev.*, Special Number, December 31, 1941. (Report on Sulzer development work on opposed-piston engines with exhaust-driven turbines geared to crankshaft; imep up to 255.)

13.741 Wallace, "Performance of Two-Stroke Compression-Ignition Engines in Combination with Compressors and Turbines," *Proc. IME* (London), *177*, 43 (1963). (Method of computation for geared-in engine-turbine systems, based on assumed component characteristics.)

13.742 Wallace, "Operating Characteristics of Compound-Engine Schemes Based on Opposed-Piston Engines and Differential Gearing," *Proc. IME* (London) *177*, 64 (1963). (Extension of previous reference to include variable gear ratios.)

13.743 Zinner, "Theoretical and Experimental Investigation of Operational Procedure Involving the Use of a Coupled Exhaust Turbine," *CIMAC International Congress*, 1962.

Gas Turbines and Diesel Engines in Parallel

13.751 "Danish Navy Frigate Has CODAG Propulsion System," *Diesel and Gas Turbine Progress*, July 1966, 36. (Combined Diesel engine and gas turbine based on P and W jet engine. Both units geared to one propeller shaft through one-way clutches.)

13.752 Kronogard, "The Volvo Dual Power Plant for Military Vehicles," *SAE Paper* 660017, Jan. 1966. (Diesel engine and gas turbine driving into a special transmission system for tanks, etc.)

13.753 "Diesel-Gas Turbine Propulsion System," *Diesel and Gas Turbine Progress*, Oct. 1966, 48. (Avco-Lycoming 2000 hp turbine, gear and control system arranged to accommodate auxiliary Diesel engine 150 to 500 hp.)

Total-Energy Systems

13.755 Sterrett, "Hospital's Total Energy System Is Expandable," *Diesel and Gas Turbine Progress*, Oct., 42 (1966). (Description of a system using Waukesha Diesel engines. Other issues of this magazine contain descriptions of similar systems based on gas turbines.)

Economics of Gas Turbines and Diesel Engines

13.760 "Diesel and Gas Engine Market," 1965–1966; "Diesel and Gas Turbine Market," 1967–1968; "Industrial and Marine Gas Turbine Engine Market," all published by Diesel and Gas Engine Catalog, P.O. Box 7406, 11225 W. Blue Mount Road, Milwaukee, Wis. (Market statistics and trends, past and projected.)

13.761 Kahle and Hung, "An Economic Assessment of Turbine Powered Industrial Vehicles," *SAE Paper* 660605, Sept. 1966. (Thorough study of costs based on 1966 prices shows turbine power is more expensive than Diesel power for a 400 hp truck. Source of Table 13–2.)

Steam Power

13.801 "Working Fluids for Power Generation Cycles," R and D Letter, May 1962, Westinghouse Electric Corp., Pittsburgh. (Curves showing relative efficiencies.)

13.802 Heldt, "Rise and Decline of the American Steam Car Industry," *Automotive and Aviation Industries*, Aug. 1, 1946, 20.

13.803 Dooley and Bell, "Description of a Modern Automotive Steam Power Plant," *SAE Preprint* No. S338, Jan. 22, 1962. (Experimental vehicle, never produced.)

13.804 Wilson, "Steam Power Plants in Aircraft," *NACA TN* 239, June 1926. (Concludes they are entirely impractical.)

The Stirling Hot-Gas Engine

The literature on this engine type is extensive. The following is a list of important references in English. See also refs. 13.00–10.01.

	Author	*Theoretical Analyses*	*Date*
13.850	Kirkly	*SAE Paper* 949B	1965
13.851	Finkelstein	*SAE Paper* 118B	1960
13.852	Wlaker and Klein	*SAE Paper* 949A	1965
13.853	Creswik	*SAE Paper* 949C	1965
13.854	Welsh *et al.*	*SAE Paper* 549C	1962
13.855	Meijer	*SAE Paper* 949E	1965
13.856	Flynn *et al.* (GMC)	*Trans. SAE 68*	1960
13.857	Heffner	*SAE Paper* 949D	1965

Electric Power

This type of power was extensively used for short-range passenger and industrial vehicles from about 1898 to 1930. For its history and technical details, see books on the history of the automobile. See also:

13.900 Papers from SAE symposium on batteries as follows:

Author	Paper	Date
Brumbaugh	*SAE Preprint* 269H	1961
Herbert and Grimes	*SAE Preprint* 269D	1961
Dittman	*SAE Preprint* 269C	1961
Lander	*SAE Preprint* 269E	1961

13.901 Riggs, "The Electric Vehicle," *SAE Preprint* 269F, 1961.
13.902 Wells, "Automobile Batteries — Selection, Service Life, New Developments," *Trans. SAE 61*, 253 (1953).
13.903 Douglas and Liebhofsky, "The Development of Fuel Batteries for the Commercial Market," *SAE Paper* 689B, 1963.

Miscellaneous Types of Power

13.910 Egli and Sherman, "Energy Sources of the Future," *Trans. SAE 72*, 306 (1964). (Review of various possibilities and their present status including thermoelectric, thermionic, magnetohydrodynamic, photoelectric, fuel-cell, and nuclear systems.)
13.911 Flynn *et al.*, "Power from Thermal Energy Storage Systems," *SAE Paper* 608B, 1962. (Attempt at fundamental evaluation of systems in a submarine environment, where air is not available. Chemical, thermal, and electrochemical alternatives. Proposed use of Stirling engine for conversion.)
13.912 Boardman, "The Space Hyperbolic Bipropellant Internal-Combustion Engine," *SAE Paper* 768A, 1963. (Design and development of a small piston engine to run on chemical reaction of "Aerozene 50," composition not specified, and N_2H_4.)
13.913 Cochran *et al.*, "Non-Air-Breathing Auxiliary Power Plants," *SAE Paper* 53T, Mar. 1959. (Basic discussion of problems of chemical and nuclear power for space vehicles.)

CHAPTER 14

Engine Testing, General
(See also ref. 11.002, Nos. 119 and 178.)

14.000 Ladenburg *et al.*, *Physical Measurements in Gas Dynamics and Combustion*, Princeton Univ. Press, Princeton, N.J., 1954.
14.011 Tuve and Domholdt, *Engineering Experimentation*, McGraw-Hill Book Company, New York, 1966. (Instrumentation and methods of measurement in

engineering. Contains much useful reference material. Approach is in some cases elementary.)

14.012 Wagner and Gorman, "Fuels, Lubricants, Engines, and Experimental Design," *Trans. SAE 71*, 684 (1963). (A mathematical approach to planning experiments with many variables.)

Engine Laboratories

14.021 Larborn, "Automation in an Engine Laboratory," *SAE Paper* 640458, 1965. (Description of new equipment and building for functional and endurance testing at Volvo plant, Sweden.)

14.022 "Ricardo," Circular issued by Ricardo and Co., Ltd., Shoreham, Sussex, England. (Brief description and photographs of buildings, equipment, and methods of one of the world's most successful engine-research laboratories.)

14.023 Taylor, "Recherches sur les Moteurs à Combustion Interne," Conférences faites à Bruxelles, les 1 et 2 juin 1949, Société Belge pour l'Etude du Petrole, Bruxelles. (Description of Sloan Laboratory and Gas-Turbine Laboratory equipment. Principal research projects, special apparatus, and instrumentation.)

14.024 Resen, "Significant Contributions of the Diesel Research Laboratory," IME (London) *Proc. Automobile Div.*, 1950–51, Pt. I, p. 7. (Excellent general article on Caterpillar development methods and experience.)

14.025 Bouchard and Oxley, "Development of a Modern Dynamometer Laboratory," *SAE Preprint* 597, 1951. (Ford laboratory, luxurious, conventional.)

14.026 "Station Nationale de Recherches," Proc. Institut Français du Pétrole, 1949. (Description of the National Engine Research Station in France.)

14.027 Henkel *et al.*, "The New Automotive Laboratory at Detroit Arsenal," *SAE Preprint* 116, June 1957. (U.S. Army laboratory; includes engine-testing equipment.)

14.028 Roensch and Farley, "Chevrolet's New Engineering Laboratory," *SAE Paper* 72F, August 1958. (Luxurious, conventional.)

14.029 "Engine and Chassis Test Facilities," *SAE Special Publication* SP-236. (Five installations described and discussed.)

Research and Test Engines
(See also refs. 11.002, No. 122, and 14.439.)

14.111 Horning, "The Cooperative Fuel-Research Committee Engine," *Trans. SAE 26*, 436 (1931). (The CFR single-cylinder variable-compression-ratio engine standardized for octane-number determinations. Widely used for basic engine research as well as for fuel testing. Manufactured and sold by Waukesha Engine Co., Waukesha, Wis.)

14.112 Bellman *et al.*, "Knock-Limited Outputs from a CFR Engine Using Internal Coolants," *NACA Wartime Report ACR* No. E5H31, E-219, Oct. 1945. (Maximum imep 1024 psi 72 kg/cm² using gasoline-air ratio 0.093, $F_R = 1.4$ plus 75% addition of a dimethylamine-water solution. Engine speed 2500 rpm, specific output 14.5 ihp per sq in. piston area, a world's record.)

14.113 Ware, "Description of the NACA Universal Test Engine and Some Test

Results," *NACA Report* 250, 1927. (Early design used on many important NACA projects.)

14.114 Pope, "The C.U.E. Cooperative Universal Engine Design for Aviation Engine Single-Cylinder Research," *Jour. SAE 48*, Jan., 33 (1941); and *SAE Paper*, June 1940. (A crankcase and auxiliary equipment designed to accommodate aircraft-type cylinders.)

14.115 Picard and Thaler, "Moteur Expérimental à Allumage Commande," *Revue de l'Institut Français du Pétrole et Annales des Combustibles Liquides, IV,* mai 1949, p. 180. (Description with illustrations of one cylinder overhead-valve domed-head engine for research work.)

14.116 Chen and Flynn, "Development of a Single-Cylinder Compression-Ignition Research Engine," *SAE Paper* 650733, 1965. (Especially designed engine for Diesel combustion-chamber work.)

14.117 "Christie Variable Compression Engine," Bulletin (1957) of Christie Machine Works, San Francisco. (High-grade 3.063 × 4.5 in. variable compression test engine.)

14.118 Ainsley and Cleveland, "The CIR Oil Test Engine," *Jour. SAE 64*, June, 46 (1956). (Cross section and description of single-cylinder, passenger-car type engine developed for Cooperative Research Council for oil testing.)

14.119 Mitchell *et al.*, "Design of a High-Compression-Ratio Test Engine for the Petroleum Industry," *SAE Paper* 260A, Nov. 1960. (GMC V-8 automobile-type engine of especially rugged design for compression ratios up to 22. Used for work of ref. 12.49 of Volume I.)

14.120 "Ricardo Variable Compression Engines," Bulletin of G. Cussons, Ltd., Manchester, England, 1930. (Describes the poppet-valve and sleeve-valve engines used in early Ricardo research. See Volume I, refs. 1.12 and 1.13.)

14.121 "Engineering Know-How in Engine Design, Part 11, 1963," *SAE Special Publication* SP-256, 1964. (Includes design, development, and application of the 17.6 cu in. test engine.)

Dynamometers, Engine

14.201 Electric dynamometers are manufactured by: General Electric Co., Apparatus Sales Div., Schenectady, N.Y.; Westinghouse Electric Corp., Box 868, Pittsburgh, Pa.; Reliance Engineering Co., Mount Vernon, N.Y.; Pohl Associates, Inc., Philadelphia, Pa.

14.202 A-c eddy-current and water-brake dynamometers are manufactured by Eaton Manufacturing Co., Dynamatic Division, Kenosha, Wis.

14.203 For manufacturers of other dynamometers in the U.S.A., see ref. 14.300.

14.204 "Construction and Operation of Brake Testing Dynamometers," Report of SAE Brake Subcommittee, No. 3, *SAE Paper* 8, Jan. 1953.

14.205 Sorenson, "Hydraulic Dynamometer Has Internal Cooling System," *Auto. Ind.*, October 1, 1950, 54. (Description of a cheap, home-made type.)

14.206 Shields, R. F., "Engine Transient Performance or Why Inertia Wheel Testing," *SAE Paper* 118, 1957. See also *Jour. SAE 65*, Oct. 1957, 63. (Interesting equipment for testing transient engine performance.)

14.207 Stanke *et al.*, "Testing with Controlled Weather Chassis Dynamometers," Ethyl Corporation Publication. See also *Jour. SAE 67*, April 1959, 63.

14.208 Moore *et al.*, "An All-Weather Chassis Dynamometer," *SAE Paper* 94E, June 1953.

14.209 MacCoull, "A Versatile Car Testing Dynamometer," *SAE Preprint* 318, March 1949. (Texas Co. installation.)

14.210 Cline, "Controlling Fleet Maintenance with Dynamometers," *SAE Quart. Trans. 4*, April 1950, 168. (Some technical information on available chassis dynamometers.)

14.211 Eipper, "Large Chassis Dynamometer," *Auto. Ind.*, March 1, 1955. (Description of Du Pont car dynamometer. Large and elaborate.)

14.212 Duff, "Precision Dynamometer for Small Motor Measurements," *Prod. Eng. 19*, May 1948, 132.

14.213 Cox, "Dynamometer Testing of High Speed Motors for Powered Aircraft Models," *Automotive and Aviation Industries 94*, May 15, 1946. (High-speed, variable frequency, electric dynamometers.)

14.214 Knudsen, "A Discussion of Present Day Dynamometers: Their Application, Operation, and Control," *Gen. Motors Eng. J. 4*, Fourth Quarter (1957). (Shows torque characteristics and applications of several types.)

Mounting of Test Engines and Dynamometers

14.250 "The Design of Vibration-Isolating Suspensions for Engine Test Set Ups," Bulletin available from Sloan Automotive Laboratories, MIT, Cambridge, Mass.

14.251 Lewis and Unholtz, "A Simplified Method for the Design of Vibration Isolating Suspensions," *Trans. ASME 69*, 813 (1947).

14.252 Crede and Walsh, "The Design of Vibration Isolating Bases for Machinery," *J. Appl. Mech. 14*, A-7 (1947).

14.253 MacDuff, "Isolation of Vibration in Spring Mounted Apparatus," *Prod. Eng. 17*, July and August 1946.

14.254 Wahl, *Mechanical Springs*, Penton Publishing Co., Cleveland, Ohio, 1944; Den Hartog, *Mechanical Vibration*, McGraw-Hill Book Company, New York, 1947.

Instruments and Equipment

(See also ref. 11.002, Nos. 143 and 178.)

14.300 "Mechanical Catalog," published annually by ASME, 29 W. 39th St., New York City, gives sources of supply of all kinds of testing and measuring equipment. Also "Electronic Buyers Guide" and Sweet's Catalog Service, "Product Design Catalog Files," published annually by McGraw-Hill Book Company, 330 W. 42nd St., New York.

14.301 Chandler, "Recent Developments in Engine Instrumentation," *SAE Preprint* 94-C, June 1953. (Ford Motor Company methods.)

14.302 Stead, "Notes on Test Methods and Some Instruments," IME (London) *Proc. Automobile Div.* 1947–48, Pt. II. (Temperature plugs, smoke meter, indicator, accelerometer, torsiograph.)

14.303 Hempson, "Instrumentation Problems of Internal-Combustion Engine

Development," IME (London), *Automobile Div. Paper*, Dec. 1961. (Valve-motion indicators, strain-gauge pressure pickups, torsiographs, temperature measurements, counters, oscillographs, etc.)

14.304 Degamo, Inc., Box 653, Chicago Heights, Ill. (Agents for several lines of European instruments including pressure transducers, crank-angle signaler, torsiographs, displacement indicators, etc.)

Instruments and Equipment for Studies of Combustion
(See also refs. 14.430–14.439.)

14.305 Warren and Hinkamp, " New Instrumentation for Engine Combustion Studies," *SAE Preprint*, Jan. 1956.

14.306 Flame photography, refs. 1.50–1.61, 1.77, 1.792, 1.83, 1.831, 2.26, 2.27, 3.11–3.192.

14.307 Schlieren photography, refs. 2.27, 2.321, and Fish and Parnham, " Focussing Schlieren Systems," Royal Aircraft Establishment, Great Britain, *Tech. Note* I.A.P. 999, Nov. 1950.

14.308 Ionization-gap equipment, refs. 1.53, 1.791, 2.324.

14.309 Spectroscopic and radiation techniques, Volume I, refs. 5.10–5.13 and Volume II, refs. 1.73, 1.75, 2.34, 2.341, 3.71–3.73.

14.310 Sound-velocity techniques, Volume I, refs. 5.15, 5.16 and Volume II, ref. 2.33.

14.311 Chemical sampling, refs. 1.72, 1.74, 2.30.

14.312 Pressure-wave measurement, refs. 2.04, 2.05, 2.061–2.093.

Force and Torque Measurements
(See also ref. 14.300.)

14.314 " MIT Hydraulic Torque Scale." Complete data and working drawings available from Sloan Automotive Laboratories, MIT, Cambridge, Mass. See also DEMA Bulletin No. 6, Jan. 3, 1949.

14.315 Bidwell and Smith, " Some Developments in Dynamometer Equipment," *SAE Paper*, March 8, 1951. (Development of MIT and General Motors Company hydraulic scales, also G.M.C. oil floating trunnion bearings.)

14.316 Markson and Williams, " Development of an Air-Operated Force-Measuring System," *Trans. ASME 70*, 271 (1948). (See also " Torque Transmitter," Republic Flowmeter Co., Chicago, Ill.)

14.317 Toledo Scale Co., Toledo 1, Ohio; Howe Scale Co., Rutland, Vermont; Buffalo Scale Co., Buffalo 13, N.Y. For others, see ref. 14.300. Gustin, " Torque Measuring Apparatus and Technique," *SAE Preprint 649*, Sept. 1951. (Waterloo Tractor Works experience.)

14.318 Bidwell and Smith, " Some Developments in Dynamometer Equipment," *SAE Paper* 598, March 1951. (Method of floating pedestal bearings to reduce friction.)

14.319 Vibro-Meter, A.G., Fribourg, Switzerland. (Load- and torque-measuring devices.)

Speed Measurements

(See also ref. 14.300 under "Tachometers.")

14.321 MIT 60-cycle flashlamp. Data available from Sloan Automotive Laboratories, MIT, Cambridge, Mass.

14.322 Stroboscopes and stroboscopic tachometers are available from General Radio Co., Concord, Mass.; Herman Sticht, Ltd., 27 Park Place, New York City; Boulin Instrument Co., Pelham, N.Y.; J. G. Biddle, Philadelphia 7, Pa.; O. Zernidrow Co., 15 Park Row, New York City; and others.

14.323 For manufacturers of centrifugal, electric, chronometric, and other types of tachometer, see ref. 14.300.

Fluid-Flow Measurements, Orifice Meters

(See also ref. 14.300 under *meters, fluid*, and Volume I, refs. A–3.0–A–3.85.)

14.401 Leary and Tsai, "Metering of Gases by Means of the ASME Square-Edged Orifice with Flanged Taps." Bulletin available from Sloan Automotive Laboratories, MIT, Cambridge, Mass. (Detailed instructions in installation and use. Formulas arranged for convenient application.)

14.402 "History of Orifice Meters and the Calibration, Construction, and Operation of Orifices for Metering," ASME, New York, 1935.

14.403 "Flow Measurement by Means of Standardized Nozzles and Orifice Plates," ASME Power Test Codes, Part V, Chapter 4, ASME, New York, 1940.

14.404 "Fluid Meters: Their Theory and Application," 4th ed., ASME, New York, 1937.

14.405 "Single-Cylinder Engine Air Flow Measuring Technique," Diesel Engine Manufacturers Association, Bulletin No. 12, February 15, 1954.

14.406 Cornelius and Caplan, "An Improved System for Control and Measurement of Air Consumption of a Single-Cylinder Engine," *SAE Paper* T12, January 1952. (Critical flow orifices to keep air flow constant automatically. Rather elaborate and expensive.)

14.407 Oppenheim and Chilton, "Pulsating-Flow Measurement — A Literature Survey," *ASME Paper* 53–A–157, 1953.

14.408 Hall, "Orifice and Flow Coefficients in Pulsating Flow," *Trans. ASME 74*, 925 (1952).

14.409 Deschere, "Basic Difficulties in Pulsating-Flow Metering," *Trans. ASME 74*, 919 (1952). (Extensive bibliography.)

14.4091 Baird and Berchtold, "The Dynamics of Pulsative Flow Through Sharp-Edged Restrictions — With Special Reference to Orifice-Metering," *Trans. ASME 74*, 1381 (1952).

Miscellaneous Types of Flowmeters for Gas

14.410 Fleming and Binder, "Study of Linear-Resistance Flowmeters," *Trans. ASME 73*, 621 (1951). (Experimental and theoretical discussion of capillary type flowmeters. Comments by Alcock of Ricardo and Co., Ltd.)

14.411 Biles and Putnam, " Use of a Consolidated Porous Medium for Measurement of Flow Rate and Viscosity of Gases at Elevated Pressures and Temperatures," *NACA TN* 2783, Sept. 1952. (Use of a consolidated porous medium as a gas-metering device and for determination of gas viscosity investigated over a moderate range of temperature and pressure.)

Fuel and Other Liquid Flow Measurement

14.412 The type used at MIT, called " Rotometer " is made by Fischer and Porter Co., Hatboro, Pa. For other manufacturers of liquid and gas flowmeters see ref. 14.300.
14.413 Shafer & Ruegg, "Liquid-Flowmeter Calibration Techniques," *ASME Paper* 57–A–70.
14.414 Wehrman *et al.*, " Measuring Rate of Fuel Injection in an Operating Engine," *SAE Preprint* 32, Jan. 1953. (General Motors' use of strain gauge in nozzle, and dynamic calibration.)

Pressure Measurements
(See also ref. 14.300 under *pressure gauges etc.* and ref. 14.439.)

14.421 Technical Memorandum on the Measurement of Small Pressure Differences with a Draft Gauge or a Micro-Manometer, mimeograph laboratory instruction dated June 7, 1955. Available from Sloan Automotive Laboratories, MIT, Cambridge, Mass.
14.422 Beadle, " Pressure and Temperature Measurement in Engine Development." From *Engineering Know-How in Engine Design*, Part 4, *SAE Special Publication* SP-143.
14.423 Biles, "A High-Pressure Differential Manometer," *Instruments*, Feb. 1951, p. 204. (U-tube manometer has been used for pressures of 1000 psi and is considered suitable for 3000 psi. Novel feature of the design is elimination from the 5/8-in. Pyrex glass tubes of all but the differential pressures. Tubing is surrounded by the lower pressure fluid which is sealed from atmosphere with 1-in. thick tempered glass plates sealed with O-rings.)

Engine Indicators

14.430 For balanced pressure indicators see Volume I, refs. 5.01–5.05 and also the following:
14.431 " P-V Conversion Machine," *DEMA Bulletin* 14, Jan. 9, 1956. (Description and diagrams of simple machine for converting p-θ diagrams to p-V diagrams and description of SLM piezoelectric indicator.)
14.432 Beadle, " Pressure and Temperature Measurement in Engine Development," *SAE Special Publication* SP-143, 1956. (Fairbanks-Morse practice. Modification and improvements to the MIT balanced-pressure indicator. Use of thermocouples for wall-temperature measurements.)
14.433 Laserson, " Construction and Calibration of a Laboratory Engine Pressure

Indicator," *Instruments*, Feb. 1951, p. 156. (The balanced diaphragm indicator is a simplified version of the MIT design, described in Volume I, ref. 5.03. Correctable error resulting from pressure necessary to bend the diaphragm was found to be 1.2 psi for a 0.003-in. diaphragm. No other appreciable errors were found at speeds between 0 and 1700 rpm and pressures up to 250 psig.)

14.434 For electric strain-gauge indicators see Volume I, refs. 5.06 and 5.091.

14.435 "SLM Pressure Indicator," *DEMA Bulletin* 14, Jan. 9, 1956. (Description with cross section of transducer. See also bulletins by the manufacturer, Kistler Instrument Corp., 8989 Sheridan Drive, Clarence, N.Y.)

14.436 For other electric indicators using crystal or capacitance elements see ref. 14.300 and Volume I, ref. 5.08.

14.437 For general discussion of various indicator types see Volume I, refs. 5.09, 5.092, 5.093.

14.438 Nagao *et al.*, " Influence of the Connecting Passage of a Low Pressure Indicator on Recording." Paper presented at 1960 Meeting, JSME.

14.439 AVL Instrument Corp (Austria), U.S. agent Degamo, Inc., Box 653, Chicago Heights, Ill. (Piezoelectric indicator units and other engine instruments. Also research and test engines.)

Temperature Measurements in Fluids

14.440 Baker *et al.*, *Temperature Measurement in Engineering*, John Wiley & Sons, New York, 1953.

14.441 Dike, *Thermoelectric Thermometry*, Leeds and Northrup Company, Philadelphia 4, Pa.

14.442 Moffatt, "Multiple Shielded High Temperature Probes — Comparison of Experimental and Calculated Errors," *SAE Paper* T13, Jan. 1952. (Airflo Instrument Co., experience.)

14.443 Allen and Hamm, "A Pyrometer for Measuring Total Temperature in Low-Density Gas Streams," *Trans. ASME 72*, 851 (1950).

14.444 Clark and Rohsenow, "A New Method for Determining the Static Temperature of High-Velocity Gas Streams," *ASME Paper* 51–SA–33. (Method of measuring the static or stream temperature of a moving gas by determining mass flow, static, and stagnation pressures at a point in the stream.)

14.445 Cesaro *et al.*, " Experimental Analysis of a Pressure-Sensitive System for Sensing Gas Temperature," *NACA TN* 2043, Feb. 1950. (Determination of high gas temperature from two flow measurements by applying the law of continuity. Comparison with several known methods.)

14.446 Vickers, " Proper Probes Keep Thermocouples Reading True," *Jour. SAE 72*, Dec. 1964, 54. (Theory, design, test results.)

14.450 Moffatt, "Multiple-Shielded High-Temperature Probes — Comparison of Experimental and Calculated Errors," *SAE Preprint* T13, Jan. 1952.

14.451 Diehl *et al.*, " Measurement of Rapidly Fluctuating Gas Temperatures," Bull. Technical University, Delft, Holland, 1956. (Use of resistance thermometer for low-speed engine work.)

14.452 Bennett *et al.*, "A Study of Exhaust Gas Temperature Measurement Methods," S.M. Thesis, Aeronautical Engineering Department, Massachusetts Institute of Technology, Cambridge, Mass., June 1947. (Plain thermocouple, well-shielded thermocouple and use of heat-transfer to cooler. Shielded thermocouple found best.)

Measurement of Cyclic Temperature in Cylinders
(See also refs. 1.40–1.42 and 2.33–2.341.)

14.461 Baker, "Preliminary Evaluation of the Thermocouple Technique for the Measurement of Instantaneous Temperatures of Unburned Gas," Columbia University report to Coordinating Research Council, New York, 1951. (Not hopeful for use of this method in engines.)

14.462 Carbon et al., "The Response of Thermocouples to Rapid Gas-Temperature Changes," *Trans. ASME 72*, 655 (1950).

Temperature Measurements of Engine Parts

14.501 Tarr, "Methods for Connection to Revolving Thermocouples," *NACA RM* E50J23a. (Refined methods for high-rotative speeds.)

14.502 Fournell and Douchet, "Temperature Measurements on Components Rotating at High Speed," *Engineers' Digest*, Vol. XIII, No. 9.

14.503 Underwood and Catlin, "An Instrument for Continuous Measurement of Piston Temperatures," *Jour. SAE 48*, Jan. 1941, 20.

14.504 Sasaki, "A New Method for Surface-Temperature Measurement," *Rev. Sci. Instr. 21*, 1 (1950). (Surface temperature of a body is equal to the temperature of the measuring junction of a thermocouple if this temperature has been so adjusted by heating or by cooling that the meter in the couple circuit does not deflect when the measuring junction comes into brief contact with the surface being measured.)

14.505 "The Measurement of Piston Temperature by Temperature Sensitive Paint," Engine Sub-Committee, Aeronautical Research Committee (London), *Report No. T.A. 4957*, 1941.

Exhaust-Gas and Cylinder-Charge Analysis
(See also Volume I, refs. 7.16, 7.162, 7.21, and Fig. 7–6.)

14.601 Hurn et al., "Application of Gas Chromatography to Analysis of Exhaust Gas," *SAE Paper* 11C, Jan. 1958. (Applies mostly to hydrocarbon residues.)

14.602 Sturgis et al., "The Application of Continuous Infrared Instruments to the Analysis of Exhaust Gas," *SAE Paper* 11B, Jan. 1958.

14.603 Gerrish et al., "The NACA Mixture Analysis and Its Application to Mixture-Distribution Measurement in Flight," *NACA TN* 1238, Mar. 1947. (Comparison with Orsat analysis.)

14.604 Spanogle and Buckley, "The NACA Combustion-Chamber Gas-Sampling Valve and Some Preliminary Test Results," *NACA TN* 454. (Description of valve and test results.)

14.605 Spindt, "Air Fuel Ratios from Exhaust Gas Analysis," *SAE Paper* 650507, May 1965. (Thorough and authoritative. Bibliography.)

14.606 Asanuma and Yanagihara, "Gas Sampling Valve for Measuring Scavenging Efficiency in High-Speed Two-Stroke Engines," *Trans. SAE 70*, 420 (1962).

Diesel Smoke Measurement
(See also refs. 3.851–3.855.)

14.701 Durant, "CRC Investigation of Diesel Smoke Measurement," *SAE Paper* 801A, Jan. 1964.
14.702 Searle, "A Sampling Smoke Meter for Automotive Diesel Use," *SAE Paper* 436C, Nov. 1961.
14.703 Durant and Eltinge, "Fuels, Engine Conditions, and Diesel Smoke," *SAE Paper* 3R, Jan. 1959.
14.704 Hunter, "Diesel Smoke Measurement," *SAE Paper* 630, Nov. 1955.

Vibration Measurements
(See also refs. 8.400, 8.490, 14.300–14.303.)

14.801 Clark, "An Inexpensive Method of Instrumenting Torsional Vibration," *SAE Paper* 652F, 1963.
14.802 Kemp *et al.*, "The Application of High-Temperature Strain Gages to the Measurement of Vibratory Stresses in Gas-Turbine Buckets," *NACA TN* 1174, April 1947. (Would be useful for high-temperature locations in reciprocating engines.)

Miscellaneous Instruments and Techniques

14.901 Manov, "The Availability of Radioisotopes and Their Application to Petroleum and Automotive Research," *SAE Paper* 688, November 1951. (By member of U.S. Atomic Energy Commission. See also ref. 11.002, No. 178.)
14.902 Meckel *et al.*, "An Apparatus for Determining Fuel Spray Characteristics," *SAE Paper* 436A, 1961. (Sprays are photographed as they are injected into a clear liquid. See also refs. 7.10–7.42.)
14.903 Friedman & Maier, Vienna, Austria. (Diesel injection-pump testing equipment.)

1984 REFERENCES FOR VOLUME 2

Since the bibliography beginning on page 637 was printed, the literature on the subject of internal-combustion engines has grown beyond the possibility of detailed listing. Publications in English that report on continuing developments include the following.

Publications of the Society of Automotive Engineers

Automotive Engineering, a monthly magazine including reviews of current practice and condensations of important papers.

SAE Reports, classified by subject matter and containing selected *SAE Papers* on various subjects. Lists of each year's reports are available annually on request.

Individual *SAE Papers*, available in limited quantities on application or at SAE meetings.

The mailing address is SAE, Warrendale, PA 15096.

Other Publications

Automotive Engineer. Monthly. Box 24, Bury St. Edmunds, Suffolk, IP326BW, England. Published bimonthly by Institute of Mechanical Engineers, Automotive Division, 1 Love Lane, London EC2V733, England.

Bulletin of Marine Engineering Society in Japan. Quarterly. Marine Engineering Society of Japan, 1-2-2 Uchisaiwai-Cho, Chiyoda-Peu, Tokyo 100, Japan.

Chilton Automotive Industries. Monthly. Chilton Way, Radnor, PA 19089. Statistical issue: April each year.

Diesel and Gas Turbine World Wide Catalog. Annual. 13555 Bishop Court, Brookfield, Wisconsin 53005-0943.

Index

763